MODERN PHYSICS

FIFTH PRINTING, DECEMBER, 1964

Library of Congress Catalog Card Number: 61–6770

Printed in the United States of America

Dedicated to

JACOB LOUIS EISBERG

Preface

This book has evolved from mimeographed notes which I have been using at the University of Minnesota as a text for a one-year advanced undergraduate course in modern physics. Its selection of material reflects some definite ideas that I have concerning such a course.

I am convinced that a modern course in modern physics should have two purposes: In addition to teaching many of the standard topics of the field, the course also should take full advantage of its unique opportunity to introduce the student to quantum mechanics at an early stage in his education, and in a way which integrates the theory into its historical development and its applications. The introduction must be elementary, but it must not be cursory because such treatment would only amplify the problems that so often arise in a first contact with quantum mechanics. This program takes time, but it pays great dividends by laying the foundation for a really mature discussion of the atom and the nucleus. A student can surely learn much more about these subjects in one year if half of it is invested in developing their theoretical bases than if all of it is spent in discussing them at an essentially empirical level.

The inclusion of more than the usual amount of quantum mechanics in this book makes it necessary that there be certain omissions. Because most institutions now offer an undergraduate course in solid state physics, I have chosen most of the omissions from material that would be covered in such a course.

Although there are items in the table of contents of this book which usually are found only in textbooks of a strictly postgraduate level, this should not be taken as a false indication of its prerequisites. I have tried to start near the beginning of each topic, and initially to employ only arguments that can be appreciated by a student well trained in elementary physics, and in mathematics through intermediate calculus. Almost all the material of the book has been presented on two or three separate occasions to a very heterogeneous group of physics, chemistry, mathematics, and engineering students, ranging from juniors to postgraduates. Their performance has shown that students at all levels, and from all specialties, are quite capable of following the material.

However, in its final form the book contains about twenty percent more material than can be covered in 90 lecture-hours. This, plus the fact that there is a gradual build-up of level in proceeding through the book, should make it adaptable to a variety of needs. A course for undergraduate students, who have had no prior contact with modern physics, might thoroughly cover the material through Chapter 8 and leave for outside reading certain topics in the remaining chapters. A course for postgraduate students, who have already studied modern physics at an elementary level, might begin with a very rapid review of the material through Chapter 5, leaving the details for outside reading, and then thoroughly cover the remaining chapters. There is also enough material for a one-semester course in introductory quantum mechanics.

I would like to thank my colleagues at the University of Minnesota, particularly Drs. W. B. Cheston, C. E. Porter, and D. R. Yennie, for reading parts of the mimeographed notes and making many useful suggestions. I would also like to thank my wife Lila for the constant aid and encouragement which made the completion of this book possible.

R. M. EISBERG

February, 1961

Contents

Introduction 1

1. What is modern physics? *1*. 2. Historical sketch, *2*.

1 The theory of relativity 3

1. The Galilean transformation and classical mechanics, *3*. 2. The Galilean transformation and electromagnetic theory, *7*. 3. The Michelson-Morley experiment, *9*. 4. Einstein's postulates, *15*. 5. Simultaneity, *16*. 6. Kinematical effects of relativity, *19*. 7. The Lorentz transformation, *24*. 8. Transformation of velocity, *27*. 9. Relativistic mechanics, *30*. 10. Transformation of momentum and energy, *36*. 11. Experimental verification of the theory, *37*. Bibliography, *39*. Exercises, *39*.

2 Thermal radiation and the origin of quantum theory 41

1. Introduction, *41*. 2. Electromagnetic radiation by accelerated charges, *42*. 3. Emission and absorption of radiation by surfaces, *46*. 4. Black body radiation, *47*. 5. Wien's law, *49*. 6. The Rayleigh-Jeans theory, *51*. 7. The Boltzmann probability distribution, *57*. 8. Comparison with experiment, *62*. 9. Planck's theory, *63*. 10. Some comments on Planck's postulate, *66*. Bibliography, *68*. Exercises, *68*.

3 Electrons and quanta 70

1. Cathode rays, *70*. 2. The e/m ratio of electrons, *72*. 3. The charge and mass of electrons, *73*. 4. Bucherer's experiment, *75*. 5. The photoelectric effect, *76*. 6. Classical theory of the photoelectric effect, *78*. 7. Quantum theory of the photoelectric effect, *79*. 8. The Compton effect, *81*. 9. The dual nature of electromagnetic radiation, *85*. Bibliography, *86*. Exercises, *86*.

4 The discovery of the atomic nucleus 87

1. Thomson's model of the atom, *87*. 2. Alpha particles, *89*. 3. The scattering of alpha particles, *91*. 4. Predictions of Thomson's model, *92*. 5. Comparison with experiment, *98*. 6. Rutherford's model of the atom, *99*. 7. Predictions of Rutherford's model, *100*. 8. Experimental verification and determination of Z, *106*. 9. The size of the nucleus, *107*. 10. A problem, *108*. Bibliography, *109*. Exercises, *109*.

5 Bohr's theory of atomic structure 110

1. Atomic spectra, *110*. 2. Bohr's postulates, *113*. 3. Bohr's theory of the one-electron atom, *115*. 4. Correction for finite nuclear mass, *122*. 5. Atomic energy states, *124*. 6. The Wilson-Sommerfeld quantization rules, *128*. 7. Sommerfeld's relativistic theory, *131*. 8. The correspondence principle, *135*. 9. A critique of the old quantum theory, *136*. Bibliography, *137*. Exercises, *138*.

6 Particles and waves 139

1. De Broglie's postulate, *139*. 2. Some properties of the pilot waves, *141*. 3. Experimental confirmation of de Broglie's postulate, *146*. 4. Interpretation of the Bohr quantization rule, *151*. 5. The uncertainty principle, *153*. 6. Some consequences of the uncertainty principle, *159*. Bibliography, *163*. Exercises, *163*.

7 Schroedinger's theory of quantum mechanics 164

1. Introduction, *164*. 2. The Schroedinger equation, *165*. 3. Interpretation of the wave function, *170*. 4. The time independent

Schroedinger equation, *176*. 5. Energy quantization in the Schroedinger theory, *178*. 6. Mathematical properties of wave functions and eigenfunctions, *184*. 7. The classical theory of transverse waves in a stretched string, *192*. 8. Expectation values and differential operators, *201*. 9. The classical limit of quantum mechanics, *206*. Bibliography, *210*. Exercises, *210*.

8 Solutions of Schroedinger's equation 212

1. The free particle, *212*. 2. Step potentials, *221*. 3. Barrier potentials, *231*. 4. Square well potentials, *239*. 5. The infinite square well, *251*. 6. The simple harmonic oscillator, *256*. Bibliography, *266*. Exercises, *266*.

9 Perturbation theory 268

1. Introduction, *268*. 2. Time independent perturbations, *268*. 3. An example, *274*. 4. The treatment of degeneracies, *278*. 5. Time dependent perturbation theory, *285*. Bibliography, *291*. Exercises, *291*.

10 One-electron atoms 293

1. Quantum mechanics for several dimensions and several particles, *293*. 2. The one-electron atom, *295*. 3. Separation and solution of the equation of relative motion, *298*. 4. Quantum numbers, eigenvalues, and degeneracies, *302*. 5. Eigenfunctions and probability densities, *304*. 6. Angular momentum operators, *313*. 7. Eigenvalue equations, *318*. 8. Angular momentum of one-electron-atom eigenfunctions, *323*. Bibliography, *325*. Exercises, *325*.

11 Magnetic moments, spin, and relativistic effects 327

1. Orbital magnetic moments, *327*. 2. Effects of an external magnetic field, *330*. 3. The Stern-Gerlach experiment and electron spin, *334*. 4. The spin-orbit interaction, *338*. 5. Total angular momentum, *346*. 6. Relativistic corrections for one-electron atoms, *353*. Bibliography, *358*. Exercises, *358*.

12 Identical particles 360

1. Quantum mechanical description of identical particles, *360*. 2. Symmetric and antisymmetric eigenfunctions, *363*. 3. The exclusion principle, *365*. 4. Additional properties of antisymmetric eigenfunctions, *369*. 5. The helium atom, *373*. 6. The Fermi gas, *381*. Bibliography, *390*. Exercises, *390*.

13 Multi-electron atoms 391

1. Introduction, *391*. 2. The Thomas-Fermi theory, *393*. 3. The Hartree theory, *396*. 4. The periodic table, *407*. 5. Excited states of atoms, *414*. 6. Alkali atoms, *418*. 7. Atoms with several optically active electrons, *425*. 8. *LS* coupling, *428*. 9. *JJ* coupling, *441*. 10. The Zeeman effect, *443*. 11. Hyperfine structure, *449*. 12. Transition rates and selection rules, *453*. 13. Lifetimes and line widths, *468*. Bibliography, *472*. Exercises, *473*.

14 X-rays 475

1. The discovery of X-rays, *475*. 2. Measurement of X-ray spectra, *476*. 3. X-ray line spectra, *483*. 4. X-ray continuum spectra, *489*. 5. The scattering of X-rays, *493*. 6. The photoelectric effect and pair production, *505*. 7. Total cross sections and attenuation coefficients, *509*. 8. Positrons and other antiparticles, *513*. Bibliography, *517*. Exercises, *517*.

15 Collision theory 518

1. Introduction, *518*. 2. Laboratory-center of mass transformations, *519*. 3. The Born approximation, *524*. 4. Some applications of the Born approximation, *530*. 5. Partial wave analysis, *534*. 6. Some applications of partial wave analysis, *546*. 7. Absorption, *554*. Bibliography, *558*. Exercises, *558*.

16 The nucleus 559

1. Introduction, *559*. 2. The composition of nuclei, *561*. 3. Nuclear sizes and the optical model, *565*. 4. Nuclear masses and

abundances, *580*. 5. The liquid drop model and the semi-empirical mass formula, *589*. 6. Magic numbers, *592*. 7. The Fermi gas model, *594*. 8. The shell model, *599*. 9. The collective model, *604*. 10. Alpha decay and fission, *609*. 11. Beta decay, *620*. 12. Gamma decay and internal conversion, *640*. 13. Properties of excited states, *655*. 14. Nuclear reactions, *662*. 15. Nuclear forces, *674*. 16. Mesons, *694*. Bibliography, *703*. Exercises, *704*.

Index **707**

Introduction

1. What Is Modern Physics?

Webster's dictionary gives as a primary definition of the word *modern*: "pertaining to the present time, or time not long past." However, when a physicist speaks of *modern physics* he is using the word in a somewhat different way. Modern physics is used to designate certain specific fields of physics. All these fields have two things in common: first, their development has taken place since, roughly speaking, the year 1900; and, second, the theories used to explain the phenomena with which these fields are concerned are startlingly different from the theories that were in existence before 1900.

The term that contrasts with modern physics is *classical physics*. Classical physics comprises fields in which the subjects studied are: Newton's theory of mechanics and the varied phenomena which can be explained in terms of that theory, Maxwell's theory of electromagnetic phenomena and its applications, thermodynamics, and the kinetic theory of gases. In modern physics the subjects studied are: the theory of relativity and associated phenomena, the quantum theories and quantum phenomena, and, in particular, the application of the relativity and quantum theories to the atom and the nucleus.

Many physicists are working today in the fields of classical physics. On the other hand, the theory of relativity, which dates from 1905, is certainly not modern in the sense of Webster's definition. We see that the words modern and classical, as used in physics, do not have quite their normal temporal significance.

In this book we shall be concerned with modern physics. However, as a point of departure, we shall first give a very brief description of the status of classical physics toward the end of the nineteenth century.

2. Historical Sketch

The astronomical observations of Tycho Brahe and their interpretation by Kepler, plus the mechanical experiments of Galileo (all of which happened within a few decades of the year 1600), were welded into an elegant yet simple theory of mechanics by Newton (1687). By the end of the nineteenth century this theory was very highly developed. It provided a successful interpretation of all mechanical phenomena known at that time and served as a basis for the kinetic theory of gases which, in turn, removed many mysteries from the theory of thermodynamics.

A great variety of facts concerning electric and magnetic fields and their interaction were discovered during the nineteenth century. All these phenomena were brought together within one framework by the beautiful theory of Maxwell (1864). This theory provided an explanation for the wave propagation of light which was in agreement with what was known at that time of both geometrical and physical optics.

In fact, almost all the experimental data known at that time could be fitted into either Newton's theory of mechanics or the electromagnetic theory of Maxwell. Physicists were beginning to feel quite self-satisfied, and it appears that the majority were of the opinion that the work of their successors would be merely to "make measurements to the next decimal place."

The turn of the present century saw the shattering of this tranquil situation by a series of quite revolutionary experimental and theoretical developments, such as the theory of relativity, which demands that we reject deeply seated intuitive ideas concerning space and time, and the quantum theories, which make similar demands on our intuitive ideas about the continuity of nature.

We shall begin our study of modern physics with the theory of relativity, not because of chronological considerations since the relativity and early quantum theories were developed concurrently, but because this order of presentation will make the discussion much more convenient. A fairly brief treatment of relativity will be followed by a similar treatment of the early quantum theories. This will lead us to quantum mechanics, which we shall study in detail and then apply to a thorough discussion of the atom and the nucleus.

CHAPTER I

The Theory of Relativity

1. The Galilean Transformation and Classical Mechanics

In classical physics, the state of any mechanical system at some time t_0 can be specified by constructing a set of coordinate axes and giving the coordinates and momenta of the various parts of the system at that time. If we know the forces acting between the various parts, Newton's laws make it possible to calculate the state of the system at any future time t in terms of its state at t_0. It is often desirable that during or after such a calculation we specify the state of the system in terms of a new set of coordinate axes, which is moving relative to the first set. The two-fold question then arises: How do we transform our description of the system from the old to the new coordinates, and what happens to the equations which govern the behavior of the system when we make the transformation? The same question often arises when treating electromagnetic systems with Maxwell's equations. This is the question with which the *theory of relativity concerns* itself.

Let us see what answer classical mechanics would give to this question. Consider the simplest possible mechanical system—one particle of mass m acted on by a force **F**.† Using rectangular coordinates to describe the system, we have the situation shown in figure (1–1).

† Boldface type will be used in this book for symbols that represent a vector quantity. The same symbol in normal type will then represent the magnitude of the vector quantity.

The set of four numbers (x, y, z, t) tells us that at time t the particle has coordinates x, y, z. Assuming that we know **F**, Newton's laws,

$$
\begin{aligned}
m \frac{d^2x}{dt^2} &= F_x \\
m \frac{d^2y}{dt^2} &= F_y \qquad (1\text{-}1) \\
m \frac{d^2z}{dt^2} &= F_z
\end{aligned}
$$

where F_x, F_y, and F_z are the x, y, and z components of the vector $\mathbf{F}$, give us a set of differential equations which govern the motion of the

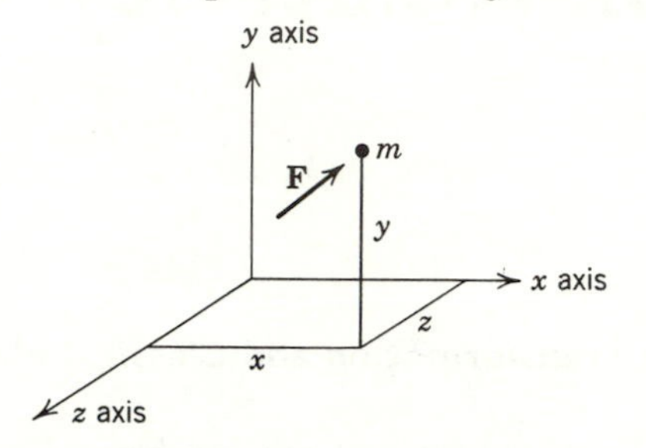

Figure I–I. A set of rectangular coordinates.

system. These equations are valid only if the frame of reference defined by the coordinates x, y, z is an *inertial frame*, i.e., a frame of reference in which a body not under the influence of forces, and initially at rest, will remain at rest. If we know the three coordinates and three components of momenta of the particle at an initial time t_0, we can evaluate the two arbitrary constants which appear in the general solution of each of the differential equations (1–1) and can make specific predictions of the location (and momentum) of the particle at any time t.

Now consider the problem of describing the system from the point of view of a new frame of reference x', y', and z', which is moving to the right with constant velocity $\mathbf{u}$ with respect to the old frame of reference, and which maintains the relative orientation between the two frames shown in figure (1–2). These two frames of reference are said to be in *uniform translation* with respect to each other. There are now two sets of four numbers (x, y, z, t) and (x', y', z', t') that can equally well be used to specify the location of the particle at any instant of time t and t', which are the times measured in the two frames of reference and defined such that $t = t' = 0$ when the two frames of reference coincide. What is the relationship between the sets of numbers (x, y, z, t) and (x', y', z', t')? Newton, almost all physicists before 1900, and most probably the reader, would certainly not hesitate to assert that

$$\begin{aligned} x' &= x - ut \\ y' &= y \\ z' &= z \\ t' &= t \end{aligned} \tag{1–2}$$

These equations are known as the *Galilean transformation* equations. They constitute the answer which classical physics gives to the first part of the question that we have posed for the case of uniform translational motion between the two frames of reference.

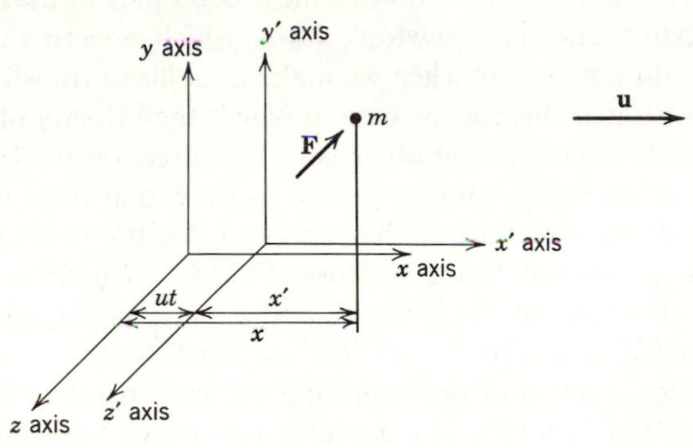

Figure 1–2. Two frames of reference in uniform translation. The x and x' axes are supposed to be collinear.

To answer the second part of the question, let us evaluate the quantity d^2x'/dt'^2 by differentiating the first of equations (1–2) with respect to t. We have

$$dx'/dt = dx/dt - u$$

and

$$d^2x'/dt^2 = d^2x/dt^2$$

Since $t' = t$,

$$d^2x'/dt'^2 = d^2x/dt^2$$

It is obvious that

$$d^2y'/dt'^2 = d^2y/dt^2 \quad \text{and} \quad d^2z'/dt'^2 = d^2z/dt^2$$

It is also true that $F_{x'} = F_x$, $F_{y'} = F_y$, $F_{z'} = F_z$ because the component of $\mathbf{F}$ in the direction of the x or x' axis is the same as seen from either frame of reference—and similarly for its other two components. Substituting for the unprimed quantities in equations (1–1), we find the equations of motion in the primed system:

$$m\frac{d^2x'}{dt'^2} = F_{x'}$$

$$m\frac{d^2y'}{dt'^2} = F_{y'} \tag{1–3}$$

$$m\frac{d^2z'}{dt'^2} = F_{z'}$$

Note that equations (1–3) have exactly the same mathematical form as equations (1–1). Thus the answer to the second part of the question is the very interesting one that Newton's laws, which govern the behavior of the system, do not change when we make a Galilean transformation. We have demonstrated this for the case in which the velocity of one frame of reference with respect to the other is in the direction of the x or x' axis. We leave it as an exercise for the reader to show that these results are true regardless of the direction of the relative velocity vector $\mathbf{u}$. The x, y, z frame was an inertial frame because $d^2x/dt^2 = d^2y/dt^2 = d^2z/dt^2 = 0$ if $\mathbf{F} = 0$. From equations (1–3) we see that x', y', z' is also an inertial frame because $d^2x'/dt'^2 = d^2y'/dt'^2 = d^2z'/dt'^2 = 0$ if $\mathbf{F} = 0$.

The physical content of the result that we have just derived is quite easy to understand and, in fact, is something with which the reader is certainly already familiar. Since Newton's laws are identical in any two inertial frames, and since the behavior of a mechanical system is governed by Newton's laws, it follows that the behavior of all mechanical systems will be identical in all inertial frames even though they are in uniform translation with respect to each other. Consider a laboratory which is located in a railroad car. With the car initially at rest with respect to the ground (assumed to be an inertial frame), let us perform a series of mechanical experiments such as measuring the period of a swinging pendulum or investigating the subject of collision mechanics by means of a serious game of billiards. If we were to perform exactly the same experiments while the car was traveling at constant speed over a very smooth and very straight stretch of track, the fact that equations (1–1) and (1–3) are identical in form requires that the experiments yield exactly the same results as we obtain when the car is at rest with respect to the ground. This is certainly in agreement with our everyday experience.

Thus we see that Newtonian mechanics predicts that all inertial frames in uniform translation with respect to each other are equivalent in so far as mechanical phenomena are concerned. A corollary of this is: No mechanical phenomena can be used to differentiate between inertial frames, so it is impossible by mechanical experiments to prove that one of the frames is in a state of absolute rest while the others move with respect to it. If the windows were covered in the railroad car moving smoothly with respect to the ground, we would not be able to detect its motion.

2. The Galilean Transformation and Electromagnetic Theory

Let us next inquire into the behavior of electromagnetic phenomena when we perform a Galilean transformation. Electromagnetic phenomena are discussed in classical physics in terms of a set of differential equations which govern their behavior (called Maxwell's equations), just as mechanics is discussed in terms of the set of differential equations which govern mechanical phenomena (called Newton's laws). We shall not actually carry through the transformation of Maxwell's equations because the calculation is quite complicated owing to the nature of these equations. However, when a Galilean transformation is performed, it is found that Maxwell's equations *do* change their mathematical form—in sharp contrast to the behavior of Newton's equations. The point is a very important one in the development of the theory of relativity, and we must discuss its physical significance.

As mentioned in the Introduction, Maxwell's equations predict the existence of electromagnetic disturbances which propagate through space in the characteristic manner of wave motion, with a propagation velocity that does not depend on the frequency of the wave motion (at least in a medium with constant index of refraction such as the vacuum). Although many different names have been given to these disturbances (radio waves for frequency $\sim 10^8$/sec, infrared radiation for frequency $\sim 10^{13}$/sec, light waves for frequency $\sim 10^{15}$/sec, X-rays for frequency $\sim 10^{19}$/sec), they are all basically the same phenomena. The nineteenth century physicists, who were very mechanistic in their outlook, felt quite sure that the propagation of waves predicted by Maxwell's equations requires the existence of a propagation medium. Just as water waves propagate through water, so, according to their argument, electromagnetic waves must propagate through a medium. This medium was given the name *ether*. The ether was obliged to have somewhat strange properties in order not to disagree with certain known facts. For instance, the ether must be massless since it was observed that electromagnetic waves such as light can travel through a vacuum. However, it must have elastic properties in order to be able to sustain the vibrations which are inherent in the idea of wave motion. Despite these difficulties, the concept of the ether was considerably more attractive than the alternative, which was to assume that electromagnetic disturbances could be propagated without the aid of a propagation medium. Existence of the ether being assumed, the disturbances would certainly be expected to propagate with a fixed velocity with respect to the ether, just as sound waves propagate with a fixed velocity with respect to the air.

It was assumed that the electromagnetic equations, in the form presented by Maxwell, were valid for the frame of reference which was at rest with respect to the ether—the so-called *ether frame*. A solution of these equations leads to an evaluation of the propagation velocity of electromagnetic waves. The result was 3.00×10^{10} cm/sec $\equiv c$, in agreement within experimental accuracy with the value for the velocity of light measured by Fizeau (1849). However, in a new frame of reference in uniform translational motion with respect to the ether, Maxwell's equations would have a mathematical form different from their mathematical form in the ether frame since, as mentioned above, a Galilean transformation changes the form of the equations. When the propagation velocity of electromagnetic waves *as seen from the new frame of reference* was evaluated by using Maxwell's equations in the form which they assume in this frame, it was found to have a value different from c. The reason is that the propagation velocity predicted by the equations depends on the form of the equations, as certainly must be the case.

Thus the application of a Galilean transformation to Maxwell's equations leads to the prediction that a measurement of the velocity of light made in the new frame of reference would yield a result different from the velocity c measured in the ether frame. It is not necessary to describe in detail the predicted relation between the velocity of light in the ether frame, the velocity of light in the new frame, and the velocity of the new frame with respect to the ether frame, since it is exactly what would be obtained from a simple vector addition calculation of the type done near the beginning of any elementary physics text. That is,

$$\mathbf{V}_{\text{light wrt. new frame}} = \mathbf{V}_{\text{light wrt. ether}} - \mathbf{V}_{\text{new frame wrt. ether}} \tag{1–4}$$

where $V_{\text{light wrt. ether}} = c$, wrt. = with respect to. The somewhat complicated calculation of the velocity of light as measured in the new frame of reference, performed by transforming Maxwell's equations and then solving them in the new frame, is in complete agreement with the very simple physical idea that light propagates in such a way that its velocity with respect to the ether frame is always c, and that its velocity with respect to any other frame can be found by doing a simple vector addition of velocities. Note that the intuitive arguments made to justify the validity of the vector addition of velocities are basically the same arguments which lie behind the Galilean transformation.

In recapitulation, the theory of physics near the end of the nineteenth century was based upon three fundamental assumptions: (a) the validity of Newton's laws, (b) the validity of Maxwell's equations, and (c) the validity of the Galilean transformation. Almost everything which could be derived from these assumptions was in good agreement with experiment in every case in which the appropriate experiment had been performed. With respect to the questions we have been discussing, these assumptions predicted that all frames of reference in uniform translation with respect to each other were equivalent as far as mechanical phenomena were concerned, if they were inertial frames, but in regard to electromagnetic phenomena they were not equivalent; there was only one frame of reference, the ether frame, in which the velocity of light was equal to c.

3. The Michelson-Morley Experiment

In 1887 Michelson and Morley carried out an experiment which proved to be of extreme importance. This experiment was designed to demonstrate the existence of the special frame of reference called the ether frame, and to determine the motion of the earth with respect to that frame.

The earth travels in a roughly circular orbit about the sun with a period (by definition) of one year, and with an orbital velocity of the order of 3×10^6 cm/sec. It would seem unrealistic to make the a priori assumption that the ether frame travels with the earth and, as we shall see later, strong arguments against this assumption were known at that time. A more reasonable working hypothesis would be to assume that the ether frame was at rest with respect to the center of mass of our solar system, or to assume that it was at rest with respect to the center of mass of the universe. In the first case the velocity of the earth with respect to the ether frame would have a magnitude of the order of 10^6 cm/sec, and furthermore the orientation of the velocity vector would change during the course of a year. (The velocity of a point on the surface of the earth, owing to its daily rotation, is negligible compared to the orbital velocity.) In the second case the velocity of the earth with respect to the ether frame would be even greater.

The basic idea of the Michelson-Morley experiment was to measure the velocity of light, as seen from a frame of reference fixed with respect to the earth, in two perpendicular directions. A moment's consideration of the classical theory, as summarized by equation (1–4), will show that the apparent velocity of light with respect to the earth would be different for light traveling in the different directions. At first glance it might be expected that the order of magnitude of the effect which would be observed in such a measurement would be

$$\frac{|\text{velocity of earth wrt. ether}|}{|\text{velocity of light wrt. ether}|} \sim \frac{3 \times 10^6}{3 \times 10^{10}} \sim 10^{-4}$$

where we have assumed that the ether frame is fixed with respect to the solar system. But, in fact, there exists a theorem due to Fresnel and Lorentz which says that it is impossible to devise an optical experiment

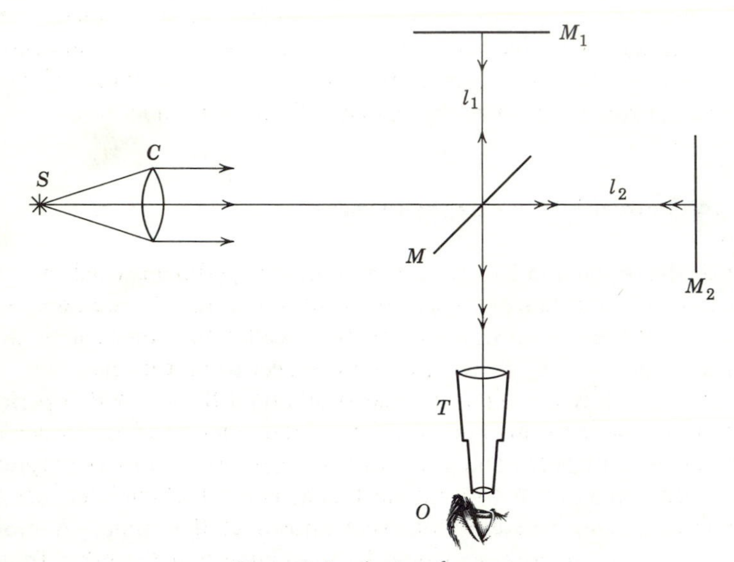

Figure 1–3. An interferometer.

where the velocity of the apparatus with respect to the ether can produce a first order effect. The biggest possible effect is a second order effect. This means that an apparatus designed to observe the difference in the velocity of light traveling in two perpendicular directions must be able to detect differences of the order of $(10^{-4})^2 = 10^{-8}$. We shall prove this theorem for a specific case in the course of our discussion of the Michelson-Morley apparatus.

The numerical estimate which we have just made indicates that an extremely sensitive apparatus would be required to observe effects due to the motion of the earth with respect to the ether, and an even more sensitive apparatus if quantitative measurements were desired. In order to do the experiment, Michelson invented an interesting device called an interferometer, shown schematically in figure (1–3).

Let us briefly consider the operation of this device. (For more details the reader should consult any text in physical optics.) Assume at first that the apparatus is at rest with respect to the ether. Light emitted by source S is collimated by lens C. Consider the central ray shown in the figure. This ray is incident at 45° upon the half-silvered mirror M. Such a mirror has the property of reflecting half the light incident upon it and transmitting the remainder. Thus the incident ray is split into ray 1 and ray 2, which travel to mirrors M_1 and M_2 and are then reflected back on themselves. Half of each of the two rays returning to M is directed into the telescope T. The intensity, due to the central ray, of the light seen by

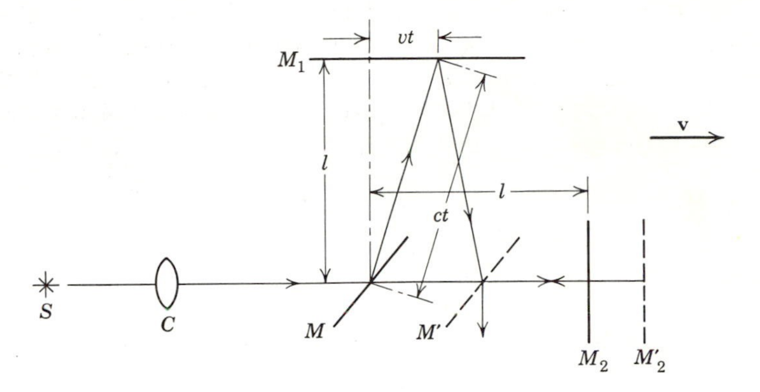

Figure 1–4. An interferometer moving through the ether.

the observer O will be a function of the phase relationship of the two recombining rays. Since the two rays are in phase when originally split by M, their relative phase upon recombining will depend only on the lengths l_1 and l_2 as follows:

$$\text{if } l_1 = l_2, l_2 \pm \frac{\lambda}{2}, l_2 \pm \frac{2\lambda}{2}, l_2 \pm \frac{3\lambda}{2}, \ldots, \quad \text{rays in phase}$$

$$\text{if } l_1 = l_2 \pm \frac{\lambda}{4}, l_2 \pm \frac{3\lambda}{4}, l_2 \pm \frac{5\lambda}{4}, \ldots, \quad \text{rays out of phase}$$

where λ is the wavelength of the light. Such a device can be used in many different ways; the standard meter is now defined in terms of the wavelength of a certain spectral line by using an interferometer.

Next, let us consider an interferometer moving, as indicated in figure (1–4), with velocity $\mathbf{v}$ with respect to the ether, where $\mathbf{v}$ is parallel to the direction of ray 2. Assume $l_1 = l_2 = l$. In the present case the incident beam of light is split by reflection from a mirror which is moving with respect to the ether. Assuming that light travels with a constant velocity in the ether frame, it is easy to show by a Huygens wavelet construction that the reflected ray of light will be thrown forward at exactly the right angle such that, after being reflected from M_1, it will be back at the axis defined by S and C just in time to meet M. This is not a crucial point because it would be possible to assume that M is a "rough" mirror which reflects light diffusely. Figure (1–4) shows the situation as seen by an observer in the ether frame; M' indicates the position of M at the instant the ray reflected from M_1 meets it; M_2' indicates the position of M_2 at the instant it reflects ray 2.

To evaluate the phase relation of the two combining rays, we calculate the time required for the two rays to travel their respective paths. As seen from the ether frame, ray 1 travels along an oblique path of length ct in a time t. During this time the apparatus moves to the right a distance vt. From the figure we see that

$$c^2t^2 = v^2t^2 + l^2$$

so

$$t = \frac{l}{c}\left(1 - \frac{v^2}{c^2}\right)^{-\frac{1}{2}}$$

Since $v^2/c^2 \ll 1$, we can use the binomial theorem and obtain

$$t = \frac{l}{c}\left(1 + \frac{1}{2}\frac{v^2}{c^2}\right)$$

The total time for ray 1 is

$$t_1 = 2t = \frac{2l}{c}\left(1 + \frac{1}{2}\frac{v^2}{c^2}\right)$$

It is easiest to calculate the time required by ray 2 from the point of view of an observer fixed with respect to the apparatus. According to the last of equations (1–2), no difficulties will arise from evaluating time intervals in different frames of reference. Equation (1–4) predicts that, for this observer, the velocity of ray 2 is $c - v$ while moving to the right, and $c + v$ while moving to the left. In each part of the journey it travels a distance l. The time for the round trip is then

$$t_2 = \frac{l}{c - v} + \frac{l}{c + v} = \frac{2l}{c}\left(1 - \frac{v^2}{c^2}\right)^{-1}$$

Using the binomial theorem, this becomes

$$t_2 = \frac{2l}{c}\left(1 + \frac{v^2}{c^2}\right)$$

The two combining rays are out of phase by an amount $\delta t/\tau$, where $\tau = \lambda/c =$ the period of the vibration for light waves of wavelength λ. This phase difference is

$$\frac{\delta t}{\tau} = \frac{(t_2 - t_1)c}{\lambda} = \frac{2l}{c}\frac{1}{2}\frac{v^2}{c^2}\frac{c}{\lambda} = \frac{l}{\lambda}\frac{v^2}{c^2} \tag{1-5}$$

In the actual experiment a measurement was made with the apparatus as shown in the figure, and then a second measurement was made after the apparatus was rotated through 90°. A moment's consideration will show that the phase difference for the apparatus in the second position is

$$\frac{\delta t'}{\tau} = -\frac{l}{\lambda}\frac{v^2}{c^2}$$

The effect of rotating the apparatus is consequently expected to be a change in the phase difference by an amount

$$\Delta \equiv \frac{\delta t}{\tau} - \frac{\delta t'}{\tau} = \frac{2l}{\lambda}\frac{v^2}{c^2} \tag{1-6}$$

[The experiment was done in this way in order to avoid the requirement that l_1 be precisely equal to l_2, as we have assumed in our calculation. It is easy to show that equation (1–6) is true if l_1 is only approximately equal to l_2 even though (1–5) would then be incorrect.]

In their experiment, Michelson and Morley used light of wavelength $\lambda = 6 \times 10^{-5}$ cm and a length $l = 10^3$ cm. If we use our estimate of $v^2/c^2 \simeq 10^{-8}$, we find that equation (1–6) gives a value for Δ of about 0.5. This means that, if the combining rays were just in phase in the first measurement, they would be just out of phase in the second measurement. An observer looking through the telescope while the apparatus was being rotated would see brightness replaced by darkness, or vice versa. To the extreme surprise of Michelson and Morley, no such effect was observed. To make sure that the null effect was not due to a fortuitous combination of velocity vectors such that their laboratory happened to be stationary with respect to the ether frame during the time their measurements were being made, Michelson and Morley made observations during various times of the day and various seasons of the year. An effect was never observed. Despite the predictions of the classical theory, the Michelson-Morley experiment showed that *the velocity of light is the same* when measured along two perpendicular axes in a reference frame which, presumably, is moving relative to the ether frame in different directions at different times of the year.

These remarkable results captured the attention of the entire community of physicists. The experiment was repeated with increased accuracy in a number of laboratories, and it was confirmed that the observed effect was zero to within the experimental accuracy of about 0.3 percent of the theoretically predicted value. Many people tried to devise explanations for the negative result.

Three separate ideas were put forth in an attempt to explain the results of the Michelson-Morley experiment and yet retain as much as possible of the physical theories then in existence. These were: the "ether drag" hypothesis, the "Lorentz contraction" hypothesis, and the so-called "emission theories."

The ether drag hypothesis assumed that the ether frame was attached to all bodies of finite mass. The hypothesis was attractive because it would certainly lead to a null result in the Michelson-Morley experiment and yet it did not involve modification of the theories of mechanics or electromagnetism. But there were two reasons why it could not be accepted. In 1853 Fizeau had measured the velocity of light in a column of rapidly flowing water. If the water were to drag the ether frame with it, the observed velocity would just be the sum of the velocity of light in stationary water and the velocity of the water. Fizeau actually did observe a change in the velocity of light, but the change was only a fraction of the velocity of the water and, furthermore, was fully accounted for by the electromagnetic theory of Maxwell, without introducing the ether drag hypothesis, as due to the motion of charges in water molecules polarized by the electromagnetic field. The second argument against this hypothesis comes from a phenomenon known in astronomy as stellar aberration. The apparent positions of stars are observed to move in circular orbits of very small angular diameter. This is a purely kinematical effect due to the motion of the earth about the sun; in fact, it is the same effect as that which causes a vertical shower of rain to appear to a moving observer as falling at an angle. From this analogy it is apparent that the astronomical effect would not be present if a light ray were to travel with constant velocity with respect to the ether frame and if that frame were fixed with respect to the earth.

The Lorentz contraction assumed that, owing to their motion relative to the ether, all material bodies contract by a factor of $(1 - v^2/c^2)^{1/2}$ in the direction of motion. This, of course, is a serious modification of the concept of a rigid body that is basic to classical mechanics. The reader would find it worth while to demonstrate for himself that, if an interferometer moving through the ether were distorted in this way, a null result would be obtained in the Michelson-Morley experiment. However, for an interferometer in which $l_1 \neq l_2$, a calculation similar to the one above, but *including* the Lorentz contraction, predicts a change in phase difference by an amount

$$\Delta = \frac{l_1 - l_2}{\lambda}\left(\frac{v^2}{c^2} - \frac{v'^2}{c^2}\right) \tag{1-6'}$$

when the velocity of the interferometer with respect to the ether changes from v to v'. Such a change in velocity could be provided by the motion of the earth in its orbit. The appropriate experiment has been done by Kennedy, who observed no effect. Thus the Lorentz contraction hypothesis is ruled out.

In an emission theory Maxwell's equations are modified in such a way that the velocity of a light wave remains associated with the velocity of its source. It is easy to see that this too would lead to a null result in the Michelson-Morley experiment, but it must be rejected because it conflicts with certain astronomical evidence concerning binary stars. Binary stars are pairs of stars which are rotating about their common center of gravity. Consider such a pair at a time when one is moving toward the earth and the other is moving away. Then, if an emission theory were valid, the velocity of light with respect to the earth from one star would be larger than that of light from the other star. This would cause them to appear to have very unusual orbits about their center of gravity. However, the observed motion of binary stars is completely accounted for by Newtonian mechanics.

None of these attempts to "patch up" the classical theories was tenable. A more fundamental change was needed.

4. Einstein's Postulates

All the evidence we have presented seems to be consistent only with the conclusion that there is no *special* frame of reference, the ether frame, with the unique property that the velocity of light in the frame is equal to c. Just as for inertial frames and mechanical phenomena, *all* frames in uniform translation with respect to each other seem to be equivalent in that the propagation velocity of light, measured in any direction, is equal to the same value, c. Einstein (1905) was willing not only to accept this evidence but also to generalize on it. In fact, he stated as a *postulate*:

The laws of electromagnetic phenomena (including the fact that the propagation velocity of light is equal to the constant value c), as well as the laws of mechanics, are the same in all inertial frames of reference, despite the fact that these frames may be in uniform translation with respect to each other. Consequently, all inertial frames are completely equivalent.

This postulate required that Einstein change either Maxwell's equations or the Galilean transformation, since the two together imply the contrary of the postulate. Guided by the failure of the emission theories, he chose not to modify Maxwell's equations. This he indicated by stating as a second *postulate*:

The velocity of light is independent of the motion of its source.

Einstein was then forced to modify the Galilean transformations. This was a very bold move. The intuitive belief in the validity of the Galilean transformation was so strong that his contemporaries had never seriously questioned it. And yet, as we shall see, the very different transformation which Einstein proposed in lieu of the Galilean one is based on realistic physical considerations whereas the Galilean transformation is grossly unrealistic. Another indication of the boldness of this move is that the arguments of section 1 show that any modification of the Galilean transformation will require some compensating modification of Newton's equations in order that the first postulate will continue to be satisfied for mechanics. We shall see later what interesting results this leads to, but first we must develop the new transformation equations.

5. Simultaneity

Consider the fourth of the Galilean transformation equations (1–2), which is

$$t' = t$$

This equation says that there is the same time scale for any two inertial frames of reference in uniform translation with respect to each other, and this is equivalent to saying that there exists a universal time scale for all such frames. Is this true? In order to find out we must realistically investigate the question of time measurement.

Let us for the moment concern ourselves with the problem of defining a time scale in a single frame of reference. Now the basic process involved in any time measurement is a measurement of simultaneity. For instance, the very common statement of the type "The train arrived at 7 o'clock," is actually an abbreviated notation for what really occurred: "The train arrived and the little hand of a *nearby* clock pointed to 7 simultaneously." No further discussion of the question of the simultaneity of events which occur at essentially the same physical location is necessary. Not so obvious is the problem of determining the simultaneity of events which occur at different locations, but this, in fact, is the key problem involved in the setting up of a time scale for a frame of reference. In order to have a time scale valid for a whole frame of reference we must have a number of clocks distributed throughout the frame so that there will everywhere be a nearby clock which can be used to measure time in its vicinity. These clocks must be synchronized; that is, we must be able to say of any two of the clocks: "The little hand of clock A and the little hand of clock B pointed to 7 simultaneously."

A number of methods for the determination of simultaneity suggest themselves. They all involve the transmission of signals between two physically separated locations. If we had at our disposal a means of transmitting information with infinite velocity, there would be no problem.

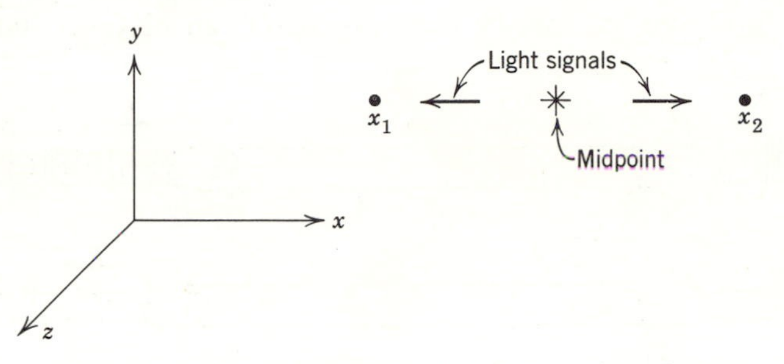

Figure 1–5. Illustrating Einstein's definition of simultaneity.

This is the point where the Galilean transformation goes wrong by tacitly assuming the existence of such a method of synchronization. In fact, no such method exists. Since we have agreed to be realistic in developing a time scale, we must use a real process to transmit the information. Electromagnetic waves, such as light signals, seem to be most appropriate for two reasons: first, their propagation velocity c is the highest known, a clearly desirable feature; second, and more fundamental, since we have postulated that the propagation velocity of light in any frame of reference will be c, it seems expedient to use the velocity of light itself in setting up our time scale. Thus we are lead to Einstein's *definition* of simultaneity:

Two instants of time t_1 and t_2, observed at two points x_1 and x_2 in a particular frame, are simultaneous if light signals simultaneously emitted from the geometrically measured midpoint between x_1 and x_2 arrive at x_1 at t_1 and at x_2 at t_2.

This is illustrated in figure (1–5).

An equivalent definition is that t_1 and t_2 are simultaneous if light signals emitted at t_1 from x_1 and at t_2 from x_2 arrive at the midpoint simultaneously. These definitions intimately mix the times t_1, t_2 and the spatial coordinates x_1, x_2. In Einstein's theory simultaneity does not have an absolute meaning, independent of the spatial coordinates, as it does in the classical theory.

A consequence of these definitions is that two events which are simultaneous when viewed from one frame of reference are in general not simultaneous when viewed from a second frame which is moving relative to the first. This can be seen by considering a simple example. Figure (1–6) illustrates the following sequence of events from the point of view of an observer O who is at rest relative to the ground. This observer has so placed two charges of dynamite at C_1 and C_2 that the distances $\overline{OC_1}$ and $\overline{OC_2}$ are equal. He can cause them to explode simultaneously by sending simultaneous light signals to C_1 and C_2 which actuate detonators. Assume he does this so that, in his frame of reference, the explosions are simultaneous with the instant that he is abreast of O', an observer stationed

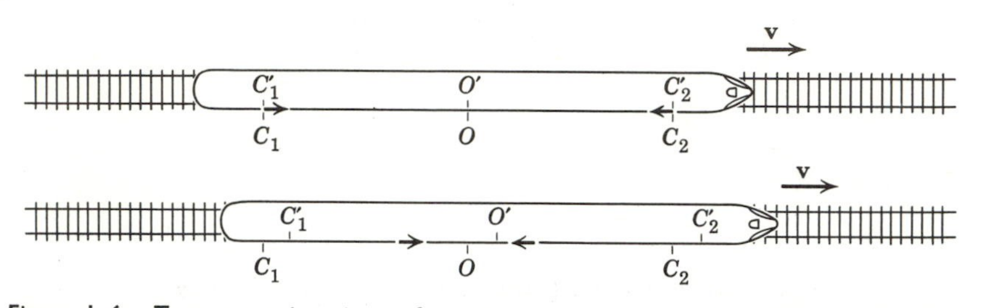

Figure 1–6. Two successive views of a train moving with uniform velocity. The small arrows indicate the light signals.

on a train which is moving by with uniform velocity $\mathbf{v}$. The explosions leave marks C_1' and C_2' on the side of the train. After the experiment O' can measure distances $\overline{O'C_1'}$ and $\overline{O'C_2'}$. He will find they are equal.† The explosions also produce flashes of light. Both O and O' record the times at which they receive the flashes. Observer O will, of course, receive the flashes simultaneously, confirming that in his frame of reference the explosions occurred simultaneously. However, O' will receive the flash which originated at C_2' before he receives the flash from C_1' because the train moves during the finite time required for the light to reach him. Since the explosions occurred at points equidistant from O' but the light signals were not received simultaneously, he concludes that the explosions were not simultaneous.

Such disagreements concerning simultaneity lead to interesting results. From the point of view of O, $\overline{C_1C_2} = \overline{C_1'C_2'}$. However, from the point of view of O', C_2' passed C_2 before C_1' passed C_1. Therefore O' must conclude that $\overline{C_1C_2} < \overline{C_1'C_2'}$. The simultaneity problem will also bring the two

† We shall see in the next paragraph that, according to O', $\overline{O'C_1'} \neq \overline{OC_1}$. However he must find $\overline{O'C_1'} = \overline{O'C_2'}$ due to the homogeneity of space.

observers into disagreement concerning the rates of clocks in their respective frames of reference. As we shall see, the nature of their disagreements about the measurement of distances and time intervals is such as to allow *both* O and O' to find a value of c for the velocity of the light pulse which came from C_1 or C_2.

6. Kinematical Effects of Relativity

Einstein's two postulates, plus his definition, uniquely determine the equations which must be used to transform the coordinates and time from

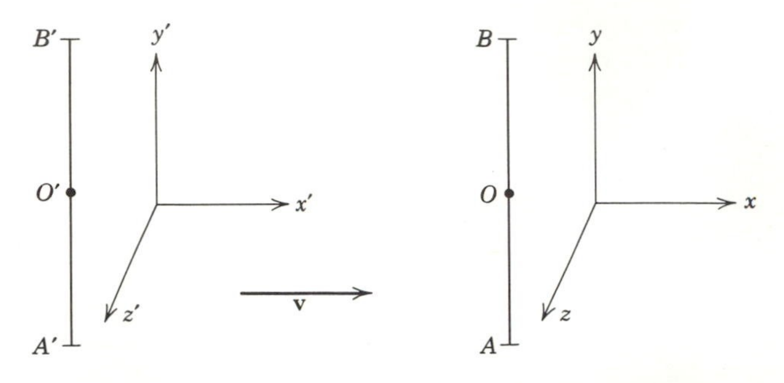

Figure 1–7. Two measuring rods in uniform translation.

one frame of reference to another. There are several ways of actually carrying out the derivation required to find these equations. We shall use here a procedure which is particularly instructive because of its emphasis on physical concepts.

(I) Comparison of Lengths Oriented Perpendicular to the Direction of Motion

Consider two observers O and O' so stationed in reference frames that O' is moving relative to O with velocity $\mathbf{v}$. As shown in figure (1–7), O and O' are each located at the center of parallel measuring rods AB and $A'B'$ which, by superposition, are known to be of identical length when at rest with respect to each other. We pose the question: What are the relative lengths of the two measuring rods, from the point of view of either observer, when they are moving with respect to each other?

As indicated in the figure, we assume that the centers of the measuring rods coincide when passing. Let O have two helpers O_A and O_B, stationed near the ends of AB, who send light signals when A' and B' cross the line defined by AB, and who also mark the location of the points at which

A' and B' cross the line. Afterwards O can compare the length $\overline{A'B'}$ with the length $\overline{AB}$. It is apparent from the symmetry of the arrangement that O will receive the signals from O_A and O_B simultaneously. Furthermore, so will O'. Therefore O' must agree that O_A and O_B made their measurements simultaneously and, consequently, must accept the results of the measurement. Let us tentatively assume that O and O' conclude that $\overline{A'B'} < \overline{AB}$. Next repeat the experiment, but now requiring that the length comparisons be carried out in the primed frame of reference. Since Einstein's first postulate requires complete symmetry between the two frames, it is apparent that O and O' must then conclude that $\overline{AB} < \overline{A'B'}$. This contradicts the previous statement. The only possible consistent conclusion is that $\overline{AB} = \overline{A'B'}$. Thus we find that the lengths of identical measuring rods, as seen by any observer, are the same independent of whether they are at rest relative to the observer or moving in a direction *perpendicular* to the direction of their orientation.

(II) Comparison of Time Intervals Measured by Clocks Which Are in Relative Motion

An observer O', moving with velocity $\mathbf{v}$ relative to observer O, wishes to compare a time interval measured by his clock with a measurement of the

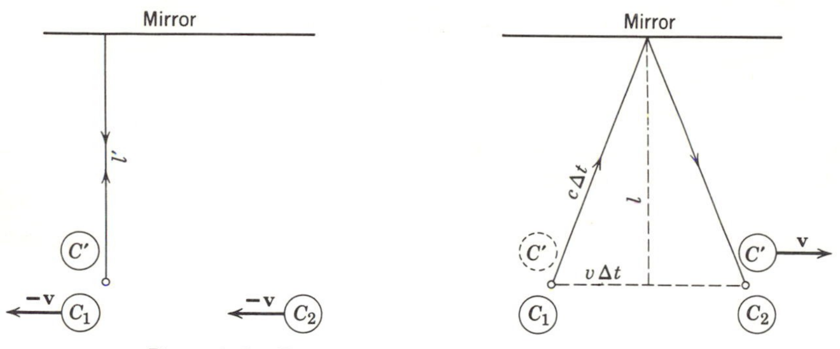

Figure 1–8. Two sets of clocks in uniform translation.

same time interval made with clocks belonging to O. We assume that they have established that, when at rest with respect to each other, all the clocks involved run at the same rate and are synchronized. It is apparent that the reading of an O' clock can be compared with the reading of an O clock that happens to be momentarily coincident with the former without any complication, even though they are moving relative to each other. Thus measurements of a time interval made with clocks in the two frames can be compared by the procedure illustrated in figure (1–8).

Observer O' sends a light signal to a mirror which reflects it back to him. Both O and O' record the time of transmission of the signal with clocks C_1 and C', respectively. The time of reception of the signal is recorded by the two observers with clocks C_2 and C'. The diagram on the left shows the sequence of events from the point of view of O', and the diagram on the right from the point of view of O.

The elapsed time between the two events measured by O' is $2\Delta t'$, where $\Delta t' = l'/c$, and where l' is the distance to the mirror as measured in his frame. The elapsed time measured by O is $2\Delta t$. From the diagram it is apparent that

$$c^2 \Delta t^2 = v^2 \Delta t^2 + l^2$$

where l is the distance to the mirror as measured by O. Solving for Δt, we have

$$\Delta t^2 = \frac{l^2}{c^2 - v^2} = \frac{l^2}{c^2} \frac{1}{1 - v^2/c^2}$$

$$\Delta t = \frac{l}{c} \frac{1}{\sqrt{1 - v^2/c^2}}$$

But argument I shows that $l = l'$. Therefore

$$\Delta t = \frac{l'}{c} \frac{1}{\sqrt{1 - v^2/c^2}} = \frac{\Delta t'}{\sqrt{1 - v^2/c^2}} \tag{1-7}$$

We see that a time interval between two events occurring at the same place in some frame of reference is *longer* by a factor $1/\sqrt{1 - v^2/c^2}$ when viewed from a frame moving relative to the first frame and, consequently, in which the two events are spatially separated. The time interval measured in the frame in which the events occurred in the same place is called a *proper time*. The effect involved is called *time dilation*.

(III) Comparison of Lengths Oriented Parallel to the Direction of Motion

Consider the experiment we have just described and imagine a measuring rod placed in the O frame with one end at clock C_1 and the other end at clock C_2. Designate by L the length of the rod in the O frame, with respect to which it is at rest. We want to evaluate L', the length of the rod as seen from the O' frame.

In this frame the rod is moving in a direction parallel to its own length. Since the velocity of O' with respect to O is $\mathbf{v}$, the velocity of O (and also of the rod) with respect to O' must be precisely $-\mathbf{v}$; otherwise there would be an inherent asymmetry between the two frames which is not allowed by Einstein's first postulate. Let t_1' be the time at which observer O' sees the front end of the rod pass, and t_2' be the time of passage of the rear end. Then the elapsed time $2\Delta t' = t_2' - t_1'$ is related to the length of the rod as measured in the O' frame, and to its velocity measured in that frame, by the equation†

$$L' = v\, 2\Delta t' \tag{1–8}$$

We must also establish an equation connecting the corresponding quantities as seen from the O frame. In this frame, C', which is moving with velocity $\mathbf{v}$, travels the distance L in a time $2\Delta t$. Thus

$$L = v\, 2\Delta t \tag{1–9}$$

From the last two equations we obtain

$$L' = \frac{L\Delta t'}{\Delta t}$$

Argument II shows that

$$\Delta t'/\Delta t = \sqrt{1 - v^2/c^2}$$

Therefore

$$L' = L\sqrt{1 - v^2/c^2} \tag{1–10}$$

We see that a measuring rod is *shorter* by a factor $\sqrt{1 - v^2/c^2}$ compared to its length in a frame in which it is at rest, when it is observed from a frame in which it is moving *parallel* to its own length. The length of the rod measured in the frame in which it is at rest is called the *proper length*. The effect is called the *Lorentz contraction* because equation (1–10) looks like the equation proposed by Lorentz (cf. section 3). In fact, there is a great distinction between these equations. In Lorentz's proposal v represented the velocity of a material body with respect to the ether frame. In the present equation v represents the relative velocity between the two arbitrary inertial frames O and O'.

It is worth while to point out that the technique used by the observer to measure the length of a moving rod in the experiment just described is different from that used by the observer in the experiment of argument I. The latter observer simultaneously marked the locations of the ends of the rod and then measured the distance between the two marks. However, it is very easy to prove that the two techniques are equivalent in that, if both can be applied to the same situation, they will give the same answer.

† Velocities measured in any given frame must, of course, be defined in terms of distances *measured in the frame* and times *measured in the frame*.

(IV) Synchronization of Clocks Which Are in Relative Motion

In section 5 we saw that observers in motion with respect to each other disagree concerning the simultaneity of two events. This will lead to a disagreement between the two observers concerning the synchronization of clocks. Our task here is to determine the quantitative nature of this disagreement.

Consider again the experiment discussed in arguments II and III with special reference to the time lapse between the two events which were timed by the clocks in both frames. We called the time interval measured in the O frame $2\Delta t$, and the time interval measured in the O' frame $2\Delta t'$, and we already know one relation between these two quantities. From equation (1–7) we have

$$2\Delta t = \frac{2\Delta t'}{\sqrt{1 - v^2/c^2}}$$

Since this equation expresses the estimate which would be made in the O frame of the time interval which was measured in the O' frame, and since the measurement was made in the O' frame with a single clock, no problem of synchronization arose in its derivation. However, if we wish to write an equation giving the estimate which would be made in the O' frame of the time interval measured in the O frame, we must account for the fact that observer O' would contend that the two clocks in the O frame were unsynchronized. But, after taking this into account, there must be the same dilation of a time interval as it is transformed from O to O' as there was when it was transformed from O' to O. Thus we put

$$2\Delta t' = \frac{2\Delta t + \delta}{\sqrt{1 - v^2/c^2}} \tag{1–11}$$

where δ is a quantity to be determined which corrects for synchronization.

This quantity can be determined by requiring that equations (1–8), (1–9), and (1–10) be satisfied. Substitute (1–8) and (1–9) into (1–11). We then have

$$\frac{L'}{v} = \frac{L/v + \delta}{\sqrt{1 - v^2/c^2}}$$

From (1–10) this becomes

$$\frac{L\sqrt{1 - v^2/c^2}}{v} = \frac{L/v + \delta}{\sqrt{1 - v^2/c^2}}$$

$$\frac{L}{v}\left(1 - \frac{v^2}{c^2}\right) = \frac{L}{v} + \delta$$

So the value of δ is given by

$$\delta = -\frac{Lv^2}{v\,c^2} = -\frac{Lv}{c^2} \qquad (1\text{–}12)$$

The minus sign means that to observer O' the second clock C_2 appears ahead of the first clock C_1.

We see again that two clocks fixed in a certain frame with a separation L, and synchronized in that frame, appear out of synchronization when viewed from a frame moving relative to the clocks. It is interesting to note that the departure from synchronization is proportional to the separation of the two clocks. Even more interesting is that the sign of the effect changes as the sign of $\mathbf{v}$ changes, since v enters linearly. Thus two observers in different frames investigating the synchronization of each other's clocks would be in *total* disagreement. Not only would each observer contend that the other observer's clocks were unsynchronized, but the two observers would even disagree on the sign of the effect. This should be compared with the effects described by equations (1–7) and (1–10), in which v enters as a squared quantity. Although in these cases there is also a disagreement between the two observers, there is a certain symmetry in their disagreement. Each observer believes that the other observer's measuring rod contracts and time scale dilates.

7. The Lorentz Transformation

We are now only a short step from our goal of deriving the coordinate transformation of the theory of relativity. Consider two observers O and O' who view the same event while moving with velocity $\mathbf{v}$ with respect to each other. Relative to their respective coordinate systems, the location and time of the event are specified by the set of numbers (x, y, z, t) or (x', y', z', t'). The desired coordinate transformation consists of the equations which allow us to calculate the primed set of numbers in terms of the unprimed set, or vice versa.

Assume that the orientation of the two coordinate systems, and of the vector $\mathbf{v}$ specifying the velocity of O' with respect to O, is as shown in figure (1–9). We define the zeros of the t and the t' scales such that clocks at the origins of the two coordinate systems read $t = t' = 0$ when the origins are coincident.

The separation of the two origins at some later time is vt, as seen by O (or vt' as seen by O'). However, the distance x' measured in the O' frame appears contracted by the factor $\sqrt{1 - v^2/c^2}$ when viewed by O. Thus O would say that the distance, parallel to $\mathbf{v}$, from his yz plane to the position of the event was $vt + x'\sqrt{1 - v^2/c^2}$. This is, by definition, the x coordinate of the event. Consequently

$$x = vt + x'\sqrt{1 - v^2/c^2}$$

Solving for x', we find

$$x' = \frac{x - vt}{\sqrt{1 - v^2/c^2}}$$

Now O believes that the time t' read on the clock fixed in the O' frame, at the point x', y', z', must be corrected by an amount $\delta' = +x'v/c^2$ in order to account for the lack of synchronization between that clock and

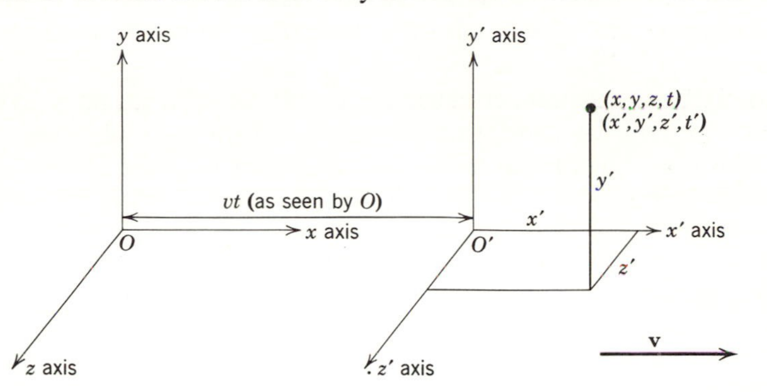

Figure 1–9. Illustrating the argument leading to the Lorentz transformation.

the clock fixed at the origin of the O' frame. (It would be instructive for the reader to convince himself that in the present case the quantity δ' is positive.) Furthermore, time intervals measured by the clock at the O' origin appear to O to be dilated by the factor $1/\sqrt{1 - v^2/c^2}$. Thus, according to O, the time t of the occurrence of the event is

$$t = \frac{t' + x'v/c^2}{\sqrt{1 - v^2/c^2}}$$

Substituting for x' from the equation above and solving for t', we find

$$t' = \frac{t - xv/c^2}{\sqrt{1 - v^2/c^2}}$$

Argument I of section 6 tells us that $y' = y$ and $z' = z$, since these distances are measured in directions perpendicular to **v**. Gathering together our results, we have

$$\begin{aligned} x' &= \frac{x - vt}{\sqrt{1 - v^2/c^2}} \\ y' &= y \\ z' &= z \\ t' &= \frac{t - xv/c^2}{\sqrt{1 - v^2/c^2}} \end{aligned} \tag{1–13}$$

These equations constitute the so-called *Lorentz transformation* which, according to Einstein, must be used for the transformation of the coordinates and time of an event from the inertial frame O into the inertial frame O'.†

An algebraic solution of equations (1–13) for x, y, z, and t produces another form of the Lorentz transformation which is useful in transforming from the primed frame to the unprimed frame. These are

$$\begin{aligned} x &= \frac{x' + vt'}{\sqrt{1 - v^2/c^2}} \\ y &= y' \\ z &= z' \\ t &= \frac{t' + x'v/c^2}{\sqrt{1 - v^2/c^2}} \end{aligned} \tag{1–13′}$$

Equations (1–13′) are identical with (1–13) except that v has been replaced by $-v$. The minus sign is merely a consequence of the fact that, for (1–13′), **v** still means the velocity of O' with respect to O. If we were to redefine **v** to mean the velocity of O with respect to O', the resulting equations would be completely identical with (1–13). We could have guessed the form of (1–13′) from the form of (1–13) without actually doing the algebra, since the observed symmetry between the two is required by Einstein's first postulate. The postulate demands that all inertial reference frames in uniform translation be equivalent. Consequently the transformation which we have derived from this postulate must be the same in going from O to O' as it is in going from O' to O.

It is immediately apparent that the Lorentz transformation reduces to the Galilean transformation when $v/c \ll 1$. This makes good physical

† Equations of exactly this form, but with v representing a velocity relative to the ether, had been proposed in connection with a classical theory of electrons by Lorentz some years before the work of Einstein.

sense. The fundamental difference between the two transformations is that in the former we realize that the signals, which are needed to synchronize the clocks of a reference frame, propagate with the finite velocity c, while in the latter we naively assume the velocity of these signals to be infinite. However, if we are required to make a coordinate transformation in a situation in which the relative velocity v is very small compared to the velocity of the synchronizing signals, it should not make any difference whether we take the latter velocity to be c or ∞. Consequently in this limit the two transformations should merge.

On the other hand, significant discrepancies arise between the predictions of the Galilean transformation and the rigorously correct Lorentz transformation when v is comparable to c. These discrepancies had not been observed in classical mechanics because experiments had not been performed at such high velocities. It is interesting to note that for $v/c > 1$ equations (1–13) are meaningless, in that real coordinates and times are transformed into imaginary ones. Thus c appears to play the role of a limiting velocity for all physical phenomena. We shall attain a better understanding of this as we continue our discussion.

To close this section, let us check the internal consistency of the Lorentz transformation with the postulates from which it was derived. Consider again figure (1–9), and imagine that at the instant $t = t' = 0$ we discharge a spark gap located at the common origin of the two frames of reference. As viewed by observer O, the resulting light pulse will travel out in all directions with the velocity c. Thus, at any instant t, the light pulse will form a spherical wave front of radius $r = ct$. The wave front will obey the equation

$$x^2 + y^2 + z^2 = r^2 \quad \text{or} \quad x^2 + y^2 + z^2 - c^2t^2 = 0 \qquad (1\text{–}14)$$

From the view point of O', according to Einstein's postulates, the light pulse must also travel in all directions with the velocity c. Thus O' must also see a spherical wave front

$$x'^2 + y'^2 + z'^2 - c^2t'^2 = 0 \qquad (1\text{–}15)$$

despite the fact that he is moving with velocity $\mathbf{v}$ relative to O. We leave it as an exercise for the reader to apply equations (1–13′) to (1–14) and obtain (1–15), thus giving the desired demonstration of consistency.

8. Transformation of Velocity

Consider the particle shown in figure (1–10), moving with velocity $\mathbf{V}$ as seen in a frame of reference O. We would like to evaluate the velocity

$\mathbf{V}'$ of the particle, as seen in the frame of reference O' which is itself moving relative to O with velocity $\mathbf{v}$.

Measured in the O frame, the velocity vector of the particle has components

$$V_x = dx/dt, \quad V_y = dy/dt, \quad V_z = dz/dt$$

The same velocity vector, as measured in the O' frame, has components

$$V_x' = dx'/dt', \quad V_y' = dy'/dt', \quad V_z' = dz'/dt'$$

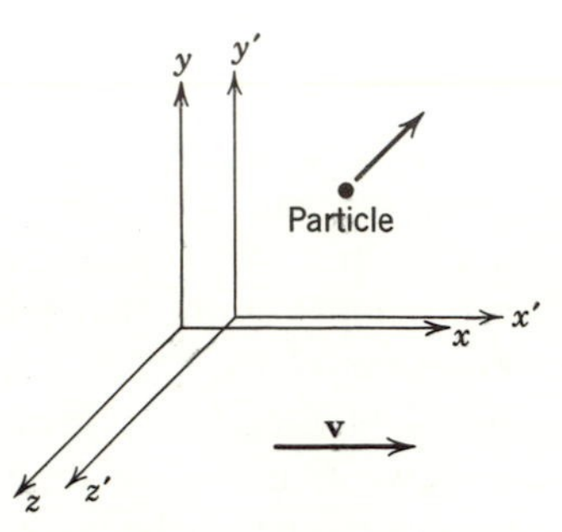

Figure 1–10. A moving particle observed from two frames of reference in uniform translation.

Now from equations (1–13) we know that the relation between the primed and the unprimed coordinates and times is

$$\begin{aligned} x' &= \frac{1}{\sqrt{1-\beta^2}}(x - vt) \\ y' &= y \\ z' &= z \\ t' &= \frac{1}{\sqrt{1-\beta^2}}\left(t - \frac{vx}{c^2}\right) \end{aligned}$$

where we have introduced the convenient notation $\beta = v/c$. Take the differential of these equations, remembering that v is a constant. This gives

$$\begin{aligned} dx' &= \frac{1}{\sqrt{1-\beta^2}}(dx - v\,dt) \\ dy' &= dy \\ dz' &= dz \\ dt' &= \frac{1}{\sqrt{1-\beta^2}}\left(dt - \frac{v\,dx}{c^2}\right) \end{aligned}$$

Then

$$V'_x = \frac{\dfrac{1}{\sqrt{1-\beta^2}}(dx - v\,dt)}{\dfrac{1}{\sqrt{1-\beta^2}}\left(dt - \dfrac{v\,dx}{c^2}\right)} = \frac{\dfrac{dx}{dt} - v}{1 - \dfrac{v}{c^2}\dfrac{dx}{dt}} = \frac{V_x - v}{1 - \dfrac{v}{c^2}V_x}$$

$$V'_y = \frac{dy}{\dfrac{1}{\sqrt{1-\beta^2}}\left(dt - \dfrac{v\,dx}{c^2}\right)} = \frac{dy/dt}{\dfrac{1}{\sqrt{1-\beta^2}}\left(1 - \dfrac{v}{c^2}\dfrac{dx}{dt}\right)} = \frac{V_y\sqrt{1-\beta^2}}{1 - \dfrac{v}{c^2}V_x} \tag{1–16}$$

and, similarly,

$$V'_z = \frac{dz}{\dfrac{1}{\sqrt{1-\beta^2}}\left(dt - \dfrac{v\,dx}{c^2}\right)} = \frac{dz/dt}{\dfrac{1}{\sqrt{1-\beta^2}}\left(1 - \dfrac{v}{c^2}\dfrac{dx}{dt}\right)} = \frac{V_z\sqrt{1-\beta^2}}{1 - \dfrac{v}{c^2}V_x}$$

These equations tell us how to transform the observed velocity from one frame of reference to another frame of reference.

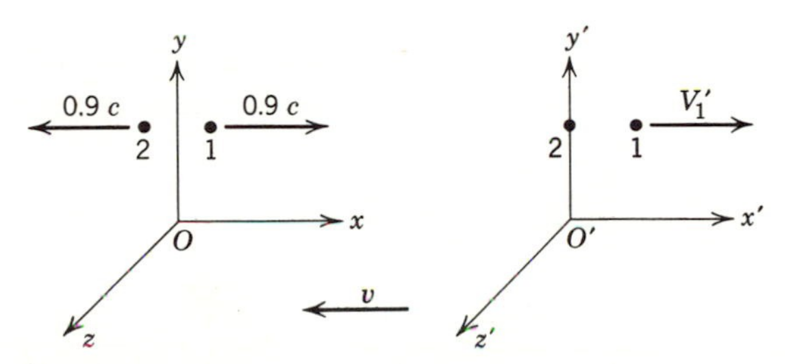

Figure 1–11. Illustrating an example of the addition of velocities.

First we note that, as V/c and v/c approach zero, equations (1–16) approach those which would be derived from the Galilean transformation. Another very interesting property of these equations is that it is impossible to choose $\mathbf{V}$ and $\mathbf{v}$ such that V', the magnitude of the velocity which is seen in the new frame, is greater than c. Consider the example illustrated in figure (1–11). As seen by O, particle 1 has velocity $0.9c$ in the direction of positive x, and particle 2 has velocity $0.9c$ in the negative x direction. To evaluate the velocity of particle 1 with respect to particle 2 we transform from the O frame to the O' frame moving in the negative x direction with velocity $v = -0.9c$, using the first of equations (1–16). We obtain

$$V'_1 = \frac{0.9c - (-0.9c)}{1 - \dfrac{(-0.9c)}{c^2}(0.9c)} = \frac{1.80c}{1.81} < c$$

It is not difficult to give a rigorous proof of this property by considering the mathematical form of equations (1–16). The velocity transformation equations demonstrate another aspect of the fact that in the theory of relativity c acts as a limiting velocity for all physical phenomena.

9. Relativistic Mechanics

It has been emphasized that Einstein's modification of the transformation equations would, in turn, necessitate a compensating modification in the laws of mechanics, so that these laws continue to satisfy the requirement of not changing form under the influence of a coordinate transformation.† In this section we develop the laws of the new mechanics, which is called *relativistic mechanics.*

It is obviously desirable to carry over into relativistic mechanics as much of classical mechanics as possible. We shall see that it is possible to preserve the classical definition of momentum,

$$\mathbf{p} = m\mathbf{u} \tag{1–17}$$

where $\mathbf{p}$ is the momentum, m the mass, and $\mathbf{u}$ the velocity of a particle, and also to preserve the classical law of conservation of momentum,

$$\left[\sum_i \mathbf{p}_i\right]_{\text{initial}} = \left[\sum_i \mathbf{p}_i\right]_{\text{final}} \tag{1–18}$$

providing we are willing to modify the classical concept of mass. It will be necessary to allow the mass of a particle to be a function of the magnitude of its velocity, i.e.,

$$m = m(u) \tag{1–19}$$

The form of this function is to be determined. However, we know a priori that we must have $m(u) \to m_0$, as $u \to 0$, where m_0 is a constant equal to the classically measured mass of the particle. This is true since, as $u \to 0$, the Lorentz transformation approaches the Galilean transformation and no modification in mechanics is necessary.

In order to evaluate the function $m(u)$, we consider the following experiment. As seen from the xyz frame indicated in figure (1–12), observers O_1 and O_2 are moving in directions parallel to the x axis with equal and opposite velocities. These observers have identical balls B_1 and B_2, each of mass m_0 as measured when they are at rest. While passing, each observer

† Just the opposite is true for the laws governing electromagnetic phenomena (Maxwell's equations), since Einstein's second postulate prohibits modification of these laws.

throws his ball toward the other with a velocity which, from his own point of view, is directed perpendicular to the x axis and is of magnitude V.

In the xyz frame, B_1 and B_2 will be seen to approach with equal velocity along parallel lines making angles $\theta_{1_i} = \theta_{2_i}$ with the x axis, and rebound at angles θ_{1_f} and θ_{2_f}. Assuming conservation of momentum and that the collision is elastic, it is easy to show that $\theta_{1_f} = \theta_{2_f}$ and that the magnitude of the velocity of the balls is the same after the collision as before. The actual value of θ_{1_f} or θ_{2_f} depends on the impact parameter d, which we assume to be such that $\theta_{1_f} = \theta_{1_i}$ as shown in the figure.

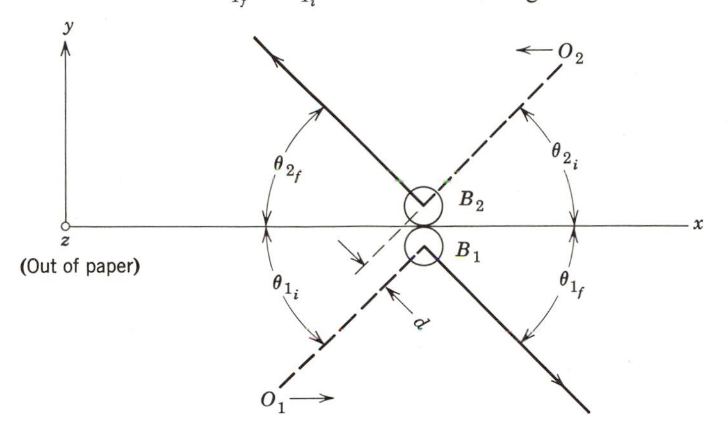

Figure 1–12. A collision between two balls of identical rest mass.

Now look at the process from the point of view of O_1, as shown in figure (1–13). Observer O_1 throws B_1 along a line parallel to his y axis with velocity V. It returns along the same line with velocity of the same magnitude but opposite sign. He sees B_2 maintain a constant x velocity which is just equal to v, the relative velocity of O_2 with respect to O_1. The component of velocity of B_2 along his y axis is observed by O_1 to change sign during the collision but to maintain a constant magnitude. To evaluate this magnitude we realize that the y component of the velocity of B_2, as seen by O_2, is V. Then we transform this to the O_1 frame with the aid of the second of equations (1–16) and obtain $V\sqrt{1 - v^2/c^2}$ for the y component of the velocity of B_2 as seen by O_1.

The y momenta of both B_1 and B_2, as measured in the O_1 frame, simply change sign during the collision. Consequently the total y momentum of the system consisting of the two balls changes sign. But the law of conservation of momentum says that the total y momentum of the system before the collision must equal the total y momentum of the system after the collision. This can only be true if the total y momentum, as seen by

O_1, is zero both before and after the collision. Evaluating the momenta according to the definition (1–17), and equating the total y momentum to zero in order to satisfy the conservation of momentum (1–18), we obtain an equation which is obviously self-contradictory if we insist that m be a constant equal to m_0. However, if we allow m to be a function of the velocity as in equation (1–19), the equation reads

$$V\,m(V) = V\sqrt{1 - v^2/c^2}\,m\left(\sqrt{V^2(1 - v^2/c^2) + v^2}\right)$$

or

$$m(V) = \sqrt{1 - v^2/c^2}\,m\left(\sqrt{V^2(1 - v^2/c^2) + v^2}\right)$$

where the argument of the function m on the right side of the equations

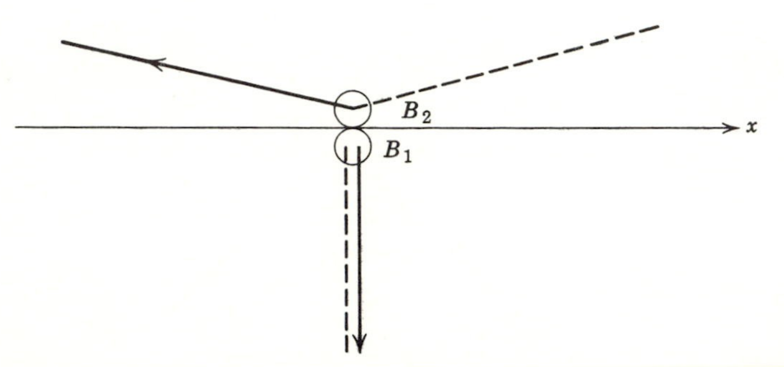

Figure 1–13. A collison between two balls of identical rest mass, as seen by observer O_1.

is simply the magnitude of the velocity of B_2 as seen by O_1. This equation is perfectly satisfactory and, in fact, enables us to evaluate the function m immediately if we take $V \to 0$. We then have

$$m(0) = m_0 = \sqrt{1 - v^2/c^2}\,m(v)$$

so

$$m(v) = m_0/\sqrt{1 - v^2/c^2} \tag{1–20}$$

[Evaluation of the functional form (1–20) from the equation for $m(V)$, by taking $V \to 0$, constitutes a short-cut but not an approximation. It is easy to show that (1–20) satisfies the equation for $m(V)$ exactly.] Thus a consistent theory of relativistic mechanics demands that the mass of a particle moving with velocity v be *larger* than its mass when at rest by a factor $1/\sqrt{1 - v^2/c^2}$. The mass $m(v)$ is called the *relativistic mass* of the particle, and m_0 is called its *rest mass*.

For the quite high velocity $v = 0.1c$, the relativistic mass is only 0.5 percent greater than the rest mass. However, with increasing v the rest mass rapidly becomes larger since $m(v) \to \infty$ as $v \to c$. It is apparent that the velocity of a particle cannot exceed c if (1–20) is valid.

Continuing our discussion of relativistic mechanics, let us consider a particle of rest mass m_0 initially at rest, but accelerated by a force of magnitude F which is applied in the x direction. The total work done on the particle in the distance x_f is

$$T = \int_0^{x_f} F\,dx = \int_0^{t_f} F \frac{dx}{dt}\,dt = \int_0^{t_f} Fv\,dt$$

In order to evaluate this we must know the relativistic form of Newton's law. We *assume* that it is

$$F = \frac{dp}{dt} = \frac{d}{dt}(mv) \tag{1–21}$$

Note that, if $m =$ constant, this is just $F = m\,dv/dt = ma$. Actually, Newton's statement of the law was that of equation (1–21) and not $F = ma$. Inserting (1–21) into the equation for T, we have

$$T = \int_0^{t_f} v \frac{dp}{dt}\,dt = \int_0^{p_f} v\,dp$$

Integrating by parts, we obtain

$$T = [vp]_0^{v_f} - \int_0^{v_f} p\,dv$$

Substituting for p from equations (1–17) and (1–20), and also expressing $v\,dv$ as $d(v^2)/2$, we have

$$T = \left[\frac{m_0 v^2}{\sqrt{1 - v^2/c^2}}\right]_0^{v_f} - \frac{m_0}{2}\int_0^{v_f} \frac{d(v^2)}{\sqrt{1 - v^2/c^2}}$$

The second term is a standard form. Integrating it, we have

$$T = m_0 c^2 \left[\frac{v^2/c^2}{\sqrt{1 - v^2/c^2}} + \sqrt{1 - v^2/c^2}\right]_0^{v_f}$$

This reduces to

$$T = m_0 c^2 \left[\frac{1}{\sqrt{1 - v^2/c^2}}\right]_0^{v_f}$$

Evaluating at the limits and then dropping the subscript f to simplify our notation, we have

$$T = \frac{m_0 c^2}{\sqrt{1 - v^2/c^2}} - m_0 c^2 \tag{1–22}$$

Now the theorem of conservation of energy implies that the total work done on the particle should equal its kinetic energy. Thus we would like to call T the kinetic energy of the particle. To check this, let $v/c \to 0$. Then

$$T = m_0c^2\left[\left(1 - \frac{v^2}{c^2}\right)^{-1/2} - 1\right] \simeq m_0c^2\left[1 + \frac{1}{2}\frac{v^2}{c^2} - 1\right]$$

$$T \simeq \frac{1}{2}\, m_0c^2\,\frac{v^2}{c^2} = \frac{1}{2}\, m_0v^2$$

This agrees with the classical expression for kinetic energy and confirms our identification of T.

Continuing with our interpretation of equation (1–22), we observe that T is a function of v which can be written as the difference between a term depending on v and a constant term, as follows:

$$T(v) = E(v) - E(0)$$

where $E(v) = m_0c^2/\sqrt{1 - v^2/c^2} = mc^2$, and also where $E(0)$ is the value of $E(v)$ for $v = 0$, i.e., $E(0) = m_0c^2$. Since T is an energy, $E(v)$ and $E(0)$ must also be energies—$E(v)$ being some energy associated with the particle when its velocity is v, and $E(0)$ some energy associated with the particle when its velocity is zero. To identify the meaning of these energies, rewrite the equation as

$$E(v) = T(v) + E(0)$$

The conclusion is inescapable. We must interpret $E(v)$ as the total energy of the particle moving with velocity v, since it is the sum of the kinetic energy $T(v)$ of the particle and an intrinsic energy $E(0)$ associated with the particle when it is at rest. The energy $E(v)$ is called the *total relativistic energy*, and $E(0)$ is called the *rest mass energy*.

We have established Einstein's well-known relation between mass and energy: *The rest mass energy of a particle is equal to c^2 times its rest mass.* That is,

$$E(0) = m_0c^2 \tag{1–23}$$

and *the total relativistic energy is equal to c^2 times its relativistic mass.* That is,

$$E(v) = mc^2 \tag{1–24}$$

Equation (1–22) tells us that the relation between total relativistic energy, kinetic energy, and rest mass energy is

$$mc^2 = T + m_0c^2 \tag{1–22$'$}$$

For situations in which the particle also has a potential energy V, it is possible to extend the argument to show that

$$mc^2 = T + V + m_0c^2 \tag{1–22″}$$

This emphasizes the fact that the relativistic mass is a direct measure of the total energy content of the particle. *Anything* that increases its energy content will increase its relativistic mass.

It is sometimes convenient to have a relation similar to (1–22′) which explicitly involves the momentum p. It can be obtained by evaluating the quantity

$$m^2c^4 - m_0^2c^4 = m_0^2c^4\left[\frac{1}{1-\beta^2} - 1\right]$$

where $\beta = v/c$. This is

$$m^2c^4 - m_0^2c^4 = m_0^2c^4\frac{\beta^2}{1-\beta^2} = \frac{m_0^2c^2v^2}{1-\beta^2} = m^2c^2v^2 = c^2p^2$$

Thus

$$E^2 = m^2c^4 = c^2p^2 + m_0^2c^4 \tag{1–25}$$

The relativistic properties of mass and energy, and the relation between the two, were derived from the Lorentz transformation and the additional assumptions we made concerning the new form of the laws of mechanics. These additional assumptions seemed reasonable since they were obviously designed to preserve as much of classical mechanics as possible. However, their ultimate justification can be found only by comparing the predictions of relativistic mechanics with experiment. We reserve this comparison for section 11, but perhaps it is appropriate to point out here that the existence of a rest mass energy m_0c^2 is not in conflict with classical physics. Since the experiments of classical physics are all such that the total rest mass involved is constant, the appropriate rest mass energies could be added to both sides of all energy balance equations without destroying their validity.

The relativistic theory of energy is, however, of more than academic interest. One important reason is that there are processes in nature in which the total rest mass of an isolated system does not remain constant. For such processes experimental evidence shows that the change in rest mass energy is exactly compensated by a change in kinetic, or potential, energy in a way that conserves the total relativistic energy of the system. This is the basis of some very practical processes for the conversion of rest mass energy into other forms of energy, as in a nuclear reactor. It also shows that in our relativistic theory we must replace the separate classical laws of conservation of mass and conservation of energy by a single comprehensive law of *conservation of total relativistic energy:*

As observed from a given frame of reference,† the total relativistic energy of an isolated system remains constant.

10. Transformation of Momentum and Energy

In certain problems, particularly those involving collision mechanics, it is convenient to transform the momentum and total relativistic energy of a particle from one frame of reference to another. In this section we derive the appropriate transformation.

Consider a particle of rest mass m_0 moving with velocity V along the positive x axis of reference frame O. In that frame it will have components of momentum

$$p_x = \frac{m_0 V}{\sqrt{1 - V^2/c^2}}, \qquad p_y = 0, \qquad p_z = 0$$

and total relativistic energy

$$E = \frac{m_0 c^2}{\sqrt{1 - V^2/c^2}}$$

Next consider an observer O' who is also moving along the positive x axis of O, but with velocity v. According to this observer, the particle would have components of momentum

$$p'_x = \frac{m_0 V'}{\sqrt{1 - V'^2/c^2}}, \qquad p'_y = 0, \qquad p'_z = 0$$

and total relativistic energy

$$E' = \frac{m_0 c^2}{\sqrt{1 - V'^2/c^2}}$$

where V' is the velocity of the particle with respect to O'.

To establish the relation between momentum and energy in the two frames, we evaluate the quantities

$$V'/\sqrt{1 - V'^2/c^2} \quad \text{and} \quad c^2/\sqrt{1 - V'^2/c^2}$$

† The same restriction applies, of course, to the energy conservation law of classical physics. For example, the total energy content of a particle will increase if we transform classically from a system in which the particle is at rest into a system in which it is moving and consequently has kinetic energy. See section 10.

in terms of V. This is done by using the velocity transformation equation (1–16), which in the present case reads

$$V' = \frac{V - v}{1 - Vv/c^2}$$

The calculation will not be reproduced here. It is perfectly straightforward but lengthy. The result is

$$\begin{aligned} p'_x &= \frac{p_x - vE/c^2}{\sqrt{1 - v^2/c^2}} \\ p'_y &= p_y \\ p'_z &= p_z \\ E' &= \frac{E - p_x v}{\sqrt{1 - v^2/c^2}} \end{aligned} \qquad (1\text{–}26)$$

Although in the derivation we assumed the velocity of the particle to be parallel to the x axis, it can be shown that equations (1–26) are valid for any orientation of its velocity vector.

11. Experimental Verification of the Theory

The theory of relativity was designed to agree with the experimental fact that the velocity of light is observed to be the same in frames of reference which are in uniform translation with respect to each other. However, in addition to achieving this, the theory *predicts* a number of new phenomena, such as length contraction, time dilation, relativistic increase in mass, and a relation between mass and energy. This is typical of a scientific theory. Also typical is the fact that the initial acceptance of the theory was only tentative even though it appeared to be based upon correct logic. The theory did not achieve full status until its predictions concerning the new phenomena had been put to the test of experiment. To close our discussion of the theory of relativity we shall briefly mention some of the experiments which confirm its predictions.

The first of these was performed in 1909 by Bucherer. It consisted of a measurement of the masses of high velocity electrons, using a fairly direct technique that will be discussed in Chapter 3. Bucherer's results are shown by the crosses in figure (1–14), some more extensive results obtained in recent years are shown by dots, and the predictions of equation (1–20) are shown by the solid line. Note that these results prove not only that equation (1–20) has the correct functional form, but also that the velocity

c, which essentially enters the theory of relativity as the limiting velocity for the transmission of information, actually is equal to the velocity of light, 3.00×10^{10} cm/sec. A number of experimental verifications of length contraction and time dilation are now known. One of the clearest of these involves measuring the lifetime of unstable particles called mesons, for mesons of various velocities. It is observed that the lifetime of a rapidly moving meson is dilated in excellent agreement with equation (1–7). These experiments will be mentioned in Chapter 16. In that chapter we shall also present an overwhelming amount of evidence proving

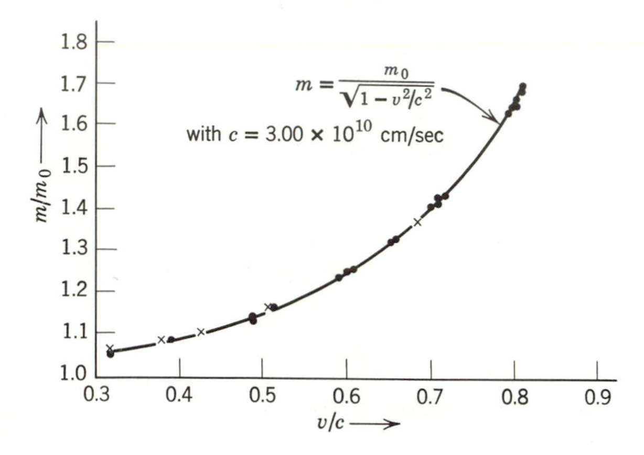

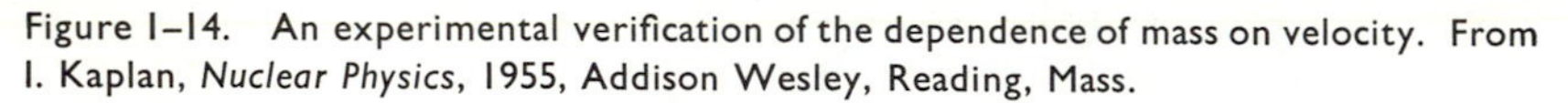

Figure 1–14. An experimental verification of the dependence of mass on velocity. From I. Kaplan, *Nuclear Physics*, 1955, Addison Wesley, Reading, Mass.

the validity of equations (1–23) and (1–24), Einstein's relations between mass and energy. The predictions of the theory of relativity have been confirmed on every point, and there is now universal agreement on its validity.

A subject that might logically be discussed at this point is the so-called *general theory of relativity*, which considers frames of reference that are *accelerating* with respect to each other. It is found that there are very interesting relations between acceleration and gravitational effects. However, the subject is such that any pretense at a serious treatment must be given in terms of tensor calculus, and therefore we shall pass it over. This will not be a handicap because the general theory has had very little interaction with the other branches of modern physics, and also because it is possible, by using various subterfuges, to employ the special theory that we have developed in certain problems involving accelerated motion. For instance, in Chapter 11 we shall make such a calculation for a case of uniform circular motion.

BIBLIOGRAPHY

Bergmann, P. G., *Introduction to the Theory of Relativity*, Prentice-Hall, Englewood Cliffs, N.J., 1942.

Leighton, R. B., *Principles of Modern Physics*, McGraw-Hill Book Co., New York, 1959.

Lorentz, H. A., A. Einstein, H. Minkowski, and H. Weyl, *The Principle of Relativity*, Dover Publications, New York, 1951.

EXERCISES

1. Write equations analogous to (1–2) for the case in which the relative velocity vector **u** is not parallel to any of the axes, and use these to show that equations (1–3) still obtain.

2. Derive equation (1–6′).

3. The distance to the farthest star in our galaxy is of the order of 10^5 light years. Explain why it is possible, in principle, for a human being to travel to this star within his lifetime, and estimate the required velocity.

4. Apply equations (1–13′) to (1–14) and obtain (1–15).

5. In frame O, particle 1 is at rest and particle 2 is moving to the right with velocity v. Now consider a frame O' which, relative to O, is moving to the right with velocity u. Find the value of u such that the two particles appear in O' to be approaching each other with equal and opposite velocities.

6. Verify that the form (1–20) for $m(v)$ satisfies exactly the equation for $m(V)$.

7. Carry through the calculation leading to (1–26).

8. In the *laboratory* (LAB) frame, particle 1 is at rest with total relativistic energy E_1, and particle 2 is moving to the right with total relativistic energy E_2 and momentum p_2. Show that the frame in which the center of the relativistic masses of the system is at rest is moving to the right with velocity

$$u = c\,\frac{cp_2}{E_1 + E_2}$$

relative to the LAB frame, and show that the total momentum of the system is zero in this *center of mass* (CM) frame. Now let the two particles have the same rest mass m_0, and let the total kinetic energy of the system in the LAB frame be T_{LAB}. Evaluate T_{CM}, the total kinetic energy of the system in the CM frame, and show that

$$T_{\text{CM}} \simeq \sqrt{2m_0c^2}\,\sqrt{T_{\text{LAB}}}$$

in the extreme relativistic limit where $T_{\text{LAB}} \gg m_0c^2$. What might be the practical consequences of this in the study of high energy nuclear reactions?

9. Consider a collision in the LAB frame between the moving particle 2 and the stationary particle 1, both of which have the same rest mass m_0. Show that the angle between the velocity vectors of the two particles after the collision is always 90° in classical mechanics, and show how this result is modified by relativistic mechanics. *Hint:* Transform to the CM frame, treat the collision, then transform back to the LAB frame.

10. An electron (charge -4.8×10^{-10} esu, mass 9.1×10^{-28} gm), initially held at rest at one plate of a plane parallel condenser, is released and allowed to fall through vacuum to the other plate under the influence of the electric field in the condenser. The separation between the plates is 10^2 cm, and the voltage across the plates is 10^7 volts ($10^7/300$ stat-volts). Calculate the time required, in the frame of reference of the condenser, for the electron to travel between its plates. Discuss the justification of using the theory developed in this chapter for the present system, which involves accelerations. *Hint:* The magnitude of the electric charge, and the magnitude of the component of the electric field parallel to the direction of motion, are both unchanged by a Lorentz transformation.

CHAPTER 2

Thermal Radiation and the Origin of Quantum Theory

1. Introduction

In the course of a successful attempt at resolving certain discrepancies between the observed energy spectrum of thermal radiation and the predictions of the classical theory, Planck was led to the idea that a system executing simple harmonic oscillations only can have energies which are integral multiples of a certain finite amount of energy (1901). A closely related idea was later applied by Einstein in explaining the photoelectric effect (1905), and by Bohr in a theory which predicted with great accuracy many of the complex features of atomic spectra (1913). The work of these three physicists, plus subsequent developments by de Broglie, Schroedinger, and Heisenberg (ca. 1925), constitutes what is known as the *quantum theory*. This theory and the theory of relativity together comprise the two most significant features of modern physics. In this chapter we begin our study of quantum theory with a discussion of thermal radiation.

The nature of the subject is such that any discussion of thermal radiation must be based upon the results of advanced treatments of several different fields of classical physics: thermodynamics, statistical mechanics, and electromagnetic theory. In many cases we shall simply quote these results and justify them with qualitative arguments since their proofs are most appropriately left to the texts which treat these subjects. However, we shall reproduce in detail that step in which Planck introduced the *quantization* of energy, and we shall be able to see exactly how its introduction resolved a serious conflict between experiment and the classical theory.

2. Electromagnetic Radiation by Accelerated Charges

The surface of every body at a temperature greater than absolute zero emits energy, in the form of thermal radiation, due to the motion of electric charges near the surface. This radiation consists of electromagnetic waves (often called infrared waves) of exactly the same nature as visible light, but of longer wavelength. As we shall be concerned on several occasions in this book with the emission of electromagnetic radiation by a moving charge, a description of the classical theory of the process is appropriate at this point. The description will be qualitative, but some of the quantitative results of the theory will be presented.

In elementary electrostatics it is shown that surrounding every stationary charge q is an electric field specified by the vector **E**. These are related by Coulomb's law

$$\mathbf{E} = \frac{q}{r^2}\left(\frac{\mathbf{r}}{r}\right) \tag{2–1}$$

where **r** is the vector from the charge to the point at which **E** is evaluated, and where $(\mathbf{r}/r)$ is a unit vector in that direction.† It is often convenient to represent the electric field in terms of *lines of force* which are constructed to be everywhere parallel to the local direction of **E** and of density (number of lines crossing a 1 cm^2 area normal to the direction of **E**) equal to the local value of E. A two dimensional representation of the lines of force surrounding a stationary charge is shown in figure (2–1). This static (i.e., time independent) electric field contains stored energy. In fact, ρ, the energy stored per unit volume, is related to the local value of E by the equation

$$\rho = E^2/8\pi \tag{2–2}$$

if the field is in a medium of unity permeability such as the vacuum. This can be verified with little difficulty by considering the case of a parallel plate condenser in which E is everywhere constant, and equating the energy required to charge the condenser to the energy stored in the electric field.

The energy stored in the field of a stationary charge is static and is not radiated away in the form of electromagnetic radiation; if this were not so, the conservation of energy would be clearly violated. It is also true that the energy stored in the field of a charge moving with uniform velocity is not radiated away but moves along with the charge, even though there

† With a few exceptions that are explicitly indicated, in this book we shall write all electromagnetic equations in the cgs-Gaussian system of units. In this system charges are expressed in esu in *all* equations.

are associated with the charge a non-static electric field **E** and also, according to Ampere's law, a non-static magnetic field **H**. It is easy to see that this must be the case by mentally transforming to a frame of reference in which the charge is stationary, realizing that it cannot radiate in that frame, and then applying the relativistic requirement that the behavior of the charge—including whether or not it radiates—cannot depend upon the frame of reference from which it is viewed. The total energy stored in the field of a uniformly moving charge is larger than if it were stationary because, when there is also a magnetic field **H**, electromagnetic theory shows that the energy density is

$$\rho = \frac{1}{8\pi}(E^2 + H^2) \tag{2–2'}$$

for unity permeability. The excess energy is supplied from the work done by the forces initially producing the motion of the charge.

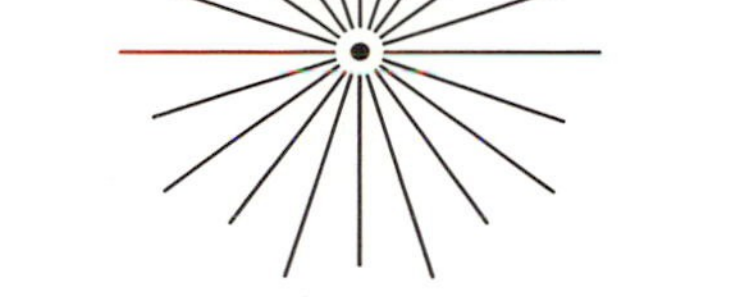

Figure 2–1. The lines of force surrounding a stationary charge.

For a charge undergoing acceleration, the non-static electric and magnetic fields cannot adjust themselves in such a way that none of the stored energy is radiated, as they can in the case of a charge moving with uniform velocity. We can see this qualitatively by considering the behavior of the electric field. In figure (2–2) we describe this field by drawing some of the lines of force surrounding a charge which was at rest at the initial instant t, suffered a constant acceleration a to the right during the interval t to t', and then continued moving with constant final velocity. The figure shows the lines of force at some later instant t'', as viewed from the frame of reference moving with the final velocity, and with the restriction that all times are measured in that frame. At small distances the lines of force are directed radially outward from the present position of the charge. At large distances they are directed radially outward from the initial position of the charge. The reason is simply that information concerning the position of the charge cannot be transmitted to distant locations with infinite velocity, but only with the velocity c. As a consequence, there are breaks in the lines of force contained between a sphere centered on the initial position and of radius $c(t'' - t)$, which is the minimum distance at which the field can "know" the acceleration started, and a sphere centered on the present position and of radius $c(t'' - t')$, which is the minimum distance at which the field can "know" that the acceleration stopped. As t'' increases, the region containing the breaks expands outward with velocity c. The electric field in this region has components which are both longitudinal and transverse to the direction of expansion, but by constructing diagrams for several values of t'' it is easy to see that the longitudinal component dies out very rapidly, and can soon be ignored,

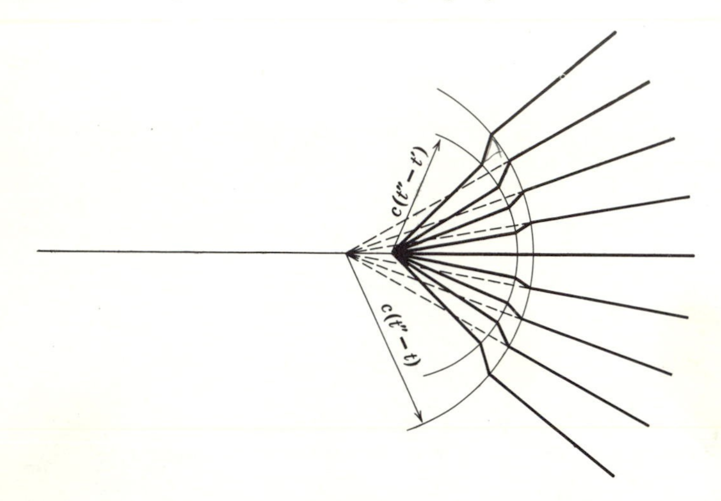

Figure 2–2. The lines of force surrounding an accelerated charge. Only some of the lines are shown.

whereas the transverse component dies out slowly. In fact, electromagnetic theory shows, by calculations based upon the same idea as our qualitative discussion, that at large distances from the region of the acceleration (large t'') the transverse electric field obeys the equation

$$E_{\perp} = \frac{qa \sin \theta}{c^2 r} \tag{2–3}$$

for unity permeability, where $r = c(t'' - t)$ is the magnitude of the vector **r** from the region at which the acceleration **a** took place to the point at which the transverse field is evaluated, and where θ is the angle between **r** and **a**. The dependence of $E_{\perp}$ on θ and r can be seen in our diagrams, and it is clear from our discussion that $E_{\perp}$ must be proportional to q and a.

The electromagnetic theory also shows that because of the acceleration there will be a transverse magnetic field $H_\perp$ moving along with $E_\perp$, and that at large distances from the region of acceleration

$$H_\perp = \frac{qa \sin \theta}{c^2 r} \tag{2–3′}$$

for unity permeability, where $r = c(t'' - t)$. These two transverse fields propagating outward with velocity c form the electromagnetic radiation emitted by the accelerated charge. For this, or any other, electromagnetic radiation propagating in a medium of unity permeability,

$$E_\perp = H_\perp$$

and the energy density of the radiation (2–2′) is

$$\rho = \frac{1}{8\pi}(E_\perp^2 + E_\perp^2) = \frac{E_\perp^2}{4\pi} \tag{2–4}$$

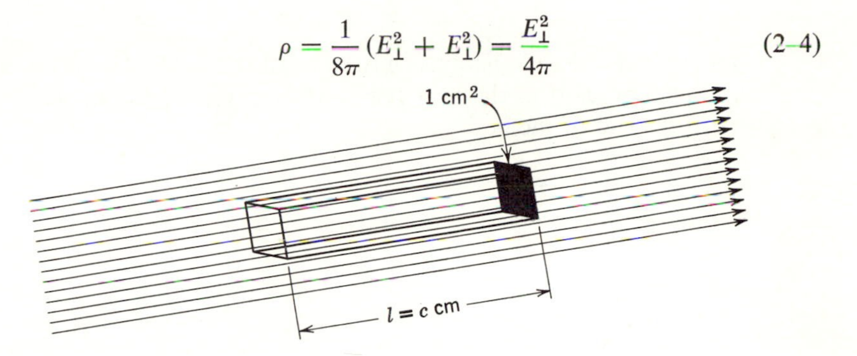

Figure 2–3. Illustrating the energy transport in a parallel stream of radiation.

However, equations (2–3) and (2–3′) are valid only if the dimensions of the region in which the acceleration took place are small compared to c times the period of its duration, as we have assumed.

The intensity I of the emitted radiation is defined as the energy which it carries per second across a 1 cm² area normal to the direction of propagation. Consider figure (2–3), which shows the radiation propagating in some direction at such a large distance from the region of acceleration that it forms an essentially parallel stream. The radiation is moving with velocity c cm/sec. In one second it moves a distance c cm. Consequently all the energy contained in the parallelepiped with a 1 cm² base normal to the direction of propagation, and with length $l = c$ cm, will be carried across the base in 1 sec. The energy contained in the parallelepiped is equal to the energy density ρ times the volume c cm³. Thus the intensity I is

$$I = \rho c \tag{2–5}$$

Using equations (2–3) and (2–4), we have

$$I(\theta) = \frac{E_\perp^2 c}{4\pi} = \frac{q^2 a^2 \sin^2 \theta}{4\pi c^3 r^2} \tag{2–6}$$

where we indicate explicitly that the intensity is a function of the direction of emission. Note that $I(\theta)$ obeys the familiar inverse square law. The total energy radiated in all directions per second, R, can be evaluated by integrating $I(\theta)$ over the area of a sphere of arbitrary radius r. That is,

$$R = \int I(\theta)\, dA = \int_0^\pi I(\theta) 2\pi r^2 \sin\theta\, d\theta$$

which yields

$$R = \frac{2}{3}\frac{q^2 a^2}{c^3} \tag{2–7}$$

for the rate of radiation of energy. Only some of the qualitative ideas of this section will be used in the next few chapters; the equations will be used toward the end of the book.

3. Emission and Absorption of Radiation by Surfaces

Let us return now to the phenomenon of thermal radiation. As we have said, every body at a temperature greater than absolute zero emits this radiation. In the classical theory, this can be pictured as the result of the accelerations of electric charges near the surface due to thermal agitation. Now, in a single acceleration lasting for a certain period of time, most of the emitted radiation has a frequency approximately equal to the reciprocal of the period, and a wavelength equal to c times the period. But, in the many different acceleration processes which give rise to the thermal radiation, a whole spectrum of wavelengths is emitted. On the basis of this picture it would be expected that the rate of emission of energy, integrated over the entire wavelength spectrum, would increase as the temperature of the surface increases because of the enhanced thermal agitation, and it would also be expected that the rate of emission of energy would be proportional to the area of the surface. This is found to be the case; an empirical equation due to Stefan (1879) states

$$I_T = \sigma e T^4 \tag{2–8}$$

The quantity I_T is the total energy emitted in all frequencies per second per cm² from a surface at absolute temperature T; e is a constant called the *emissivity* which ranges from 0 to 1 depending on the nature of the emitting surface; and $\sigma = 0.567 \times 10^{-4}$ erg-cm^{-2}-deg^{-4}-sec^{-1} is the Stefan-Boltzmann constant. The energy emitted as thermal radiation is supplied by the thermal agitation energy.

We shall also be concerned with the absorption of thermal radiation by a surface. In this process, energy is removed from incident thermal radiation, through its action on electric charges, and ends up as thermal agitation energy. There is an interesting relation between the efficiency of a surface as an emitter of thermal radiation, as measured by e, and its efficiency as an absorber. The latter property is measured by a constant called the *absorptivity* a that is defined as the ratio of the total thermal energy absorbed by the surface to the total thermal energy incident upon it. The relation between e and a, which is a theorem due to Kirchhoff (1895), is simply

$$e = a \tag{2–9}$$

This was proved in a thermodynamic argument, independent of any detailed assumptions about the emission and absorption processes, by considering the thermal equilibrium of surfaces of different nature which are exchanging thermal energy by emission and absorption. It has also been verified experimentally.

4. Black Body Radiation

Consider a body with a surface that has the property of absorbing all the radiation incident upon it, i.e., $a = 1$. Such a body is called a *black body*. A number of objects found in nature are quite good approximations to a black body. One example would be any object coated with a diffuse layer of black pigment; another example will be given shortly. Since the absorptivity of a black body is equal to unity, it follows from Kirchhoff's law that its emissivity is also. A black body, which is the most efficient absorber, is also the most efficient radiator. Applying the condition $e = 1$ to equation (2–8), we see that the radiant power emitted per cm^2 is the same for all black bodies which are at the same temperature. This strongly suggests that the other properties of the thermal radiation emitted by black bodies, such as the spectral distribution of the radiation, also depend only on the temperature of the black body but not on its detailed nature. It can be proved by thermodynamic arguments, involving two black bodies in equilibrium, that this is indeed the case. It is evident that the universal properties of the thermal radiation emitted by black bodies makes them of particular theoretical interest.

The spectral distribution of black body radiation is specified by the quantity $I_T(\lambda)$, which is so defined that $I_T(\lambda)\,d\lambda$ is equal to the energy emitted per second, in radiation of wavelength in the interval λ to $\lambda + d\lambda$, from 1 cm^2 of a surface at temperature T. The names Lummer and Pringsheim are associated with the earliest accurate measurements of this quantity (1899). The measurements were made with an instrument essentially similar to the prism spectrometers used in measuring optical spectra, except that special materials had to be used in order to have the lenses, prisms, etc., transparent to the long wavelength thermal radiation.

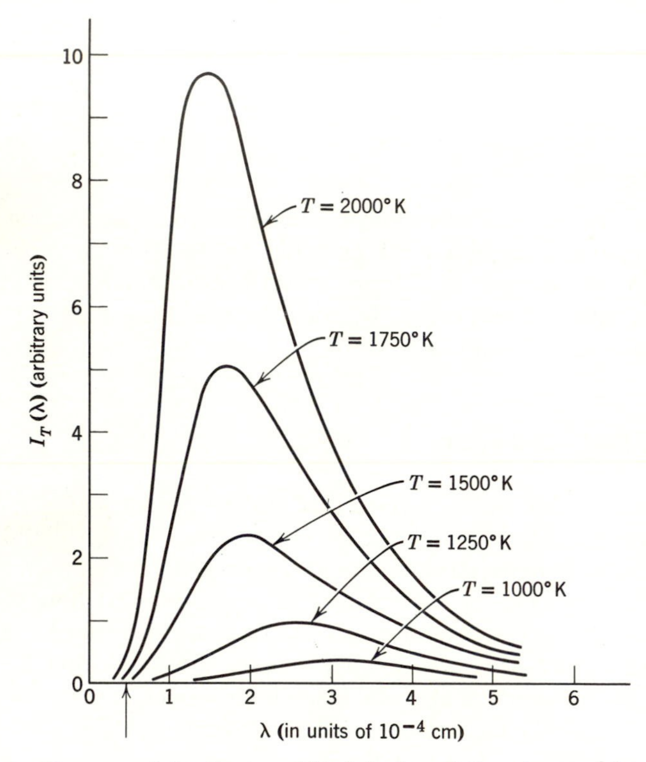

Figure 2–4. The spectral distribution of black body radiation at several temperatures.

The observed dependence of $I_T(\lambda)$ on λ and T is indicated in figure (2–4). The arrow on the abscissa indicates the wavelength at which the eye has its maximum response (green light).

We see that, for any fixed λ, $I_T(\lambda)$ increases with increasing T. The integral of $I_T(\lambda)$ over all λ is, of course, just equal to the quantity I_T previously defined. This integral, which is equal to the area under the curves, does increase with the fourth power of T, in agreement with equation (2–8). Figure (2–4) also shows that the spectrum shifts toward shorter wavelengths as T increases. A quantitative inspection of the figure will demonstrate the validity of the equation

$$\lambda_{\max} \propto 1/T \tag{2–10}$$

where $\lambda_{\max}$ is the λ at which $I_T(\lambda)$ has its maximum value for a particular T. All these results are in agreement with the everyday experience that bodies emit more heat as their temperature increases, and that with increasing temperature their "color" shifts from dull red to blue white (i.e., increasingly more radiant energy is emitted in the region of short wavelengths).

Now consider an object containing a cavity which is connected to the outside by a small hole. Radiation incident upon the hole from the outside enters the cavity and is reflected back and forth by the walls of the cavity, eventually being absorbed on these walls. If the area of the hole is very small compared with the area of the inner surface of the cavity, a negligible amount of the incident radiation will be reflected back to the hole. Then all the radiation incident upon the hole is absorbed. For the hole, $a = 1$ and therefore the hole must have the properties of the surface of a black body.

Next assume that the walls of the cavity are uniformly heated to a temperature T. The walls will emit thermal radiation which will fill the cavity. The small fraction of this radiation incident from the inside upon the hole will pass through the hole. Thus the hole will act as an emitter of thermal radiation. Since the hole must have the properties of the surface of a black body, the radiation emitted by the hole must have a black body spectrum. But, as the hole is merely sampling the thermal radiation present inside the cavity, it is clear that the radiation in the cavity must also have a black body spectrum. In fact, it will have a black body spectrum characteristic of the temperature T of the walls since this is the only temperature defined for the system. The spectrum emitted by the hole in the cavity is specified in terms of an energy flux $I_T(\lambda)$, but it is convenient to specify the spectrum of the radiation inside the cavity in terms of an energy density $\rho_T(\lambda)$ which is defined such that $\rho_T(\lambda)\, d\lambda$ is the energy contained in 1 cm^3 of the cavity in the wavelength interval λ to $\lambda + d\lambda$. It is apparent from the discussion above that $\rho_T(\lambda)$ is proportional to $I_T(\lambda)$, with a proportionality constant that does not depend on λ or T. This statement, as well as all the other statements of the last two paragraphs, can, of course, be proved rigorously.

5. Wien's Law

Many of the properties of black body radiation can be understood from the classical theory of thermodynamics. In 1884 Boltzmann produced a theoretical derivation of the equation giving I_T, the total energy emitted per cm^2 per second, for a black body radiator (equation 2–8 for the case $e = 1$). In this derivation he considered a cavity with reflecting walls in the form of a cylinder with a movable piston and filled with thermal radiation at temperature T. Thermal radiation, in common with all other electromagnetic radiation, can be shown to exert a pressure proportional to its energy density.† Taking this system through a cycle of expansion and

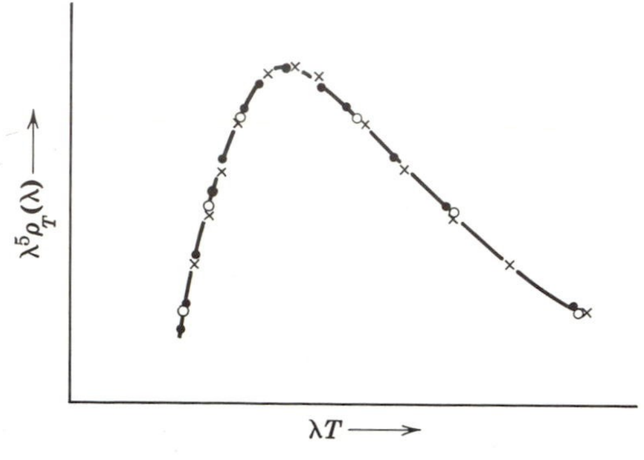

Figure 2–5. An experimental verification of Wien's law. ●, $T = 1646°$K; ×, $T = 1449°$K; ○, $T = 1259°$K. From F. K. Richtmyer, E. H. Kennard, and T. Lauritsen, *Introduction to Modern Physics*, 5th ed., McGraw-Hill Book Co., New York, 1955.

compression, called in thermodynamics a Carnot cycle, Boltzmann obtained a relation between the work done by the pressure of the radiation, and its temperature. This relation leads to the desired result since the pressure can be expressed in terms of the energy density, which can in turn be expressed in terms of I_T.

In the process of expansion or compression of such a cavity filled with radiation, the wavelength of any spectral component of the radiation will be changed as the result of a Doppler shift upon reflection from the moving piston. From a detailed consideration of this fact, Wien (1893) was able to derive a general functional form for the spectral distribution of black body radiation known as *Wien's law*:

$$\rho_T(\lambda) = \frac{f(\lambda T)}{\lambda^5} \tag{2–11}$$

where $f(\lambda T)$ is some function of the product of the wavelength and the temperature, whose form is not specified by Wien's derivation. This equation is in excellent agreement with the experimental data, as can be seen from figure (2–5). This figure plots λ^5 times $\rho_T(\lambda)$ as a function of

† In this respect it behaves just like a gas.

λT, using 1259°, 1449°, and 1646° data of Lummer and Pringsheim for $I_T(\lambda)$ [which is proportional to $\rho_T(\lambda)$]. We see that all sets of data fall on the same smooth curve, confirming the prediction of equation (2–11) that $\lambda^5\rho_T(\lambda)$ is equal to a universal function of the variable λT. It is apparent that Wien's law agrees with the empirical fact expressed by equation (2–10).

6. The Rayleigh-Jeans Theory

Wien's thermodynamical derivation performed the useful role of allowing all black body spectra measured at different temperatures to be discussed theoretically in terms of the single function $f(\lambda T)$. But the derivation did not evaluate the form of this function. It is typical of a thermodynamical argument to show that certain relations must obtain between the variables which describe some physical system, but not to provide a complete theory of the behavior of the system. The reason is that such arguments are based on general principles that apply to all physical systems, but do not involve the details of the composition of the particular system in question. A theory which would be able to evaluate the function $f(\lambda T)$ must take into account some of the detailed properties of a black body.

At the turn of the present century several attempts were made to formulate such a theory. In one of them the behavior of accelerating electric charges in the walls of a black body cavity was considered. These charges were assumed to be the source of the black body radiation. It was also assumed that each charge undergoes simple harmonic oscillation, of some particular frequency, about its equilibrium position. According to electromagnetic theory, each charge then radiates at a frequency precisely equal to its particular oscillation frequency. In thermal equilibrium the energy of a particular frequency, or wavelength, component of the black body radiation must be proportional to the average energy of the corresponding charged oscillator because the radiation and the oscillator are constantly exchanging energy. From detailed considerations of the absorption and emission properties of a charged oscillator, the proportionality constant was evaluated. It contains the square of the frequency. Next, certain results of the theory of statistical mechanics (to be described later) were used to evaluate the average energy of the charged oscillators in terms of the temperature of the walls in which they are located. From these results the spectral distribution of the black body radiation, and therefore the function $f(\lambda T)$, is immediately obtained. We shall not pursue this theory further. Instead, we shall turn to the theory of Rayleigh and Jeans which leads to the same results, and which has a sounder basis because it does not require such a detailed picture of the origin of the black body radiation. In fact, it has been shown that the form of $f(\lambda T)$ obtained by Rayleigh and Jeans is a necessary consequence of the theories of classical physics.

Consider a cavity with metallic walls at temperature T. The walls emit thermal electromagnetic radiation. In thermal equilibrium this radiation has a black body spectrum characteristic of the temperature T. Furthermore, as we shall show, in the steady state attained at equilibrium the electromagnetic radiation inside the cavity must exist in the form of standing waves with nodes at the metallic surfaces. Rayleigh and Jeans calculated the number of standing waves with nodes at the surfaces of the cavity in the wavelength interval λ to $\lambda + d\lambda$. They then calculated the average energy of these waves. The energy depends on the temperature T and is evaluated from the theory of statistical mechanics. The number of standing waves in the wavelength interval times the average energy of the waves divided by the volume of the cavity is equal to the average energy per cm^3 in the wavelength interval λ to $\lambda + d\lambda$, which is just $\rho_T(\lambda)$. From this $f(\lambda T)$ is evaluated by using equation (2–11).

Let us assume that the metallic walled cavity filled with electromagnetic radiation is in the form of a perfect cube of edge length a, as shown in figure (2–6). Then the radiation reflecting back and forth between the walls can be analyzed into three components along the three mutually perpendicular directions defined by the edges of the cavity. As the opposing walls are exactly parallel to each other, the three components of the radiation do not mix, and we may treat them separately. Consider the x component and the metallic wall at $x = 0$. All the radiation of this component which is incident upon the wall is reflected by it, and the incident and reflected waves combine to form a standing wave. Now, since electromagnetic radiation is a transverse vibration with the electric field vector **E** perpendicular to the propagation direction, and since the propagation direction for this component is perpendicular to the wall in question, its **E** is parallel to the wall. But a metallic wall cannot support an electric field parallel to its surface, as currents can always flow in such a way as to neutralize the electric field. Therefore **E** for this component must always vanish at the wall—that is, the standing wave associated with the x component of the radiation must have a node at $x = 0$. The standing wave must also have a node at $x = a$ because there can be no parallel electric field in the corresponding wall. Furthermore, similar conditions apply to the other two components; the standing wave associated with the y component must have nodes at $y = 0$ and $y = a$, and the standing wave associated with the z component must have nodes at $z = 0$ and $z = a$.

These conditions put a limitation on the possible wavelengths of the electromagnetic radiation contained in the cavity.

To determine this limitation, consider radiation of wavelength λ and frequency $\nu = c/\lambda$, propagating in the direction defined by the three angles α, β, γ, as shown in figure (2–7). The radiation must be a standing wave since all three of its components are standing waves. We have indicated the locations of some of the fixed nodes of this standing wave by a set of planes perpendicular to the propagation direction α, β, γ. The

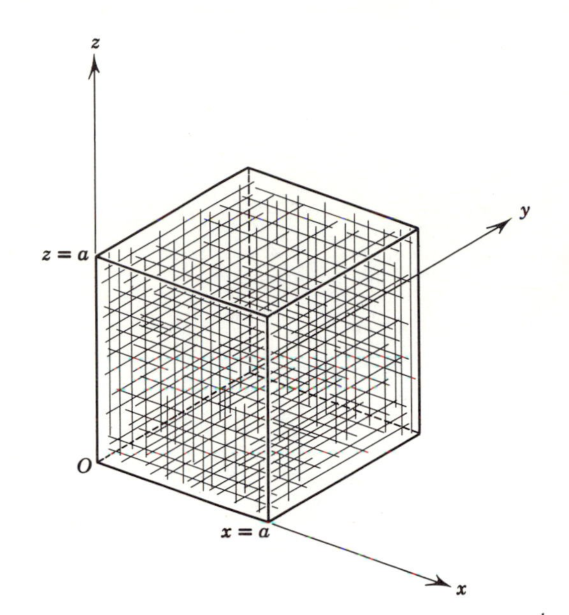

Figure 2–6. A cubical cavity containing electromagnetic radiation.

distance between these nodal planes of the radiation is just $\lambda/2$, where λ is its wavelength. We have also indicated the locations at the three axes of the nodes of the three components. The distances between these nodes are

$$\frac{\lambda_x}{2} = \frac{\lambda}{2\cos\alpha}$$

$$\frac{\lambda_y}{2} = \frac{\lambda}{2\cos\beta} \qquad (2\text{–}12)$$

$$\frac{\lambda_z}{2} = \frac{\lambda}{2\cos\gamma}$$

Let us write expressions for the magnitudes at the three axes of the electric fields of the three components. They are

$$
\begin{aligned}
E(x, t) &= A \sin(2\pi x/\lambda_x) \sin(2\pi \nu t) \\
E(y, t) &= B \sin(2\pi y/\lambda_y) \sin(2\pi \nu t) \\
E(z, t) &= C \sin(2\pi z/\lambda_z) \sin(2\pi \nu t)
\end{aligned}
\tag{2–13}
$$

The expression for the x component represents a wave with a maximum amplitude A, with a spatial variation $\sin(2\pi x/\lambda_x)$, and which is oscillating

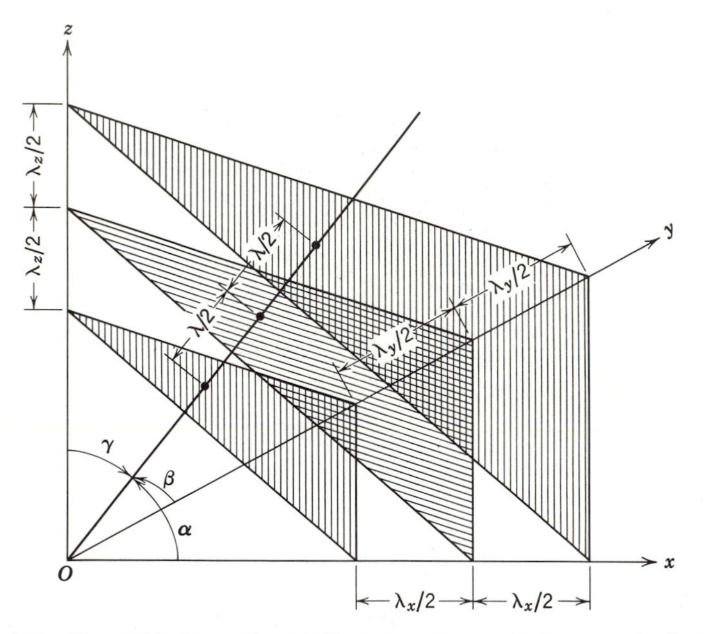

Figure 2–7. The nodal planes of a standing wave propagating in a certain direction in a cubical cavity.

with frequency ν. As $\sin(2\pi x/\lambda_x)$ vanishes for $2x/\lambda_x = 0, 1, 2, 3, \ldots$, the wave is a standing wave of wavelength λ_x because it has fixed nodes separated by the distance $\Delta x = \lambda_x/2$. The expressions for the y and z components represent standing waves of maximum amplitudes B and C and wavelengths λ_y and λ_z, but all three component standing waves oscillate with the frequency ν of the radiation. Note that these expressions automatically satisfy the requirement that the x component have a node at $x = 0$, the y component have a node at $y = 0$, and the z component have a node at $z = 0$. To make them also satisfy the requirement that the x component have a node at $x = a$, the y component have a node at $y = a$, and the z component have a node at $z = a$, set

$$
\begin{aligned}
2x/\lambda_x &= n_x, \qquad \text{for } x = a \\
2y/\lambda_y &= n_y, \qquad \text{for } y = a \\
2z/\lambda_z &= n_z, \qquad \text{for } z = a
\end{aligned}
\tag{2–14}
$$

where $n_x = 0, 1, 2, 3, \ldots$; $n_y = 0, 1, 2, 3, \ldots$; $n_z = 0, 1, 2, 3, \ldots$. Using equations (2–12) these conditions become

$$\frac{2a}{\lambda}\cos\alpha = n_x, \quad \frac{2a}{\lambda}\cos\beta = n_y, \quad \frac{2a}{\lambda}\cos\gamma = n_z$$

Squaring both sides of these equations and adding, we obtain

$$(2a/\lambda)^2(\cos^2\alpha + \cos^2\beta + \cos^2\gamma) = n_x^2 + n_y^2 + n_z^2$$

But the angles α, β, γ have the property

$$\cos^2\alpha + \cos^2\beta + \cos^2\gamma = 1$$

Thus

$$2a/\lambda = \sqrt{n_x^2 + n_y^2 + n_z^2}$$

where n_x, n_y, n_z take on all possible integral values. This equation describes the limitation on the possible wavelengths of the electromagnetic radiation contained in the cavity.

It is convenient to continue the discussion in terms of the allowed frequencies instead of the allowed wavelengths. They are

$$\nu = \frac{c}{\lambda} = \frac{c}{2a}\sqrt{n_x^2 + n_y^2 + n_z^2} \tag{2–15}$$

Now we shall count the number of allowed frequencies in a given frequency interval by constructing a uniform cubic lattice in one octant of a rectangular coordinate system in such a way that the three coordinates of each point of the lattice are equal to a possible set of the three integers n_x, n_y, n_z. See figure (2–8). By construction, each lattice point corresponds to an allowed frequency. Furthermore, $N(\nu)\,d\nu$, the number of allowed frequencies between ν and $\nu + d\nu$, is equal to $N(r)\,dr$, the number of points contained between concentric shells of radii r and $r + dr$, where

$$r = \sqrt{n_x^2 + n_y^2 + n_z^2}$$

From equation (2–15) this is

$$r = \frac{2a}{c}\nu \tag{2–16}$$

Since $N(r)\,dr$ is equal to the volume enclosed by the shells times the density of lattice points, and since, by construction, the density is unity, $N(r)\,dr$ is simply

$$N(r)\,dr = \frac{1}{8}\,4\pi r^2\,dr = \frac{\pi r^2\,dr}{2}$$

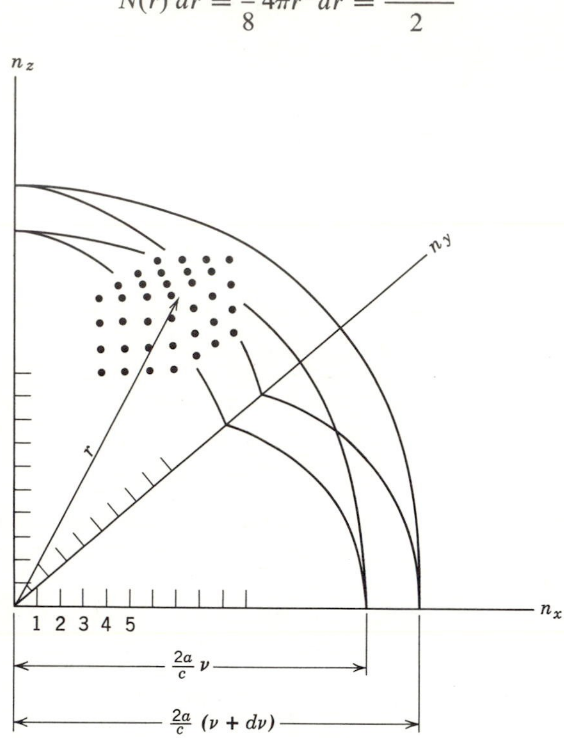

Figure 2–8. The rectangular coordinate system used in counting the number of allowed frequencies in a cubical cavity. Only a few of the points are shown.

Setting this equal to $N(\nu)\,d\nu$, and evaluating $r^2\,dr$ from equation (2–16), we have

$$N(\nu)\,d\nu = \frac{\pi}{2}\left(\frac{2a}{c}\right)^3 \nu^2\,d\nu$$

This completes the calculation except that we must multiply these results by a factor of 2 because, for each of the allowed frequencies we have enumerated, there are actually two independent waves corresponding to the two possible states of polarization of electromagnetic radiation.† Thus

† The fact that electromagnetic radiation in the visible range (light) can have either of two independent circular polarizations is discussed in almost any elementary physics text.

the number of allowed frequencies in the frequency interval ν to $\nu + d\nu$ is

$$N(\nu)\, d\nu = \frac{8\pi a^3 \nu^2\, d\nu}{c^3} \tag{2–17}$$

It can be shown that $N(\nu)\, d\nu$ is independent of the assumed *shape* of the cavity and depends only on its volume, a^3.

7. The Boltzmann Probability Distribution

The next step in the Rayleigh-Jeans calculation is the evaluation of the average energy contained in each standing wave of frequency ν. According to classical physics, the *particular energy* of some wave can have *any value* from zero to infinity; the actual value is proportional to the square of its average amplitude (cf. equation 2–4). But, if we have a system containing a large number of physical entities of the same kind which are in thermal equilibrium with each other at a temperature T, such as the system of standing waves in equilibrium in the black body cavity, the classical theory of statistical mechanics demands that the energies of these entities be distributed according to a definite probability distribution whose form is specified by T. As the *average energy* is determined by the probability distribution, it must have a *definite value* that depends on T.

It is not difficult to see the essential ideas underlying these results of statistical mechanics, if a sufficiently simple system is considered. Imagine a system consisting of entities of the same kind which contain energy. An example would be a set of identical coil springs which are vibrating. Assume that the system is isolated so that its total energy content is constant, and assume also that the entities can exchange energy with each other through some mechanism so that the system is in thermal equilibrium. Finally, assume that there are only four of these entities in the system, that the total energy of any entity is restricted to the values $\varepsilon = 0$, $\Delta\varepsilon$, $2\Delta\varepsilon$, $3\Delta\varepsilon$, $4\Delta\varepsilon$, $5\Delta\varepsilon, \ldots$, and that the total energy of the system (which now must be some integral multiple of $\Delta\varepsilon$) has the value $3\Delta\varepsilon$. The last assumptions are made strictly for the purpose of simplifying the calculation we are about to perform. After the calculation we shall imagine letting $\Delta\varepsilon$ approach zero so that ε may have any value, but at the same time keeping the total energy of the system constant. We shall also imagine letting the number of entities in the system become large.

Since the entities can exchange energy with each other, all possible divisions of the total energy $3\Delta\varepsilon$ between the four entities can occur. The only possible divisions are indicated in the diagrams of figure (2–9). Consider the diagram labeled $i = 1$. This corresponds to divisions of total energy in which $\varepsilon = 0$ for three entities and $\varepsilon = 3\Delta\varepsilon$ for the fourth. But there are actually four such divisions, since any one of the four entities can be the one in the energy state $\varepsilon = 3\Delta\varepsilon$. This is indicated in the column marked "number of duplicate divisions." The same is true for the divisions of the type $i = 3$. The reader should verify for himself that there are twelve duplicate divisions of the type $i = 2$. In evaluating this number, any rearrangement of entities between different energy states is to be

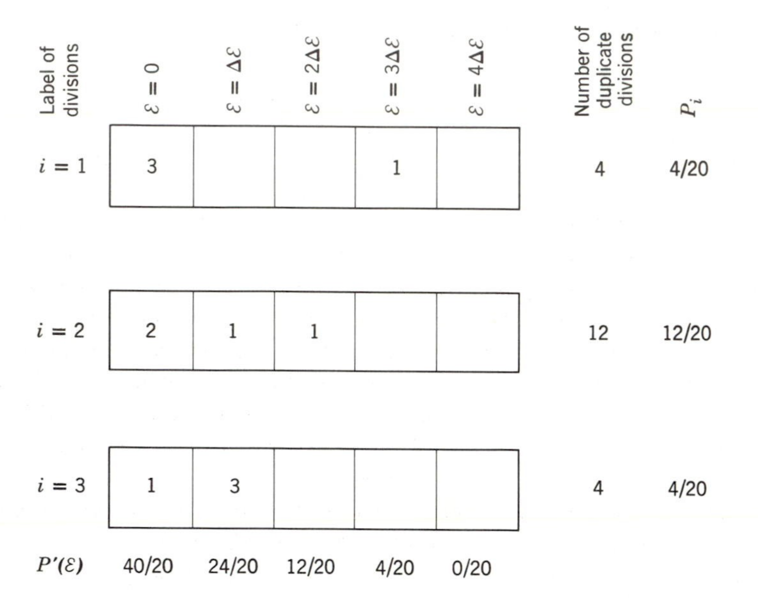

Figure 2–9. Illustrating a simple calculation leading to the Boltzmann probability distribution.

counted as a new division since the entities could presumably be distinguished experimentally when in different energy states even though they are of the same kind, but any rearrangement of entities in the same energy state is not counted because entities in the same energy state could not be distinguished.† We now make the final *assumption*: *All possible divisions of the total energy occur with the same probability.* Then the probability that the divisions of a given type will occur is proportional to the number of duplicate divisions of that type, and the relative probability P_i is just equal to that number divided by the total number of duplicate divisions. The P_i are evaluated in the column marked "P_i." Next let us calculate

† *Hint:* The total number of rearrangements (permutations) of the four entities is $4! \equiv 4 \times 3 \times 2 \times 1$. But rearrangements of the two entities within the same energy state do not count, so this number must be divided by 2!

$P'(\varepsilon)$, the probability of finding an entity in the energy state ε. Consider $\varepsilon = 0$. For divisions of the type $i = 1$ there are three entities in this state, and the relative probability P_i that these divisions occur is 4/20; for $i = 2$ there are two entities, and P_i is 12/20; for $i = 3$ there is one entity, and P_i is 4/20. Thus $P'(0)$, the probability of finding an entity in this state, is $3 \times 4/20 + 2 \times 12/20 + 1 \times 4/20 = 40/20$. The values of $P'(\varepsilon)$ calculated in the same way for the other values of ε are given in the row marked

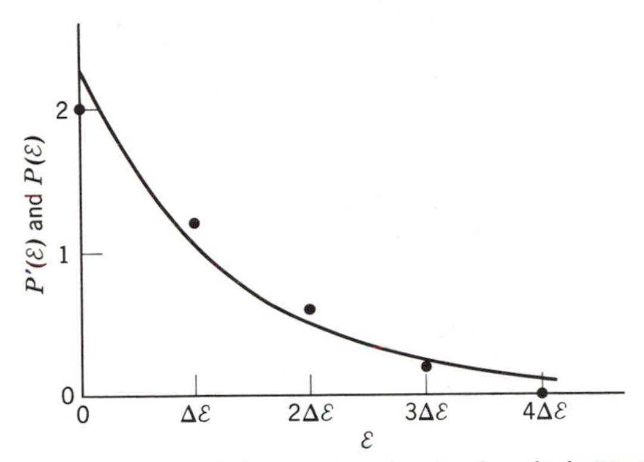

Figure 2–10. A comparison of the results of a simple calculation and the Boltzmann probability distribution.

"$P'(\varepsilon)$." They are also plotted as points in figure (2–10). The solid line is the exponential function

$$P(\varepsilon) = Ae^{-\varepsilon/\varepsilon_0} \tag{2–18}$$

where A and ε_0 are constants. (In the figure the constants have been adjusted to give the best fit to the points representing the results of our calculation.)

Now we must imagine letting $\Delta\varepsilon$ go to zero—and at the same time changing the condition on the total energy to read that it is equal to $N\,\Delta\varepsilon$, with N increasing as rapidly as $\Delta\varepsilon$ decreases, so that the total energy remains constant. The result of this step is that the calculated function $P'(\varepsilon)$ becomes defined for values of ε which are closer and closer together. In the limit, the energy ε becomes a continuous variable as classical physics demands it must, and the probability distribution $P'(\varepsilon)$ becomes a continuous function. Finally, when the number of entities in the system is allowed to become large, this function is found to be identical with the $P(\varepsilon)$ of equation (2–18).

The quantity $P(\varepsilon)$ specifies the probability of finding an entity in one of its energy states between ε and $\varepsilon + d\varepsilon$ for a system containing a large number of entities of the same kind in thermal equilibrium with each other. The probability is simply $P(\varepsilon)\,d\varepsilon$. It also specifies the average energy of one of the entities, which is

$$\bar{\varepsilon} = \frac{\int_0^\infty \varepsilon P(\varepsilon)\,d\varepsilon}{\int_0^\infty P(\varepsilon)\,d\varepsilon} \tag{2-19}$$

The integrand in the numerator is the energy ε weighted by the probability that an entity will have this energy.† The denominator is the probability of finding an entity with any energy; it is just the total number of entities in the system. Evaluating $P(\varepsilon)$ from equation (2–18),

$$P(\varepsilon) = Ae^{-\varepsilon/\varepsilon_0} = Ae^{-\alpha\varepsilon}, \qquad \alpha \equiv 1/\varepsilon_0$$

we find

$$\bar{\varepsilon} = \frac{\int_0^\infty A\varepsilon e^{-\alpha\varepsilon}\,d\varepsilon}{\int_0^\infty Ae^{-\alpha\varepsilon}\,d\varepsilon} = \frac{\int_0^\infty \varepsilon e^{-\alpha\varepsilon}\,d\varepsilon}{\int_0^\infty e^{-\alpha\varepsilon}\,d\varepsilon}$$

Now

$$-\frac{d}{d\alpha}\ln\int_0^\infty e^{-\alpha\varepsilon}\,d\varepsilon = \frac{-\int_0^\infty \frac{d}{d\alpha} e^{-\alpha\varepsilon}\,d\varepsilon}{\int_0^\infty e^{-\alpha\varepsilon}\,d\varepsilon} = \frac{\int_0^\infty \varepsilon e^{-\alpha\varepsilon}\,d\varepsilon}{\int_0^\infty e^{-\alpha\varepsilon}\,d\varepsilon}$$

Thus

$$\bar{\varepsilon} = -\frac{d}{d\alpha}\ln\int_0^\infty e^{-\alpha\varepsilon}\,d\varepsilon$$

Evaluating the integral, we have

$$\bar{\varepsilon} = -\frac{d}{d\alpha}\ln\frac{1}{\alpha} = -\alpha\frac{d\alpha^{-1}}{d\alpha} = \alpha\alpha^{-2} = \frac{1}{\alpha}$$

Therefore

$$\bar{\varepsilon} = \varepsilon_0 \tag{2-20}$$

By treating a system containing two different kinds of entities in equilibrium, it can be proved that for a given temperature the constant ε_0 must be independent of the nature of the entities of which the system is comprised. Consider then the system containing a set of identical coil

† We assume that the number of energy states in the interval ε to $\varepsilon + d\varepsilon$ is independent of ε. The assumption is valid for a system of simple harmonic oscillators to which we shall apply the equation.

springs which can exchange energy by some process or other, and which are in equilibrium at temperature T. Assume that the lengths of the springs can vibrate about their quiescent lengths, but that they cannot otherwise move. Then one "coordinate" per spring (its length) is sufficient to describe the system at any instant, and the system is said to have one degree of freedom per entity. Now it is shown in almost any elementary physics text that, for a system of gas molecules moving in thermal equilibrium at temperature T, the average *kinetic* energy per degree of freedom is $kT/2$, where $k = 1.38 \times 10^{-16}$ erg-deg^{-1} is *Boltzmann's constant*. This is the *law of equipartition of energy*, and it is true for any *classical* system. Thus, for our classical system of springs, the average kinetic energy of vibration of one of the springs is $kT/2$. But, for a vibrating spring, or any entity in which the single "coordinate" executes simple harmonic oscillations, the average total energy is twice the average kinetic energy. The reader may easily prove this for a pendulum. Therefore, for one of the springs of our system, the average *total* energy $\bar{\varepsilon}$ is

$$\bar{\varepsilon} = kT \tag{2-21}$$

With equations (2–20) and (2–21), we find

$$\varepsilon_0 = kT \tag{2-22}$$

This determines the value of ε_0, and equation (2–18) becomes

$$P(\varepsilon) = Ae^{-\varepsilon/kT} \tag{2-23}$$

This is the famous *Boltzmann probability distribution.*

These equations should apply directly to a system of standing waves in equilibrium in a black body cavity because each of the entities of this system also has a single "coordinate" (its amplitude) which executes simple harmonic oscillations.† In the Rayleigh-Jeans calculation for this system we are interested only in equation (2–21). It could have been written directly from the law of equipartition of energy, but later we shall need (2–23) and the calculation preceding (2–20).

† Equations (2–22) and (2–23) also apply to a system of moving gas molecules; however, equation (2–21) is replaced by $\bar{\varepsilon} = \frac{3}{2}kT$ as the average total energy equals the average kinetic energy, but there are 3 degrees of freedom. There is no inconsistency here since equation (2–20) is also changed. It is replaced by $\bar{\varepsilon} = \frac{3}{2}\varepsilon_0$ because a factor proportional to $\varepsilon^{1/2}$ must be included in the integrands of equation (2–19) to account for the fact that the number of energy states in the interval ε to $\varepsilon + d\varepsilon$ is proportional to $\varepsilon^{1/2}$ (cf. equation 12–37). The product of this factor and $P(\varepsilon)$ gives the Maxwellian energy distribution of gas molecules, $A\varepsilon^{1/2}e^{-\varepsilon/kT}$

8. Comparison with Experiment

The energy per cm^3 in the frequency interval ν to $\nu + d\nu$ of the black body spectrum of a cavity at temperature T is just the product of the average energy per standing wave (equation 2–21) and the number of standing waves in the frequency interval (equation 2–17) divided by the volume of the cavity. That is,

$$\rho_T(\nu)\, d\nu = \frac{8\pi\nu^2 kT}{c^3}\, d\nu \tag{2–24}$$

To make a comparison with our earlier discussion, we must express the energy density in terms of the wavelength λ. This can be done by using the relation

$$\nu = c/\lambda$$

Thus

$$d\nu = -\frac{c}{\lambda^2}\, d\lambda \tag{2–25}$$

Consider a certain interval of the spectrum, which can be defined in terms of frequency as ν to $\nu + d\nu$, or equally well in terms of wavelength as λ to $\lambda + d\lambda$. For this interval the following relation obtains:

$$\rho_T(\lambda)\, d\lambda = -\rho_T(\nu)\, d\nu \tag{2–26}$$

because both sides are equal to the energy content per cm^3 in that interval of the spectrum. [The minus sign compensates for the fact that, according to equation (2–25), $d\nu$ and $d\lambda$ are of the opposite sign.] Combining equations (2–25) and (2–26), we have

$$\rho_T(\lambda)\, d\lambda = \rho_T(\nu)\, \frac{c}{\lambda^2}\, d\lambda \tag{2–27}$$

Using this to transform equation (2–24) to the variable λ, we find

$$\rho_T(\lambda)\, d\lambda = \frac{8\pi k}{\lambda^5}\, \lambda T\, d\lambda \tag{2–28}$$

This is the Rayleigh-Jeans spectral distribution of black body radiation. We note that it has the form required by Wien's law (equation 2–11) since it can be written $\rho_T(\lambda)\, d\lambda = f(\lambda T)/\lambda^5$, where $f(\lambda T) = 8\pi k\lambda T$. However, as is seen in figure (2–11), the Rayleigh-Jeans spectrum is in complete disagreement with the experimental data; it predicts the wrong form for the function $f(\lambda T)$.

In the limit of long wavelengths, the Rayleigh-Jeans spectrum approaches the experimental results. But, as $\lambda \to 0$, their theory predicts that $\rho_T(\lambda) \to \infty$, whereas experiment shows that $\rho_T(\lambda)$ always remains finite and is equal to zero at zero wavelength. The unrealistic behavior of the Rayleigh-Jeans distribution at short wavelengths is known in physics as the "ultraviolet catastrophe." This terminology is suggestive of the importance

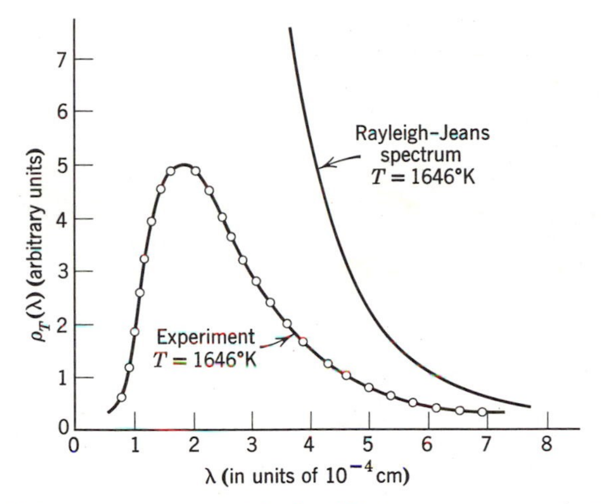

Figure 2–11. A comparison of the Rayleigh-Jeans spectrum and experiment.

of the failure of the Rayleigh-Jeans theory. Let us evaluate the total energy per cm³ for the Rayleigh-Jeans spectrum:

$$\int_0^\infty \rho_T(\lambda)\, d\lambda = 8\pi kT \int_0^\infty \frac{d\lambda}{\lambda^4} = \frac{8\pi}{3}\, kT \lim_{\lambda \to 0} \left(\frac{1}{\lambda^3}\right)$$

This goes to infinity unless $T = 0$. We have here an even more impressive indication of the complete failure of the Rayleigh-Jeans spectrum. Yet this spectrum is, as we have mentioned earlier, a necessary consequence of the theories of classical physics.

9. Planck's Theory

The discrepancy between experiment and theory was resolved in 1901 by Planck—but only at the expense of introducing a postulate which was not only new, but also drastically at variance with certain concepts of classical physics. Planck's *postulate* may be stated as follows:

Any physical entity whose single "coordinate" executes simple harmonic oscillations (i.e., is a sinusoidal function of time)† can possess only total energies ε which satisfy the relation

$$\varepsilon = nh\nu, \qquad n = 0, 1, 2, 3, \ldots \tag{2–29}$$

where ν is the frequency of the oscillation and h is a universal constant.

An *energy level diagram* provides a convenient way of illustrating the behavior of an entity governed by this postulate, and it is also useful in contrasting this behavior with what would be predicted by classical

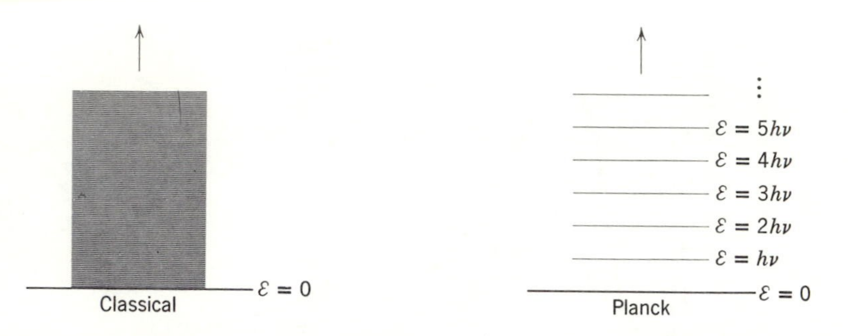

Figure 2–12. Energy level diagrams for a classical simple harmonic oscillator and a simple harmonic oscillator obeying Planck's postulate.

physics. See figure (2–12). In such diagrams we indicate each of the possible energy states of the entity with a horizontal line. The distance from the line to the zero energy line is proportional to the total energy to which it corresponds. Since the entity may have any energy from zero to infinity according to classical physics, the classical energy level diagram consists of a continuum of lines extending from zero up. However, the entity executing simple harmonic oscillations can have only one of the discrete total energies $\varepsilon = nh\nu$, $n = 0, 1, 2, 3, \ldots$, if it obeys Planck's postulate. This is indicated by the discrete set of lines in its energy level diagram. The energy of the entity obeying Planck's postulate is said to be *quantized*, the allowed energy states are called *quantum states*, and the integer n is called the *quantum number*.

Let us recalculate $\rho_T(\lambda)$ under the assumption that Planck's postulate is obeyed by the electromagnetic standing waves whose amplitudes are, according to equation (2–13), executing simple harmonic oscillation with

† The word "coordinate" is used in its general sense to mean any quantity which describes the instantaneous condition of the entity. Examples are: the length of a coil spring, the angular position of a pendulum bob, the amplitude of a wave. All these examples happen also to be sinusoidal functions of time.

frequency ν. No modification will be required in the calculation of $N(\nu)\,d\nu$ since at no point in that calculation was it necessary to specify the energy of the standing waves. However, the evaluation of $\bar{\varepsilon}$ involved integrals in which the energy of the standing waves, ε, was the variable of integration. This was proper in the original calculation since classically the energy can assume any value. But, under Planck's postulate, ε can only have the discrete values $nh\nu$ where $n = 0, 1, 2, 3, 4, \ldots$. Thus we must recalculate $\bar{\varepsilon}$, replacing all integrals over ε by summations over this variable. Then

$$\bar{\varepsilon} = \frac{\sum_{n=0}^{\infty} \varepsilon P(\varepsilon)}{\sum_{n=0}^{\infty} P(\varepsilon)} = \frac{\sum_{n=0}^{\infty} A\varepsilon e^{-\varepsilon/kT}}{\sum_{n=0}^{\infty} Ae^{-\varepsilon/kT}} = \frac{\sum_{n=0}^{\infty} nh\nu e^{-\alpha nh\nu}}{\sum_{n=0}^{\infty} e^{-\alpha nh\nu}}, \qquad \alpha = \frac{1}{kT}$$

where we have assumed, as did Planck, that the Boltzmann probability distribution still applies. In a manner completely analogous to that employed in the calculation preceding equation (2–20), we can show this yields

$$\bar{\varepsilon} = -\frac{d}{d\alpha} \ln \sum_{n=0}^{\infty} e^{-\alpha nh\nu}$$

Now

$$\sum_{n=0}^{\infty} e^{-\alpha nh\nu} = 1 + e^{-\alpha h\nu} + e^{-2\alpha h\nu} + e^{-3\alpha h\nu} + \cdots$$

or

$$\sum_{n=0}^{\infty} e^{-\alpha nh\nu} = 1 + x + x^2 + x^3 + \cdots$$

where $x = e^{-\alpha h\nu}$. But

$$(1 - x)^{-1} = 1 + x + x^2 + x^3 + \cdots$$

Therefore

$$\sum_{n=0}^{\infty} e^{-\alpha nh\nu} = (1 - e^{-\alpha h\nu})^{-1}$$

and

$$\bar{\varepsilon} = -\frac{d}{d\alpha} \ln (1 - e^{-\alpha h\nu})^{-1}$$

$$\bar{\varepsilon} = -(1 - e^{-\alpha h\nu})(-1)(1 - e^{-\alpha h\nu})^{-2}[-(-h\nu)e^{-\alpha h\nu}]$$

$$\bar{\varepsilon} = \frac{h\nu e^{-\alpha h\nu}}{1 - e^{-\alpha h\nu}} = \frac{h\nu}{e^{\alpha h\nu} - 1} = \frac{h\nu}{e^{h\nu/kT} - 1} \tag{2–30}$$

Evaluating $\rho_T(\nu)\,d\nu$ as before, using equation (2–17), we have

$$\rho_T(\nu)\,d\nu = \frac{\bar{\varepsilon} N(\nu)\,d\nu}{a^3} = \frac{8\pi\nu^2}{c^3} \frac{h\nu}{e^{h\nu/kT} - 1}\,d\nu \tag{2–31}$$

Transforming to the variable λ with equation (2–25), we obtain

$$\rho_T(\lambda)\,d\lambda = \frac{8\pi hc}{\lambda^5}\frac{1}{e^{hc/k\lambda T} - 1}\,d\lambda \tag{2–32}$$

This is the black body spectral distribution derived by Planck. It satisfies Wien's law with $f(\lambda T) = 8\pi hc/(e^{hc/k\lambda T} - 1)$. This function involves the constant h whose numerical value is to be determined by fitting the

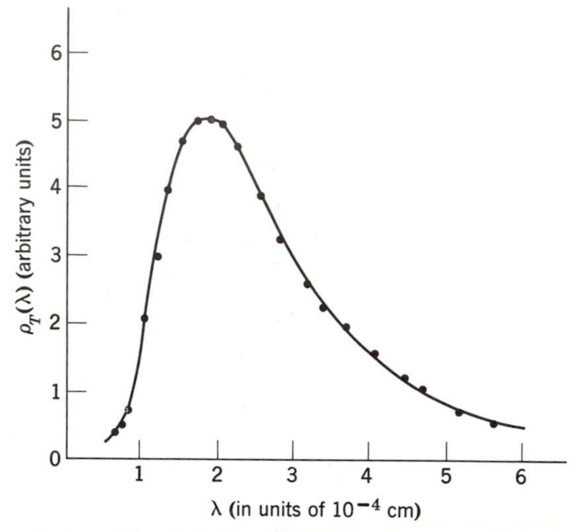

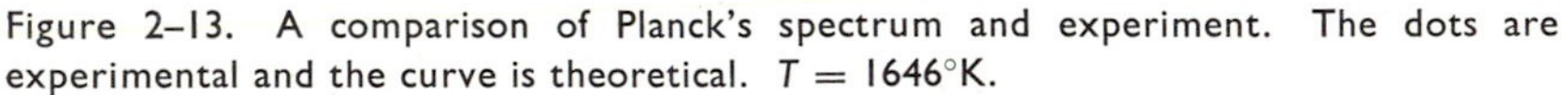

Figure 2–13. A comparison of Planck's spectrum and experiment. The dots are experimental and the curve is theoretical. $T = 1646°$K.

theoretical spectrum to the experimental data. Such a fit is shown in figure (2–13). The value providing the best fit to the data is

$$h = 6.63 \times 10^{-27} \text{ erg-sec}$$

This constant is known as *Planck's constant*. The figure shows that Planck's theoretical spectrum is in extremely good agreement with the experimental spectrum for $T = 1646°$ K. Since it satisfies Wien's law, we know that it will agree equally well with experiment at any other temperature.

10. Some Comments on Planck's Postulate

It is worth while to give a qualitative explanation of how Planck's postulate, which in terms of the mathematics only has the effect of replacing integrals by sums, is able to produce such a remarkable change in the results of the calculation. The failure of the Rayleigh-Jeans theory is essentially due to the rapid increase in $N(\nu)\,d\nu$ as ν becomes large. Since the classical law of equipartition of energy requires that $\bar{\varepsilon}$ be independent of ν, the spectrum itself will also increase rapidly with increasing ν (or decreasing λ). Planck's postulate has the effect of putting a cut-off into $\bar{\varepsilon}$ so that it does not obey the classical law. This keeps the spectrum bounded and, in fact, brings the spectrum to zero as $\nu \rightarrow \infty$ (or as $\lambda \rightarrow 0$). This behavior of $\bar{\varepsilon}$ can be seen directly from equation (2–30). The decrease in $\bar{\varepsilon}$ as $\nu \rightarrow \infty$ comes about in the following way. Under Planck's postulate, the energy of the standing waves is either $\varepsilon = 0$, or $\varepsilon = h\nu$, or $\varepsilon = 2h\nu$, etc. If we now consider the Boltzmann probability distribution (equation 2–23), we see that for ν large enough that $h\nu \gg kT$ the probability that a standing wave will have an energy other than $\varepsilon = 0$ becomes negligible. Consequently, when we evaluate $\bar{\varepsilon}$ for standing waves whose frequency ν is large, we find $\bar{\varepsilon} \simeq 0$. On the other hand, for the very low frequency standing waves in which $h\nu \ll kT$, the quantum states are so close together that they behave as if they were continuously distributed, as in the classical theory. Thus in the low frequency limit Planck's value of $\bar{\varepsilon}$ approaches the classical value of $\bar{\varepsilon} = kT$, as is also apparent from equation (2–30).

In its original form, Planck's postulate was not so far reaching as it is in the form we have given. Planck's initial work was done with the theory, described in section 6, in which the energy in a particular frequency component of the black body radiation is related to the energy of a (charged) particle in the wall of the black body cavity oscillating sinusoidally at the same frequency, and he postulated only that the energy of the oscillating particle is quantized. After the development of the Rayleigh-Jeans theory it became apparent that it is equivalent to quantize directly the oscillating electromagnetic waves, and the postulate was broadened to include any entity whose single "coordinate" oscillates sinusoidally. Planck's first step consisted in finding, by trial and error, an empirical equation which agreed with the experimental spectra. Then he set about finding what modifications of the classical theory were required to yield this equation. In considering the derivation of the Boltzmann probability distribution $P(\varepsilon)$, which we described in terms of a simple example in section 7, he was led to try holding constant the quantity that we called $\Delta\varepsilon$ in our example, instead of taking it to zero as required by classical theory. Although $P(\varepsilon)$ then has values only for discrete ε, these values are not changed. However $\bar{\varepsilon}$ is changed, as we have seen, and its new value depends on $\Delta\varepsilon$. Planck soon found that the desired spectral distribution is obtained if $\Delta\varepsilon$ is proportional to the oscillation frequency ν. Then the possible energies $\varepsilon = n\,\Delta\varepsilon$ become $\varepsilon = nh\nu$, as stated in his postulate.

The success of Planck's postulate in leading to a theoretical black body spectrum which agrees with experiment requires that we tentatively accept its validity, at least until such time as it may be proved to lead to conclusions which disagree with experiment. It may have occurred to the reader that there are physical systems whose behavior seems to be obviously in disagreement with the postulate. For instance, an ordinary pendulum executes simple harmonic oscillations, and yet this system certainly appears to be capable of possessing a continuous range of energies. Before we accept this argument, it would be prudent to make some simple numerical calculations concerning such a system. Consider a pendulum consisting of a 1-gm weight suspended from a 10-cm string. The oscillation frequency of this pendulum is about 1.6/sec. The energy of the pendulum depends on the amplitude of the oscillations. Assume the amplitude to be such that the string in its extreme position makes an angle of 0.1 radian with the vertical; then the energy E is approximately 50 ergs. If the energy of the pendulum is quantized, any decrease in the energy of the pendulum, due for instance to frictional effects, will take place in discontinuous jumps of magnitude $\Delta E = h\nu$. This is not observed. But, if we consider that $\Delta E = h\nu = 6.63 \times 10^{-27} \times 1.6 \simeq 10^{-26}$ ergs, whereas $E \simeq 50$ ergs, it is apparent that even the most sensitive experimental equipment would be totally incapable of resolving the discontinuous nature of the energy decrease.

No evidence, either positive or negative, concerning the validity of Planck's postulate is to be found from experiments involving a pendulum. The same is true of all other macroscopic mechanical systems. Only when we consider systems in which ν is so large and/or E is so small that $h\nu$ is of the order of E are we in a position to test Planck's postulate. One example is, of course, the high frequency standing waves in black body radiation. Other examples will be considered in the following chapters.

BIBLIOGRAPHY

Mayer, J. E., and M. G. Mayer, *Statistical Mechanics*, John Wiley and Sons, New York, 1940.

Peck, E. R., *Electricity and Magnetism*, McGraw-Hill Book Co., New York, 1953.

Richtmyer, F. K., E. H. Kennard, and T. Lauritsen, *Introduction to Modern Physics*, McGraw-Hill Book Co., New York, 1955.

EXERCISES

1. Verify (2–2) by using the procedure suggested immediately below that equation.

2. When the sun is directly overhead, the thermal energy incident upon the earth is about 1.4×10^6 ergs-cm^{-2}-sec^{-1}. The diameter of the sun is about 1.6×10^{11} cm, and the distance from the sun to the earth is about 1.3×10^{13} cm. Assuming that the sun radiates like a black body, use equation (2–8) to estimate its surface temperature.

3. Repeat the calculation of section 7 for a system with six entities and total energy $6\ \Delta\varepsilon$. Compare the results with equation (2–18), as in figure (2–10).

4. Prove the fact, used in obtaining equation (2–21), that the total energy of a simple harmonic oscillator is twice its average kinetic energy. Do this for the case of a pendulum.

5. Integrate equation (2–32) over all λ, and show that the T dependence of the results agrees with equation (2–8).

6. To a certain approximation, an atom of a crystal executes simple harmonic oscillations about its equilibrium position because it possesses thermal agitation energy and experiences an approximately linear restoring force. Use the bulk compressibility of some crystal to estimate the strength of the restoring force and, from this, the oscillation frequency. Use equation (2–21) to estimate the oscillation energy at room temperature. From these results determine the approximate value of the quantum number that describes the motion of this microscopic simple harmonic oscillator, and compare it with the approximate value of the quantum number that describes the motion of the macroscopic simple harmonic oscillator discussed in section 10.

CHAPTER **3**

Electrons and Quanta

1. Cathode Rays

In this chapter we shall primarily be concerned with two phenomena, the *photoelectric effect* and the *Compton effect*, the interpretation of which played a crucial role in the development of the quantum theory. These phenomena involve the interaction of electromagnetic radiation with

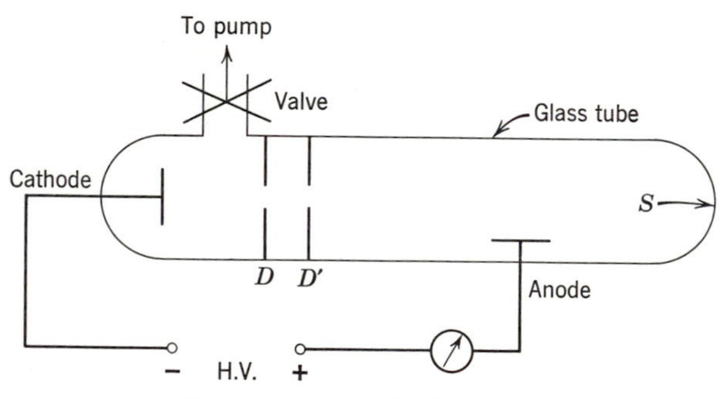

Figure 3–1. A cathode ray tube.

matter—specifically with the *electrons* contained in matter. Consequently we shall first briefly describe the discovery of electrons and the quantitative measurement of their charge and mass.

The conduction of electricity through rarefied gases was a popular subject of investigation in the second half of the nineteenth century. A typical experimental arrangement is shown in figure (3–1). A high voltage applied between the negative electrode, or cathode, and the positive electrode, or anode, causes current to flow through the tube which is filled with some gas (such as air) at a low pressure. The current produces a striking pattern of layers of glowing gas which alternate with dark layers. The exact nature of the pattern depends on the pressure of the gas as well as on its composition and the magnitude of the voltage. A complicated set of phenomena, involving ionization of the gas, is responsible for the conduction of current and for the light emitted by the gas. We shall not discuss these processes here. When the tube is pumped to very low pressures, in the region of 0.01 mm of Hg or less, a quite different phenomenon develops. The gas ceases to glow (for one reason, because there is very little gas left). Nevertheless, the ammeter indicates that a current is still flowing in the tube. Another indication that something is still happening in the tube is the presence of a luminous spot S on the inner surface of the glass wall at the end of the tube. The size of the spot and its position on the end wall of the tube are observed to depend on the size and location of the holes in diaphragms D and D' in such a way as to imply that the luminous spot is caused by rays of particles which leave from the cathode and travel in straight lines, finally impinging upon the wall. Additional supporting evidence is obtained from the observation that a solid object placed to the right of D' will cast a sharp shadow on the luminous spot at the end of the tube. The position of the luminous spot can be influenced by the presence of a strong electric or magnetic field in the region to the right of D', showing that the particles traveling from the cathode to the wall of the tube are electrically charged. The sense of the deflection of the spot is such as to indicate that the charge of the particles is negative.† This was confirmed by an experiment in which an insulated metallic chamber was placed inside the tube at S; the chamber was observed to collect a negative charge. These particles, which travel in ray-like paths from the cathode, were called *cathode ray particles*, or sometimes simply *cathode rays*. The apparatus shown in figure (3–1) is called a *cathode ray tube*. It is now known that the cathode ray particles are emitted from the cathode as a result of its bombardment by positive ions of the gas contained in the tube.

† The reader may wonder why the negatively charged particles travel in straight lines in the region to the right of D', instead of being attracted to the positively charged anode. The reason is that most of the voltage drop across the tube occurs near the cathode. Thus the charged particles are accelerated in the strong electric field near the cathode, and then they continue moving in straight lines. The electric field near the anode is too weak to produce an appreciable deflection.

2. The e/m Ratio of Electrons

In 1897 Thomson made accurate measurements of e/m, the ratio of the charge of the cathode ray particles to their mass. This provided the key to their identity. Thomson's apparatus is shown in figure (3–2). Cathode ray particles of (negative) charge e and mass m travel from the cathode C toward the anode A, which is pierced with a small hole and

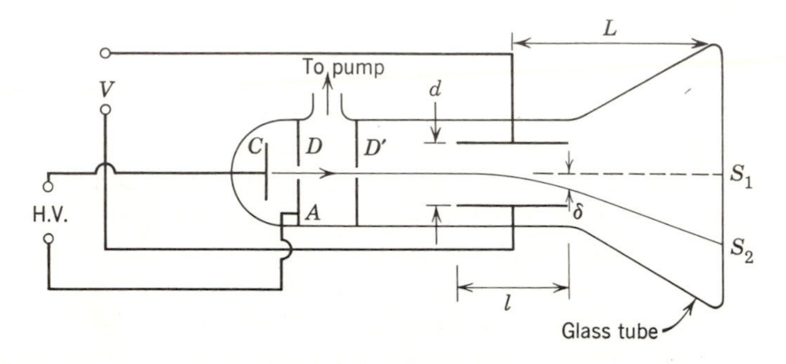

Figure 3–2. An apparatus used to measure the e/m ratio for electrons.

also acts as the first of two diaphragms, D and D', that define the particles to a narrow beam. The particles then pass between two parallel metal plates of length l and separation d. Finally they strike the end of the tube, producing a luminous spot at S_1. The application of a voltage V across the two plates creates a force $F = eV/d$ which acts on the particles in a direction perpendicular to their velocity v. This force produces a transverse acceleration $a = (e/m)(V/d)$. The acceleration lasts for the time $t = l/v$ during which the particles are passing between the plates. Upon emerging, their transverse deflection is $\delta = \frac{1}{2}at^2 = \frac{1}{2}(e/m)(V/d)(l/v)^2$. The deflection $\overline{S_1S_2}$ of the luminous spot is amplified by a factor which, for small deflections, is almost exactly $2L/l$, where L is the distance from the center of the plates to the end of the tube. Thus

$$\overline{S_1S_2} = \frac{e}{m}\frac{V}{d}\frac{L}{l}\left(\frac{l}{v}\right)^2 \tag{3–1}$$

If the deflection $\overline{S_1S_2}$ is measured and the velocity v of the particles is known, equation (3–1) allows the evaluation of e/m.

To determine v, Thomson made a separate measurement employing an electrostatic force due to a voltage V across the two plates and also a magnetic force resulting from the interaction of the moving charged particles with a magnetic field H applied over the region between the plates and in a direction perpendicular to the plane of the figure (3–2). With the correct orientation of the magnetic field, the electrostatic and magnetic forces will be opposed; in fact, for a certain value of the magnetic field strength the forces can be made to cancel out exactly. For this value of H the following relation obtains:

$$eV/d = Hev/c \tag{3–2}$$

and the luminous spot suffers no deflection. The left side of this equation is the electrostatic force, and the right side the magnetic force. From equations (3–1) and (3–2), e/m can be evaluated in terms of the quantities measured in the experiment. The currently accepted value is

$$\begin{aligned} e/m &= 5.27 \times 10^{17} \text{ esu/gm} \\ &= 1.76 \times 10^{8} \text{ coulombs/gm} \end{aligned}$$

The results of the experiment were observed to be independent of both the composition of the cathode and the nature of the residual gas in the tube. This showed that the value of e/m stated above is characteristic of the cathode ray particles themselves and that all cathode ray particles have the same value of e/m.

The ratio of the charge to the mass of cathode ray particles is very much larger than the value of that ratio for ionized atoms. The latter quantity is measured in experiments involving the electrolysis of solutions containing ionized atoms; an electrolytic cell is operated for a certain period of time and the total charge carried by the ionized atoms, as well as the total mass of the atoms deposited at the electrode, is measured. The largest atomic value of e/m is that for an ionized hydrogen atom, but e/m for a cathode ray particle is 1836 times greater than e/m for an ionized hydrogen atom. Either the charge of a cathode ray particle is very much larger than the charge of an ionized atom, or the mass of a cathode ray particle is very much smaller than an atomic mass.

Thomson made the assumption that the second alternative was correct. In fact, he assumed that the charge of the cathode ray particles is exactly equal in magnitude to the charge of a singly ionized atom. This unit of charge is called the *electronic charge*. The cathode ray particles, in modern terminology, are called *electrons*.

3. The Charge and Mass of Electrons

The magnitude of the electronic charge was determined in a series of experiments performed by Townsend and Thomson (ca. 1900) and by Millikan (1909). These experiments were based on the observation that very small drops of liquid often pick up electric charges. If a voltage V is applied between two horizontal parallel plates located above and below the region containing the charged droplets, a droplet can be suspended against the force of gravity when

$$q\frac{V}{d} = Mg \tag{3–3}$$

where d is the separation of the parallel plates, q is the charge on the droplet, M is its mass, and $g = 980$ cm-sec^{-2}. In order to determine M, the voltage is removed, and the droplet falls under the influence of gravity until it reaches terminal velocity because the frictional drag F_v becomes equal to the gravitational force Mg. According to *Stokes' law*, $F_v = 6\pi\eta av$, where a is the radius of the droplet, v is its terminal velocity, and η is the coefficient of viscosity of the medium (air) through which the droplet is falling. Thus at terminal velocity we have the relation

$$Mg = 6\pi\eta av \tag{3–4}$$

The radius of the droplet can be expressed in terms of its mass and density from the equation $M = \frac{4}{3}\pi a^3\rho$. Putting this into equation (3–4), we can solve for M in terms of the terminal velocity v. The latter quantity is measured by observing with the aid of a microscope the distance through which the droplet falls in a given time. Knowing M, we find q from equation (3–3).

The results of a large number of measurements on the charge of different droplets could all be expressed as

$$q = n \times 4.80 \times 10^{-10} \text{ esu} \tag{3–5}$$

where n is an integer, and where we quote currently accepted values. That is, it was observed that the charge of a droplet was always some integral multiple of 4.80×10^{-10} esu. It is reasonable to interpret this as meaning that there exists a basic unit of charge of magnitude

$$\begin{aligned} e &= 4.80 \times 10^{-10} \text{ esu} \\ &= 1.60 \times 10^{-19} \text{ coulombs} \end{aligned}$$

since no droplet was ever found to have a charge less than this amount.

Thomson and his colleagues assumed that the electronic charge was equal to this basic unit of charge. Then, knowing the value of e/m for an electron, they were able to evaluate the mass of an electron. The value they obtained was not far from the currently accepted value,

$$m = 9.11 \times 10^{-28} \text{ gm}$$

There is now an overwhelming body of evidence proving the correctness of the assumption that the electronic charge is equal to 4.80×10^{-10} esu. For instance, it is possible to evaluate the electron mass from the equation

$$m = m_{H^+} \frac{(e/m)_{H^+}}{e/m} = m_{H^+} \frac{1}{1836}$$

where $(e/m)_{H^+}$ represents the charge to mass ratio for an ionized hydrogen atom and m_{H^+} represents its mass. The latter quantity is equal to the chemical atomic weight of hydrogen divided by Avogadro's number.† Thus

$$m = \frac{1.008}{6.02 \times 10^{23}} \frac{1}{1836} = 9.11 \times 10^{-28} \text{ gm}$$

This provides an independent check on the value of the electron mass determined from e/m and the value of e and thereby confirms the correctness of the value of e given above.

In recapitulation, the electron is a particle having a negative charge the magnitude of which is equal to the magnitude of the charge of a singly ionized atom, and a mass which is smaller than the mass of an ionized hydrogen atom by a factor of 1836.

4. Bucherer's Experiment

This is an appropriate place to describe the work of Bucherer (1909) which provided the first experimental verification of the relativistic dependence of mass on velocity: $m = m_0/\sqrt{1 - v^2/c^2}$ (equation 1–20). This experiment consisted of a series of measurements of the e/m ratio of high velocity electrons. The reader will recall that the theory of relativity does not alter the laws of electromagnetic phenomena; the charge of an electron, e, is a constant independent of its velocity in the theory of relativity just as it is in the classical theory. Consequently the relativistic theory predicts

$$\frac{e}{m} = \frac{e}{m_0}\left(1 - \frac{v^2}{c^2}\right)^{1/2} = \left(\frac{e}{m}\right)_{v/c\to 0}\left(1 - \frac{v^2}{c^2}\right)^{1/2} \qquad (3\text{–}6)$$

where $(e/m)_{v/c\to 0}$ is the value of e/m for electrons of velocities small compared to c. The quantity $(e/m)_{v/c\to 0}$ is what was measured in Thomson's experiments, since the velocity of the electrons in the cathode ray tubes which he used is very small compared to c. In order to obtain electrons

† We are ignoring the 0.05 percent difference between the mass of an ionized hydrogen atom and that of a neutral hydrogen atom.

with velocities comparable to c, Bucherer used the electrons which are spontaneously emitted from the naturally radioactive elements (cf. section 2, Chapter 4). These electrons are emitted with a continuous spectrum of velocities which extend, in some cases, to velocities as high as $0.99c$. Bucherer's apparatus was essentially similar to that used by Thomson. Electrons of a certain velocity were first selected from the continuous spectrum by a combination of electric and magnetic fields. These electrons were then deflected in a magnetic field, and the e/m ratio was determined from the magnitude of the deflection. The results of the experiment, which are shown in figure (1–14), are in excellent agreement with equation (3–6) over the entire range of velocities investigated.

5. The Photoelectric Effect

In a cathode ray tube, electrons are emitted from a metallic electrode (the cathode) as a result of the bombardment of the electrode by positive ions of the gas contained in the tube. Another process in which electrons are emitted from the surface of a metal was discovered by Hertz in 1887. A later form of his apparatus is shown in figure (3–3). The glass tube

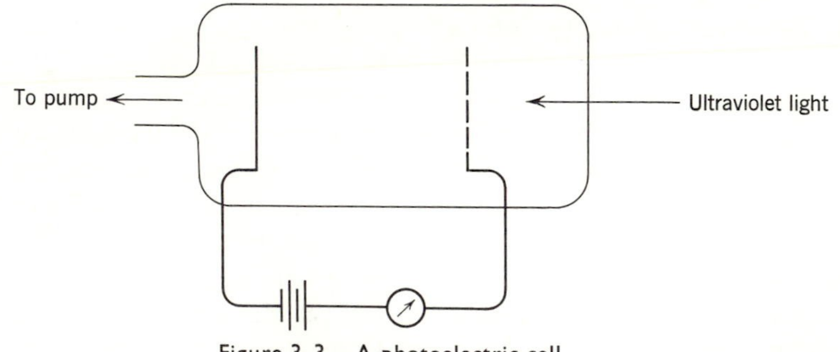

Figure 3–3. A photoelectric cell.

contains a polished metal electrode called a *photocathode*, and a second electrode in the form of a perforated metal plate. The two electrodes are maintained at a potential difference of a few volts (normally with the second electrode positive with respect to the photocathode). When *ultraviolet light* (frequency $\nu \sim 10^{16}$ sec^{-1}) passes through the perforated second electrode and is incident upon the inner surface of the photocathode, a current is observed to flow through the tube. This phenomenon is called the *photoelectric effect*. The effect persists even when the tube is evacuated to a very low pressure, implying that gaseous ions are not the carriers of the current. Experiments in which a magnetic field was applied to the region between the photocathode and the second electrode indicated that the current consisted of the flow of negatively charged particles. Thomson's discovery of the existence of electrons suggested that the negatively charged particles of the photoelectric effect could also be electrons which were liberated from the photocathode as a result of its bombardment by ultraviolet light. This hypothesis was confirmed in 1900 by Lenard, who measured the e/m ratio of the photoelectric particles and showed that it was the same as that for electrons.

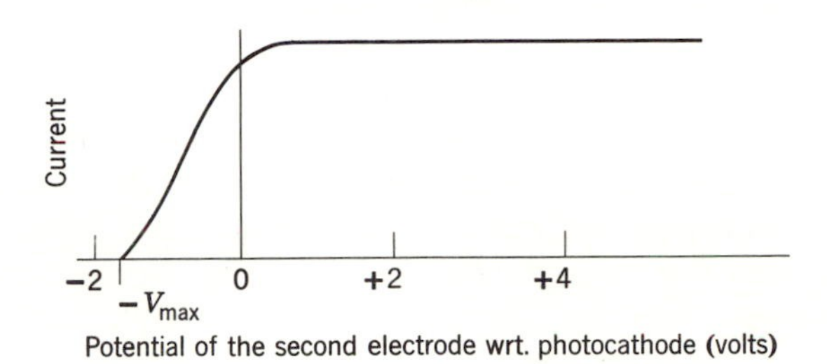

Figure 3–4. The voltage dependence of the current liberated in a photoelectric cell.

The experiments of Lenard cleared up the question of the identity of the photoelectric particles, but they also demonstrated some properties of the photoelectric effect which, as we shall see, were very difficult to understand in terms of the theories of classical physics. Lenard measured the current reaching the second electrode as a function of the potential between that electrode and the photocathode, keeping all other parameters fixed. His data are represented in figure (3–4). An interesting feature of these data is that there is still some current reaching the second electrode when the potential V of that electrode is negative with respect to the photocathode. A potential of this sign will, of course, tend to repel the negatively charged photoelectrons. This must mean that the photoelectrons are ejected from the photocathode with non-negligible kinetic energy. The shape of the curve of current versus potential difference indicates that not all photoelectrons have the same kinetic energy; but the well-defined endpoint at $V = -V_{max}$ shows that there is a well-defined maximum kinetic energy $E_{max} = eV_{max}$, where e is the magnitude of the electronic charge. It was soon suggested that the photoelectrons of maximum energy were emitted from the surface of the photocathode, whereas lower energy photoelectrons originated inside the surface and lost kinetic energy in the process of reaching the surface. This suggestion implies that E_{max} is a good measure of the energy given to an electron in the photoelectric process. Lenard made a series of measurements, of the type indicated in figure (3–4), as a function of the intensity of the incident ultraviolet light. He found that the photoelectric current for positive values of V (for which photoelectrons of all kinetic energies will reach the second electrode) was directly proportional to the light intensity. However the cut-off potential, $-V_{max}$, was observed to be independent of the light intensity. Thus the energy E_{max} acquired by the photoelectrons does not depend on the intensity of the incident light.

6. Classical Theory of the Photoelectric Effect

Let us consider what the classical theory would predict about the photoelectric effect. Thomson's experiments showed that metals contain electrons, although it was not really known whether the electrons were free to move throughout the metal or were bound to atoms. In either case, the mechanism for the ejection of these electrons by the absorption of incident light would involve an interaction between the electric field of the electromagnetic light waves and the electric charge of the electrons. The oscillating electric field acting on the charged electrons would set them into vibration with an amplitude that would be proportional to the amplitude of the electric field. Now, it is easy to show that the average kinetic energy of any vibrating body is proportional to the square of the amplitude of its vibration. Furthermore, the electromagnetic theory shows that the amplitude of the oscillating electric field is proportional to the square root of the light intensity (cf. equation 2–6). Thus we have

$$\text{(av. kin. energy)} \propto \text{(amp. of vib.)}^2 \propto \text{(amp. of elect. fld.)}^2 \propto (\sqrt{\text{light int.}})^2$$

That is, the average kinetic energy of the vibrating electrons would be proportional to the intensity of the incident light. This makes it difficult to understand why the energy acquired by the photoelectrons was observed experimentally to be independent of the light intensity.

An even more serious problem, which arises in any attempt to explain the photoelectric effect in terms of the classical wave theory of light, is that according to such theories a very long time is required for an electron to absorb enough energy. The problem can be illustrated by a simple calculation.

This calculation requires an estimate of the radius of an atom. The estimate can be obtained from a knowledge of the macroscopic density of an element, ρ, its atomic weight A, and Avogadro's number N. Consider a typical element, silver: $\rho \simeq 10$ gm-cm^{-3}, $A \simeq 100$, also $N \simeq 6 \times 10^{23}$ gm^{-1}. Now the mass of an atom is $A/N \simeq 100/6 \times 10^{23}$ gm. The volume of an atom is $V = \frac{4}{3}\pi r^3$, where r is its radius. Assuming that the atoms in the solid are closely packed, the density of an atom will be approximately equal to the density of the solid. So

$$\rho \simeq \frac{\text{mass/atom}}{\text{volume/atom}} \quad \text{or} \quad 10 \text{ gm-cm}^{-3} \simeq \frac{100/6 \times 10^{23} \text{ gm}^{-1}}{\frac{4}{3}\pi r^3}$$

from which we obtain $r \sim 10^{-8}$ cm. This result can be confirmed in several ways. For instance, from an analysis of the measured diffusion rate of a gas composed of atoms, in terms of the kinetic theory of gases, it is also found that $r \sim 10^{-8}$ cm.

Now let us calculate the time required for photoelectrons to absorb the energy $E_{\max} = eV_{\max}$ which is observed experimentally to be of the order of 1.6×10^{-19} coulombs $\times$ 1 volt $\simeq 10^{-19}$ joules $= 10^{-12}$ ergs. Let us assume that a source which emits energy in the form of ultraviolet light, at the rate of 1 watt = 1 joule-sec^{-1} = 10^7 erg-sec^{-1}, is located 1 meter from the photoelectric tube. If the light is emitted with spherical symmetry, the energy flux per cm^2 at the tube is equal to the energy emitted per second times the ratio of 1 cm^2 to the area of a sphere of radius 1 meter; that is, 10^7 erg-sec^{-1} $\times$ $1/4\pi(10^2)^2$ cm^2 $\simeq 10^2$ erg-cm^{-2}-sec^{-1}. Let us assume that the electrons emitted in the photoelectric effect are bound to atoms and that they are somehow able to absorb all the energy incident upon the atom to which they are bound. The radius of an atom is of the order of 10^{-8} cm. Thus its cross sectional area is of the order of 10^{-16} cm^2, and the rate at which energy is incident upon this area is approximately 10^2 erg-cm^{-2}-sec^{-1} $\times$ 10^{-16} cm^2 $= 10^{-14}$ erg-sec^{-1}. If we were to assume that the electrons were not bound to atoms, their rate of absorption of energy would presumably be less. Finally we can estimate that the time required for a photoelectron to absorb the observed 10^{-12} ergs is about 10^2 sec. This calculation, which is based on the assumption characteristic of the wave theory—that the light energy is uniformly distributed over spherical wave fronts spreading out from the source—predicts a delay between the time at which the light source was turned on and the emission of the first photoelectrons of about a minute. No such time delay was observed by Lenard. In fact, experiments performed in 1928 by Lawrence and Beams, using a light source many orders of magnitude *weaker* than we assumed in the above calculation, set an upper limit on the time delay of about 10^{-9} sec!

7. Quantum Theory of the Photoelectric Effect

In 1905 Einstein enunciated a quantum theory of the photoelectric effect which was closely related to the quantum theory of black body radiation. He reasoned that Planck's requirement that the energy content of the electromagnetic waves of frequency ν in a radiant source (e.g., the ultraviolet light source) can only be 0, or $h\nu$, or $2h\nu$, . . . , or $nh\nu$, . . . implies that in the process of going from energy state $nh\nu$ to energy state $(n - 1)h\nu$ the source would emit a burst of electromagnetic energy of energy content $h\nu$.

Einstein assumed that such a burst of emitted energy is initially localized in a small volume of space; and furthermore that it remains localized as it moves away from the source with velocity c, instead of spreading out in the manner characteristic of moving waves. He assumed that the energy content ε of such a bundle, or quantum *of energy, is related to its frequency ν by the equation*

$$\varepsilon = h\nu \tag{3-7}$$

He also assumed that in the photoelectric process one of the quanta is completely absorbed by an electron in the photocathode.

This would impart to the electron an energy equal to $h\nu$. If this energy is greater than the quantity ΔE, the electron can escape from the photocathode, where ΔE is the energy which the electron must expend in reaching the surface of the photocathode plus the energy W required to overcome the attractive forces which we know must be present at the surface since electrons do not normally escape. The kinetic energy of the electron after escaping from the photocathode will be equal to $E = h\nu - \Delta E$. For an electron originating at the surface, ΔE will be just equal to W, and E will have its maximum value

$$E_{\text{max}} = h\nu - W \tag{3-8}$$

where W is a constant which depends on the composition of the photocathode.

According to this theory, the rate of emission of photoelectrons will be proportional to the flux of quanta incident upon the photocathode, which in turn will be proportional to the intensity of the incident electromagnetic radiation. This is in agreement with Lenard's observation that the photoelectric current is proportional to the intensity. As can be seen from equation (3–8), the theory also agrees with the observation that E_{max} does not depend on the intensity. Furthermore, the theory removes the difficulty concerning the time required for the photoelectrons to receive enough energy. This difficulty arose in the wave theory because the energy of the source was spread uniformly over the entire wave front. In the quantum theory the energy is concentrated in the quanta. Within a very short time after the source is turned on, a quantum will strike the photocathode where it will be absorbed by one of the electrons, and the ejection of that electron will result.

Einstein's theory, as summarized by equation (3–8), makes the prediction that the maximum kinetic energy of the photoelectrons should be a linear function of the frequency of the incident electromagnetic radiation. This prediction was tested in 1916 by Millikan, who measured E_{max} for radiation in the frequency range 6 to 12 $\times$ 10^{14} sec^{-1}. The nature of his data is indicated in figure (3–5). The experiment shows that E_{max} actually is a linear function of the frequency ν. According to equation (3–8), the

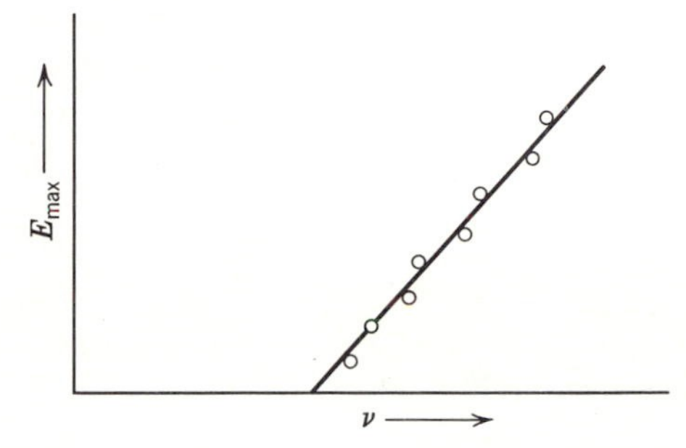

Figure 3–5. The frequency dependence of the maximum kinetic energy of photoelectrons.

slope of the straight line fitting the data should be equal to Planck's constant h, and its intercept on the ν axis should be equal to W/h. Thus the photoelectric experiments, with the aid of Einstein's theory, could be used to determine the numerical value of h. The value of h which was obtained agreed to within better than 0.5 percent with the value of that constant which had been determined by the completely independent method of fitting Planck's theory to the experimental black body spectrum. This finding represented a great triumph for the quantum theories.

8. The Compton Effect

In 1923 Compton discovered that, when a beam of X-rays of well-defined wavelength λ_0 is scattered through an angle θ by sending the radiation through a metallic foil, the scattered radiation contains a component of well-defined wavelength λ_1 which is longer than λ_0. This phenomenon is called the *Compton effect*. As we shall see, it provides extremely convincing evidence for the existence of quanta. Although a description of the production of X-rays and the measurement of their wavelength will not be presented until Chapter 14, it is desirable to discuss the Compton effect at this time.

X-rays consist of electromagnetic radiation of wavelength which is shorter than that of visible light by a factor of the order of 10^4. This is proved by experiments which show that X-rays can be diffracted and polarized just like light. In Compton's experiment the wavelength λ_0 of the incident radiation was 0.708×10^{-8} cm. The wavelength λ_1 was observed to depend on the angle of scattering θ, but not on the material comprising the foil. A typical set of data is shown in figure (3–6) for

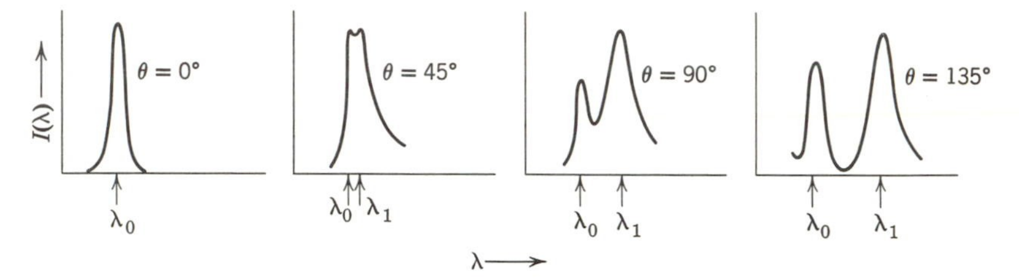

Figure 3–6. Wavelength spectra of quanta scattered at various angles from a carbon foil. From A. H. Compton, *Phys. Rev.*, **22,** 409 (1923).

$\theta = 0°$, 45°, 90°, and 135°. These curves are wavelength spectra of the scattered radiation. The abscissas are a measure of the wavelength, and the ordinates are a measure of the intensity per unit wavelength interval. At all angles the scattered radiation is observed to contain a component of wavelength equal to the wavelength λ_0 of the incident radiation. The reason for the presence of this component will be discussed in Chapter 14. Here we shall be concerned only with the component in the scattered radiation of wavelength λ_1. The wavelength of this component is observed always to be greater than the wavelength of the incident radiation, and to increase with increasing angle of scattering. At $\theta = 90°$, $\lambda_1 - \lambda_0 = 0.024 \times 10^{-8}$ cm. In terms of the frequency of the scattered radiation, $\nu_1 = c/\lambda_1$, we see that ν_1 is always less than the frequency of the incident radiation ν_0, and decreases with increasing angle of scattering.

This suggested to Compton that, since ν_1 is proportional to the energy E_1 of the quanta, which according to Einstein are associated with electromagnetic radiation of this frequency, the dependence of E_1 on θ is qualitatively similar to the angular dependence of the energy of a particle which has been scattered from another particle. Pursuing this, he developed a simple but very successful theory.

Consider a quantum in the incident beam of electromagnetic radiation. For radiation of frequency ν, the energy of the quantum is

$$E = h\nu \tag{3–9}$$

Taking the idea of a quantum as a localized bundle of energy quite literally, we shall consider it to be a particle of energy E and momentum p. Such a particle must, however, have certain quite specialized properties. Consider equations (1–20) and (1–24), which give the total relativistic energy of a particle in terms of its rest mass m_0 and its velocity v:

$$E = m_0c^2/\sqrt{1 - v^2/c^2}$$

Since the velocity of a quantum equals c, and since its energy content $E = h\nu$ is finite, it is apparent that the rest mass of a quantum must be zero. Thus the quantum can be considered to be a particle of zero rest mass, and of total relativistic energy E which is entirely kinetic. The momentum of a quantum can be evaluated from the general relation (1–25) between the total relativistic energy E, momentum p, and rest mass m_0. This is

$$E^2 = c^2p^2 + (m_0c^2)^2$$

For a quantum the second term on the right is zero, and we have

$$p = E/c = h\nu/c = h/\lambda \tag{3–10}$$

where λ is the wavelength of the electromagnetic radiation associated with the quantum. It is quite interesting to note that Maxwell's classical wave theory of electromagnetic radiation leads to an equation $p = E/c$, where p represents the momentum content per unit volume of the radiation and E represents the energy content per unit volume.

Now the frequency ν of the scattered radiation was observed to be independent of the material in the foil. This implies that the scattering does not involve entire atoms. Compton assumed that the scattering was due to collisions between the quanta and individual electrons in the foil. He also assumed that the electrons participating in this scattering process are free and initially stationary. Some a priori justification of this assumption can be found from considering the fact that the energy of an X-ray quantum is several orders of magnitude greater than the energy of an ultraviolet quantum, and from our discussion of the photoelectric effect it is apparent that the energy of an ultraviolet quantum is comparable to the energy with which an electron is bound in a metal. Consider, then, a collision between a quantum and a free stationary electron, as in figure (3–7). In the diagram on the left, a quantum of total relativistic energy E_0 and momentum p_0 is incident on a stationary electron of rest mass energy m_0c^2. In the diagram on the right, the quantum is scattered at an angle θ and moves off with total relativistic energy E_1 and momentum p_1, while the electron recoils at an angle ϕ with kinetic energy T and momentum p. Compton applied the conservation of momentum and total relativistic energy to this collision problem. Momentum conservation requires

$$p_0 = p_1 \cos\theta + p\cos\phi \tag{3–11}$$

and

$$p_1 \sin\theta = p\sin\phi \tag{3–12}$$

Squaring these equations, we obtain

$$(p_0 - p_1\cos\theta)^2 = p^2\cos^2\phi$$

and

$$p_1^2 \sin^2\theta = p^2\sin^2\phi$$

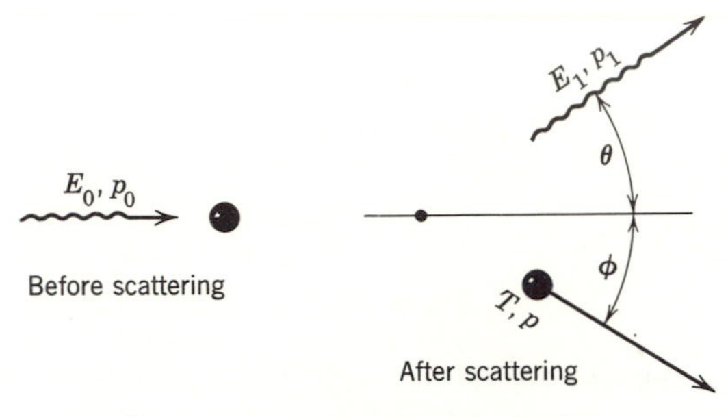

Figure 3–7. A collision between a quantum and a free stationary electron.

Adding, we find

$$p_0^2 + p_1^2 - 2p_0p_1\cos\theta = p^2 \tag{3–13}$$

Conservation of total relativistic energy requires

$$E_0 + m_0c^2 = E_1 + T + m_0c^2 \tag{3–14}$$

Thus

$$E_0 - E_1 = T$$

According to equation (3–10), this is

$$c(p_0 - p_1) = T \tag{3–15}$$

According to equations (1–22′) and (1–25),

$$(T + m_0c^2)^2 = c^2p^2 + (m_0c^2)^2$$

So

$$T^2 + 2Tm_0c^2 = c^2p^2$$

or

$$T^2/c^2 + 2Tm_0 = p^2 \tag{3–16}$$

Evaluating p^2 from equation (3–13), and T from (3–15), we have

$$(p_0 - p_1)^2 + 2m_0c(p_0 - p_1) = p_0^2 + p_1^2 - 2p_0p_1\cos\theta$$

which reduces to

$$m_0c(p_0 - p_1) = p_0p_1(1 - \cos\theta)$$

or

$$\left(\frac{1}{p_1} - \frac{1}{p_0}\right) = \frac{1}{m_0c}(1 - \cos\theta)$$

Multiplying through by h and applying equation (3–10), we obtain

$$\lambda_1 - \lambda_0 = \lambda_C(1 - \cos\theta) \tag{3–17}$$

where

$$\lambda_C \equiv \frac{h}{m_0c} = 0.02426 \times 10^{-8}\ \text{cm}$$

is the so-called *Compton wavelength.* Equation (3–17) is the *Compton equation.* It predicts that the increase in wavelength $(\lambda_1 - \lambda_0)$ of the scattered electromagnetic radiation depends only on the scattering angle θ and the universal constant h/m_0c and is independent of the wavelength of the incident radiation. The equation was verified quantitatively by the experiments of Compton and others. The recoil electrons were observed experimentally by Bothe and by Wilson (1923). Bothe and Geiger (1925) showed that, in the scattering of a single quantum by an electron, the scattered quantum and the recoil electron appear simultaneously. The energy of the recoil electrons were measured by Bless (1927) and found to be in agreement with the predictions of the theory. These experiments, and their interpretation in terms of the theory, provide even stronger evidence for the existence of quanta than does the photoelectric effect.

9. The Dual Nature of Electromagnetic Radiation

Einstein's quantum theory of the photoelectric effect was very successful in explaining a phenomenon which simply could not be understood on the basis of the theories of classical physics. Nevertheless, many physicists were very reluctant to accept it because it was completely contrary to the well-established wave nature of electromagnetic radiation. After the discovery and interpretation of the Compton effect, the existence of quanta could no longer be questioned. As a result physics was in what seemed at the time to be a very uncomfortable situation. On the one hand, electromagnetic radiation manifests certain properties, such as diffraction, which could only be explained in terms of wave motion. On the other hand, some of its properties, such as those evident in the photoelectric and Compton effects, could only be explained in terms of quanta which, being *localized*, have essentially the characteristics of particles. Electromagnetic radiation appeared to possess a split personality, sometimes behaving as if it were composed of waves and sometimes behaving as if it were composed of particles.

The experimental facts concerning the properties of electromagnetic radiation, as well as the interpretations of these properties in terms of the existence of *both* wave aspects and particle aspects, remain essentially unchanged today. However, as a consequence of the broader point of view provided by the development of the theory of quantum mechanics, the attitude of physicists concerning this situation is now very different from their initial attitude. The duality evident in the wave-particle aspects of electromagnetic radiation is no longer considered unusual because it is now known to be a general characteristic of all physical entities. Furthermore, this duality is no longer considered to represent a problem, since it is possible to reconcile the existence of both aspects with the aid of the theory of quantum mechanics. This theory will be discussed in detail later.

BIBLIOGRAPHY

Born, M., *Atomic Physics*, Blackie and Son, London, 1957.

Richtmyer, F. K., E. H. Kennard, and T. Lauritsen, *Introduction to Modern Physics*, McGraw-Hill Book Co., New York, 1955.

EXERCISES

1. Electrons are emitted at thermal velocities from a heated cathode, are accelerated through a voltage drop V, and then move in a circle of radius R through a magnetic field of strength H perpendicular to their direction of motion. Derive an expression relating V, H, and R to e/m. Design equipment, based upon this expression, with which it would be possible to make a 1 percent measurement of e/m.

2. Prove the statement, made in section 6, that the average kinetic energy of a vibrating system is proportional to the square of the amplitude of its vibration. Do this for the case of a pendulum.

3. Derive an expression for the kinetic energy of the recoiling electron in the Compton scattering process.

4. Derive the Compton equation for the case $\theta = 180°$ by transforming to the CM frame, treating the collision, and then transforming back to the LAB frame.

5. List the experimental evidence, mentioned in section 9, for the properties of electromagnetic radiation that can be explained only by wave motion.

CHAPTER **4**

The Discovery of the Atomic Nucleus

1. Thomson's Model of the Atom

That electrons are emitted from the metallic cathode of a cathode ray or photoelectric tube implies that the atoms of which the cathode is comprised contain electrons. If so, it is reasonable to assume that all atoms contain electrons. This assumption is attractive because it leads to the simple picture of a positively ionized atom as an atom from which one or more electrons have been removed. It agrees with the experimental observation that the charge of a singly ionized atom is equal in magnitude to the charge of a single electron, or that the charge of a doubly ionized atom is equal to the magnitude of the charge of two electrons, etc. Additional evidence for the existence of electrons in atoms was soon obtained from the experiments of Barkla and others (1909) concerning the scattering of X-rays by atoms. These experiments will be discussed in Chapter 14, but it is appropriate to mention at this point that the experiments provided an estimate of Z, the number of electrons in an atom. It was found that Z is roughly equal to $A/2$, where A is the chemical atomic weight of the atom in question. Another set of experiments which provided a measure of the number of electrons in an atom will be described in this chapter.

Since atoms are normally neutral, they must also contain positive charge equal in magnitude to the negative charge carried by their normal complement of electrons. Thus a neutral atom has a negative charge of magnitude

Ze, where *e* is the electronic charge, and also a positive charge of the same magnitude. That the mass of an electron is very small compared to the mass of even the lightest atom implies that most of the mass of the atom must be associated with the positive charge.

All these considerations naturally led to the question of the distribution of the positive and negative charges within the atom. Thomson proposed a tentative description, or *model*, of the composition of an atom, according to which the negatively charged electrons were located within a continuous

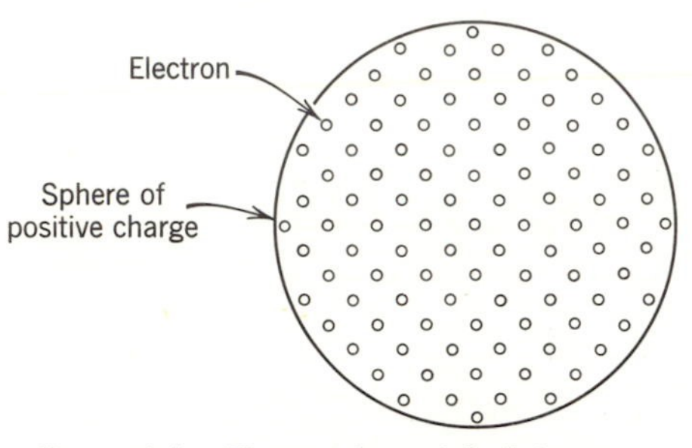

Figure 4–1. Thomson's model of the atom.

distribution of positive charge. The positive charge distribution was assumed to be spherical in shape with a radius of the order of 10^{-8} cm, which was known to be the order of magnitude of the radius of an atom. Owing to their mutual repulsion, the electrons would be uniformly distributed through the sphere of positive charge. This is illustrated in figure (4–1). In an atom in its lowest possible energy state, the electrons would be fixed at their equilibrium positions. In excited atoms (e.g., atoms in a material at high temperature), the electrons would vibrate about their equilibrium positions. Since the electromagnetic theory predicts that an accelerated charged body, such as a vibrating electron, emits electromagnetic radiation, it was possible to understand qualitatively the emission of such radiation by excited atoms on the basis of Thomson's model. However, a calculation of the spectrum of radiation which would be emitted showed that the model did not appear to be able to lead to quantitative agreement with the experimentally observed spectra.

Really conclusive proof of the inadequacy of Thomson's model was obtained in 1911 by Rutherford from the analysis of certain experiments involving the scattering of *alpha particles* by atoms. Rutherford's analysis showed that, instead of being spread throughout the atom, the positive charge is concentrated in a very small region, or *nucleus*, at the center of the atom. Rutherford's work comprised one of the most important developments in the subject of atomic physics, as well as the foundation of the subject of nuclear physics.

2. Alpha Particles

Alpha particles are doubly ionized helium atoms which are spontaneously emitted at very high velocities from several of the heaviest elements, such as U and Ra. This phenomenon will be discussed in detail later in this book; here we shall present only a few pertinent points which are required as background for a description of the experiments which led Rutherford to the discovery of the nucleus.

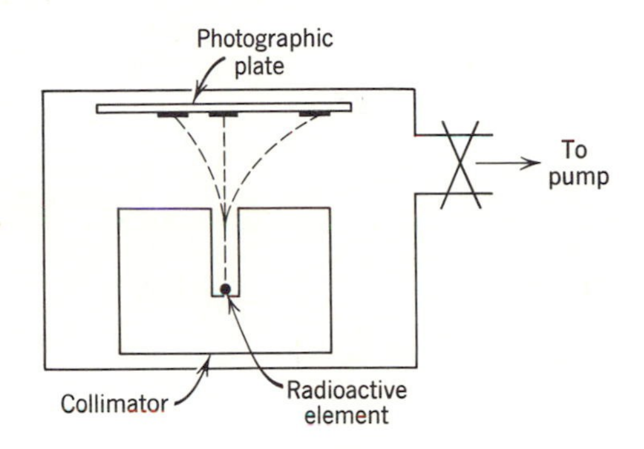

Figure 4–2. An apparatus used to study the types of radiation emitted by radioactive elements.

In 1896–98 Becquerel and Mme. Curie discovered that several of the heaviest elements spontaneously emit penetrating radiation capable of blackening a photographic plate. Soon the radiation emitted by these so-called *radioactive* elements was being investigated in many laboratories. With apparatus of the type indicated in figure (4–2), it was shown that the radiation consists of three separate components. Radiation is emitted from a radioactive element at the bottom of a channel in a lead block which collimates the radiation into a well-defined beam. With no magnetic field, a photographic plate exposed to the radiation shows, upon development, a darkened spot at a point in line with the collimating channel. When a magnetic field is applied to the region between the collimator and the plate, in a direction perpendicular to the plane of the figure, three darkened spots are seen on the plate after it is developed. The two spots which are deviated from the zero field position must be due to charged particles emitted by the radioactive element, and which are deflected, by the application of the magnetic field, in a direction depending on the sign of the charge. The undeviated spot is presumably due to uncharged radiation upon which the magnetic field would have no influence. The particles of which the positively charged component of the radiation are comprised are called *alpha particles*. The negatively charged particles are called *beta particles*, and the uncharged radiation is given the name *gamma radiation*.

In the experiments just described the chamber containing the collimator and plate is evacuated. If the experiment is repeated with the chamber at atmospheric pressure, the spot due to the alpha particles does not appear. Apparently a few centimeters of air is sufficient to stop the alpha particles.

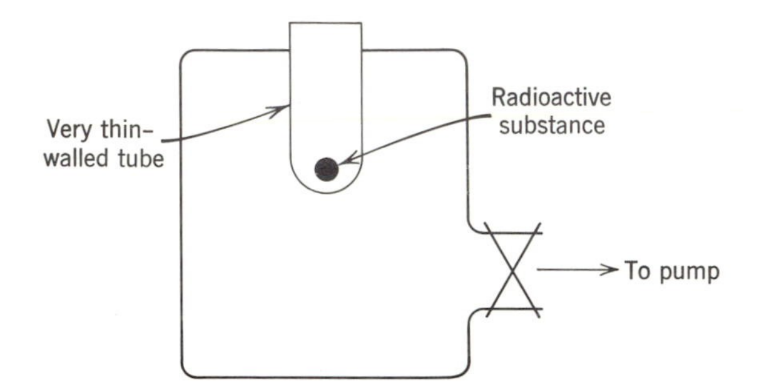

Figure 4–3. An apparatus used to prove that alpha particles are doubly ionized helium atoms.

Similar experiments establish that a foil of some dense substance several millimeters thick will stop the beta particles from reaching the plate, but that the intensity of the spot due to the gamma radiation is significantly decreased only when several centimeters of lead are placed between the collimator and the plate. We know now that the beta particles are high energy electrons, whereas the gamma radiation consists of electromagnetic radiation of the same nature as light, but of very much higher frequency.

The nature of the alpha particles was established in a series of measurements performed by Rutherford. Using apparatus basically similar to that used by Thomson, he measured the charge to mass ratio of the alpha particles as well as their velocity. These measurements showed that the alpha particles were emitted with a well-defined velocity in the range 1.4×10^9 to 2.2×10^9 cm/sec, the exact value depending upon the radioactive element from which they were emitted. Note that these velocities are roughly one-twentieth the velocity of light. The charge to mass ratio of the alpha particles was found to be one-half the value of that ratio for singly ionized hydrogen atoms. Thus, if their charge was equal in magnitude to a single electronic charge, the alpha particles would have a mass twice the mass of hydrogen atoms. Alternatively, if the alpha particles were assumed to have a charge twice the magnitude of an electronic charge, their mass would be four times the mass of hydrogen atoms (i.e., equal to the mass of helium atoms). Rutherford was inclined toward the latter interpretation since it would mean that alpha particles were simply doubly ionized helium atoms.

By a simple experiment, using the apparatus shown in figure (4–3), Rutherford proved that this is indeed the case. A very thin-walled glass tube containing a radioactive substance was surrounded by a glass enclosure which was initially evacuated. Some of the alpha particles emitted by the radioactive substance were able to penetrate the thin-walled tube and enter the outer enclosure. After a time, a sensitive test of the contents of the outer enclosure showed that it contained a detectable amount of ordinary helium gas. This finding confirmed the argument that alpha particles are doubly ionized helium atoms—which can pick up two electrons and become neutral helium atoms during or after their passage through the thin-walled tubing.

3. The Scattering of Alpha Particles

Rutherford and his collaborators performed many experiments in the course of a program designed to determine the properties of alpha particles and of their interaction with matter. By far the most interesting of these were the experiments concerning the scattering of alpha particles upon

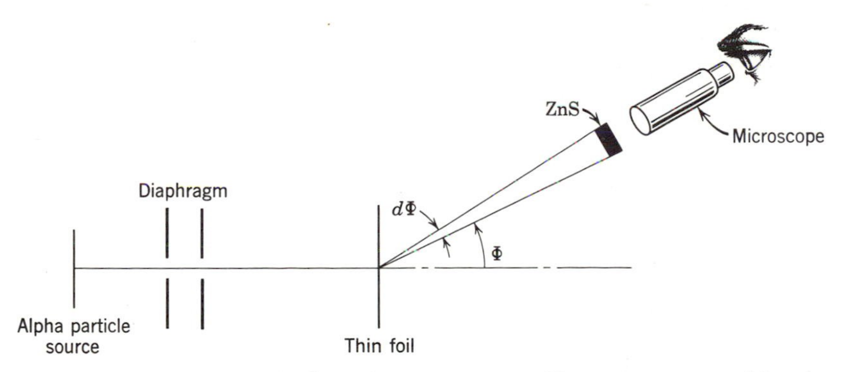

Figure 4–4. An alpha particle scattering experiment. The region traversed by the alpha particles is evacuated.

passing through thin foils of various substances. A typical experimental arrangement is shown in figure (4–4). The radioactive source emits alpha particles which are collimated into a narrow parallel beam by the two diaphragms. The beam is incident upon a foil of some substance, usually a metal. The foil is so thin that the alpha particles pass completely through with only a small decrease in their velocity. However, in the process of traversing the foil, each alpha particle experiences many small deflections due to the Coulomb force acting between its charge and the positive and negative charges of the atoms of the foil. Since the deflection of an alpha particle in passing through a single atom depends on the details of its trajectory through the atom, it is apparent that the net deflection in passing through the entire foil is different for different alpha particles in the beam. As a result, the beam emerging from the far side of the foil is divergent. A quantitative measure of its divergent character is obtained by measuring the number of alpha particles scattered into each angular range Φ to $\Phi + d\Phi$. This is accomplished with the aid of an alpha particle detector consisting of a layer of the crystalline compound ZnS and a microscope. ZnS has the useful property of producing a small flash of light when struck by an alpha particle. If observed by a microscope, the flash due to the incidence of a *single* alpha particle can be distinguished.† In the experiment an observer counts the number of light flashes produced per unit time as a function of the angular position of the detector.

4. Predictions of Thomson's Model

The deflection of an alpha particle in following a given trajectory through an atom can be calculated from Coulomb's law and the laws of mechanics, if the distribution of charges in the atom is given. Thus, for the charge distribution assumed in Thomson's model of the atom, it is possible to calculate the expected results of the experiment just described. It is

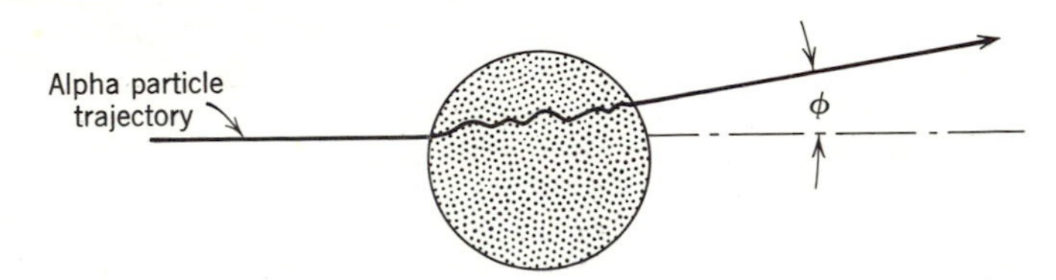

Figure 4–5. An alpha particle passing through a Thomson's model atom.

apparent from the discussion above that the calculation will not be able to predict the behavior of a single alpha particle, but only the *average* behavior of a large number of alpha particles. The average behavior, however, is precisely what is measured in the experiment.

Consider the traversal of an alpha particle through a single atom of the type discussed in Thomson's model; this is indicated schematically in figure (4–5). We shall estimate the order of magnitude of ϕ_{max}, the maximum possible deflection angle of the alpha particle. The calculation will

† This effect is used in the radium dials commonly found on wrist watches. Individual light flashes can be seen by a dark-adapted eye with the aid of a magnifying glass.

use classical mechanics, which is justifiable since $v/c \simeq 1/20$. The deflection is assumed to be due to the Coulomb forces acting between the charged alpha particle and the charges of the various parts of the atom. According to Coulomb's law, the force due to some region of the atom at a distance r from the instantaneous position of the alpha particle, and containing charge dq, is

$$d\mathbf{F} = \frac{2e\,dq}{r^2}\left(\frac{\mathbf{r}}{r}\right) \tag{4–1}$$

where $(\mathbf{r}/r)$ is a vector of unit length directed from the location of the charge dq to the position of the alpha particle, and where $2e$ is the charge of the alpha particle.

Let us begin by evaluating the maximum deflection to be expected from the forces due to the electrons in the atom. At first glance, equation (4–1) would seem to imply that the electrons could give very large deflections to the alpha particle since $d\mathbf{F} \rightarrow \infty$ as $r \rightarrow 0$, so that the alpha particle would experience a very large force and consequently suffer a large deflection if it happened to pass very close to an electron. This is actually not the case because the mass of an electron is so very small compared to the mass of the alpha particle. We leave it as an exercise for the reader to show that, in the collision of a very massive body initially moving at velocity v, with a stationary body of small mass, the velocity of the light body after the collision cannot be greater than $2v$.† This result follows strictly from the conservation of momentum and energy and does not depend on the nature of the force between the two bodies. Consequently the maximum momentum which can be transferred to an electron is $p = m(2v)$, where m is the mass of an electron. This must be equal to the momentum lost by the alpha particle in the collision. Thus Δp_α, the maximum change in the momentum of the alpha particle due to a collision with an electron, is equal to $2mv$. To evaluate the *order of magnitude* of the maximum deflection angle ϕ_{max}, assume that the momentum change is directed perpendicular to the initial momentum p of the alpha particle as illustrated in figure (4–6). Then we have $\phi_{\text{max}} \sim \Delta p_\alpha/p_\alpha \sim 2mv/Mv$, where M is the mass of an alpha particle which is equal to $4 \times 1836m$. Consequently

$$\phi_{\text{max}} \sim \frac{2m}{4(1836)m} = \frac{2}{4(1836)} \sim 10^{-4} \text{ radians}$$

This is the order of magnitude of the maximum deflection angle due to a collision with a single electron in the atom; but it is clear that the

† We assume the binding energy of the electron to the atom to be small compared to the energy of the alpha particle. Then in the collision the electron will be essentially free.

maximum value of the total deflection angle (due to all the collisions with electrons in the atom) will be of the same order since the chance of having a very close collision with more than one or two of the electrons is negligible.

To estimate $\phi_{\max}$ due to the forces arising from the positive charge, we integrate equation (4–1) over the positive charge distribution to obtain

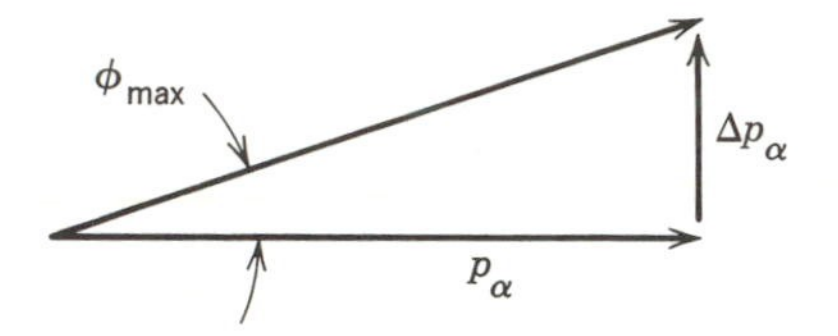

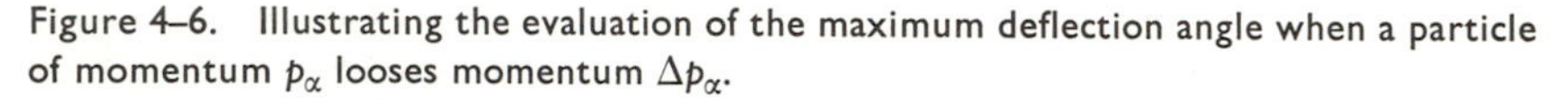

Figure 4–6. Illustrating the evaluation of the maximum deflection angle when a particle of momentum p_α looses momentum Δp_α.

the total force acting on the alpha particle at some instant when it is inside the atom. This gives

$$\mathbf{F} = \int \frac{2e\,dq}{r^2}\left(\frac{\mathbf{r}}{r}\right) \tag{4–2}$$

where r is the distance from the position of the alpha particle at that instant to a volume element containing the positive charge dq, and $(\mathbf{r}/r)$ is a vector of unit length directed from the volume element to the position of the alpha particle. It is clear that in evaluating the integral there will be cancellation from volume elements symmetrically disposed with respect to the position of the alpha particle. However, to estimate the *order of magnitude* of the maximum total force, we ignore such cancellation and take

$$F_{\max} \sim \int \frac{2e}{r^2}\,dq$$

The order of magnitude of the integral can be evaluated by writing it as

$$F_{\max} \sim 2e\overline{\left(\frac{1}{r^2}\right)}\int dq = 2eq\overline{\left(\frac{1}{r^2}\right)}$$

where $\overline{(1/r^2)}$ is the value of $1/r^2$ averaged over a region of space of atomic dimensions, and $q = Ze$ is the total positive charge of the atom. Now $\overline{(1/r^2)}$ must be of the order of $1/R^2$, where R is the radius of an atom. Consequently

$$F_{\max} \sim 2Ze^2/R^2 \tag{4–3}$$

To estimate the maximum deflection angle $\phi_{\max}$, we evaluate Δp_α the momentum given to the alpha particle by the force F_{max} acting for a time of the order of Δt, the transit time through the atom. The time Δt is approximately equal to R/v, where v is the velocity of the alpha particle, so

$$\Delta p_\alpha \sim F_{\text{max}}\,\Delta t \sim 2Ze^2/Rv$$

and we have

$$\phi_{\text{max}} \sim \Delta p_\alpha / p_\alpha \sim 2Ze^2/RMv^2 \tag{4–4}$$

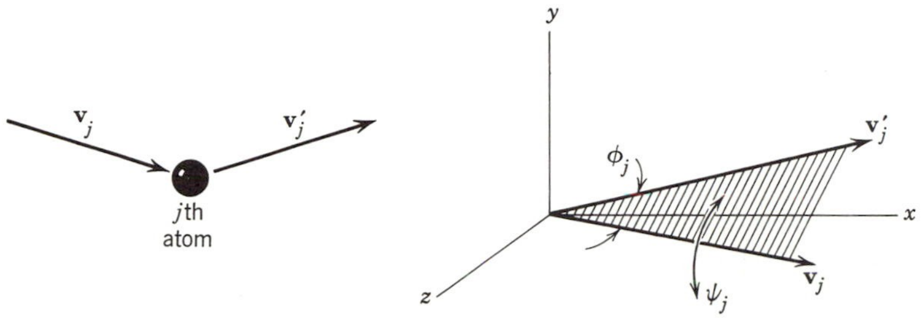

Figure 4–7. The velocity vectors before and after an alpha particle traverses the jth atom.

For atoms containing the largest positive charge, $Z \simeq 100$. The other quantities entering into equation (4–4) have the values

$$e \simeq 5 \times 10^{-10} \text{ esu}$$
$$M \simeq 8 \times 10^{-24} \text{ gm}$$
$$v \simeq 2 \times 10^{-9} \text{ cm/sec}$$
$$R \simeq 10^{-8} \text{ cm}$$

Consequently

$$\phi_{\text{max}} \sim \frac{2 \times 10^2 \times 2 \times 10^{-19}}{10^{-8} \times 8 \times 10^{-24} \times 4 \times 10^{18}} \sim 10^{-4} \text{ radians}$$

We see that the positive charge can produce only a very small deflection because, since the charge $+Ze$ is distributed uniformly over a sphere of radius $\sim 10^{-8}$ cm, the very high velocity alpha particles never get close enough to a large enough quantity of charge that they experience a Coulomb force of the intensity required to produce a large deflection.

Now that we know something about the angle through which an alpha particle is deflected in the traversal of a single atom, let us see what we can learn about the angle through which the particle is deflected in the process of traversing the large number of atoms which lie along its path through the foil. Clearly, summing the results of the individual deflections is involved, and in this sum we must recognize the three dimensional character of the individual deflections. Consider figure (4–7), which indicates typical orientations of $\mathbf{v}_j$ and $\mathbf{v}_j'$, the velocity vectors before and after traversing the jth atom, and a coordinate system useful in describing these vectors. The coordinate system is chosen such that the x axis is in the direction of the beam of alpha particles incident upon the foil. Because the change in the momentum vector of the alpha particle in any single deflection is extremely small, the angles between all the velocity vectors and the initial velocity vector remain small and, for any j, both $\mathbf{v}_j$ and $\mathbf{v}_j'$ are nearly

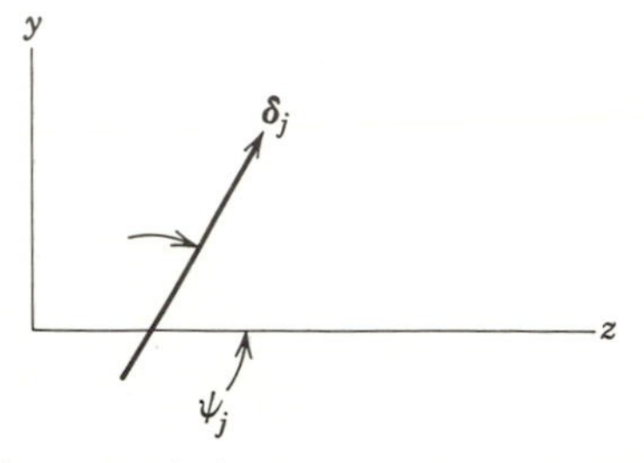

Figure 4–8. A two dimensional diagram representing the *j*th deflection of an alpha particle.

perpendicular to the yz plane. Consequently, the nature of the deflection can be specified by the angle ϕ_j measured between the two vectors in the plane which they define, and the angle ψ_j between that plane and the z axis. This allows us to represent the deflection in terms of a two dimensional diagram consisting of the y and z axes and a vector $\boldsymbol{\delta}_j$, as shown in figure (4–8). The length of the vector is equal to the magnitude of the angle ϕ_j, and the angle between the vector and the z axis is equal to the angle ψ_j. A moment's consideration will show that the relative orientation between the velocity vector after the Nth deflection and the initial velocity vector can be described in the same representation by a vector $\boldsymbol{\delta}$ where

$$\boldsymbol{\delta} = \sum_{j=1}^{N} \boldsymbol{\delta}_j$$

Thus the entire deflection process can be represented by a two dimensional diagram, a typical example of which is shown in figure (4–9). The net angle of deflection resulting from N individual deflections is equal to the length of the vector $\boldsymbol{\delta}$. Except for the constraint, $\phi_j \lesssim 10^{-4}$ radians, the magnitude of the individual deflection angles ϕ_j, as well as the orientation of the deflection planes specified by ψ_j, depends in an essentially fortuitous manner on both the exact trajectory of the alpha particle through the atom and on the exact location of the charges within the atom at the time the particle passes through. Thus the vectors $\boldsymbol{\delta}_j$ will have random directions and a distribution of lengths, as indicated in the figure.

The random nature of the $\boldsymbol{\delta}_j$, and the fact that there are a large number of them, suggest that we apply the mathematical theory of statistics to the problem of determining the length of the vector $\boldsymbol{\delta}$. Now the diagram in figure (4–9) represents the deflections suffered by one alpha particle in its traversal of the foil. If we were to observe the traversals of a number of alpha particles and make similar diagrams for each of them, we could evaluate two characteristic properties of these diagrams: the *average*

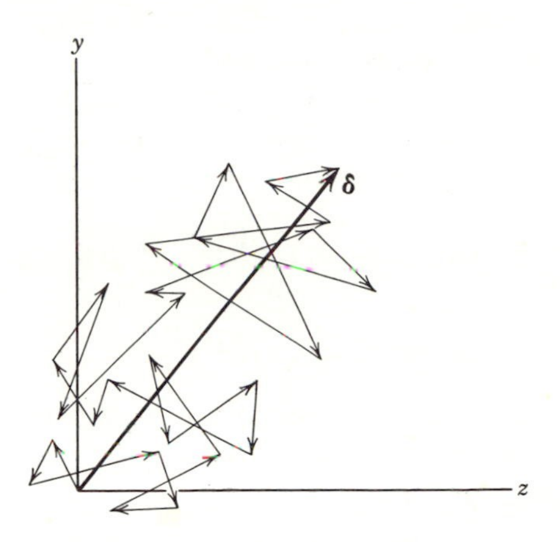

Figure 4–9. A two dimensional diagram representing the entire deflection process experienced by an alpha particle in passing through a foil.

length of the vector $\boldsymbol{\delta}$, and the *distribution* of the lengths of this vector. The theory of statistics makes predictions concerning both of these properties. These predictions are particularly easy to obtain because we have formulated the problem in such a way that it is identical with a well-known statistical problem called the "random walk."† According to the theory $(\overline{\delta^2})^{1/2}$, the root-mean-square average length of the net deflection vector $\boldsymbol{\delta}$, is related to the total number of individual deflections N and to $(\overline{\delta_j^2})^{1/2}$, the root-mean-square average length of the individual deflection vectors $\boldsymbol{\delta}_j$, by the equation

$$(\overline{\delta^2})^{1/2} = \sqrt{N}(\overline{\delta_j^2})^{1/2}$$

It is interesting to note that the average length of the vector $\boldsymbol{\delta}$ increases only as the square root of N because of the random directions of the $\boldsymbol{\delta}_j$. (If, on the other hand, the $\boldsymbol{\delta}_j$ were all pointing in the same direction, which is just the opposite of being randomly directed, the average length of $\boldsymbol{\delta}$

† See any textbook on the theory of statistics.

would clearly be proportional to the first power of N.) The theoretical predictions of the distribution of the lengths of the vector $\boldsymbol{\delta}$ can be stated as

$$\mathcal{N}(\delta)\, d\delta = \frac{2\mathcal{N}\delta}{\overline{\delta^2}}\, e^{-\delta^2/\overline{\delta^2}}\, d\delta$$

The quantity $\mathcal{N}(\delta)\, d\delta$ is the number of cases in which the length of the net deflection vector is within the range of values δ to $\delta + d\delta$; $\overline{\delta^2}$ is the average value of the square of the length of the net deflection vectors; and $\mathcal{N}$ is the total number of cases considered.

Since the lengths of the vectors $\boldsymbol{\delta}$ and $\boldsymbol{\delta}_j$ are equal, respectively, to the net deflection angle Φ and the individual deflection angle ϕ_j, the equations above can be immediately written in terms of the angles:

$$\mathcal{N}(\Phi)\, d\Phi = \frac{2\mathcal{N}\Phi}{\overline{\Phi^2}}\, e^{-\Phi^2/\overline{\Phi^2}}\, d\Phi \tag{4-5}$$

$$(\overline{\Phi^2})^{1/2} = \sqrt{N}(\overline{\phi_j^2})^{1/2} \tag{4-6}$$

The quantity $N(\Phi)\, d\Phi$ is the number of alpha particles scattered within the angular range Φ to $\Phi + d\Phi$; $\overline{\Phi^2}$ is the mean square scattering angle; $\overline{\phi_j^2}$ is the mean square scattering angle in a deflection from a single atom; $\mathcal{N}$ is the number of alpha particles passing through the foil; and N is the number of atoms traversed by an alpha particle in its passage through the foil. These equations, plus the restriction

$$\phi_j \lesssim 10^{-4} \text{ radians} \tag{4-7}$$

that the angle of deflection in passing through a single atom is very small, comprise the predictions of Thomson's model of the atom for the scattering of alpha particles traversing a thin foil.

5. Comparison with Experiment

These predictions were tested by Rutherford and his collaborators. In a typical experiment (Geiger and Marsden, 1909) alpha particles were scattered by an Au foil 10^{-4} cm thick. The average scattering angle $(\overline{\Phi^2})^{1/2}$ was found to be about $1° \simeq 2 \times 10^{-2}$ radians, and more than 99 percent of the alpha particles were scattered at angles less than about 3°. Now, the number of atoms traversed by the alpha particle is approximately equal to the thickness of the foil divided by the diameter of an atom. Thus $N \simeq 10^{-4}\text{ cm}/10^{-8}\text{ cm} = 10^4$. Knowing N and $(\overline{\Phi^2})^{1/2}$, we can use equation (4–6) to evaluate the average deflection angle in traversing a single atom. We obtain

$$(\overline{\phi_j^2})^{1/2} = \frac{(\overline{\Phi^2})^{1/2}}{\sqrt{N}} \simeq \frac{2 \times 10^{-2}}{10^2} = 2 \times 10^{-4} \text{ radians}$$

This is in reasonable agreement with the restriction stated in equation (4–7). Accurate measurements of $\mathcal{N}(\Phi)\, d\Phi$ were difficult to obtain because of the small angles involved. However, the available data were in agreement with equation (4–5), using 1° for $(\overline{\Phi^2})^{1/2}$, for angles less than 2° or 3°.

On the other hand, the angular distribution of the small number of particles scattered at angles large compared to 1° was observed to be in marked disagreement with equation (4–5). This equation predicts that $\mathcal{N}(\Phi)\, d\Phi$ will decrease *very* rapidly with increasing Φ. For instance, if we evaluate the predicted fraction of alpha particles scattered at angles greater than 90°, we find

$$\frac{\mathcal{N}(\Phi > 90^\circ)}{\mathcal{N}} = \frac{\int_{90^\circ}^{180^\circ} \mathcal{N}(\Phi)\, d\Phi}{\mathcal{N}}$$

Using the experimental value of 1° for $(\overline{\Phi^2})^{1/2}$, we obtain the impressively small number $\mathcal{N}(\Phi > 90^\circ)/\mathcal{N} = e^{-(90)^2} \simeq 10^{-3500}$. But experimentally it was found that $\mathcal{N}(\Phi > 90^\circ)/\mathcal{N} \simeq 10^{-4}$. In general, the number of scattered alpha particles was observed to be very much larger than the predicted number for all scattering angles greater than a few degrees. The existence of a small, but non-zero, probability for scattering at large angles was totally inexplicable in terms of Thomson's model of the atom, which could only lead to the *multiple small angle scattering* process described in section 4.

6. Rutherford's Model of the Atom

Experiments utilizing foils of various thicknesses showed that the number of large angle scatterings was proportional to N, the number of atoms traversed by the alpha particle. This is just the dependence on N that would arise if there was a small probability that an alpha particle could be scattered through a large angle in traversing a *single* atom. Since this cannot happen in Thomson's model of the atom, Rutherford proposed a new model (1911).

In Rutherford's model of the structure of the atom, all the positive charge of the atom, and consequently essentially all its mass, is assumed to be concentrated in a small region called the nucleus.

From symmetry considerations the nucleus was assumed to be located at the center of the atom, but its exact location played no role in Rutherford's work.

The arguments of section 4 indicate that a large angle scattering would indeed be possible in the traversal of a single atom of such structure if the alpha particle happened to pass very near the nucleus, provided that the dimensions of the nucleus are small enough. We can get a preliminary idea of the dimensions required from equation (4–4) since, as a review of its derivation will demonstrate, it can be used to estimate the scattering angle of an alpha particle passing near (or through) the nucleus if we interpret R to be the nuclear radius. In the computation directly below equation (4–4), we used $R \sim 10^{-8}$ cm and obtained $\phi \sim 10^{-4}$ radians. Consequently, to obtain a large angle scattering, $\phi \sim 1$ radian, the value of R must be $\sim 10^4$ times smaller; that is, $R \sim 10^{-12}$ cm. As will be indicated later in this chapter, and discussed in detail in Chapter 16, 10^{-12} cm is, in fact, a good estimate of the radius of the atomic nucleus.

7. Predictions of Rutherford's Model

Rutherford made a detailed calculation of the angular distribution to be expected for the scattering of alpha particles from atoms of the type proposed in his model. The calculation was concerned only with scattering at angles greater than several degrees. Consequently, as we have seen above, the scattering due to the atomic electrons can be ignored. The scattering is then due to the repulsive Coulomb force acting between the positively charged alpha particle and the positively charged nucleus. Furthermore, the calculation considered only the scattering from heavy atoms and thus permitted the assumption that the mass of the nucleus is so large compared to the mass of the alpha particle that the nucleus does not recoil (remains fixed in space) during the scattering process. It was also assumed that the alpha particle does not actually penetrate the nuclear region, so that the alpha particle and the nucleus act like point charges as far as the Coulomb force is concerned. We shall see later that all these assumptions are quite valid except for the scattering of alpha particles from the lighter nuclei. The calculation employs classical mechanics since $v/c \sim 1/20$.

Figure (4–10) illustrates the scattering of a particle, of charge $+ze$ and mass M, in passing near a nucleus of charge $+Ze$. The nucleus is fixed at the origin of the coordinate system. When the particle is very far from the nucleus, the Coulomb force on it will vanish, and so the particle approaches the nucleus along a straight line with constant velocity v. After the scattering, the particle will move off finally along a straight line with constant velocity v'. The position of the particle is specified by the radial coordinate r and the polar angle θ, which is measured from an axis drawn parallel to the initial direction of motion. The perpendicular

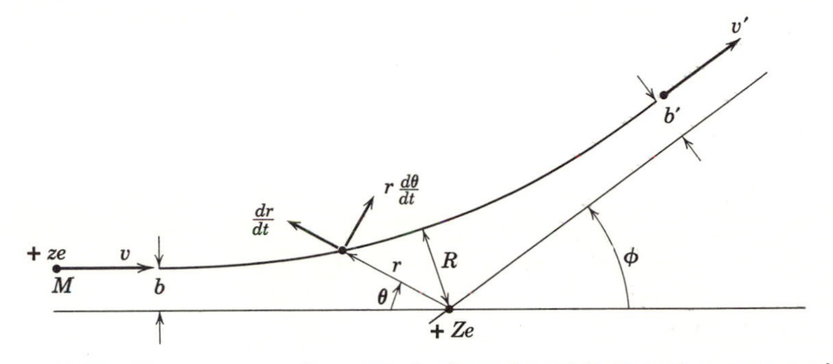

Figure 4–10. The scattering of a positively charged particle in passing near a nucleus.

distance from that axis to the line of initial motion is specified by the impact parameter b. The scattering angle ϕ is just the angle between the axis and a line drawn through the origin parallel to the line of final motion, and the perpendicular distance between these two lines is b'. We resolve the velocity of the particle at any point along its trajectory into a radial component of magnitude dr/dt and a tangential component of magnitude $r\,(d\theta/dt)$.

Now, the force acting on the particle is always in the radial direction. Consequently its angular momentum $Mr^2\,(d\theta/dt)$ has the constant value L. That is,

$$Mr^2 \frac{d\theta}{dt} = L \tag{4–8}$$

Specifically, the initial angular momentum is equal to the final angular momentum, or

$$Mvb = Mv'b' = L \tag{4–8$'$}$$

Of course, the kinetic energy of the particle does not remain constant during the scattering. However, the initial kinetic energy must be equal to the final kinetic energy since the nucleus is assumed to remain stationary. Thus

$$\tfrac{1}{2}Mv^2 = \tfrac{1}{2}Mv'^2$$

From the last two equations we see that $v = v'$ and $b = b'$, as indicated in figure (4–10).

Next, applying Newton's law to the radial component of motion, we have

$$\frac{zZe^2}{r^2} = M\left[\frac{d^2r}{dt^2} - r\left(\frac{d\theta}{dt}\right)^2\right] \tag{4–9}$$

The left-hand term is the Coulomb force. The first term in the bracket is the radial acceleration due to the change in r, and the second term is the (radially directed) centrifugal acceleration. To effect the fastest solution of this equation we transform from the coordinates r, θ to the coordinates u, θ, where

$$r = 1/u \tag{4–10}$$

Then

$$\frac{dr}{dt} = \frac{dr}{d\theta}\frac{d\theta}{dt} = \frac{dr}{du}\frac{du}{d\theta}\frac{d\theta}{dt}$$

$$\frac{dr}{dt} = -\frac{1}{u^2}\frac{du}{d\theta}\frac{Lu^2}{M} = -\frac{L}{M}\frac{du}{d\theta}$$

$$\frac{d^2r}{dt^2} = \frac{d}{d\theta}\left[\frac{dr}{dt}\right]\frac{d\theta}{dt} = -\frac{L}{M}\frac{d^2u}{d\theta^2}\frac{Lu^2}{M}$$

$$\frac{d^2r}{dt^2} = -\frac{L^2u^2}{M^2}\frac{d^2u}{d\theta^2}$$

Substituting into equation (4–9), we have

$$-\frac{L^2u^2}{M^2}\frac{d^2u}{d\theta^2} - \frac{1}{u}\left(\frac{Lu^2}{M}\right)^2 = \frac{zZe^2u^2}{M}$$

$$\frac{d^2u}{d\theta^2} + u = -\frac{zZe^2M}{L^2} = -\frac{zZe^2M}{M^2v^2b^2} \tag{4–11}$$

The constant D, defined by

$$D \equiv \frac{zZe^2}{\frac{1}{2}Mv^2} \tag{4–12}$$

is a convenient parameter equal to the distance of closest approach to the nucleus in a *head-on collision* ($b = 0$), since D is the distance at which the potential energy zZe^2/D is equal to the initial kinetic energy $\frac{1}{2}Mv^2$; at this point the particle would come to a stop and then reverse its direction of motion. In terms of this parameter, equation (4–11) reads

$$d^2u/d\theta^2 + u = -D/2b^2 \tag{4–13}$$

This is a second order differential equation (i.e., involving a second derivative) for u as a function of θ. The *general solution* to this equation is

$$u = A \cos \theta + B \sin \theta - D/2b^2 \tag{4–14}$$

which contains two arbitrary constants, A and B. We may prove that equation (4–14) is, in fact, the solution to (4–13) by evaluating

$$du/d\theta = -A \sin \theta + B \cos \theta$$

and

$$d^2u/d\theta^2 = -A \cos \theta - B \sin \theta$$

and substituting into (4–13). This gives

$$-A \cos \theta - B \sin \theta + A \cos \theta + B \sin \theta - D/2b^2 \equiv -D/2b^2, \quad \text{Q.E.D.}$$

To determine A and B, we require that equation (4–14) conform to the *initial conditions:* $\theta \rightarrow 0$ as $r \rightarrow \infty$, and $dr/dt \rightarrow -v$ as $r \rightarrow \infty$. Thus

$$u = \frac{1}{r} = 0 = A \cos 0 + B \sin 0 - \frac{D}{2b^2}$$

or

$$A = \frac{D}{2b^2}$$

and

$$\frac{dr}{dt} = -\frac{L}{M}\frac{du}{d\theta} = -v = -\frac{L}{M}(-A \sin 0 + B \cos 0)$$

or

$$B = \frac{Mv}{L} = \frac{Mv}{Mvb} = \frac{1}{b}$$

Therefore the *particular solution* of the differential equation pertinent to the problem at hand is

$$u = \frac{D}{2b^2} \cos \theta + \frac{1}{b} \sin \theta - \frac{D}{2b^2}$$

or

$$\frac{1}{r} = \frac{1}{b} \sin \theta + \frac{D}{2b^2} (\cos \theta - 1) \tag{4–15}$$

This is the equation of a hyperbola in polar coordinates. We see that the trajectory of the particle moving under the influence of a repulsive inverse square Coulomb force is one of the conic sections. The trajectory of a particle moving under an attractive inverse square force (e.g., a satellite in the gravitational field of the earth) could be obtained from the derivation above by simply reversing the sign of the term on the left side of equation (4–9). The trajectory would then turn out to be one of the conic sections—exactly which one depending on the values of the parameters.

From the equation (4–15) of the trajectory of the particle, we may evaluate the scattering angle ϕ by letting $r \to \infty$. Then

$$0 = \frac{1}{b} \sin \theta' + \frac{D}{2b^2} (\cos \theta' - 1)$$

One solution to this equation is $\theta' = 0$. This is the polar angle for large r before scattering. The interesting solution is

$$\frac{2b}{D} = \frac{1 - \cos \theta'}{\sin \theta'} = \tan \frac{\theta'}{2}$$

which gives the polar angle for large r after scattering. It is apparent from figure (4–10) that for this solution $\phi = \pi - \theta'$. Consequently

$$\tan \left(\frac{\pi - \phi}{2} \right) = \frac{2b}{D}$$

which is equivalent to

$$\cot \frac{\phi}{2} = \frac{2b}{D} \tag{4–16}$$

It will also be useful to evaluate R, the distance of closest approach of the particle to the center of the nucleus (the origin). From figure (4–10) it is apparent that the radial coordinate will assume this value when the polar angle is equal to $\theta'/2 = (\pi - \phi)/2$. Evaluating equation (4–15) for this angle, and using (4–16), we obtain

$$R = \frac{D}{2} \left[1 + \frac{1}{\sin (\phi/2)} \right] \tag{4–17}$$

It is easy to verify that $R \to D$ as $\phi \to \pi$, and that $R \to b$ as $\phi \to 0$, as would be expected.

We see from equation (4–16) that, in the scattering of an alpha particle by a single nucleus, if the impact parameter is in the range b to $b + db$ the scattering angle is in the range ϕ to $\phi + d\phi$, where the relation between b and ϕ is given by the equation. Thus the problem of calculating the number of alpha particles scattered in the angular range ϕ to $\phi + d\phi$ in traversing the entire foil is equivalent to the problem of calculating the number which are incident, with impact parameter from b to $b + db$, upon the nuclei in the foil. Let the number of nuclei (i.e., the number of atoms) per cm^3 in the foil be ρ, and let the foil be t cm thick. Consider a segment of the foil with a cross-sectional area of 1 cm^2, as shown in figure

(4–11). This diagram indicates a ring, of inner radius b and outer radius $b + db$, drawn around an axis passing through each nucleus. The area of each ring is $2\pi b\, db$. The number of such rings in the segment is ρt, where t is numerically equal to the volume of the segment. The probability $P(b)\, db$ that an alpha particle will pass through one of these rings is equal

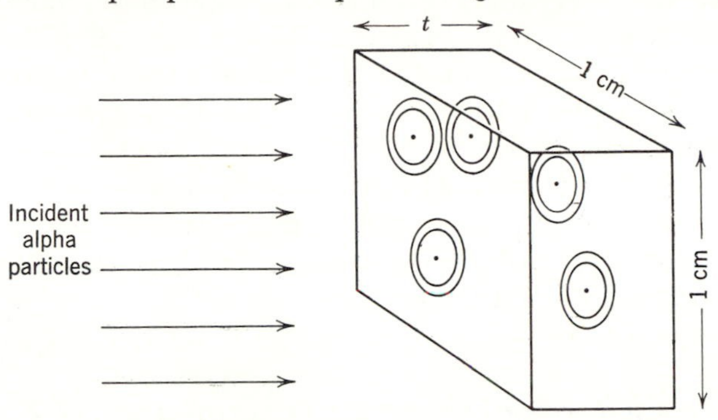

Figure 4–11. A beam of alpha particles incident upon a 1 cm² foil. In actuality the rings are very much smaller and there are very many more of them.

to the total area obscured by the rings, as seen by the incident alpha particles, divided by the total area of the segment. That is,

$$P(b)\, db = \rho t 2\pi b\, db$$

Now

$$b = \frac{D}{2} \cot \frac{\phi}{2}$$

so

$$db = -\frac{D}{2} \frac{\frac{1}{2} d\phi}{\sin^2 (\phi/2)}$$

and

$$b\, db = -\frac{D^2}{8} \frac{\cos (\phi/2)\, d\phi}{\sin^3 (\phi/2)} = -\frac{D^2}{16} \frac{\sin \phi\, d\phi}{\sin^4 (\phi/2)}$$

Thus

$$P(b)\, db = -\frac{\pi}{8} \rho t D^2 \sin \phi \frac{d\phi}{\sin^4 (\phi/2)}$$

But $-P(b)\, db$ is equal to the probability that the incident particles will be scattered in the angular range ϕ to $\phi + d\phi$. (The minus sign compensates for the fact that db and $d\phi$ are of the opposite sign.) Using the notation employed in equation (4–5), this is

$$\frac{\mathscr{N}(\Phi)\, d\Phi}{\mathscr{N}} = -P(b)\, db = \frac{\pi}{8} \rho t D^2 \frac{\sin \Phi\, d\Phi}{\sin^4 (\Phi/2)}$$

Evaluating D, we have

$$\mathcal{N}(\Phi)\,d\Phi = \mathcal{N}\,\frac{\pi}{8}\,\rho t\left(\frac{zZe^2}{\frac{1}{2}Mv^2}\right)^2 \frac{\sin\Phi\,d\Phi}{\sin^4(\Phi/2)} \tag{4–18}$$

This is the angular distribution derived by Rutherford for the scattering of alpha particles in passing through a foil composed of atoms with the nuclear structure proposed by his model. The quantity $\mathcal{N}$ is the number of alpha particles incident on the foil of thickness t and ρ nuclei per cm^3; $\mathcal{N}(\Phi)\,d\Phi$ is the number scattered at angles from Φ to $\Phi + d\Phi$; ze is the charge of the alpha particle; M is its mass; v is its velocity; and Ze is the charge of the nuclei. We note that, although the angular factor in this equation decreases rapidly with increasing angle, nevertheless the decrease is very much less rapid than the decrease predicted by the multiple small angle scattering angular distribution (equation 4–5).

8. Experimental Verification and the Determination of Z

Detailed experimental tests of equation (4–18) were performed by Geiger and Marsden within a few months after its derivation (1911). They investigated the following points.

1. The angular dependence was tested, using foils of Ag and Au, over the angular range 5° to 150°. Although $\mathcal{N}(\Phi)\,d\Phi$ varies by a factor of about 10^5 over this range, the experimental data remained proportional to the theoretical angular distribution to within a few percent.
2. The quantity $\mathcal{N}(\Phi)\,d\Phi$ was found indeed to be proportional to t for a range of thickness of about 10 for all the elements investigated.
3. Equation (4–18) predicts that the number of scattered alpha particles will be inversely proportional to the square of their kinetic energy, $\frac{1}{2}mv^2$. This was tested by using alpha particles from several different radioactive elements. The predicted energy dependence was observed over an energy variation of about a factor of 3.
4. Finally, the equation predicts $\mathcal{N}(\Phi)\,d\Phi$ to be proportional to $(Ze)^2$, the square of the nuclear charge. At the time, Z was not known for the various atoms, except that, as mentioned in section 1, X-ray scattering experiments had shown that Z was roughly equal to $A/2$, where A is the chemical atomic weight of the atom. The validity of equation (4–18) being assumed, the experimental data could be used to determine Z since everything else was known. The resulting values agreed with the law $Z \sim A/2$. Furthermore it was observed that, to within the accuracy of the experiment, the measured values of Z were equal to the chemical atomic number of the atoms in question, i.e., equal to the ordering number

of the atoms in the periodic table. This implies that the first atom, H, in the periodic table contains 1 electron, the second atom, He, contains 2 electrons, the third atom, Li, contains 3 electrons, etc., since Z is the number of electrons in the atom. This was shortly confirmed by very accurate determinations of Z using X-ray techniques, which will be described in Chapter 14.

9. The Size of the Nucleus

By evaluating R, the distance of closest approach, Rutherford was able to place limits on the size of the nucleus. As an example, in the scattering of Ra alpha particles ($v = 1.6 \times 10^9$ cm/sec) from Cu ($A = 63, Z = 29$) it was observed that the Rutherford scattering law was obeyed to the largest scattering angle investigated, $\Phi = 180°$. From equation (4–17) we see that R takes on its smallest value at that angle. So

$$R_{180°} = D = \frac{zZe^2}{\frac{1}{2}Mv^2}$$

$$R_{180°} = \frac{2 \times 29 \times (4.8 \times 10^{-10})^2}{\frac{1}{2}(4 \times 1.67 \times 10^{-24}) \times (1.6 \times 10^9)^2}$$

$$R_{180°} \simeq 1.7 \times 10^{-12} \text{ cm}$$

Rutherford's derivation depends on the assumption that the force acting on the alpha particle is always strictly a Coulomb force between two point charges. This would not be true if the particle penetrated the nuclear region at its point of closest approach. The obvious implication is that the radius of the Cu nucleus is $\leqslant 1.7 \times 10^{-12}$ cm.†

The equation above shows that R decreases as Z decreases. The question arises: How much can R decrease before $R <$ nuclear radius? Departures were actually observed in the scattering from the very light (low Z) nuclei. Part of this was due to a violation, for the very light nuclei, of the assumption that the nuclear mass is large compared to the alpha particle mass. However, deviations remained even after the finite nuclear mass was taken into account in the theory. Figure (4–12) shows some data for the scattering of alpha particles, of various velocities, at a fixed large

† We may also conclude that the use of classical mechanics to calculate the angular distribution for scattering from a Coulomb field is valid, even for collisions in which R is as small as 10^{-12} cm. In Chapter 15 we shall find that quantum theory verifies this for the case of a Coulomb force, but that for any other force classical mechanics would give only an approximation in such collisions.

angle from an Al foil. The ordinate is the ratio of the observed number of scattered particles to the number predicted by the Rutherford theory (corrected for the finite nuclear mass). The abscissa is the distance of closest approach calculated from equation (4–17). These data imply that the radius of the Al nucleus is about 10^{-12} cm.

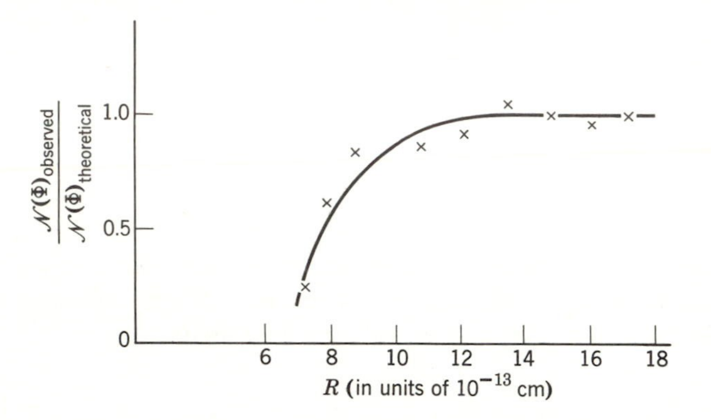

Figure 4–12. Some data obtained in the scattering of alpha particles from aluminum.

10. A Problem

The detailed experimental verification of the predictions of the nuclear model of the atom left little room for doubt concerning the validity of the model. However, the question of the *stability* of such an atom presented a serious problem.

If we assume that the electrons in the atom are stationary, it is easy to see that there exists no stable arrangement of the electrons surrounding the nucleus which would prevent the electrons from falling into the nucleus under the influence of its Coulomb attraction. We cannot allow the atom to collapse, because then its radius would be of the order of a nuclear radius, which is four orders of magnitude smaller than what we know the radius of an atom to be.

At first glance it appears that we can simply allow the electrons to circulate about the nucleus in orbits similar to the orbits of the planets circulating about the sun. Such a system can be stable mechanically because the centrifugal force can be made to balance the Coulomb attraction just as it balances the gravitational attraction in the planetary system. But a difficulty arises in trying to carry over this idea from the planetary system to the atomic system. The problem is that the charged electrons would be constantly accelerating in their motion around the nucleus and, according to the classical electromagnetic theory, all accelerating charged bodies radiate energy in the form of electromagnetic radiation. The energy would be emitted at the expense of the mechanical energy of the electron, and the electron would spiral into the nucleus. Again we have an atom which would rapidly collapse to nuclear dimensions. Furthermore, the continuous spectrum of the radiation which would be emitted is not in agreement with the discrete spectrum which is known to be emitted by atoms.

BIBLIOGRAPHY

Evans, R. D., *The Atomic Nucleus*, McGraw-Hill Book Co., New York, 1955.

Rutherford, E., J. Chadwick, and C. D. Ellis, *Radiations from Radioactive Substances*, Cambridge University Press, Cambridge, England, 1930.

EXERCISES

1. Use classical mechanics to prove the statement, made in section 4, that in a collision of a very massive body, initially moving at velocity v, with a free stationary body of small mass, the velocity of the light body after the collision can be no greater than $2v$. *Hint:* Make a Galilean transformation to the CM system, treat the collision, then transform back to the LAB system.

2. Derive equation (4–8) from Newton's law.

3. Adapt the calculation of section 7 to the case of a particle moving under an attractive inverse square force. Use the results to discuss the motion of a satellite.

4. Verify from equation (4–17) that $R \to D$ as $\phi \to \pi$, and that $R \to b$ as $\phi \to 0$. Give physical arguments explaining these results.

5. Adapt the calculation of section 7 to the case of an electron moving in a hydrogen atom, under the assumption that it radiates energy at the rate given by equation (2–7) because it is accelerated. If the radius of its orbit is initially 10^{-8} cm, how long would it take for it to spiral into the nucleus of radius 10^{-12} cm?

CHAPTER **5**

Bohr's Theory of Atomic Structure

1. Atomic Spectra

The difficult problem of the stability of atoms actually led to a simple theory of atomic structure. A key feature of this very successful theory, proposed in 1913 by Bohr, was the prediction of the spectrum of electromagnetic radiation emitted by certain atoms. Because of this prediction, and because we have mentioned atomic spectra on two occasions in the previous chapter, it is necessary at this point to describe some of the less complex features of such spectra. A detailed discussion of atomic spectra is, however, best deferred until after we have developed the theory of quantum mechanics.

A typical apparatus used in the measurement of atomic spectra is indicated in figure (5–1). The source consists of an electric discharge passing through a region containing a monatomic gas. Owing to collisions with electrons, and with each other, some of the atoms in the discharge are put into a state in which their total energy is greater than it is in a normal atom. In returning to their normal energy state, the atoms give up their excess energy by emitting electromagnetic radiation. The radiation is collimated by the slit, and then it passes through a prism (or diffraction grating), consequently breaking up into its spectrum, which is recorded on the photographic plate.

The nature of the observed spectra is indicated by the front view of the photographic plate. In contrast to the continuous spectrum of electromagnetic radiation emitted, for instance, from the surface of solids at high temperature, *the electromagnetic radiation emitted by a free atom is concentrated at a number of discrete wavelengths.* Each of these wavelength components is called a *line* because of the line which it produces on the photographic plate. Upon investigating the spectra emitted from different kinds of atoms, it is observed that each kind of atom has its own characteristic spectrum, i.e., characteristic set of wavelengths at which the lines of the spectrum are found. This feature is of great practical importance

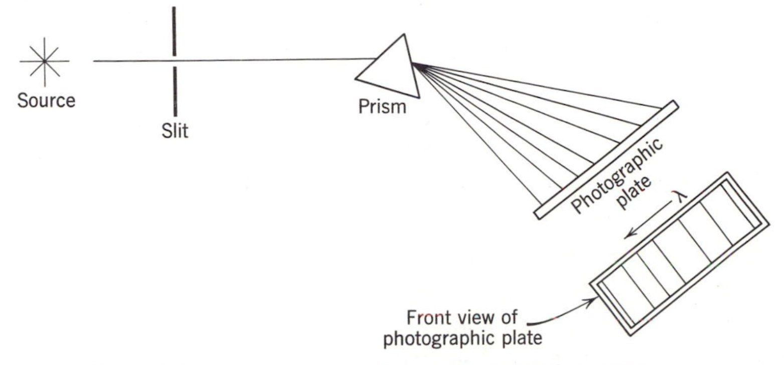

Figure 5–1. An apparatus used to measure atomic spectra.

because it makes *spectroscopy* a very useful addition to the usual techniques of chemical analysis. Primarily as a result of this motivation, considerable effort was devoted in the nineteenth century to the accurate measurement of atomic spectra. And, in fact, considerable effort was required because the spectra consist of many hundreds of lines and in general are very complicated.

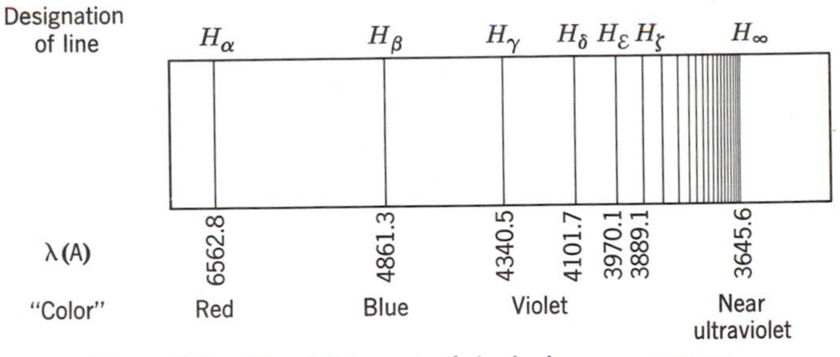

Figure 5–2. The visible part of the hydrogen spectrum.

However, the spectrum of hydrogen is relatively simple. This is perhaps not surprising since hydrogen, which contains just one electron, is itself the simplest atom. The H spectrum is of considerable interest, not only because of its simplicity, but also because of historical and theoretical reasons which will become apparent later. Figure (5–2) represents that part of the H spectrum which falls approximately within the wavelength range of visible light. The wavelength λ of the lines is measured in units of 1 Angstrom = 1 A = 10^{-8} cm; the unit is named after a physicist who was responsible for some of the first precise spectroscopic measurements. We see that the spacing, in wavelengths, between adjacent lines of the spectrum continuously decreases with decreasing wavelength of the lines, so that the *series of lines* converges to the so-called *series limit* at 3645.6 A. The short wavelength lines, including the series limit, are hard to observe experimentally because of their close spacing and because they are in the ultraviolet.

The obvious regularity of the H spectrum tempted several people to look for an empirical formula which would represent the wavelength of the lines. Such a formula was discovered in 1885 by Balmer. He found that the simple equation

$$\lambda = 3646 \frac{n^2}{n^2 - 4} \quad \text{(in A units)} \tag{5-1}$$

where $n = 3$ for H_α, $n = 4$ for H_β, $n = 5$ for H_γ etc., was able to predict the wavelength of the first nine lines of the series, which were all that were known at the time, to better than 1 part in 1000. This discovery initiated a search for similar empirical formulas that would apply to series of lines which can sometimes be identified in the complicated distribution of lines that constitute the spectra of other elements. Most of this work was done by Rydberg (ca. 1890), who found it convenient to deal with the *wave number*,

$$k \equiv 1/\lambda \tag{5-2}$$

of the lines, instead of their wavelength. In terms of this quantity, the Balmer formula can be rewritten

$$k = R_{\mathrm{H}}(1/2^2 - 1/n^2), \qquad n = 3, 4, 5, \ldots \tag{5-3}$$

where R_{H} is the so-called *Rydberg constant* for hydrogen. From recent spectroscopic data, its value is known to be

$$R_{\mathrm{H}} = 109677.576 \pm 0.012 \text{ cm}^{-1} \tag{5-4}$$

This is indicative of the accuracy possible in spectroscopic measurements.

Formulas of this type were found for a number of series. For instance, we now know of the existence of five series of lines in the H spectrum, as shown in table (5–1). For alkali elements the series formulas are of the same general structure:

$$k = R\left(\frac{1}{(m - a)^2} - \frac{1}{(n - b)^2}\right) \tag{5-5}$$

where R is the Rydberg constant for the particular element, a and b are

TABLE (5–1)

Names, Wavelength Ranges, and Formulas for the Hydrogen Series

Lyman	Ultraviolet	$k = R_H\left(\frac{1}{1^2} - \frac{1}{n^2}\right)$,	$n = 2, 3, 4, \ldots$
Balmer	Near ultraviolet and visible	$k = R_H\left(\frac{1}{2^2} - \frac{1}{n^2}\right)$,	$n = 3, 4, 5, \ldots$
Paschen	Infrared	$k = R_H\left(\frac{1}{3^2} - \frac{1}{n^2}\right)$,	$n = 4, 5, 6, \ldots$
Brackett	Infrared	$k = R_H\left(\frac{1}{4^2} - \frac{1}{n^2}\right)$,	$n = 5, 6, 7, \ldots$
Pfund	Infrared	$k = R_H\left(\frac{1}{5^2} - \frac{1}{n^2}\right)$,	$n = 6, 7, 8, \ldots$

constants for the particular series, m is an integer which is fixed for the particular series, and n is a variable integer. To within about 0.05 percent the Rydberg constant has the same value for all elements, although it does show a very slight systematic increase with increasing atomic weight.

We have been discussing the *emission spectrum* of an atom. A closely related property is the *absorption spectrum.* This may be measured with apparatus similar to that shown in figure (5–1) except that a source emitting a continuous spectrum is used and a glass-walled cell, containing the monatomic gas to be investigated, is inserted somewhere between the source and the prism. After exposure and development, the photographic plate is found to be darkened everywhere except for a number of unexposed lines. These lines represent a set of discrete wavelength components which were missing from the otherwise continuous spectrum incident upon the prism, and which must have been absorbed by the atoms in the gas cell. It is observed that for every line in the absorption spectrum of an element there is a corresponding (same wavelength) line in its emission spectrum. However, the reverse is not true. Only certain emission lines show up in the absorption spectrum. For hydrogen gas, normally only lines corresponding to the Lyman series appear in the absorption spectrum; but, when the gas is at very high temperatures, e.g., at the surface of a star, lines corresponding to the Balmer series are found.

2. Bohr's Postulates

All these features of atomic spectra, and many more which we have not discussed, must be explained by any successful theory of atomic structure. Furthermore, the very great precision which characterizes spectroscopic measurements imposes particularly severe requirements on the accuracy with which such a theory must be able to predict the quantitative features of the spectra.

Nevertheless, in 1913 Bohr developed a theory which was in accurate quantitative agreement with certain of the spectroscopic data (e.g., the hydrogen spectrum). This theory had the additional attraction that the mathematics involved was very easy to understand. However, the postulates upon which it was based were not so transparent. These postulates were:

1. *An electron in an atom moves in a circular orbit about the nucleus under the influence of the Coulomb attraction between the electron and the nucleus, and obeying the laws of classical mechanics.*

2. *But, instead of the infinity of orbits which would be possible in classical mechanics, it is only possible for an electron to move in an orbit for which its orbital angular momentum L is an integral multiple of Planck's constant h, divided by 2π.*

3. *Despite the fact that it is constantly accelerating, an electron moving in such an allowed orbit does not radiate electromagnetic energy. Thus its total energy E remains constant.*

4. *Electromagnetic radiation is emitted if an electron, initially moving in an orbit of total energy E_i, discontinuously changes its motion so that it moves in an orbit of total energy E_f. The frequency of the emitted radiation ν is equal to the quantity $(E_i - E_f)$ divided by Planck's constant h.*

The first postulate bases Bohr's theory on the existence of the atomic nucleus. It embodies some of the ideas concerning the stability of a nuclear atom discussed at the end of the preceding chapter. The second postulate introduces quantization. The reader should note the difference between Bohr's quantization of the *orbital angular momentum* of an atomic electron moving under the influence of an *inverse square (Coulomb) force*,

$$L = nh/2\pi, \qquad n = 1, 2, 3, \ldots \tag{5-6}$$

and Planck's quantization of the *total energy* of a particle, such as an electron, executing simple harmonic motion under the influence of a *harmonic restoring force:* $E = nh\nu$, $n = 0, 1, 2, \ldots$. We shall see in the next section that the quantization of the orbital angular momentum of the atomic electron does lead to the quantization of its total energy, but with an energy quantization equation which is different from Planck's equation. The third postulate removes the problem of the stability of an electron moving in a circular orbit, due to of the emission of the electromagnetic radiation required of the electron by classical theory, by simply postulating that this particular feature of the classical theory is not valid for the case of an atomic electron. The postulate was based on the fact that atoms are *observed* by *experiment* to be stable—even though this is not *predicted* by the *classical theory*. The fourth postulate,

$$\nu = \frac{E_i - E_f}{h} \tag{5–7}$$

is obviously closely related to Einstein's postulate (equation 3–7) that the frequency of a quantum of electromagnetic radiation is equal to the energy carried by the quantum divided by Planck's constant.

These postulates do a thorough job of mixing classical and non-classical physics. The electron moving in a circular orbit is assumed to obey classical mechanics, and yet the non-classical idea of quantization of orbital angular momentum is included. The electron is assumed to obey one feature of classical electromagnetic theory (Coulomb's law), and yet not to obey another feature (emission of radiation by an accelerated charged body). However, we should not be surprised if the laws of classical physics, which are based on our experience with macroscopic systems, are not completely valid when dealing with microscopic systems such as the atom.

3. Bohr's Theory of the One-Electron Atom

The justification of Bohr's postulates, or of any set of postulates, can be found only by comparing the predictions that can be derived from the postulates with the results of experiment. In this section we derive some of these predictions and compare them with the data of section 1.

Consider an atom consisting of a nucleus of charge $+Ze$ and mass M, and a single electron of charge $-e$ and mass m. For a neutral hydrogen atom $Z = 1$, for a singly ionized helium atom $Z = 2$, for a doubly ionized lithium atom $Z = 3$, etc. We assume that the electron rotates in a circular orbit about the nucleus. Initially we suppose the mass of the electron to be completely negligible compared to the mass of the nucleus, and consequently assume that the nucleus remains fixed in space. Thus we consider at first the situation depicted in figure (5–3). The condition of mechanical stability of the electron is

$$\frac{Ze^2}{r^2} = m\frac{v^2}{r} \tag{5–8}$$

where v is the velocity of the electron in its orbit, and r is the radius of the orbit. The left side of equation (5–8) is the Coulomb force acting on the electron, and the right side is the centrifugal force required to keep it in its circular orbit. Now, the orbital angular momentum of the electron

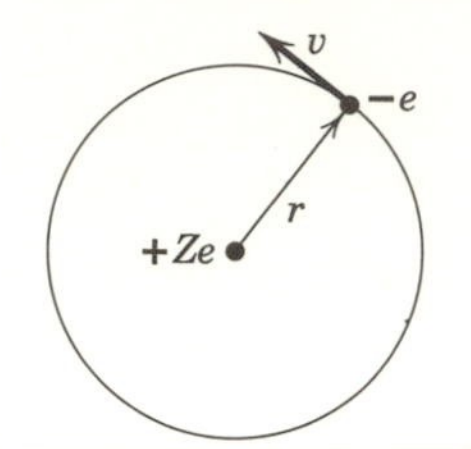

Figure 5–3. An electron moving in a circular orbit about a fixed nucleus.

must be a constant because the force acting on the electron is entirely in the radial direction. It is equal to

$$L = mvr \tag{5–9}$$

Applying the quantization condition (5–6) to L, we have

$$mvr = nh/2\pi, \qquad n = 1, 2, 3, \ldots$$

or, in terms of the convenient symbol

$$\hbar \equiv h/2\pi \tag{5–10}$$

$$mvr = n\hbar, \qquad n = 1, 2, 3, \ldots \tag{5–11}$$

Solving for v and substituting into equation (5–8), we obtain

$$Ze^2 = mrv^2 = mrn^2\hbar^2/m^2r^2 = n^2\hbar^2/mr$$

So

$$r = n^2\hbar^2/mZe^2, \qquad n = 1, 2, 3, \ldots \tag{5–12}$$

and

$$v = n\hbar mZe^2/mn^2\hbar^2 = Ze^2/n\hbar, \qquad n = 1, 2, 3, \ldots \tag{5–13}$$

The application of the angular momentum quantization condition has restricted the possible circular orbits to those of radii given by equation (5–12). Note that these radii are proportional to the square of the quantum number n. Let us evaluate the radius of the smallest orbit ($n = 1$) for a hydrogen atom ($Z = 1$). Inserting the known values of $\hbar$, m, and e, we obtain $r = 5.3 \times 10^{-9}$ cm. It is shown below that the electron has its minimum total energy when in the orbit corresponding to $n = 1$. Consequently we may interpret the radius of this orbit as a measure of the radius of a hydrogen atom in its normal state. It is in good agreement with the estimate, mentioned previously, that the order of magnitude of an atomic radius is 10^{-8} cm. Evaluating the orbital velocity of an electron in the smallest orbit of a hydrogen atom from equation (5–13), we find $v = 2.2 \times 10^8$ cm/sec.† It is apparent from the equation that this is the largest velocity possible for an electron in a hydrogen atom. The fact that this velocity is less than 1 percent of the velocity of light is the justification for using classical mechanics instead of relativistic mechanics in the Bohr theory. On the other hand, equation (5–13) shows that for large values of Z the electron velocity becomes relativistic. The theory could not be applied in such cases.

Next we calculate the total energy of an atomic electron moving in one of the allowed orbits. Let us define the potential energy to be zero when the electron is infinitely distant from the nucleus. Then the potential energy V at any finite distance r can be obtained by integrating the energy imparted to the electron by the Coulomb force acting from ∞ to r. Thus

$$V = \int_{\infty}^{r} \frac{Ze^2}{r^2}\, dr = -\frac{Ze^2}{r}$$

The potential energy is negative because the Coulomb force is attractive. The kinetic energy of the electron, T, can be evaluated, with the aid of equation (5–8), to be

$$T = \tfrac{1}{2}mv^2 = Ze^2/2r$$

The total energy of the electron, E, is then

$$E = T + V = -Ze^2/2r = -T$$

Using equation (5–12), we have

$$E = -Ze^2mZe^2/2n^2\hbar^2 = -mZ^2e^4/2n^2\hbar^2, \qquad n = 1, 2, 3, 4, \ldots \qquad (5\text{–}14)$$

We see that the quantization of the orbital angular momentum of the electron leads to a quantization of its total energy. The information contained in equation (5–14) is presented as an energy level diagram in figure (5–4). The energy of each level, as evaluated from equation (5–14), is shown on the left in terms of ergs and also in terms of electron volts (ev), a unit of energy which will be explained in section 5, and the quantum number of the level is shown on the right. The diagram is so constructed that the distance from any level to the zero energy level is proportional to the energy of that level. Note that the lowest (most negative) allowed value of total energy occurs for the smallest quantum number $n = 1$. As n increases, the total energy of the quantum state becomes less negative, with E approaching zero as n approaches infinity. Since the state of lowest total energy is, of course, the most stable state for the electron, we see

† Equation (5–13) explains why Bohr could not allow the quantum number n ever to assume the value $n = 0$, as it may in Planck's quantization equation.

that the normal state of the electron in a one-electron atom is the state for which $n = 1$.

Next we calculate the frequency ν of the electromagnetic radiation emitted when the electron makes a transition from the quantum state n_i to the quantum state n_f, that is, when an electron initially moving in an orbit characterized by the quantum number n_i discontinuously changes its

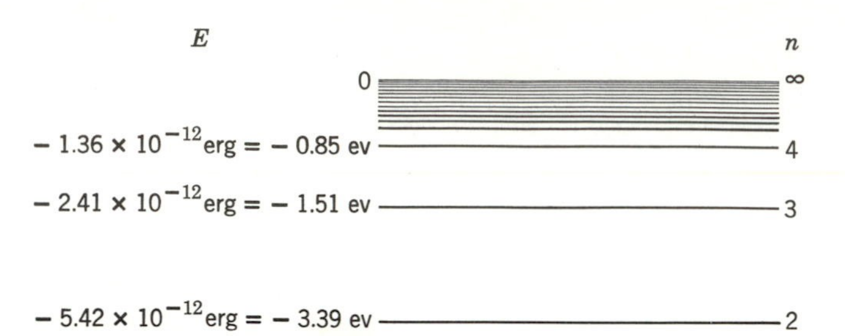

Figure 5–4. An energy level diagram for the hydrogen atom.

motion so that it moves in an orbit characterized by quantum number n_f. Using Bohr's fourth postulate (equation 5–7), and equation (5–14), we have

$$\nu = \frac{E_i - E_f}{h} = +\frac{mZ^2e^4}{4\pi\hbar^3}\left(\frac{1}{n_f^2} - \frac{1}{n_i^2}\right)$$

In terms of the wave number $k = 1/\lambda = \nu/c$, this is

$$k = \frac{me^4}{4\pi c\hbar^3}\, Z^2\left(\frac{1}{n_f^2} - \frac{1}{n_i^2}\right)$$

or

$$k = R_\infty Z^2\left(\frac{1}{n_f^2} - \frac{1}{n_i^2}\right), \qquad R_\infty \equiv \frac{me^4}{4\pi c\hbar^3} \tag{5–15}$$

where n_i and n_f are integers.

The essential predictions of the Bohr theory are contained in equations (5–14) and (5–15). Let us first discuss the emission of electromagnetic radiation by a one-electron Bohr atom in terms of these equations.

1. The normal state of the atom will be the state in which the electron has the lowest energy, i.e., the state $n = 1$. This is called the *ground state.*

2. In an electric discharge the atom receives energy due to collisions, etc. This means that the electron must make a transition to a state of higher energy, or *excited state*, in which $n > 1$.

3. Obeying the common tendency of all physical systems, the atom will emit its excess energy and return to the ground state. This is accomplished by a series of transitions in which the electron drops to states of successively lower energy, finally reaching the ground state. In each transition electromagnetic radiation is emitted with a wave number which depends on the energy lost by the electron, i.e., upon the initial and final quantum numbers. In a typical case, the electron might be excited into state $n = 7$ and drop successively through the states $n = 4$ and $n = 3$ to the ground state $n = 1$. Three lines of the atomic spectrum are emitted with wave numbers given by equation (5–15) for $n_i = 7$ and $n_f = 4$, $n_i = 4$ and $n_f = 3$, and $n_i = 3$ and $n_f = 1$.

4. In the very large number of excitation and de-excitation processes which take place during a measurement of an atomic spectrum, all possible transitions occur and the complete spectrum is emitted. The wave numbers of the set of lines which constitute the spectrum are given by equation (5–15), where we allow n_i and n_f to take on all possible integral values subject only to the restriction that $n_i > n_f$.

For hydrogen ($Z = 1$) let us consider the subset of spectral lines which arises from transitions in which $n_f = 2$. According to equation (5–15) the wave numbers for these lines are

$$k = R_\infty(1/n_f^2 - 1/n_i^2), \qquad n_f = 2,\ n_i > n_f$$

or

$$k = R_\infty(1/2^2 - 1/n^2), \qquad n = 3, 4, 5, 6, \ldots$$

This is identical with the series formula for the Balmer series of the hydrogen spectrum (equation 5–3), if R_∞ is equal to R_H. According to the Bohr theory, $R_\infty = me^4/4\pi c\hbar^3$. Although the numerical values of some of the quantities entering into this equation were not very accurately known at the time, Bohr evaluated R_∞ in terms of these quantities and found that the resulting number was in quite good agreement with the experimental value of R_H. In the next section we shall make a detailed comparison, using recent data, between the experimental value of R_H and the prediction of Bohr's theory, and show that the two agree almost perfectly.

According to the Bohr theory, each of the five known series of the hydrogen spectrum arises from a subset of transitions in which the electron goes to a certain final quantum state n_f. For the Lyman series $n_f = 1$; for the Balmer $n_f = 2$; for the Paschen $n_f = 3$; for the Brackett $n_f = 4$; and for the Pfund $n_f = 5$. These series are conveniently illustrated in terms of the energy level diagram of figure (5–5). The transition giving rise to a particular line of a series is indicated in this diagram by an arrow

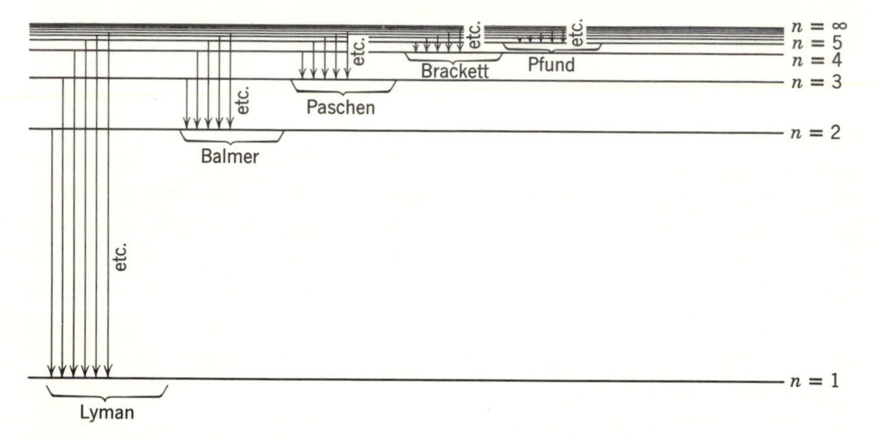

Figure 5–5. Illustrating the production of the five known series of the hydrogen spectrum.

going from the initial quantum state n_i to the final quantum state n_f. Only the arrows corresponding to the first few lines of each series are shown. Since the distance between any two energy levels in such a diagram is proportional to the difference between the energy of the two levels, and since equation (5–7) states that the frequency ν or wave number k is proportional to the energy difference, the length of any arrow is proportional to the frequency or wave number of the corresponding spectral line.

The wave numbers of the lines of all these series are fitted very accurately by equation (5–15) by using the appropriate value of n_f. This was a great triumph for Bohr's theory. The success of the theory was particularly impressive because the Lyman, Brackett, and Pfund series had not been discovered at the time the theory was developed by Bohr. The existence of these series was predicted by the theory, and the series were soon observed experimentally by the gentlemen they are named after.

The theory worked equally well when applied to the case of one-electron atoms with $Z = 2$, i.e., singly ionized helium atoms He^+. Such atoms can be produced by passing a particularly violent electric discharge (a spark) through normal helium gas. They make their presence apparent by emitting a simpler spectrum than that emitted by normal helium atoms. In fact, the atomic spectrum of He^+ is exactly the same as the hydrogen spectrum except that the wave numbers of all the lines are almost exactly four times as great. This is explained very easily, in terms of the Bohr theory, by setting $Z^2 = 4$ in equation (5–15).

The properties of the absorption spectrum of one-electron atoms are also easy to understand in terms of the Bohr theory. Since the atomic electron must have a total energy exactly equal to the energy of one of the allowed energy states, the atom can only absorb discrete amounts of energy from the incident electromagnetic radiation. This fact leads to the idea that we consider the incident radiation to be a beam of quanta, and that only those quanta can be absorbed whose frequency is given by $E = h\nu$, where E is one of the discrete amounts of energy which can be absorbed by the atom. The process of absorbing electromagnetic radiation is then just the inverse of the normal emission process, and the lines of the absorption spectrum will have exactly the same wave numbers as the lines of the emission spectrum. Normally the atom is always initially in the ground state $n = 1$, so that only absorption processes from $n = 1$ to $n > 1$ can occur. Thus only the absorption lines which correspond (for hydrogen) to the Lyman series will normally be observed. However, if the gas containing the absorbing atoms is at a very high temperature, then, owing to collisions, some of the atoms will initially be in the first excited state $n = 2$, and absorption lines corresponding to the Balmer series will be observed. The required temperature can be estimated from the Boltzmann probability distribution (equation 2–23), which shows that the ratio of the probability of finding an atom in the state $n = 2$ to the probability of finding it in the ground state $n = 1$ is

$$\frac{P_{n=2}}{P_{n=1}} = \frac{e^{-E_{n=2}/kT}}{e^{-E_{n=1}/kT}}$$

where k is Boltzmann's constant $= 1.38 \times 10^{-16}$ ergs/°K. For hydrogen the energies of these two states are given in the energy level diagram (figure 5–4): $E_{n=1} = -21.7 \times 10^{-12}$ ergs, $E_{n=2} = -5.42 \times 10^{-12}$ ergs. Thus

$$\frac{P_{n=2}}{P_{n=1}} = e^{\frac{-(-5.42+21.7)\times 10^{-12}\text{ ergs}}{(1.38\times 10^{-16}\text{ ergs/°K})T}} = e^{\frac{-1.18\times 10^{5}\text{ °K}}{T}}$$

It is apparent that a significant fraction of the hydrogen atoms will initially be in the $n = 2$ state only when T is of the order of, or greater than, 10^5 °K. As mentioned before, Balmer absorption lines are actually observed in the hydrogen gas of some stellar atmospheres. Thus we have one way of estimating the temperature of the surface of a star.

4. Correction for Finite Nuclear Mass

We have assumed the mass of the atomic nucleus to be infinitely large compared to the mass of the atomic electron, so that the nucleus remains fixed in space. This is a good approximation even for hydrogen, which contains the lightest nucleus, since the mass of that nucleus is about 2000 times larger than the electron mass. However, the spectroscopic data are

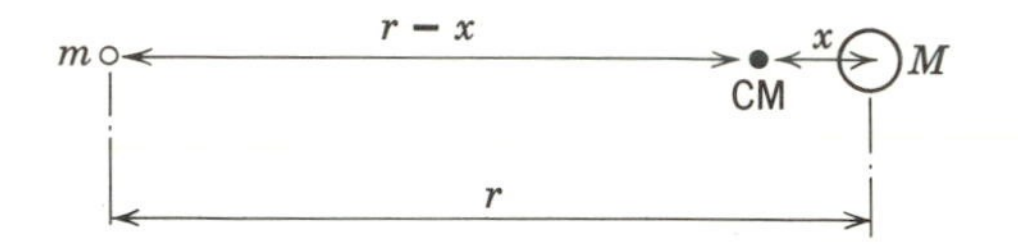

Figure 5–6. The center of mass of an electron-nucleus system.

so very accurate that before we make a detailed numerical comparison of these data with the Bohr theory we must take into account the fact that the nuclear mass is actually finite.

Consider the electron-nucleus system shown in figure (5–6). The electron of mass m and the nucleus of mass M must both move so that the *center of mass* (CM) remains fixed in space. Let r be the distance between the nucleus and the electron. Then the distance x from the nucleus to the CM is determined by the equation $m(r - x) = Mx$; so

$$x = \frac{mr}{m + M} \quad \text{and} \quad r - x = \frac{Mr}{m + M} \tag{5–16}$$

If the electron rotates about the CM with angular velocity ω, the nucleus must also rotate about the CM with the same angular velocity. Let us calculate the total orbital angular momentum L of such a system. We have

$$L = m(r - x)^2\omega + Mx^2\omega = \frac{mM^2r^2\omega}{(m + M)^2} + \frac{Mm^2r^2\omega}{(m + M)^2}$$

So

$$L = \frac{mM(m + M)}{(m + M)^2}\, r^2\omega = \frac{mM}{m + M}\, r^2\omega \tag{5–17}$$

As $M/m \to \infty$, $L \to (mM/M)r^2\omega$ and $x \to 0$. Thus, for an infinitely heavy nucleus the electron rotates about the stationary nucleus with orbital angular momentum $L = mr^2\omega$. This is just what we had before in equation (5–9), since $r\omega$ is equal to the orbital velocity v.

To handle the case at hand, Bohr modified his second postulate to read:

Instead of the infinity of orbits which would be possible in classical mechanics, it is only possible for an electron to move in an orbit for which the total orbital angular momentum of the atom, L, is an integral multiple of Planck's constant h, divided by 2π.

This obviously reduces to the original statement of the postulate for an infinitely heavy nucleus discussed in the previous section. Applying this postulate to equation (5–17), we have

$$\frac{mM}{m+M} r^2\omega = \frac{mM}{m+M} vr = n\hbar, \qquad n = 1, 2, 3, \ldots$$

Comparing this with equation (5–11), we see that the net effect has been to replace the mass of the electron, m, by the so-called *reduced mass*,

$$\mu = \frac{mM}{m+M} \tag{5–18}$$

which is slightly less than m by the factor $1/(1 + m/M)$.

It is not difficult to go through the rest of Bohr's derivation for the case of finite nuclear mass. It is found that all the equations are identical with those derived before except that the electron mass m is replaced by the reduced electron mass μ. In particular, the formula for the wave numbers of the spectral lines becomes

$$k = R_M Z^2\left(\frac{1}{n_f^2} - \frac{1}{n_i^2}\right), \qquad R_M \equiv \frac{mM}{m+M}\frac{e^4}{4\pi c\hbar^3} \tag{5–19}$$

where R_M is the *Rydberg constant for a nucleus of mass M*. As $M/m \to \infty$, it is apparent that $R_M \to R_\infty$, the *Rydberg constant for an infinitely heavy nucleus* which appears in equation (5–15). In general, the Rydberg constant R_M is less than R_∞ by the factor $1/(1 + m/M)$. For the most extreme case of hydrogen, $M/m = 1836$ (see Chapter 3) and R_H is less than R_∞ by about 1 part in 2000.

If we evaluate R_H from equation (5–19), using the currently accepted values of the quantities m, M, e, c, and $\hbar$, we find $R_H = 109681$ cm^{-1}. Comparing this with the experimental value of R_H given in equation (5–4), we see that the Bohr theory, corrected for finite nuclear mass, agrees with the spectroscopic data to within 3 parts in 100,000!

Before closing this section, it is worth while to present another demonstration of the numerical accuracy of both the spectroscopic data and the Bohr theory. This involves a comparison of the spectrum of hydrogen, H, and the spectrum of singly ionized helium, He^+. As mentioned above, the lines of these two spectra are observed to differ in wave number by a factor F which has a value very close to 4. According to equation (5–19), the factor should be

$$F = Z^2 \frac{\dfrac{mM_{\mathrm{He}^+}}{m + M_{\mathrm{He}^+}}}{\dfrac{mM_{\mathrm{H}}}{m + M_{\mathrm{H}}}} = 4 \frac{M_{\mathrm{He}^+}}{M_{\mathrm{H}}} \frac{(m + M_{\mathrm{H}})}{(m + M_{\mathrm{He}^+})}$$

The mass of the helium nucleus is just the mass of the helium atom less two electron masses.† That is,

$$M_{\mathrm{He}^+} = M_{\mathrm{He\ atom}} - 2m$$

and the mass of the hydrogen nucleus is equal to the mass of the hydrogen atom less one electron mass.† So

$$M_{\mathrm{H}} = M_{\mathrm{H\ atom}} - m$$

From the study of chemistry we have an accurate value of the ratio

$$\frac{M_{\mathrm{He\ atom}}}{M_{\mathrm{H\ atom}}} = \frac{4.004}{1.0081}$$

The four equations above, plus the experimental value of the factor F, allow an evaluation of M_{H}/m. The result is $M_{\mathrm{H}}/m = 1837$, in very good agreement with the value 1836 obtained in the experiments discussed in Chapter 3.

5. Atomic Energy States

The Bohr theory predicts that the total energy of an atomic electron is quantized. For example, equation (5–14) gives the allowed energy values for the electron in a one-electron atom. Although we have not attempted to derive similar expressions for the electrons in a multi-electron atom, it is clear that the total energy of each of the electrons will also be quantized and, consequently, that the total energy content of the atom must be restricted to certain discrete values. Thus the theory predicts that the atomic energy states are quantized. The correctness of this prediction was confirmed in a simple experiment performed by Franck and Hertz in 1914, less than one year after the announcement of Bohr's theory.

Apparatus of the type used by these investigators is indicated in figure (5–7). Electrons are emitted thermally at low energy from the heated

† We ignore the relativistic contribution to the atomic mass, which arises from the fact that the electrons have a negative total energy of the order of 10^{-11} ergs, since this mass $M = E/c^2 \simeq 10^{-11}/10^{21} = 10^{-32}$ gm is small compared to an electron mass.

cathode C.† They are accelerated to the anode A by a potential V applied between the two electrodes. Some of the electrons pass through holes in A and travel to plate P, providing their kinetic energy upon leaving A is enough to overcome a small retarding potential V_r applied between P and A. The entire tube is filled at a low pressure with a gas or vapor of the atoms to be investigated. The experiment involves measuring the electron current reaching P (indicated by the current I flowing through the meter) as a function of the accelerating voltage V.

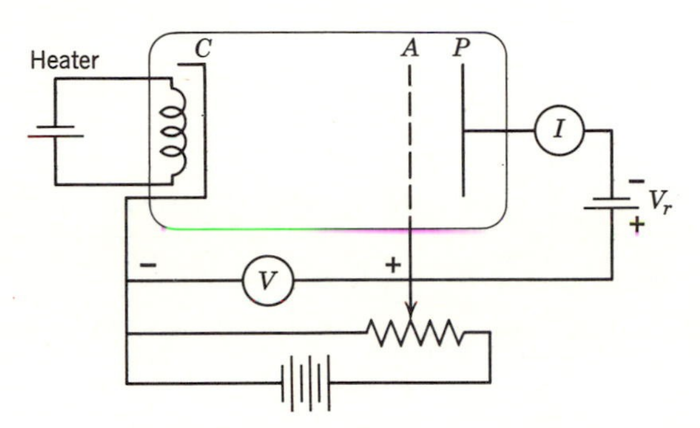

Figure 5–7. An apparatus used to prove that atomic energy states are quantized.

The first experiment was performed with the tube containing Hg vapor. The results are indicated in figure (5–8). At low accelerating voltage, the current I is observed to increase with increasing voltage V. When V reaches 4.9 volts, the current abruptly drops. This was interpreted as indicating that some interaction between the electrons and the Hg atoms suddenly commences when the electrons attain a kinetic energy of 4.9 ev $= 7.85 \times 10^{-12}$ ergs, as a result of falling through a potential difference of 4.9 volts.‡ Apparently a significant fraction of the electrons of this energy excite the Hg atoms and in so doing entirely lose their kinetic energy. If V is only slightly more than 4.9 volts, the excitation process must occur just in front of the anode A, and after the process the electrons

† By raising the temperature of the cathode to the point at which the thermal agitation kinetic energy of its electrons is comparable to the energy W required to overcome the attractive forces acting on the electrons at the surface of the cathode, some of the electrons will be spontaneously emitted.

‡ An electron falling unimpeded through a potential difference of 1 volt attains a kinetic energy $E = eV = 1.60 \times 10^{-19}$ coulomb $\times$ 1 volt $= 1.60 \times 10^{-19}$ joules $=$ 1.60×10^{-12} ergs. In the present case the electron energy is $4.9 \times 1.60 \times 10^{-12}$ ergs $=$ 7.85×10^{-12} ergs. For a very convenient unit of energy, we define the *electron volt* (*ev*) such that 1 ev $= 1.60 \times 10^{-12}$ ergs. This unit is used extensively in atomic and nuclear physics. The kinetic energy of the electron discussed above can be specified as 4.9 ev or as 7.85×10^{-12} ergs.

cannot gain enough kinetic energy in falling toward A to overcome the retarding potential V_r and reach plate P. At somewhat larger V, the electrons can gain enough kinetic energy after the excitation process to overcome V_r and reach P. The sharpness of the break in the curve indicates that electrons of energy less than 4.9 ev are *not able* to transfer their energy to an Hg atom. This interpretation is consistent with the existence of discrete energy states for the Hg atom, as predicted by the Bohr theory.

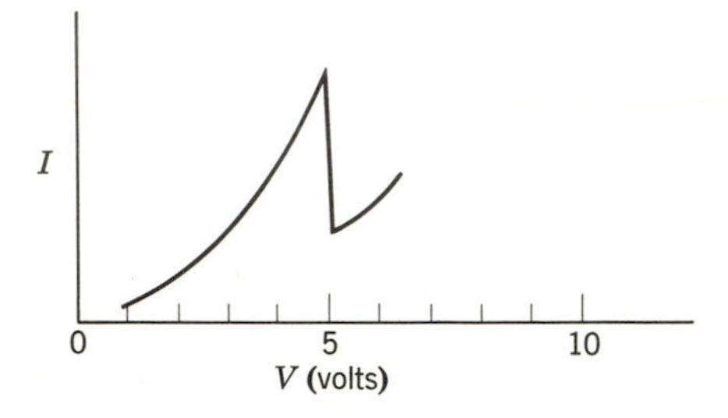

Figure 5–8. The voltage dependence of the current measured in the Franck-Hertz experiment.

Assuming the first excited state of Hg to be 4.9 ev higher in energy than the ground state, an Hg atom would simply not be able to accept energy from the bombarding electrons unless these electrons had at least 4.9 ev. Now, if the separation between the ground state and the first excited state is actually 4.9 ev = 7.85×10^{-12} ergs, there should be a line in the Hg spectrum of wave number†

$$k = \frac{\nu}{c} = \frac{E}{ch} = \frac{7.85 \times 10^{-12}}{3.00 \times 10^{10} \times 6.62 \times 10^{-27}} = 3.95 \times 10^4 \text{ cm}^{-1}$$

corresponding to the transition from the first excited state to the ground state. Franck and Hertz found a line in the spectrum of very nearly this wave number. Furthermore, they found that when the energy of the bombarding electrons is less than 4.9 ev no spectral lines at all are emitted from the Hg vapor in the tube, and when the energy is not more than a few electron volts greater than this value only the single line $k = 3.95 \times 10^4$ cm^{-1} is seen in the spectrum. All these observations are precisely what would be expected if the energy of the Hg atom is quantized. The Franck-Hertz experiment provided a striking confirmation of the predictions of the Bohr theory. It also provided a method for the direct measurement of the energy differences between the quantum states of an

† In evaluating this we use $E = 7.85 \times 10^{-12}$ ergs since we have expressed all other quantities in the cgs system and want the answer in that system.

atom—the answers appear on the dial of a voltmeter! When the curve of I versus V is extended to higher voltages, additional breaks are found. Some are due to electrons exciting the first excited state of the atoms on several separate occasions in their trip from C to A; but some are due to excitation of the higher excited states and, from the position of these breaks, the energy differences between the higher excited states and the ground state can be directly measured.

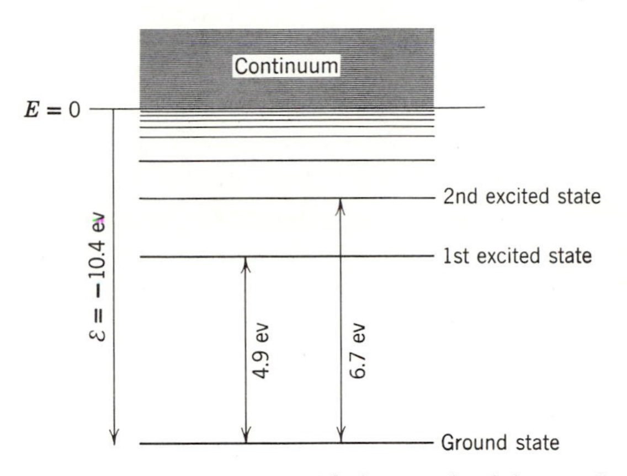

Figure 5–9. A considerably simplified energy level diagram for mercury.

Another experimental method of determining the separations between the energy states of an atom is to measure its atomic spectrum and then empirically construct a set of energy states which would lead to such a spectrum. In practice this is often quite difficult to do since the set of lines constituting the spectrum, as well as the set of energy states, is often very complicated. However, in common with all spectroscopic techniques, it is a very accurate method. In all cases in which determinations of the separations between the energy states of a certain atom have been made using both this technique and the Franck-Hertz technique, the results have been found to be in excellent agreement.

In order to illustrate the preceding discussion, we show in figure (5–9) a considerably simplified representation of the energy states of Hg in terms of an energy level diagram. The separations between the ground state and the first and second excited states are known, from the Franck-Hertz experiment, to be 4.9 ev and 6.7 ev. These numbers can be confirmed, and in fact determined with much higher accuracy, by measuring the wave numbers of the two spectral lines corresponding to transitions of an electron in the Hg atom from these two states to the ground state. The energy $\varepsilon = -10.4$ ev, of the ground state relative to a state of zero total energy, is not determined by the Franck-Hertz experiment. However, it can be found by measuring the wave number of the line corresponding to a transition of an atomic electron from a state of zero total energy to the ground state. This is the series limit of the series terminating on the ground state. The energy ε can also be measured by measuring the energy which must be supplied to an Hg atom in order to send one of its electrons from the ground state to a state of zero total energy. Since an electron of zero total energy is no longer bound to the atom, ε is the energy required to ionize the atom and is therefore called the *ionization energy*.

Lying above the highest discrete state at $E = 0$ are the energy states of the system consisting of an unbound electron plus an ionized Hg atom. *The total energy of an unbound electron* (*a free electron with* $E > 0$) *is not quantized.* Thus any energy $E > 0$ is possible for the electron, and the energy states form a continuum. The electron can be excited from its ground state to a continuum state if the Hg atom receives an energy greater than 10.4 ev. Conversely, it is possible for an ionized Hg atom to capture a free electron into one of the quantized energy states of the neutral atom. In this process, radiation of wave number greater than the series limit corresponding to that state will be emitted. The exact value of the wave number depends on the initial energy E of the free electron. Since E can have any value, the spectrum of Hg should have a continuum extending beyond every series limit in the direction of increasing wave number. This can actually be seen experimentally, although with some difficulty. These comments concerning the continuum of energy states for $E > 0$, and its consequences, have been made in reference to the Hg atom, but they are equally true for all atoms.

6. The Wilson-Sommerfeld Quantization Rules

The success of the Bohr theory, as measured by its agreement with experiment, was certainly very striking. But it only accentuated the mysterious nature of the postulates on which the theory was based. One of the biggest of these mysteries was the question of the relation between Bohr's quantization of the angular momentum of an electron moving in a circular orbit and Planck's quantization of the total energy of an entity, such as an electron, executing simple harmonic motion. In 1916 some light was shed upon this by Wilson and Sommerfeld, who enunciated a set of rules for the quantization of any physical system for which the coordinates are periodic functions of time. These rules included both the Planck and the Bohr quantization as special cases. They were also of considerable use in broadening the range of applicability of the quantum theory. These rules can be stated as follows:

For any physical system in which the coordinates are periodic functions of time, there exists a quantum condition for each coordinate. These quantum conditions are

$$\oint p_q \, dq = n_q h \tag{5–20}$$

where q is one of the coordinates, p_q is the momentum associated with that coordinate, n_q is a quantum number which takes on integral values, and $\oint$ means that the integration is taken over one period of the coordinate q.

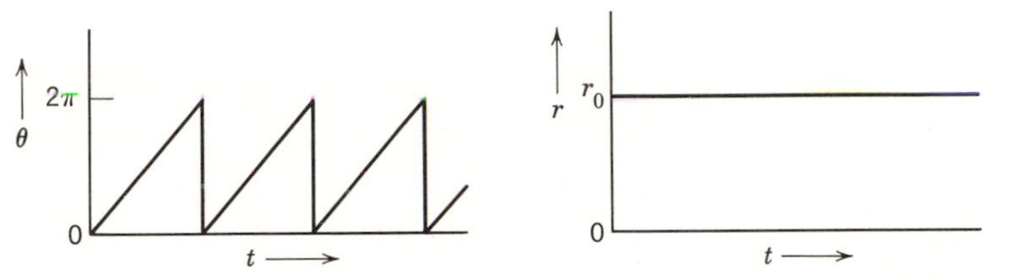

Figure 5–10. The time dependence of the coordinates of an electron in a Bohr orbit.

The meaning of these rules can best be illustrated in terms of some specific examples.

Consider a particle of mass m moving with constant angular velocity ω in a circular orbit of radius r_0 (an atomic electron in a Bohr orbit). The position of the particle can be specified by the polar coordinates r and θ. The behavior of these two coordinates is shown in figure (5–10) as functions of time. They are both periodic functions of time, if we consider $r = r_0$ to be a limiting case of this behavior. The momentum associated with the angular coordinate θ is the angular momentum $L = mr^2 \, d\theta/dt$. The momentum associated with the radial coordinate r is the radial momentum $p_r = m \, dr/dt$. In the present case $r = r_0$, $dr/dt = 0$, and $d\theta/dt = \omega$, a constant. Thus $L = mr_0^2\omega$ and $p_r = 0$. We do not need to apply equation (5–20) to the radial coordinate r in the limiting case in which the coordinate is constant. The application of the equation to the angular coordinate θ is easy to carry through for the present example. We have $q = \theta$ and $p_q = L$, a constant. Write $n_q = n$; then

$$\oint p_q \, dq = \oint L \, d\theta = L \oint d\theta = L \int_0^{2\pi} d\theta = 2\pi L$$

So the condition

$$\oint p_q \, dq = n_q h$$

becomes

$$2\pi L = nh$$

or

$$L = nh/2\pi \equiv n\hbar$$

which is identical with Bohr's quantization law.

Next let us apply the Wilson-Sommerfeld quantization rules to a particle of mass m executing simple harmonic motion with frequency ν. The

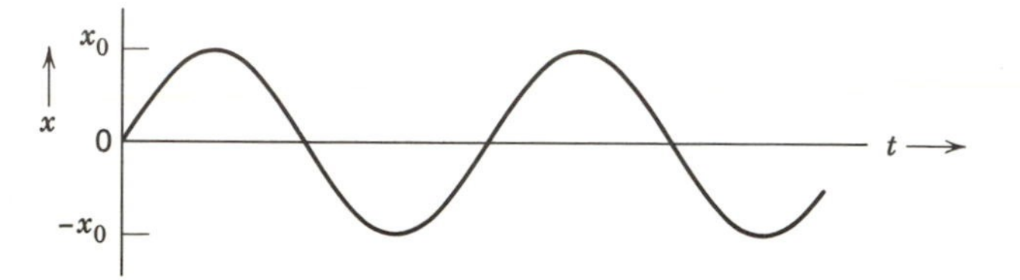

Figure 5–11. The time dependence of the coordinate of a simple harmonic oscillator.

position of the particle can be specified by the single linear coordinate x. The behavior of this coordinate with time is illustrated in figure (5–11) and can be expressed as

$$x = x_0 \sin 2\pi\nu t = x_0 \sin \omega t, \qquad \omega \equiv 2\pi\nu$$

where x_0 is the amplitude of the oscillation. The momentum of the particle is

$$p = m\frac{dx}{dt} = mx_0\omega \cos \omega t$$

In this case

$$\oint p_q\, dq = \oint p\, dx = \oint mx_0\omega \cos \omega t\, dx$$

To evaluate this integral it is convenient to express $\cos \omega t$ in terms of x:

$$x^2 = x_0^2 \sin^2 \omega t = x_0^2(1 - \cos^2 \omega t)$$

Then

$$\cos^2 \omega t = \frac{x_0^2 - x^2}{x_0^2}, \qquad \cos \omega t = \frac{\sqrt{x_0^2 - x^2}}{x_0}$$

and

$$\oint p\, dx = mx_0\omega \oint \frac{\sqrt{x_0^2 - x^2}}{x_0}\, dx = m\omega\, 4\int_0^{x_0} \sqrt{x_0^2 - x^2}\, dx$$

The last step depends upon the fact that

$$\oint = \int_0^{x_0} + \int_{x_0}^{0} + \int_0^{-x_0} + \int_{-x_0}^{0} = 4\int_0^{x_0}$$

because the integrand is an even function of x. This gives

$$\oint p\,dx = m\omega\,4\left[\frac{x\sqrt{x_0^2 - x^2}}{2} + \frac{x_0^2}{2}\sin^{-1}\frac{x}{x_0}\right]_0^{x_0}$$

$$\oint p\,dx = m\omega\,2(x_0^2 \sin^{-1} 1 - x_0^2 \sin^{-1} 0)$$

$$\oint p\,dx = m x_0^2\,\omega\pi$$

We would like to write this in terms of the total energy E of the particle; it is easy to do if we recall that the total energy of a harmonic oscillator is equal to its kinetic energy when $x = 0$. So

$$E = \frac{1}{2}\,m\left(\frac{dx}{dt}\right)^2_{t=0} = \frac{1}{2}\,m(x_0\omega \cos \omega t)^2_{t=0} = \frac{1}{2}\,m x_0^2\omega^2$$

since $x = 0$ when $t = 0$. Thus we have

$$\oint p\,dx = \frac{1}{2}\,m x_0^2\omega^2\,\frac{2\pi}{\omega} = \frac{E}{\nu}$$

and

$$E/\nu = n_q h \equiv nh, \qquad E = nh\nu$$

which is identical with Planck's quantization law.

7. Sommerfeld's Relativistic Theory

One of the important applications of the Wilson-Sommerfeld quantization rules was to the case of a hydrogen atom in which it was assumed that the electron could move in elliptical orbits. This was done by Sommerfeld in an attempt to explain the *fine structure* of the hydrogen spectrum. The fine structure is a splitting of the spectral lines, into several distinct components, which is found in all atomic spectra. It can be observed only by using equipment of very high resolution since the separation, in terms of wave number, between adjacent components of a single spectral line is of the order of 10^{-4} times the separation between adjacent lines. According to the Bohr theory, this must mean that what we had thought was a single energy state of the hydrogen atom actually consists of several states which are very close together in energy.

Sommerfeld first evaluated the size and shape of the allowed elliptical orbits, as well as the total energy of an electron moving in such an orbit, using the formulas of *classical mechanics*. Describing the motion in terms of the polar coordinates r and θ, he applied the two quantum conditions

$$\oint L\, d\theta = n_\theta h$$

$$\oint p_r\, dr = n_r h$$

The first condition yields the same restriction on the orbital angular momentum,

$$L = n_\theta \hbar, \qquad n_\theta = 1, 2, 3, \ldots$$

that it does for the circular orbit theory. The second condition (which was not applicable in the limiting case of purely circular orbits) leads to the following relation between L and a/b, the ratio of the semimajor axis to the semiminor axis of the ellipse:

$$L(a/b - 1) = n_r \hbar, \qquad n_r = 0, 1, 2, 3, \ldots$$

By applying the condition of mechanical stability analogous to equation (5–8), a third equation is obtained. From these equations Sommerfeld evaluated a and b, which give the size and shape of the elliptical orbits, and also the total energy E of an electron in such an orbit. The results are

$$a = \frac{n^2 \hbar^2}{\mu Z e^2}$$

$$b = a \frac{n_\theta}{n} \tag{5–21}$$

$$E = -\frac{\mu Z^2 e^4}{2 n^2 \hbar^2}$$

where μ is the reduced mass of the electron, and where the quantum number n is defined by

$$n \equiv n_\theta + n_r$$

Since $n_\theta = 1, 2, 3, \ldots$ and $n_r = 0, 1, 2, 3, \ldots$, n can take on the values

$$n = 1, 2, 3, 4, \ldots$$

For a given value of n, n_θ can assume only the values

$$n_\theta = 1, 2, 3, \ldots, n$$

The integer n is called the *principal quantum number*, and n_θ is called the *azimuthal quantum number*.

The second of equations (5–21) shows that the shape of the orbit (the ratio of the semimajor to the semiminor axes) is determined by the ratio of n_θ to n. For $n_\theta = n$, the orbits are circles of radius a. Note that the equation giving a in terms of n is identical with equation (5–12), the equation giving the radius of the circular Bohr orbits.† Figure (5–12) shows,

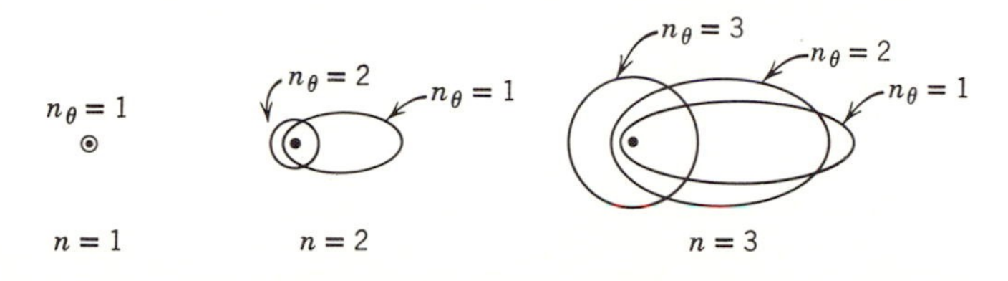

Figure 5–12. Some elliptical Bohr-Sommerfeld orbits.

to scale, the possible orbits corresponding to the first three values of the principal quantum number. Corresponding to each value of the principal quantum number n there are n different allowed orbits. One of these, the circular orbit, is just the orbit described by the original Bohr theory. The others are elliptical. But, despite the very different paths followed by an electron moving in the different possible orbits for a given n, the third of equations (5–21) tells us that the total energy of the electron is the same. The total energy of the electron depends only on n. The several orbits characterized by a common value of n are said to be *degenerate*.

This degeneracy in the total energy of an electron, following the orbits of very different shape but common n, is the result of a very delicate balance between potential and kinetic energy which is characteristic of treating an inverse square law of force by the methods of classical mechanics. Sommerfeld removed the degeneracy by treating the problem with the theory of *relativistic mechanics*. In the discussion following equation (5–13) we showed that, for an electron in a hydrogen atom, $v/c \simeq 10^{-2}$ or less. Thus we would expect the relativistic corrections to the total energy, due to the relativistic variation of the electron mass, which will be of the order of $(v/c)^2$, to be only of the order of 10^{-4}. However, this is just the order of magnitude of the splitting in the energy states of hydrogen that would be needed to explain the fine structure of the hydrogen spectrum. The actual size of the correction depends on the average velocity of the electron which, in turn, depends on the ellipticity of the orbit. After a calculation which is much too tedious to reproduce here, Sommerfeld

† Remember that equation (5–12) will have m replaced by μ if proper account is taken of the finite nuclear mass.

showed that the total energy of an electron in an orbit characterized by the quantum numbers n and n_θ is equal to

$$E = -\frac{\mu Z^2 e^4}{2n^2\hbar^2}\left[1 + \frac{\alpha^2 Z^2}{n}\left(\frac{1}{n_\theta} - \frac{3}{4n}\right)\right] \tag{5-22}$$

The quantity $\alpha \equiv e^2/\hbar c$ is a pure number called the *fine structure constant.* The numerical value of this constant is

$$\alpha \equiv e^2/\hbar c = 7.297 \times 10^{-3} \simeq 1/137 \tag{5-23}$$

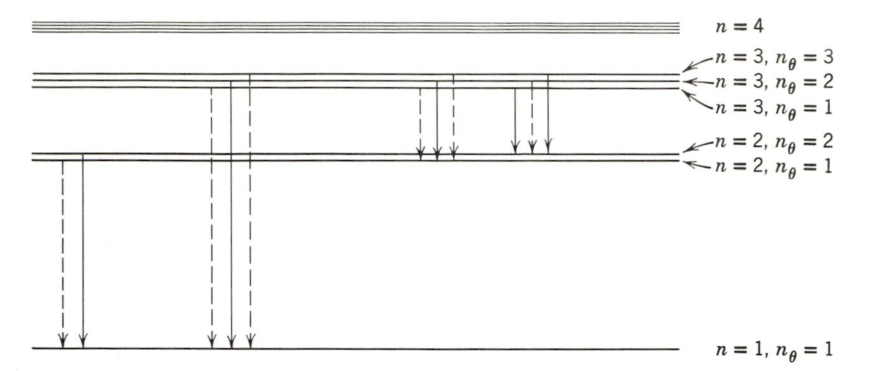

Figure 5–13. The fine structure splitting of some energy levels of the hydrogen atom. The splitting is greatly exaggerated. The possible transitions between the levels are indicated by solid arrows.

In figure (5–13) we represent the first few energy states of the hydrogen atom in terms of an energy level diagram. The separation between the several levels with a common value of n has been greatly exaggerated for the sake of clarity. Arrows indicate transitions between the various energy states which produce the lines of the atomic spectrum. Lines corresponding to the transitions represented by the solid arrows are observed in the hydrogen spectrum. The wave numbers of these lines are in very good agreement with the predictions derived from equation (5–22). However, the lines corresponding to the transitions represented by dashed arrows are not found in the spectrum. The transitions concerned do not take place. Inspection of the figure will demonstrate that transitions only occur if

$$n_{\theta_i} - n_{\theta_f} = \pm 1 \tag{5-24}$$

This is called a *selection rule.*

8. The Correspondence Principle

The reason for the existence of selection rules is not contained in the theory we have described so far. However, with the aid of an auxiliary postulate known as the *correspondence principle*, a theoretical justification of these rules could sometimes be found. This principle, enunciated by Bohr in 1923, consists of two parts:

1. *The predictions of the quantum theory for the behavior of any physical system must correspond to the predictions of classical physics in the limit in which the quantum numbers specifying the state of the system become very large.*
2. *A selection rule holds true over the entire range of the quantum number concerned. Thus any selection rules which are necessary to obtain the required correspondence in the classical limit (large n) also apply in the quantum limit (small n).*

Concerning the first part, it is obvious that the quantum theory must correspond to the classical theory in the limit in which the system behaves classically. The only question is: Where is the classical limit? Bohr's assumption is that the classical limit is always to be found in the limit of large quantum numbers. In making this assumption he was guided by certain evidence available at the time. For instance, the classical Rayleigh-Jeans theory of the black body spectrum agrees with experiment in the limit of small ν. Since Planck's quantum theory agrees with experiment everywhere, we see that correspondence between the quantum and classical theories is found, in this case, in the limit of small ν. But it is easy to see from equation (2–30) that as ν becomes small the average value $\bar{n}$, of the quantum number specifying the energy state of black body electromagnetic waves of frequency ν, will become large.† The second part of the correspondence principle was purely an assumption, but certainly a reasonable one.

Let us illustrate the correspondence principle by applying it to a simple harmonic oscillator, such as a pendulum oscillating at frequency ν. One of the predictions of the quantum theory for this system is that the allowed energy states are given by the equation $E = nh\nu$. In the discussion of section 10, Chapter 2, we have seen that, in the limit of large n, this prediction is not in disagreement with what we actually know about the energy states of a classical pendulum. In this sense, the quantum and classical theories do correspond for $n \to \infty$ in so far as the energy states of a pendulum are concerned. Next assume that the pendulum bob carries

† $\varepsilon = nh\nu$, so $\bar{\varepsilon} = \bar{n}h\nu$. Equation (2–30) shows that $\bar{\varepsilon} \to kT$ as $\nu \to 0$, so in this limit $\bar{n}h\nu = kT$, which is a constant. Thus $\bar{n} \to \infty$ as $\nu \to 0$.

an electric charge, so that we can compare the predictions of the two theories concerning the emission and absorption of electromagnetic radiation by such a system. Classically the system would emit radiation due to the accelerated motion of the charge, and the frequency of the emitted radiation would be exactly ν. According to the quantum theory, radiation is emitted as a result of the system making a transition from quantum state n_i to quantum state n_f. The energy emitted in such a transition is equal to $E_i - E_f = (n_i - n_f)h\nu$. This energy is carried away by a quantum of frequency $(E_i - E_f)/h = (n_i - n_f)\nu$. Thus, in order to obtain correspondence between the classical and quantum predictions of the frequency of the emitted radiation, we must require that the selection rule $n_i - n_f = 1$ be valid in the classical limit. A similar argument concerning the absorption of radiation by the charged pendulum shows that in the classical limit there is also the possibility of a transition in which $n_i - n_f = -1$. The validity of these selection rules in the quantum limit can be tested by investigating the spectrum of radiation emitted by a vibrating diatomic molecule. The vibrational energy states for such a system are just those of a simple harmonic oscillator, since the force which leads to the equilibrium separation of the two atoms has the same form as a harmonic restoring force. From the vibrational spectrum it can be determined that the selection rule $n_i - n_f = \pm 1$ actually is in operation in the limit of small quantum numbers.

A number of selection rules were discovered empirically in the analysis of atomic and molecular spectra. It was often possible to understand these selection rules in terms of a correspondence principle argument, but sometimes ambiguities arose.

9. A Critique of the Old Quantum Theory

In Chapters 2, 3, and 5, we have described the high points in the development of what is now referred to as the *old quantum theory*. In many respects this theory was very successful—even more so than may be apparent to the reader since we have not mentioned a number of successful applications of the theory to phenomena, such as the heat capacity of solids at low temperatures, which were inexplicable in terms of the classical theories. However, the old quantum theory was certainly not free of criticism. To complete our discussion of this theory we must indicate some of its undesirable aspects.

1. The theory only tells us how to treat systems which are periodic, but there are many systems of physical interest which are not periodic.

2. Although it does tell us how to calculate the energies of the allowed states of a system, and the frequency of the quanta emitted or absorbed when the system makes a transition between allowed states, the theory does not tell us how to calculate the *rate* at which such transitions take place.

3. The theory is only really applicable to a one-electron atom. The alkali elements (Li, Na, K, Rb, Cs) can be treated approximately, but only because they are in many respects similar to a one-electron atom. The theory fails badly when applied to the neutral He atom, which contains two electrons.

4. Finally we might mention the subjective criticism that the entire theory seems somehow to lack coherence—to be intellectually unsatisfying.

That some of these objections are really of a very fundamental nature was realized by everyone concerned, and much effort was expended in attempts to develop a quantum theory which would be free of these and other objections. The effort was well rewarded. In 1925, following an idea proposed in the preceding year by de Broglie, Schroedinger developed his theory of *quantum mechanics*. This theory is now almost universally considered completely acceptable.

As we shall see, the Schroedinger theory is in many respects very different from the old quantum theory. For instance, the picture of atomic structure provided by the theory of quantum mechanics is the antithesis of the picture, used in the old quantum theory, of electrons moving in well-defined orbits. Nevertheless, the old quantum theory is still often employed as a first approximation to the more accurate description of quantum phenomena provided by quantum mechanics. The reasons are that the old quantum theory is often capable of giving numerically correct results with mathematical procedures which are considerably less complicated than those used in quantum mechanics, and that the old quantum theory is often helpful in visualizing processes which are difficult to visualize in terms of the somewhat abstract language of the theory of quantum mechanics. Therefore the reader would be well advised to become thoroughly familiar with the content of the old quantum theory, particularly the Bohr theory.

BIBLIOGRAPHY

Born, M., *Atomic Physics*, Blackie and Son, London, 1957.

Ruark, A. E., and H. C. Urey, *Atoms, Molecules, and Quanta*, McGraw-Hill Book Co., New York, 1930.

Richtmyer, F. K., E. H. Kennard, and T. Lauritsen, *Introduction to Modern Physics*, McGraw-Hill Book Co., New York, 1955.

EXERCISES

1. Carry through the details of the Bohr calculation for a nucleus of finite mass and obtain equation (5–19).

2. Consider a particle of mass m moving back and forth along the x axis between two points $x = 0$ and $x = a$ from which it rebounds elastically. Apply the Wilson-Sommerfeld quantization rules to find the predicted possible values of the total energy of the particle.

3. Consider a body rotating about an axis. Apply the Wilson-Sommerfeld quantization rules, and show that the possible values of its total energy are predicted to be

$$E = \hbar^2 n^2/2I, \qquad n = 0, 1, 2, 3, \ldots$$

where I is its moment of inertia about the axis of rotation.

4. Apply the Wilson-Sommerfeld quantization rules to the results of exercise 3, Chapter 4, and obtain equations (5–21).

5. Prove the statement, made in section 7, that for small v/c the error in the classical expression $E = p^2/2m + V$ is smaller than E by a factor of the order of $(v/c)^2$.

6. Derive a relation connecting the frequency of the electromagnetic radiation, emitted in a transition between two states of a Bohr atom, and the orbital frequencies of the electron in these states. Study this relation in the limit of large quantum numbers, and comment on its correspondence with the predictions of classical physics.

CHAPTER 6

Particles and Waves

1. De Broglie's Postulate

In his doctoral dissertation (1924) de Broglie put forth a simple but extremely important idea which initiated the development of the theory of quantum mechanics. De Broglie's line of thought ran something like the following: In classical physics electromagnetic radiation had been considered to be purely a wave propagation phenomenon. However the work of Einstein and Compton has demonstrated that under certain circumstances it displays properties characteristic of particles (quanta). This being the case, could it also be true that physical entities which we normally think of as particles (electrons, alpha particles, billiard balls, etc.) will under certain circumstances display properties characteristic of waves?

The particle-like aspects of electromagnetic radiation become apparent when the interaction of the radiation with matter is investigated, while the wave-like aspects become apparent when the manner in which the radiation propagates is investigated. It is immaterial whether the situation is described by saying that electromagnetic radiation actually consists of waves which upon interacting with matter are able to manifest a particle-like behavior, or by saying that it actually consists of particles whose motion is governed by the wave propagation properties of certain associated waves. In fact, it is somewhat naive even to imply that there is a choice to be made. However, by temporarily adopting the latter statement, de Broglie was guided by analogy to look for the proposed wave-like aspects of particles in terms of some wave propagation character in their motion. Thus he was led to investigate the idea that the motion of a particle is governed by the wave propagation properties of certain *pilot waves* (to use his terminology) which are associated with the particle.

At the time of de Broglie's proposal, wave-like behavior in the motion of a particle had never been observed although the motion of particles had been extensively investigated. But this does not prove anything because

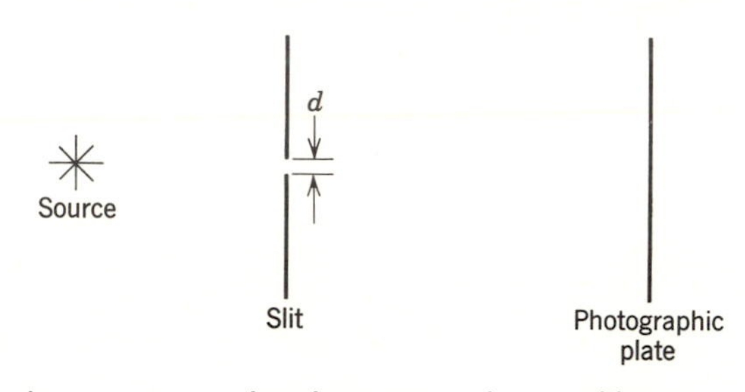

Figure 6–1. An apparatus used to demonstrate the wave-like nature of electromagnetic radiation.

it might be that the wavelength of the pilot waves is so small, compared to the dimensions of the physical systems which had been used to investigate particle motion, that the wave-like properties would not have been observable. The meaning of this statement can be illustrated by considering a single-slit diffraction experiment typical of those which can be used to demonstrate the wave-like nature of electromagnetic radiation. The experiment is indicated in figure (6–1). It is known that the results of such an experiment depend in a drastic manner on the ratio of the wavelength of the radiation, λ, to the width of the slit, d. For $\lambda \gg d$, a diffraction pattern characteristic of wave phenomena is recorded on the photographic plate. However, for $\lambda \ll d$ a quite sharply defined image of the slit is observed. Under these circumstances the radiation betrays little sign of its wave-like character, and its propagation can be very adequately described in terms of the laws of ray optics, a particle-like theory.

It is clear that de Broglie's first task was to specify some procedure for determining the wavelength of the pilot waves. He did this by making the reasonable assumption that the wavelength of pilot waves, and also their frequency, could be evaluated from the same equations which are used to evaluate these parameters for electromagnetic radiation. Consider a beam of electromagnetic radiation containing quanta of total relativistic energy E. According to Einstein's equation (3–7), the frequency ν of the radiation is

$$\nu = E/h \tag{6–1}$$

The wavelength λ of the radiation can be evaluated from the usual relation between λ, ν, and w, the velocity of propagation of the wave. This is

$$\lambda = w/\nu \tag{6–2}$$

For electromagnetic radiation $w = c$, so

$$\lambda = c/\nu = hc/E \tag{6–3}$$

According to equation (3–10), $E/c = p$, the momentum of the quantum. Thus λ can also be evaluated from the equation

$$\lambda = h/p \tag{6–4}$$

De Broglie postulated that the wavelength λ and the frequency ν of the pilot waves associated with a particle of momentum p and total relativistic energy E are given by the equations

$$\left.\begin{aligned}\lambda &= h/p\\ \nu &= E/h\end{aligned}\right\} \tag{6–5}$$

and that the motion of the particle is governed by the wave propagation properties of the pilot waves.

His reason for choosing equations analogous to (6–4) and (6–1), instead of to (6–3) and (6–1), will become apparent shortly.

2. Some Properties of the Pilot Waves

Let us calculate the propagation velocity w of the pilot waves associated with a particle, using equation (6–2) and evaluating λ and ν from equations (6–5). This gives

$$w = \nu\lambda = \frac{E}{h}\frac{h}{p} = \frac{E}{p} \tag{6–6}$$

Using equation (1–25) to evaluate the total relativistic energy of the particle, we have

$$w = \frac{\sqrt{c^2p^2 + (m_0c^2)^2}}{p} = \frac{c\sqrt{p^2 + (m_0c)^2}}{p}$$

or

$$w = c\sqrt{1 + (m_0c/p)^2} \tag{6–7}$$

Note that w is greater than c. This seems, at first, quite disturbing because the velocity v of the particle must necessarily be less than c, and it would appear that the particle could not keep up with its own pilot waves.

Actually there is no difficulty, as the following argument shows. Imagine that a free particle (experiencing no forces) and its associated pilot waves are moving in such a way that they can be described by the single spatial coordinate x. Assume that we have distributed at a large number of points along the x axis a set of hypothetical instruments which are capable of measuring the intensity of the pilot waves. At a certain time t_0, we record the readings of these instruments. The results of the experiment

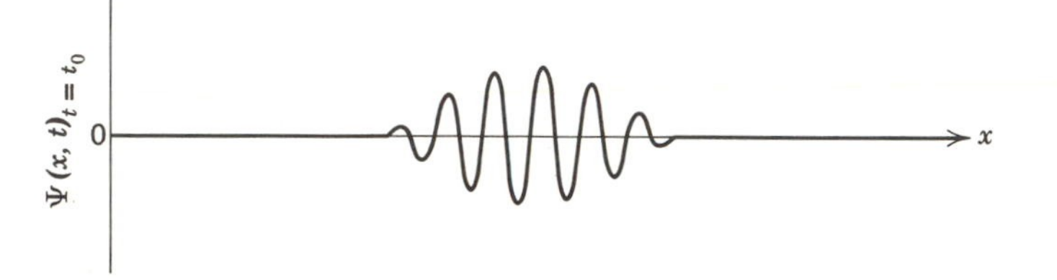

Figure 6–2. A group of pilot waves.

could be presented as a plot of the instantaneous value of the pilot waves, which we designate by the symbol $\Psi(x, t)_{t=t_0}$, as a function of x. Although we know little about the pilot waves at present, it is evident that the plot must look *qualitatively* like the one shown in figure (6–2). The amplitude of the pilot waves must be modulated in such a way that their value is non-zero only over a finite region of space in the vicinity of the particle. This is necessary because the pilot waves must somehow be associated spatially with the particle whose motion they control. The pilot waves form a *group* of waves and, as a function of time, the group must move along the x axis with the same velocity as the particle. The reader may recall, from his study of wave motion in elementary physics, that for such a moving group of waves it is necessary to distinguish between the velocity g of the group and the velocity w of the individual oscillations of the waves. Furthermore, g is in general less than w. This is encouraging, but of course we must prove that g is equal to the velocity of the particle.

To do this, we develop a relation between g and the quantities ν and λ comparable to the relation (6–2) between w and these two quantities. We start by considering the simplest type of wave motion, a sinusoidal wave of frequency ν and wavelength λ, which is of constant amplitude from $-\infty$ to $+\infty$, but which is moving with uniform velocity in the direction of increasing x. Such a wave can be represented by the function

$$\Psi(x, t) = \sin 2\pi\left(\frac{x}{\lambda} - \nu t\right)$$

or

$$\Psi(x, t) = \sin 2\pi(kx - \nu t), \qquad \text{where } k \equiv 1/\lambda \tag{6–8}$$

That this does represent the wave can be seen from the following considerations:

1. Holding x fixed, we see that the function oscillates in time sinusoidally with frequency ν.
2. Holding t fixed, we see that the function has a sinusoidal dependence on x, with wavelength λ or wave number k.
3. The zeros of the function, which correspond to the nodes of the wave it represents, are found at positions x_n for which

$$2\pi(kx_n - \nu t) = \pi n, \qquad n = 0, \pm 1, \pm 2, \ldots$$

or

$$x_n = \frac{n}{2k} + \frac{\nu}{k}t$$

Thus these nodes, and in fact all points on the wave, are moving in the direction of increasing x with velocity

$$w = dx_n/dt$$

which is equal to

$$w = \nu/k$$

Note that this is identical with equation (6–2) since $k = 1/\lambda$.

Next we discuss the case in which the amplitude of the waves is modulated to form a group. We can obtain mathematically *one* group of waves moving in the direction of increasing x [something similar to the group of pilot waves pictured in figure (6–2)] by adding together an *infinitely large* number of waves of the form (6–8), each with infinitesimally differing frequencies ν and wave numbers k. However, the mathematical techniques become a little involved and, for the purposes of the present argument, it will suffice to consider what happens when we add together only two such waves. Thus we take

$$\Psi(x, t) = \Psi_1(x, t) + \Psi_2(x, t) \qquad (6\text{–}9)$$

where

$$\Psi_1(x, t) = \sin 2\pi[kx - \nu t]$$

$$\Psi_2(x, t) = \sin 2\pi[(k + dk)x - (\nu + d\nu)t]$$

Now

$$\sin A + \sin B = 2 \cos \tfrac{1}{2}(A - B) \sin \tfrac{1}{2}(A + B)$$

Applying this to the case at hand, we have

$$\Psi(x, t) = 2 \cos 2\pi\left[\frac{dk}{2}x - \frac{d\nu}{2}t\right] \sin 2\pi\left[\frac{(2k + dk)}{2}x - \frac{(2\nu + d\nu)}{2}t\right]$$

Since $d\nu \ll 2\nu$ and $dk \ll 2k$, this is

$$\Psi(x, t) \cong 2 \cos 2\pi\left(\frac{dk}{2}\,x - \frac{d\nu}{2}\,t\right) \sin 2\pi(kx - \nu t) \qquad (6\text{–}10)$$

A plot of $\Psi(x, t)$ as a function of x for a fixed value of $t = t_0$ is shown in figure (6–3). The second term of $\Psi(x, t)$ is a wave of the same form as equation (6–8), but this wave is modulated by the first term so that the

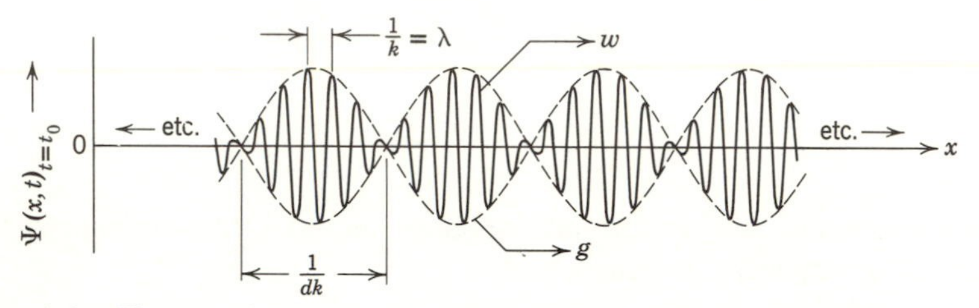

Figure 6–3. The sum of two sinusoidal waves of slightly different frequencies and wave numbers.

oscillations of $\Psi(x, t)$ fall within an envelope of periodically varying amplitude. *Two* waves of slightly different frequency and wave number alternately interfere and reinforce in such a way as to produce an *infinite* succession of groups. These groups, and the individual waves which they contain, are both moving in the direction of increasing x. The velocity w of the individual waves can be evaluated by considering the second term of $\Psi(x, t)$, and the velocity g of the groups can be evaluated from the first term. Proceeding as in 3 above, we find

$$w = \nu/k$$

$$g = \frac{d\nu/2}{dk/2} = \frac{d\nu}{dk} \qquad (6\text{–}11)$$

It can be shown that for an *infinitely large* number of waves combining to form *one* moving group, the dependence of the wave velocity w, and the group velocity g, on ν, k, and $d\nu/dk$ is exactly the same as for the simple case we have considered. Equations (6–11) have a general validity.

Finally we are in a position to calculate the group velocity g of the group of pilot waves associated with the moving particle. From equations (6–5), we have

$$\nu = E/h \quad \text{and} \quad k \equiv 1/\lambda = p/h$$

So

$$d\nu = dE/h \quad \text{and} \quad dk = dp/h$$

Thus the group velocity is

$$g = d\nu/dk = dE/dp$$

According to equation (1–25),

$$E^2 = c^2p^2 + (m_0c^2)^2$$

So

$$2E\,dE = c^2 2p\,dp$$

$$\frac{dE}{dp} = c^2 \frac{p}{E}$$

Therefore

$$g = c^2 \frac{p}{E} \tag{6–12}$$

According to equations (1–24) and (1–17),

$$E = mc^2 \quad \text{and} \quad p = mv$$

where $m = m_0/\sqrt{1 - v^2/c^2}$, the total relativistic mass, and v is the velocity of the particle. Thus we have the satisfying result that

$$g = c^2 \frac{mv}{mc^2} = v \tag{6–13}$$

The velocity of the group of pilot waves is just equal to the velocity of the particle whose motion they govern, and de Broglie's postulate is internally consistent.

It is easy to see why de Broglie did not take a pair of equations

$$k \equiv 1/\lambda = E/hc$$

$$\nu = E/h$$

analogous to equations (6–3) and (6–1) to specify the frequency and wave number of the pilot waves associated with a particle. Applying (6–11), these equations lead to a wave velocity

$$w = \nu/k = c$$

and to a group velocity

$$g = \frac{d\nu}{dk} = \frac{dE/h}{dE/hc} = c$$

This is correct for a quantum which moves with a velocity which is always equal to c, but certainly not correct for a particle which moves with a velocity v which is necessarily different from c.

From equations (6–6) and (6–12) we find the following relation between the wave velocity w and the group velocity g:

$$w = c^2/g$$

According to equation (6–13), this can be written in terms of the velocity of the particle v:

$$w = c^2/v \tag{6–14}$$

Since w is larger than g, the individual waves are constantly moving through the group from the rear to the front. The same situation occurs in a group of water waves.

3. Experimental Confirmation of de Broglie's Postulate

As mentioned in section 1, comparison with the propagation characteristics of electromagnetic radiation predicts that wave-like behavior in the motion of a particle will only be apparent when the *de Broglie wavelength* $\lambda = h/p$ is of the order of, or larger than, the characteristic dimensions of the system used to investigate the motion. Since Planck's constant h has the very small value 6.62×10^{-27} erg-sec, it is immediately apparent that, unless the momentum p is also very small, λ will be so small as to preclude any hope of observing the postulated effects. Let us evaluate p and λ for some specific cases.

Consider a very small dust particle of radius $r = 10^{-4}$ cm, density $\rho = 10$, moving with the low velocity $v = 1$ cm-sec^{-1}. Since $v \ll c$, we may evaluate p from the classical formula

$$p = mv = \tfrac{4}{3}\pi r^3 \rho v \simeq 4 \times 10^{-11} \text{ gm-cm-sec}^{-1}$$

The de Broglie wavelength of this particle is

$$\lambda = \frac{h}{p} = \frac{6.6 \times 10^{-27}}{4 \times 10^{-11}} \simeq 1.6 \times 10^{-16} \text{ cm}$$

This is extremely small compared to the dimensions of *any* physical system, although we have taken the momentum of the particle to be very much less than that of the particles typically discussed in classical mechanics. Thus we would not expect to be able to either prove or disprove the validity of de Broglie's postulate by investigating the motion of macroscopic particles.

Next consider an electron with the relatively low kinetic energy $T = 10$ ev $= 1.6 \times 10^{-11}$ ergs. This is, for example, approximately equal to the kinetic energy of an electron in a hydrogen atom. The velocity of the electron is small compared to c, so we may evaluate its momentum from the classical equation

$$p = \sqrt{2mT} = \sqrt{2 \times 9.1 \times 10^{-28} \times 1.6 \times 10^{-11}}$$

$$p \simeq 1.7 \times 10^{-19} \text{ gm-cm-sec}^{-1}$$

The corresponding de Broglie wavelength is

$$\lambda = \frac{6.6 \times 10^{-27}}{1.7 \times 10^{-19}} \simeq 3.9 \times 10^{-8} \text{ cm}$$

Although this is still a very small wavelength, it is of the order of the size of an atom and, consequently, of the interatomic spacing of the

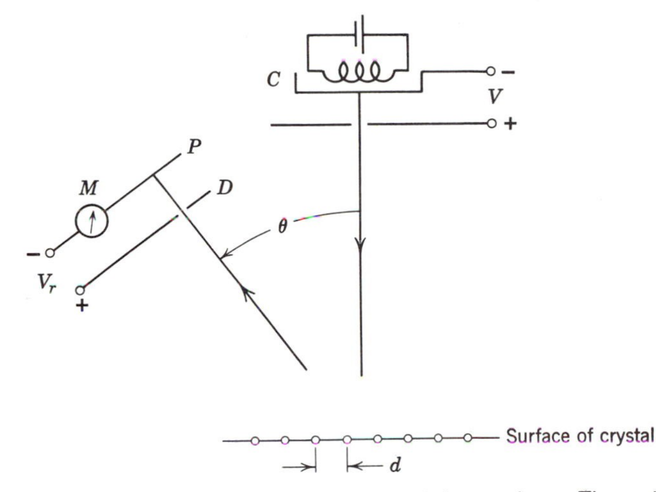

Figure 6–4. An apparatus used to verify de Broglie's postulate. The region traversed by the electrons is evacuated.

atoms in a crystal. This suggests the possibility of looking for diffraction effects in the reflection or transmission of a beam of electrons by a crystal. Such effects were first found by Davisson and Germer.

In the Davisson-Germer experiment (1927), illustrated schematically in figure (6–4), a parallel beam of monoenergetic low energy electrons was produced by accelerating electrons, thermally emitted from a cathode C, through a voltage drop V. This beam was incident normal to the surface of a Ni crystal. Some of the electrons were scattered back from the surface of the crystal. The number scattered at angle θ (defined in the figure) was measured with a detector consisting of a plate P behind a diaphragm D. A retarding voltage V_r, slightly less than V, allowed only electrons scattered from the crystal with little energy loss to reach P and produce a current indicated by meter M. Low energy electrons lose energy rapidly in traversing a solid, so the detected electrons must have been scattered essentially from the surface of the crystal. The periodic distribution of the atoms of that surface is indicated in the figure. For a Ni crystal the repetition length d of the periodic structure is 2.15×10^{-8} cm,† which is comparable to the de Broglie wavelength of the incident electrons. Since the essential feature of a diffraction grating is its periodicity, the pattern of electrons scattered from the crystal surface should show

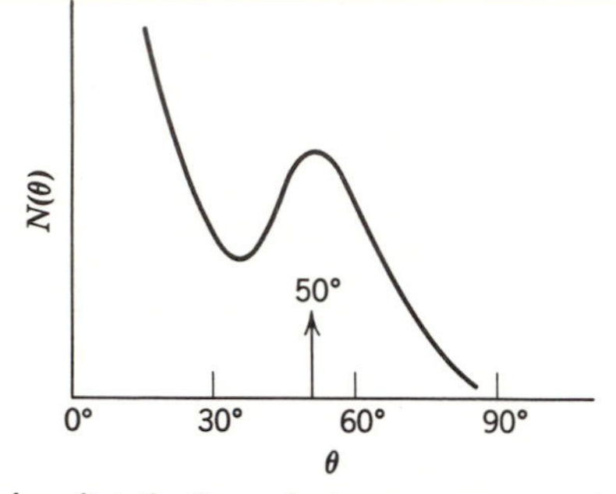

Figure 6–5. The angular distribution of electrons scattered from a nickel crystal. $V = 54.0$ volts.

diffraction effects if the de Broglie postulate is correct. Plotted in figure (6–5) is the form of Davisson and Germer's data for the number of scattered electrons $N(\theta)$ as a function of θ, measured for an accelerating voltage V of 54.0 volts. The angular distribution $N(\theta)$ becomes very large for small θ; that is, most of the scattered electrons are specularly reflected (angle of reflection = angle of incidence). This is not particularly informative because such behavior is characteristic of both particle motion and wave motion. However, at an angle of 50° a peak is observed in $N(\theta)$. The existence of this peak proves the qualitative validity of the de Broglie postulate because *such a peak can only be explained as a constructive interference of waves scattered by the periodically placed atoms of the crystal surface.* This is not an interference between waves associated with one electron and waves associated with another. Instead, it is an interference between waves, associated with a single electron, that have been scattered from various parts of the crystal. This conclusion can be demonstrated experimentally by making measurements with an electron

† This is known from measurements of the diffraction pattern produced in the scattering of X-rays of known wavelength by a Ni crystal. Such measurements will be described in Chapter 14.

beam of such low intensity that the electrons go through the apparatus one at a time, and by showing that the pattern of the scattered electrons remains the same.

The data of Davisson and Germer can also be used to show that de Broglie's postulate is quantitatively correct. If we evaluate the de Broglie wavelength of a 54 ev electron from the equation $\lambda = h/p$, as above, we find

$$\lambda = 1.67 \times 10^{-8}\ \text{cm} \equiv 1.67\ \text{A}$$

The wavelength of the waves which are interfering constructively can also be evaluated from the well-known *grating equation*,†

$$n\lambda = d \sin \theta, \qquad n = 1, 2, 3, \ldots \tag{6–15}$$

Assuming that the peak corresponds to a first order diffraction ($n = 1$), we have

$$\lambda = 2.15 \times 10^{-8}\ \text{cm} \times \sin 50^\circ = 1.65\ \text{A}$$

The two values of λ agree to within the accuracy of the experiment. Comparable agreement was found in the data obtained at a number of different accelerating voltages. At somewhat higher voltages a second order peak ($n = 2$) was also seen.

Diffraction effects in transmission through crystals were observed by G. P. Thomson. His experiment (1928) involved sending a collimated beam of electrons through a polycrystalline foil. The foil thickness was only about 10^{-5} cm, but even so it was necessary to use an electron energy of about 10^4 ev so that the energy loss in traversing the foil would not be too large. In passing through the randomly oriented crystals of the foil, the electrons will find certain ones for which the set of scattering angles $\theta_1, \theta_2, \theta_3, \ldots$ will satisfy the equation analogous to (6–15) which applies to this case. The electrons will be strongly scattered from these crystals at the angles indicated. The angles are all small since λ/d is small ($\lambda \simeq 5 \times 10^{-2}$ A); however, by placing a photographic plate some distance behind the scattering foil a series of concentric rings corresponding to the allowed scattering angles can be detected. Typical results are shown in figure (6–6). The photograph on the left shows a ring pattern resulting from the scattering of electrons by a thin layer of Au crystals. The photograph on the right shows the completely similar pattern observed in the scattering of electromagnetic radiation (X-rays) by a thin layer of ZrO_2 crystals. This comparison gives another convincing demonstration of wave-like behavior in the motion of electrons. Furthermore the values of the scattering angles measured from the electron ring pattern agree with

† See any textbook on elementary physics.

those of the scattering angles calculated from the wavelength predicted by the de Broglie postulate.

Several years after the initial experiments with electrons, Estermann, Frisch, and Stern demonstrated the existence of diffraction effects in the scattering of a beam of He atoms from the surface of a LiF crystal. The beam was obtained from an enclosure heated to a temperature of 400°K. Emerging from a channel communicating with the interior of the

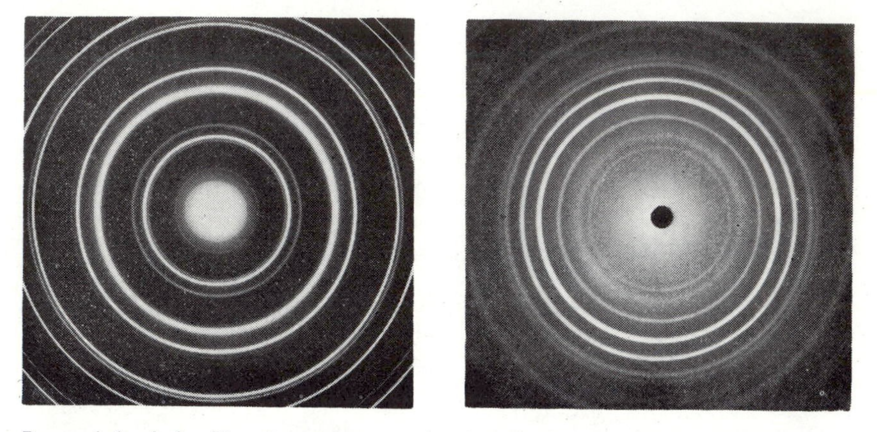

Figure 6–6. Left: The ring pattern produced in the scattering of electrons by gold crystals. Right: The ring pattern produced in the scattering of X-rays by zirconium oxide crystals. From U. Fano and L. Fano, *Basic Physics of Atoms and Molecules*, John Wiley and Sons, New York, 1959.

enclosure was a beam of atoms of average energy $\bar{E} = \frac{3}{2}kT$, where k is Boltzmann's constant. The de Broglie wavelength of an atom of average energy was

$$\lambda = \frac{h}{p} = \frac{h}{\sqrt{2ME}} = \frac{h}{\sqrt{3MkT}}$$

$$= \frac{6.6 \times 10^{-27}}{\sqrt{3 \times 6.7 \times 10^{-24} \times 1.4 \times 10^{-16} \times 4 \times 10^2}} \simeq 0.62 \text{ A}$$

The scattered atoms were detected by a very sensitive pressure gauge. The pattern of scattered He atoms was found to be similar to that observed by Davisson and Germer for the scattering of electrons. The angular location of the diffraction peak was in good agreement with what would be expected for the wavelength predicted by de Broglie's postulate.

We have described three of the original experiments verifying that there are wave propagation effects in the motion of particles, as well as verifying the wavelength predicted by de Broglie's postulate. Since then very many more examples of these effects have been observed experimentally, and *the validity of de Broglie's postulate has been undeniably confirmed.*

4. Interpretation of the Bohr Quantization Rule

We have seen above that the de Broglie wavelength of an electron of kinetic energy comparable to that of an electron in an H atom is about 4×10^{-8} cm. According to the Bohr theory, the radius of the ground state orbit of the H atom is about 0.5×10^{-8} cm. Since $\lambda \gg r$, we would expect the wave propagation properties of the pilot waves to play an important part in determining the motion of the electron in the atom. However, there is a significant difference between the type of particle motion we have previously discussed in this chapter and the motion of an electron in an atom. In the examples above the particles travel their trajectories just once; they are *unbound* particles, and their associated pilot waves are *traveling waves* (moving nodes). An electron in an atom is *bound* to the atomic nucleus. It travels repeatedly through the same orbit. Consequently its associated pilot waves would be expected to be *standing waves* (fixed nodes).

Standing waves are one of the most striking phenomena of wave propagation. In 1924 de Broglie showed how a simple consideration of their properties could lead to an interpretation of the mysterious Bohr angular momentum quantization rule. According to equation (5–11), this rule can be written

$$mvr = pr = nh/2\pi, \qquad n = 1, 2, 3, \ldots$$

where p is the linear momentum of an electron in an allowed orbit of radius r. If we substitute into this equation the expression

$$p = h/\lambda$$

for p in terms of the corresponding de Broglie wavelength, the equation becomes

$$hr/\lambda = nh/2\pi$$

This is

$$2\pi r = n\lambda, \qquad n = 1, 2, 3, \ldots \tag{6–16}$$

Thus the allowed orbits are those in which the circumference of the orbit can contain exactly an integral number of de Broglie wavelengths. If we think of an electron traveling repeatedly around the same orbit, it becomes apparent that equation (6–16) is just the necessary condition that the pilot waves associated with the particle in a particular traversal of the orbit will combine coherently with the pilot waves associated with the previous traversals to set up a standing wave. However, if this equation is violated, then in a large number of traversals the pilot waves will interfere with each other in such a way that their average intensity will vanish. Since the intensity of the pilot waves must be some sort of a measure of where the particle is located, we interpret this as meaning that an electron could not be found in such an orbit. Figure (6–7) is an attempt to illustrate

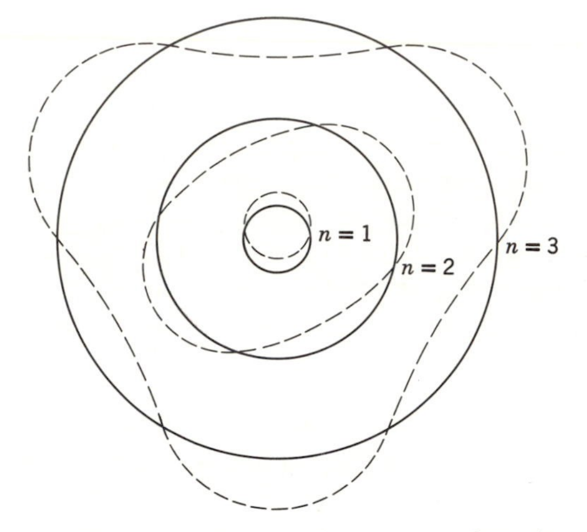

Figure 6–7. A schematic representation of the patterns of standing waves set up in the first three allowed Bohr orbits.

the $\Psi_{t=t_0}$, at some particular value of t_0, for the standing waves set up in the first three allowed Bohr orbits. At some other value of t_0 the magnitude and sign of the amplitudes of these oscillatory patterns might be different, but the locations of the nodes would be the same since the nodes of any standing wave are fixed.

It is not difficult to show that *the requirement that the pilot waves associated with a particle undergoing any sort of periodic motion be a set of standing waves is equivalent to the requirement that the motion of the particle satisfy the Wilson-Sommerfeld quantization rules* (*5–20*). We shall see in the following chapters that the properties of standing waves also play an important role in the Schroedinger theory of quantum mechanics. For instance, the time independent features of the standing waves associated with an electron in one of its allowed states in an atom will make it possible to understand why the motion described by the standing wave does not cause the electron to emit electromagnetic radiation.

5. The Uncertainty Principle

An inescapable result of de Broglie's standing wave description of an electron in a Bohr orbit is that the position of the electron, and the orientation of the vector indicating its linear momentum, cannot be exactly specified at a given instant of time since the azimuthal symmetry of the pattern of standing waves, pictured in figure (6–7) at the instant $t = t_0$, indicates that at that instant the electron in a particular orbit could be located essentially anywhere in the orbit. This is in contrast to the exact specification of position and linear momentum which is possible in Bohr's particle-like description of the motion of an atomic electron, just as it is possible in Newton's description of the motion of a planet. This uncertainty in the position and momentum of a particle at any instant is, as we shall see, a general feature of de Broglie's description of the motion of a particle in terms of the propagation of its associated pilot waves.

Let us consider a freely moving particle and its associated group, or groups, of pilot waves. We take first the mathematically simplified case of the pilot waves, described by equation (6–9), obtained by combining two unmodulated waves of slightly different k and ν. This is

$$\Psi = \sin 2\pi[kx - \nu t] + \sin 2\pi[(k + \Delta k)x - (\nu + \Delta\nu)t] \quad (6\text{–}17)$$

where we use Δk and $\Delta\nu$ instead of dk and $d\nu$. This function is plotted, at the instant $t = t_0$, in figure (6–3). It represents an infinite succession of groups of waves which are traveling in the direction of increasing x. Now, in consideration of the connection that we have made on several previous occasions between the intensity of the pilot waves and the location of the particle, it is apparent that we should interpret this particular pilot wave as indicating that the particle with which it is associated has equal probability of being located within any one of the groups at the time $t = t_0$. Considering momentarily a single group, our interpretation of Ψ indicates that even the location of the particle within that group is undetermined to within a distance comparable to the length Δx of the group. From equation (6–10), or from figure (6–3), we obtain the following relation between Δx and the difference Δk between the wave numbers of the two constituent waves:

$$\Delta x \, \Delta k = 1 \quad (6\text{–}18)$$

A similar relation can be obtained for the case, mentioned in section 2, in which an infinitely large number of unmodulated waves, each with infinitesimally different k and ν, combine to form a single traveling group.

Such a group is pictured at $t = t_0$ in figure (6–8). A single group is formed by adjusting the phases of all the unmodulated waves so that at one value of x (the center of the group) they are all in phase and combine to give large net values of $\Psi_{t=t_0}$. Proceeding away from this value of x, in either direction, the unmodulated waves begin to get out of phase with each other because their wave numbers differ. Beyond certain points they are completely out of phase, and never again get back into phase. Everywhere in these regions there are, on the average, as many unmodulated waves with positive values as with negative values, and the waves combine to give a $\Psi_{t=t_0}$ whose value is zero at all x. It is clear that, the larger the range

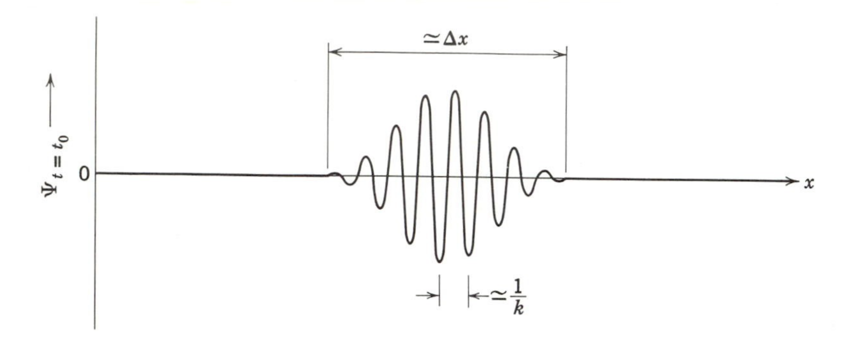

Figure 6–8. A group of pilot waves.

of the wave numbers of the unmodulated waves, the smaller is the distance to the points at which they become completely out of phase. In fact, the distance Δx between these two points is just inversely proportional to the range Δk of the wave numbers, as in equation (6–18). However, the proportionality constant is different. From the theory of Fourier analysis,† it can be shown that a combination of unmodulated waves with wave numbers covering the range Δk will form a group of length Δx, where

$$\Delta x \, \Delta k \simeq 1/2\pi \tag{6–19}$$

The exact value of the constant depends on the relative amplitudes of the different unmodulated waves. The relative amplitudes also determine the exact shape of the group.

Equation (6–19) describes a property which is common to *all* wave motion. Let us now consider the pilot waves which govern the motion of a particle moving along the x axis. For these waves de Broglie postulates

$$k = 1/\lambda = p_x/h$$

† The "Fourier integral" is used to sum the contributions from the infinitely large number of unmodulated waves with infinitesimally differing wave numbers.

where p_x is the momentum of the particle (the x signifies that the momentum is in the x direction). We then have

$$\Delta k = \Delta p_x/h \tag{6-20}$$

where Δk is the range of wave numbers involved in the composition of a group of pilot waves having, at some instant, a length Δx. Now, we may say that Δk represents the uncertainty in the wave number which should be associated with the group at that instant. Therefore we may also say that Δp_x represents the uncertainty in the momentum which should be associated with the particle at the same instant. Furthermore, the discussion above indicates that, for pilot waves, the length Δx of the group represents the uncertainty in the instantaneous position of the particle. Thus from equations (6–19) and (6–20) we have the following relation between the uncertainties in the instantaneous values of the position and of the momentum of the particle:

$$\Delta x \, \Delta p_x \simeq \hbar, \qquad \hbar \equiv h/2\pi \tag{6-21}$$

This is one statement of the *uncertainty principle*, which was first enunciated by Heisenberg in the year 1927.

This relation sets a fundamental limit on the ultimate precision with which we can *simultaneously* know both the x coordinate of a particle and its momentum in the x direction. We may arbitrarily decrease the uncertainty in one of these quantities, but only at the expense of increasing the uncertainty in the other. For example, consider a particle whose associated pilot waves form a group of length ten times less than the group pictured in figure (6–8). The instantaneous uncertainty in the position of the particle is then less by a factor of 10. However, equation (6–19) says that the range of wave numbers present in the new group will be increased by a factor of 10. Consequently the uncertainty in the momentum of the particle at that instant will also be increased by a factor of 10, and the product $\Delta x \, \Delta p_x$ remains approximately equal to $\hbar$. If we demand that the uncertainty in our knowledge of the position of the particle at some instant approach zero, then equation (6–21) requires that the uncertainty in our knowledge of the momentum of the particle at that instant approach infinity. Going to the other limit, we see that if at any time the momentum of the particle is precisely known ($\Delta p_x = 0$), then at that time the position must be completely unknown ($\Delta x \to \infty$). A pilot wave associated with a particle in this situation is given in equation (6–8). This is

$$\Psi = \sin 2\pi(kx - \nu t) \tag{6-22}$$

Here p_x always has the precise value $p_x = hk$, but at any time the position of the particle is entirely unknown since $\Psi_{t=t_0}$ extends from $-\infty$ to $+\infty$.

It should be pointed out that the uncertainty principle gives a *lower limit* to the product $\Delta x \, \Delta p_x$. That is, it is possible to find a group of waves for which the relation corresponding to equation (6–19) is $\Delta x \, \Delta k = \gamma$, where $\gamma > 1/2\pi$. For a particle whose pilot waves form such a group, the product of the instantaneous uncertainties in the position and momentum of the particle would be larger than $\hbar$. As a limiting example of this, consider a particle whose associated pilot wave is represented by the function

$$\Psi = \sin 2\pi[kx - \nu t] + \sin 2\pi[(k + \Delta k)x - (\nu + \Delta\nu)t] \qquad (6\text{–}23)$$

given in equation (6–9), with a large value of Δk. For this particle $\Delta x \rightarrow \infty$, and $\Delta p_x = h \, \Delta k$ will also be large.† A situation in which $\Delta x \, \Delta p_x$ is significantly larger than $\hbar$ means simply that the experimentalist has not achieved the ultimate limit of precision set by the uncertainty principle. In fact, it is very difficult to obtain $\Delta x \, \Delta p_x$ comparable to $\hbar$ in an actual measurement because of the very small size of this constant. This is why the uncertainty principle was never noticed in the experiments of classical mechanics. Because of the very small size of $\hbar \sim 10^{-27}$ erg-sec, the uncertainty principle is not important in classical mechanics; but it is extremely important when we are dealing with the very small distances and momenta involved in atomic and nuclear systems.

If we are considering a three dimensional problem then, according to Heisenberg:

There are three independently operating uncertainty relations, one for each coordinate. For rectangular coordinates x, y, z and corresponding momentum components p_x, p_y, p_z these relations are

$$\begin{aligned} \Delta x \, \Delta p_x &\gtrsim \hbar \\ \Delta y \, \Delta p_y &\gtrsim \hbar \\ \Delta z \, \Delta p_z &\gtrsim \hbar \end{aligned} \qquad (6\text{–}24)$$

If some of the coordinates are angular coordinates, then for each coordinate θ and corresponding angular momentum L_θ the uncertainty relation is

$$\Delta\theta \, \Delta L_\theta \gtrsim \hbar \qquad (6\text{–}25)$$

† It may seem that the use of functions (6–22) and (6–23) in these examples weakens the arguments of the second paragraph of section 2, but this is not actually so. Consider a 1 cm long beam of alpha particles which are being scattered by an atom. From the point of view of the atom, and in terms of distances of the order of the atomic radius R, the x position of some particular particle in the beam may be entirely unknown; $\Delta x \gg R$. If so, (6–22) or (6–23), depending on the accuracy with which the momentum is known, could provide a *good approximation in the vicinity of the atom* to the pilot waves associated with a particle of the beam. However, Δx is really 1 cm and not infinity, and the arguments of section 2 remain valid.

The symbol $\gtrsim$ has been used to indicate specifically that the uncertainty principle specifies only a lower limit. Note that there is no restriction at all on products such as $\Delta x \, \Delta p_y$.

The "physical reason" for the existence of a limit in the maximum precision with which we can simultaneously know the position and the momentum of a particle is quite easy to understand. What happens is that a measurement performed to find the value of one of these quantities

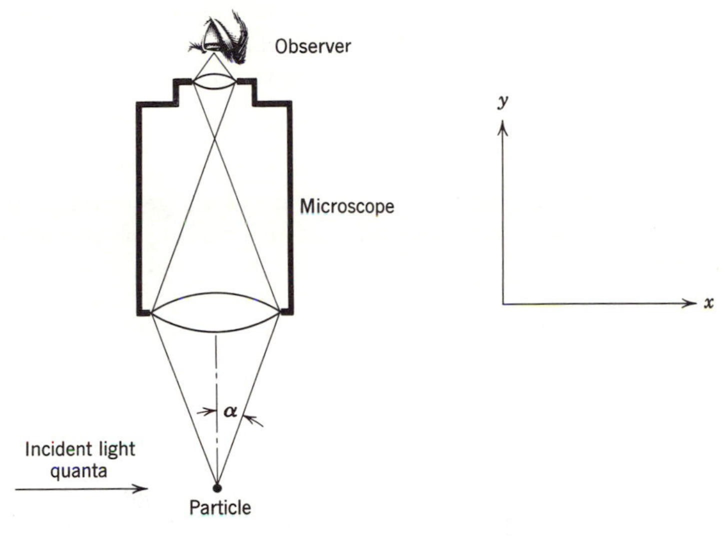

Figure 6–9. Bohr's microscope experiment.

will always disturb the particle in such a way as to leave the value of the other quantity uncertain. A good example of this was pointed out by Bohr in 1928. Consider a measurement, illustrated in figure (6–9), made to determine the instantaneous location of a particle by means of a microscope. In such a measurement the particle must be illuminated, because it is actually the light quanta scattered by the particle that the microscopist sees. The resolving power of the microscope determines the ultimate accuracy with which the particle can be located. The resolving power is known to be approximately $\lambda/\sin \alpha$, where λ is the wavelength of the scattered light and α is the half-angle subtended by the objective lens of the microscope. So the accuracy of the measurement is

$$\Delta x \simeq \frac{\lambda}{\sin \alpha}$$

Let us assume that the microscopist needs to see only one scattered light quantum in order to complete his measurement. The magnitude of the momentum of that quantum is $p = h/\lambda$. But the quantum may have been scattered anywhere within the angle $\pm\alpha$, since there is no way of measuring the exact scattering angle. Consequently the x component of the momentum of the quantum after scattering is uncertain by an amount

$$2p \sin \alpha = \frac{2h \sin \alpha}{\lambda}$$

As the x component of the momentum of the quantum can be known exactly before the scattering (there being no need to know its x coordinate), conservation of momentum requires that the particle receive an amount of recoil momentum in the x direction which is uncertain by the same amount. Thus

$$\Delta p_x \simeq \frac{2h \sin \alpha}{\lambda}$$

and the product of the uncertainties in the knowledge of the x position and the x component of momentum of the particle at the instant of measurement is

$$\Delta x \, \Delta p_x \simeq 2h > h$$

By using light of shorter wavelength, the microscopist can increase the accuracy of the position measurement, but this will lead to an increase in the uncertainty of the momentum of the particle.

We see that the state of affairs described by the uncertainty principle is a direct result of quantization. The quantization of electromagnetic radiation requires that either at least one "unit" of light (a quantum of momentum $p = h/\lambda$) must be scattered from the particle, or else absolutely no light at all. If it were not for this requirement, there would be no reason in principle why the microscopist would not be able to see the particle with good resolution using an illumination of arbitrarily small momentum content. The scattered light quantum provides the interaction which must exist between the measuring instrument and the particle. This interaction disturbs the particle in a way which is uncontrollable and unpredictable. As a result, the coordinates and momentum of the particle cannot be completely known after the measurement. The relation $\Delta x \, \Delta p_x \gtrsim \hbar$ shows that *Planck's constant is a measure of the (minimum) magnitude of this uncontrollable disturbance.*

There is another form of the uncertainty principle which again follows from the mathematical properties of waves plus the de Broglie postulate, in just the same way that equation (6–21) follows from (6–19) and (6–20). From the theory of Fourier analysis it can be shown that the frequency range $\Delta\nu$, which must be involved in the composition of a single group of waves of length Δx and group velocity g, is given by the relation

$$\Delta\tau \, \Delta\nu \gtrsim 1/2\pi \qquad (6\text{–}26)$$

where $\Delta\tau = \Delta x/g$ is the time required for the group to pass any given point; i.e., $\Delta\tau$ is the duration of the pulse of waves.†

Now consider a group of pilot waves of duration $\Delta\tau$. This quantity would be a measure of the uncertainty in the time at which the particle, whose motion is governed by the waves, would pass by a given point. Furthermore, the frequency of the pilot waves is related to the energy of the particle by de Broglie's postulate

$$\nu = E/h$$

Consequently

$$\Delta\nu = \Delta E/h \tag{6–27}$$

The uncertainty $\Delta\nu$ in the frequency of the waves leads to an uncertainty ΔE in the energy associated with the particle, and that uncertainty is related to the uncertainty in the time associated with the particle by the relation

$$\Delta\tau\, \Delta E \gtrsim \hbar \tag{6–28}$$

Heisenberg's interpretation of this uncertainty relation was, however, much broader than that implied by the discussion above. He stated:

A measurement of the energy of a particle (or any system) performed during a time interval $\Delta\tau$ must be uncertain by an amount ΔE, where the relation between these two quantities is $\Delta\tau\ \Delta E \gtrsim \hbar$.

We shall discuss several examples of this uncertainty relation in subsequent chapters.

6. Some Consequences of the Uncertainty Principle

Consider a particle moving freely along the x axis. Assume that at the instant $t = 0$ the position of the particle is uncertain by the amount Δx_0. Let us calculate the uncertainty in the position of the particle at some later time t. According to equations (6–24), the momentum of the particle at $t = 0$ is uncertain by the amount $\Delta p_x \sim \hbar/\Delta x_0$. Consequently the velocity

† The property of wave motion described by equation (6–26) makes it necessary to use very high frequencies in the transmission of television programs. The signal from a television station consists of pulses of duration $\Delta\tau \sim 10^{-6}$ sec. Thus the range of frequencies which must be present is $\Delta\nu \sim 10^6$ sec^{-1}, and the entire broadcast band ($\nu \simeq 0.5 \times 10^6$ sec^{-1} to $\nu \simeq 1.5 \times 10^6$ sec^{-1}) would be able to accommodate only a single television "channel." At the frequencies used in television transmission ($\nu \sim 10^8$ sec^{-1}) many channels can fit into a reasonable portion of the spectrum.

of the particle at that instant is uncertain by the amount $\Delta v = \Delta p_x/m \sim \hbar/m\,\Delta x_0$, and the distance traveled by the particle in the time t is uncertain by the amount

$$\Delta x = t\,\Delta v \sim \frac{\hbar t}{m\,\Delta x_0} \tag{6–29}$$

If by a measurement at $t = 0$ we have localized the particle within the range Δx_0, then in a measurement of its position at time t the particle could be found anywhere within the range Δx. Since Δx is inversely proportional to Δx_0, we see that, the more carefully we localize the particle at the initial instant, the less we shall know about its final position. Furthermore, this uncertainty increases linearly with t.† The behavior of a system, involving distances and momenta small enough that the uncertainty principle is important, is in complete contrast to the behavior of a system described by classical mechanics.

In classical mechanics, if the position and momentum of each particle in an isolated system are known exactly at any instant, the exact behavior of the particles of the system can be predicted for all future time. The uncertainty principle shows us that it is impossible to do this for systems involving small distances and momenta because it is impossible to know, with the required accuracy, the instantaneous positions and momenta of the particles of such a system. As a result, we shall be able to make predictions only of the *probable* behavior of these particles. This idea has already crept into our discussion through the interpretation of the intensity of pilot waves, at a particular position, as some sort of measure of the probability that the associated particle is located at that position. This interpretation will be put into a quantitative form in the next chapter. In the same chapter we shall develop the differential equation (Schroedinger's equation) to which the pilot waves are a solution. In discussing this equation we shall see that the behavior of the pilot waves *is* exactly predictable. But the pilot waves only give information about the *probable* behavior of the associated particle.

The uncertainty principle provides us with a deep insight into our ability to describe the behavior of small scale physical entities. It also helps us to reconcile conflicts which appear to arise from the dual (particle-wave) nature of these entities. Let us consider, as an example, a double-slit

† This statement implies that the group of pilot waves associated with the particle spreads in time, so that its length at any instant is essentially the Δx of equation (6–29). This can be understood as a consequence of the fact that the unmodulated waves which form the group have a range of wave velocities Δw because they have a range of wave numbers Δk and a range of frequencies $\Delta\nu$. As these waves move with different velocities, they cannot maintain the phase relations required to form the group and, in time, the group spreads.

diffraction experiment, using the photoelectric effect as a detector. This is shown in figure (6–10). The source emits electromagnetic radiation. From a study of the energy and time distribution of the photoelectrons ejected from the photocathode, we conclude that the radiation must possess particle-like aspects. However, when we measure the number of photoelectrons ejected as a function of the y coordinate, the periodic pattern which we observe (characteristic double-slit intensity pattern) forces us to conclude that the radiation must also possess wave-like aspects. If we think of the radiation as quanta whose motion is governed

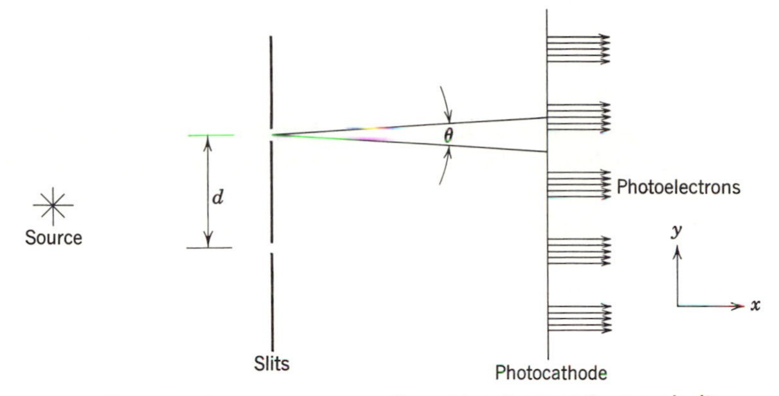

Figure 6–10. An experiment illustrating the particle-wave duality.

by the wave propagation properties of certain associated waves, then we are faced with the following problem: Each quantum must pass through either one slit or the other; if so, how can its motion in the region to the right of the slits be influenced by the interaction of its associated waves with a slit through which it did not pass?

Let us analyze this problem in the spirit of Einstein's analysis of the Galilean transformation equations. The statement that a particular quantum must pass through either one slit or the other is really meaningless unless some provision is made to determine through which slit the quantum actually passed. One way of doing this would be to place a number of very small particles in the region immediately to the right of the slits. After its passage through a slit, the quantum would strike one of the particles, causing it to recoil. By observing the recoiling particle, we could tell through which slit the quantum passed, provided the uncertainty in the y coordinate of the particle after the collision is small compared to the distance between the slits; that is, provided that

$$\Delta y \ll d$$

Now, in colliding the quantum exchanges some y momentum with the particle. The exact amount exchanged depends on details of the collision which we have no way of determining. Thus the y momentum exchange is uncertain by an amount which we call Δp_y. In order not to destroy the double-slit diffraction pattern produced in the passage of a number of quanta through the system, Δp_y must always obey the restriction

$$\Delta p_y/p_x \ll \theta$$

where θ is the angle between adjacent minima and maxima of the pattern, and p_x is the x component of the momentum of the quantum. If this condition is not satisfied, a quantum could receive so much y momentum in the collision that it would end up in what had been a minimum of the diffraction pattern. Since we can assume that the y momentum of the particle was precisely known before the collision because there was no need to know its y position at that time, and since momentum must be conserved in the collision, Δp_y is also equal to the uncertainty in the y momentum of the particle after the collision. Now it is known that the angle θ is related to the wavelength of the radiation and to the distance between the slits by the equation

$$\theta = \lambda/2d$$

Expressing this in terms of the momentum of the quantum by means of equation (6–5), we have

$$\theta = h/2dp_x$$

Combining this with the second inequality above gives

$$\Delta p_y/p_x \ll \theta = h/2dp_x$$

so

$$\Delta p_y \ll h/2d$$

Multiplying this into the first inequality, we have

$$\Delta y \, \Delta p_y \ll h/2$$

This is a condition on the uncertainty in the y position and y momentum of the particle which must always be satisfied in order that we may be able to determine through which slit each quantum passed without destroying the double-slit diffraction pattern. But the uncertainty principle tells us that this condition cannot be satisfied. If we do demonstrate that each quantum actually did pass through one slit or the other, we shall no longer observe the double-slit diffraction pattern. The uncertainty principle shows us that the problem we have drawn attention to is actually illusory. In Chapter 8 we shall see additional examples of the way in which the uncertainty principle resolves apparent conflicts between the wave-like and particle-like aspects of the same entity.

BIBLIOGRAPHY

Bohm, D., *Quantum Theory*, Prentice-Hall, Englewood Cliffs, N.J., 1951.
Bohr, N., *Atomic Physics and Human Knowledge*, John Wiley and Sons, New York, 1958.

EXERCISES

1. Add graphically six unmodulated waves of 1 cm amplitudes and wavelengths $\lambda = 1.5$, 1.6, 1.7, 1.8, 1.9, and 2.0 cm, adjusting the phases so that, at $x = 0$, $\Psi(x, t)_{t=t_0} = +1$ cm for all the waves. Carry out the addition to $x = \pm 10$ cm, and compare the length Δx of the group formed, and the range Δk of the wave numbers used, with equation (6–19). Note that, for a finite number of components with finite differences in wave numbers, a single group is not obtained because there are always many values of x at which all the components will be in phase again.

2. Find the possible values of the momentum of the particle, treated in exercise 2, Chapter 5, from the possible values of energy predicted by the Wilson-Sommerfeld quantization rules. Use these values to evaluate the predicted possible wavelengths of the pilot waves. Show that, with the exception of the lowest energy state where the momentum is zero, it is just for these wavelengths that the waves can combine to form standing waves, as stated in section 4.

3. Discuss the exception mentioned in exercise 2 in terms of the uncertainty principle.

4. Consider a single-slit diffraction experiment for particles, similar to that illustrated in figure (6–1) for electromagnetic radiation. First treat the diffraction of the pilot waves, and show by a Huygens wavelet construction that the angular width of the diffraction pattern is $\theta \sim \lambda/d$. Next treat the deflection of the particles by using the uncertainty principle to estimate the transverse momentum introduced in passing through the slit by the measurement of transverse position that the slit provides, and show that the deflection angle is $\theta \sim \lambda/d$. Comment on why both treatments lead to the same predictions.

5. A boy on top of a ladder of height H is dropping marbles of mass m to the floor and is trying to hit a crack in the floor. To aim, he is using equipment of the highest possible precision. Show that, despite his great care, the marbles will miss the crack by an average distance of the order of

$$(\hbar/m)^{1/2}(H/g)^{1/4}$$

where g is the acceleration due to gravity. Using reasonable values of H and m, evaluate this distance.

CHAPTER 7

Schroedinger's Theory of Quantum Mechanics

1. Introduction

In this chapter we begin our development of Schroedinger's theory of quantum mechanics. This theory, and its applications to a number of typical problems, will occupy our attention for much of the remainder of the book. Furthermore, the discussion of atomic and nuclear physics which follows the discussion of quantum mechanics will be largely based on the Schroedinger theory. For these reasons our treatment of the theory, particularly of the mathematical details, will be considerably more thorough than the treatments we have given previously. This is necessary because it is often easier to think in terms of the mathematics of quantum mechanics than it is to think in terms of the somewhat elusive physical pictures which the mathematics represents—and the mathematics is sufficiently involved that a cursory treatment could not help being confusing. However, we shall, as usual, be perfectly willing to sacrifice rigor for clarity when it seems advisable.

Before embarking on Schroedinger's theory, we should mention that an alternative approach was developed at about the same time by Heisenberg. His theory did not discuss pilot waves. Instead he dealt only with *dynamical quantities* such as x, p_x, E, which were represented by *matrices*. The quantum aspects were introduced into the theory via the uncertainty principle, and applied as commutation conditions on the matrices. The uncertainty principle is equivalent to the de Broglie postulate. Consequently the Heisenberg and Schroedinger theories are actually identical in content, although very different in form. Since the Schroedinger theory is better adapted to an introductory treatment of the subject, we consider here only that theory.

2. The Schroedinger Equation

It should be apparent to the reader that de Broglie's postulate is essentially correct. It should also be apparent that the postulate does not provide us with a complete theory for the behavior of a particle, but only the first step toward such a theory. The postulate says that the motion of a particle is governed by the propagation of its associated pilot waves, but it does not tell us the way in which these waves propagate. Only for the simple case of a free particle have we been able to learn some of the features of the propagation of pilot waves.† To handle the case of a particle moving under the influence of forces, we must have an equation which will tell how the pilot waves propagate under these more general circumstances. In addition, we must also have a quantitative connection between the pilot waves and the associated particle; that is, we must know exactly how these waves "govern" the motion of the particle. The required propagation equation for pilot waves, called the *Schroedinger equation*, was developed in 1925, and the quantitative connection between these waves and the associated particle was developed in the following year.

With several exceptions, Schroedinger directly followed de Broglie's ideas. He did not use the picturesque term "pilot waves" but instead denoted both the waves, and the mathematical function $\Psi(x, t)$ which represents them, by the term *wave function*. We shall adopt this terminology ourselves. A more serious change was that Schroedinger attempted only to develop a theory which would be valid in the non-relativistic range of velocities, although, as we have seen in section 2, Chapter 6, the de Broglie postulate was consistent with the theory of relativity. Schroedinger adopted de Broglie's two equations

$$\lambda = h/p \tag{7–1}$$

and

$$\nu = E/h \tag{7–2}$$

† We know that the velocity of a group of pilot waves associated with a free particle is equal to the velocity of the particle (equation 6–13); and also that a group spreads with increasing time, its length Δx being proportional to t (equation 6–29).

but he did not adopt de Broglie's definition of E as the total relativistic energy. Now, from the theory presented in Chapter 1 (cf. equation 1–22″), it is apparent that, in the case in which there is a potential energy V, the total relativistic energy approaches the expression

$$E_{\text{total relativistic}} \rightarrow p^2/2m_0 + V + m_0c^2$$

in the non-relativistic limit $v/c \rightarrow 0$. In his non-relativistic theory Schroedinger did not bother to carry along the constant m_0c^2, and he took for E the classical definition of total energy

$$E = p^2/2m + V \tag{7–3}$$

where $m = m_0$, the rest mass. We see from equation (7–2) that a change in the definition of E will change the value of ν. But recall that the experiments described in the previous chapter were all such as to check the validity of equation (7–1), not equation (7–2). We shall see later that the actual value of ν is of no significance in Schroedinger's theory. However, one point should be investigated immediately. Assuming equation (7–3), can we still obtain the necessary result that the group velocity g of the wave function for a free particle is equal to the velocity v of the particle? Take

$$\nu = E/h = p^2/2mh + V/h$$

and

$$k \equiv 1/\lambda = p/h$$

For a free particle the potential energy V is a constant (cf. 3 below), so

$$d\nu = \frac{2p\,dp}{2mh}$$

Also we have

$$dk = dp/h$$

From equation (6–11),

$$g = d\nu/dk$$

Thus we obtain the acceptable result

$$g = \frac{p\,dp}{mh}\frac{h}{dp} = \frac{p}{m} = v$$

We begin our development of the Schroedinger equation by listing three requirements which we know a priori it must satisfy.

1. It must be consistent with equations (7–1), (7–2), and (7–3).
2. It must be linear in $\Psi(x, t)$. That is, if $\Psi_1(x, t)$ and $\Psi_2(x, t)$ are solutions to the equation, a linear combination of these functions $\Psi(x, t) = a_1\Psi_1(x, t) + a_2\Psi_2(x, t)$ must also be a solution for any values of the

constants a_1 and a_2. This requirement ensures that we shall be able to add together wave functions to produce the interference phenomena which are observed, for example, in the Davisson-Germer experiments. In the previous chapter we have already assumed that the wave functions can be added (cf. equation 6–9).

3. The potential energy V is, in general, a function of x and t. In the special case $V(x, t) = V_0$, a constant, the momentum of the particle will be constant since $dp/dt = F$, and the force

$$F = -\frac{\partial V(x, t)}{\partial x} = -\frac{\partial V_0}{\partial x} = 0.$$

Also the total energy E is constant in this case. This is the situation of a free particle with constant values of $k = p/h$ and $\nu = E/h$ discussed in the previous chapter. For this case we require the Schroedinger equation to have oscillatory traveling wave solutions of constant wave number and frequency, similar to the wave functions used in that chapter.

Rewriting equations (7–1) and (7–2) in terms of the convenient parameters,

$$K \equiv 2\pi k, \qquad \omega \equiv 2\pi\nu \tag{7–4}$$

we have

$$p = \hbar K, \qquad E = \hbar\omega \tag{7–5}$$

Combining equations (7–3) and (7–5), we obtain

$$\hbar^2 K^2/2m + V(x, t) = \hbar\omega \tag{7–6}$$

To satisfy requirement 1 the Schroedinger equation must be consistent with equation (7–6).

Requirement 2 demands that every term in the equation be linear in $\Psi(x, t)$. Consequently every term must contain either $\Psi(x, t)$ itself, or a derivative of $\Psi(x, t)$ such as†

$$\frac{\partial\Psi(x, t)}{\partial x}, \quad \frac{\partial^2\Psi(x, t)}{\partial x^2}, \ldots \quad \text{or} \quad \frac{\partial\Psi(x, t)}{\partial t}, \quad \frac{\partial^2\Psi(x, t)}{\partial t^2}, \ldots$$

The equation cannot contain terms independent of $\Psi(x, t)$, or terms such as $[\Psi(x, t)]^2$.

We shall use requirement 3 as an aid in finding the form of the Schroedinger equation. We take one of the free particle wave functions $\Psi_f(x, t)$,

† A derivative of Ψ, such as $\partial\Psi/\partial x$, is linear in Ψ because if

$$\Psi = a_1\Psi_1 + a_2\Psi_2$$

then

$$\frac{\partial\Psi}{\partial x} = \frac{\partial}{\partial x}(a_1\Psi_1 + a_2\Psi_2) = a_1\frac{\partial\Psi_1}{\partial x} + a_2\frac{\partial\Psi_2}{\partial x}$$

and then find an equation involving only terms in this function and its derivatives which is consistent with (7–6). This equation will give us a differential equation which has the free particle wave function as a solution when $V(x, t) = V_0$, a constant. We shall then postulate that, even for the general case in which the potential $V(x, t)$ is not a constant, the solutions of this differential equation are the correct wave functions to be associated with a particle moving under the influence of such a potential.

Let us take the simplest free particle wave function discussed in the previous chapter, $\Psi_f = \sin 2\pi(kx - \nu t)$. In terms of K and ω, this is

$$\Psi_f = \sin (Kx - \omega t) \tag{7–7}$$

Evaluate some of its derivatives, remembering that K and ω are constants because we have assumed $V(x, t) = V_0$. We find

$$\begin{aligned} \frac{\partial \Psi_f}{\partial x} &= K \cos (Kx - \omega t), & \frac{\partial^2 \Psi_f}{\partial x^2} &= -K^2 \sin (Kx - \omega t) \\ \frac{\partial \Psi_f}{\partial t} &= -\omega \cos (Kx - \omega t), & \frac{\partial^2 \Psi_f}{\partial t^2} &= -\omega^2 \sin (Kx - \omega t) \end{aligned} \tag{7–8}$$

We must find an equation consistent with

$$\frac{\hbar^2}{2m} K^2 + V_0 = \hbar\omega \tag{7–9}$$

The fact that differentiating twice with respect to x brings in a factor of K^2 and differentiating once with respect to t brings in a factor of ω suggests that we try the equation

$$\alpha \frac{\partial^2 \Psi_f}{\partial x^2} + V_0 \Psi_f = \beta \frac{\partial \Psi_f}{\partial t} \tag{7–10}$$

where α and β are constants to be determined. Inserting (7–7) and (7–8) into this equation, we find

$$-\alpha \sin (Kx - \omega t) K^2 + \sin (Kx - \omega t) V_0 = -\beta \cos (Kx - \omega t)\omega \tag{7–11}$$

Even though the constants α and β are at our disposal, we obviously cannot make this equation agree with (7–9) for all possible values of x and t, and so this attempt has failed.

The observation that the differentiation process changes $\sin (Kx - \omega t)$ into $\cos (Kx - \omega t)$, and vice versa, suggests that we try again to construct the differential equation by starting with a combination of two functions of the form (7–7), one being a sine and the other a cosine. That is, we try

$$\Psi_f = \cos (Kx - \omega t) + \gamma \sin (Kx - \omega t) \tag{7–12}$$

where γ is a constant to be determined that gives us some freedom which, it might be guessed, we shall need. Evaluating some derivatives, we find

$$\frac{\partial \Psi_f}{\partial x} = -K \sin (Kx - \omega t) + K\gamma \cos (Kx - \omega t)$$

$$\frac{\partial^2 \Psi_f}{\partial x^2} = -K^2 \cos (Kx - \omega t) - K^2\gamma \sin (Kx - \omega t) \qquad (7\text{–}13)$$

$$\frac{\partial \Psi_f}{\partial t} = \omega \sin (Kx - \omega t) - \omega\gamma \cos (Kx - \omega t)$$

We again assume the form of the differential equation to be given by equation (7–10) and substitute these derivatives into that equation. This gives

$$-\alpha K^2 \cos (Kx - \omega t) - \alpha K^2\gamma \sin (Kx - \omega t) + V_0 \cos (Kx - \omega t) + V_0\gamma \sin (Kx - \omega t) = \beta\omega \sin (Kx - \omega t) - \beta\omega\gamma \cos (Kx - \omega t)$$

$$[-\alpha K^2 + V_0 + \beta\omega\gamma] \cos (Kx - \omega t) + [-\alpha K^2\gamma + V_0\gamma - \beta\omega] \sin (Kx - \omega t) = 0$$

If this is to be true for all values of x and t, the coefficients of both the cosine and the sine must vanish. That is,

$$-\alpha K^2 + V_0 = -\beta\gamma\omega \qquad (7\text{–}14)$$

$$-\alpha K^2 + V_0 = \frac{\beta}{\gamma}\,\omega \qquad (7\text{–}15)$$

This looks encouraging; we have three equations to satisfy, (7–9), (7–14), (7–15), and three free constants, α, β, γ. Subtracting (7–15) from (7–14) gives

$$0 = -\beta\gamma\omega - \frac{\beta}{\gamma}\,\omega$$

$$\gamma = -1/\gamma$$

$$\gamma^2 = -1$$

$$\gamma = \pm\sqrt{-1} \equiv \pm i \qquad (7\text{–}16)$$

Substituting this into equation (7–14) yields

$$-\alpha K^2 + V_0 = \mp i\beta\omega$$

Comparing this equation with (7–9), we see that

$$\alpha = -\hbar^2/2m$$

$$\mp i\beta = \hbar, \quad \beta = \mp\hbar/i = \pm i\hbar \qquad (7\text{–}17)$$

There are two possible sets of solutions, depending on whether we take $\gamma = +i$ or $\gamma = -i$. Following common practice, we take $\gamma = +i$; then $\beta = +i\hbar$, and the differential equation (7–10) is

$$-\frac{\hbar^2}{2m}\frac{\partial^2\Psi_f}{\partial x^2} + V_0\Psi_f = i\hbar\frac{\partial\Psi_f}{\partial t} \tag{7–18}$$

This equation is consistent with the three a priori requirements.

Equation (7–18) has been obtained for the special case of a constant potential $V(x, t) = V_0$, starting from a free particle wave function Ψ_f similar to those used in the previous chapter for this case. We now assume that, for the general case in which $V(x, t)$ is not a constant, the differential equation controlling the propagation of the wave function is of the same form as (7–18). That is, we *postulate* the equation to be

$$-\frac{\hbar^2}{2m}\frac{\partial^2\Psi(x, t)}{\partial x^2} + V(x, t)\,\Psi(x, t) = i\hbar\frac{\partial\Psi(x, t)}{\partial t} \tag{7–19}$$

This is the famous *Schroedinger equation.*

It is a second order partial differential equation (i.e., involving second partial derivatives) for Ψ as a function of x and t. In many respects it is similar to other partial differential equations which arise in the theories of classical physics.† However, there is one feature of the Schroedinger equation which sets it off in a very striking way from the equations of classical physics. The Schroedinger equation contains the imaginary number i. As a consequence, its solutions are necessarily *complex* functions of x and t. We have already seen one example of this; we were forced to take a complex function

$$\Psi_f = \cos(Kx - \omega t) + i\sin(Kx - \omega t) \tag{7–20}$$

for the case of the free particle.

3. Interpretation of the Wave Function

The wave function $\Psi(x, t)$ is an inherently complex function. We did not know this in Chapter 6 when we developed some of the qualitative properties of the free particle wave functions. However, each of the arguments of that chapter could be stated again, always replacing the phrase "wave function" by, for example, the phrase "real part of the wave

† See the equation (7–58) for the propagation of transverse waves in a stretched string.

function."† In every case the argument would remain valid and the conclusion would remain the same. Recall also that in the discussion of the difference between group and wave velocities, in section 2, Chapter 6, we described the results of a hypothetical measurement of the value of the wave function for various coordinates x at a certain instant t_0. This was done simply to provide something concrete to think about. If we were to repeat that discussion, we would now have sufficient acquaintance with the idea of a wave function to render such an artifice unnecessary. We would also realize that the hypothetical instruments for measuring the value of a wave function cannot possibly exist, since the value of an inherently complex quantity cannot be measured with an actual physical instrument.

The fact that wave functions are complex functions should not be considered a weak point of the quantum mechanical theory. Actually, it is a desirable feature because it makes it immediately obvious that we should not attempt to attribute to wave functions a physical existence in the same sense that water waves have a physical existence. That is, we should not try to answer, or even pose, the questions: Exactly what is waving, and what is it waving in? The reader will recall that consideration of just such questions concerning the nature of electromagnetic waves led nineteenth century physicists to the fallacious concept of the ether. As the wave functions are complex, there is no temptation to make the same mistake again. Instead, it is apparent from the outset that the wave functions are computational devices which have an existence—or at least a significance—only in the context of the Schroedinger theory of which they are a part. This point is emphasized by the fact that wave functions never enter in the Heisenberg theory. Yet that theory is completely equivalent, in terms of its end results, to the Schroedinger theory. These comments should not be construed as implying that the wave functions are lacking in physical interest. We shall see in this section, and in section 8, that a wave function actually contains all the information which the uncertainty principle allows us to know about the associated particle.

In order to obtain this information we must have a quantitative connection between $\Psi(x, t)$ and the dynamical quantities describing the associated particle. Now, it has been indicated in the previous chapter that there must be a relation between some measure of the intensity of $\Psi(x, t)$, in the region of the coordinate x at the time t, and the probability per unit length, or *probability density* $P(x, t)$, of finding the associated particle in that region at that instant. It is apparent that $\Psi(x, t)$ cannot simply be set equal to $P(x, t)$ because the probability density is a real

† The real part of the wave function $\Psi(x, t)$ is the function $R(x, t)$ defined by equation (7–22).

quantity whereas the wave function is complex. However, a way to make this association was proposed in 1926 by Born in the form of the following *postulate:*

If, at the instant t, a measurement is made to locate the particle associated with the wave function $\Psi(x, t)$, then the probability $P(x, t)\,dx$ that the particle will be found at a coordinate between x and $x + dx$ is

$$P(x, t)\,dx = \Psi^*(x, t)\Psi(x, t)\,dx \tag{7-21}$$

The symbol $\Psi^*(x, t)$ represents the *complex conjugate* of $\Psi(x, t)$. To illustrate the meaning of this term, consider the complex function

$$\Psi(x, t) = R(x, t) + i\,I(x, t) \tag{7-22}$$

where $R(x, t)$ and $I(x, t)$ are real functions. (A complex function can always be decomposed in this manner.) Then the complex conjugate of this function is *defined* as

$$\Psi^*(x, t) = R(x, t) - i\,I(x, t) \tag{7-23}$$

An example we shall often come across is

$$\Psi(x, t) = e^{i(Kx-\omega t)} \tag{7-24}$$

This provides a convenient way of writing the free particle wave function (7-20) because $e^{iz} = \cos z + i \sin z$.† From this relation and the definition (7-23), it is easy to show that

$$\Psi^*(x, t) = e^{-i(Kx-\omega t)} \tag{7-25}$$

Now let us evaluate $\Psi^*\Psi$ from equations (7-22) and (7-23). This gives

$$\Psi^*\Psi = (R - iI)(R + iI)$$
$$\Psi^*\Psi = R^2 - i^2 I = R^2 + I^2$$

since $i^2 = -1$. Thus we see that $\Psi^*\Psi$ is *always* a real function, and Born is not inconsistent in equating $\Psi^*(x, t)\Psi(x, t)$ to $P(x, t)$.

Of course, this does not prove that these two quantities must be equated because there are other functions of $\Psi(x, t)$, such as the absolute value $|\Psi(x, t)|$, which are also always real. Some justification for Born's choice of $\Psi^*(x, t)\Psi(x, t)$ to give the probability density can be found

† This equality can easily be proved from a consideration of the series expansions of these functions:

$$e^{iz} = 1 + \frac{(iz)}{1!} + \frac{(iz)^2}{2!} + \frac{(iz)^3}{3!} + \frac{(iz)^4}{4!} + \cdots$$

$$\cos z = 1 - \frac{z^2}{2!} + \frac{z^4}{4!} - \frac{z^6}{6!} + \cdots$$

$$\sin z = z - \frac{z^3}{3!} + \frac{z^5}{5!} - \frac{z^7}{7!} + \cdots$$

from the following argument. Consider Ψ and Ψ^* which are, respectively, solutions to the Schroedinger equation,

$$-\frac{\hbar^2}{2m}\frac{\partial^2\Psi}{\partial x^2} + V\Psi = i\hbar\frac{\partial\Psi}{\partial t} \tag{7–26}$$

and the complex conjugate of the Schroedinger equation,†

$$-\frac{\hbar^2}{2m}\frac{\partial^2\Psi^*}{\partial x^2} + V\Psi^* = -i\hbar\frac{\partial\Psi^*}{\partial t} \tag{7–26$'$}$$

Multiply equation (7–26) by Ψ^* and (7–26′) by Ψ and subtract. This gives

$$-\frac{\hbar^2}{2m}\left(\Psi^*\frac{\partial^2\Psi}{\partial x^2} - \Psi\frac{\partial^2\Psi^*}{\partial x^2}\right) = i\hbar\left(\Psi^*\frac{\partial\Psi}{\partial t} + \Psi\frac{\partial\Psi^*}{\partial t}\right)$$

which reduces to

$$-\frac{\hbar^2}{2m}\left(\Psi^*\frac{\partial^2\Psi}{\partial x^2} - \Psi\frac{\partial^2\Psi^*}{\partial x^2}\right) = i\hbar\frac{\partial}{\partial t}\Psi^*\Psi$$

and further to

$$-\frac{\hbar^2}{2m}\frac{\partial}{\partial x}\left(\Psi^*\frac{\partial\Psi}{\partial x} - \Psi\frac{\partial\Psi^*}{\partial x}\right) = i\hbar\frac{\partial}{\partial t}\Psi^*\Psi$$

Integrating both sides over x between the limits x_1 to x_2, we find

$$-\frac{\hbar^2}{2m}\int_{x_1}^{x_2}\frac{\partial}{\partial x}\left(\Psi^*\frac{\partial\Psi}{\partial x} - \Psi\frac{\partial\Psi^*}{\partial x}\right)dx = i\hbar\int_{x_1}^{x_2}\frac{\partial}{\partial t}\Psi^*\Psi\,dx$$

which gives

$$+\frac{i\hbar}{2m}\left[\Psi^*\frac{\partial\Psi}{\partial x} - \Psi\frac{\partial\Psi^*}{\partial x}\right]_{x_1}^{x_2} = \frac{\partial}{\partial t}\int_{x_1}^{x_2}\Psi^*\Psi\,dx \tag{7–27}$$

† Take the complex conjugate of every term of equation (7–26). This gives

$$\left(-\frac{\hbar^2}{2m}\frac{\partial^2\Psi}{\partial x^2}\right)^* + (V\Psi)^* = \left(i\hbar\frac{\partial\Psi}{\partial t}\right)^*$$

It is easy to show from the definition (7–23) that the complex conjugate of a product is equal to the product of the complex conjugates. Thus we have

$$\left(-\frac{\hbar^2}{2m}\right)^*\left(\frac{\partial^2\Psi}{\partial x^2}\right)^* + V^*\Psi^* = (i\hbar)^*\left(\frac{\partial\Psi}{\partial t}\right)^*$$

According to equation (7–23), the complex conjugate of a purely real quantity, such as $(-\hbar^2/2m)$ or V, is equal to that quantity. The complex conjugate of a purely imaginary quantity, such as $(i\hbar)$, is equal to the negative of the quantity. Furthermore, it is easy to show from (7–23) and the mathematical definition of a derivative that the complex conjugate of the derivative of a function is equal to the derivative of the complex conjugate of the function. From these properties we obtain immediately equation (7–26′).

Now let us take for Ψ the free particle wave function (7–24). Then

$$\Psi = e^{i(Kx-\omega t)} \quad \text{and} \quad \Psi^* = e^{-i(Kx-\omega t)}$$

Differentiation yields

$$\frac{\partial \Psi}{\partial x} = iKe^{i(Kx-\omega t)} = iK\Psi$$

and

$$\frac{\partial \Psi^*}{\partial x} = -iKe^{-i(Kx-\omega t)} = -iK\Psi^*$$

Substituting into equation (7–27), we have

$$-\frac{\hbar}{m}\Big[K\Psi^*\Psi\Big]_{x_1}^{x_2} = \frac{\partial}{\partial t}\int_{x_1}^{x_2} \Psi^*\Psi\, dx$$

From equation (7–5), $\hbar K/m = p/m = v$, where v is the velocity of the particle. Therefore

$$(v\Psi^*\Psi)_{x=x_1} - (v\Psi^*\Psi)_{x=x_2} = \frac{\partial}{\partial t}\int_{x_1}^{x_2} \Psi^*\Psi\, dx \qquad (7\text{–}28)$$

This is really not a very interesting equation for the case of a free particle, where $V(x, t) =$ constant, because $\Psi^*\Psi = e^{-i(Kx-\omega t)}e^{i(Kx-\omega t)} = 1$, $v =$ constant, and the equation amounts to $0 = 0$. However, if we consider a case in which $V(x, t)$ is a very slowly varying function of x and t (almost a constant), then, at least for a restricted range of the x axis, the wave function can be written

$$\Psi = Ae^{i(Kx-\omega t)}$$

where A, K, and ω are very slowly varying functions of x and t. In this case $\Psi^*\Psi = A^2$, and also v is not constant, but we shall still obtain equation (7–28) as we can ignore derivatives of A, K, and ω in evaluating $\partial\Psi/\partial x$ and $\partial\Psi^*/\partial x$. Under these circumstances we can use equation (7–28) to confirm the identification of $\Psi^*\Psi$ with the probability density by comparing this equation with the one dimensional *conservation equation* of flowing fluids,

$$(v\rho)_{x=x_1} - (v\rho)_{x=x_2} = \frac{\partial}{\partial t}\int_{x_1}^{x_2} \rho\, dx \qquad (7\text{–}29)$$

or

$$(S)_{x=x_1} - (S)_{x=x_2} = \frac{\partial}{\partial t}\int_{x_1}^{x_2} \rho\, dx \qquad (7\text{–}29')$$

This equation is illustrated schematically in figure (7–1). The quantity ρ is the mass density of a fluid, and v is its velocity. The quantity $v\rho = S$ is the mass flux of the fluid, which is the mass crossing a given point x

per unit time, and the left side of the equation is equal to the mass of fluid flowing into a region per unit time at x_1, less the mass flowing out per unit time at x_2. The integral is the total mass contained in the region x_1 to x_2, so the right side is just the change in mass in that region per unit time. Thus the equation simply states that in the region x_1 to x_2 the fluid is neither created nor destroyed—it is conserved. We show in the next paragraph that the same kind of conservation equation must apply to the quantum mechanical probability. By comparing equations (7–28) and (7–29), we see that this will be true if we accept Born's postulate.

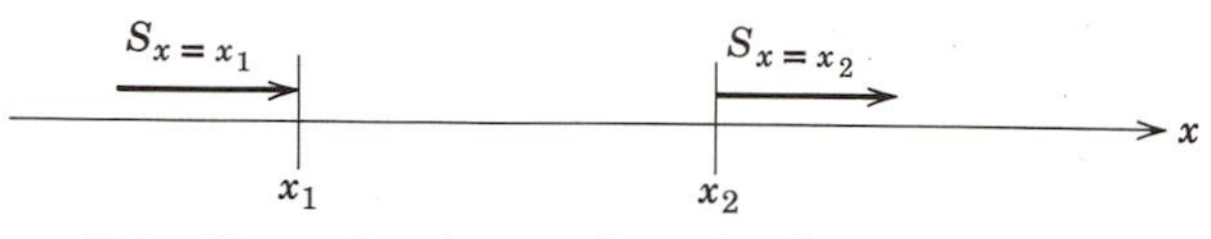

Figure 7–1. Illustrating the one dimensional conservation equation.

The reason why we must have conservation of probability is easy to see. Consider the integral $\int_{-\infty}^{\infty} \Psi^*(x, t)\,\Psi(x, t)\,dx$. This is just the total probability that the particle associated with the wave function $\Psi(x, t)$ be somewhere between $-\infty$ and $+\infty$ at the time t. It is clear then that we must have

$$\int_{-\infty}^{\infty} \Psi^*(x, t)\,\Psi(x, t)\,dx = 1 \tag{7–30}$$

and that this must be true for all t. Thus the total probability must be conserved. Unless we are willing to make some very special assumptions, this can only be achieved by conserving probability in every region x_1 to x_2. In other words, we must have a conservation equation of the form (7–29).

Before leaving this argument, we take the opportunity to obtain from it the formula for the *probability flux* $S(x, t)$. This quantity is the probability per unit time that the particle associated with the wave function $\Psi(x, t)$ will cross the point x, in the direction of increasing values of that coordinate. By comparing equations (7–28) and (7–29′) we see that, in the case where $V(x, t)$ is a slowly varying function, $S(x, t)$ is just $v\Psi^*(x, t)\,\Psi(x, t)$. An expression for $S(x, t)$ of general validity can be obtained by considering equation (7–27), which was derived without any approximations. If we *postulate* the probability flux to be

$$S(x, t) = -\frac{i\hbar}{2m}\left[\Psi^*(x, t)\,\frac{\partial \Psi(x, t)}{\partial x} - \Psi(x, t)\,\frac{\partial \Psi^*(x, t)}{\partial x}\right] \tag{7–31}$$

then equation (7–27) becomes a quantum mechanical probability conservation equation of general validity. We take this as justification for the postulate. It is easy to show that $S(x, t)$ is always real.

4. The Time Independent Schroedinger Equation

The Schroedinger equation (7–19) is a partial differential equation for Ψ as a function of x and t. A standard technique for the solution of such an equation is to look for solutions $\Psi(x, t)$ which are products of a function of x times a function of t, that is, solutions of the form

$$\Psi(x, t) = \psi(x)\,\phi(t) \tag{7–32}$$

We shall see that such solutions exist if the *potential energy is a function of x alone*. We shall also see that this procedure reduces the problem of solving the partial differential equation to the problem of solving two ordinary differential equations.

Let us substitute (7–32) into (7–19), assuming $V(x, t) = V(x)$. This gives

$$-\frac{\hbar^2}{2m}\frac{\partial^2}{\partial x^2}\psi(x)\,\phi(t) + V(x)\,\psi(x)\,\phi(t) = i\hbar\frac{\partial}{\partial t}\psi(x)\,\phi(t)$$

Now

$$\frac{\partial^2}{\partial x^2}\psi(x)\,\phi(t) = \phi(t)\frac{\partial^2}{\partial x^2}\psi(x) = \phi(t)\frac{d^2\psi(x)}{dx^2}$$

the notation $\partial^2\psi(x)/\partial x^2$ being redundant with $d^2\psi(x)/dx^2$ since ψ is a function of x only. Similarly we have

$$\frac{\partial}{\partial t}\psi(x)\,\phi(t) = \psi(x)\frac{d\phi(t)}{dt}$$

and the equation becomes

$$-\frac{\hbar^2}{2m}\phi(t)\frac{d^2\psi(x)}{dx^2} + V(x)\,\psi(x)\,\phi(t) = i\hbar\psi(x)\frac{d\phi(t)}{dt}$$

Dividing both sides by $\psi(x)\,\phi(t)$, we have

$$\frac{1}{\psi(x)}\left\{-\frac{\hbar^2}{2m}\frac{d^2\psi(x)}{dx^2} + V(x)\,\psi(x)\right\} = i\hbar\frac{1}{\phi(t)}\frac{d\phi(t)}{dt}$$

The left side of the equation depends only on the variable x; the right side depends only on the variable t. Since x and t are independent variables, both sides must be equal to a quantity which depends on neither x nor t in order that the equality of the left and right sides of the equation can be true for arbitrary values of the variables. Thus both sides must be equal to the same *separation constant* C. That is,

$$\frac{1}{\psi(x)}\left\{-\frac{\hbar^2}{2m}\frac{d^2\psi(x)}{dx^2} + V(x)\,\psi(x)\right\} = C \tag{7–33}$$

and

$$i\hbar \frac{1}{\phi(t)} \frac{d\phi(t)}{dt} = C \tag{7–33$'$}$$

Equation (7–33′) is a simple first order ordinary differential equation for ϕ as a function of t. It has the solution

$$\phi(t) = e^{-iCt/\hbar} \tag{7–34}$$

This solution can be verified by differentiating and substituting into equation (7–33′):

$$\frac{d\phi(t)}{dt} = -\frac{iC}{\hbar} e^{-iCt/\hbar} = -\frac{iC}{\hbar} \phi(t)$$

which shows that

$$i\hbar \frac{1}{\phi(t)} \frac{d\phi(t)}{dt} = i\hbar \frac{1}{\phi(t)} \left[-\frac{iC}{\hbar} \phi(t) \right] \equiv C; \quad \text{Q.E.D.}$$

The function $\phi(t)$ is a complex oscillatory function of time,

$$\phi(t) = e^{-iCt/\hbar} = \cos(Ct/\hbar) - i \sin(Ct/\hbar)$$

with frequency ν given by $2\pi\nu = C/\hbar$. Thus $\nu = C/2\pi\hbar = C/h$. But, according to equations (7–2) and (7–3), we must have $\nu = E/h$, where E is the total energy of the particle, because $\phi(t)$ gives $\Psi(x, t)$ its time dependence. Thus C must be equal to the total energy E. Knowing this, we may write equation (7–33) as

$$-\frac{\hbar^2}{2m} \frac{d^2\psi(x)}{dx^2} + V(x)\, \psi(x) = E\psi(x) \tag{7–35}$$

Setting $C = E$ in equation (7–34), we can write the solution to the Schroedinger equation as

$$\Psi(x, t) = \psi(x) e^{-iEt/\hbar} \tag{7–36}$$

where $\psi(x)$ is a solution to equation (7–35), which is called the *time independent Schroedinger equation.*

It is an ordinary second order differential equation for ψ as a function of x. It contains no imaginary numbers and, consequently, its solutions $\psi(x)$ are not *necessarily* complex functions.† In section 7 we shall show

† Of course, they can be complex if we wish, because the differential equation is linear in $\psi(x)$ so we can always take a solution which is the sum of a real solution plus i times a second real solution, provided the two solutions correspond to the same value of E. This is proved in the footnote on page 189. An example is the wave function (7–24), which is a solution for the potential $V(x, t) = 0$. This can be written as

$$\Psi(x, t) = \psi(x)\, \phi(t), \quad \text{where} \quad \phi(t) = e^{-i\omega t} = e^{-iEt/\hbar}$$

and

$$\psi(x) = e^{iKx} = \cos Kx + i \sin Kx$$

that this equation is very closely related to the time independent differential equation for classical wave motion. The functions $\psi(x)$ are called the *eigenfunctions*. The reader is cautioned to keep clearly in mind the difference between the eigenfunctions $\psi(x)$ and the wave functions $\Psi(x, t)$, and also the difference between the time independent Schroedinger equation and the Schroedinger equation.

5. Energy Quantization in the Schroedinger Theory

Energy quantization appears in a very natural way in the Schroedinger theory. In this section we present, in some detail, a qualitative description of how this comes about. A number of quantitative examples will be presented in Chapter 8.

Consider a particle of mass m moving under the influence of a force $F(x, t)$, Corresponding to this force is a potential energy $V(x, t)$. The two are related by the equation

$$F(x, t) = -\frac{\partial V(x, t)}{\partial x} \tag{7–37}$$

To determine the motion of the particle under the influence of this potential from the Schroedinger theory, we must solve the Schroedinger equation with the potential $V(x, t)$. Now consider the class of problems for which the potential is a function of x only. (The more complicated case of a time dependent potential will be discussed in a subsequent chapter.) For a time independent potential, we know that there are solutions of the form $\Psi(x, t) = \psi(x)\,\phi(t)$. Since $\phi(t)$ has already been evaluated (equation 7–34), the problem of finding these solutions reduces to the problem of finding the solutions $\psi(x)$ to the time independent Schroedinger equation for the potential $V(x)$ pertinent to the problem. The behavior of the particle is then specified by the functions $\Psi(x, t)$ through the probability postulates (7–21) and (7–31). Energy quantization arises because, in solving the time independent Schroedinger equation, we shall find that *acceptable* solutions exist only for certain values of the total energy E.

To be an acceptable solution, an eigenfunction $\psi(x)$ and its derivative $d\psi(x)/dx$ must obey the following restrictions for all values of x:

1a. $\psi(x)$ *must be finite.* 1b. $\dfrac{d\psi(x)}{dx}$ *must be finite.*

(7–38)

2a. $\psi(x)$ *must be continuous.* 2b. $\dfrac{d\psi(x)}{dx}$ *must be continuous.*

If $\psi(x)$ or $d\psi(x)/dx$ were to violate requirement 1a or 1b, then

$$\Psi(x, t) = e^{-iEt/\hbar}\,\psi(x) \quad \text{and} \quad \frac{\partial \Psi(x, t)}{\partial x} = e^{-iEt/\hbar}\frac{d\psi(x)}{dx}$$

would also. This would mean that the probability density

$$P(x, t) = \Psi^*(x, t)\,\Psi(x, t)$$

and/or the probability flux

$$S(x, t) = \frac{-i\hbar}{2m}\left[\Psi^*(x, t)\frac{\partial \Psi(x, t)}{\partial x} - \Psi(x, t)\frac{\partial \Psi^*(x, t)}{\partial x}\right]$$

were not well defined—that is, did not have a finite and definite value for all values of the variable x.† This cannot be allowed since $P(x, t)$ and

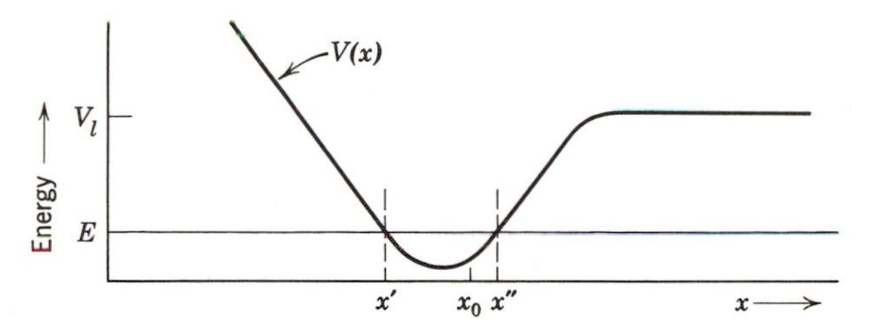

Figure 7–2. An energy diagram showing the relation between potential and total energies in a typical time independent Schroedinger equation.

$S(x, t)$ describe the behavior of a real physical particle and must be well defined. In order to have requirement 1b, we must also have requirement 2a. The need for requirement 2b can be demonstrated by considering the time independent Schroedinger equation (7–35), which we write

$$\frac{d^2\psi(x)}{dx^2} = \frac{2m}{\hbar^2}\,[V(x) - E]\,\psi(x) \tag{7–39}$$

For finite $V(x)$, E, and $\psi(x)$, we see that $d^2\psi(x)/dx^2$ must be finite. This demands that we impose requirement 2b.

Keeping in mind these restrictions on the acceptable form of $\psi(x)$, let us mentally solve the time independent Schroedinger equation for a potential energy $V(x)$ of the form indicated in figure (7–2). By "mentally solve the equation" we mean that we imagine going through exactly the same type of integration procedure which would be used in finding

† All the eigenfunctions we shall deal with in the next few chapters are *single valued*, so the requirement that $P(x, t)$ and $S(x, t)$ have a *definite* value for all values of x is automatically satisfied. However, for the H atom eigenfunctions treated in Chapter 10 it will be necessary to consider this requirement explicitly.

solutions to the differential equation with the aid of a high speed electronic computing machine. This technique is used in situations in which the form of the function $V(x)$ is such that the standard mathematical methods for obtaining an explicit integration of the differential equation (which will be presented in the following chapters) are not applicable.

In order to integrate the differential equation (7–39), we must know the value of the constant total energy E of the particle. But E is not known initially—all we know is the potential $V(x)$. Consequently we must arbitrarily choose a value for this constant, put this value in the equation, and see if we can find an acceptable solution. The value of E which we choose is indicated on the plot in figure (7–2) by the horizontal line: Energy $= E$. We note that, for the value of E chosen, there are two intersections of the line Energy $= E$ and the line Energy $= V(x)$, which divide the x axis into three regions: $x < x'$, $x' \leqslant x \leqslant x''$, $x > x''$. In the first and third regions the quantity $[V(x) - E]$ is positive; in the second region this quantity is negative.

In integrating the differential equation on a computing machine we must start at some point x_0, which we shall take such that $x' < x_0 < x''$, and assume an initial value for the function $\psi(x)$ at that point. Since the equation is linear in $\psi(x)$, the value we choose for $[\psi(x)]_{x_0}$ is immaterial. Let us choose $[\psi(x)]_{x_0} = +1$.† We must also assume an initial value of the derivative $[d\psi(x)/dx]_{x_0}$ at that point. Then, knowing the values of the function and its derivative at the point x_0, we can calculate their values at the nearby point x_1 as follows. For $(x_1 - x_0)$ very small, $[\psi(x)]_{x_1}$ is given simply by

$$[\psi(x)]_{x_1} - [\psi(x)]_{x_0} = \left[\frac{d\psi(x)}{dx}\right]_{x_0} (x_1 - x_0)$$

A similar formula for $[d\psi(x)/dx]_{x_1}$ can be found by using the Schroedinger equation (7–39), which is

$$\frac{d^2\psi(x)}{dx^2} = \frac{2m}{\hbar^2}[V(x) - E]\psi(x)$$

or

$$d\left[\frac{d\psi(x)}{dx}\right] = \frac{2m}{\hbar^2}[V(x) - E]\psi(x)\,dx$$

So

$$\left[\frac{d\psi(x)}{dx}\right]_{x_1} - \left[\frac{d\psi(x)}{dx}\right]_{x_0} = \frac{2m}{\hbar^2}[V(x) - E]_{x_0}[\psi(x)]_{x_0}(x_1 - x_0)$$

provided $(x_1 - x_0)$ is very small.

† As $\psi(x)$ is not necessarily a complex function, we simplify this argument by assuming that it is real.

Repeating the procedure from the starting point x_1, we can find $\psi(x)$ and $d\psi(x)/dx$ at x_2. Continuing in this manner, we can trace out the function and its derivative over the entire x axis. Now note that in the region $x' \leqslant x \leqslant x''$, where $[V(x) - E] < 0$, the sign of the change in the derivative for increasing values of x is opposite to the sign of the function itself. This means that the function is concave downwards in that region if the value of the function is positive, and concave upwards if its value is

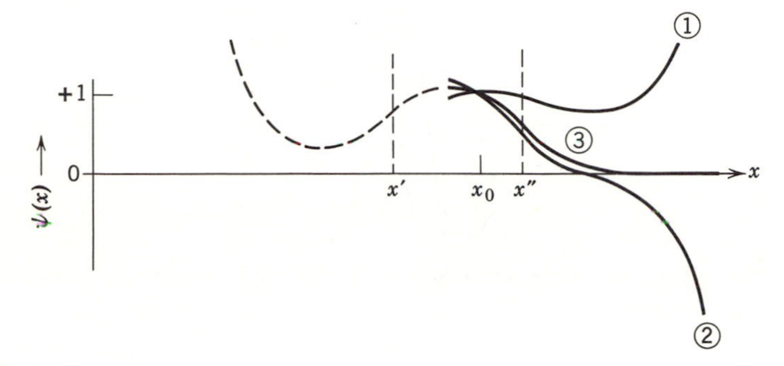

Figure 7–3. Illustrating three unsuccessful attempts at numerical integration of a typical time independent Schroedinger equation.

negative. In the regions $x < x'$ and $x'' < x$, where $[V(x) - E] > 0$, the function is concave upwards if the value of the function is positive, and concave downwards if the value is negative.

On curve 1 of figure (7–3) we record the results of a calculation, starting at x_0, which traced ψ and $d\psi/dx$ in the direction of increasing x. For $x < x''$ the function was concave downwards, but on passing x'' it became concave upwards. Although the derivative was negative at $x = x''$, it soon became zero, and then positive. The function ψ then started increasing, and matters rapidly went from bad to worse because, according to Schroedinger's equation, the rate of change of the derivative, i.e. $d^2\psi/dx^2$, is proportional to ψ. In this calculation we found $\psi \rightarrow \infty$ as $x \rightarrow \infty$. Since this is not acceptable behavior for an eigenfunction, we made a second attempt in which we tried to avoid this divergent behavior by changing our choice of the initial value of the derivative $[d\psi/dx]_{x_0}$. The results are indicated in curve 2. We were not quite successful because ψ became negative in the region where the function is concave downwards if its value is negative. The difficulty in obtaining an acceptable eigenfunction should now be apparent. It should also be apparent that, by making exactly the right choice of $[d\psi/dx]_{x_0}$, we shall be able to find a function whose acceptable behavior with increasing x is as shown in curve 3.

This ψ asymptotically approaches the x axis. The closer it gets the less concave upwards it becomes, so it never turns up.

In figure (7–3) we also indicate with a dashed curve the results of extending the function of curve 3 in the direction of decreasing x. From the discussion above we must expect that the function will, in general, diverge when we extend it to small x. We cannot adjust $[d\psi/dx]_{x_0}$, as that would disturb the situation at large x. Nor can we attempt to prevent divergence at both large

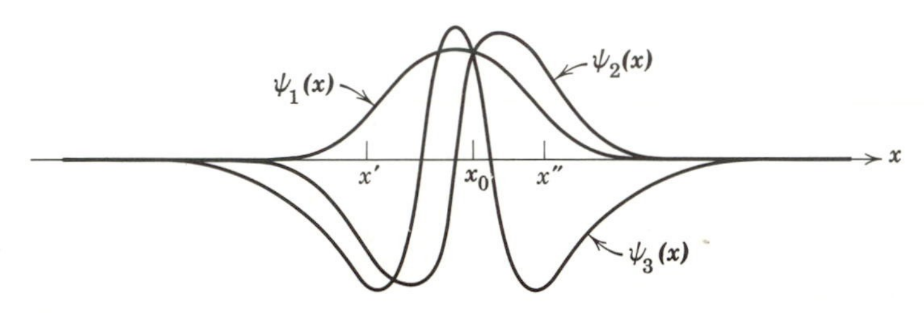

Figure 7–4. The first three acceptable solutions of a typical time independent Schroedinger equation.

and small x by joining two functions of different derivative at $x = x_0$. This is ruled out by the requirement that the derivative of an eigenfunction be everywhere continuous. We are forced to conclude that, for the particular value of the total energy E which we initially chose, there is no acceptable solution to the time independent Schroedinger equation. The relation between a function ψ and its second derivative $d^2\psi/dx^2$, imposed by the equation for that value of E, is such that the function must diverge at either large x or small x (or both).

However, by repeating this procedure for many different choices of the energy E, we shall eventually find a value E_1 for which the time independent Schroedinger equation has an acceptable solution, $\psi_1(x)$. In fact, there will, in general, be a number of allowed values of total energy, E_1, E_2, $E_3, \ldots,$ for which the time independent Schroedinger equation has acceptable solutions $\psi_1(x)$, $\psi_2(x)$, $\psi_3(x), \ldots$. In figure (7–4) we indicate the form of the first three acceptable solutions. The behavior of $\psi_1(x)$ for both small and large x is the same as the behavior of the function shown in curve 3 of figure (7–3) for large x. For $x > x_0$, the behavior of $\psi_2(x)$ is essentially similar to the behavior of $\psi_1(x)$. But, since its second derivative is relatively larger in magnitude, $\psi_2(x)$ crosses the axis at some value of x less than x_0 but greater than x'. When this happens, the sign of the second derivative reverses and the function becomes concave upwards. At $x = x'$ the second derivative reverses again and, for $x < x'$, the function asymptotically approaches the axis.

From figure (7–4), and from the time independent Schroedinger equation,

$$\frac{d^2\psi(x)}{dx^2} = \frac{2m}{\hbar^2}\,[V(x) - E]\,\psi(x)$$

we can see that the allowed energy E_2 is greater than the allowed energy E_1. Consider the point x_0 where both $\psi_1(x)$ and $\psi_2(x)$ have the same value. It is apparent from figure (7–4) that at this point

$$\left|\frac{d^2\psi_2(x)}{dx^2}\right| > \left|\frac{d^2\psi_1(x)}{dx^2}\right|$$

Thus

$$|V(x) - E_2| > |V(x) - E_1|, \quad \text{and} \quad E_2 > E_1$$

From a similar argument we can show that $E_3 > E_2$. It is also apparent that the quantities $(E_2 - E_1)$, $(E_3 - E_2)$, etc., are not infinitesimals since, for example, the quantity

$$\left[\left|\frac{d^2\psi_2(x)}{dx^2}\right| - \left|\frac{d^2\psi_1(x)}{dx^2}\right|\right]_{x_0}$$

is not an infinitesimal. Thus the allowed values of energy are well separated and form a *discrete* set of energies. For a particle moving under the influence of a time independent potential $V(x)$, acceptable solutions to the Schroedinger equation exist only if the total energy of the particle is quantized to be precisely equal to one of the discrete set of energies $E_1, E_2, E_3, \ldots$.

This statement is true as long as the relation between the potential energy $V(x)$ and the total energy E is essentially as shown in figure (7–2); that is, as long as there exists an x' and an x'' such that $[V(x) - E] > 0$ for all $x < x'$ and all $x > x''$. For a potential of the type shown in that figure, in which $V(x)$ has a finite limiting value V_l as x becomes very large, there is generally room only for a finite number of discrete allowed energy values which satisfy that condition.† This is illustrated in figure (7–5). For $E > V_l$, the situation changes. There are now only two regions of the x axis: $x < x'$ and $x > x'$. In the second region $[V(x) - E]$ will be negative for all values of x out to infinity. But, when $[V(x) - E]$ is negative, $\psi(x)$ is concave downwards if its value is positive, and concave upwards if its value is negative. It always tends to return to the axis and

† Potentials which approach the limiting value V_l very slowly can be an exception, because the separation between the allowed energies E_n and E_{n+1} becomes very small as $V_l - E_n$ becomes very small. In such a case it is possible for there to be an infinite number of discrete values of E_n which satisfy the condition $V_l > E_n$. An important example is the Coulomb potential $V(r) \propto -r^{-1}$ which very slowly approaches the limiting value $V_l = 0$.

is, therefore, an oscillatory function. Consequently there will be no problem of $\psi(x)$ diverging for large values of x. Since we can always make $\psi(x)$ asymptotically approach the axis for small values of x by a proper choice of $[d\psi(x)/dx]_{x_0}$, we shall be able to find an acceptable eigenfunction for *any* value of $E > V_l$. Thus the allowed energy values for $E > V_l$ are continuously distributed. They are said to form a *continuum*. It is evident

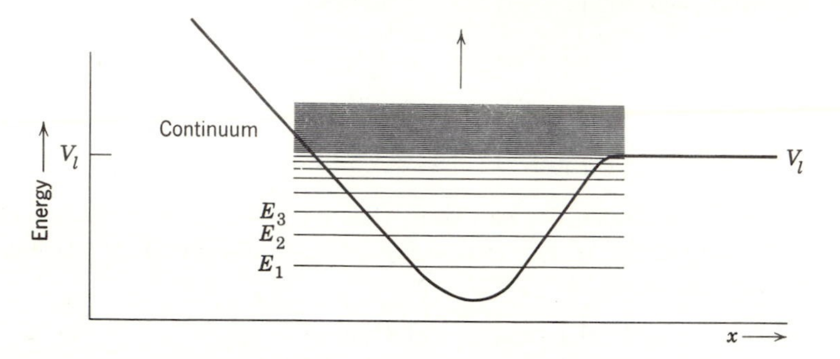

Figure 7–5. The discrete and continuum values of allowed energy for a typical time independent Schroedinger equation.

that, if the potential $V(x)$ is limited for small values of x, or for both large and small values of x, then the allowed energy values will form a continuum for all energies greater than the smallest limit.

We note that when the relation between the potential energy $V(x)$ and the total energy E is such that classically the particle would be *bound* to a finite region of the x axis, in order to satisfy the classical requirement that the total energy E can never be less than the potential energy $V(x)$, then according to the Schroedinger theory the total energy of the particle can only be one of a *discrete* set of energies. When the relation between $V(x)$ and E is such that classically the particle is *unbound* and could move in an infinite region of the x axis, then according to the Schroedinger theory the total energy can have any value and the allowed energies form a *continuum*. All the conclusions we have obtained from "mentally solving" the time independent Schroedinger equation will be verified in the following chapters by explicit integrations of this equation.

6. Mathematical Properties of the Wave Functions and Eigenfunctions

(a) We have found that for a particular time independent potential $V(x)$ acceptable solutions to the Schroedinger equation exist only for certain values of the energy, which we list in order of increasing energy as

$$E_1, E_2, \ldots, E_n, \ldots \tag{7–40}$$

These energies are called the *eigenvalues* of the potential. The eigenvalues early in the list are discrete. However, unless the potential $V(x)$ increases without limit for both very large and very small values of x, the eigenvalues become continuous beyond a certain point. Corresponding to each eigenvalue is an *eigenfunction*

$$\psi_1(x), \psi_2(x), \ldots, \psi_n(x), \ldots \tag{7–41}$$

which is a solution to the time independent Schroedinger equation for the potential $V(x)$. For each eigenvalue there is also a corresponding *wave function*

$$\Psi_1(x, t), \Psi_2(x, t), \ldots, \Psi_n(x, t), \ldots \tag{7–42}$$

which is a solution to the Schroedinger equation for the potential. According to equation (7–36), we know that these wave functions are

$$e^{-iE_1t/\hbar}\psi_1(x),\ e^{-iE_2t/\hbar}\psi_2(x), \ldots, e^{-iE_nt/\hbar}\psi_n(x), \ldots \tag{7–43}$$

The index n, which takes on integral values from 1 to ∞, and which is used to designate a particular eigenvalue and its corresponding eigenfunction and wave function, is called the *quantum number*.

(b) Each of the wave functions $\Psi_n(x, t)$ is a particular solution to the Schroedinger equation for the potential $V(x)$. Since that equation is linear in the wave functions, any linear combination of these functions is also a solution. Let us check this for the simple case

$$\Psi(x, t) = a_1\Psi_1(x, t) + a_2\Psi_2(x, t)$$

Substituting this function into the Schroedinger equation,

$$-\frac{\hbar^2}{2m}\frac{\partial^2\Psi}{\partial x^2} + V\Psi - i\hbar\frac{\partial\Psi}{\partial t} = 0$$

we have

$$-\frac{\hbar^2}{2m}\left(a_1\frac{\partial^2\Psi_1}{\partial x^2} + a_2\frac{\partial^2\Psi_2}{\partial x^2}\right) + V(a_1\Psi_1 + a_2\Psi_2) - i\hbar\left(a_1\frac{\partial\Psi_1}{\partial t} + a_2\frac{\partial\Psi_2}{\partial t}\right) = 0$$

or

$$a_1\left[-\frac{\hbar^2}{2m}\frac{\partial^2\Psi_1}{\partial x^2} + V\Psi_1 - i\hbar\frac{\partial\Psi_1}{\partial t}\right] + a_2\left[-\frac{\hbar^2}{2m}\frac{\partial^2\Psi_2}{\partial x^2} + V\Psi_2 - i\hbar\frac{\partial\Psi_2}{\partial t}\right] = 0$$

Since Ψ_1 and Ψ_2 satisfy the Schroedinger equation, both brackets vanish. Thus

$$a_1\left[-\frac{\hbar^2}{2m}\frac{\partial^2\Psi_1}{\partial x^2} + V\Psi_1 - i\hbar\frac{\partial\Psi_1}{\partial t}\right] + a_2\left[-\frac{\hbar^2}{2m}\frac{\partial^2\Psi_2}{\partial x^2} + V\Psi_2 - i\hbar\frac{\partial\Psi_2}{\partial t}\right] \equiv 0; \qquad \text{Q.E.D.}$$

It is obvious that this argument may be extended to show that the function

$$\Psi(x, t) = \sum_{n=1}^{\infty} a_n \Psi_n(x, t) \tag{7–44}$$

where the a_n are arbitrary constants, is a solution to the Schroedinger equation. In fact, it is the most general form of the solution for the potential $V(x)$.

(c) Let us calculate the probability density $\Psi^*(x, t)\Psi(x, t)$ for the general case of a particle whose associated wave function is of the form of the wave function (7–44). We write it

$$\Psi(x, t) = \sum_{n=1}^{\infty} a_n e^{-iE_nt/\hbar}\psi_n(x) \tag{7–45}$$

Then the complex conjugate is

$$\Psi^*(x, t) = \sum_{l=1}^{\infty} a_l^* e^{+iE_lt/\hbar}\psi_l^*(x) \tag{7–45$'$}$$

We have used a different symbol to designate the index of summation for the two series to remind ourselves that these two series are independent and that their summation indices are independent. Multiplying the two series together, we have†

$$\Psi^*(x, t)\Psi(x, t) = \sum_n a_n^* a_n \psi_n^*(x)\,\psi_n(x) + \sum_{l}\sum_{\substack{n \\ l\neq n}} a_l^* a_n \psi_l^*(x)\,\psi_n(x) e^{-i(E_n - E_l)t/\hbar} \tag{7–46}$$

† For the benefit of readers not proficient in handling series, we illustrate this by writing out a few terms.

$$\begin{aligned}
&[a_1^* e^{iE_1t/\hbar}\psi_1^* + a_2^* e^{iE_2t/\hbar}\psi_2^* + a_3^* e^{iE_3t/\hbar}\,\psi_3^* + \cdots] \\
&\quad\times [a_1 e^{-iE_1t/\hbar}\psi_1 + a_2 e^{-iE_2t/\hbar}\psi_2 + a_3 e^{-iE_3t/\hbar}\psi_3 + \cdots] \\
&= [a_1^* a_1 \psi_1^*\psi_1 + a_2^* a_2 \psi_2^*\psi_2 + a_3^* a_3 \psi_3^*\psi_3 + \cdots] \\
&\quad + [(a_1^* a_2 e^{-i(E_2-E_1)t/\hbar}\psi_1^*\psi_2 + a_1^* a_3 e^{-i(E_3-E_1)t/\hbar}\psi_1^*\psi_3 + \cdots) \\
&\quad + (a_2^* a_1 e^{-i(E_1-E_2)t/\hbar}\psi_2^*\psi_1 + a_2^* a_3 e^{-i(E_3-E_2)t/\hbar}\psi_2^*\psi_3 + \cdots) \\
&\quad + (a_3^* a_1 e^{-i(E_1-E_3)t/\hbar}\psi_3^*\psi_1 + a_3^* a_2 e^{-i(E_2-E_3)t/\hbar}\psi_3^*\psi_2 + \cdots) \\
&\quad + (\qquad\quad) + (\qquad\quad) + \cdots]
\end{aligned}$$

We see that in general the probability density depends on time. Thus the relative probabilities of the various possible results of a measurement made to locate the particle depend on the time at which the measurement is made.

(d) Next consider the special case of a particle whose associated wave function is one of the wave functions $\Psi_n(x, t)$ corresponding to a single eigenvalue E_n. Then

$$\Psi_n^*(x, t)\,\Psi_n(x, t) = e^{iE_nt/\hbar}e^{-iE_nt/\hbar}\psi_n^*(x)\,\psi_n(x) = \psi_n^*(x)\,\psi_n(x) \quad (7\text{–}47)$$

and we see that the probability density is independent of time. Even though the wave function is time dependent, the relative probabilities of the possible results of a measurement of the location of the particle do not depend on the time at which the measurement is made. In such a case the particle is said to be in a stationary state, or *eigenstate*, of the potential $V(x)$.†

(e) According to equation (7–30), the probability interpretation of $\Psi^*(x, t)\,\Psi(x, t)$ demands that any wave function $\Psi(x, t)$ must be *normalized* so that it satisfies the relation

$$\int_{-\infty}^{\infty} \Psi^*(x, t)\,\Psi(x, t)\,dx = 1 \quad (7\text{–}48)$$

If we consider the general form of the wave function

$$\Psi(x, t) = \sum_{n=1}^{\infty} a_n \Psi_n(x, t)$$

it is apparent that normalization of the wave function can be achieved, at least at any given instant of time, by a proper choice of the arbitrary constants a_n.‡ Of course, equation (7–48) must be satisfied at all times. Later we shall derive a time independent relation between the constants a_n which, when satisfied, will effect the normalization of $\Psi(x, t)$ at all times.

(f) Consider the special case of a particle whose wave function $\Psi(x, t)$ is simply one of the eigenstate wave functions $\Psi_n(x, t)$. Then we have from equation (7–47)

$$\Psi^*(x, t)\,\Psi(x, t) = \psi_n^*(x)\,\psi_n(x)$$

† In section 12, Chapter 13, it will be seen that an atom does not emit electromagnetic radiation when the atomic electrons are in eigenstates because its charge distribution, which is proportional to $\Psi^*\Psi$, is time independent.

‡ There are certain difficulties connected with normalizing the wave functions for an unbound particle, for instance the free particle wave function $\Psi(x, t) = e^{i(Kx-\omega t)}$. However, these difficulties are not fundamental; they will be discussed in the first section of the next chapter.

In this case equation (7–48) requires that

$$\int_{-\infty}^{\infty} \psi_n^*(x)\,\psi_n(x)\,dx = 1 \tag{7–49}$$

Thus the eigenfunctions must also be normalized. This can always be accomplished as follows. Assume that we have obtained an eigenfunction $\psi_n'(x)$ by solving the time independent Schroedinger equation. We evaluate the integral

$$\int_{-\infty}^{\infty} \psi_n'^*(x)\,\psi_n'(x)\,dx = A, \quad \text{a constant}$$

Then, if we construct a new function $\psi_n(x) = A^{-1/2}\psi_n'(x)$, we see that $\int_{-\infty}^{\infty} \psi_n^*(x)\,\psi_n(x)\,dx = 1$, and that $\psi_n(x)$ is a normalized eigenfunction.

(g) The set of eigenfunctions of the time independent Schroedinger equation for a particular potential $V(x)$ have an interesting and very useful property expressed by the equation

$$\int_{-\infty}^{\infty} \psi_l^*(x)\,\psi_n(x)\,dx = 0, \qquad l \neq n \tag{7–50}$$

The integral, over all x, of the product of one eigenfunction of the set times the complex conjugate of a different eigenfunction of the set always vanishes. Because of this property eigenfunctions are said to be *orthogonal.* The proof of equation (7–50) is not difficult. Consider two different eigenfunctions $\psi_n(x)$ and $\psi_l(x)$ which are, respectively, solutions to the time independent Schroedinger equations

$$-\frac{\hbar^2}{2m}\frac{d^2\psi_n}{dx^2} + V\psi_n = E_n\psi_n \tag{7–51}$$

and

$$-\frac{\hbar^2}{2m}\frac{d^2\psi_l}{dx^2} + V\psi_l = E_l\psi_l \tag{7–51′}$$

Taking the complex conjugate of the second equation, we obtain the equation for $\psi_l^*(x)$,

$$-\frac{\hbar^2}{2m}\frac{d^2\psi_l^*}{dx^2} + V\psi_l^* = E_l\psi_l^* \tag{7–52}$$

since everything in that equation is real except possibly the eigenfunction. Then we multiply (7–51) by ψ_l^* and (7–52) by ψ_n and subtract. This gives

$$-\frac{\hbar^2}{2m}\left(\psi_l^*\frac{d^2\psi_n}{dx^2} - \psi_n\frac{d^2\psi_l^*}{dx^2}\right) + V\psi_l^*\psi_n - V\psi_n\psi_l^* = (E_n - E_l)\psi_l^*\psi_n$$

Next we integrate with respect to x from $-\infty$ to $+\infty$, and find

$$\frac{2m}{\hbar^2}(E_l - E_n)\int_{-\infty}^{\infty}\psi_l^*\psi_n\,dx = \int_{-\infty}^{\infty}\left(\psi_l^*\frac{d^2\psi_n}{dx^2} - \psi_n\frac{d^2\psi_l^*}{dx^2}\right)dx$$

which reduces to

$$\frac{2m}{\hbar^2}(E_l - E_n)\int_{-\infty}^{\infty}\psi_l^*\psi_n\,dx = \int_{-\infty}^{\infty}\frac{d}{dx}\left(\psi_l^*\frac{d\psi_n}{dx} - \psi_n\frac{d\psi_l^*}{dx}\right)dx$$

Integrating the right side gives

$$\frac{2m}{\hbar^2}(E_l - E_n)\int_{-\infty}^{\infty}\psi_l^*\psi_n\,dx = \left[\psi_l^*\frac{d\psi_n}{dx} - \psi_n\frac{d\psi_l^*}{dx}\right]_{-\infty}^{\infty} \qquad (7\text{–}53)$$

If the eigenvalues E_n and E_l are discrete, thus corresponding to eigenfunctions $\psi_n(x)$ and $\psi_l(x)$ representing a bound particle, then we have seen in the previous section that $\psi_n(x)$ and $\psi_l(x)$ not only remain finite for very large and very small values of x, but actually go to zero at these limits. In this case the right side of equation (7–53) vanishes at both limits and, since $(E_l - E_n) \neq 0$, we obtain the desired proof. For eigenvalues in the continuum, the corresponding eigenfunctions remain finite for very large or very small values of x (or both). However, it is still possible to arrange things in such a way that the right side of equation (7–53) vanishes.† Sometimes it happens that $E_n = E_l$ for $l \neq n$. That is, the eigenvalues corresponding to two different eigenfunctions are numerically equal. In this case the two different eigenstates of the particle are said to be *degenerate*. (Recall the degeneracy discussed in the Sommerfeld treatment of the hydrogen atom.) It is, however, easy to show that degenerate eigenfunctions can always be made orthogonal.‡

† This is associated with the question of normalizing the wave functions for an unbound particle; it is discussed in the first section of the next chapter.

‡ Consider ψ_n and ψ_l which are solutions to the time independent Schroedinger equation for the potential $V(x)$ and for the same eigenvalue $E_n = E_l = E$. Then the linear combination

$$\psi_a = a_n\psi_n + a_l\psi_l$$

is also a solution. To check, substitute into the equation. We have

$$-\frac{\hbar^2}{2m}\frac{d^2}{dx^2}(a_n\psi_n + a_l\psi_l) + V(a_n\psi_n + a_l\psi_l) - E(a_n\psi_n + a_l\psi_l) = 0$$

But this is

$$a_n\left(-\frac{\hbar^2}{2m}\frac{d^2\psi_n}{dx^2} + V\psi_n - E_n\psi_n\right) + a_l\left(-\frac{\hbar^2}{2m}\frac{d^2\psi_l}{dx^2} + V\psi_l - E_l\psi_l\right) \equiv 0; \quad \text{Q.E.D.}$$

(Incidentally this makes it obvious that a linear combination of eigenfunctions is not a

(h) The normality and orthogonality of the set of eigenfunctions for a particular potential $V(x)$ can be summarized in the single equation

$$\int_{-\infty}^{\infty} \psi_l^*(x)\, \psi_n(x)\, dx = \begin{matrix} 1, & l = n \\ 0, & l \neq n \end{matrix} \tag{7–54}$$

These properties often allow a considerable simplification of quantum mechanical calculations.

(i) We now derive the relation between the arbitrary constants a_n of the general form for a wave function

$$\Psi(x, t) = \sum_{n=1}^{\infty} a_n \Psi_n(x, t)$$

which must be satisfied in order that the wave function satisfies the normalization requirement

$$\int_{-\infty}^{\infty} \Psi^*(x, t)\, \Psi(x, t)\, dx = 1$$

Integrating both sides of equation (7–46) over all x, we have

$$\int_{-\infty}^{\infty} \Psi^*(x, t)\, \Psi(x, t)\, dx = \sum_n a_n^* a_n \int_{-\infty}^{\infty} \psi_n^*(x)\, \psi_n(x)\, dx$$
$$+ \sum_{\substack{l \\ l \neq n}} \sum_n a_l^* a_n e^{-i(E_n - E_l)t/\hbar} \int_{-\infty}^{\infty} \psi_l^*(x)\, \psi_n(x)\, dx$$

---

solution to the time independent Schroedinger equation in the normal case in which $E_n \neq E_l$.) Now choose

$$\frac{a_n}{a_l} = -\frac{\int_{-\infty}^{\infty} \psi_n^* \psi_l \, dx}{\int_{-\infty}^{\infty} \psi_n^* \psi_n \, dx}$$

Then ψ_a is orthogonal to ψ_n since

$$\int_{-\infty}^{\infty} \psi_n^* \psi_a \, dx = a_n \int_{-\infty}^{\infty} \psi_n^* \psi_n \, dx + a_l \int_{-\infty}^{\infty} \psi_n^* \psi_l \, dx$$

$$= a_l \int_{-\infty}^{\infty} \psi_n^* \psi_n \, dx \left[\frac{a_n}{a_l} + \frac{\int_{-\infty}^{\infty} \psi_n^* \psi_l \, dx}{\int_{-\infty}^{\infty} \psi_n^* \psi_n \, dx} \right] = 0$$

By applying the condition that ψ_a be normalized, the values of a_n and a_l can be completely determined. The pair of orthogonal degenerate eigenfunctions ψ_n and ψ_a are used instead of the original pair, ψ_n and ψ_l.

Since the double sum contains only terms for which $l \neq n$, we may apply equation (7–54) ($l \neq n$) to each integral appearing in that sum. We may also apply equation (7–54) ($l = n$) to each integral in the single sum. Then we obtain

$$\int_{-\infty}^{\infty} \Psi^*(x, t)\,\Psi(x, t)\,dx = \sum_{n=1}^{\infty} a_n^* a_n$$

Thus normalization of the wave function is achieved if we have

$$\sum_{n=1}^{\infty} a_n^* a_n = 1 \tag{7–55}$$

Since this condition does not involve time, we see that the wave function will always remain normalized.

(j) Consider a general wave function $\Psi(x, t)$ which is a solution to the Schroedinger equation for the potential $V(x)$. If we know the form of this wave function for any particular time, say $t = 0$, then it is easy (at least formally) to evaluate explicitly the wave function at any other time t. At the time $t = 0$, we write the function $\Psi(x, t)$ as a sum of the eigenstate wave functions $\Psi_n(x, t)$ of the Schroedinger equation for the potential $V(x)$. That is, we take equation (7–44), as evaluated in (7–45), at the instant $t = 0$. This gives

$$\Psi(x, 0) = \sum_{n=1}^{\infty} a_n \psi_n(x) \tag{7–56}$$

(This can be thought of as a generalized type of Fourier analysis.) Multiplying both sides by the complex conjugate of one of the eigenfunctions, and integrating over all x, we obtain

$$\int_{-\infty}^{\infty} \psi_l^*(x)\,\Psi(x, 0)\,dx = \sum_{n=1}^{\infty} a_n \int_{-\infty}^{\infty} \psi_l^*(x)\,\psi_n(x)\,dx$$

From equation (7–54) we see that in every term of the sum the integral will vanish, except for the term $n = l$ in which the integral equals unity. Thus we have

$$a_l = \int_{-\infty}^{\infty} \psi_l^*(x)\,\Psi(x, 0)\,dx$$

This tells us the value of the constant a_l. The constant a_n can be evaluated from the same equation if we replace l by n throughout. Furthermore, the value of a definite integral does not depend on what we call the variable of integration. Let us then replace x by x', and write

$$a_n = \int_{-\infty}^{\infty} \psi_n^*(x')\,\Psi(x', 0)\,dx'$$

To evaluate the wave function $\Psi(x, t)$ for an arbitrary value of t, we insert these values of a_n in equation (7–45) and obtain

$$\Psi(x, t) = \sum_{n=1}^{\infty}\left[\int_{-\infty}^{\infty} \psi_n^*(x')\,\Psi(x', 0)\,dx'\right]e^{-iE_nt/\hbar}\psi_n(x) \qquad (7\text{–}57)$$

7. The Classical Theory of Transverse Waves in a Stretched String

In this section we shall develop the classical theory of transverse waves in a stretched string. At each step we shall compare and contrast the classical wave theory with the quantum mechanical wave theory that we

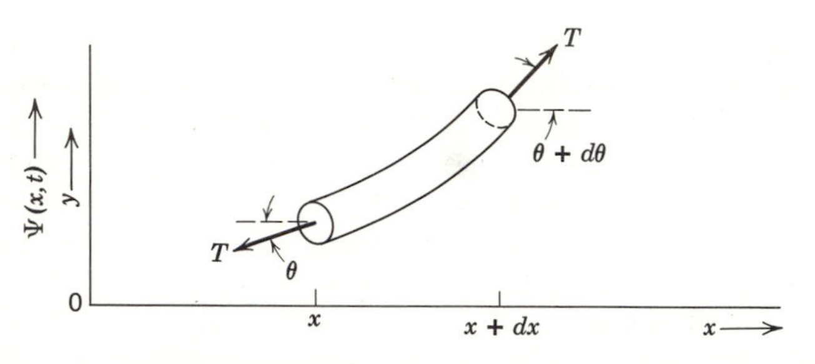

Figure 7–6. A small element of a string at an instant when a transverse wave is passing.

have developed in the previous sections. As we shall see, the mathematical properties of these two theories are often very similar. Consequently we shall be able to illustrate the significance of much of the quantum mechanical wave theory in terms of the familiar phenomena described by the classical wave theory.

Consider a string which is under tension. If the string is given an up and down displacement at one end, the displacement will travel down the string in the manner characteristic of a moving wave. The propagation of this wave can be completely described by the wave function $\Psi(x, t)$, the value of which is numerically equal to the transverse displacement at time t of a point on the string at longitudinal coordinate x. We note that in the classical theory the wave function is a real, measurable function with an immediate physical interpretation. The quantum mechanical wave function is a complex function and is, therefore, not measurable. Furthermore, a physical interpretation exists only for the product of the quantum mechanical wave function times its complex conjugate.

To discuss the propagation of either the classical or the quantum mechanical waves, we must know the differential equation which the wave function obeys. The quantum mechanical wave equation cannot be derived from classical physics and is obtained, essentially, by postulate. The classical wave equation can be derived directly from Newton's laws as follows.

Consider a small element of the string located between the coordinates x and $x + dx$. In figure (7–6) we show that element at an instant t when the displacement is passing the point x. The figure can also be thought of as part of a plot of the function $\Psi(x, t)$ versus x, at the instant t. Now let us impose the restriction: $\theta \ll 1$. Our classical wave equation will apply only to waves on the string of small amplitude. If $\theta \ll 1$,

$$\sin \theta = \tan \theta = \theta$$

$$\cos \theta = 1$$

length of element of string $= dx$

the tension T is a constant

The net force on the element in the x direction is then

$$T \cos (\theta + d\theta) - T \cos \theta = T - T = 0$$

and the net force in the y direction is

$$T \sin (\theta + d\theta) - T \sin \theta = T(\theta + d\theta) - T\theta = T\, d\theta$$

Now $\theta = \tan \theta = \partial\Psi(x, t)/\partial x$, which is the derivative of the function $\Psi(x, t)$ with respect to x, evaluated at the point x and the time t. Thus

$$d\theta = \frac{\partial}{\partial x} \frac{\partial \Psi(x, t)}{\partial x}\, dx$$

which is the change in the derivative, per unit change in x, times the change in x. Therefore

$$T\, d\theta = T \frac{\partial^2 \Psi(x, t)}{\partial x^2}\, dx$$

According to Newton's laws, this net force in the y direction must equal the mass of the element of string times its acceleration in the y direction. The mass is $\rho\, dx$, where ρ is the mass per unit length of the string. Since there is no net x force, the element moves only in the y direction and its acceleration is just $\partial^2\Psi(x, t)/\partial t^2$. Consequently we have

$$T \frac{\partial^2 \Psi(x, t)}{\partial x^2}\, dx = \rho \frac{\partial^2 \Psi(x, t)}{\partial t^2}\, dx$$

or

$$\frac{\partial^2 \Psi(x, t)}{\partial x^2} = \frac{\rho}{T} \frac{\partial^2 \Psi(x, t)}{\partial t^2} \qquad (7\text{–}58)$$

This is the classical wave equation. It governs the propagation of waves (of small amplitude) in a string. Essentially the same equation governs the propagation of electromagnetic waves (of any amplitude) through space. The classical wave equation relates the second space derivative of $\Psi(x, t)$ to its second time derivative. All the constants in the equation are real. The quantum mechanical wave equation

$$-\frac{\hbar^2}{2m}\frac{\partial^2\Psi(x, t)}{\partial x^2} + V(x, t)\,\Psi(x, t) = i\hbar\frac{\partial\Psi(x, t)}{\partial t}$$

relates the second space derivative of $\Psi(x, t)$ to its first time derivative, and to the function itself. The constants in the quantum mechanical wave equation are partly real and partly imaginary. Nevertheless, the procedures involved in solving these two partial differential equations are very similar. Let us consider an example of the solution of the classical wave equation.

A string of length a is fixed at both ends and under tension T. The string is given a small but arbitrary initial displacement, and we are required to evaluate its subsequent motion. We define an x axis extending along the length of the string with an origin at its center. Then we must solve the wave equation, subject to the conditions

$$\Psi(x, t) = 0, \qquad \text{for } x = \pm\, a/2, \text{ for all } t \tag{7–59}$$

which state that the ends of the string are fixed. Similar conditions obtain for the quantum mechanical wave function when it describes the motion of a particle which is bound to a certain region of space. Such quantum mechanical wave functions always go to zero at both large and small x.

It is obviously appropriate in this problem to look for solutions of the form

$$\Psi(x, t) = \psi(x)\,\phi(t)$$

Substituting this into equation (7–58), we obtain

$$\phi(t)\frac{d^2\psi(x)}{dx^2} = \frac{\rho}{T}\,\psi(x)\frac{d^2\phi(t)}{dt^2}$$

Dividing both sides by $\psi(x)\,\phi(t)$, we have

$$\frac{1}{\psi(x)}\frac{d^2\psi(x)}{dx^2} = \frac{\rho}{T}\frac{1}{\phi(t)}\frac{d^2\phi(t)}{dt^2} = -K^2$$

Since x and t are independent variables, both sides of this equation must be equal to the same separation constant, which we call $-K^2$. Thus we have the two second order ordinary differential equations

$$\frac{d^2\psi(x)}{dx^2} + K^2\psi(x) = 0 \tag{7–60}$$

and

$$\frac{d^2\phi(t)}{dt^2} + K^2 \frac{T}{\rho} \phi(t) = 0 \tag{7–61}$$

Consider equation (7–60). It is identical in form with the time independent Schroedinger equation for the case $V(x) = 0$.† The general solution of the equation is

$$\psi(x) = A \sin Kx + B \cos Kx \tag{7–62}$$

where A and B are arbitrary constants. To prove this, evaluate

$$\frac{d\psi(x)}{dx} = AK \cos Kx - BK \sin Kx$$

and

$$\frac{d^2\psi(x)}{dx^2} = -AK^2 \sin Kx - BK^2 \cos Kx = -K^2\psi(x)$$

and substitute into equation (7–60). This gives

$$-K^2\psi(x) + K^2\psi(x) \equiv 0; \quad \text{Q.E.D.}$$

To determine the allowed values of the separation constant, we apply the conditions (7–59), which read

$$\Psi(\pm a/2, t) = \psi(\pm a/2)\,\phi(t) = 0, \qquad \text{for all } t$$

So

$$\psi(\pm a/2) = 0 \tag{7–63}$$

Then we have

$$A \sin \frac{Ka}{2} + B \cos \frac{Ka}{2} = 0$$

and

$$-A \sin \frac{Ka}{2} + B \cos \frac{Ka}{2} = 0$$

Substracting and adding, we obtain

$$A \sin \frac{Ka}{2} = 0$$

$$B \cos \frac{Ka}{2} = 0$$

Both of these equations must be satisfied.

† The analogy is even closer than this. If we consider a string of mass per unit length $\rho(x)$ which is a function of x, we shall obtain an equation identical in form with the time independent Schroedinger equation for a general potential $V(x)$.

Now there is no value of K for which both $\sin(Ka/2)$ and $\cos(Ka/2)$ vanish simultaneously. However, we can satisfy these equations *either* by choosing K such that $\cos(Ka/2)$ vanishes and setting A equal to zero, *or* by choosing K such that $\sin(Ka/2)$ vanishes and setting B equal to zero. That is, we take

$$A = 0;\ \cos\frac{Ka}{2} = 0, \text{ which implies } \frac{Ka}{2} = \frac{n\pi}{2} \text{ with } n = 1, 3, 5, \ldots$$

or (7–64)

$$B = 0;\ \sin\frac{Ka}{2} = 0, \text{ which implies } \frac{Ka}{2} = \frac{n\pi}{2} \text{ with } n = 2, 4, 6, \ldots.$$

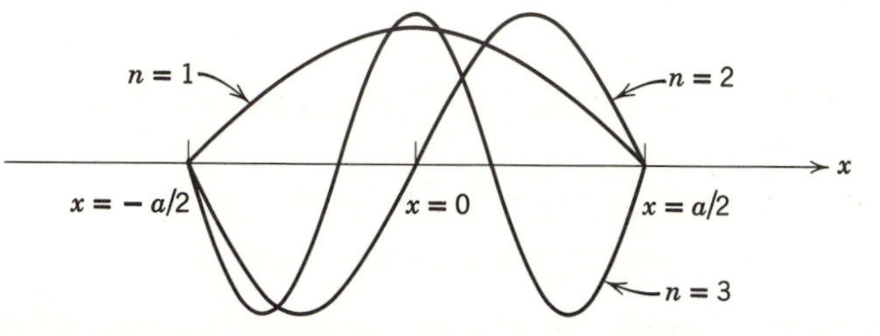

Figure 7–7. The first three eigenfunctions for a string fixed at both ends.

Thus there are two classes of solutions. For the *first class*,

$$\psi_n(x) = B\cos\frac{n\pi x}{a}, \qquad n = 1, 3, 5, \ldots$$

For the *second class*, (7–65)

$$\psi_n(x) = A\sin\frac{n\pi x}{a}, \qquad n = 2, 4, 6, \ldots$$

These are the eigenfunctions for the string of length a fixed at both ends. In figure (7–7) we plot the first three eigenfunctions. Note the essential similarity of these functions to the three quantum mechanical eigenfunctions for a bound particle plotted in figure (7–4). Also note the difference in their physical significance. In the classical theory, $\psi_n(x)$ gives the spatial dependence of the amplitude of the vibration. In the quantum mechanical theory, $\psi_n^*(x)\,\psi_n(x)$ gives the spatial dependence of the probable location of the associated particle.

Next consider equation (7–61). By comparison with equations (7–60) and (7–62), it is apparent that the general solution to this equation is

$$\phi(t) = \alpha\sin(K\sqrt{T/\rho}\,t) + \beta\cos(K\sqrt{T/\rho}\,t) \qquad (7–66)$$

where α and β are arbitrary constants. By proper choice of the zero of time, we may set $\alpha = 0$, and simplify equation (7–66) to

$$\phi(t) = \beta \cos (K\sqrt{T/\rho}\, t) \tag{7–66$'$}$$

Since equation (7–61) involves no complex constants, it has a real solution (7–66′). The corresponding equation in the quantum mechanical case (7–33′) contains the imaginary number i, and its solution (7–34) is necessarily complex.

Combining equations (7–59), (7–64), (7–65), and (7–66′), we find the wave functions for a string of length a, density per unit length ρ, and tension T, which is fixed at both ends:

$$\Psi_n(x, t) = a_n \cos (2\pi\nu_n t)\, \psi_n(x) \tag{7–67}$$

where

$$\psi_n(x) = \begin{cases} \cos \dfrac{n\pi x}{a}, & n = 1, 3, 5, \ldots \\[2ex] \sin \dfrac{n\pi x}{a}, & n = 2, 4, 6, \ldots \end{cases}$$

and where

$$\nu_n = \frac{K}{2\pi}\sqrt{\frac{T}{\rho}} = \frac{n}{2a}\sqrt{\frac{T}{\rho}}$$

The a_n are arbitrary constants, and the ν_n are the vibration frequencies of the string. This discrete set of frequencies constitutes the familiar "fundamental," "second harmonic," "third harmonic," etc., produced by any stringed musical instrument. In the classical theory the frequency is quantized. In the quantum theory the energy is quantized, but the energy is related to the frequency by the equation $E = h\nu$.

An interesting difference between the two theories is that in the classical theory the wave functions are not required to satisfy the normalization condition (7–48); the integral $\int_{-\infty}^{\infty} \Psi^*\Psi\, dx$ can assume any value because the value of the constants a_n, which determine the amplitude of the vibrations, can assume any value that is not too large to violate the restriction $\theta \ll 1$. However, the eigenfunctions of the classical theory of a vibrating string are orthogonal. The orthogonality relation reads

$$\int_{-a/2}^{a/2} \psi_l(x)\, \psi_n(x)\, dx = 0, \qquad l \neq n \tag{7–68}$$

This is completely equivalent to equation (7–50) because, in the present case, the range of x is only from $-a/2$ to $+a/2$, and because $\psi_l^*(x) = \psi_l(x)$. The reader should verify equation (7–68) by actually evaluating some of the integrals.

The general form of the classical wave function is an arbitrary linear combination of the functions (7–67). That is,

$$\Psi(x, t) = \sum_{n=1}^{\infty} a_n \cos(2\pi\nu_n t)\, \psi_n(x) \tag{7–69}$$

Normally, the form of the x dependence of the function $\Psi(x, t)$, at some particular time t, depends on the value of t. This is analogous to the situation described by equation (7–46). However, if we set the string into

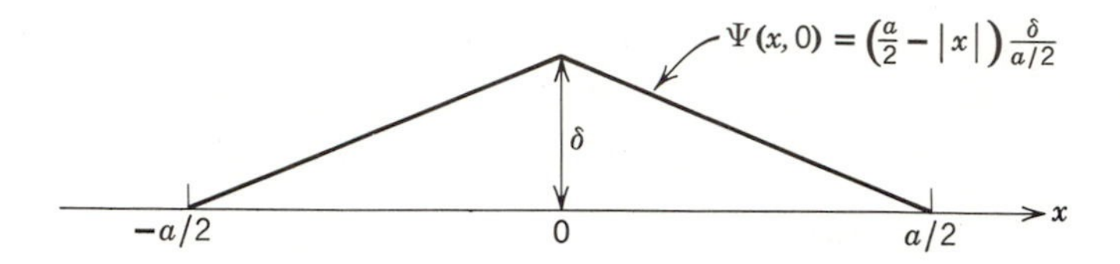

Figure 7–8. A string plucked at its center.

vibration by giving each segment an initial displacement which is in the form of one of the eigenfunctions, the form of the x dependence of $\Psi(x, t)$ will be constant in time. For instance, let us give the string a displacement at $t = 0$ of the form $\Psi(x, 0) = C \cos(\pi x/a)$. Then, in equation (7–69), we have

$$C \cos \frac{\pi x}{a} = \sum_{n=1}^{\infty} a_n \psi_n(x)$$

In this case all $a_n = 0$, except $a_1 = C$. Since the a_n are time independent, the wave function at any instant of time is given by

$$\Psi(x, t) = C \cos\left(2\pi \frac{1}{2a}\sqrt{\frac{T}{\rho}}\, t\right) \cos \frac{\pi x}{a}$$

This situation is analogous to a quantum mechanical eigenstate.

Consider a string which, at the instant $t = 0$, is given a displacement of a form different from that of an eigenfunction; say we pluck the string at its center, so that $\Psi(x, 0) = (a/2 - |x|)\, \delta/(a/2)$. This is shown in figure (7–8). To determine the subsequent motion of the string, we use its initial displacement to evaluate the coefficients a_n. At $t = 0$, equation (7–69) reads

$$\Psi(x, 0) = \sum_{n=1}^{\infty} a_n\, \psi_n(x) \tag{7–70}$$

To evaluate a particular constant, say a_l, multiply both sides of this equation by $\psi_l(x)$ and integrate over the range of x. This gives

$$\int_{-a/2}^{a/2} \psi_l(x)\,\Psi(x, 0)\,dx = \sum_{n=1}^{\infty} a_n \int_{-a/2}^{a/2} \psi_l(x)\,\psi_n(x)\,dx$$

Because of the orthogonality condition (7–68), every term of the summation vanishes except the term for $n = l$. We consequently have

$$a_l = \frac{\displaystyle\int_{-a/2}^{a/2} \psi_l(x)\,\Psi(x, 0)\,dx}{\displaystyle\int_{-a/2}^{a/2} [\psi_l(x)]^2\,dx} \tag{7–71}$$

which is

$$a_l = \frac{\displaystyle\int_{-a/2}^{a/2} \psi_l(x)\left(\frac{a}{2} - |x|\right)\frac{\delta}{a/2}\,dx}{\displaystyle\int_{-a/2}^{a/2} [\psi_l(x)]^2\,dx}$$

Evaluating the eigenfunctions from equation (7–67), and performing the integrations, we find

$$a_n = \begin{matrix} 8\delta/\pi^2 n^2, & n = 1, 3, 5, \ldots \\ 0, & n = 2, 4, 6, \ldots \end{matrix}$$

where we have replaced l by n. The coefficients of $\psi_n(x)$ in the series (7–70) are non-zero only for odd n because of the symmetry of $\Psi(x, 0)$. The series is

$$\Psi(x, 0) = \sum_{n=1,3,5,\ldots} \frac{8\delta}{\pi^2 n^2}\,\psi_n(x)$$

From equation (7–67) this is

$$\Psi(x, 0) = \sum_{n=1,3,5,\ldots} \frac{8\delta}{\pi^2 n^2} \cos\frac{n\pi x}{a}$$

The coefficients of the cosine functions decrease rapidly with increasing n, and the series rapidly converges. Thus a good approximation to $\Psi(x, 0)$ can be obtained by taking only the first few terms of the series. This is illustrated in figure (7–9) by plotting each of the first three terms of the series, their sum, and also the function

$$\Psi(x, 0) = \left(\frac{a}{2} - |x|\right)\frac{\delta}{a/2}$$

to which the series converges.

What we have done is to write the initial displacement $\Psi(x, 0)$ in terms of a series of the eigenfunctions $\psi_n(x)$. [As the $\psi_n(x)$ happen to be sines and cosines, this amounts to making a Fourier analysis of $\Psi(x, 0)$.] This allows us to evaluate the displacement of the string at any instant of time by simply inserting, into the series for the general form of the wave function (7–69), the values of a_n which we have determined. That is,

$$\Psi(x, t) = \sum_{n=1,3,5,\ldots} \frac{8\delta}{\pi^2 n^2} \cos(2\pi\nu_n t) \cos\frac{n\pi x}{a}$$

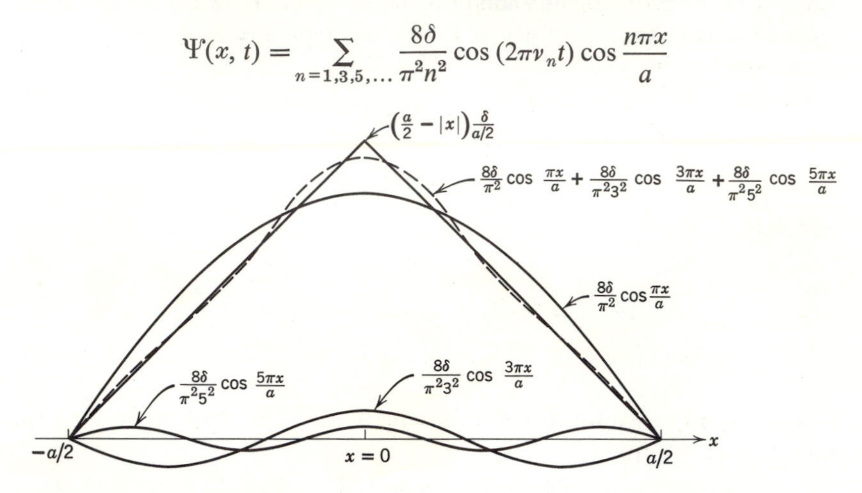

Figure 7–9. Illustrating the Fourier analysis of a string plucked at its center.

A good approximation to $\Psi(x, t)$ can be obtained from the first few terms of the series. The physical significance of this series can be interpreted as follows. At $t = 0$, we simultaneously set into vibration a number of different eigenstates. As time progresses, these eigenstates vibrate independently, each at its own characteristic frequency ν_n. Then at any instant the net displacement at x can be obtained by summing the instantaneous displacements of the eigenstates at that value of x.

Using the general expression (7–71) for the a_n, we can write a general solution to the problem of a string with fixed ends that is vibrating as a result of an arbitrary given initial displacement $\Psi(x, 0)$. The solution is

$$\Psi(x, t) = \sum_{n=1}^{\infty} \frac{\int_{-a/2}^{a/2} \psi_n(x')\,\Psi(x', 0)\,dx'}{\int_{-a/2}^{a/2} [\psi_n(x')]^2\,dx'} \cos(2\pi\nu_n t)\,\psi_n(x)$$

This equation is completely analogous to equation (7–57), which gives the quantum mechanical wave function for a particle moving in a certain potential in terms of a given initial form of the wave function.

8. Expectation Values and Differential Operators

In the previous section we compared some of the mathematical properties of the classical wave theory with certain similar properties which we have developed for the quantum mechanical wave theory. We return now to the task of continuing the development of the latter theory.

Consider a particle and its associated wave function $\Psi(x, t)$. In a measurement of the position of the particle, there would be a finite probability of finding it at any x coordinate in the interval x to $x + dx$, as long as the wave function is non-zero in that interval. In general, the wave function is non-zero over a certain range of the x axis. Thus we are generally not able to state that the x coordinate of the particle has a certain definite value. However, it is possible to specify some sort of an average position of the particle in the following way. Let us imagine making a measurement of the position of the particle at the instant t. The probability of finding it between x and $x + dx$ is, according to the postulate (7–21),

$$P(x, t)\, dx = \Psi^*(x, t)\Psi(x, t)\, dx \qquad (7\text{–}72)$$

Imagine repeating this measurement a number of times, always at the same value of t, and recording the observed values of x.† We can then use the average of the observed values to characterize the position of the particle at the instant t. This we call $\bar{x}$, the *expectation value* of the x coordinate of the particle at the instant t. It is evident that $\bar{x}$ will be given by

$$\bar{x} = \int_{-\infty}^{\infty} xP(x, t)\, dx$$

The integrand is just the value of the x coordinate weighted by the probability of observing that value and, upon integrating, we obtain the average of the observed values. From equation (7–72) this can be written‡

$$\bar{x} = \int_{-\infty}^{\infty} \Psi^*(x, t)\, x\Psi(x, t)\, dx \qquad (7\text{–}73)$$

It is easy to see that an expression of the same form as (7–73) would be appropriate for the evaluation of the expectation value of any function of x. That is,

$$\overline{x^2} = \int_{-\infty}^{\infty} \Psi^*(x, t)\, x^2\Psi(x, t)\, dx$$

† We can visualize doing this either by making a number of measurements on identical particles in identical environments, each particle being in the state associated with the wave function $\Psi(x, t)$, or by making a number of repeated measurements on the same particle, always with the particle in the state associated with the wave function $\Psi(x, t)$.

‡ The terms of the integrand are written in this order to preserve symmetry with a notation which will be used later.

and

$$\overline{f(x)} = \int_{-\infty}^{\infty} \Psi^*(x, t) f(x) \Psi(x, t)\, dx$$

where $f(x)$ is any function of x. Even for a function which explicitly depends on the time, such as the potential energy $V(x, t)$, we may still write

$$\overline{V(x, t)} = \int_{-\infty}^{\infty} \Psi^*(x, t)\, V(x, t) \Psi(x, t)\, dx \tag{7-74}$$

because all measurements made to evaluate $V(x, t)$ are made at the same value of t, and so the arguments of the previous paragraph would still hold.

The coordinate x and the potential energy $V(x, t)$ are two examples of the *dynamical quantities* which can be used to characterize the state of a particle. Examples of other dynamical quantities are the momentum p and the total energy E. The expectation value of these quantities is always given by the same type of expression. For instance, for the momentum p the expression is

$$\bar{p} = \int_{-\infty}^{\infty} \Psi^*(x, t)\, p\Psi(x, t)\, dx \tag{7-75}$$

In order to evaluate the integral in equation (7–75), the integrand $\Psi^*(x, t)\, p\Psi(x, t)$ must be expressed as a function of the variables x and t. In a problem in classical mechanics, p can always be written as a function of the variables x and/or t after the problem has been solved. For instance, for a particle moving in a time independent potential, p can be written as a function of x since, after the classical problem has been solved, the momentum is precisely known at every point of the trajectory. But the uncertainty principle shows us that in quantum mechanics this cannot be done because p and x cannot be simultaneously known with complete precision. We must find some other way of expressing the integrand in terms of x and t.

A clue can be found by considering the free particle wave function (7–24), which is

$$\Psi(x, t) = e^{i(Kx - \omega t)}$$

Differentiate with respect to x. This gives

$$\frac{\partial \Psi(x, t)}{\partial x} = iKe^{i(Kx - \omega t)} = iK\Psi(x, t)$$

Since $K = p/\hbar$,

$$\frac{\partial \Psi(x, t)}{\partial x} = i\frac{p}{\hbar}\Psi(x, t)$$

which can be written

$$p[\Psi(x, t)] = -i\hbar \frac{\partial}{\partial x} [\Psi(x, t)]$$

This indicates that there is an association between the dynamical quantity p and the *differential operator* $-i\hbar\, \partial/\partial x$. A similar association between the dynamical quantity E and the differential operator $i\hbar\, \partial/\partial t$ can be found by differentiating the free particle wave function with respect to t:

$$\frac{\partial \Psi(x, t)}{\partial t} = -i\omega e^{i(Kx - \omega t)} = -i\omega \Psi(x, t)$$

Since $\omega = E/\hbar$, this can be written

$$E[\Psi(x, t)] = i\hbar \frac{\partial}{\partial t} [\Psi(x, t)]$$

It may seem that these associations apply only to the case of a free particle wave function, but this is actually not so. There are several ways of showing this; perhaps the most convincing one is the following. Consider the equation (7–3) defining the classical total energy E:

$$p^2/2m + V(x, t) = E$$

Let us replace the dynamical quantities p and E by their associated differential operators. Then we have

$$\frac{1}{2m}\left(-i\hbar \frac{\partial}{\partial x}\right)^2 + V(x, t) = i\hbar \frac{\partial}{\partial t}$$

or

$$-\frac{\hbar^2}{2m} \frac{\partial^2}{\partial x^2} + V(x, t) = i\hbar \frac{\partial}{\partial t} \tag{7–76}$$

This is an operator equation. To obtain a differential equation, we multiply the equation into the equality

$$\Psi(x, t) = \Psi(x, t)$$

and obtain

$$-\frac{\hbar^2}{2m} \frac{\partial^2 \Psi(x, t)}{\partial x^2} + V(x, t)\, \Psi(x, t) = i\hbar \frac{\partial \Psi(x, t)}{\partial t}$$

This is just the Schroedinger equation (7–19). We see that *postulating the associations*

$$\begin{aligned} p &\leftrightarrow -i\hbar \frac{\partial}{\partial x} \\ E &\leftrightarrow i\hbar \frac{\partial}{\partial t} \end{aligned} \tag{7–77}$$

is equivalent to postulating the Schroedinger equation. This is essentially the procedure originally followed by Schroedinger. The procedure provides us with a powerful method for obtaining Schroedinger equations for more complicated cases than we have discussed in this chapter. We shall use it in Chapter 10 to obtain Schroedinger equations for three dimensional problems, and for problems involving more than one particle.

Let us use (7–77) in the equation (7–75) for the expectation value of the momentum. Then

$$\bar{p} = \int_{-\infty}^{\infty} \Psi^*(x, t)\, p\Psi(x, t)\, dx$$

becomes

$$\bar{p} = \int_{-\infty}^{\infty} \Psi^*(x, t)\left(-i\hbar \frac{\partial}{\partial x}\right) \Psi(x, t)\, dx$$

or

$$\bar{p} = -i\hbar \int_{-\infty}^{\infty} \Psi^*(x, t) \frac{\partial \Psi(x, t)}{\partial x}\, dx \qquad (7\text{–}78)$$

We thus obtain an expression which can be immediately evaluated if we know $\Psi(x, t)$.† Similarly, we can evaluate the expectation value of the energy as follows:

$$\bar{E} = \int_{-\infty}^{\infty} \Psi^*(x, t)\, E\, \Psi(x, t)\, dx$$

$$\bar{E} = \int_{-\infty}^{\infty} \Psi^*(x, t)\left(i\hbar \frac{\partial}{\partial t}\right) \Psi(x, t)\, dx$$

$$\bar{E} = i\hbar \int_{-\infty}^{\infty} \Psi^*(x, t) \frac{\partial \Psi(x, t)}{\partial t}\, dx \qquad (7\text{–}79)$$

From equation (7–76), this can also be written

$$\bar{E} = \int_{-\infty}^{\infty} \Psi^*(x, t)\left(-\frac{\hbar^2}{2m} \frac{\partial^2}{\partial x^2} + V(x, t)\right) \Psi(x, t)\, dx$$

† We see now the reason for the ordering of the terms in the integrand of $\bar{p}$. It would not be possible to have

$$\bar{p} = -i\hbar \int_{-\infty}^{\infty} \Psi^*(x, t)\Psi(x, t) \frac{\partial}{\partial x}\, dx$$

since this is meaningless. Nor would it be possible to have

$$\bar{p} = -i\hbar \int_{-\infty}^{\infty} \frac{\partial}{\partial x} \Psi^*(x, t)\Psi(x, t)\, dx = -i\hbar\, [\Psi^*(x, t)\Psi(x, t)]_{-\infty}^{\infty}$$

because this always vanishes.

In general, we can obtain the expectation value of any dynamical quantity—that is, a quantity which is a function of x, p, and possibly t—by evaluating the integral

$$\overline{f(x, p, t)} = \int_{-\infty}^{\infty} \Psi^*(x, t) f_{\text{op}}\left(x, -i\hbar \frac{\partial}{\partial x}, t\right) \Psi(x, t)\, dx \qquad (7\text{–}80)$$

where the operator $f_{\text{op}}(x, -i\hbar\, \partial/\partial x, t)$ is obtained from the function $f(x, p, t)$ specifying the dynamical quantity, by everywhere replacing p by $-i\hbar\, \partial/\partial x$.†

We see that the wave function $\Psi(x, t)$ contains more information than just the probability density

$$P(x, t) = \Psi^*(x, t)\Psi(x, t)$$

and the probability flux

$$S(x, t) = -\frac{i\hbar}{2m}\left[\Psi^*(x, t)\frac{\partial \Psi(x, t)}{\partial x} - \Psi(x, t)\frac{\partial \Psi^*(x, t)}{\partial x}\right]$$

The wave function also contains, through equation (7–80), the expectation values of the coordinate x, the potential energy $V(x, t)$, the momentum p, the total energy E, and, in general, of any dynamical quantity $f(x, p, t)$. In fact, the wave function $\Psi(x, t)$ contains all the information which, in consideration of the uncertainty principle, we can ever expect to learn about the particle associated with that wave function.

Let us calculate the expectation value $\bar{E}$ of the total energy of a particle whose associated wave function is of the general form

$$\Psi(x, t) = \sum_{n=1}^{\infty} a_n \Psi_n(x, t) = \sum_{n=1}^{\infty} a_n e^{-iE_n t/\hbar}\, \psi_n(x) \qquad (7\text{–}81)$$

According to equation (7–79),

$$\bar{E} = \int_{-\infty}^{\infty} \Psi^*(x, t) i\hbar \frac{\partial \Psi(x, t)}{\partial t}\, dx$$

$$\bar{E} = \int_{-\infty}^{\infty} \sum_{l=1}^{\infty} a_l^* e^{iE_l t/\hbar}\, \psi_l^*(x) i\hbar \sum_{n=1}^{\infty} a_n \left(-\frac{iE_n}{\hbar}\right) e^{-iE_n t/\hbar} \psi_n(x)\, dx$$

$$\bar{E} = \sum_{n=1}^{\infty}\sum_{l=1}^{\infty} E_n a_l^* a_n e^{i(E_l - E_n)t/\hbar} \int_{-\infty}^{\infty} \psi_l^*(x)\, \psi_n(x)\, dx$$

From equation (7–54) we see that the integral vanishes except when

† Cf. exercise 8 at end of this chapter.

$l = n$, for which the integral equals unity. So the summation over l contributes only the single term $l = n$, and we have

$$\bar{E} = \sum_{n=1}^{\infty} E_n a_n^* a_n e^{i(E_n - E_n)t/\hbar}$$

or

$$\bar{E} = \sum_{n=1}^{\infty} E_n a_n^* a_n \tag{7–82}$$

Since the total energy of the particle is quantized and can only be equal to one of the eigenvalues E_n, any single measurement of this quantity can only yield one of these values. However, as the wave function (7–81) is a sum of a number of different wave functions $\Psi_n(x, t)$, each corresponding to a different eigenvalue, any one of these eigenvalues could be found in the measurement. In light of this fact, it is apparent that we should interpret equation (7–82) as meaning that *the quantity $a_n^* a_n$ is equal to the probability that a single measurement of the total energy will yield the value E_n; i.e., it is equal to the probability that the particle would be found in the eigenstate n.* Then that equation will just give the average of the values found in a large number of measurements of the energy of a particle in a state described by the wave function (7–81), which is the definition of the expectation value of the total energy. In general, $\bar{E}$ will not happen to be equal to one of the eigenvalues. However, if the particle is known to be in single eigenstate, say $\Psi_k(x, t)$, then in equation (7–81) $a_n = 1$ if $n = k$, and $a_n = 0$ otherwise. In this case equation (7–82) gives $\bar{E} = E_k$, as would be expected.

9. The Classical Limit of Quantum Mechanics

Consider a particle moving under the influence of a potential $V(x, t)$. In the limit in which the characteristic distances and momenta involved in describing the motion of the particle are so large that the uncertainty principle can be ignored, we shall prove that the predictions of quantum mechanics coincide with Newton's laws:†

$$\frac{dp}{dt} = -\frac{\partial V(x, t)}{\partial x} \tag{7–83}$$

where

$$p = m\frac{dx}{dt}$$

† The first equation is simply $dp/dt = F$, where F is the force. Cf. equation (7–37).

Such a proof was first given by Ehrenfest in 1927. It is considerably more involved than showing that the laws of relativistic mechanics approach the laws of Newtonian mechanics in the limit in which the characteristic velocities are small compared to the velocity of light. However, it is obviously of the greatest importance to be sure that quantum mechanics agrees with classical mechanics in the appropriate limit.

In Chapter 6 we saw that, for large characteristic distances and momenta, it is possible to associate with the particle a wave function $\Psi(x, t)$ in the form of a group involving uncertainties in these quantities which will always be relatively small. Now for any wave function, and in particular for $\Psi(x, t)$, the expectation value of the coordinate x is

$$\bar{x} = \int_{-\infty}^{\infty} \Psi^* x \Psi \, dx$$

Evaluating the time derivative of this quantity, we obtain

$$\frac{d\bar{x}}{dt} = \frac{d}{dt} \int_{-\infty}^{\infty} \Psi^* x \Psi \, dx = \int_{-\infty}^{\infty} \left(\Psi^* x \frac{\partial \Psi}{\partial t} + \frac{\partial \Psi^*}{\partial t} x \Psi \right) dx$$

Substituting for $\partial\Psi/\partial t$ and $\partial\Psi^*/\partial t$ from the Schroedinger equations (7–26) and (7–26′) gives

$$\frac{d\bar{x}}{dt} = -\frac{i}{\hbar} \int_{-\infty}^{\infty} \left[\Psi^* x \left(-\frac{\hbar^2}{2m} \frac{\partial^2 \Psi}{\partial x^2} + V\Psi \right) - \left(-\frac{\hbar^2}{2m} \frac{\partial^2 \Psi^*}{\partial x^2} + V\Psi^* \right) x \Psi \right] dx$$

The terms involving V cancel out, and we have

$$\frac{d\bar{x}}{dt} = \frac{i\hbar}{2m} \int_{-\infty}^{\infty} \left(\Psi^* x \frac{\partial^2 \Psi}{\partial x^2} - \frac{\partial^2 \Psi^*}{\partial x^2} x \Psi \right) dx$$

Separate out the second term of the integrand, and integrate it by parts as follows:

$$\int_{-\infty}^{\infty} \underbrace{x\Psi}_{u} \underbrace{\frac{\partial^2 \Psi^*}{\partial x^2} dx}_{dv} = \left[\underbrace{x\Psi}_{u} \underbrace{\frac{\partial \Psi^*}{\partial x}}_{v} \right]_{-\infty}^{\infty} - \int_{-\infty}^{\infty} \underbrace{\frac{\partial \Psi^*}{\partial x}}_{v} \underbrace{\frac{\partial (x\Psi)}{\partial x} dx}_{du}$$

Since Ψ must be in the form of a group of limited spatial extent in order that the uncertainty in the x coordinate be relatively small, both the wave function and its derivatives vanish as $x \to \pm\infty$. Consequently the integrated term is equal to zero, and we have

$$\int_{-\infty}^{\infty} x\Psi \frac{\partial^2 \Psi^*}{\partial x^2} dx = -\int_{-\infty}^{\infty} \frac{\partial (x\Psi)}{\partial x} \frac{\partial \Psi^*}{\partial x} dx$$

Integrating by parts again,

$$\int_{-\infty}^{\infty} x\Psi \frac{\partial^2 \Psi^*}{\partial x^2}\, dx = -\left[\frac{\partial(x\Psi)}{\partial x}\Psi^*\right]_{-\infty}^{\infty} + \int_{-\infty}^{\infty} \Psi^* \frac{\partial^2(x\Psi)}{\partial x^2}\, dx$$

Again this reduces to

$$\int_{-\infty}^{\infty} x\Psi \frac{\partial^2 \Psi^*}{\partial x^2}\, dx = \int_{-\infty}^{\infty} \Psi^* \frac{\partial^2(x\Psi)}{\partial x^2}\, dx$$

Putting this back into the equation for $d\bar{x}/dt$, we have

$$\frac{d\bar{x}}{dt} = \frac{i\hbar}{2m}\int_{-\infty}^{\infty} \Psi^*\left[x\frac{\partial^2\Psi}{\partial x^2} - \frac{\partial^2(x\Psi)}{\partial x^2}\right] dx$$

Consider the bracket in the integrand. It can be written

$$x\frac{\partial^2\Psi}{\partial x^2} - \frac{\partial^2(x\Psi)}{\partial x^2} = x\frac{\partial^2\Psi}{\partial x^2} - \frac{\partial}{\partial x}\left(x\frac{\partial\Psi}{\partial x} + \Psi\right)$$

$$= x\frac{\partial^2\Psi}{\partial x^2} - x\frac{\partial^2\Psi}{\partial x^2} - \frac{\partial\Psi}{\partial x} - \frac{\partial\Psi}{\partial x}$$

$$= -2\frac{\partial\Psi}{\partial x}$$

Consequently

$$\frac{d\bar{x}}{dt} = -\frac{i\hbar}{m}\int_{-\infty}^{\infty} \Psi^* \frac{\partial\Psi}{\partial x}\, dx$$

According to equation (7–78), the expectation value of the momentum of the particle is

$$\bar{p} = -i\hbar\int_{-\infty}^{\infty} \Psi^* \frac{\partial\Psi}{\partial x}\, dx \tag{7–84}$$

Thus we see that

$$\frac{d\bar{x}}{dt} = \frac{\bar{p}}{m}$$

or

$$\bar{p} = m\frac{d\bar{x}}{dt} \tag{7–85}$$

Next let us evaluate the time derivative of $\bar{p}$. Differentiating equation (7–84), we have

$$\frac{d\bar{p}}{dt} = -i\hbar\int_{-\infty}^{\infty}\left(\frac{\partial\Psi^*}{\partial t}\frac{\partial\Psi}{\partial x} + \Psi^*\frac{\partial^2\Psi}{\partial x\,\partial t}\right) dx$$

Again we substitute for $\partial\Psi/\partial t$ and $\partial\Psi^*/\partial t$ from the Schroedinger equation.

Then

$$\frac{d\bar{p}}{dt} = -\frac{\hbar^2}{2m}\int_{-\infty}^{\infty}\left(\frac{\partial^2\Psi^*}{\partial x^2}\frac{\partial\Psi}{\partial x} - \Psi^*\frac{\partial^3\Psi}{\partial x^3}\right)dx$$
$$+\int_{-\infty}^{\infty}\left(V\Psi^*\frac{\partial\Psi}{\partial x} - \Psi^*\frac{\partial(V\Psi)}{\partial x}\right)dx$$

This is

$$\frac{d\bar{p}}{dt} = -\frac{\hbar^2}{2m}\int_{-\infty}^{\infty}\frac{\partial}{\partial x}\left(\frac{\partial\Psi^*}{\partial x}\frac{\partial\Psi}{\partial x} - \Psi^*\frac{\partial^2\Psi}{\partial x^2}\right)dx$$
$$-\int_{-\infty}^{\infty}\Psi^*\frac{\partial V}{\partial x}\Psi\,dx$$

Integrating the first term on the right, we have

$$\frac{d\bar{p}}{dt} = -\frac{\hbar^2}{2m}\left[\frac{\partial\Psi^*}{\partial x}\frac{\partial\Psi}{\partial x} - \Psi^*\frac{\partial^2\Psi}{\partial x}\right]_{-\infty}^{\infty} - \int_{-\infty}^{\infty}\Psi^*\frac{\partial V}{\partial x}\Psi\,dx$$

Then we see that this term vanishes because of the behavior of the wave function as $x \to \pm\infty$. Thus

$$\frac{d\bar{p}}{dt} = -\int_{-\infty}^{\infty}\Psi^*\left(\frac{\partial V}{\partial x}\right)\Psi\,dx \tag{7–86}$$

Now $\partial V/\partial x \equiv \partial V(x, t)/\partial x$ is simply some function of x and t. Consequently, on comparing equation (7–86) with the general equation for the expectation value of a dynamical quantity (7–80), we see that

$$\frac{d\bar{p}}{dt} = -\frac{\overline{\partial V(x, t)}}{\partial x} \tag{7–87}$$

where $\overline{\partial V(x, t)/\partial x}$ is the expectation value of the spatial derivative of the potential energy.

Equations (7–85) and (7–87) are identical with Newton's laws (7–83), except that they involve expectation values. In the limit in which the distances and momenta involved in describing the motion of the particle are very large compared to the uncertainties in these quantities, the dynamical quantities all have precise values and it is no longer necessary to distinguish between $\bar{x}$ and x, $\bar{p}$ and p, or $\overline{\partial V/\partial x}$ and $\partial V/\partial x$. Then the equations we have derived from the theory of quantum mechanics become completely identical with the laws of Newtonian mechanics. Quantum mechanics does indeed agree with Newtonian mechanics in the limit in which the uncertainty principle is not important. However, we shall see in the next chapter that, when we are not at this limit, quantum mechanics predicts a number of extremely interesting phenomena which have no counterpart in Newtonian mechanics.

BIBLIOGRAPHY

Bohm, D., *Quantum Theory*, Prentice-Hall, Englewood Cliffs, N.J., 1951.

Pauling, L., and E. B. Wilson, *Introduction to Quantum Mechanics*, McGraw-Hill Book Co., New York, 1935.

Schiff, L. I., *Quantum Mechanics*, McGraw-Hill Book Co., New York, 1955.

Sherwin, C. W., *Introduction to Quantum Mechanics*, Henry Holt and Co., New York, 1959.

EXERCISES

1. Show that the probability flux (equation 7–31) is always real. *Hint:* A quantity is real if it equals its own complex conjugate.

2. Consider a particle of mass m in the state of lowest energy E_1 of the potential

$$V(x) = \begin{matrix} V_0, & x < -a/2, \quad x > a/2 \\ 0, & -a/2 < x < a/2 \end{matrix}$$

where m, V_0, and a are such that $E_1 = 0.5V_0$. Make a sketch of the form of the eigenfunction $\psi_1(x)$ for this particle. Now assume that V_0 is increased by a factor of 3. What effect does this have on the shape of $\psi_1(x)$ in the regions where $V(x) > E_1$? What effect does this have on the shape of $\psi_1(x)$ in the region where $V(x) < E_1$? What effect does this have on E_1? These questions are to be answered qualitatively by a direct consideration of the behavior of the time independent Schroedinger equations for the potentials concerned.

3. Take a potential of the form indicated in exercise 2, except that $\sqrt{mV_0a^2/2\hbar^2} = 4$, and use the formulas for $[\psi(x)]_{x_{j+1}} - [\psi(x)]_{x_j}$ and $[d\psi(x)/dx]_{x_{j+1}} - [d\psi(x)/dx]_{x_j}$, developed in section 5, to make a numerical integration of the time independent Schroedinger equation and find the first eigenfunction $\psi_1(x)$ and the first eigenvalue E_1. Compare these with the results of the analytical treatment of this potential presented in section 4, Chapter 8. *Hint:* Start the integration at $x = 0$, where it is apparent from symmetry that $d\psi_1(x)/dx = 0$, and take $(x_{j+1} - x_j) = 0.1a/2$.

4. Use numerical integration to find the first eigenvalue E_1 for the potential

$$V(x) = \begin{matrix} \infty, & x < -a/2, \quad x > a/2 \\ 0, & -a/2 < x < -a/4, \quad a/4 < x < a/2 \\ v_0, & -a/4 < x < a/4 \end{matrix}$$

where

$$v_0 = \pi^2\hbar^2/8ma^2$$

Employ the conditions $\psi_1(x) = 0$, $x \leqslant -a/2$, $x \geqslant a/2$, and $\psi_1(x)$ need not be continuous at $x = \pm a/2$, which are justified in section 5, Chapter 8, for a potential of this type. One of the exercises of that chapter consists of an analytical treatment of this potential.

5. Derive the classical wave equation for a string of variable density per unit length $\rho(x)$. Separate it into two ordinary differential equations, and compare the equation in x with the time independent Schroedinger equation for the general potential $V(x)$.

6. Verify equation (7–68) by evaluating some of the integrals.

7. Set up and solve the problem of the motion of a string fixed at both ends, which is plucked at a point at distance b from its center. Check the solution by seeing that when $b \to 0$ it reduces to the solution of the problem treated in section 7.

8. In calculating the expectation value of the product of position times momentum, an ambiguity arises because it is not apparent which of the two expressions

$$\overline{xp} = \int_{-\infty}^{\infty} \Psi^* x \left(-i\hbar \frac{\partial}{\partial x} \right) \Psi \, dx$$

$$\overline{xp} = \int_{-\infty}^{\infty} \Psi^* \left(-i\hbar \frac{\partial}{\partial x} \right) x \Psi \, dx$$

should be used. Show that neither of these is acceptable because they violate the obvious requirement that $\overline{xp}$ be real. Then show that the expression

$$\overline{xp} = \int_{-\infty}^{\infty} \Psi^* \left[\frac{x\left(-i\hbar \dfrac{\partial}{\partial x} \right) + \left(-i\hbar \dfrac{\partial}{\partial x} \right) x}{2} \right] \Psi \, dx$$

is acceptable because it does satisfy this requirement. *Hint:* Feel free to integrate by parts.

CHAPTER **8**

Solutions of Schroedinger's Equation

1. The Free Particle

In this chapter we shall demonstrate some of the features of the quantum mechanical theory and, in particular, some of the interesting phenomena predicted by the theory, by solving Schroedinger's equation for several specific forms of the potential energy $V(x, t)$ and finding the eigenfunctions and eigenvalues.

Let us begin by considering the simplest case: $V(x, t) =$ constant. For a particle under the influence of such a potential the force $F(x, t) = -\partial V(x, t)/\partial x$ will vanish; it is a free particle. Since this is true regardless of the value of the constant, we do not lose generality by defining the potential energy such that $V(x, t) = 0$. We know that in classical mechanics a free particle may be either at rest or moving with constant momentum p. In either case its total energy E is a constant.

To find the behavior predicted by quantum mechanics for a free particle, we solve the Schroedinger equation for $V(x, t) = 0$. Since this potential energy is not a function of t, the problem reduces to the problem of solving the time independent Schroedinger equation. For $V(x) = 0$ that equation reads

$$-\frac{\hbar^2}{2m}\frac{d^2\psi(x)}{dx^2} = E\psi(x) \tag{8-1}$$

The solutions to the equation are the eigenfunctions $\psi(x)$. The wave functions are

$$\Psi(x, t) = e^{-iEt/\hbar}\,\psi(x) \tag{8-2}$$

where E is the total energy of the particle. From the discussion of section 5 of the last chapter, we know that *an acceptable solution to equation* (8-1) *exists for any value of* $E \geqslant 0$.

A solution to the differential equation (8–1) can be written

$$\psi(x) = Ae^{iKx} + Be^{-iKx} \tag{8–3}$$

where A and B are arbitrary constants, and where $K = \sqrt{2mE}/\hbar$. The validity of this solution is demonstrated by evaluating

$$\frac{d^2\psi(x)}{dx^2} = i^2K^2Ae^{iKx} + i^2K^2Be^{-iKx} = -K^2\psi(x) = -\frac{2mE}{\hbar^2}\psi(x)$$

and substituting into equation (8–1):

$$-\frac{\hbar^2}{2m}\left(-\frac{2mE}{\hbar^2}\right)\psi(x) \equiv E\psi(x); \quad \text{Q.E.D.}$$

Since there are two arbitrary constants, equation (8–3) is the general form of the solution to the second order ordinary differential equation (8–1). Thus the wave function for a free particle may be written

$$\Psi(x, t) = e^{-iEt/\hbar}\psi(x) \tag{8–4}$$

where

$$\psi(x) = Ae^{iKx} + Be^{-iKx} \tag{8–4'}$$

and

$$K = \sqrt{2mE}/\hbar$$

We consider first the case in which one of the constants, say B, is equal to zero. Then the wave function is

$$\Psi(x, t) = Ae^{-iEt/\hbar}e^{iKx} = Ae^{i(Kx - Et/\hbar)} \tag{8–5}$$

which can also be written

$$\Psi(x, t) = A\cos(Kx - Et/\hbar) + iA\sin(Kx - Et/\hbar)$$

This is a *traveling wave*, oscillating at angular frequency $\omega = E/\hbar$. Its nodes move to the right with velocity $w = \nu/k \equiv \omega/K = E/\hbar K$. Let us evaluate its probability density $P(x, t)$ and its probability flux $S(x, t)$. These are

$$P(x, t) = \Psi^*\Psi = A^*e^{-i(Kx - Et/\hbar)}Ae^{i(Kx - Et/\hbar)}$$

$$P(x, t) = A^*A \tag{8–6}$$

and

$$S(x, t) = -\frac{i\hbar}{2m}\left(\Psi^*\frac{\partial \Psi}{\partial x} - \Psi\frac{\partial \Psi^*}{\partial x}\right)$$

$$S(x, t) = -\frac{i\hbar}{2m}[A^*e^{-i(Kx-Et/\hbar)}AiKe^{i(Kx-Et/\hbar)}$$

$$- Ae^{i(Kx-Et/\hbar)}A^*(-iK)e^{-i(Kx-Et/\hbar)}]$$

$$S(x, t) = -\frac{i\hbar}{2m}A^*A2iK = \frac{\hbar K}{m}A^*A \tag{8–7}$$

According to the de Broglie postulate (equation 7–5),

$$\hbar K = p = mv$$

where v is the velocity of the particle. Thus the probability flux is

$$S(x, t) = vA^*A = vP(x, t) \tag{8–8}$$

which is the same as the familiar classical relation between flux, velocity, and density. For the wave function (8–5), both the probability density and the probability flux are constant in time and independent of position. Furthermore, the wave function involves only the single momentum $p = \hbar K$ and the single energy E. Consequently we realize that this wave function is associated with a particle which is traveling along the x axis with perfectly known and constant momentum p, but with completely unknown x coordinate. This is the physical situation of a particle of momentum that is known to be exactly p, and moving somewhere in an infinitely long *beam* of particles. For the wave function obtained when the constant A is set equal to zero, which is

$$\Psi(x, t) = Be^{-i(Kx+Et/\hbar)} \tag{8–9}$$

we find

$$P(x, t) = B^*B \tag{8–10}$$

and

$$S(x, t) = -\frac{\hbar K}{m}B^*B \tag{8–11}$$

The sign of the probability flux $S(x, t)$ indicates, of course, the direction of motion of the particle. (In three dimensions the probability flux is a vector quantity.) The particle associated with wave function (8–5) is moving in a beam in the direction of increasing x, and the particle associated with wave function (8–9) is moving in a beam in the direction of decreasing x.

Let us evaluate the integral of the probability density for the wave function (8–5). This is

$$\int_{-\infty}^{\infty} \Psi^*\Psi \, dx = \int_{-\infty}^{\infty} A^*A \, dx = A^*A \int_{-\infty}^{\infty} dx$$

We see that it diverges unless $A = 0$. This is the difficulty of normalizing the wave function of an unbound particle, which we have already come across in the previous chapter. But the difficulty is not really fundamental. This wave function represents the highly idealized situation of a particle moving in a beam of infinite length and whose x coordinate is, consequently, completely unknown. But in any real physical situation this is not the case; instead it is known that the particle would certainly not be found outside some region of the x axis of finite length. In an experiment the beam is actually always of finite length. A realistic wave function describing the actual situation would be of essentially constant amplitude over the length of the beam but would vanish at very large and very small values of x. It would be in the form of a group of length Δx which is long, but finite. Furthermore, we know from the uncertainty principle that the realistic wave function must involve not only the single momentum $p = \hbar K$ but also a distribution of momenta spread over the range $\Delta p = \hbar \, \Delta K \sim \hbar/\Delta x$ which is centered about the momentum $p = \hbar K$. The wave function (8–5) can be thought of as representing the behavior of the realistic wave function in the limit $\Delta x \to \infty$. There is never any difficulty in normalizing the wave function as long as Δx is finite. In the limit a difficulty does arise in the sense that the wave function can be normalized only if the multiplicative constant A vanishes. Even so, the limiting form of the wave function can be used to calculate quantities which do not depend on the value of the multiplicative constant. In subsequent sections we shall perform a number of calculations using wave functions of the limiting form; but they will always be used in such a way that a physical significance is attached only to the *ratio* of the multiplicative constant of one wave function to the multiplicative constant of another wave function. This procedure bypasses the normalization difficulty.

In certain situations it is necessary to use a wave function in the form of a group of finite length. Such a wave function can be constructed by adding together a number of wave functions of the form (8–5) if the group is moving in the direction of increasing x, or of the form (8–9) otherwise, each with a different value of K and E. That is, we take

$$\Psi(x, t) = \sum_{n=1}^{\infty} A_n \Psi_n(x, t) \tag{8–12}$$

where

$$\Psi_n(x, t) = e^{-iE_n t/\hbar} e^{iK_n x}$$

and where

$$K_n = \sqrt{2mE_n}/\hbar$$

Since each of the $\Psi_n(x, t)$ is a solution to the Schroedinger equation, equation (7–44) shows that (8–12) will also be a solution. The values of the constants A_n can be calculated as in equation (7–57) in terms of the form of the wave function at some particular instant of time, say $t = 0$. As an example, consider a free particle moving in the potential $V(x) = 0$, in the direction of increasing x, where at $t = 0$ the particle was known to have an x coordinate within the range Δx centered about the expectation value $\bar{x}$ and momentum in the range Δp centered about the expectation value $\bar{p}$. Furthermore, let the product of Δx and Δp be equal to $\hbar$, the minimum allowed by the uncertainty principle. Then the normalized wave function for this particle at any time t can be shown to be†

$$\Psi(x, t) = \sum_{n=1}^{\infty} \left[\int_{-\infty}^{\infty} \psi_n(x') \Psi(x', 0) \, dx' \right] e^{-iE_n t/\hbar} \psi_n(x) \tag{8–13}$$

where

$$\Psi(x', 0) = [2\pi (\Delta x)^2]^{-1/4} e^{-\left[\frac{(x'-\bar{x})^2}{4(\Delta x)^2} - \frac{i\bar{p}x'}{\hbar}\right]}$$

and

$$\psi_n(x) = e^{iK_n x}$$

[For a particle moving with the same initial conditions in a more general potential $V(x)$, the expression for $\Psi(x, t)$ is essentially the same except that the eigenfunctions $\psi_n(x)$ for that potential must be used.] We present equation (8–13) strictly for the sake of interest. Fortunately it is almost always possible to formulate quantum mechanical problems in such a way that the wave function for a free particle moving in the potential $V(x) = 0$ can be represented by the wave function (8–5) or (8–9), or by a simple combination of the two, instead of something as complicated as equation (8–13). [The same is true for a particle moving in a more general potential $V(x)$.] Only such problems will be considered in this book.

Consider the two wave functions, (8–5)

$$\Psi(x, t) = Ae^{-iEt/\hbar} e^{iKx}$$

and (8–9)

$$\Psi(x, t) = Be^{-iEt/\hbar} e^{-iKx}$$

which are solutions to the Schroedinger equation for a free particle. Choose the ratio of the arbitrary constants such that $A = -B$, and add the two wave functions. The result is a new wave function,

$$\Psi(x, t) = Ae^{-iEt/\hbar}(e^{iKx} - e^{-iKx})$$

† To be correct, (8–13), and also (8–12), must be written as integrals since K_n is a continuous variable for the case of a free particle. Equation (8–13) has been written here as a sum to bring out its analogy with equation (7–57).

which can be written

$$\Psi(x, t) = Ae^{-iEt/\hbar}[\cos Kx + i \sin Kx - \cos(-Kx) - i \sin(-Kx)]$$

$$\Psi(x, t) = 2Aie^{-iEt/\hbar} \sin Kx$$

$$\Psi(x, t) = A'e^{-iEt/\hbar} \sin Kx \tag{8–14}$$

where A' is an arbitrary constant which is equal to $2i$ times the arbitrary constant A. Next choose the ratio of the constants such that $A = B$, and add the two wave functions. The result is a second new wave function,

$$\Psi(x, t) = Be^{-iEt/\hbar}(e^{iKx} + e^{-iKx})$$

which can be written

$$\Psi(x, t) = Be^{-iEt/\hbar}[\cos Kx + i \sin Kx + \cos(-Kx) + i \sin(-Kx)]$$

$$\Psi(x, t) = 2Be^{-iEt/\hbar} \cos Kx$$

$$\Psi(x, t) = B'e^{-iEt/\hbar} \cos Kx \tag{8–15}$$

where B' is an arbitrary constant equal to 2 times B, and which is independent of A'. Since the Schroedinger equation is a linear differential equation, the wave functions (8–14) and (8–15) are also solutions, and so is their sum,

$$\Psi(x, t) = e^{-iEt/\hbar}(A' \sin Kx + B' \cos Kx)$$

which can be written

$$\Psi(x, t) = e^{-iEt/\hbar}\psi(x) \tag{8–16}$$

where

$$\psi(x) = A' \sin Kx + B' \cos Kx \tag{8–16$'$}$$

and

$$K = \sqrt{2mE}/\hbar$$

We recognize from the form of equation (8–16) that (8–16′) must also be a solution to the time independent Schroedinger equation (8–1) for a free particle. Since it contains two arbitrary constants, it is the general form of the solution—just as general as (8–4′). The solution (8–16′) could have been written down directly as was (8–4′). However, the procedure we have followed has the advantage of emphasizing the relation between the two general forms of the solution.

Furthermore, it helps us to interpret the meaning of the wave function (8–16). First of all, we note that the wave function was obtained by adding and subtracting two traveling waves of equal amplitudes and must, therefore, be a *standing wave*. This, of course, can be seen directly from the form of (8–16), since the nodes of $\Psi(x, t)$ are fixed in space at values of x for which $A' \sin Kx + B' \cos Kx = 0$. Next we recall that each of the two traveling waves have been associated with a particle of precisely known momentum moving in an infinitely long beam, with the momentum being equal in magnitude but oppositely directed in the two cases. Consequently we must associate the standing wave (8–16) with a particle which is moving in an infinitely long beam in such a way that the *magnitude* of its momentum p is precisely known but the *direction* of the momentum is not known. That is, the wave function is to be associated with a particle which could be moving in the direction of either increasing x, or decreasing x, with momentum of magnitude precisely equal to p, and with a completely unknown x coordinate.

Let us check our interpretation of equation (8–16) by evaluating the probability density $P(x, t)$ and the probability flux $S(x, t)$. For simplicity, set one of the constants, say B', equal to zero. Then

$$\Psi(x, t) = A'e^{-iEt/\hbar} \sin Kx \tag{8–17}$$

and

$$P(x, t) = \Psi^*\Psi = A'^*e^{iEt/\hbar} \sin Kx\, A'e^{-iEt/\hbar} \sin Kx$$

$$P(x, t) = A'^*A' \sin^2 Kx \tag{8–18}$$

and also

$$S(x, t) = -\frac{i\hbar}{2m}\left(\Psi^*\frac{\partial\Psi}{\partial x} - \Psi\frac{\partial\Psi^*}{\partial x}\right)$$

$$S(x, t) = -\frac{i\hbar}{2m}[A'^*e^{iEt/\hbar} \sin Kx\, A'e^{-iEt/\hbar} K \cos Kx$$
$$- A'e^{-iEt/\hbar} \sin Kx A'^*e^{iEt/\hbar} K \cos Kx]$$

$$S(x, t) = 0 \tag{8–19}$$

The fact that the probability flux vanishes is certainly in agreement with our interpretation of $\Psi(x, t)$. According to that interpretation, the probability per unit time that the particle associated with the wave function will cross the point x in the direction of increasing x will be equal to the probability per unit time that it will cross in the direction of decreasing x. These will balance each other out, and the total probability flux $S(x, t)$ will be equal to zero. Next consider the probability density $P(x, t)$, which is plotted in figure (8–1). Also plotted in the same figure is the constant probability density which would be expected classically for a particle in the situation which we have described as being associated with the wave function (8–17). The fact that the quantum mechanical $P(x, t)$ extends to infinitely large and small values of x agrees with our interpretation of the wave function and with the classical behavior of the particle, but the presence of points along the x axis at which the particle would never be found is in contrast to what would be expected classically. This is a manifestation of the wave-like aspects of a particle in the situation described by this wave function. Such behavior is not observed classically because the distance $d = \pi/K = h/2p$ is only one-half the de Broglie wavelength of the particle. In the classical limit this distance becomes extremely small, and the $P(x, t)$ pattern is so compressed in the x direction that only the average (constant) behavior of $P(x, t)$ can be resolved experimentally.

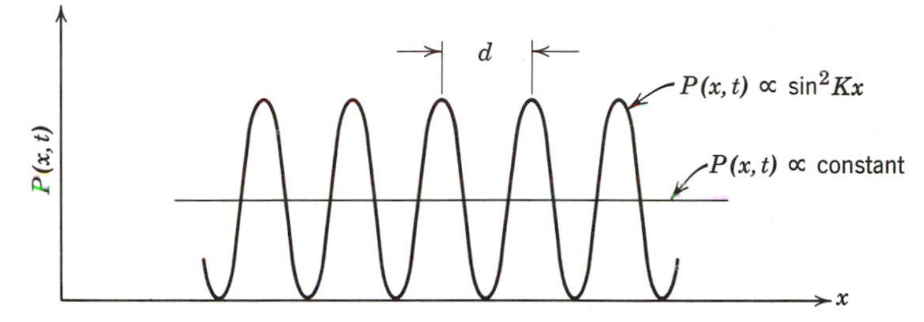

Figure 8–1. The probability density for a free particle standing wave function.

For the standing wave function (8–16) there will be the same problem concerning normalization as there is for the traveling wave function. For the standing wave there is a particularly simple way to remove the difficulty. The normalization problem arises because of the unrealistic way the wave function extends to infinitely large and small values of the coordinate x. Consequently the problem can be removed by completely suppressing the wave function outside the large, but finite, region $-L/2 \leqslant x \leqslant L/2$. We do this by constructing a more realistic wave function which is the function (8–16) inside the region, and zero outside that region. That is, we take

$$\Psi(x, t) = e^{-iEt/\hbar}\psi(x) \tag{8–20}$$

where

$$\psi(x) = \begin{matrix} A' \sin Kx + B' \cos Kx, & -L/2 \leqslant x \leqslant L/2 \\ 0, \quad x \leqslant -L/2, & x \geqslant L/2 \end{matrix} \tag{8–20′}$$

It is clear that there will be no difficulty in normalizing the wave function (8–20). Furthermore, the eigenfunctions now vanish as $x \rightarrow \pm \infty$, and it is immediately apparent from equation (7–53) that there will be no difficulty in proving that such eigenfunctions are orthogonal. We shall see in section 5 that the procedure indicated in equation (8–20′) is mathematically equivalent to saying that the potential energy $V(x)$ becomes infinitely large for $x \leqslant -L/2$ and $x \geqslant L/2$. This is as if the region of space $-L/2 \leqslant x \leqslant L/2$ were enclosed in a (one dimensional) box with completely impenetrable walls. The procedure is known as *box normalization.* As long as the walls of the box are very far away from the region of interest, their presence will have no physical effect of importance, but they will remove the mathematical difficulties associated with normality and orthogonality for a free particle. The values of the constants A' and B' required to normalize the wave function will depend on the length of the box, L; however, these constants will not enter into the results of a calculation of any quantity of physical interest. Box normalization will also change the eigenvalues E from continuous to discrete. But we shall see that the separation between adjacent eigenvalues is proportional to L^{-2}. With a very large value of L the eigenfunctions will, for all practical purposes, remain continuous.

Normalization of the traveling wave function (8–4) in a box with impenetrable walls is not possible because, as will be shown in section 5, the wave function must always vanish at the walls of the box. This boundary condition can be satisfied by a standing wave with fixed nodes, but it cannot be satisfied by a traveling wave with moving nodes. However, we shall see in Chapter 15 that a form of box normalization can be used for the traveling wave function, if we assume that the walls of the box are penetrable so that the wave function need not vanish at the walls. In this form of box normalization we impose the so-called *periodic boundary conditions* by demanding that at any instant the wave function and its spatial derivative have the same values at both walls. It is immediately apparent from equation (7–53) that with these boundary conditions the eigenfunctions are orthogonal. These boundary conditions also lead to the convenient result that both forms of box normalization produce the same discreteness of the eigenvalues. With normalization in a box of penetrable walls, we are interested only in what happens in the box, and we restrict the domain of the variable x to the region between its walls. The wave function can then be normalized. The value of the normalization constant depends on the length of the box, but this always drops out of all calculations of physically measurable quantities.

The situation relative to the problem of normality and orthogonality for the free particle wave functions and eigenfunctions, both traveling wave and standing wave, may be summarized briefly by saying that it is usually possible simply to ignore the problem. The mathematical consistency of the theory is ensured by the fact that a wave function in the form of a group *may* be used, or that one of the forms of box normalization *may* be employed. In practice, it is not normally necessary actually to do either, and the wave functions (8–4) and (8–16) can be used "as is."

2. Step Potentials

In the next few sections we shall calculate the quantum mechanical motion of a particle whose potential energy can be represented by a function $V(x)$ which has a different constant value in each of several adjacent ranges of the x axis, and which changes discontinuously in going

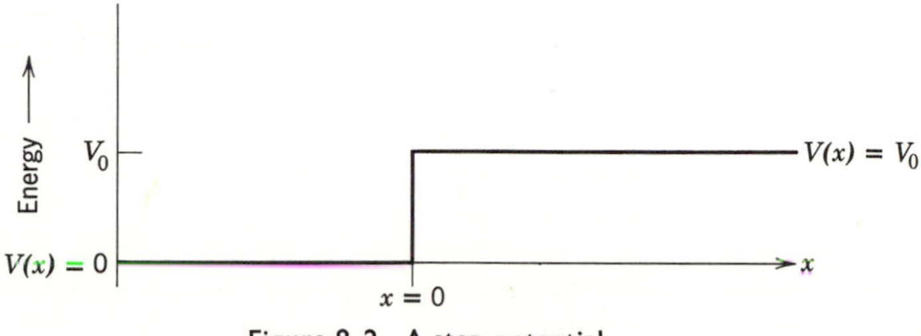

Figure 8–2. A step potential.

from one range to the next. Of course, the situations represented by potentials which are discontinuous functions of x do not really exist in nature. However, such potentials are often used in quantum mechanics to approximate real situations because, being constant in each range, they are easy to treat mathematically. With such potentials we shall be able to illustrate a number of characteristic quantum mechanical phenomena.

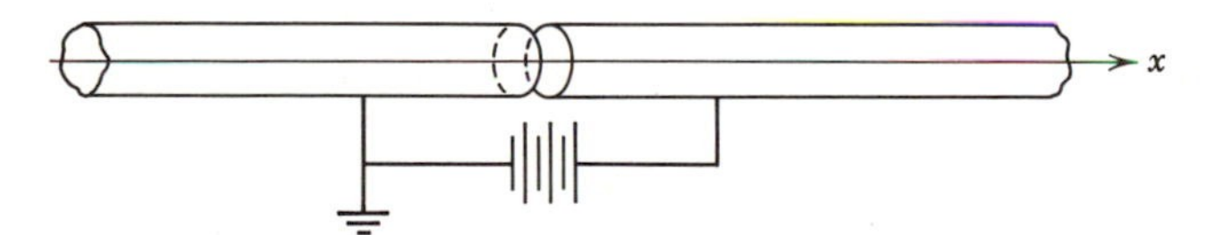

Figure 8–3. A physical system approximating a step potential.

Let us consider first the simplest example, a *step potential*, as shown in figure (8–2). We choose the origin of the x axis to be at the step. Then $V(x)$ can be written

$$V(x) = \begin{matrix} V_0, & x > 0 \\ 0, & x < 0 \end{matrix} \tag{8–21}$$

where V_0 is a constant. We may think of $V(x)$ as a limiting representation of the potential energy function for a charged particle moving along the axis of a system of two electrodes which are held at different voltages, as illustrated in figure (8–3).

Assume that a particle of mass m and total energy E is in the region $x < 0$ and is moving toward the point at which $V(x)$ changes. According to classical mechanics the particle will move freely in that region until it reaches $x = 0$, where it is subjected to an impulsive force $F(x) = -\partial V(x)/\partial x$ acting in the direction of decreasing x. The subsequent motion of the particle depends, classically, on the relation between E and V_0; this is also true in quantum mechanics. We take first the case:

$E < V_0$

This is illustrated in figure (8–4). Since the total energy E is a constant,

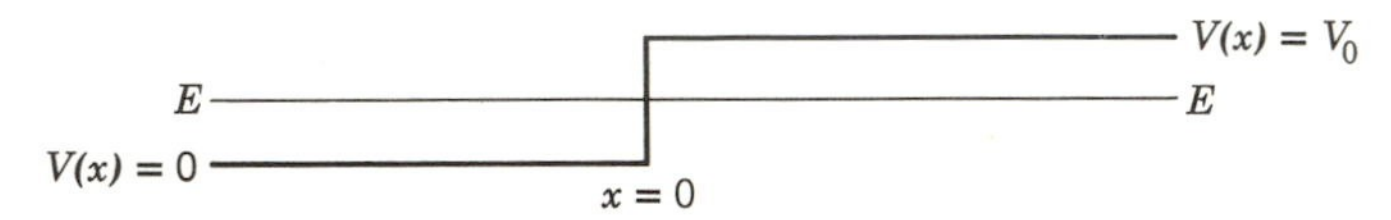

Figure 8–4. The case in which classical mechanics predicts that the particle is reflected at the discontinuity of the potential.

classical mechanics says that the particle cannot enter the region $x > 0$, as in that region

$$E = p^2/2m + V(x) < V(x)$$

or

$$p^2/2m < 0$$

The impulsive force will change the momentum of the particle in such a way that it will exactly reverse its motion, traveling off in the direction of decreasing x with momentum equal in magnitude (so that the total energy remains constant) but opposite in direction to its initial momentum.

To determine the motion of the particle quantum mechanically, we must find the wave function which is a solution, for the total energy $E < V_0$, to the Schroedinger equation for the potential (8–21). We know from our discussion of section 5, Chapter 7, that *acceptable solutions to the Schroedinger equation for the potential $V(x)$ exist for any value of $E \geqslant 0$.* Since we are dealing with a time independent potential, the actual problem is to solve the time independent Schroedinger equation and find the eigenfunction. With this potential the x axis breaks up into two regions. In one region the eigenfunction is a solution to the simple time independent Schroedinger equation

$$-\frac{\hbar^2}{2m}\frac{d^2\psi(x)}{dx^2} = E\psi(x), \qquad x < 0 \tag{8–22}$$

In the other region it is a solution to the simple time independent Schroedinger equation

$$-\frac{\hbar^2}{2m}\frac{d^2\psi(x)}{dx^2} + V_0\psi(x) = E\psi(x), \qquad x > 0 \tag{8–23}$$

The two equations are solved separately. Then an eigenfunction valid for the entire range of x is constructed by joining the two solutions together at $x = 0$ in such a way as to satisfy the conditions (7–38), which require that $\psi(x)$ and $d\psi(x)/dx$ be everywhere finite and continuous.

Consider equation (8–22). This is precisely the time independent Schroedinger equation (8–1) for a free particle. The general solution to this equation can be written either in the traveling wave form (8–4′) or in the standing wave form (8–16′). Since the two are mathematically equivalent, either may be used. However, by using the form most appropriate to the problem at hand, it is often possible to simplify the calculation or the interpretation of the results. When discussing the motion of an unbound particle it is usually most convenient to use the traveling wave form. If the particle is bound, the standing wave form is usually the most convenient. For the problem at hand we take

$$\psi(x) = Ae^{iK_1x} + Be^{-iK_1x}, \qquad x < 0 \tag{8–24}$$

where

$$K_1 = \sqrt{2mE}/\hbar$$

Next consider equation (8–23). The general solution to this equation is

$$\psi(x) = Ce^{K_2x} + De^{-K_2x}, \qquad x > 0 \tag{8–25}$$

where

$$K_2 = \sqrt{2m(V_0 - E)}/\hbar \quad \text{and} \quad E < V_0$$

To verify this, calculate

$$\begin{aligned}\frac{d^2\psi(x)}{dx^2} &= CK_2^2e^{K_2x} + D(-K_2)^2e^{-K_2x} \\ &= K_2^2\psi(x) = \frac{2m(V_0 - E)}{\hbar^2}\,\psi(x)\end{aligned}$$

and substitute into equation (8–23). This gives

$$-\frac{\hbar^2}{2m}\frac{2m}{\hbar^2}(V_0 - E)\psi(x) + V_0\psi(x) \equiv E\psi(x); \quad \text{Q.E.D.}$$

The arbitrary constants A, B, C, and D must be so chosen that the eigenfunction satisfies the conditions (7–38). Consider first the behavior of the eigenfunction in the limit $x \rightarrow +\infty$. In this region of the x axis the solution to the time independent Schroedinger equation is given by equation (8–25). It will, in general, diverge as $x \rightarrow +\infty$ because of the first term. In order to prevent this, we must set the coefficient of the first term equal to zero. Thus

$$C = 0 \tag{8–26}$$

Now consider the eigenfunction at the point $x = 0$. At this point the solutions (8–24) and (8–25) must join in such a way that $\psi(x)$ and $d\psi(x)/dx$ are continuous. Continuity of $\psi(x)$ is obtained if the relation

$$D(e^{-K_2 x})_{x=0} = A(e^{iK_1 x})_{x=0} + B(e^{-iK_1 x})_{x=0}$$

is satisfied. This yields

$$D = A + B \tag{8–27}$$

Continuity of the derivative of the solutions

$$\frac{d\psi(x)}{dx} = -K_2 D e^{-K_2 x}, \qquad x > 0$$

and

$$\frac{d\psi(x)}{dx} = iK_1 A e^{iK_1 x} - iK_1 B e^{-iK_1 x}, \qquad x < 0$$

is obtained if the relation

$$-K_2 D(e^{-K_2 x})_{x=0} = iK_1 A(e^{iK_1 x})_{x=0} - iK_1 B(e^{-iK_1 x})_{x=0}$$

is satisfied. This yields

$$\frac{iK_2}{K_1} D = A - B \tag{8–28}$$

Adding equations (8–27) and (8–28) gives

$$A = \frac{D}{2}\left(1 + \frac{iK_2}{K_1}\right) \tag{8–29}$$

Subtracting gives

$$B = \frac{D}{2}\left(1 - \frac{iK_2}{K_1}\right) \tag{8–30}$$

Thus the eigenfunction for this potential, and for the energy E, is

$$\psi(x) = \begin{matrix} \frac{D}{2}(1 + iK_2/K_1)e^{iK_1 x} + \frac{D}{2}(1 - iK_2/K_1)e^{-iK_1 x}, & x \leqslant 0 \\ De^{-K_2 x}, & x \geqslant 0 \end{matrix} \tag{8–31}$$

The arbitrary constant D has not been determined because we have not normalized the eigenfunction. This could be done by the techniques described in section 1, but it is not necessary.

The wave function is

$$\Psi(x, t) = \begin{matrix} Ae^{-iEt/\hbar}\, e^{iK_1 x} + Be^{-iEt/\hbar}\, e^{-iK_1 x}, & x \leqslant 0 \\ De^{-iEt/\hbar}\, e^{-K_2 x}, & x \geqslant 0 \end{matrix} \tag{8–32}$$

In the region $x < 0$ it consists of a wave traveling in the direction of increasing x, plus a wave traveling in the direction of decreasing x; in the region $x > 0$ it consists of a standing wave. In interpreting this wave function, it is useful to calculate the probability flux $S(x, t)$ in the region $x < 0$. By comparison with equations (8–7) and (8–11) and the wave functions from which these expressions were derived, it is apparent that in the present case

$$S(x, t) = vA^*A - vB^*B, \qquad x < 0 \tag{8–33}$$

where $v = \hbar K_1/m$, the magnitude of the velocity of the particle. The first term, which comes from the first term in the wave function, is a probability flux flowing in the direction of increasing x. The second term, which comes from the second term in the wave function, is a probability flux flowing in the opposite direction. Consequently, we associate the first term in $\Psi(x, t)$ or $S(x, t)$, in the region $x < 0$, with the incidence of the particle upon the point at which the potential energy changes, and the second term with the reflection of the particle from the change in potential. Let us calculate the ratio of the intensity of the reflected probability flux to the intensity of the incident probability flux. We obtain

$$R = \frac{vB^*B}{vA^*A} = \frac{B^*B}{A^*A} = \frac{(1 - iK_2/K_1)^*(1 - iK_2/K_1)}{(1 + iK_2/K_1)^*(1 + iK_2/K_1)}$$

$$= \frac{(1 + iK/_2K_1)(1 - iK_2/K_1)}{(1 - iK_2/K_1)(1 + iK_2/K_1)} = 1$$

The quantity B^*B/A^*A is also the ratio of the intensity of the reflected traveling wave to the intensity of the incident traveling wave in the region $x < 0$. That these ratios are unity means that a particle incident upon the change in potential, with total energy $E < V_0$, has unity probability of being reflected. This is in complete agreement with the classical predictions.

Consider the eigenfunction (8–31). Writing $e^{iK_1x} = \cos K_1x + i \sin K_1x$, etc., it is easy to show that the eigenfunction can be expressed as

$$\psi(x) = \begin{matrix} D \cos K_1x - D\dfrac{K_2}{K_1} \sin K_1x, & x \leqslant 0 \\ De^{-K_2x}, & x \geqslant 0 \end{matrix} \tag{8–34}$$

The wave function corresponding to this eigenfunction is actually a standing wave for all x. In this problem the incident and reflected traveling waves combine to form a standing wave because they are of equal intensity. Let us plot (8–34), which is a real function of x if we take D real. The plot is shown in figure (8–5). Here we find a feature which is in sharp contrast to the classical predictions. Although in the region $x > 0$ the probability density

$$P(x, t) = \Psi^*(x, t)\,\Psi(x, t) = D^* D e^{-2K_2 x} \tag{8–35}$$

decreases rapidly with increasing x, there *is* a finite probability of finding the particle in this region. In classical mechanics it would be absolutely impossible to find the particle in the region $x > 0$ because there the total

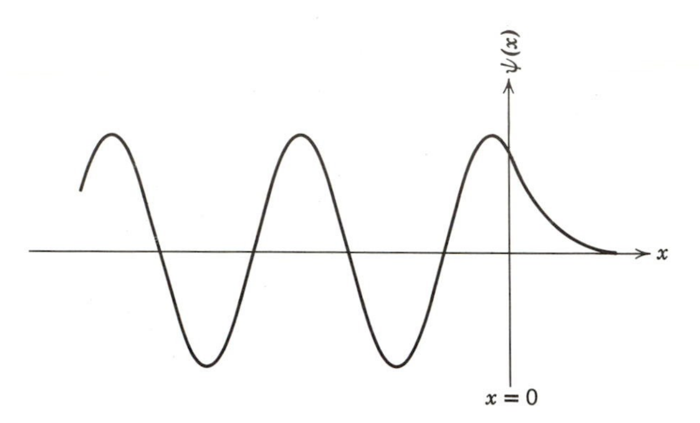

Figure 8–5. An eigenfunction for a step potential in the case in which classical mechanics predicts that the particle is reflected.

energy is less than the potential energy. Penetration of the classically excluded region is one of the more striking predictions of quantum mechanics.

We shall discuss later certain experiments which confirm this prediction. Here we would like to point out that equation (8–35) is not in disagreement with the experiments of classical mechanics. It is apparent from that equation that probability of finding the particle with a coordinate $x > 0$ is only appreciable in a region starting at $x = 0$ and of length Δx of the order of $1/K_2$. Thus

$$\Delta x \sim 1/K_2 = \hbar/\sqrt{2m(V_0 - E)}$$

In the classical limit the product of m and $(V_0 - E)$ is so large, compared to $\hbar^2$, that Δx is immeasurably small.

Furthermore, the uncertainty principle demonstrates that the wave-like properties manifested by an entity in penetrating the classically excluded region are not really in conflict with its particle-like properties. Consider an experiment capable of showing that the particle was located somewhere in the region $x > 0$. Since $P(x, t)$ for $x > 0$ is appreciable only in a region of length Δx, the experiment would amount to localizing the particle within that region. In doing this, the experiment would lead to an uncertainty Δp in the momentum of the particle, where

$$\Delta p \sim \hbar/\Delta x \sim \sqrt{2m(V_0 - E)}$$

Consequently the energy of the particle would be uncertain by an amount

$$\Delta E = (\Delta p)^2/2m \sim V_0 - E$$

and it would no longer be possible to say that the total energy of the particle E was definitely less than the potential energy V_0.

Next let us consider the motion of a particle in a step potential when:

$E > V_0$

This is illustrated in figure (8–6). Classically, a particle of total energy E traveling in the region $x < 0$, in the direction of increasing x, will suffer

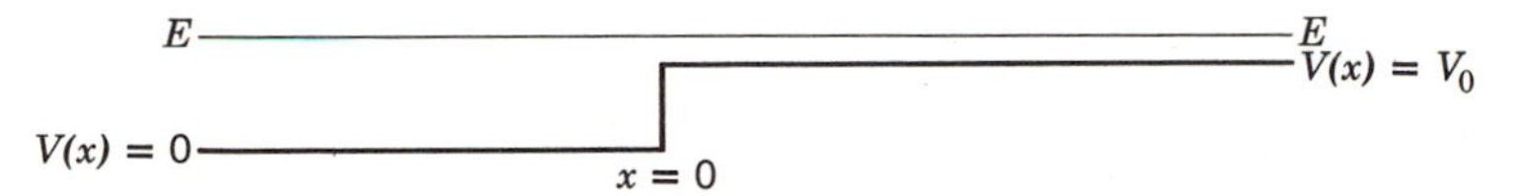

Figure 8–6. The case in which classical mechanics predicts that the particle is not reflected at the discontinuity of the potential.

an impulsive retarding force $F(x) = -\partial V(x)/\partial x$ at the point $x = 0$. But the impulse will only slow the particle and it will enter the region $x > 0$, continuing its motion in the direction of increasing x. Its total energy E remains constant; its momentum in the region $x < 0$ is p_1, where $E = p_1^2/2m$; its momentum in the region $x > 0$ is p_2, where $E = p_2^2/2m + V_0$.

Quantum mechanically, the motion of the particle is described by the wave function $\Psi(x, t) = e^{-iEt/\hbar}\psi(x)$, where the eigenfunction $\psi(x)$ is a solution to

$$-\frac{\hbar^2}{2m}\frac{d^2\psi(x)}{dx^2} = E\psi(x), \qquad x < 0 \qquad (8\text{–}36)$$

and

$$-\frac{\hbar^2}{2m}\frac{d^2\psi(x)}{dx^2} = (E - V_0)\psi(x), \qquad x > 0 \qquad (8\text{–}37)$$

and where the eigenfunction satisfies the conditions (7–38) at the point $x = 0$. The first equation describes the motion of a free particle of momentum p_1. The traveling wave form of its general solution is

$$\psi(x) = Ae^{iK_1x} + Be^{-iK_1x}, \qquad x < 0 \qquad (8\text{–}38)$$

where

$$K_1 = \sqrt{2mE}/\hbar = p_1/\hbar$$

The second equation describes the motion of a free particle of momentum p_2. The traveling wave form of its general solution is

$$\psi(x) = Ce^{iK_2x} + De^{-iK_2x}, \qquad x > 0 \tag{8–39}$$

where

$$K_2 = \sqrt{2m(E - V_0)}/\hbar = p_2/\hbar \quad \text{and} \quad E > V_0$$

The wave function specified by equations (8–38) and (8–39) consists of traveling waves of de Broglie wavelength $\lambda_1 = h/p_1 = 2\pi/K_1$ in the region $x < 0$, and of longer de Broglie wavelength $\lambda_2 = h/p_2 = 2\pi/K_2$ in the region $x > 0$.

Classically, the particle has unity probability of passing the point $x = 0$ and entering the region $x > 0$. This is not true quantum mechanically. Because of the wave-like properties of the particle, there is a certain probability that the particle will be reflected at the point $x = 0$, where there is a discontinuous change in the de Broglie wavelength. Thus we need to take both terms of the general solution (8–38) to describe the incident and reflected traveling waves in the region $x < 0$. However, we do not need to take the second term of the general solution (8–39). This term describes a wave traveling in the direction of decreasing x in the region $x > 0$. Since the particle is incident in the direction of increasing x, such a wave could only arise from a reflection at some point with a large positive x coordinate. As there is nothing out there to cause a reflection, we know that there is only a transmitted traveling wave in the region $x > 0$, and so we set the arbitrary constant

$$D = 0 \tag{8–40}$$

The arbitrary constants A, B, and C must be chosen to make $\psi(x)$ and $d\psi(x)/dx$ continuous at $x = 0$. The first requirement is satisfied if

$$A(e^{iK_1x})_{x=0} + B(e^{-iK_1x})_{x=0} = C(e^{iK_2x})_{x=0}$$

or

$$A + B = C \tag{8–41}$$

The second requirement is satisfied if

$$iK_1A(e^{iK_1x})_{x=0} - iK_1B(e^{-iK_1x})_{x=0} = iK_2C(e^{iK_2x})_{x=0}$$

or

$$K_1(A - B) = K_2C \tag{8–42}$$

From equations (8–41) and (8–42) we find

$$B = \frac{K_1 - K_2}{K_1 + K_2}A$$

and (8–43)

$$C = \frac{2K_1}{K_1 + K_2}A$$

Thus the eigenfunction is

$$\psi(x) = \begin{cases} Ae^{iK_1x} + A\dfrac{K_1 - K_2}{K_1 + K_2}e^{-iK_1x}, & x \leqslant 0 \\ A\dfrac{2K_1}{K_1 + K_2}e^{iK_2x}, & x \geqslant 0 \end{cases} \tag{8–44}$$

The arbitrary constant A can be chosen to satisfy the normalization condition, but we do not bother to do this. It is clear that an eigenfunction satisfying the two continuity conditions and the normalization condition could not be found if we had initially set the coefficient B of the reflected wave equal to zero, since then we would have only two arbitrary constants to satisfy three conditions. We shall not plot the complex function (8–44).

Writing down the wave function, and evaluating the probability flux at some point $x < 0$, we find

$$S(x, t) = v_1A^*A - v_1B^*B$$

where

$$v_1 = \hbar K_1/m = p_1/m$$

The first term is the incident flux, and the second term is the reflected flux. The transmitted probability flux can be evaluated by calculating $S(x, t)$ at some point $x > 0$. This gives

$$S(x, t) = v_2C^*C$$

where

$$v_2 = \hbar K_2/m = p_2/m$$

The ratio of the intensity of the reflected flux to the intensity of the incident flux is the probability that the particle will be reflected back into the region $x < 0$. This is

$$R = \frac{v_1B^*B}{v_1A^*A} = \frac{(K_1 - K_2)^2}{(K_1 + K_2)^2} \tag{8–45}$$

The ratio of the intensity of the transmitted flux to the intensity of the incident flux is the probability that the particle will be transmitted into the region $x > 0$. This is

$$T = \frac{v_2C^*C}{v_1A^*A} = \frac{K_2}{K_1}\frac{(2K_1)^2}{(K_1 + K_2)^2}$$

$$T = \frac{4K_1K_2}{(K_1 + K_2)^2} \tag{8–46}$$

It is easy to show that

$$R + T = 1 \tag{8–47}$$

The probability flux incident upon the potential discontinuity is split into a transmitted flux and a reflected flux, but from equation (8–47) we see that the total probability flux is conserved. This statement says that either the particle will be reflected or it will be transmitted, but it will not vanish. Of course, the particle is never split at the discontinuity. In any particular trial the particle will go one way or the other. For a large number of trials, the average probability of going in the direction of decreasing x is R, and the average probability of going in the direction of increasing x is T. Note that R and T are both unchanged if K_1 and K_2 are exchanged. A moment's consideration will show that this means the same values of R and T would be obtained if the particle were incident upon the potential discontinuity in the direction of decreasing x from the region $x > 0$. The wave function describing the motion of the particle, and consequently the probability flux, is partially reflected simply because there is a *discontinuity* in $V(x)$, and not because $V(x)$ becomes larger in the direction in which the particle is going.

It is interesting to investigate the value of R for $E \simeq V_0$, and also for $E \gg V_0$. In the first case, $K_2 = \sqrt{2m(E - V_0)}/\hbar \to 0$, while $K_1 = \sqrt{2mE}/\hbar \to \sqrt{2mV_0}/\hbar$. Thus

$$R = \left(\frac{K_1 - K_2}{K_1 + K_2}\right)^2 \to \left(\frac{K_1}{K_1}\right)^2 = 1$$

and the probability that the particle is reflected approaches unity. In the second case, $K_2 \to \sqrt{2mE}/\hbar = K_1$. Thus $R \to 0$ and the probability that the particle is reflected becomes negligible. However, the case $E \gg V_0$ is not necessarily the case of the classical limit at which we know there will be no reflection. This becomes apparent if we evaluate equation (8–45) in terms of E and V_0, giving

$$R = \left(\frac{1 - \sqrt{1 - V_0/E}}{1 + \sqrt{1 - V_0/E}}\right)^2 \qquad (8\text{–}48)$$

Planck's constant does not enter into the equation; the value of R depends only on the ratio of V_0 to E. Keeping this ratio constant, equation (8–48) predicts that there will be reflection even when V_0 and E are so large that the predictions of classical mechanics should be valid. This seems paradoxical until we realize that, in the limit of large energies, the assumption that the change in $V(x)$ is perfectly sharp can no longer be even an approximation to a real physical situation. The point is that the reflection is the result of an abrupt change in the de Broglie wavelength; "abrupt" means that the wavelength changes by an appreciable amount in a distance of 1 wavelength. If the change in wavelength in 1 wavelength becomes very small, because the change in the potential in that distance is very small, the reflection becomes negligible. This can be demonstrated by solving the Schroedinger equation for a potential of the form shown in figure (8–7). These statements, which follow from the characteristic properties of wave motion, should be familiar to the reader from his study of the behavior of water waves or light waves. In the classical limit the

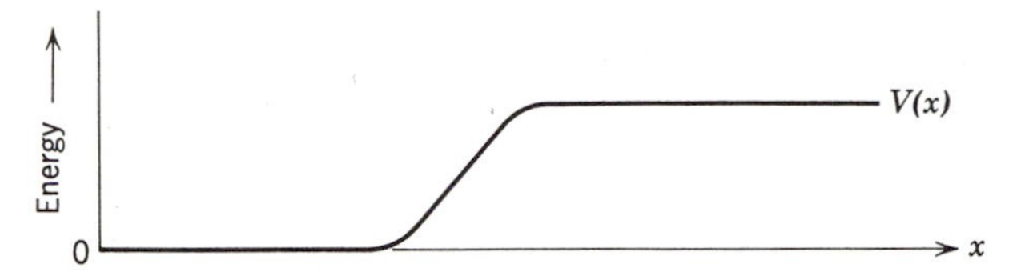

Figure 8–7. A smoothed step potential.

de Broglie wavelength is so very short that any physically realistic potential $V(x)$ changes only by a negligible amount in 1 wavelength, and no reflection results. However, for particles in atomic or nuclear systems, the de Broglie wavelength can be long relative to the distance in which the potential changes significantly, and reflection phenomena can become very important.

3. Barrier Potentials

We consider next a *barrier potential*, illustrated in figure (8–8). This potential is

$$V(x) = \begin{matrix} V_0, & 0 < x < a \\ 0, & x < 0, x > a \end{matrix} \qquad (8\text{–}49)$$

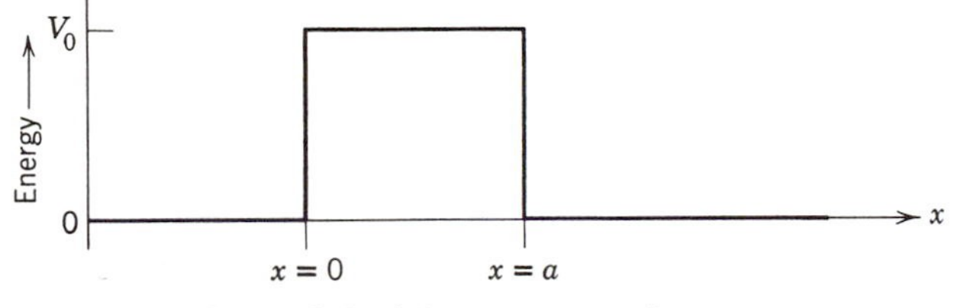

Figure 8–8. A barrier potential.

Classically, a particle of total energy E in the region $x < 0$, which is incident upon the barrier in the direction of increasing x, will have unity probability of being reflected if $E < V_0$, and unity probability of being transmitted if $E > V_0$. Neither of these conclusions obtains in quantum mechanics.

For this potential *acceptable solutions to the time independent Schroedinger equation exist for all* $E \geqslant 0$. Also the equation breaks into three separate equations for the three regions: $x < 0$, $0 < x < a$, and $x > a$. In the first and third regions the equation is that of a free particle of total energy E. The traveling wave solutions are

$$\begin{aligned} \psi(x) &= Ae^{iK_{\text{I}}x} + Be^{-iK_{\text{I}}x}, \qquad x < 0 \\ \psi(x) &= Ce^{iK_{\text{I}}x} + De^{-iK_{\text{I}}x}, \qquad x > a \end{aligned} \tag{8-50}$$

where

$$K_{\text{I}} = \sqrt{2mE}/\hbar$$

In the second region the form of the equation and of the solutions depends on whether $E < V_0$ or $E > V_0$. Both of these cases have been discussed in the previous section. In the first case,

$$\psi(x) = Fe^{-K_{\text{II}}x} + Ge^{K_{\text{II}}x}, \qquad 0 < x < a \tag{8-51}$$

where

$$K_{\text{II}} = \sqrt{2m(V_0 - E)}/\hbar \quad \text{and} \quad E < V_0$$

In the second case,

$$\psi(x) = Fe^{iK_{\text{III}}x} + Ge^{-iK_{\text{III}}x}, \qquad 0 < x < a \tag{8-52}$$

where

$$K_{\text{III}} = \sqrt{2m(E - V_0)}/\hbar \quad \text{and} \quad E > V_0$$

In evaluating the motion of a particle incident in the direction of increasing x, we may set

$$D = 0$$

since we know that there can only be a transmitted wave in the region $x > a$. However, in the present situation we cannot set $G = 0$ in equation (8–51) since x remains finite in the region $0 < x < a$. Nor can we set $G = 0$ in equation (8–52) since there will be a reflection from the discontinuity at $x = a$.

Consider first the case:

$E < V_0$

In matching $\psi(x)$ and $d\psi(x)/dx$ at the points $x = 0$ and $x = a$, four equations in the constants A, B, C, F, and G will be obtained. We leave it as an exercise for the reader to set up these equations and use them to evaluate B, C, F, and G in terms of A. The value of A would be determined by the normalization condition, if it were applied. The form of the eigenfunction obtained is indicated in figure (8–9). This figure is meant to be only an indication, since the exact form of $\psi(x)$ in the region $0 < x < a$, as well as the ratio of its amplitude in the region $x > a$ to its amplitude in the region $x < 0$, depends on the values of E, V_0, a, and m. Furthermore, $\psi(x)$ is actually a complex function. [For example, $\psi(x) = Ce^{iK_1x}$ for $x > a$.] In the figure we have plotted the real part of $\psi(x)$.

The most interesting result of the calculation is the ratio T of the intensity of the probability flux transmitted into the region $x > a$ to the intensity of the incident probability flux. The value of T is

$$T = \frac{v_1 C^* C}{v_1 A^* A} = \left[1 + \frac{\sinh^2 K_{II} a}{(4E/V_0)(1 - E/V_0)}\right]^{-1}, \qquad E < V_0$$

which is

$$T = \left[1 + \frac{\sinh^2\left(\sqrt{\frac{2mV_0a^2}{\hbar^2}(1 - E/V_0)}\right)}{(4E/V_0)(1 - E/V_0)}\right]^{-1}, \qquad E < V_0 \qquad (8\text{–}53)$$

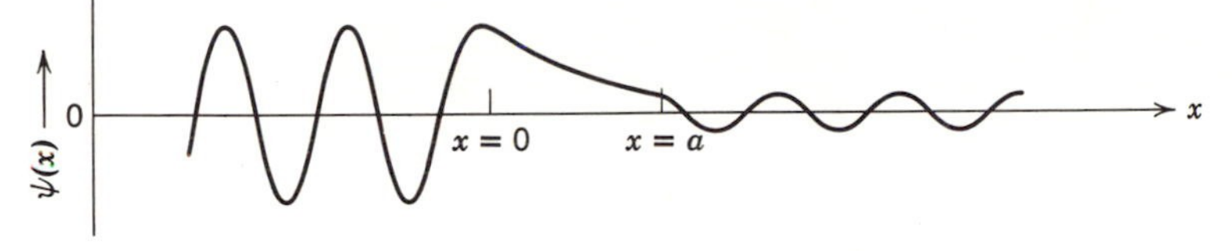

Figure 8–9. The real part of a barrier penetration eigenfunction.

If the argument of the hyperbolic sine is large compared to 1, T will have a very small value which is given approximately by

$$T \simeq 16 \frac{E}{V_0}(1 - E/V_0)e^{-2K_{II}a} \qquad (8\text{–}54)$$

These equations make the remarkable (from the point of view of classical mechanics) prediction that a particle of mass m and total energy E, which is incident on a potential barrier of height $V_0 > E$ and finite thickness a, actually has a certain probablity T of penetrating the barrier and appearing on the other side. This phenomenon is called *barrier penetration.* Of course, T is vanishingly small in the classical limit because in that limit the quantity $2mV_0a^2/\hbar^2$, which is a measure of the "opacity" of the barrier, is extremely large.

Next consider the case:

$E > V_0$

In this case the wave function is a traveling wave in all three regions, but of longer wavelength in the region $0 < x < a$. Evaluation of the constants B, C, F, and G by the application of the continuity conditions at $x = 0$ and $x = a$, leads to the value

$$T = \frac{v_1 C^* C}{v_1 A^* A} = \left[1 + \frac{\sin^2 K_{III} a}{(4E/V_0)(E/V_0 - 1)}\right]^{-1}, \qquad E > V_0$$

which is

$$T = \left[1 + \frac{\sin^2 \left(\sqrt{\frac{2mV_0 a^2}{\hbar^2}(E/V_0 - 1)}\right)}{(4E/V_0)(E/V_0 - 1)}\right]^{-1}, \qquad E > V_0 \quad (8\text{–}55)$$

The forms of the functions (8–53) and (8–55) depend on the value of $2mV_0a^2/\hbar^2$. Figure (8–10) shows a plot of T as a function of E/V_0 for

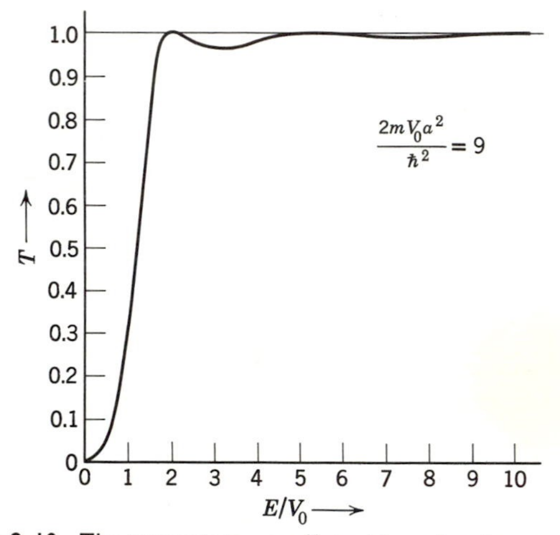

Figure 8–10. The transmission coefficient for a barrier potential.

an electron incident upon a barrier of height $V_0 = 10$ eV and thickness $a = 1.85 \times 10^{-8}$ cm; $(2mV_0a/\hbar^2 = 9)$. This is typical of the values for the potentials encountered, for instance, by an electron moving through the atoms of a crystal. For $E/V_0 \ll 1$, T is very small. But, when E/V_0 is only somewhat smaller than unity, T is not at all negligible. It is apparent that barrier penetration can be very important in atomic systems. For $E/V_0 > 1$, T is in general less than 1, owing to reflection at the discontinuities in the potential. However, from equation (8–55) it can be seen that $T = 1$ whenever $K_{III}a = \pi, 2\pi, 3\pi, \ldots$. This is simply the condition that the length a of the barrier region is equal to an integral or half-integral number of wavelengths $\lambda_{III} = 2\pi/K_{III}$ in that region. For this particular barrier, electrons of energy $E \simeq 21$ ev, 53 ev, etc., satisfy the condition $K_{III}a = \pi$, 2π, etc., and so pass into the region $x > a$ without any reflection. This effect is a result of destructive interference between reflections at $x = 0$ and $x = a$. It is closely related to the *Ramsauer effect* observed in the scattering of electrons by atoms, which will be discussed in Chapter 15.

Barrier penetration is a manifestation of wave-like behavior in the motion of a particle. The same phenomenon is also observed in other types of wave motion. We have seen that the time independent differential equation governing the propagation of waves in a string is essentially the same as the time independent Schroedinger equation. This is also true of the time independent differential equation for electromagnetic radiation. For radiation of frequency ν propagating through a medium of index of refraction μ this equation is

$$\frac{d^2\psi(x)}{dx^2} + \left(\frac{2\pi\nu}{c}\mu\right)^2 \psi(x) = 0 \tag{8-56}$$

When we compare this with equation (7–35), it is apparent that the time independent Schroedinger equation is identical with that for the propagation of light through a medium in which the index of refraction is the following function of position:

$$\mu(x) = \frac{c}{2\pi\nu}\sqrt{\frac{2m}{\hbar^2}[E - V(x)]} \tag{8-57}$$

In fact, there are optical phenomena which are exactly analogous to each of the quantum mechanical phenomena that arise in considering the motion of an unbound particle. For instance, an optical phenomenon, completely analogous to the quantum mechanical phenomenon of unity transmission coefficient for barriers of length equal to an integral or half-integral number of wavelengths, is used in the "coating" of lenses to obtain very high light transmission. In particular, the situation $E < V(x)$ encountered in quantum mechanical barrier penetration is equivalent to the optical situation for an imaginary index of refraction. This occurs in the case of total internal reflection.

Consider a ray of light incident upon a glass to air interface at an angle greater than the critical angle θ_c. This situation, which is called *total internal reflection*, is illustrated in figure (8–11). A detailed treatment of this process in terms of electromagnetic theory shows that the index of refraction, measured along the line *ABC*, is real in the region *AB* but imaginary in the region *BC*. Furthermore, it shows that there are electromagnetic vibrations in the region *BC* of exactly the same form as the exponentially decreasing standing wave of equation (8–34) for the region $E < V(x)$. The flux of energy is zero in this electromagnetic standing wave, just as the flux of probability is zero in the quantum mechanical standing wave. However, if a second block of glass is placed near enough to the first block to be in the region in which the intensity of the electromagnetic vibrations is still appreciable, these vibrations are picked up and they propagate through the second block, and the electromagnetic vibrations in the air gap now carry a flux of energy through to the second block.

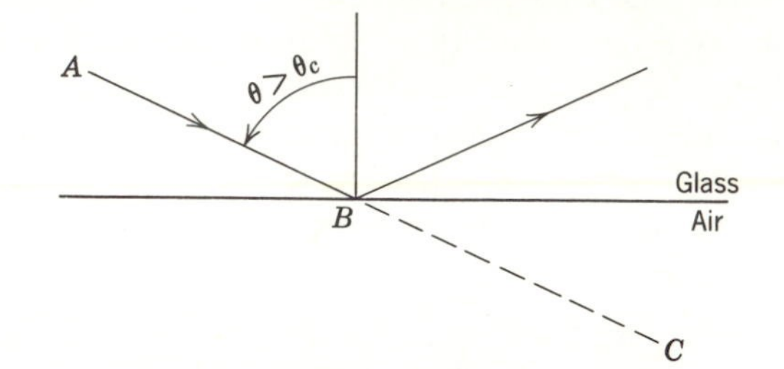

Figure 8–11. Illustrating total internal reflection.

This situation, which is called *frustrated total internal reflection*, is illustrated in figure (8–12). Essentially the same thing happens in the quantum mechanical case when the region in which $E < V(x)$ is reduced from infinite thickness (step potential) to finite thickness (barrier potential). The transmission of light through an air gap, at an angle of incidence greater than θ_c, was first observed by Newton (ca. 1700). The equation

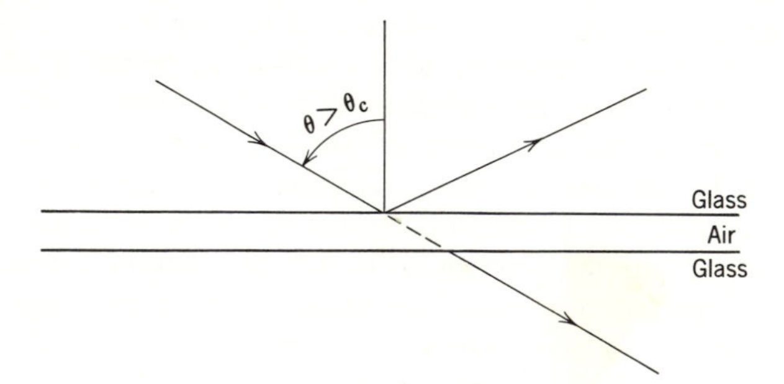

Figure 8–12. Illustrating frustrated total internal reflection.

relating the intensity of the transmitted beam to the thickness of the air gap, and the other parameters, is identical in form with equation (8–53), and it has been verified experimentally.

Barrier penetration of particles was first discussed in connection with a long-standing problem concerning the emission of alpha particles by radioactive elements. As a typical example, consider a nucleus of the radioactive element U^{238}. The potential energy $V(r)$ of an alpha particle at a distance r from the nucleus had been investigated by means of Rutherford scattering experiments. Using as a probe the 8.8×10^6 ev = 8.8 Mev† alpha particles emitted from the radioactive element Po^{212}, it was observed that the probability of scattering agreed with the predictions of Rutherford's formula over the entire angular range. From this it was concluded that, for the U^{238} nucleus, $V(r)$ exactly follows the Coulomb law $V(r) = zZe^2/r$ at least in to a distance $r' = 3.0 \times 10^{-12}$ cm at which $V(r') = 8.8$ Mev. It was also known from the experiments involving the scattering of alpha particles from the nuclei of light atoms that $V(r)$ must

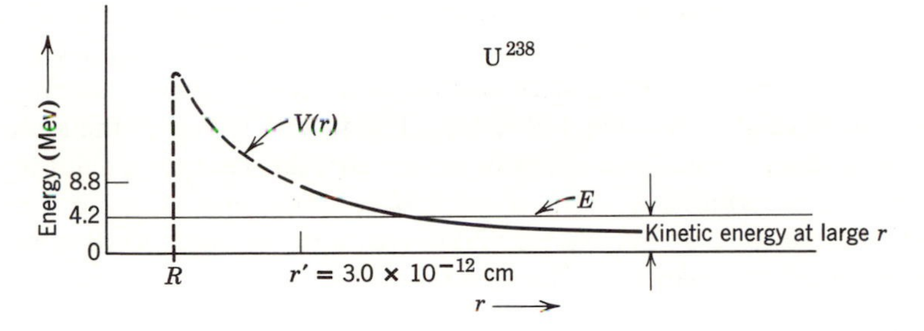

Figure 8–13. An energy diagram showing the relation between potential and total energies in alpha particle emission.

eventually depart from a $1/r$ law when $r < R$, the nuclear radius, although at that time the exact value of R was not known for the heavy radioactive nuclei. Furthermore, since alpha particles are (very infrequently) emitted by U^{238} nuclei, it was assumed that they must exist inside such nuclei—to which they are normally bound by the potential $V(r)$. From these arguments it was concluded that the form of $V(r)$ in the region $r < r'$ must be qualitatively as depicted in figure (8–13). (This assumption has been verified by recent experiments involving the scattering of alpha particles of energies high enough to investigate the potential over the entire range of r.) It was also known that the kinetic energy of alpha particles emitted by U^{238} was 4.2 Mev. The kinetic energy was, of course, measured at a very large distance from the nucleus, at which $V(r) = 0$ and the kinetic energy equals the total energy. This value of the constant total energy is also shown in figure (8–13). From the point of view of classical mechanics, this situation is completely paradoxical. An alpha particle of total energy E is initially in the region $r < R$. This region is separated from the rest of space by a potential barrier of height which is known to be at least twice E.

† 1 Mev = 10^6 ev = 1.602×10^{-6} ergs. This unit of energy is used extensively in nuclear physics.

Yet it is observed that on occasion the alpha particle penetrates the barrier and moves off to large values of r.

In 1928 Gamow, Condon, and Gurney treated alpha particle emission as a quantum mechanical barrier penetration problem. They assumed that $V(r) = zZe^2/r$ for $r > R$, and $V(r) < E$ for $r < R$, as shown in figure (8–13). Equation (8–54) was used to evaluate the transmission factor T since $K_{II}a \gg 1$. In fact, this quantity is so large that the exponential completely dominates the behavior of T, and it was sufficient to take

$$T \simeq e^{-2K_{II}a} = e^{-2\sqrt{(2m/\hbar^2)(V_0-E)}\,a}$$

This expression was derived for a rectangular barrier of height V_0 and width a, but in the limit that the expression is valid it can be applied to the barrier $V(r)$ by considering it to consist of a set of adjacent rectangular barriers of height $V(r_i)$ and width Δr_i. The reason is that in the limit reflections from the discontinuities in the rectangular barriers do not make important contributions to the transmission factor, and its value is completely dominated by the exponential decrease in the eigenfunction in the regions within the barriers. The transmission factor for the ith rectangular barrier is

$$T_i = e^{-2\sqrt{(2m/\hbar^2)[V(r_i)-E]}\,\Delta r_i}$$

If reflections can be ignored so that there is no interaction between the individual barriers, the total transmission factor will be the product of the individual transmission factors because independent probabilities are multiplicative. Then

$$\begin{aligned}
T &= e^{-2\sqrt{(2m/\hbar^2)[V(r_1)-E]}\,\Delta r_1}\,e^{-2\sqrt{(2m/\hbar^2)[V(r_2)-E]}\,\Delta r_2}\,e^{-2\sqrt{(2m/\hbar^2)V[(r_3)-E]}\,\Delta r_3}\cdots \\
&= e^{-2\left(\sqrt{(2m/\hbar^2)[V(r_1)-E]}\,\Delta r_1+\sqrt{(2m/\hbar^2)[V(r_2)-E]}\,\Delta r_2+\sqrt{(2m/\hbar^2)[V(r_3)-E]}\,\Delta r_3+\cdots\right)} \\
&= e^{-2\sum_i\sqrt{(2m/\hbar^2)[V(r_i)-E]}\,\Delta r_i}
\end{aligned}$$

Letting the Δr_i become very small, this becomes

$$T = e^{-2\int\sqrt{(2m/\hbar^2)[V(r)-E]}\,dr} \tag{8–58}$$

A rigorous treatment of this problem shows that this procedure is justified, in that it gives the correct answer, if T is very small, as it is for the case of alpha particle emission.

The quantity T gives the probability that in one trial an alpha particle will penetrate the barrier. The number of trials per second can be estimated as

$$N \simeq v/2R \tag{8–59}$$

if it is assumed that an alpha particle is bouncing back and forth with velocity v inside the nucleus of diameter $2R$. Then the probability per second that the nucleus will emit an alpha particle, which is conventionally designated by the symbol λ, is

$$\lambda \simeq \frac{v}{2R} e^{-2\int_R^\infty \sqrt{(2m/\hbar^2)[(zZe^2/r)-E]}\,dr} \tag{8-60}$$

In applying this expression to a particular radioactive nucleus, all the quantities in this expression except v and R are known. Assuming v to be comparable to the velocity of the alpha particle after emission (i.e., $\frac{1}{2}mv^2 = E$), then λ is a function only of the nuclear radius R. Using $R \simeq 9 \times 10^{-13}$ cm, which was certainly in line with the values obtained from Rutherford's analysis of the alpha particle scattering experiments from light elements, Gamow, Condon, and Gurney obtained values of λ which were in good agreement with those experimentally measured, although they vary over a tremendously large range. For U^{238}, $\lambda = 5 \times 10^{-18}\ \text{sec}^{-1}$. For Po^{212}, $\lambda = 2 \times 10^6\ \text{sec}^{-1}$. This variation is due primarily to the variation, from one radioactive element to the next, of the energy E of the emitted alpha particles. The height and breadth of the barrier (Z and R) do not change significantly for nuclei in the limited range of the periodic table in which the radioactive elements are found.

The successful application of the theory of quantum mechanics to the process of alpha particle emission provides one of the most convincing verifications of the theory. Furthermore, this process gives a striking illustration of the wave-particle duality. On either side of the barrier the alpha "particle" could be localized and shown to behave like a particle. In penetrating the barrier it behaves purely like a wave. Any attempt to localize it in the barrier would, according to the uncertainty principle, necessarily introduce such a large uncertainty in the energy that it would no longer be possible to say definitely that it was not going over the top of the barrier.

4. Square Well Potentials

In this section we shall discuss the simplest potential which is capable of binding a particle to a limited region of space, a *square well potential*, indicated in figure (8–14). This potential is

$$V(x) = \begin{cases} V_0, & x < -a/2, \quad x > +a/2 \\ 0, & -a/2 < x < +a/2 \end{cases} \tag{8-61}$$

Consider first the case:

$E > V_0$ (traveling waves)

Classically, a particle under the influence of this potential with total energy $E > V_0$ would be able to move over the entire range of the x axis. Upon entering the region $-a/2 < x < +a/2$, it would receive an accelerating impulse and travel through that region with momentum higher than its initial momentum. Upon leaving, it would receive a compensating decelerating impulse. Its final momentum would equal its initial momentum. We shall not develop the traveling wave solutions to the quantum

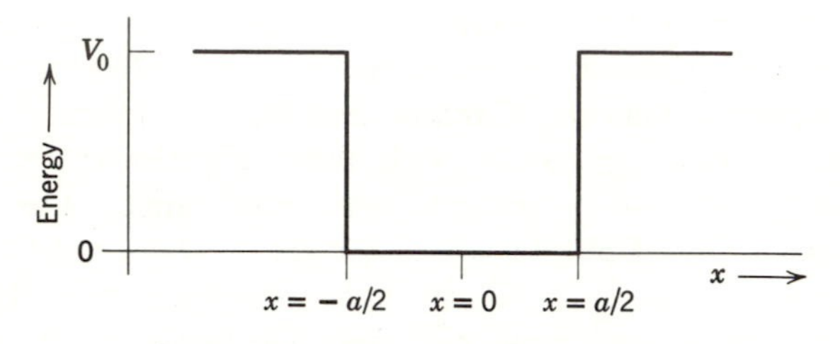

Figure 8–14. A square well potential.

mechanical problem, *which exist for all* $E > V_0$, since they are essentially the same as those for a barrier potential with $E > V_0$. In fact, the transmission factor T can be obtained directly from equation (8–55) if the terms are appropriately redefined. This factor is, as before, less than unity except for energies at which a is just equal to an integral or half-integral number of de Broglie wavelengths in the region $-a/2 < x < +a/2$. We return later to discuss the standing wave solutions for the square well potential with $E > V_0$.

We consider instead the case:

$E < V_0$ (standing waves)

If the particle has total energy $E < V_0$, then classically it could only be in the region $-a/2 < x < +a/2$. The particle would be bound to that region and would bounce back and forth between the ends of the region with momentum of constant magnitude but alternating direction. Classically, any value of the total energy E is possible for a bound state. In section 5, Chapter 7, we have seen qualitatively that in quantum mechanics *only certain values of the total energy E are possible.* In this section we shall evaluate quantitatively the discrete eigenvalues, and the corresponding eigenfunctions, for a square well potential.

Such a potential is very often used in quantum mechanics to represent a situation in which a particle moves in a restricted region of space under the influence of forces which hold it in that region. Although this simplified potential loses some details of the motion, it retains the essential feature of binding the particle by forces of a certain strength to a region of a certain size. As an example, many aspects of the motion of a nuclear particle in a nucleus can be explained by assuming that the particle is bound in a square well potential with a depth of about 50 Mev and linear dimensions of about 10^{-12} cm.

The description of the classical motion of particle bound by the square well suggests that it would be most appropriate to look for solutions to the Schroedinger equation in the form of standing waves. Thus we take, as a solution to the time independent Schroedinger equation in the region $-a/2 < x < +a/2$, in which $V(x) = 0$, the free particle standing wave eigenfunction (8–16′). This is

$$\psi(x) = A \sin K_{\text{I}} x + B \cos K_{\text{I}} x, \qquad -a/2 < x < +a/2 \tag{8–62}$$

where

$$K_{\text{I}} = \sqrt{2mE}/\hbar$$

In the regions $x < -a/2$ and $x > +a/2$ the time independent Schroedinger equation is the same as equation (8–23). According to equation (8–25), the solutions are

$$\psi(x) = Ce^{K_{\text{II}} x} + De^{-K_{\text{II}} x}, \qquad x < -a/2 \tag{8–63}$$

and

$$\psi(x) = Fe^{K_{\text{II}} x} + Ge^{-K_{\text{II}} x}, \qquad x > +a/2 \tag{8–64}$$

where

$$K_{\text{II}} = \sqrt{2m(V_0 - E)}/\hbar \quad \text{and} \quad E < V_0$$

To determine the arbitrary constants first impose the requirement that the eigenfunctions remain finite for all x. Consider equation (8–63) in the limit $x \to -\infty$. It is apparent that this requirement demands

$$D = 0 \tag{8–65}$$

Similarly, it is necessary to set

$$F = 0 \tag{8–66}$$

in order that equation (8–64) remain finite in the limit $x \to +\infty$. Next impose the requirement that the eigenfunctions and their first derivatives be continuous at $x = -a/2$ and $x = +a/2$. Four equations are obtained. They are

$$-A \sin (K_{\text{I}} a/2) + B \cos (K_{\text{I}} a/2) = Ce^{-K_{\text{II}} a/2} \tag{8–67}$$

$$AK_{\text{I}} \cos (K_{\text{I}} a/2) + BK_{\text{I}} \sin (K_{\text{I}} a/2) = CK_{\text{II}} e^{-K_{\text{II}} a/2} \tag{8–68}$$

$$A \sin (K_{\text{I}} a/2) + B \cos (K_{\text{I}} a/2) = Ge^{-K_{\text{II}} a/2} \tag{8–69}$$

$$AK_{\text{I}} \cos (K_{\text{I}} a/2) - BK_{\text{I}} \sin (K_{\text{I}} a/2) = -GK_{\text{II}} e^{-K_{\text{II}} a/2} \tag{8–70}$$

Subtracting equation (8–67) from (8–69) yields

$$2A \sin (K_{\text{I}}a/2) = (G - C)e^{-K_{\text{II}}a/2} \tag{8–71}$$

Adding equation (8–67) to (8–69) yields

$$2B \cos (K_{\text{I}}a/2) = (G + C)e^{-K_{\text{II}}a/2} \tag{8–72}$$

Subtracting equation (8–70) from (8–68) yields

$$2BK_{\text{I}} \sin (K_{\text{I}}a/2) = (G + C)K_{\text{II}}e^{-K_{\text{II}}a/2} \tag{8–73}$$

Adding equation (8–70) to (8–68) yields

$$2AK_{\text{I}} \cos (K_{\text{I}}a/2) = -(G - C)K_{\text{II}}e^{-K_{\text{II}}a/2} \tag{8–74}$$

Provided $B \neq 0$ and $(G + C) \neq 0$, we may divide equation (8–73) by (8–72) and obtain

$$K_{\text{I}} \tan (K_{\text{I}}a/2) = K_{\text{II}} \tag{8–75}$$
$$\text{if} \quad B \neq 0 \quad \text{and} \quad (G + C) \neq 0$$

Provided $A \neq 0$ and $(G - C) \neq 0$, we may divide equation (8–74) by (8–71) and obtain

$$K_{\text{I}} \cot (K_{\text{I}}a/2) = -K_{\text{II}} \tag{8–76}$$
$$\text{if} \quad A \neq 0 \quad \text{and} \quad (G - C) \neq 0$$

It is easy to see that both equations (8–75) and (8–76) cannot be satisfied simultaneously. If they could, the equation obtained by adding these two,

$$K_{\text{I}} \tan (K_{\text{I}}a/2) + K_{\text{I}} \cot (K_{\text{I}}a/2) = 0$$

would be valid. Multiply through by $\tan (K_{\text{I}}a/2)$. Then the equation becomes

$$K_{\text{I}} \tan^2 (K_{\text{I}}a/2) + K_{\text{I}} = 0$$

or

$$\tan^2 (K_{\text{I}}a/2) = -1$$

But this cannot be valid as both K_{I} and $a/2$ are real. Thus it is only possible *either* to satisfy (8–75) but not (8–76) *or* to satisfy (8–76) but not (8–75). Just as for the eigenfunctions of the vibrating string, the eigenfunctions of the square well potential form two classes. For the *first class*,

$$\begin{aligned} K_{\text{I}} \tan (K_{\text{I}}a/2) &= K_{\text{II}} \\ A &= 0 \\ G - C &= 0 \end{aligned} \tag{8–77}$$

Then equation (8–69) reads

$$B \cos (K_{\text{I}}a/2) = Ge^{-K_{\text{II}}a/2}$$
$$G = B \cos (K_{\text{I}}a/2)e^{K_{\text{II}}a/2} = C$$

and the eigenfunctions are

$$\psi(x) = \begin{cases} [B \cos (K_{\text{I}}a/2)e^{K_{\text{II}}a/2}]e^{K_{\text{II}}x}, & x < -a/2 \\ [B] \cos (K_{\text{I}}x), & -a/2 < x < a/2 \\ [B \cos (K_{\text{I}}a/2)e^{K_{\text{II}}a/2}]e^{-K_{\text{II}}x}, & x > a/2 \end{cases} \qquad (8\text{–}77')$$

For the *second class*,

$$K_{\text{I}} \cot (K_{\text{I}}a/2) = -K_{\text{II}}$$
$$B = 0 \qquad (8\text{–}78)$$
$$G + C = 0$$

Then equation (8–69) reads

$$A \sin (K_{\text{I}}a/2) = Ge^{-K_{\text{II}}a/2}$$
$$G = A \sin (K_{\text{I}}a/2)e^{K_{\text{II}}a/2} = -C$$

and the eigenfunctions are

$$\psi(x) = \begin{cases} [-A \sin (K_{\text{I}}a/2)e^{K_{\text{II}}a/2}]e^{K_{\text{II}}x}, & x < -a/2 \\ [A] \sin (K_{\text{I}}x), & -a/2 < x < a/2 \\ [A \sin (K_{\text{I}}a/2)e^{K_{\text{II}}a/2}]e^{-K_{\text{II}}x}, & x > a/2 \end{cases} \qquad (8\text{–}78')$$

Consider the first of equations (8–77). Evaluating K_{I} and K_{II}, and multiplying through by $a/2$, the equation becomes

$$\sqrt{mEa^2/2\hbar^2} \tan \left(\sqrt{mEa^2/2\hbar^2}\right) = \sqrt{m(V_0 - E)a^2/2\hbar^2} \qquad (8\text{–}79)$$

For a given particle of mass m and a given potential well of depth V_0 and width a, this is an equation in the single unknown E. Its solutions are the allowed values of the total energy of the particle—the eigenvalues for eigenfunctions of the first class. Solutions of this transcendental equation can be obtained only by numerical or graphical methods. We present a simple graphical method which will illustrate the important features of the equation. Let us make the change of variable

$$\varepsilon \equiv \sqrt{mEa^2/2\hbar^2} \qquad (8\text{–}80)$$

so the equation becomes

$$\varepsilon \tan \varepsilon = \sqrt{mV_0a^2/2\hbar^2 - \varepsilon^2} \qquad (8\text{–}81)$$

If we plot the function

$$p(\varepsilon) = \varepsilon \tan \varepsilon$$

and the function

$$q(\varepsilon) = \sqrt{mV_0a^2/2\hbar^2 - \varepsilon^2}$$

the intersections specify values of ε which are solutions to equation (8–81). Such a plot is shown in figure (8–15). The function $p(\varepsilon)$ has zeros at $\varepsilon = 0, \pi, 2\pi, \ldots$ and has asymptotes at $\varepsilon = \pi/2, 3\pi/2, 5\pi/2, \ldots$. The function $q(\varepsilon)$ is a quarter-circle of radius $\sqrt{mV_0a^2/2\hbar^2}$. It is clear from the figure that the number of solutions which exist for equation (8–81) depends on the radius of the quarter-circle. Each solution gives an eigenvalue for $E < V_0$ corresponding to an eigenfunction of the first

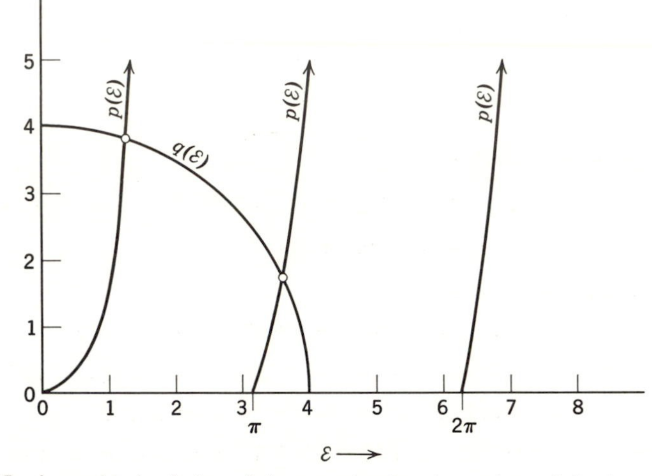

Figure 8–15. A graphical solution of the equation for eigenvalues of the first class of a particular square well potential. [Solution of $\varepsilon \tan \varepsilon = \sqrt{mV_0a^2/2\hbar^2 - \varepsilon^2}$ or $p(\varepsilon) = q(\varepsilon)$.]

class. There exists one such eigenvalue if $\sqrt{mV_0a^2/2\hbar^2} < \pi$; two if $\pi \leqslant \sqrt{mV_0a^2/2\hbar^2} < 2\pi$; three if $2\pi \leqslant \sqrt{mV_0a^2/2\hbar^2} < 3\pi$; etc. The case $\sqrt{mV_0a^2/2\hbar^2} = 4$ is illustrated in the figure. For this case there are two solutions: $\varepsilon \simeq 1.25$ and $\varepsilon \simeq 3.60$. From equation (8–80), the eigenvalues are

$$E = \varepsilon^2 \frac{2\hbar^2}{ma^2} = \varepsilon^2 \frac{2\hbar^2}{mV_0a^2} V_0 \simeq \left(\frac{1.25}{4}\right)^2 V_0 \simeq 0.098 V_0$$

and

$$E = \varepsilon^2 \frac{2\hbar^2}{mV_0a^2} V_0 \simeq \left(\frac{3.60}{4}\right)^2 V_0 \simeq 0.81 V_0$$

The eigenvalues corresponding to eigenfunctions of the second class are found from the solutions of an analogous equation obtained from (8–78), which is

$$-\varepsilon \cot \varepsilon = \sqrt{mV_0a^2/2\hbar^2 - \varepsilon^2} \tag{8–82}$$

Figure (8–16) illustrates the solution of this equation. It is apparent that there will be no eigenvalues for $E < V_0$ corresponding to eigenfunctions of the second class if $\sqrt{mV_0a^2/2\hbar^2} < \pi/2$; there will be one if $\pi/2 \leqslant \sqrt{mV_0a^2/2\hbar^2} < 3\pi/2$; two if $3\pi/2 \leqslant \sqrt{mV_0a^2/2\hbar^2} < 5\pi/2$; etc. The figure illustrates the case $\sqrt{mV_0a^2/2\hbar^2} = 4$. The single solution to equation (8–82) is $\varepsilon \simeq 2.45$, and the eigenvalue is

$$E = \varepsilon^2 \frac{2\hbar^2}{mV_0a^2} V_0 \simeq \left(\frac{2.45}{4}\right)^2 V_0 \simeq 0.37V_0$$

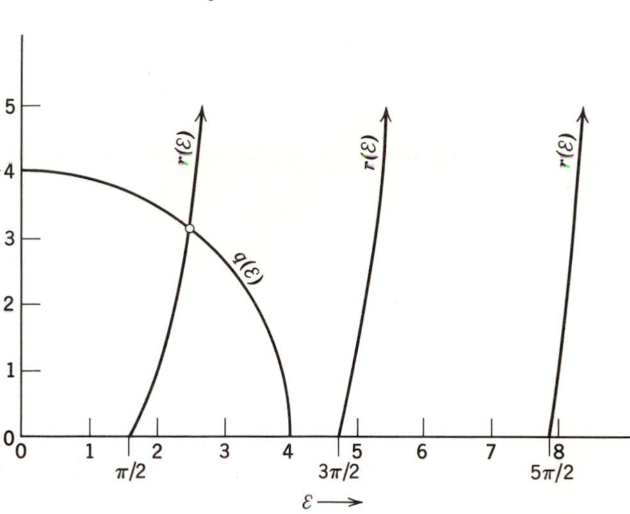

Figure 8–16. A graphical solution of the equation for eigenvalues of the second class of a particular square well potential. [Solution of $-\varepsilon \cot \varepsilon = \sqrt{mV_0a^2/2\hbar^2 - \varepsilon^2}$ or $r(\varepsilon) = q(\varepsilon)$.]

We see that for a given potential well there are only a restricted number of allowed values of total energy E for $E < V_0$. These are the discrete eigenvalues for the bound states of the particle. On the other hand, we know that any value of E is allowed for $E > V_0$; the eigenvalues for the unbound states form a continuum. For a potential well which is very shallow or very narrow or both, only a single eigenstate of the first class will be bound. With increasing values of $\sqrt{mV_0a^2/2\hbar^2}$ an eigenstate of the second class will be bound. For even larger values of this parameter an additional eigenstate of the first class will be bound. Next, an additional eigenstate of the second class will be bound, etc. As an example consider the case $\sqrt{mV_0a^2/2\hbar^2} = 4$. The potential and the discrete and continuum eigenvalues are illustrated to scale in figure (8–17). We have used the quantum numbers $n = 1, 2, 3, 4, 5, \ldots$ to label the eigenvalues in order of increasing energy. For this potential only the first three eigenstates are bound.

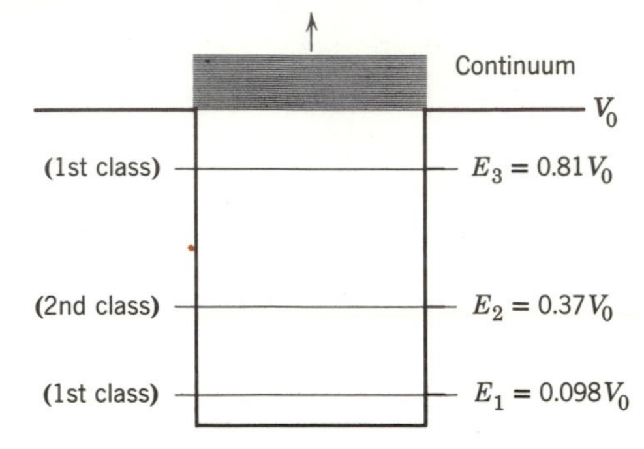

Figure 8–17. The eigenvalues of a particular square well potential.

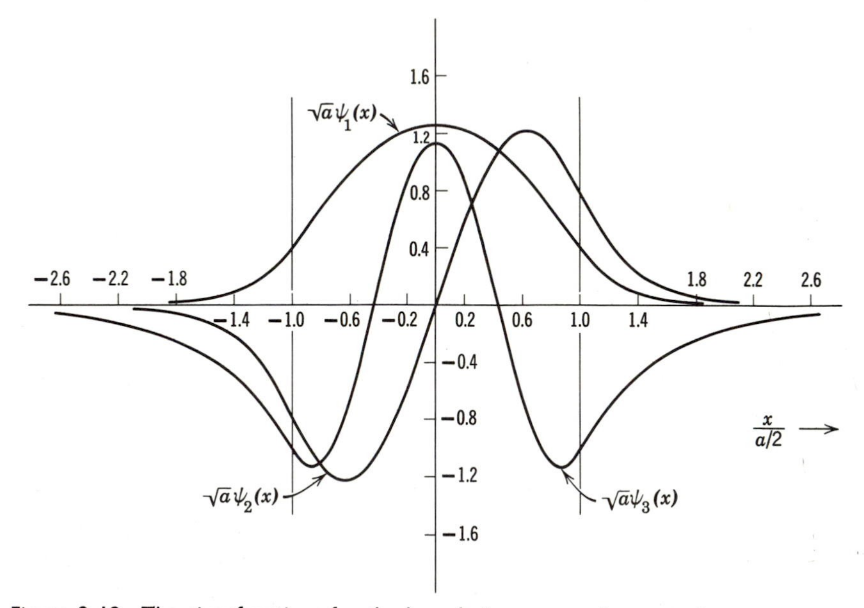

Figure 8–18. The eigenfunctions for the bound eigenstates of a particular square well potential.

From the solutions ε, of equations (8–81) and (8–82) for a given value of $\sqrt{mV_0a^2/2\hbar^2}$, the explicit forms of the eigenfunctions (8–77′) and (8–78′) may be evaluated. The required relations are

$$K_{\text{I}} \frac{a}{2} = \varepsilon \quad \text{and} \quad K_{\text{II}} \frac{a}{2} = \sqrt{mV_0a^2/2\hbar^2 - \varepsilon^2} \tag{8–83}$$

The value of the constant A or B must be adjusted so that each eigenfunction satisfies the normalization condition (7–49). For the case $\sqrt{mV_0a^2/2\hbar^2} = 4$, the three normalized eigenfunctions corresponding to the eigenvalues E_1, E_2, and E_3 are

$$\psi_1(x) = \begin{cases} 17.9 \dfrac{1}{\sqrt{a}} e^{3.80 \frac{x}{a/2}}, & x \leqslant -a/2 \\ 1.26 \dfrac{1}{\sqrt{a}} \cos\left(1.25 \dfrac{x}{a/2}\right), & -a/2 \leqslant x \leqslant a/2 \\ 17.9 \dfrac{1}{\sqrt{a}} e^{-3.80 \frac{x}{a/2}}, & x \geqslant a/2 \end{cases}$$

$$\psi_2(x) = \begin{cases} -18.6 \dfrac{1}{\sqrt{a}} e^{3.16 \frac{x}{a/2}}, & x \leqslant -a/2 \\ 1.23 \dfrac{1}{\sqrt{a}} \sin\left(2.45 \dfrac{x}{a/2}\right), & -a/2 \leqslant x \leqslant a/2 \\ 18.6 \dfrac{1}{\sqrt{a}} e^{-3.16 \frac{x}{a/2}}, & x \geqslant a/2 \end{cases}$$

$$\psi_3(x) = \begin{cases} -5.80 \dfrac{1}{\sqrt{a}} e^{1.74 \frac{x}{a/2}}, & x \leqslant -a/2 \\ 1.13 \dfrac{1}{\sqrt{a}} \cos\left(3.60 \dfrac{x}{a/2}\right), & -a/2 \leqslant x \leqslant a/2 \\ -5.80 \dfrac{1}{\sqrt{a}} e^{-1.74 \frac{x}{a/2}}, & x \geqslant a/2 \end{cases}$$

The eigenfunctions, multiplied by $\sqrt{a}$, are plotted in figure (8–18) as a function of $x/(a/2)$.

This figure makes quite apparent the essential difference between the two classes of eigenfunctions. The eigenfunctions of the first class, $\psi_1(x)$ and $\psi_3(x)$, are *even functions* of x; that is,

$$\psi(-x) = +\psi(x) \tag{8–84}$$

In quantum mechanics these functions are said to be of *even parity*. The eigenfunction of the second class, $\psi_2(x)$, is an *odd function* of x; that is

$$\psi(-x) = -\psi(x) \tag{8–85}$$

and is said to be of *odd parity*. Inspection of equations (8–77′) and (8–78′) will demonstrate that, for any value of $\sqrt{mV_0a^2/2\hbar^2}$, the functions of the first class are all of even parity and the functions of the second class are all of odd parity. Furthermore, the continuum standing wave eigenfunctions also form two classes, one of even parity and the other of odd parity. That all the standing wave eigenfunctions have a *definite parity*, either even or odd, is a direct result of the fact that we have so chosen the origin of the x axis that the square well potential $V(x)$ is an even function of x.† For such a choice it is intuitively obvious that any measurable quantity describing the motion of the particle in a bound eigenstate (or in a standing wave continuum eigenstate) will be an even function of x. We note that this is true for the probability density function $P(x, t)$ for both even parity and odd parity eigenstates since

$$P(-x, t) = \psi^*(-x)\,\psi(-x) = [\pm\psi^*(x)][\pm\psi(x)] = \psi^*(x)\,\psi(x) = P(x, t) \tag{8–86}$$

This is not true for the wave function itself in the case of an odd parity eigenstate; such a wave function is an odd function of x. Of course, this is not a contradiction as the wave function is not measurable. The traveling wave eigenfunctions do not have a definite parity since they are linear combinations of standing wave eigenfunctions of opposite parities. For a one dimensional problem the fact that the standing wave eigenfunctions have a definite parity, if $V(-x) = V(x)$, is of importance largely because it leads to a simplification in the mathematics of certain quantum mechanical calculations. But in a three dimensional problem this property takes on a deeper significance that will become apparent later.

The probability densities multiplied by a, for the three bound eigenstates of $\sqrt{mV_0a^2/2\hbar^2} = 4$, are plotted in figure (8–19) as a function of $x/(a/2)$. Also illustrated in the figure is the normalized probability density which would be predicted classically for a bound particle bouncing back and forth between $-a/2$ and $+a/2$. Since classically the particle would spend an equal amount of time in any element of the x axis in that region, it would be equally likely to be found in any such element. The quantum mechanical probability densities show that, in each eigenstate, there is a finite probability of finding the particle in the classically excluded region. This effect is largest for $n = 3$, which corresponds to the smallest value of $(V_0 - E_n)$. Also there are points in the classically allowed region at which quantum mechanical probability densities show that there is zero probability of finding the particle.

For a potential in which $\sqrt{mV_0a^2/2\hbar^2}$ is large enough that there are many bound eigenstates, the quantum mechanical probability density

† In figure (8–18), if we redefine the origin of the x axis to be, for instance, at the point $x = -a/2$, it becomes apparent that the eigenfunctions will no longer have a definite parity.

oscillates more and more rapidly with increasing values of n. For such a potential, consider $a\,\psi_n^*(x)\,\psi_n(x)$, where n is large but not so large as to violate the condition

$$\sqrt{mV_0a^2/2\hbar^2 - mE_na^2/2\hbar^2} = K_{\rm II}a/2 \gg 1 \tag{8–87}$$

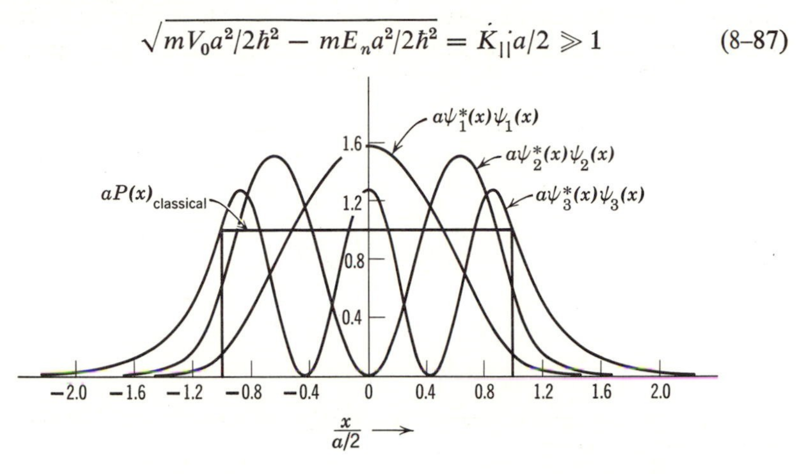

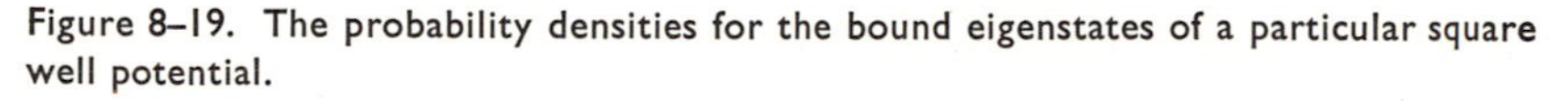

Figure 8–19. The probability densities for the bound eigenstates of a particular square well potential.

The probability density has the form indicated in figure (8–20). The function $a\,\psi_n^*(x)\,\psi_n(x)$ oscillates rapidly, and because of condition (8–87) there is a negligible chance of finding the particle in the classically excluded

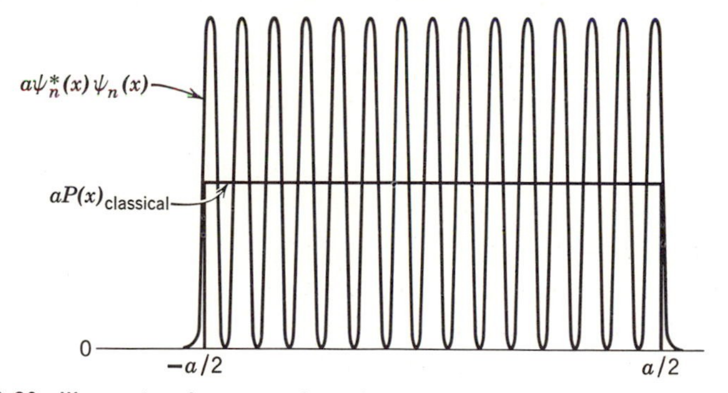

Figure 8–20. Illustrating the approach to the classical limit of the probability density for a square well potential.

region. In the limit $n \to \infty$, the oscillations in $a\psi_n^*(x)\,\psi_n(x)$ are so rapid that only the average behavior can be observed experimentally and the results of the quantum mechanical theory approach those of classical mechanics.

It is interesting to note that in this case the classical limit of the quantum mechanical theory is found in the limit of large quantum numbers, as predicted by the correspondence principle of the old quantum theory. However, it should also be noted that it is necessary to apply the restriction (8–87), which was not anticipated by the correspondence principle. For

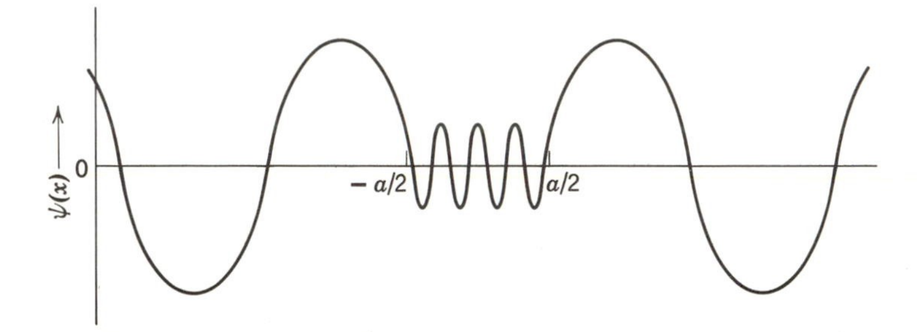

Figure 8–21. An even parity unbound eigenfunction for a square well potential and a typical value of total energy.

unbound eigenstates, we have seen that the approach to the classical limit has little relation to the technique of letting $n \to \infty$.

Finally, let us consider the case:

$E > V_0$ (standing waves)

Any value of total energy $E > V_0$ is allowed, but the features of the standing wave eigenfunctions are not the same for all values of E. We shall not bother actually to solve the time independent Schroedinger equation for this case (it is easier to do than for $E < V_0$) but shall only present certain qualitative arguments and results.

In each of the three regions of the x axis the eigenfunctions are sinusoidal standing waves. In the region $-a/2 < x < +a/2$, the momentum of the particle is $p_{\text{I}} = \sqrt{2mE}$, so the wavelength of the eigenfunctions is $\lambda_{\text{I}} = h/\sqrt{2mE}$. In the regions $x < -a/2$ and $x > +a/2$, the momentum is $p_{\text{II}} = \sqrt{2m(E - V_0)}$, and the wavelength is $\lambda_{\text{II}} = h/\sqrt{2m(E - V_0)}$. Figure (8–21) depicts the qualitative form of an even parity eigenfunction for a typical value of E. It is easy to see that, because $\lambda_{\text{II}} > \lambda_{\text{I}}$, it is possible to join smoothly the sinusoidal solutions of the three regions only if the amplitude inside the region $-a/2 < x < +a/2$ is less than it is outside. However, if the value of E is such that exactly an integral number of wavelengths fit in the region $-a/2$ to $+a/2$, the slope of the interior solution is zero at the boundaries and the interior and exterior solutions have the same amplitude. This is shown in figure (8–22), which depicts the eigenfunction for a value of E slightly larger than for the previous figure. For odd parity eigenfunctions, the amplitude in the interior region is less than the amplitude in the exterior regions except for values of E such that exactly a half-integral number of wavelengths fit in the region. Considering both classes of eigenfunctions, we see that the probability

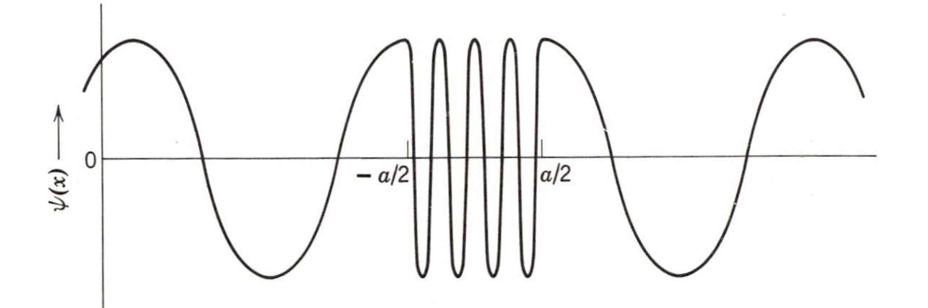

Figure 8–22. An even parity unbound eigenfunction for a square well potential when the total energy is at resonance.

of finding the particle in the interior region, which is equal to

$$\int_{-a/2}^{a/2} \psi^*(x)\,\psi(x)\,dx$$

is generally less than the probability of finding it in an exterior region of the same length. This is analogous to the classical statement that the particle moves faster in the interior region and so it would be less likely to find the particle in that region. However, when the length of the region is equal to an integral or half-integral number of wavelengths in the region, the probability of finding it in the interior region equals that of finding it in a comparable exterior region. This is precisely the condition (which arose in the discussion of the traveling wave eigenfunctions) that the probability T will be unity for the transmission of the particle from the region $x < -a/2$ to the region $x > +a/2$. These so-called *resonances* lead to some interesting situations in atomic and nuclear physics.

5. The Infinite Square Well

An important special case of a square well potential is found in the limit $V_0 \rightarrow \infty$, which is illustrated in figure (8–23). This potential is

$$V(x) = \begin{matrix} 0, & -a/2 < x < a/2 \\ \infty, & x < -a/2, \quad x > a/2 \end{matrix} \tag{8–88}$$

We shall see that it is easy to find simple closed-form expressions for the eigenvalues and eigenfunctions of such a potential. For small values of the quantum number n, these eigenvalues and eigenfunctions can often be used to approximate the corresponding (same n) eigenvalues and eigenfunctions of a square well potential with large but finite V_0. An infinite

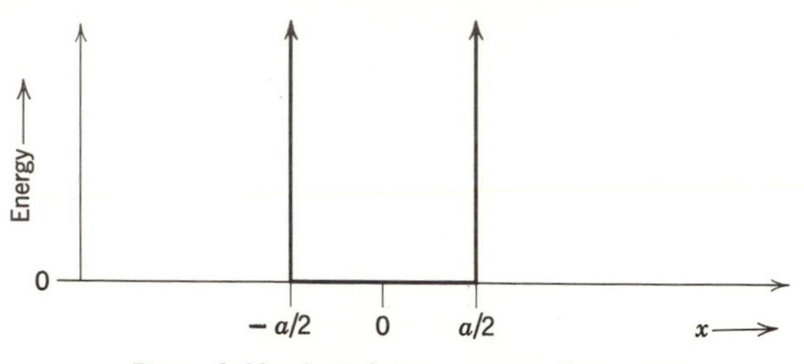

Figure 8–23. An infinite square well potential.

square well potential is also used in discussing the quantum mechanical properties of a system of gas molecules (or other particles) which are strictly confined within a box of certain dimensions. In addition, such a potential is used in the technique for the impenetrable walled box normalization of free particle eigenfunctions.

Let us consider the effect on the eigenfunctions of letting $V_0 \to \infty$. Shown in figure (8–24) is the eigenfunction for $n = 1$ in a finite square well potential. As $V_0 \to \infty$, E_1 will change; but it will change very slowly compared to the change in V_0. Thus the coefficient of x in the exponential will become infinitely large and the value of $\psi_1(x)$ will decrease infinitely

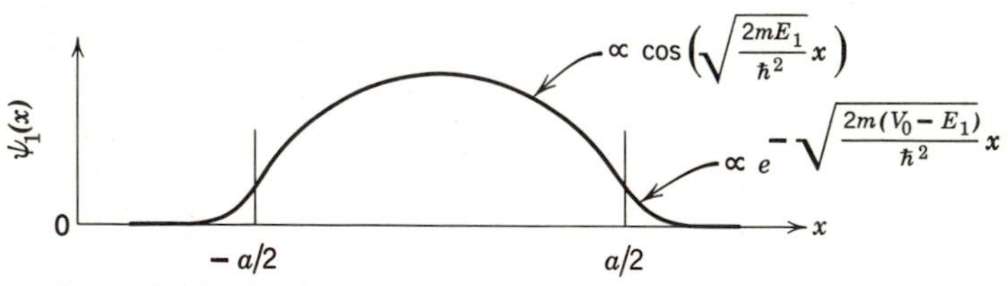

Figure 8–24. The first eigenfunction for a finite square well potential.

rapidly with increasing x for $x > a/2$. In the limit, $\psi_1(x)$ must be zero for all $x > a/2$. Then the requirement that an eigenfunction be everywhere continuous demands that $\psi_1(x) = 0$ at $x = a/2$. For an infinite square well potential, $\psi_1(x)$ must have the form shown in figure (8–25). It is apparent that this argument holds for all the eigenfunctions. That is,

$$\psi_n(x) = 0, \qquad x \leqslant -a/2, \qquad x \geqslant a/2 \tag{8–89}$$

for all n. The condition (8–89) can only be satisfied by violating the requirement of (7–38) that the derivative $d\psi_n(x)/dx$ of an eigenfunction be continuous everywhere. However, if the reader will review the argument which was presented to justify this requirement, he will find that $d\psi_n(x)/dx$ must be continuous only when V_0 is finite.

For the infinite square well potential, we must solve the time independent Schroedinger equation

$$\frac{d^2\psi_n(x)}{dx^2} + \frac{2m}{\hbar^2} E_n\psi_n(x) = 0 \tag{8–90}$$

subject to the condition (8–89). Comparing with the differential equation

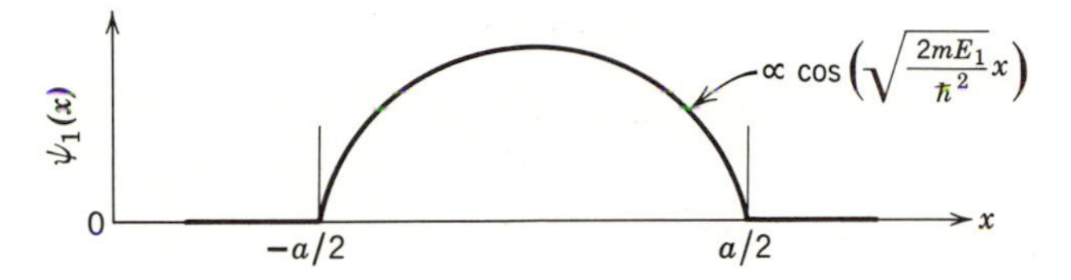

Figure 8–25. The first eigenfunction for an infinite square well potential.

(7–60) and the condition (7–63), we see that the present problem is mathematically identical with the problem of a vibrating string fixed at both ends. The solution of the present problem could be obtained directly from equation (7–65) by making the appropriate substitutions. It could also be obtained from the equations of the preceding section by letting $V_0 \to \infty$. However, we shall solve the problem again in order to illustrate how the knowledge that the eigenfunctions must have a definite parity, because $V(-x) = V(x)$, can be used to simplify the calculation.

From equation (8–16′) the standing wave form of the general solution to (8–90) is

$$\psi_n(x) = A_n \sin(K_n x) + B_n \cos(K_n x), \qquad -a/2 \leqslant x \leqslant a/2$$

where

$$K_n = \sqrt{2mE_n}/\hbar$$

Since $\cos(K_n x)$ is an even function of x and $\sin(K_n x)$ is an odd function of x, the even parity eigenfunctions are

$$\psi_n(x) = B_n \cos(K_n x), \qquad -a/2 \leqslant x \leqslant a/2 \tag{8–91}$$

and the odd parity eigenfunctions are

$$\psi_n(x) = A_n \sin(K_n x), \qquad -a/2 \leqslant x \leqslant a/2 \tag{8–92}$$

In both cases the values of K_n must be such as to satisfy condition (8–89). Because of the evenness or oddness of the eigenfunctions, it is only necessary to apply condition (8–89) at $x = +a/2$. From equation (8–91) we have

$$0 = B_n \cos(K_n a/2)$$

This demands

$$K_n a/2 = \pi/2, 3\pi/2, 5\pi/2, \ldots$$

which is

$$K_n = n\pi/a, \qquad n = 1, 3, 5, \ldots \tag{8–93}$$

Thus the even parity eigenfunctions are those for quantum numbers $n = 1, 3, 5, \ldots$. Next apply condition (8–89) at $x = +a/2$ to (8–92) and obtain

$$0 = A_n \sin(K_n a/2)$$

This demands

$$K_n a/2 = \pi, 2\pi, 3\pi, \ldots$$

which is

$$K_n = n\pi/a, \qquad n = 2, 4, 6, \ldots\dagger \tag{8–94}$$

The odd parity eigenfunctions are those for quantum numbers $n = 2, 4, 6, \ldots$. To normalize the eigenfunctions, evaluate the integral

$$\int_{-\infty}^{\infty} \psi_n^*(x)\, \psi_n(x)\, dx$$

For even parity eigenfunctions this is

$$B_n^2 \int_{-a/2}^{a/2} \cos^2(K_n x)\, dx = B_n^2 \frac{a}{2}$$

Since this must equal unity, we have

$$B_n = \sqrt{2/a}$$

Similarly it is found that

$$A_n = \sqrt{2/a}$$

Thus the normalized eigenfunctions are

$$\psi_n(x) = \begin{cases} \sqrt{2/a}\cos(n\pi x/a), & n = 1, 3, 5, \ldots, \quad -a/2 \leqslant x \leqslant a/2 \\ \sqrt{2/a}\sin(n\pi x/a), & n = 2, 4, 6, \ldots, \quad -a/2 \leqslant x \leqslant a/2 \\ 0, & x \leqslant -a/2, \quad x \geqslant a/2 \end{cases} \tag{8–95}$$

† The value $n = 0$ satisfies condition (8–89), but it leads to the physically uninteresting solution $\psi_0(x) = 0$ for all x.

It is hardly necessary to plot $\psi_n(x)$ or $\psi_n^*(x)\,\psi_n(x)$. Except for the fact that there is no penetration of the classically excluded region for an infinite square well potential, the eigenfunctions and probability densities are similar to those for a finite square well potential. The approach to the classical limit as $n \rightarrow \infty$ is the same as for a finite square well except that there is no auxiliary condition on n comparable to (8–87).

From equations (8–93) and (8–94), the eigenvalues are

$$E_n = K_n^2\hbar^2/2m = \pi^2\hbar^2 n^2/2ma^2, \qquad n = 1, 2, 3, 4, 5, \ldots \qquad (8\text{–}96)$$

The potential $V(x)$ and the first few eigenvalues are illustrated in figure (8–26). Of course, all the eigenvalues are discrete for an infinite square

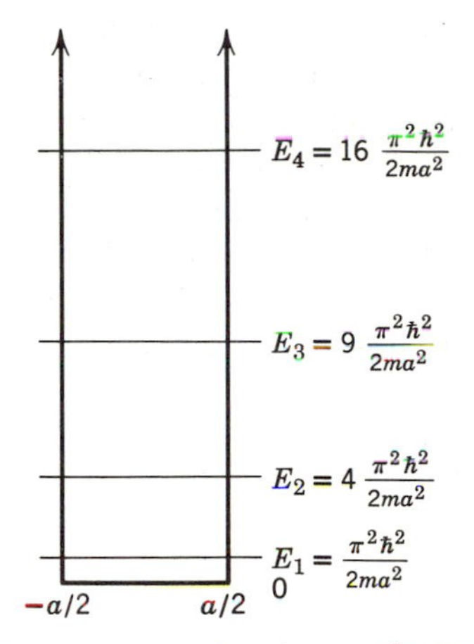

Figure 8–26. The first few eigenvalues for an infinite square well potential.

well potential since the particle is bound for any finite eigenvalue. Of particular interest is the first eigenvalue,

$$E_1 = \pi^2\hbar^2/2ma^2 \qquad (8\text{–}97)$$

This is called the *zero point energy*. It is the lowest possible total energy the particle can have if it is bound by the infinite square well potential to the region $-a/2 \leqslant x \leqslant +a/2$. The particle *cannot* have zero total energy. This is a manifestation of the uncertainty principle. If the particle is bound by the potential, then we know its x coordinate to within an uncertainty $\Delta x = a$. Consequently the uncertainty in its x momentum must be $\Delta p_x \gtrsim \hbar/\Delta x = \hbar/a$. The uncertainty principle cannot allow the particle to be bound by the potential with zero total energy since that would mean the uncertainty in its x momentum would be zero. For the eigenstate E_1, the magnitude of the momentum is $p_1 = \sqrt{2mE_1} = \pi\hbar/a$.

Since the particle whose motion is described by the standing wave eigenfunction can be moving in either direction, the actual value of the x momentum is uncertain by an amount $\Delta p_x = 2p_1 = 2\pi\hbar/a$. This is in agreement with the lower limit set by the uncertainty principle. The zero point energy is responsible for several interesting phenomena, particularly in connection with the behavior of matter at very low temperatures.

6. The Simple Harmonic Oscillator

We have discussed most of the interesting cases of potentials which are discontinuous functions of position with constant values in adjacent regions. Let us now consider potentials which are continuous functions of position. It turns out that there are only a limited number of such potentials for which it is possible to obtain a solution to the Schroedinger equation in closed form. But, fortunately, these potentials include some of the most important cases: for example, the Coulomb potential $V(r) \propto r^{-1}$, which will be discussed in Chapter 10, and the simple harmonic oscillator potential $V(x) \propto x^2$, which is discussed in this section. Techniques for handling the more recalcitrant continuous potentials will be presented in the next chapter.

The simple harmonic oscillator potential is of great importance because it can be used to describe the broad class of problems in which a particle is executing small vibrations about a position of stable equilibrium. At a position of stable equilibrium, $V(x)$ must have a minimum. For any $V(x)$ which is continuous, the shape of the curve in the region near the minimum, which is all that is of interest for small vibrations, can be well approximated by a parabola. Thus, if we choose the zeros of the x axis and the energy axis to be at the minimum, we can write

$$V(x) = Cx^2/2 \tag{8–98}$$

where C is a constant. This potential is illustrated in figure (8–27). A particle moving under the influence of such a potential experiences a simple harmonic restoring force $F = -\partial V/\partial x = -Cx$. The theory of classical mechanics predicts that, if the particle is displaced by an amount x_0 from the equilibrium position, and then let go, it will oscillate sinusoidally about the equilibrium position with frequency

$$\nu = \frac{1}{2\pi}\sqrt{\frac{C}{m}} \tag{8–99}$$

where m is the mass of the particle. According to that theory, the total energy E of the particle is proportional to $x_0{}^2$ and can have any value, since x_0 is arbitrary. However, the old quantum theory (Planck's postulate) predicts that E can have only one of the values:

$$E_n = nh\nu, \qquad n = 0, 1, 2, 3, \ldots \tag{8–100}$$

What are the predictions of quantum mechanics?

Let us specify the simple harmonic oscillator potential (8–98) in terms of the classical oscillation frequency (8–99). From these two equations we have

$$V(x) = 2\pi^2 m\nu^2 x^2 \tag{8–101}$$

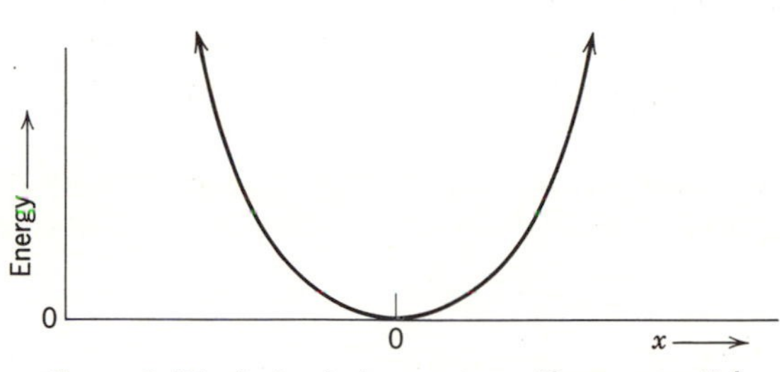

Figure 8–27. A simple harmonic oscillator potential.

The corresponding time independent Schroedinger equation is

$$-\frac{\hbar^2}{2m}\frac{d^2\psi}{dx^2} + 2\pi^2 m\nu^2 x^2\psi = E\psi$$

or

$$\frac{d^2\psi}{dx^2} + \left[\frac{2mE}{\hbar^2} - \left(\frac{2\pi m\nu}{\hbar}\right)^2 x^2\right]\psi = 0 \tag{8–102}$$

Define

$$\alpha \equiv 2\pi m\nu/\hbar, \qquad \beta \equiv 2mE/\hbar^2 \tag{8–103}$$

Then the equation can be written

$$d^2\psi/dx^2 + (\beta - \alpha^2x^2)\psi = 0 \tag{8–104}$$

It is convenient to express this in terms of a new variable,

$$\xi \equiv \sqrt{\alpha}\, x \tag{8–105}$$

Then

$$\frac{d\psi}{dx} = \frac{d\psi}{d\xi}\frac{d\xi}{dx} = \sqrt{\alpha}\,\frac{d\psi}{d\xi}$$

$$\frac{d^2\psi}{dx^2} = \frac{d}{d\xi}\left(\frac{d\psi}{dx}\right)\frac{d\xi}{dx} = \alpha\,\frac{d^2\psi}{d\xi^2}$$

and the time independent Schroedinger equation becomes

$$\alpha \frac{d^2\psi}{d\xi^2} + (\beta - \alpha\xi^2)\psi = 0$$

or

$$\frac{d^2\psi}{d\xi^2} + \left(\frac{\beta}{\alpha} - \xi^2\right)\psi = 0 \tag{8–106}$$

We must find solutions $\psi(\xi)$ which are continuous, and finite, for all ξ from $-\infty$ to $+\infty$. The first condition will be automatically satisfied by the solutions we shall obtain. However, it will be necessary to take explicit consideration of the requirement that $\psi(\xi)$ remain finite as $|\xi| \to \infty$. For this purpose it is useful first to consider the form of $\psi(\xi)$ for very large values of $|\xi|$.

For any finite value of the total energy E, the quantity β/α becomes negligible compared to ξ^2 for very large values of $|\xi|$. Thus we may write

$$d^2\psi/d\xi^2 = \xi^2\psi, \qquad |\xi| \to \infty \tag{8–107}$$

The general solution to this equation is

$$\psi = Ae^{\xi^2/2} + Be^{-\xi^2/2} \tag{8–108}$$

To check the solution, calculate

$$\begin{aligned} d\psi/d\xi &= A\xi e^{\xi^2/2} + B(-\xi)e^{-\xi^2/2} \\ d^2\psi/d\xi^2 &= A\xi^2 e^{\xi^2/2} + Ae^{\xi^2/2} + B\xi^2 e^{-\xi^2/2} - Be^{-\xi^2/2} \\ &= A(\xi^2 + 1)e^{\xi^2/2} + B(\xi^2 - 1)e^{-\xi^2/2} \end{aligned}$$

But for $|\xi| \to \infty$ this is essentially

$$d^2\psi/d\xi^2 = A\xi^2 e^{\xi^2/2} + B\xi^2 e^{-\xi^2/2}$$

or

$$d^2\psi/d\xi^2 = \xi^2(Ae^{\xi^2/2} + Be^{-\xi^2/2}) \equiv \xi^2\psi; \quad \text{Q.E.D.}$$

Now apply the condition that the eigenfunctions must remain finite as $|\xi| \to \infty$. It is apparent from equation (8–108) that this demands we set

$$A = 0$$

Thus the form of the eigenfunctions for very large $|\xi|$ must be

$$\psi(\xi) = Be^{-\xi^2/2}, \qquad |\xi| \to \infty \tag{8–109}$$

This suggests looking for solutions to equation (8–106), which will be valid for all ξ, in the form

$$\psi(\xi) = e^{-\xi^2/2}H(\xi) \tag{8–110}$$

The $H(\xi)$ are functions which must, at $|\xi| \rightarrow \infty$, be slowly varying compared to $e^{-\xi^2/2}$. To evaluate these functions, calculate

$$\frac{d\psi}{d\xi} = -\xi e^{-\xi^2/2}H + e^{-\xi^2/2}\frac{dH}{d\xi}$$

$$\frac{d^2\psi}{d\xi^2} = -e^{-\xi^2/2}H + \xi^2 e^{-\xi^2/2}H - \xi e^{-\xi^2/2}\frac{dH}{d\xi} - \xi e^{-\xi^2/2}\frac{dH}{d\xi} + e^{-\xi^2/2}\frac{d^2H}{d\xi^2}$$

$$\frac{d^2\psi}{d\xi^2} = e^{-\xi^2/2}\left[-H + \xi^2 H - 2\xi\frac{dH}{d\xi} + \frac{d^2H}{d\xi^2}\right]$$

Then substitute ψ and $d^2\psi/d\xi^2$ into equation (8–106). This gives

$$e^{-\xi^2/2}\left[-H + \xi^2 H - 2\xi\frac{dH}{d\xi} + \frac{d^2H}{d\xi^2}\right] + \frac{\beta}{\alpha}e^{-\xi^2/2}H - \xi^2 e^{-\xi^2/2}H = 0$$

Dividing by $e^{-\xi^2/2}$, and canceling the terms involving $\xi^2 H$, we have

$$\frac{d^2H}{d\xi^2} - 2\xi\frac{dH}{d\xi} + \left(\frac{\beta}{\alpha} - 1\right)H = 0 \qquad \text{(8–111)}$$

This equation determines the functions $H(\xi)$.

Let us recapitulate. We started with the time independent Schroedinger equation (8–106). As it stands, this equation cannot be solved in closed form. However, by writing the solutions to the equation as products of the function $e^{-\xi^2/2}$, which is the form of the solution for $|\xi| \rightarrow \infty$, times the functions $H(\xi)$, we transform the problem to one of solving equation (8–111). This equation is soluable in closed form. In fact, it is identical with an equation called the *Hermite* differential equation whose solutions are functions which are well known in the mathematical literature. However, we shall solve it ourselves.

To do this we must make use of the most general procedure available for the purpose of obtaining an explicit solution to a differential equation. This consists in assuming a solution in the form of a power series in the independent variable. That is, we assume

$$H(\xi) = \sum_{k=0}^{\infty} a_k \xi^k \equiv a_0 + a_1\xi + a_2\xi^2 + a_3\xi^3 + \cdots \qquad \text{(8–112)}$$

The coefficients a_0, a_1, a_2, . . . are then determined by substituting equation (8–112) into (8–111) and demanding that the resulting equation be satisfied for any value of ξ. Calculate

$$\frac{dH}{d\xi} = \sum_{k=1}^{\infty} k a_k \xi^{k-1} \equiv 1a_1 + 2a_2\xi + 3a_3\xi^2 + \cdots$$

and

$$\frac{d^2H}{d\xi^2} = \sum_{k=2}^{\infty}(k-1)ka_k\xi^{k-2} \equiv 1\cdot 2a_2 + 2\cdot 3a_3\xi + 3\cdot 4a_4\xi^2 + \cdots$$

Then substitute into the differential equation. This gives

$$\begin{aligned}
&1\cdot 2a_2 + 2\cdot 3a_3\xi + 3\cdot 4a_4\xi^2 + 4\cdot 5a_5\xi^3 + \cdots \\
&\quad - 2\cdot 1a_1\xi - 2\cdot 2a_2\xi^2 - 2\cdot 3a_3\xi^3 - \cdots \\
&+ (\beta/\alpha - 1)a_0 + (\beta/\alpha - 1)a_1\xi + (\beta/\alpha - 1)a_2\xi^2 + (\beta/\alpha - 1)a_3\xi^3 + \cdots = 0
\end{aligned}$$

Since this is to be true for all values of ξ, the coefficients of each power of ξ must vanish individually so that the validity of the equation will not depend on the value of ξ. Gathering the coefficients together and equating them to zero, we have

$$\begin{aligned}
&\xi^0: && 1\cdot 2a_2 + (\beta/\alpha - 1)a_0 = 0 \\
&\xi^1: && 2\cdot 3a_3 + (\beta/\alpha - 1 - 2\cdot 1)a_1 = 0 \\
&\xi^2: && 3\cdot 4a_4 + (\beta/\alpha - 1 - 2\cdot 2)a_2 = 0 \\
&\xi^3: && 4\cdot 5a_5 + (\beta/\alpha - 1 - 2\cdot 3)a_3 = 0
\end{aligned}$$

For the kth power of ξ the relation is

$$\xi^k: \qquad (k+1)(k+2)a_{k+2} + \left(\frac{\beta}{\alpha} - 1 - 2k\right)a_k = 0$$

so

$$a_{k+2} = -\frac{\beta/\alpha - 1 - 2k}{(k+1)(k+2)}\,a_k \tag{8–113}$$

This is called the *recursion relation*. It allows us to calculate successively the coefficients $a_2, a_4, a_6, \ldots$ in terms of a_0, and the coefficients $a_3, a_5, a_7, \ldots$ in terms of a_1. The coefficients a_0 and a_1 are not specified by equation (8–113). But this is as it should be; the second order differential equation for $H(\xi)$ should have a general solution containing two arbitrary constants. Thus the general solution splits up into two independent series, which we write as

$$\begin{aligned}
H(\xi) = a_0&\left(1 + \frac{a_2}{a_0}\xi^2 + \frac{a_4}{a_2}\frac{a_2}{a_0}\xi^4 + \frac{a_6}{a_4}\frac{a_4}{a_2}\frac{a_2}{a_0}\xi^6 + \cdots\right) \\
&+ a_1\left(\xi + \frac{a_3}{a_1}\xi^3 + \frac{a_5}{a_3}\frac{a_3}{a_1}\xi^5 + \frac{a_7}{a_5}\frac{a_5}{a_3}\frac{a_3}{a_1}\xi^7 + \cdots\right)
\end{aligned} \tag{8–114}$$

The ratios a_{k+2}/a_k are given by the recursion relation. The first series is an even function of ξ, and the second series is an odd function of that variable.

For an arbitrary value of β/α, both the even and the odd series will contain an infinite number of terms. As we shall see, this will not lead to acceptable eigenfunctions. Consider either series and evaluate the ratio of the coefficients of successive powers of ξ for large k. This is

$$\frac{a_{k+2}}{a_k} = -\frac{(\beta/\alpha - 1 - 2k)}{(k+1)(k+2)} \simeq \frac{2k}{k^2} = \frac{2}{k}$$

Let us compare this with the ratio for the series expansion of the function e^{ξ^2}, which is

$$e^{\xi^2} = 1 + \xi^2 + \frac{\xi^4}{2!} + \frac{\xi^6}{3!} + \cdots + \frac{\xi^k}{(k/2)!} + \frac{\xi^{k+2}}{(k/2+1)!} + \cdots$$

For large k, the ratio of the coefficients of successive powers of ξ is

$$\frac{1/(k/2+1)!}{1/(k/2)!} = \frac{(k/2)!}{(k/2+1)!} = \frac{(k/2)!}{(k/2+1)(k/2)!} = \frac{1}{k/2+1} \simeq \frac{1}{k/2} = \frac{2}{k}$$

The two ratios are the same. This means that the terms of high power in ξ in the series for e^{ξ^2} can only differ from the corresponding terms in the even series of $H(\xi)$ by a constant K. They can only differ from the terms in the odd series of $H(\xi)$ by ξ times another constant K'. But, for $|\xi| \to \infty$, the terms of low power in ξ are not important in determining the value of any of these series. Consequently we conclude that

$$H(\xi) = a_0 K e^{\xi^2} + a_1 K' \xi e^{\xi^2}, \qquad |\xi| \to \infty$$

Now, according to equation (8–110), the solutions to the time independent Schroedinger equation are

$$\psi(\xi) = e^{-\xi^2/2} H(\xi)$$

Thus, if the series of $H(\xi)$ contain an infinite number of terms, the behavior of these solutions for $|\xi| \to \infty$ is

$$e^{-\xi^2/2} H(\xi) = a_0 K e^{\xi^2/2} + a_1 K' \xi e^{\xi^2/2}, \qquad |\xi| \to \infty$$

But this diverges as $|\xi| \to \infty$, which is not acceptable behavior for an eigenfunction.

However, acceptable eigenfunctions can be obtained for certain values of β/α. Set *either* the arbitrary constant a_0 *or* the arbitrary constant a_1 equal to zero. Force the remaining series of $H(\xi)$ to *terminate* by setting

$$\beta/\alpha = 2n + 1 \tag{8–115}$$

where

$$n = 1, 3, 5, \ldots \quad \text{if } a_0 = 0$$
$$n = 0, 2, 4, 6, \ldots \quad \text{if } a_1 = 0$$

It is clear from equation (8–113) that such a choice of β/α will cause the series to terminate at the nth term since we shall have for $k = n$

$$a_{n+2} = -\frac{(\beta/\alpha - 1 - 2n)}{(n+1)(n+2)} a_n = -\frac{(2n + 1 - 1 - 2n)}{(n+1)(n+2)} a_n = 0 \cdot a_n = 0$$

The coefficients $a_{n+4}, a_{n+6}, a_{n+8}, \ldots$ will also be zero since they are proportional to a_{n+2}. The resulting solutions $H_n(\xi)$ are polynomials of order n, called *Hermite polynomials.* For polynomial solutions to the Hermite equation, the corresponding eigenfunctions

$$\psi(\xi) = e^{-\xi^2/2} H_n(\xi) \tag{8–116}$$

will always have the acceptable behavior of going to zero as $|\xi| \to \infty$, because, for large $|\xi|$, the function $e^{-\xi^2/2}$ varies so much more rapidly than the polynomial $H_n(\xi)$ that it completely dominates the behavior of the eigenfunctions. In fact,

$$\psi(\xi) \propto e^{-\xi^2/2}, \qquad |\xi| \to \infty$$

in agreement with equation (8–109).

Acceptable solutions to the Schroedinger equation exist only for the values of β/α specified by equation (8–115). According to (8–103), this quantity is

$$\frac{\beta}{\alpha} = \frac{2mE}{\hbar^2} \frac{\hbar}{2\pi m\nu} = \frac{2E}{h\nu}$$

Thus equation (8–115) specifies the allowed values of the total energy E, which are

$$2E/h\nu = 2n + 1$$

or

$$E_n = (n + \tfrac{1}{2})h\nu, \qquad n = 0, 1, 2, 3, \ldots \tag{8–117}$$

where ν is the classical oscillation frequency of the particle, and where we have used the quantum number n to label the eigenvalues. The potential and the eigenvalues are illustrated in figure (8–28). All eigenvalues are discrete for the simple harmonic oscillator potential, since the particle is bound for any finite eigenvalue. Comparing equation (8–117) with the energy quantization equation postulated by Planck (8–100), we see that in quantum mechanics all the eigenvalues are shifted up by an amount $\frac{1}{2}h\nu$. As a consequence, the minimum possible total energy for a particle bound to the potential is $E_0 = \frac{1}{2}h\nu$. This is the zero point energy for this potential, the existence of which is required by the uncertainty principle. It is interesting to note that Planck's postulated quantization of the simple harmonic oscillator was actually in error by the additive constant $\frac{1}{2}h\nu$. This constant canceled out in most applications of the Planck postulate because they involved only calculating the difference between two eigenvalues (cf. Einstein's arguments leading to the idea of quanta). However, discrepancies were noticed when attempts were made to explain certain low temperature phenomena in terms of the old quantum theory.

Each Hermite polynomial $H_n(\xi)$ can be evaluated from equation (8–114)

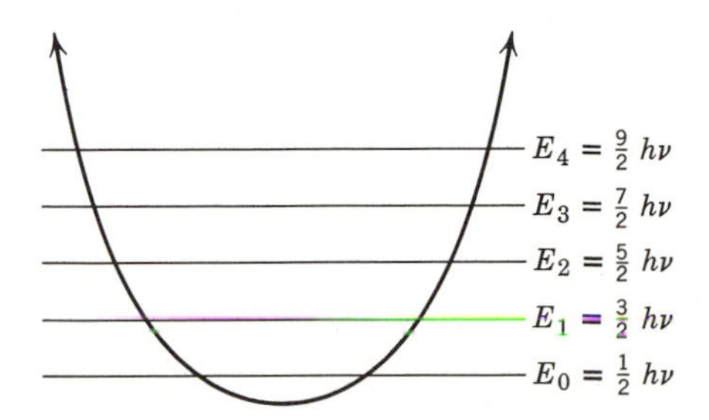

Figure 8–28. The first few eigenvalues for a simple harmonic oscillator potential.

by calculating the coefficients from (8–113) with β/α given by (8–115) for that value of n. The first few are listed below.

$$
\begin{aligned}
H_0(\xi) &= 1 \\
H_1(\xi) &= 2\xi \\
H_2(\xi) &= 2 - 4\xi^2 \\
H_3(\xi) &= 12\xi - 8\xi^3 \\
H_4(\xi) &= 12 - 48\xi^2 + 16\xi^4 \\
H_5(\xi) &= 120\xi - 160\xi^3 + 32\xi^5 \\
H_6(\xi) &= 120 - 720\xi^2 + 480\xi^4 - 64\xi^6
\end{aligned}
\tag{8–118}
$$

In each case the arbitrary constant a_0 or a_1 has been so chosen that the normalized eigenfunctions are given by

$$
\psi_n(\xi) = \left(\frac{1}{\sqrt{\pi}\,2^n n!}\right)^{1/2} e^{-\xi^2/2} H_n(\xi) \tag{8–119}
$$

The eigenfunctions can be written in terms of x by making the substitution $\xi = \sqrt{\alpha}\,x$, but it is usually more convenient to leave them in terms of ξ. Since $e^{-\xi^2/2}$ is an even function, the parity of the eigenfunctions is the same as that of the Hermite polynomials: even for even n, and odd for odd n. The eigenfunction corresponding to the eigenvalue of lowest energy is of even parity, just as in the case of a square potential well.

The first few eigenfunctions are plotted in figure (8–29). The vertical lines indicate the limits of the classically allowed region in each case. In figure (8–30) we show the quantum mechanical probability density for $n = 10$, and also the probability density which would be predicted classically for a

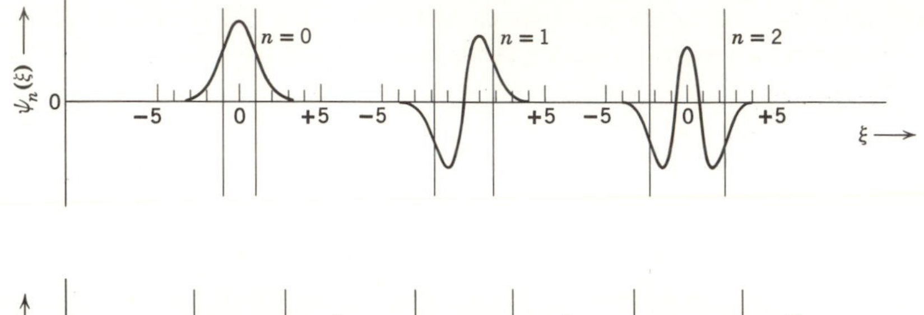

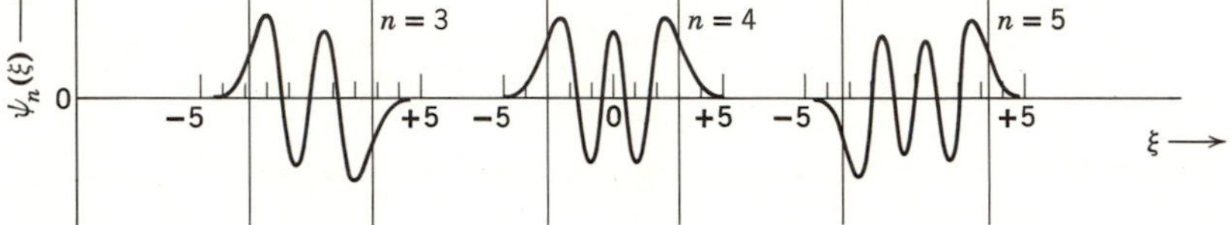

Figure 8–29. The first few eigenfunctions for a simple harmonic oscillator potential.

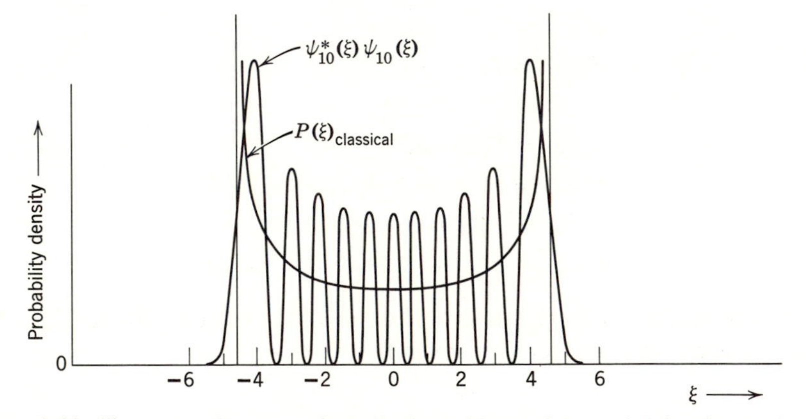

Figure 8–30. Illustrating the approach to the classical limit of the probability density for a simple harmonic oscillator potential. From L. Pauling and E. B. Wilson, *Introduction to Quantum Mechanics*, McGraw-Hill Book Co., New York, 1935.

particle of total energy equal to E_{10}. The velocity of a particle classically executing simple harmonic oscillations goes to zero at the extremes of its motion. Consequently the particle spends most of the time near the limits of the classically allowed region, and the classical probability of finding the particle in an element of the ξ axis behaves as shown. The figure makes quite apparent the nature of the correspondence between the quantum mechanical and classical probability densities in the limit $n \to \infty$.

To close our discussion of the simple harmonic oscillator, let us calculate the expectation values $\bar{T}$ and $\bar{V}$ of the kinetic and potential energies of a particle in the state $n = 0$. The wave function is

$$\Psi_0(\xi, t) = \psi_0(\xi)e^{-iE_0t/\hbar} = \frac{1}{\pi^{1/4}}\, e^{-\xi^2/2}\, e^{-iE_0t/\hbar}$$

Now

$$T = \frac{p^2}{2m}$$

Thus, according to equation (7–80),

$$\bar{T} = \int_{-\infty}^{\infty} \Psi_0^* \left[\frac{1}{2m}\left(-i\hbar\frac{\partial}{\partial x}\right)^2\right] \Psi_0 \, d\xi$$

Change to a derivative with respect to ξ:

$$\frac{\partial}{\partial x} = \sqrt{\alpha}\,\frac{\partial}{\partial\sqrt{\alpha}x} = \sqrt{\alpha}\,\frac{\partial}{\partial \xi}$$

Then

$$\bar{T} = -\frac{\hbar^2\alpha}{2m}\int_{-\infty}^{\infty} \Psi_0^* \frac{\partial^2 \Psi_0}{\partial \xi^2}\, d\xi$$

The derivative is

$$\frac{\partial^2 \Psi_0}{\partial \xi^2} = \frac{1}{\pi^{1/4}}\,(\xi^2 e^{-\xi^2/2} - e^{-\xi^2/2})e^{-iE_0t/\hbar}$$

The complex conjugate is

$$\Psi_0^* = \frac{1}{\pi^{1/4}}\, e^{-\xi^2/2} e^{+iE_0t/\hbar}$$

So

$$\bar{T} = \frac{\hbar^2\alpha}{2m\sqrt{\pi}}\int_{-\infty}^{\infty} e^{-\xi^2/2}(e^{-\xi^2/2} - \xi^2 e^{-\xi^2/2})\, d\xi$$

$$\bar{T} = \frac{\hbar^2\alpha}{m\sqrt{\pi}}\left[\int_0^{\infty} e^{-\xi^2}\, d\xi - \int_0^{\infty} \xi^2 e^{-\xi^2}\, d\xi\right]$$

The definite integrals can be found in any table. They give

$$\bar{T} = \frac{\hbar^2\alpha}{m\sqrt{\pi}}\left[\frac{\sqrt{\pi}}{2} - \frac{\sqrt{\pi}}{4}\right] = \frac{\hbar^2\alpha}{4m}$$

Evaluating α, we have

$$\bar{T} = \frac{\hbar^2 2\pi m\nu}{4m\hbar} = \frac{h\nu}{4}$$

According to equation (8–117) this is

$$\bar{T} = E_0/2$$

From (8–101), (8–103), and (8–105) we may write the potential energy as

$$V = \hbar\pi\nu\xi^2$$

Then, according to (7–80),

$$\bar{V} = \int_{-\infty}^{\infty} \Psi_0^*[\hbar\pi\nu\xi^2]\Psi_0 \, d\xi$$

$$\bar{V} = \frac{\hbar\pi\nu}{\sqrt{\pi}} \int_{-\infty}^{\infty} e^{-\xi^2/2}\xi^2 e^{-\xi^2/2} \, d\xi$$

$$\bar{V} = 2\nu\hbar\sqrt{\pi}\int_0^{\infty} \xi^2 e^{-\xi^2} \, d\xi = 2\nu\hbar\pi/4 = h\nu/4 = E_0/2$$

It can be shown that the relation

$$\bar{T} = \bar{V} = E_n/2 \qquad (8\text{–}120)$$

is true for all values of the quantum number. Furthermore, the same relation is true classically if $\bar{T}$ and $\bar{V}$ are understood to represent the values of T and V averaged over one period of the oscillation.

BIBLIOGRAPHY

Bohm, D., *Quantum Theory*, Prentice-Hall, Englewood Cliffs, N.J., 1951.

Pauling, L., and E. B. Wilson, *Introduction to Quantum Mechanics*, McGraw-Hill Book Co., New York, 1935.

Schiff, L. I., *Quantum Mechanics*, McGraw-Hill Book Co., New York, 1955.

EXERCISES

1. Derive equation (8–53) and verify (8–54).

2. Derive equation (8–55).

3. Consider the potential

$$V(x) = \begin{cases} 0, & x < 0 \\ V_1, & 0 < x < a \\ V_2, & x > a \end{cases}$$

where

$$0 < V_1 < V_2$$

and a particle of total energy $E > V_2$ approaching $x = 0$ in the direction of increasing x. Show that its probability of continuing into the region $x > a$ is a maximum if a is an integral or half-integral number of de Broglie wavelengths in the region $0 < x < a$.

4. Obtain the transmission factor for a square well potential by redefining the terms of equation (8–55), as suggested near the beginning of section 4.

5. Evaluate quantitatively the standing wave solutions for the square well potential of section 4 in the case $E > V_0$. Use these solutions to verify the qualitative conclusions reached at the end of that section.

6. Obtain an analytical solution to the time independent Schroedinger equation for the potential of exercise 4, Chapter 7, and determine the first eigenvalue E_1. Compare with the results of the numerical integration performed in that exercise.

7. Evaluate several of the $H_n(\xi)$ by using equations (8–113), (8–114), and (8–115). Check these against the forms presented in (8–118), and also verify that several of the eigenfunctions (8–119) are properly normalized.

8. Verify equation (8–120) for $n = 1$.

CHAPTER 9

Perturbation Theory

1. Introduction

The technique which we employed in solving the Schroedinger equation for the simple harmonic oscillator potential will not, in general, be of use in the case of a potential of arbitrary form $V(x)$. What happens is that the recursion relation is found to involve more than two coefficients, making it impossible to find solutions to the differential equation in closed form. In such cases the equation can always be solved by numerical integration in the manner described in section 5, Chapter 7. This method can be tedious if the calculation must be done "manually." But, with the aid of modern high speed computing machines, numerical integration is completely feasible and is widely used. In addition, there are approximation techniques that are very useful for treating certain potentials for which the Schroedinger equation cannot be solved in closed form. The study of one of these techniques forms the subject of *perturbation theory*, to which this chapter is devoted.

2. Time Independent Perturbations

Consider a potential $V'(x)$, for which it is either difficult or impossible to solve the Schroedinger equation in closed form, but which can be decomposed as follows:

$$V'(x) = V(x) + v(x) \tag{9-1}$$

where $V(x)$ is a potential for which the Schroedinger equation has been solved, and where $v(x)$ is a potential that is small compared to $V(x)$. We shall develop expressions from which it will be easy to obtain good approximations to the eigenvalues and eigenfunctions of the *perturbed potential* $V'(x)$, in terms of the *perturbation* $v(x)$ and the known eigenvalues and eigenfunctions of the *unperturbed potential* $V(x)$. An example of equation (9–1) is illustrated in figure (9–1). The potential $V'(x)$ has been decomposed into a square well potential $V(x)$, plus a perturbation $v(x)$ which is small compared to $V(x)$.

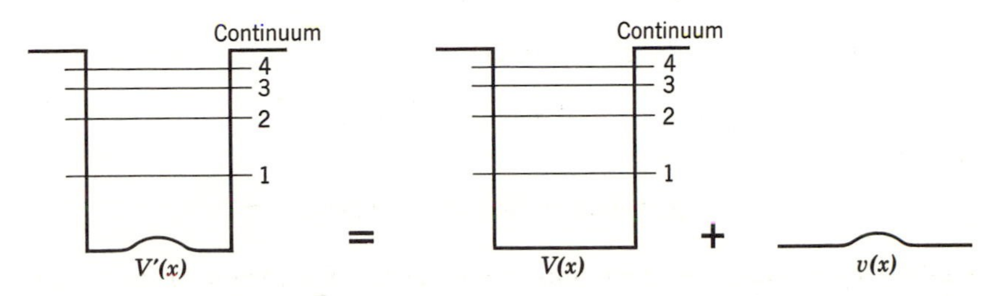

Figure 9–1. Illustrating the decomposition of a perturbed potential into an unperturbed potential plus a perturbation.

On several occasions in Chapter 7 we wrote a function of x as a sum of a certain set of eigenfunctions (cf. equations 7–56 and 7–70). Let us do this for some particular perturbed eigenfunction $\psi_n'(x)$. That is, we write†

$$\psi'_n(x) = \sum_l a_{nl}\psi_l(x) \tag{9–2}$$

The $\psi_l(x)$ are the unperturbed eigenfunctions, and the coefficients a_{nl} specify how much of each of the $\psi_l(x)$ is contained in $\psi_n'(x)$. The summation runs over all the values of the quantum number l, including those in the continuum. The unperturbed eigenfunctions are solutions to the time independent Schroedinger equation for the potential V, which is

$$-\frac{\hbar^2}{2m}\frac{d^2\psi_l}{dx^2} + V\psi_l = E_l\psi_l \tag{9–3}$$

The perturbed eigenfunctions are solutions to the same equation for the potential V', which is

$$-\frac{\hbar^2}{2m}\frac{d^2\psi'_n}{dx^2} + V'\psi'_n = E'_n\psi'_n$$

† It can be shown that every set of eigenfunctions $\psi_l(x)$ forms a "complete set" in terms of which almost any function of x can be expanded (i.e., written as in equation 9–2).

Using equation (9–1) this can be written

$$-\frac{\hbar^2}{2m}\frac{d^2\psi'_n}{dx^2} + V\psi'_n + v\psi'_n = E'_n\psi'_n \tag{9–4}$$

where E_l and E'_n are, respectively, the unperturbed and perturbed eigenvalues. Now substitute equation (9–2) into (9–4):

$$\sum_l a_{nl}\left[-\frac{\hbar^2}{2m}\frac{d^2\psi_l}{dx^2} + V\psi_l\right] + \sum_l a_{nl}v\psi_l = \sum_l a_{nl}E'_n\psi_l$$

According to equation (9–3) the bracket is equal to $E_l\psi_l$. Thus we have

$$\sum_l a_{nl}E_l\psi_l + \sum_l a_{nl}v\psi_l = \sum_l a_{nl}E'_n\psi_l$$

or

$$\sum_l a_{nl}(E'_n - E_l)\psi_l = \sum_l a_{nl}v\psi_l$$

Multiplying through by the complex conjugate of a certain unperturbed eigenfunction ψ_m, and integrating over all x, we have

$$\sum_l a_{nl}(E'_n - E_l)\int_{-\infty}^{\infty}\psi_m^*\psi_l\,dx = \sum_l a_{nl}\int_{-\infty}^{\infty}\psi_m^*v\psi_l\,dx$$

The unperturbed eigenfunctions are, of course, orthogonal. Assuming that they have also been normalized,† equation (7–54) shows that the integral on the left will be equal to zero if $l \neq m$, and equal to unity if $l = m$. Thus there will be only one non-vanishing term in the summation on the left side of the equation, and

$$a_{nm}(E'_n - E_m) = \sum_l a_{nl}\int_{-\infty}^{\infty}\psi_m^*v\psi_l\,dx \tag{9–5}$$

Let us define the symbol

$$v_{ml} \equiv \int_{-\infty}^{\infty}\psi_m^*(x)\,v(x)\,\psi_l(x)\,dx \tag{9–6}$$

Then equation (9–5) can be written

$$a_{nm}(E'_n - E_m) = \sum_l a_{nl}v_{ml} \tag{9–7}$$

This equation is exact, but it is not very useful. In order to obtain one that is useful, we shall employ the condition that the perturbation $v(x)$ is small compared to the unperturbed potential $V(x)$, so that the perturbed

† This involves box normalization for the continuum eigenfunctions. In doing this, the continuum eigenvalues actually become discrete (although very closely spaced). This removes any difficulty of interpreting the summation $\sum_l$ for the continuum values of l.

potential $V'(x)$ differs only slightly from $V(x)$. For such a situation it is reasonable to assume that the perturbed eigenfunctions will differ only slightly from the unperturbed eigenfunctions. In terms of equation (9–2), this means that we *assume*

$$a_{nl} \begin{matrix} \ll 1, & l \neq n \\ \simeq 1, & l = n \end{matrix} \tag{9–8}$$

If we also *require* that $v(x)$ be small compared to the eigenvalues of $V(x)$, it is clear that the v_{ml} must then *all* be small compared to the unperturbed eigenvalues because, according to (9–6), these quantities are just certain averages of $v(x)$. Now let us divide both sides of equation (9–7) by the unperturbed eigenvalue E_m. We have

$$a_{nm} \frac{(E_n' - E_m)}{E_m} = \sum_l a_{nl} \frac{v_{ml}}{E_m}$$

Every term in the summation, except the term $l = n$, is the product of two small quantities a_{nl} and v_{ml}/E_m. We shall neglect such terms, keeping only the term for $l = n$. Then we have

$$a_{nm} \frac{(E_n' - E_m)}{E_m} \simeq a_{nn} \frac{v_{mn}}{E_m}$$

or

$$a_{nm}(E_n' - E_m) \simeq a_{nn} v_{mn} \tag{9–9}$$

Now take $m = n$. We obtain

$$a_{nn}(E_n' - E_n) \simeq a_{nn} v_{nn}$$

so

$$E_n' - E_n \simeq v_{nn} \tag{9–10}$$

If we take $m \neq n$, we obtain

$$a_{nm} \simeq a_{nn} \frac{v_{mn}}{E_n' - E_m}$$

Setting $a_{nn} = 1$ because of (9–8), this becomes

$$a_{nm} \simeq \frac{v_{mn}}{E_n' - E_m} \tag{9–11}$$

Using (9–10) to evaluate E_n', we find that

$$a_{nm} \simeq \frac{v_{mn}}{E_n - E_m + v_{nn}} = \frac{v_{mn}}{(E_n - E_m)\left(1 + \dfrac{v_{nn}}{E_n - E_m}\right)}$$

$$= \frac{v_{mn}}{E_n - E_m}\left(1 + \frac{v_{nn}}{E_n - E_m}\right)^{-1} \simeq \frac{v_{mn}}{E_n - E_m}\left(1 - \frac{v_{nn}}{E_n - E_m}\right)$$

We have taken the first term in the binomial expansion of 1 plus the quantity $v_{nn}/(E_n - E_m)$. Next we shall drop the term involving the product of the two quantities $v_{mn}/(E_n - E_m)$ and $v_{nn}/(E_n - E_m)$. The validity of these two steps depends on the *additional requirement* that $v(x)$ be small even compared to the difference between E_n and any other eigenvalue E_m which enters into our calculations. We have finally

$$a_{nm} \simeq \frac{v_{mn}}{E_n - E_m} \tag{9–12}$$

Equations (9–10) and (9–12) are the expressions which provide good approximations to the eigenvalues and eigenfunctions of the perturbed potential $V'(x)$. Consider (9–10), and evaluate v_{nn} from (9–6). This yields

$$E'_n - E_n \simeq v_{nn} = \int_{-\infty}^{\infty} \psi_n^*(x)\, v(x)\, \psi_n(x)\, dx \tag{9–13}$$

This gives an approximation to the nth perturbed eigenvalue in terms of the nth unperturbed eigenvalue and a certain integral involving the corresponding unperturbed eigenfunction and the perturbation $v(x)$. This integral is the expectation value of $v(x)$ for the nth unperturbed eigenstate. To see this, consider (7–74), with $V(x, t) = v(x)$ and $\Psi(x, t) = \Psi_n(x, t) = e^{-iE_nt/\hbar}\, \psi_n(x)$. The equation reads

$$\overline{v(x)} = \int_{-\infty}^{\infty} e^{iE_nt/\hbar}\, \psi_n^*(x)\, v(x)\, e^{-iE_nt/\hbar}\, \psi_n(x)\, dx$$

or

$$\overline{v(x)} = \int_{-\infty}^{\infty} \psi_n^*(x)\, v(x)\, \psi_n(x)\, dx$$

Thus perturbation theory gives the very reasonable result that the shift in the energy of the nth eigenvalue, due to the presence of the perturbing potential $v(x)$, is approximately equal to the value of $v(x)$ averaged over the nth unperturbed eigenstate with a weighting factor equal to the probability density $\psi_n^*(x)\, \psi_n(x)$ for that eigenstate.

Next consider (9–11), and evaluate the symbol v_{mn}. This yields

$$a_{nm} \simeq \frac{1}{E_n - E_m} \int_{-\infty}^{\infty} \psi_m^*(x)\, v(x)\, \psi_n(x)\, dx, \qquad m \neq n \tag{9–14}$$

This equation gives the approximate value of the coefficients a_{nm} which specify how much of each of the unperturbed eigenfunctions $\psi_m(x)$ is mixed in with the dominant unperturbed eigenfunction $\psi_n(x)$ to form the perturbed eigenfunction $\psi_n'(x)$. Then in the series (9–2), with l replaced by m,

$$\psi'_n(x) = \sum_m a_{nm}\psi_m(x) \tag{9–15}$$

we may use (9–14) to evaluate all the coefficients except a_{nn}. From (9–8) we know $a_{nn} \simeq 1$. Its exact value can be determined by requiring that $\psi_n'(x)$ be normalized. Note that a_{nm} is proportional to $1/(E_n - E_m)$. Thus the perturbation $v(x)$ will mix in with the unperturbed eigenfunction $\psi_n(x)$ only a negligibly small amount of any unperturbed eigenfunction $\psi_m(x)$ whose eigenvalue E_m is very different from the eigenvalue E_n. This has the important consequence that a good approximation to the series (9–15) may be obtained by taking only the term for $m = n$, plus a few terms for m not very different from n. The coefficient a_{nm} is also proportional to the quantity

$$v_{mn} = \int_{-\infty}^{\infty} \psi_m^*(x)\, v(x)\, \psi_n(x)\, dx$$

This is a certain average of $v(x)$, with a weighting factor $\psi_m^*(x)\,\psi_n(x)$ which depends on the eigenfunction for the mth unperturbed eigenstate as well as the eigenfunction for the nth unperturbed eigenstate.

The quantities v_{mn}, for $m = n$ as well as $m \neq n$, are called *the matrix elements of the perturbation v taken between the state n and the state m.* This terminology is used because in advanced treatments of quantum mechanics it is convenient to consider a matrix in which each element is one of the quantities v_{mn}. Such a matrix,

$$\begin{bmatrix} v_{11} & v_{12} & v_{13} & \cdot & \cdot & \cdot & v_{1n} & \cdot & \cdot & \cdot \\ v_{21} & v_{22} & v_{23} & \cdot & \cdot & \cdot & v_{2n} & \cdot & \cdot & \cdot \\ v_{31} & v_{32} & v_{33} & \cdot & \cdot & \cdot & \cdot & \cdot & \cdot & \cdot \\ \cdot & \cdot & \cdot & \cdot & \cdot & \cdot & \cdot & \cdot & \cdot & \cdot \\ \cdot & \cdot & \cdot & \cdot & \cdot & \cdot & \cdot & \cdot & \cdot & \cdot \\ \cdot & \cdot & \cdot & \cdot & \cdot & \cdot & \cdot & \cdot & \cdot & \cdot \\ v_{m1} & v_{m2} & \cdot & \cdot & \cdot & \cdot & \cdot & \cdot & \cdot & \cdot \\ \cdot & \cdot & \cdot & \cdot & \cdot & \cdot & \cdot & \cdot & \cdot & \cdot \\ \cdot & \cdot & \cdot & \cdot & \cdot & \cdot & \cdot & \cdot & \cdot & \cdot \\ \cdot & \cdot & \cdot & \cdot & \cdot & \cdot & \cdot & \cdot & \cdot & \cdot \end{bmatrix}$$

contains all possible information concerning the application of a perturbation $v(x)$ to a system whose unperturbed eigenfunctions are $\psi_1(x)$, $\psi_2(x)$, $\psi_3(x)$, $\psi_4(x)$, ...

3. An Example

Let us illustrate the use of (9–13) and (9–14) by doing a simple perturbation calculation. We shall evaluate the first eigenvalue and eigenfunction for the potential indicated in figure (9–2) and specified by the equation

$$V'(x) = \begin{cases} \delta \dfrac{|x|}{a/2}, & -\dfrac{a}{2} < x < \dfrac{a}{2} \\ \infty, & x < -\dfrac{a}{2},\ x > \dfrac{a}{2} \end{cases} \tag{9–16}$$

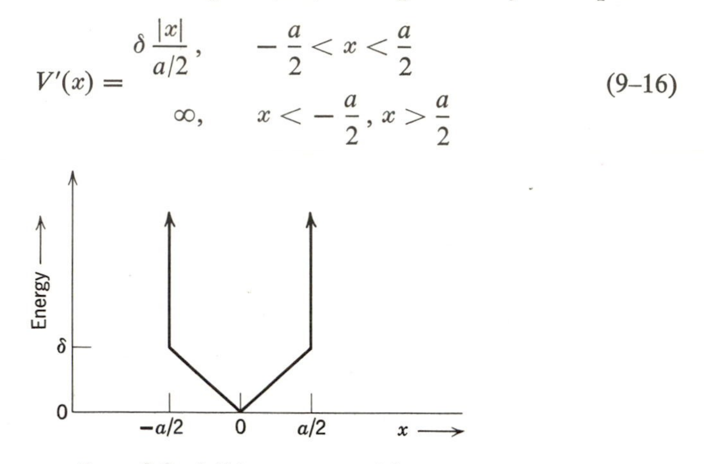

Figure 9–2. A V-bottom potential.

We consider this as the sum of an unperturbed potential,

$$V(x) = \begin{cases} 0, & -a/2 < x < a/2 \\ \infty, & x < -a/2,\ x > a/2 \end{cases}$$

which is an infinite square well, plus a perturbation

$$v(x) = \delta \frac{|x|}{a/2}$$

According to (8–95), the normalized unperturbed eigenfunctions are

$$\psi_m(x) = \begin{cases} \sqrt{2/a}\cos(m\pi x/a), & m = 1, 3, 5, \ldots \\ \sqrt{2/a}\sin(m\pi x/a), & m = 2, 4, 6, \ldots \end{cases}$$

According to (8–96), the unperturbed eigenvalues are

$$E_m = \pi^2\hbar^2 m^2/2Ma^2, \qquad m = 1, 2, 3, 4, \ldots$$

where we use M for the mass of the particle. If δ is small compared to the first eigenvalue,

$$E_1 = \pi^2\hbar^2/2Ma^2$$

the perturbation technique should be applicable.

To evaluate $\psi_1'(x)$, take $n = 1$ in (9–14). This gives

$$a_{1m} = \frac{1}{E_1 - E_m}\int_{-\infty}^{\infty} \psi_m^*(x)\, v(x)\, \psi_1(x)\, dx, \qquad m \neq 1$$

which is

$$a_{1m} = \begin{cases} \dfrac{8M\delta}{\pi^2\hbar^2}\dfrac{1}{(1 - m^2)}\displaystyle\int_{-a/2}^{a/2} \cos\left(\frac{m\pi x}{a}\right)|x| \cos\left(\frac{\pi x}{a}\right) dx, & m = 3, 5, 7, \ldots \\ \dfrac{8M\delta}{\pi^2\hbar^2}\dfrac{1}{(1 - m^2)}\displaystyle\int_{-a/2}^{a/2} \sin\left(\frac{m\pi x}{a}\right)|x| \cos\left(\frac{\pi x}{a}\right) dx, & m = 2, 4, 6, \ldots \end{cases}$$

For $m = 2, 4, 6, \ldots$ the integrand is an odd function of x. Since the integral is taken over a range symmetrical about $x = 0$, the integral will vanish. Thus we have

$$a_{1m} = 0, \qquad m = 2, 4, 6, \ldots$$

For $m = 3, 5, 7, \ldots$ the integral is an even function of x; it gives

$$a_{1m} = \frac{16M\delta}{\pi^2\hbar^2}\frac{1}{(1 - m^2)}\int_0^{a/2} \cos\left(\frac{m\pi x}{a}\right) x \cos\left(\frac{\pi x}{a}\right) dx, \qquad m = 3, 5, 7, \ldots$$

Let $Z = \pi x/a$; then this becomes

$$a_{1m} = \frac{8}{\pi^2}\frac{\delta}{E_1}\frac{1}{(1 - m^2)}\int_0^{\pi/2} \cos(mZ)\, Z \cos Z\, dZ$$

where we have introduced the convenient dimensionless ratio $\delta/E_1 = 2Ma^2\delta/\pi^2\hbar^2$. The integral can be evaluated easily by writing $\cos(mZ) = \frac{1}{2}(e^{+imZ} + e^{-imZ})$, etc. The result is

$$a_{1m} = \frac{8}{\pi^2}\frac{\delta}{E_1}\frac{1}{(1 - m^2)}\left\{\frac{\cos[(m + 1)\pi/2] - 1}{2(m + 1)^2} + \frac{\cos[(m - 1)\pi/2] - 1}{2(m - 1)^2}\right\},$$

$$m = 3, 5, 7, \ldots$$

The first few non-vanishing coefficients have the values

$$a_{13} = \frac{1}{32}\frac{8}{\pi^2}\frac{\delta}{E_1}$$

$$a_{15} = \frac{1}{864}\frac{8}{\pi^2}\frac{\delta}{E_1}$$

$$a_{17} = \frac{1}{1728}\frac{8}{\pi^2}\frac{\delta}{E_1}$$

$$a_{19} = \frac{1}{8000}\frac{8}{\pi^2}\frac{\delta}{E_1}$$

It is not surprising that $a_{1m} = 0$ for $m = 2, 4, 6, \ldots$. The perturbed potential $V'(x)$ is symmetrical about the origin, and so its first eigenstate must be of even parity. Consequently there can be no odd parity unperturbed eigenfunctions mixed into the first perturbed eigenfunction, and the odd parity unperturbed eigenfunctions are precisely those for $m = 2, 4, 6, \ldots$. The perturbed eigenfunction $\psi_1'(x)$ is obtained by substituting the a_{1m} in the series (9–15). Since the a_{1m} decrease rapidly with increasing m [owing partly to the $1/(E_1 - E_m)$ term and partly to the v_{m1} term], it is apparent that we can get a very good approximation to the series by taking only the terms for $m = 1$ and $m = 3$. Thus

$$\psi_1'(x) \simeq a_{11}\psi_1(x) + \frac{1}{32}\frac{8}{\pi^2}\frac{\delta}{E_1}\psi_3(x) \tag{9–17}$$

Finally the coefficient a_{11} must be adjusted so that $\psi_1'(x)$ is normalized, but we leave this as an exercise for the reader.

Figure (9–3) illustrates equation (9–17). The relative amount of $\psi_3(x)$ has been exaggerated for the sake of clarity. Fixing our attention on $\psi_1'(x)$ and $\psi_1(x)$, we see that the second derivative of the perturbed eigenfunction is relatively small near the ends of the region $-a/2$ to $+a/2$, and relatively large near the center, compared to the second derivative of the unperturbed eigenfunction. Consideration of the form of the time independent Schroedinger equation for the perturbed and unperturbed potentials will make it clear why this happens.

Next let us evaluate $E_1' - E_1$. Taking $n = 1$ in equation (9–13), and inserting the appropriate unperturbed eigenfunction, we have

$$E_1' - E_1 = \frac{4\delta}{a^2}\int_{-a/2}^{a/2} |x| \cos^2\left(\frac{\pi x}{a}\right) dx$$

$$E_1' - E_1 = \frac{8\delta}{a^2}\int_{0}^{a/2} x \cos^2\left(\frac{\pi x}{a}\right) dx$$

$$E_1' - E_1 = \frac{8\delta}{\pi^2}\int_{0}^{\pi/2} Z \cos^2 Z \, dZ$$

$$E_1' - E_1 = \frac{8\delta}{\pi^2}\left(\frac{\pi^2}{16} - \frac{1}{4}\right)$$

which is

$$E_1' - E_1 = 0.297\delta$$

Figure (9–4) shows the perturbed eigenvalue E_1' in terms of the dimensionless ratio $(E_1' - E_1)/E_1$, plotted as a function of the dimensionless ratio δ/E_1. Perturbation theory predicts the straight line of slope 0.297. The points are the correct answer. They were calculated from the eigenvalues E_1' obtained by an exact (numerical integration) solution of the

Schroedinger equation for the four potentials $V'(x)$ corresponding to the values of δ/E_1 indicated. The shift in the energy of the first eigenvalue, as

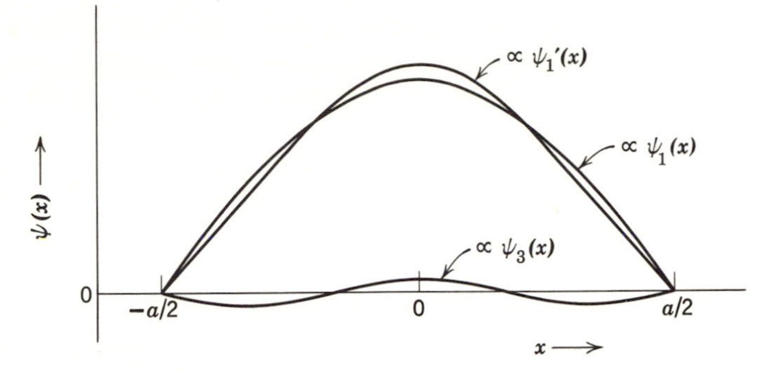

Figure 9–3. Illustrating the composition of the first eigenfunction for a V-bottom potential.

predicted by perturbation theory, is seen to be in error by about 10 percent for $\delta/E_1 \simeq 0.9$, which corresponds to $(E_1' - E_1)/E_1 \simeq 0.25$. For $(E_1' - E_1)/E_1 \simeq 0.05$, the error is about 0.5 percent. Now it is apparent that the error in the perturbation theory we have developed is of the order of the square of a small quantity since, throughout the development, the

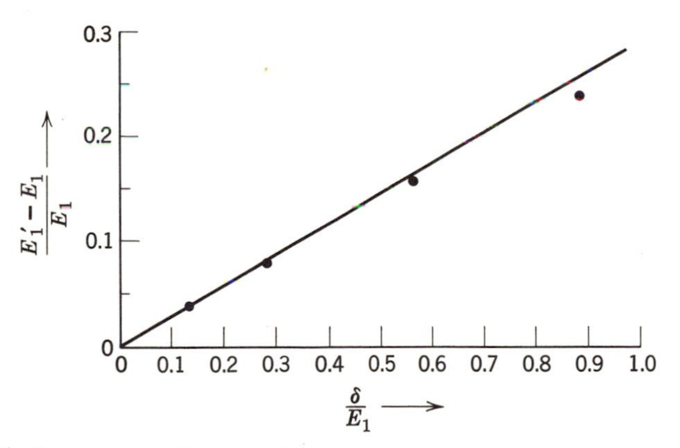

Figure 9–4. A comparison between the first eigenvalues for several V-bottom potentials obtained from perturbation theory and from exact solutions of the Schroedinger equation.

squares of small quantities were always neglected. The numbers just quoted indicate that, in the present case, an approximate measure of the size of this small quantity is the ratio $(E_1' - E_1)/E_1$. Note also that the eigenvalue E_1' calculated by perturbation theory is always too large. It can be shown that this is true for any form of the perturbation $v(x)$, and it is easy to see why it happens. Perturbation theory uses the unperturbed eigenvalue $\psi_1(x)$ to evaluate $E_1' - E_1 = \int_{-\infty}^{\infty} \psi_1^*(x)\, v(x)\, \psi_1(x)\, dx$. Comparing the plots of $\psi_1(x)$ and of $\psi_1'(x)$, we see that this procedure gives too much weight to the values of $v(x)$ near the ends of the region. But near the ends of the region $v(x)$ is largest, and therefore the contribution of $v(x)$ to the perturbed eigenvalue E_1' is overestimated.

A comparison of the exact form of the eigenfunction $\psi_1'(x)$ of the potential $V'(x)$ with the form (9–17) predicted by perturbation theory shows that the error in the coefficient a_{13} is also of the order of the square of the quantity $(E_1' - E_1)/E_1$.

If more accurate estimates of E_n' and $\psi_n'(x)$ are needed, it is possible to extend the perturbation theory to obtain expressions in which the error is of the order of the cube, or even of a higher power, of the appropriate small quantity. However, in practice equations (9–13) and (9–14) are normally adequate.

4. The Treatment of Degeneracies

Consider the case of two different unperturbed eigenfunctions, which we label $\psi_1(x)$ and $\psi_2(x)$, whose corresponding unperturbed eigenvalues E_1 and E_2 happen to be exactly equal. These eigenfunctions are said to be *degenerate*. We shall see later that a number of important examples of this situation actually arise in the study of atomic and nuclear physics. For instance, many of the eigenfunctions are degenerate for an electron bound in the $1/r$ Coulomb potential of a hydrogen atom. When eigenfunctions are degenerate, we shall often be interested in studying the effect of a small perturbation which changes the potential in such a way as to remove the degeneracy.

However, to apply perturbation theory in a case involving degenerate eigenfunctions, we must make certain modifications in the theory. This need is clearly indicated by equation (9–11), which makes the prediction $a_{12} \sim 1$ and $a_{21} \sim 1$ for the case $E_1 = E_2$.† This really tells us only that the theory we have developed breaks down in this case. But it also provides some clue to the nature of the difficulty by showing that, when $E_1 = E_2$, the assumption $a_{nl} \ll 1$ for $n \neq l$ (equation 9–8) is not consistent with the results obtained from that assumption in the cases $n = 1, 2$ and $l = 1, 2$.

† Equation (9–11) states $a_{12} \simeq v_{21}/(E_1' - E_2)$. Taking $E_1 = E_2$, and using (9–10), this becomes $a_{12} \simeq v_{21}/(E_1' - E_1) \simeq v_{21}/v_{11} \sim 1$, in general. Note that this result does not depend on the "additional requirement" that $v(x)$ be small compared to the difference between two unperturbed eigenvalues.

The difficulty is resolved when we realize that there is certainly no a priori basis for the assumption that $a_{12} \ll 1$ and $a_{21} \ll 1$, when $\psi_1(x)$ and $\psi_2(x)$ correspond to eigenvalues which are exactly equal. Under such circumstances it might very well be that, in contrast to the assumption, a small perturbation could have a big effect and thoroughly mix up the two degenerate eigenfunctions.

To account for this situation we modify the procedure as follows. First we consider only the mixing, due to the presence of the perturbation $v(x)$, of the two degenerate unperturbed eigenfunctions $\psi_1(x)$ and $\psi_2(x)$ with each other. In doing this we ignore the mixing with $\psi_1(x)$ and $\psi_2(x)$ of any of the non-degenerate unperturbed eigenfunctions $\psi_3(x)$, $\psi_4(x)$, $\psi_5(x)$, ... From this calculation we shall find the eigenfunctions

$$\begin{aligned} \psi_1^0(x) &= a_{11}\psi_1(x) + a_{12}\psi_2(x) \\ \psi_2^0(x) &= a_{21}\psi_1(x) + a_{22}\psi_2(x) \end{aligned} \tag{9–18}$$

and the eigenvalues

$$E_1^0, \quad E_2^0$$

In all but exceptional cases it will turn out that $E_1^0 \neq E_2^0$, i.e., *the perturbation removes the degeneracy*. The eigenfunctions ψ_1^0 and ψ_2^0 are in error by the first powers of small quantities (the a_{nl} for $n = 1, 2$ and $l = 3, 4, 5, \ldots$, which have been neglected), but the eigenvalues E_1^0 and E_2^0 are in error only by the squares of small quantities. Since it is usually necessary only to know the eigenvalues, the calculation is often completed when E_1^0 and E_2^0 are obtained. However, if it is also necessary to know the eigenfunctions, we may take the set of non-degenerate eigenfunctions and eigenvalues

$$\begin{gathered} \psi_1^0(x),\ \psi_2^0(x),\ \psi_3^0(x),\ \psi_4^0(x), \ldots \\ E_1^0,\ E_2^0,\ E_3^0,\ E_4^0, \ldots \end{gathered} \tag{9–19}$$

where

$$\begin{aligned} \psi_3^0(x) &= \psi_3(x) \quad \text{and} \quad E_3^0 = E_3 \\ \psi_4^0(x) &= \psi_4(x) \quad \text{and} \quad E_4^0 = E_4 \\ &\ \ \vdots \end{aligned}$$

and apply the normal non-degenerate perturbation theory to obtain the set of eigenfunctions and eigenvalues

$$\begin{gathered} \psi_1'(x),\ \psi_2'(x),\ \psi_3'(x),\ \psi_4'(x), \ldots \\ E_1',\ E_2',\ E_3',\ E_4', \ldots \end{gathered} \tag{9–20}$$

where

$$E_1' = E_1^0 \quad \text{and} \quad E_2' = E_2^0$$

All these eigenfunctions and eigenvalues are in error only by the square of a small quantity.

Now let us study the calculation which evaluates ψ_1^0, ψ_2^0, E_1^0, and E_2^0. Assume that $\psi_1(x)$ and $\psi_2(x)$ are orthogonal to each other, as well as to all the other $\psi_3(x)$, $\psi_4(x)$, $\psi_5(x)$ . . . .† Assume also that all the eigenfunctions have been normalized. Then it is apparent that we may write an equation identical with (9–7), but E_n' replaced by E_n^0. That is,

$$a_{nm}(E_n^0 - E_m) = \sum_l a_{nl} v_{ml}$$

Now set all $a_{nl} = 0$ except for $n = 1, 2$ and $l = 1, 2$. Two equations are obtained, one for $m = 1$ and one for $m = 2$:

$$a_{n1}(E_n^0 - E_1) = \sum_{l=1,2} a_{nl} v_{1l}, \qquad n = 1, 2$$

$$a_{n2}(E_n^0 - E_2) = \sum_{l=1,2} a_{nl} v_{2l}, \qquad n = 1, 2$$

Evaluating the summations explicitly, these equations give

$$a_{n1}(E_n^0 - E) = a_{n1} v_{11} + a_{n2} v_{12}, \qquad n = 1, 2 \tag{9–21}$$

$$a_{n2}(E_n^0 - E) = a_{n1} v_{21} + a_{n2} v_{22}, \qquad n = 1, 2 \tag{9–22}$$

where we have set $E_1 = E_2 = E$. Write these equations in the form

$$(E_n^0 - E - v_{11})a_{n1} + (-v_{12})a_{n2} = 0, \qquad n = 1, 2$$

$$(-v_{21})a_{n1} + (E_n^0 - E - v_{22})a_{n2} = 0, \qquad n = 1, 2$$

Then it becomes apparent that they constitute a pair of homogeneous linear equations in the unknowns a_{n1} and a_{n2}. In order that a solution exist for such a set of equations, it is necessary that the determinant of the coefficients of the unknowns be equal to zero. That is, we must have

$$\begin{pmatrix} (E_n^0 - E - v_{11}) & -v_{12} \\ -v_{21} & (E_n^0 - E - v_{22}) \end{pmatrix} = 0, \qquad n = 1, 2 \tag{9–23}$$

Evaluating the determinant, we find

$$(E_n^0 - E - v_{11})(E_n^0 - E - v_{22}) - v_{21} v_{12} = 0, \qquad n = 1, 2 \tag{9–24}$$

In many cases of physical interest it turns out that $v_{11} = v_{22}$, and that v_{12} and v_{21} are real. To simplify the discussion, we shall assume that both hold true. Since the definition (9–6) shows that in all cases $v_{21} = v_{12}^*$, we shall then have $v_{21} = v_{12}$, as well as $v_{11} = v_{22}$, and equation (9–24) becomes

$$(E_n^0 - E - v_{11})^2 - v_{12}^2 = 0, \qquad n = 1, 2$$

† The footnote on page 189 proves that this is possible.

This is a quadratic in E_n^0. It has two solutions,

$$E_n^0 - E - v_{11} = \pm v_{12}, \qquad n = 1, 2$$

one for each value of n. We arbitrarily take the + sign for $n = 1$ and the − sign for $n = 2$. Then

$$\begin{aligned} E_1^0 &= E + v_{11} + v_{12} \\ E_2^0 &= E + v_{11} - v_{12} \end{aligned} \tag{9–25}$$

Now consider equation (9–21) with $n = 1$. It is

$$a_{11}(E_1^0 - E) = a_{11}v_{11} + a_{12}v_{12}$$

Substitute E_1^0 from (9–25) and transpose. The result is

$$a_{11}(v_{11} + v_{12} - v_{11}) = a_{12}v_{12}$$

Therefore

$$a_{11} = a_{12}$$

Thus the perturbed eigenfunction (9–18), corresponding to perturbed eigenvalue E_1^0, is equal to

$$\psi_1^0(x) = a_{11}[\psi_1(x) + \psi_2(x)]$$

The coefficient a_{11} must be adjusted to normalize $\psi_1^0(x)$. It is easy to see that this requires $a_{11} = 1/\sqrt{2}$. Then

$$\psi_1^0(x) = \frac{1}{\sqrt{2}}[\psi_1(x) + \psi_2(x)] \tag{9–26}$$

and we also have

$$E_1^0 = E + v_{11} + v_{12} \tag{9–26$'$}$$

Consider equation (9–21) with $n = 2$. It is

$$a_{21}(E_2^0 - E) = a_{21}v_{11} + a_{22}v_{12}$$

Evaluating E_2^0, we have

$$a_{21}(v_{11} - v_{12} - v_{11}) = a_{22}v_{12}$$

So

$$-a_{21} = a_{22}$$

and

$$\psi_2^0(x) = a_{22}[\psi_1(x) - \psi_2(x)]$$

Normalization gives

$$\psi_2^0(x) = \frac{1}{\sqrt{2}}[\psi_1(x) - \psi_2(x)] \tag{9–27}$$

and we also have

$$E_2^0 = E + v_{11} - v_{12} \tag{9–27$'$}$$

Except for the unusual case in which the matrix element v_{12} is equal to zero, equations (9–26′) and (9–27′) show that the perturbation $v(x)$ has removed the degeneracy. The two perturbed eigenvalues E_1^0 and E_2^0 are separated by the energy

$$E_1^0 - E_2^0 = 2v_{12} \tag{9–28}$$

Then $\psi_1^0(x)$, $\psi_2^0(x)$, E_1^0 and E_2^0, plus the original non-degenerate eigenfunctions $\psi_3(x)$, $\psi_4(x)$, . . . and eigenvalues E_3, E_4, . . . , provide the completely non-degenerate set of eigenfunctions and eigenvalues (9–19) to which we can apply the normal perturbation theory and obtain (9–20). Note that equations (9–26) and (9–27) certainly confirm our suspicion that even an arbitrarily small perturbation will produce a large mixing of the degenerate eigenfunctions $\psi_1(x)$ and $\psi_2(x)$, because both $\psi_1^0(x)$ and $\psi_2^0(x)$ contain equal amounts of these eigenfunctions, independent of the magnitude of $v(x)$.

For a case in which $v_{12} \neq v_{21}$, and/or $v_{11} \neq v_{22}$, the procedure is the same in essence but the details are more complicated. For a case in which there are α degenerate eigenfunctions, the procedure is the same except that equation (9–23) becomes a determinant of order α. This leads to an equation analogous to (9–24) of order α in E_n^0. For large values of α, such an equation can be very difficult to solve, but it is often possible to perform certain manipulations on the determinant so that the equation for E_n^0 comes to a form for which its roots are easy to find.

Furthermore, it is often possible to eliminate the entire procedure of degenerate perturbation theory, and use instead non-degenerate perturbation theory, if physical arguments can be used to find the particular linear combinations of degenerate eigenfunctions that do not mix under the application of the perturbation. Consider the problem that we have been treating, and assume that ψ_1^0 and ψ_2^0 are known a priori. Then E_1^0 and E_2^0, which are usually what we are really interested in finding, can be evaluated simply by applying the equation (9–13) of non-degenerate perturbation theory. That is, equation (9–13) applied to ψ_1^0 gives

$$E_1^0 - E = \int \psi_1^{0*} v \psi_1^0 \, dx = \int \frac{1}{\sqrt{2}} [\psi_1^* + \psi_2^*] v \frac{1}{\sqrt{2}} [\psi_1 + \psi_2] \, dx$$

$$= \frac{1}{2}\left[\int \psi_1^* v \psi_1 \, dx + \int \psi_1^* v \psi_2 \, dx + \int \psi_2^* v \psi_1 \, dx + \int \psi_2^* v \psi_2 \, dx\right]$$

$$= \tfrac{1}{2}[v_{11} + v_{12} + v_{21} + v_{22}]$$

$$= v_{11} + v_{12}$$

in agreement with equations (9–25), and the same equation applied to ψ_2^0 gives

$$E_2^0 - E = \int \psi_2^{0*} v \psi_2^0 \, dx = \int \frac{1}{\sqrt{2}} [\psi_1^* - \psi_2^*] v \frac{1}{\sqrt{2}} [\psi_1 - \psi_2] \, dx$$

$$= \frac{1}{2}\left[\int \psi_1^* v \psi_1 \, dx - \int \psi_1^* v \psi_2 \, dx - \int \psi_2^* v \psi_1 \, dx + \int \psi_2^* v \psi_2 \, dx\right]$$

$$= \tfrac{1}{2}[v_{11} - v_{12} - v_{21} + v_{22}]$$

$$= v_{11} - v_{12}$$

also in agreement with equations (9–25).

Non-degenerate perturbation theory can be applied to the particular linear combinations ψ_1^0 and ψ_2^0 of the degenerate eigenfunctions ψ_1 and ψ_2, because those particular combinations are not mixed by the application of the perturbation. We can see this by evaluating the integrals that appear in the coefficients which, according to the equation (9–14) of non-degenerate perturbation theory, determine the mixing. For instance,

$$\int \psi_1^{0*} v \psi_2^0 \, dx = \int \frac{1}{\sqrt{2}} [\psi_1^* + \psi_2^*] v \frac{1}{\sqrt{2}} [\psi_1 - \psi_2] \, dx$$

$$= \frac{1}{2}\left[\int \psi_1^* v \psi_1 \, dx - \int \psi_1^* v \psi_2 \, dx + \int \psi_2^* v \psi_1 \, dx - \int \psi_2^* v \psi_2 \, dx\right]$$

$$= \tfrac{1}{2}[v_{11} - v_{12} + v_{21} - v_{22}] = 0$$

Similarly,

$$\int \psi_2^{0*} v \psi_1^0 \, dx = 0$$

So the perturbation does not mix ψ_1^0 and ψ_2^0, and *non-degenerate perturbation theory can be applied directly to these particular linear combinations, even though they are degenerate before the application of the perturbation.*

In subsequent chapters we shall consider a number of examples of the physical arguments which are used in quantum mechanical systems to find the particular linear combinations of degenerate eigenfunctions that are not mixed by the application of a perturbation. (Cf. Chapters 11, 12, and 13.) We consider here an example of a physical argument which can be used in a classical system. In one of the higher frequency modes of a circular drum head, the drum head vibrates with a nodal line lying along a diameter. This mode is degenerate because the same frequency is obtained for all orientations of the nodal line. But there is only a two-fold degeneracy because there are only two independent vibrations—the vibrations whose nodal lines are perpendicular. These independent degenerate vibrations are indicated in figure (9–5). All other vibrations at this frequency can be obtained by linear combinations of these two. In particular, other sets of two independent degenerate vibrations, with perpendicular nodal lines of different orientation, can be obtained by appropriate linear combinations. In the absence of a perturbation, all these sets are equivalent.

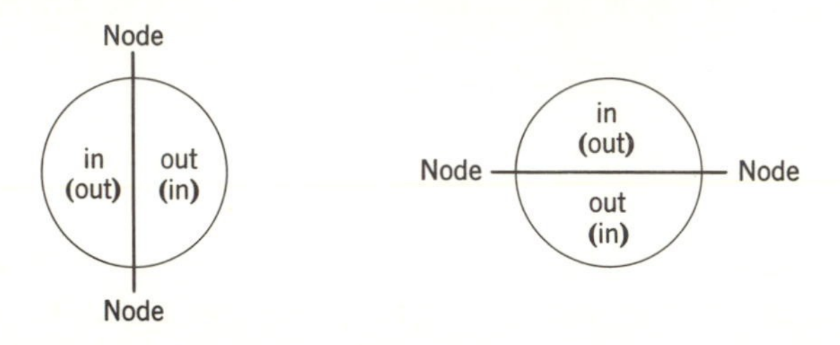

Figure 9–5. Two independent degenerate vibrations of a circular drum head.

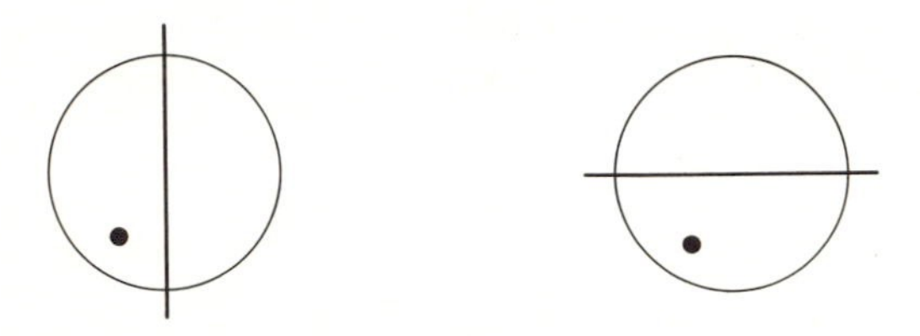

Figure 9–6. Applying a perturbation to a circular drum head.

Figure 9–7. The results of applying a perturbation to a circular drum head.

Now imagine applying a perturbation by fixing a small weight to the drum head at some position other than its center, as indicated in figure (9–6). Because of the asymmetry introduced by the perturbation, the two previously independent vibrations are mixed together to form two new vibrations. The forms of these new vibrations, as well as their frequencies, can be found by a calculation using the procedure of degenerate classical perturbation theory. The vibrations which are found are indicated in figure (9–7). It is also found that the perturbation removes the degeneracy because the weight lies along the nodal line for one vibration and therefore has no effect on the frequency of that vibration, while it does effect the frequency of the other vibration.

After gaining some experience with these problems, it is possible to tell from a priori physical arguments what the form of the perturbed vibrations must be. This makes it possible to eliminate the complicated procedure of degenerate classical perturbation theory by taking the set of independent degenerate unperturbed vibrations to be the particular set for which one nodal line runs through the weight. Then the application of the perturbation does not mix the vibrations because they have the same form both before and after its application, and the non-degenerate classical perturbation theory can be used. This considerably simplifies the calculation of the frequency shift produced by the perturbation, which is the quantity that is usually of interest.

5. Time Dependent Perturbation Theory

We now develop a theory for treating perturbations which are functions of both position and time. This is an important case for several reasons, one being that time dependent perturbation theory provides the only useful method for solving the Schroedinger equation for a time dependent potential $V(x, t)$.† Thus we consider a time dependent potential $V'(x, t)$ which can be decomposed as follows:

$$V'(x, t) = V(x) + v(x, t) \tag{9–29}$$

where $V(x)$ is a time independent unperturbed potential and $v(x, t)$ is a small time dependent perturbation. The solutions to the Schroedinger equation for $V(x)$ are the set of unperturbed wave functions

$$\Psi_n(x, t) = e^{-iE_nt/\hbar}\psi_n(x) \tag{9–30}$$

where the E_n and $\psi_n(x)$ are the unperturbed eigenvalues and eigenfunctions. Assume that a solution to the Schroedinger equation for $V'(x, t)$ can be written

$$\Psi'(x, t) = \sum_n a_n(t)\Psi_n(x, t) \tag{9–31}$$

where the coefficients $a_n(t)$ are functions of time. Different solutions will have different sets of coefficients, but here we shall not use a second subscript to indicate this explicitly. Substitute (9–31) into the equation

$$-\frac{\hbar^2}{2M}\frac{\partial^2\Psi'}{\partial x^2} + V'\Psi' - i\hbar\frac{\partial\Psi'}{\partial t} = 0$$

† One exception is a time dependent potential of the form $V(x, t) = V_1(x) + V_2(t)$. For this form only, the Schroedinger equation can be separated in the manner of section 4, Chapter 7, by assuming a solution $\Psi(x, t) = \psi(x)\,\phi(t)$.

which it is supposed to satisfy. This gives

$$\sum_n a_n\left[-\frac{\hbar^2}{2M}\frac{\partial^2\Psi_n}{\partial x^2} + V\Psi_n - i\hbar\frac{\partial\Psi_n}{\partial t}\right] + \sum_n a_n v\Psi_n - i\hbar\sum_n \frac{da_n}{dt}\Psi_n = 0$$

The bracket vanishes because the Ψ_n are solutions to the Schroedinger equation for the potential V. Multiply the remaining terms by the complex conjugate of some particular unperturbed wave function $\Psi_m = e^{-iE_m t/\hbar}\,\psi_m$, and integrate over all x. Then, evaluating the Ψ_n, we have

$$\sum_n a_n e^{-i(E_n-E_m)t/\hbar}\int_{-\infty}^{\infty}\psi_m^* v\psi_n\,dx = i\hbar\sum_n \frac{da_n}{dt}\,e^{-i(E_n-E_m)t/\hbar}\int_{-\infty}^{\infty}\psi_m^*\psi_n\,dx$$

Since the ψ_n are orthogonal and normalized,

$$\frac{da_m(t)}{dt} = -\frac{i}{\hbar}\sum_n a_n(t)e^{-i(E_n-E_m)t/\hbar}\,v_{mn} \tag{9–32}$$

where we have extended the definition of the matrix element v_{mn} (equation 9–6) to include time dependent perturbations. This is a set of coupled first order ordinary differential equations, one for each m, which determine the $a_n(t)$. The details of the solution of these equations depend on the details of the particular problem at hand. We consider here a simple but illustrative case.

Assume a perturbation of the form

$$v(x, t) = \begin{matrix} 0, & t < 0 \\ v(x), & t > 0 \end{matrix} \tag{9–33}$$

This is a perturbation $v(x)$ which is "switched on" at $t = 0$. For this case the set of unperturbed wave functions (9–30) are exact solutions for $t < 0$. Next assume that the wave function for the particle is known to be equal to a single one of these wave functions, say $\Psi_k(x, t)$, for $t < 0$. This amounts to assuming that the total energy of the particle is known to be precisely E_k for $t < 0$. This does not conflict with the uncertainty principle

$$\Delta\tau\,\Delta E \gtrsim \hbar \tag{9–34}$$

because in the infinite time before $t = 0$ it would be possible to measure the energy of the particle with perfect precision. In terms of equation (9–31), this assumption provides the following set of initial conditions for the $a_n(t)$ at $t = 0$.

$$a_n(0) = \begin{matrix} 0, & n \neq k \\ 1, & n = k \end{matrix} \tag{9–35}$$

[We assume that the $a_n(t)$ do not change discontinuously at $t = 0$. This assumption will be justified by the results of the calculation.] We would like to find the perturbed wave function $\Psi'(x, t)$ for the particle at a time $t > 0$. To do this we shall evaluate the $a_n(t)$ for $t > 0$.

Let us require that the perturbation $v(x)$ be small enough, *or* that the time t be short enough, that

$$a_n(t) \begin{matrix} \ll 1, & n \neq k, \\ \simeq 1, & n = k, \end{matrix} \qquad t > 0 \tag{9–36}$$

Then we may neglect all terms in the right side of equation (9–32) except for $n = k$. This gives

$$\frac{da_m(t)}{dt} \simeq -\frac{i}{\hbar} a_k(t) e^{-i(E_k - E_m)t/\hbar} v_{mk} \tag{9–37}$$

To evaluate $a_k(t)$, set $m = k$. Then

$$\frac{da_k(t)}{dt} \simeq -\frac{i}{\hbar} a_k(t) v_{kk}$$

or

$$\frac{da_k(t)}{a_k(t)} \simeq -\frac{i}{\hbar} v_{kk}\, dt$$

Integrate both sides from 0 to $t' > 0$, remembering that the v_{mk} are all independent of t for $t > 0$. This gives

$$\left[\ln a_k(t)\right]_0^{t'} \simeq \left[-\frac{i}{\hbar} v_{kk} t\right]_0^{t'}$$

which is

$$\ln\left[\frac{a_k(t')}{a_k(0)}\right] \simeq -\frac{i}{\hbar} v_{kk} t', \qquad t' > 0$$

According to (9–35), $a_k(0) = 1$. So we find

$$a_k(t) \simeq e^{-iv_{kk}t/\hbar}, \qquad t > 0 \tag{9–38}$$

where we have dropped the primes to simplify the notation.

Next evaluate the $a_n(t)$, $n \neq k$, by setting $m = n$ in (9–37) and by making the additional approximation that $a_k(t) = 1$. We have

$$\frac{da_n(t)}{dt} \simeq -\frac{i}{\hbar} e^{-i(E_k - E_n)t/\hbar} v_{nk}, \qquad n \neq k$$

or

$$da_n(t) \simeq -\frac{i}{\hbar} v_{nk} e^{-i(E_k - E_n)t/\hbar}\, dt, \qquad n \neq k$$

Integrate from 0 to $t' > 0$ to obtain

$$\left[a_n(t)\right]_0^{t'} \simeq \left[\frac{v_{nk}}{E_k - E_n} e^{-i(E_k - E_n)t/\hbar}\right]_0^{t'}$$

From (9–35), $a_n(0) = 0$, so

$$a_n(t) \simeq \frac{v_{nk}}{E_k - E_n} \left[e^{-i(E_k - E_n)t/\hbar} - 1\right], \qquad n \neq k \tag{9–39}$$

where we have again dropped the primes. Evaluating $\Psi'(x, t)$ from (9–31), (9–38), and (9–39), we find

$$\Psi'(x, t) \simeq e^{-i(E_k + v_{kk})t/\hbar} \psi_k(x) + \sum_{\substack{n \\ n \neq k}} \frac{v_{nk}}{E_k - E_n} \left[e^{-i(E_k - E_n)t/\hbar} - 1\right] e^{-iE_n t/\hbar} \psi_n(x) \tag{9–40}$$

Note that the energy $E_k + v_{kk}$ appearing in the exponential of the first term is exactly the perturbed energy $E_k' = E_k + v_{kk}$, which would be predicted for a completely time independent perturbation equal to $v(x)$.

It is of interest to consider the quantity $a_n^*(t)\, a_n(t)$. This real function of t is the square of the magnitude of the coefficient $a_n(t)$. Multiplying (9–39) into its complex conjugate, we find

$$a_n^*(t)\, a_n(t) \simeq \frac{v_{nk}^* v_{nk}}{\hbar^2} \frac{\sin^2\left[\dfrac{(E_n - E_k)}{2\hbar} t\right]}{\left(\dfrac{E_n - E_k}{2\hbar}\right)^2} \tag{9–41}$$

This quantity oscillates in time between zero and $4v_{nk}^* v_{nk}/(E_n - E_k)^2$, with frequency $\nu = (E_n - E_k)/h$. We plot in figure (9–8) the factor $\sin^2 [(E_n - E_k)t/2\hbar]/[(E_n - E_k)/2\hbar]^2$ as a function of $(E_n - E_k)/2\hbar$ for fixed t. Now the wave function describing the particle initially contained only the eigenstate wave function $\Psi_k(x, t)$. The perturbation $v(x, t)$ has the effect of mixing in contributions from other eigenstates over a whole range of the quantum number n. However, we see that the most important contributions come from those n which correspond to eigenvalues E_n lying within a range centered about E_k and of width ΔE, where

$$\Delta E/2\hbar \sim \pi/t$$

or

$$\Delta E \sim 2\pi\hbar/t \tag{9–42}$$

In section 8, Chapter 7, we showed that the value of $a_n^*(t)\, a_n(t)$ at any instant t is equal to the probability of finding the particle in the eigenstate

n at that instant. (The argument actually considered the a_n to be constants, but inspection will demonstrate that it is equally valid when the a_n are functions of t.) Thus at any time t there is a certain probability of finding the particle in final eigenstate n which is different from the initial eigenstate k, and with total energy E_n different from the initial total energy E_k. This appears to be a violation of the law of conservation of energy by an amount $E_n - E_k$, which may be large compared to the energy v_{kk} supplied

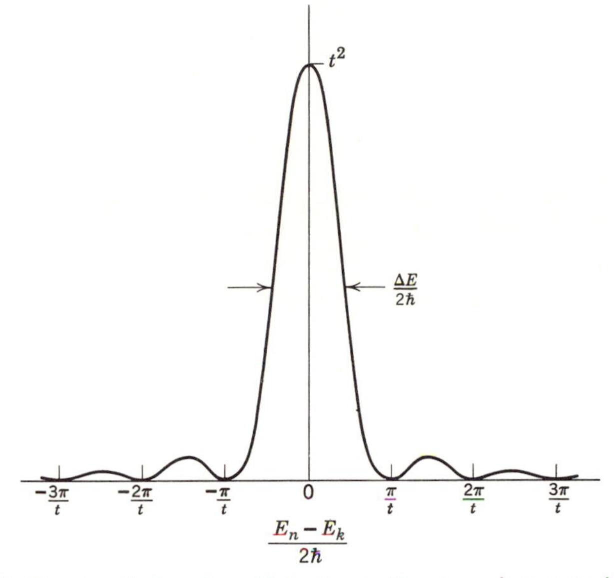

Figure 9–8. The plot of a function which arises in time dependent perturbation theory.

by the perturbation. However, in the time interval 0 to t the probability of finding the particle with energy E_n is important only when $E_n - E_k$ is at most equal to about ΔE, where t and ΔE are related by equation (9–42). According to the uncertainty principle (9–34) any measurement of the total energy of the particle which is carried out in this time interval must be uncertain by an amount of the order of $\hbar/t$, which is comparable to ΔE. This removes the difficulty and provides an example of this statement of the uncertainty principle.

Consider equation (9–41) for small values of $t > 0$. The equation says that the probability of finding the particle in a particular eigenstate n is proportional to the square of t. This statement is in contrast to the linear dependence on t that might be expected intuitively. However, physical intuition is always based on our experience with systems in the classical limit. In that limit the resolution of any experimental apparatus is so large compared to the separation of the eigenvalues, or even to the width of the range ΔE, that it is not possible to measure $a_n^*(t)\, a_n(t)$ for a single value of n. All that can be measured classically is the total probability of finding that the particle has made a transition from the initial eigenstate k to some other final eigenstate n. We express this in terms of the *transition probability* P_k, which is defined as

$$P_k \equiv \sum_{\substack{n \\ n \neq k}} a_n^*(t)\, a_n(t) \tag{9–43}$$

To evaluate this, we assume that there are a large number of closely spaced final eigenstates in the range ΔE; the number of final eigenstates dN_n per energy interval dE_n is the *density of final states* $\rho_n = dN_n/dE_n$. Then the summation over n can be approximated by an integral over dN_n. That is,

$$P_k \simeq \int_{-\infty}^{\infty} a_n^*(t)\, a_n(t)\, dN_n = \int_{-\infty}^{\infty} a_n^*(t)\, a_n(t) \frac{dN_n}{dE_n}\, dE_n$$

Evaluating $a_n^*(t)\, a_n(t)$ from (9–41), we have

$$P_k \simeq \frac{1}{\hbar^2} \int_{-\infty}^{\infty} v_{nk}^* v_{nk}\, \rho_n \frac{\sin^2\left[\dfrac{(E_n - E_k)}{2\hbar}\, t\right]}{\left(\dfrac{E_n - E_k}{2\hbar}\right)^2} dE_n$$

Owing to the factor $\sin^2[(E_n - E_k)t/2\hbar]/[(E_n - E_k)/2\hbar]^2$, most of the contribution to the integral comes from the range ΔE. If we assume that the matrix element v_{nk} and the density of final states ρ_n are both slowly varying functions of n in that range, we can write

$$P_k \simeq \frac{v_{nk}^* v_{nk}}{\hbar^2}\, \rho_n \int_{-\infty}^{\infty} \frac{\sin^2\left[\dfrac{(E_n - E_k)}{2\hbar}\, t\right]}{\left(\dfrac{E_n - E_k}{2\hbar}\right)^2} dE_n$$

The quantum number n now refers to a typical final eigenstate in the neighborhood of the initial eigenstate k. Let $Z = (E_n - E_k)t/2\hbar$; then

$$P_k \simeq \frac{v_{nk}^* v_{nk}}{\hbar^2}\, \rho_n\, 2\hbar t \int_{-\infty}^{\infty} \frac{\sin^2 Z}{Z^2}\, dZ$$

which gives

$$P_k \simeq \frac{2\pi}{\hbar}\, v_{nk}^* v_{nk} \rho_n t \tag{9–44}$$

The transition probability is proportional to t, as expected. The *transition rate* $R_k \equiv dP_k/dt$ is independent of t, since

$$R_k \simeq \frac{2\pi}{\hbar} v_{nk}^* v_{nk} \rho_n \tag{9–45}$$

This important formula is often called *Golden Rule No. 2.* It is very widely used because it is of very general applicability.† In any situation in which transitions are made to an essentially continuous range of final states under the influence of a constant perturbation, the transition rate can be evaluated from this formula. Note that we have here a good example of the use of quantum mechanics in the evaluation of transition rates. The ability to do this is one of its most important advantages over the old quantum theory.

BIBLIOGRAPHY

Bohm, D., *Quantum Theory*, Prentice-Hall, Englewood Cliffs, N.J., 1951.

Pauling, L., and E. B. Wilson, *Introduction to Quantum Mechanics*, McGraw-Hill Book Co., New York, 1935.

Schiff, L. I., *Quantum Mechanics*, McGraw-Hill Book Co., New York, 1955.

Sherwin, C. W., *Introduction to Quantum Mechanics*, Henry Holt and Co., New York, 1959.

EXERCISES

1. Use perturbation theory to calculate the first eigenvalue E_1 and the first eigenfunction $\psi_1(x)$ of the potential

$$V(x) = \begin{cases} \infty, & x < -a/2,\ x > a/2 \\ \delta \dfrac{x}{a/2}, & -a/2 < x < a/2 \end{cases}$$

where δ is small relative to E_1. Compare with the results obtained in the calculation of section 3.

2. Use perturbation theory to calculate the first eigenvalue E_1 for the potential of exercise 4, Chapter 7. Compare with the results of the numerical integration performed in that exercise, and also with the results of the analytical treatment of exercise 6, Chapter 8.

3. Except for certain pathological cases, no degeneracies arise in problems involving one particle moving in one dimension. In order to obtain a simple

† We shall use it in Chapters 15 and 16, and use a closely related formula in Chapter 13.

example of the application of degenerate perturbation theory, consider one particle moving in the two dimensional infinite square well potential

$$V(x, y) = \begin{cases} \infty, & x < -a/2, x > a/2, y < -a/2, y > a/2 \\ 0, & -a/2 < x < a/2, \ -a/2 < y < a/2 \end{cases}$$

After consulting the first section of the next chapter, set up the time independent Schroedinger equation for this potential. Separate this partial differential equation into two ordinary differential equations by the usual method, making use of the fact that $V(x, y)$ can be written $V(x) + V(y)$. Since these equations, and the conditions on ψ at the edges of the well, have the same form as for a one dimensional infinite square well, their solutions can be written immediately. Note that there are degeneracies in almost all the eigenstates. Now consider the application of the perturbation

$$v(x, y) = \begin{cases} \delta, & x > a/4, \ y > a/3 \\ 0, & \text{otherwise} \end{cases}$$

Make a routine degenerate perturbation theory calculation to investigate the effect of this perturbation on the first two degenerate eigenstates.

4. Take the linear combinations of the unperturbed degenerate eigenfunctions found in exercise 3, and show that if these linear combinations are used in non-degenerate perturbation theory the energy shifts which are obtained are the same as those obtained in that exercise. Show also that in non-degenerate perturbation theory these linear combinations of the unperturbed eigenfunctions are not mixed by the perturbation, even though they are degenerate before the perturbation is applied.

5. At $t < 0$, an electron is known to be in the $n = 1$ eigenstate of a one dimensional infinite square well potential which extends from $x = -a/2$ to $x = a/2$. At $t = 0$, a uniform electric field E is applied in the direction of increasing x. The electric field is left on for a short time τ and then removed. Use time dependent perturbation theory to calculate the probability that the electron will be in the $n = 2$, 3, or 4 eigenstates at $t > \tau$. Make plots of these probabilities as a function of τ. *Hint:* Some of the results of exercise 1 can be used.

CHAPTER 10

One-Electron Atoms

1. Quantum Mechanics for Several Dimensions and Several Particles

Up to this point we have restricted our treatment of quantum mechanics to a single particle moving in a one dimensional space. Now, in preparation for the discussion of atoms, we must generalize to the case of a system of particles moving in a three dimensional space.

Let us consider N particles whose positions are defined by a set of rectangular coordinates, x, y, z. We first develop the Schroedinger equation for this case, using the same technique as in section 8, Chapter 7. We write the classical expression defining the total energy of the system:

$$\sum_{j=1}^{N} \frac{1}{2m_j}(p_{x_j}^2 + p_{y_j}^2 + p_{z_j}^2) + V(x_1, y_1, z_1, \ldots, x_j, y_j, z_j, \ldots, x_N, y_N, z_N, t) = E \quad (10\text{–}1)$$

The quantity p_{x_j} is the x component of the momentum of the jth particle, p_{y_j} and p_{z_j} are its y and z components, and m_j is the mass of that particle. Thus the jth term of the summation is the kinetic energy of the jth particle, and the value of the sum is the kinetic energy of the system. The potential energy of the system depends, in general, on the coordinates of each particle, as well as on the time t. The quantities x_j, y_j, z_j are the three coordinates of the jth particle. Now we replace the dynamical quantities $p_{x_1}, \ldots, p_{z_N}$, and E, by their associated differential operators (cf. 7–77). This gives

$$\sum_{j=1}^{N} -\frac{\hbar^2}{2m_j}\left(\frac{\partial^2}{\partial x_j^2} + \frac{\partial^2}{\partial y_j^2} + \frac{\partial^2}{\partial z_j^2}\right) + V(x_1, \ldots, z_N, t) = i\hbar\frac{\partial}{\partial t} \quad (10\text{–}2)$$

Multiplying this operator equation into the equality

$$\Psi(x_1, \ldots, z_N, t) = \Psi(x_1, \ldots, z_N, t) \tag{10–3}$$

we obtain the Schroedinger equation for the system,

$$\sum_{j=1}^{N} -\frac{\hbar^2}{2m_j}\nabla_j^2\Psi + V\Psi = i\hbar\frac{\partial\Psi}{\partial t} \tag{10–4}$$

where we use the symbol

$$\nabla_j^2 \equiv \frac{\partial^2}{\partial x_j^2} + \frac{\partial^2}{\partial y_j^2} + \frac{\partial^2}{\partial z_j^2} \tag{10–5}$$

which is called the *Laplacian* operator in rectangular coordinates of the jth particle.

The wave function for the system, $\Psi(x_1, \ldots, z_N, t)$, is a function of $3N + 1$ variables—3 spatial coordinates for each of the N particles, plus the time. The significance of this wave function is specified by a natural extension of the interpretive postulates (7–21), (7–31), and (7–80). For instance, we *postulate* that:

If, at the instant t, a measurement is made to locate the particles in the system associated with the wave function $\Psi(x_1, \ldots, z_N, t)$, the probability $P(x_1, \ldots, z_N, t)\, dx_1 \cdots dz_N$ that the x coordinate of the first particle will be found between x_1 and $x_1 + dx_1$, etc., and finally that the z coordinate of the Nth particle will be found between z_N and $z_N + dz_N$, is

$$P(x_1, \ldots, z_N, t)\, dx_1 \cdots dz_N = \Psi^*(x_1, \ldots, z_N, t)\, \Psi(x_1, \ldots, z_N, t)\, dx_1 \cdots dz_N$$

Essentially all the properties of the Schroedinger equation (10–4) and the wave function (10–3) can be obtained by obvious extensions of the properties developed in Chapters 7, 8, and 9. As an example, it is easy to show by the usual method that, when the potential V does not depend on time, there are solutions to the Schroedinger equation which can be written

$$\Psi(x_1, \ldots, z_N, t) = \psi(x_1, \ldots, z_N)\, \phi(t) \tag{10–6}$$

where

$$\phi(t) = e^{-iEt/\hbar} \tag{10–7}$$

and where $\psi(x_1, \ldots, z_N)$ is a solution to the time independent Schroedinger equation

$$\sum_{j=1}^{N} -\frac{\hbar^2}{2m_j}\nabla_j^2\psi + V\psi = E\psi \tag{10–8}$$

Note that the time independent Schroedinger equation is, in the general case we are now discussing, a partial differential equation. Also note that the wave function $\Psi(x_1, \ldots, z_N, t)$ is defined in a non-physical space of more than three dimensions. This makes it quite clear that the wave function is only a computational device.

2. The One-Electron Atom

The procedures involved in solving the time independent Schroedinger equation (10–8) are best illustrated by taking a specific example. In this section we shall begin our treatment of one of the simplest—the one-electron atom. This is also perhaps the most interesting example because of the unique importance of the one-electron atom in atomic theory. The reader will recall that Bohr's one-electron atom is the most significant feature of the old quantum theory. We shall see that the role of the one-electron atom is equally important in the theory of quantum mechanics. Above and beyond its own intrinsic interest, the theory of this simple system is fundamental because it forms the foundation of the quantum mechanical treatment of all multi-electron atoms.

The one-electron atom is a three dimensional system containing two particles, a nucleus and an electron, moving under the influence of a mutual Coulomb attraction and bound together by that attraction. Let the mass of the nucleus be m_1 and its charge be $+Ze$ ($Z = 1$ for neutral hydrogen, $Z = 2$ for singly ionized helium, etc.). Designate the mass of the electron by m_2 and its charge by $-e$. Then the (Coulomb) potential energy of the system is

$$V(x_1, \ldots, z_2) = -\frac{Ze^2}{\sqrt{(x_2 - x_1)^2 + (y_2 - y_1)^2 + (z_2 - z_1)^2}} \tag{10–9}$$

where x_1, y_1, z_1 are the rectangular coördinates of the nucleus and x_2, y_2, z_2 are the rectangular coordinates of the electron. The denominator of equation (10–9) is the distance between the two particles. The time independent Schroedinger equation for this case is

$$-\frac{\hbar^2}{2m_1}\left(\frac{\partial^2\psi_T}{\partial x_1^2} + \frac{\partial^2\psi_T}{\partial y_1^2} + \frac{\partial^2\psi_T}{\partial z_1^2}\right) - \frac{\hbar^2}{2m_2}\left(\frac{\partial^2\psi_T}{\partial x_2^2} + \frac{\partial^2\psi_T}{\partial y_2^2} + \frac{\partial^2\psi_T}{\partial z_2^2}\right) + V(x_1, \ldots, z_2)\psi_T = E_T\psi_T \tag{10–10}$$

The reason for using the subscript T will become apparent later.

By making a change of variables, this equation goes over into a form in which it can be separated into two parts. One part represents the translational motion of the center of mass of the system, and the other part represents the motion of the two particles relative to each other. The new variables are

$$x = \frac{m_1x_1 + m_2x_2}{m_1 + m_2}$$

$$y = \frac{m_1y_1 + m_2y_2}{m_1 + m_2} \tag{10–11}$$

$$z = \frac{m_1z_1 + m_2z_2}{m_1 + m_2}$$

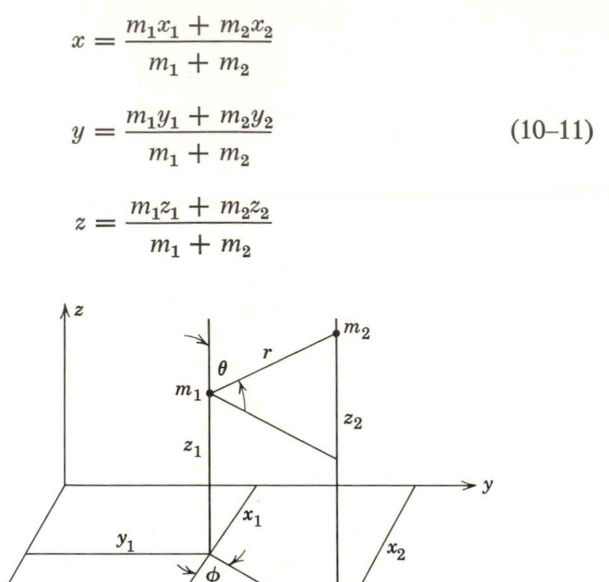

Figure 10–1. The one-electron atom described in rectangular and spherical coordinates.

the rectangular coordinates of the center of mass of the system, and

$$r, \theta, \phi$$

the spherical coordinates of the electron relative to the nucleus.

These coordinates are illustrated in figure (10–1). From this figure it is easy to see that r, θ, ϕ are related to $x_1, \ldots, z_2$ by the equations

$$r \sin\theta \cos\phi = x_2 - x_1$$

$$r \sin\theta \sin\phi = y_2 - y_1 \tag{10–12}$$

$$r \cos\theta = z_2 - z_1$$

Using equations (10–11) and (10–12), equation (10–10) can be transformed to the new variables. This is actually quite a tedious task, and so we present here only the results. They are

$$-\frac{\hbar^2}{2(m_1+m_2)}\left(\frac{\partial^2\psi_{\mathrm{T}}}{\partial x^2}+\frac{\partial^2\psi_{\mathrm{T}}}{\partial y^2}+\frac{\partial^2\psi_{\mathrm{T}}}{\partial z^2}\right)-\frac{\hbar^2}{2\mu}\left\{\frac{1}{r^2}\frac{\partial}{\partial r}\left(r^2\frac{\partial\psi_{\mathrm{T}}}{\partial r}\right)\right.$$

$$\left.+\frac{1}{r^2\sin^2\theta}\frac{\partial^2\psi_{\mathrm{T}}}{\partial\phi^2}+\frac{1}{r^2\sin\theta}\frac{\partial}{\partial\theta}\left(\sin\theta\frac{\partial\psi_{\mathrm{T}}}{\partial\theta}\right)\right\}+V(r)\psi_{\mathrm{T}}=E_{\mathrm{T}}\psi_{\mathrm{T}} \qquad (10\text{–}13)$$

where $$\mu \equiv \frac{m_1 m_2}{m_1+m_2} \quad \text{and} \quad V(r) = -\frac{Ze^2}{r}$$

The quantity μ is the reduced mass.

The separation of this time independent Schroedinger equation is effected by looking for solutions of the form

$$\psi_{\mathrm{T}}(x, y, z, r, \theta, \phi) = \psi_{\mathrm{CM}}(x, y, z)\,\psi(r, \theta, \phi) \qquad (10\text{–}14)$$

Substitute this into equation (10–13), and divide through by $\psi_{\mathrm{T}} = \psi_{\mathrm{CM}}\psi$. This gives

$$\left[-\frac{\hbar^2}{2(m_1+m_2)}\frac{1}{\psi_{\mathrm{CM}}}\left(\frac{\partial^2\psi_{\mathrm{CM}}}{\partial x^2}+\frac{\partial^2\psi_{\mathrm{CM}}}{\partial y^2}+\frac{\partial^2\psi_{\mathrm{CM}}}{\partial z^2}\right)\right]$$

$$+\left[-\frac{\hbar^2}{2\mu}\frac{1}{\psi}\left\{\frac{1}{r^2}\frac{\partial}{\partial r}\left(r^2\frac{\partial\psi}{\partial r}\right)\right.\right.$$

$$\left.\left.+\frac{1}{r^2\sin^2\theta}\frac{\partial^2\psi}{\partial\phi^2}+\frac{1}{r^2\sin\theta}\frac{\partial}{\partial\theta}\left(\sin\theta\frac{\partial\psi}{\partial\theta}\right)\right\}+V(r)\right]=E_{\mathrm{T}}$$

The first term on the left is a function only of x, y, z and the second term is a function only of r, θ, ϕ. Since the sum of the two terms is equal to a constant E_{T}, even though the variables x, y, z are independent of the variables r, θ, ϕ, it is clear that each term must separately be equal to a constant. Thus we obtain two equations,

$$-\frac{\hbar^2}{2(m_1+m_2)}\frac{1}{\psi_{\mathrm{CM}}}\left(\frac{\partial^2\psi_{\mathrm{CM}}}{\partial x^2}+\frac{\partial^2\psi_{\mathrm{CM}}}{\partial y^2}+\frac{\partial^2\psi_{\mathrm{CM}}}{\partial z^2}\right)=E_{\mathrm{CM}}$$

and

$$-\frac{\hbar^2}{2\mu}\frac{1}{\psi}\left\{\frac{1}{r^2}\frac{\partial}{\partial r}\left(r^2\frac{\partial\psi}{\partial r}\right)+\frac{1}{r^2\sin^2\theta}\frac{\partial^2\psi}{\partial\phi^2}+\frac{1}{r^2\sin\theta}\frac{\partial}{\partial\theta}\left(\sin\theta\frac{\partial\psi}{\partial\theta}\right)\right\}+V(r)=E$$

where

$$E_{\mathrm{CM}} + E = E_{\mathrm{T}}$$

Consider the first equation, which we write

$$-\frac{\hbar^2}{2(m_1 + m_2)}\left(\frac{\partial^2 \psi_{\mathrm{CM}}}{\partial x^2} + \frac{\partial^2 \psi_{\mathrm{CM}}}{\partial y^2} + \frac{\partial^2 \psi_{\mathrm{CM}}}{\partial z^2}\right) = E_{\mathrm{CM}}\psi_{\mathrm{CM}}$$

This is the time independent Schroedinger equation for a particle of mass $m_1 + m_2$, whose coordinates are those of the center of mass of the atom, and which is moving with total energy E_{CM} in a region of zero potential. The equation describes the translational motion of the atom. Consider next the second equation, which we write

$$-\frac{\hbar^2}{2\mu}\left\{\frac{1}{r^2}\frac{\partial}{\partial r}\left(r^2\frac{\partial \psi}{\partial r}\right) + \frac{1}{r^2 \sin^2\theta}\frac{\partial^2 \psi}{\partial \phi^2} + \frac{1}{r^2 \sin\theta}\frac{\partial}{\partial \theta}\left(\sin\theta\frac{\partial \psi}{\partial \theta}\right)\right\} + V(r)\psi = E\psi \quad (10\text{–}15)$$

The curly bracket is $\nabla^2\psi$, where ∇^2 is the Laplacian operator in the spherical coordinates r, θ, ϕ. This equation is the time independent Schroedinger equation describing the motion of the electron relative to the nucleus. Since $E = E_{\mathrm{T}} - E_{\mathrm{CM}}$, where E_{T} is the total energy of the atom and E_{CM} is its energy of translational motion, E is the total energy of relative motion—that is, the total energy of the atom in a frame of reference in which the center of mass is stationary. Equation (10–15) is formally identical with the time independent Schroedinger equation for an electron of mass μ, instead of mass m_1, moving with coordinates r, θ, ϕ about an infinitely massive nucleus which, consequently, remains stationary (cf. section 4, Chapter 5).

It is apparent that in equation (10–14) we have decomposed the total eigenfunction ψ_{T} into the product of an eigenfunction ψ_{CM}, specifying the motion of the center of mass of the atom, times an eigenfunction ψ, specifying the motion of the electron relative to the nucleus. *This decomposition is possible whenever the potential V depends only on the coordinates of one particle relative to the other.*

3. Separation and Solution of the Equation of Relative Motion

The translational motion of the atom is of no interest to us here. Consequently, we shall assume that we are viewing the atom from a frame of reference in which the center of mass is at rest. Then we need consider only the time independent Schroedinger equation (10–15). For the case of a Coulomb potential $V = -Ze^2/r$ and, in fact, *whenever the potential V is any function of the radial coordinate r only, it is possible to separate* *the equation into three ordinary differential equations.* To do this, we look for solutions of the form

$$\psi(r, \theta, \phi) = R(r)\Theta(\theta)\Phi(\phi) \qquad (10\text{–}16)$$

Inserting this into equation (10–15) and dividing through by $-R\Theta\Phi\hbar^2/2\mu$, we obtain

$$\frac{1}{r^2R}\frac{d}{dr}\left(r^2\frac{dR}{dr}\right) + \frac{1}{r^2\sin^2\theta\Phi}\frac{d^2\Phi}{d\phi^2} + \frac{1}{r^2\sin\theta\Theta}\frac{d}{d\theta}\left(\sin\theta\frac{d\Theta}{d\theta}\right) + \frac{2\mu}{\hbar^2}[E - V(r)] = 0$$

Next multiply through by $r^2 \sin^2\theta$ and transpose. This gives

$$\frac{1}{\Phi}\frac{d^2\Phi}{d\phi^2} = -\frac{\sin^2\theta}{R}\frac{d}{dr}\left(r^2\frac{dR}{dr}\right) - \frac{\sin\theta}{\Theta}\frac{d}{d\theta}\left(\sin\theta\frac{d\Theta}{d\theta}\right) - \frac{2\mu}{\hbar^2}r^2\sin^2\theta[E - V(r)]$$

Since the left side depends only on ϕ and the right side only on r and θ, both must equal a constant, which we shall call $-m^2$. Thus we obtain two equations,

$$\frac{d^2\Phi}{d\phi^2} = -m^2\Phi \qquad (10\text{–}17)$$

and

$$-\frac{1}{R}\frac{d}{dr}\left(r^2\frac{dR}{dr}\right) - \frac{1}{\sin\theta\Theta}\frac{d}{d\theta}\left(\sin\theta\frac{d\Theta}{d\theta}\right) - \frac{2\mu}{\hbar^2}r^2[E - V(r)] = -\frac{m^2}{\sin^2\theta}$$

By transposing, we rewrite the latter as

$$\frac{1}{R}\frac{d}{dr}\left(r^2\frac{dR}{dr}\right) + \frac{2\mu r^2}{\hbar^2}[E - V(r)] = \frac{m^2}{\sin^2\theta} - \frac{1}{\sin\theta\Theta}\frac{d}{d\theta}\left(\sin\theta\frac{d\Theta}{d\theta}\right)$$

According to the usual argument, both sides of the equation must equal a constant α. Consequently,

$$-\frac{1}{\sin\theta}\frac{d}{d\theta}\left(\sin\theta\frac{d\Theta}{d\theta}\right) + \frac{m^2\Theta}{\sin^2\theta} = \alpha\Theta \qquad (10\text{–}18)$$

and

$$\frac{1}{r^2}\frac{d}{dr}\left(r^2\frac{dR}{dr}\right) + \frac{2\mu}{\hbar^2}[E - V(r)]R = \alpha\frac{R}{r^2} \qquad (10\text{–}19)$$

The problem has been reduced to that of solving the three ordinary differential equations (10–17), (10–18), and (10–19). We shall find that equation (10–17) has acceptable solutions only for certain values of m. Using these values of m in equation (10–18), it turns out that the equation has acceptable solutions only for certain values of α. With these values of α in (10–19), the equation is found to have acceptable solutions only for certain values of the total energy E. Thus the energy is quantized.

Consider equation (10–17). The reader can easily verify that it has a solution

$$\Phi(\phi) = e^{im\phi}$$

Now we must, for the first time, explicitly consider the requirement that the probability density and probability flux have definite values, i.e., be *single-valued*, for all values of their variables. This must be the case because they are measurable quantities describing the behavior of a real physical particle. We may ensure that it will be so by demanding that *the eigenfunction ψ must be single-valued.* This requires the function $\Phi(\phi)$ to be single-valued, and it must be considered explicitly because the azimuthal angles 0 and 2π are actually the same angle. Thus we must be sure that

$$\Phi(0) = \Phi(2\pi)$$

This is

$$e^{im0} = e^{im2\pi}$$

or

$$1 = \cos m2\pi + i \sin m2\pi$$

This is true only if

$$|m| = 0, 1, 2, 3, \ldots \tag{10–20}$$

that is, if m has integral values.† Thus the set of functions which are acceptable solutions to equation (10–17) are

$$\Phi_m(\phi) = e^{im\phi} \tag{10–21}$$

where m assumes the values specified by (10–20). The *quantum number m* is used as a subscript to identify the form of a particular solution.

In working with equation (10–18), it is convenient to change variables from θ to ξ, where

$$\xi = \cos\theta \tag{10–22}$$

† In a more rigorous treatment it is not necessary to demand that ψ be single-valued, but only that the measurable quantity $\psi^*\psi$ be single-valued. It is evident that this would admit either integral or half-integral values of m. However, in the present context the half-integral values can be ruled out because they lead to solutions of the $\Theta(\theta)$ equation for which the probability flux has a physically impossible behavior. On the other hand, half-integral values of a quantum corresponding to m do arise in connection with functions, related to the $\Phi(\phi)$, that are used to describe "spin." This will be discussed in the next chapter.

Then the equation reads

$$\frac{d}{d\xi}\left\{(1-\xi^2)\frac{d\Theta(\xi)}{d\xi}\right\}+\left\{\alpha-\frac{m^2}{1-\xi^2}\right\}\Theta(\xi)=0 \tag{10–23}$$

This differential equation is not directly solvable by the power series method because the recursion relation involves more than two terms (cf. section 6, Chapter 8). But, by considering the behavior of the differential equation at $\xi = \pm 1$, the extreme values of the variable, it is found that a power series solution can be obtained for the function $F(\xi)$ defined by

$$\Theta(\xi) = (1-\xi^2)^{|m|/2}\,F(\xi)$$

After solving the differential equation for $F(\xi)$, it is found that solutions for $\Theta(\xi)$, which have the acceptable behavior of being everywhere finite, exist only when $F(\xi)$ is a polynomial. This happens only when α has the value

$$\alpha = l(l+1) \tag{10–24}$$

where l is one of the integers

$$l = |m|,\ |m|+1,\ |m|+2,\ |m|+3,\ \ldots \tag{10–25}$$

Corresponding to each value of the *quantum number l* there is an acceptable solution $F_l(\xi)$. To identify the particular solutions to equation (10–18), which are written

$$\Theta_{lm}(\xi) = (1-\xi^2)^{|m|/2}\,F_l(\xi) \tag{10–26}$$

it is necessary to use as a label the two quantum numbers l and m. The functions $\Theta_{lm}(\xi)$ are known in the mathematical literature as *associated Legendre functions*. Some examples of these functions will be presented in section 5.

Setting $\alpha = l(l+1)$ in equation (10–19), we have

$$\frac{1}{r^2}\frac{d}{dr}\left(r^2\frac{dR}{dr}\right)+\left\{-\frac{l(l+1)}{r^2}+\frac{2\mu}{\hbar^2}[E-V(r)]\right\}R=0 \tag{10–27}$$

In terms of the new variable

$$\rho = 2\beta r \tag{10–28}$$

where

$$\beta^2 = -\frac{2\mu E}{\hbar^2} \tag{10–29}$$

and also using

$$\gamma = \frac{\mu Z e^2}{\hbar^2 \beta} \tag{10–30}$$

the equation becomes

$$\frac{1}{\rho^2}\frac{d}{d\rho}\left(\rho^2\frac{dR(\rho)}{d\rho}\right)+\left\{-\frac{1}{4}-\frac{l(l+1)}{\rho^2}+\frac{\gamma}{\rho}\right\}R(\rho)=0 \tag{10–31}$$

Again it turns out that the differential equation is not directly amenable to a power series solution. However, by considering the behavior of the equation for $\rho \rightarrow \infty$, it is found that a series solution can be obtained for the function $G(\rho)$ defined by

$$R(\rho) = e^{-\rho/2}\rho^l G(\rho)$$

In order that $G(\rho)$ be a polynomial, which is necessary to prevent $R(\rho)$ from diverging as $\rho \rightarrow \infty$, it is found that γ must have the value

$$\gamma = n \qquad (10\text{–}32)$$

where n is one of the integers

$$n = l + 1, l + 2, l + 3, \ldots \qquad (10\text{–}33)$$

The functions that are acceptable solutions to equation (10–19), which are written

$$R_{nl}(\rho) = e^{-\rho/2}\rho^l G_n(\rho) \qquad (10\text{–}34)$$

carry as labels the *quantum number n* and the quantum number l. Examples of these *associated Laguerre functions* are given in section 5.†

The reader will note that the procedures involved in solving equations (10–18) and (10–19) are essentially the same as those which were employed in the solution of the simple harmonic oscillator equation. Thus he should not have too much difficulty in filling in the details himself, using the outline presented above.

4. Quantum Numbers, Eigenvalues, and Degeneracy

According to equation (10–16), the eigenfunctions are

$$\psi_{nlm}(r, \theta, \phi) = R_{nl}(r)\, \Theta_{lm}(\theta)\, \Phi_m(\phi) \qquad (10\text{–}35)$$

All three quantum numbers are required to identify the eigenfunctions because their mathematical form depends on the values of each of the quantum numbers. That three quantum numbers arise is a consequence of the fact that the time independent Schroedinger equation contains three

† Note that for a given set of quantum numbers there is only one acceptable solution $R_{nl}(\rho)$ to equation (10–31), even though this is a second order differential equation and therefore has two particular solutions. The reason is that one of the particular solutions is not acceptable as it becomes infinite within the range of the variable ρ. The same situation obtains concerning the single acceptable solution $\Theta_{lm}(\xi)$ to (10–23).

independent variables. Gathering together the conditions which the quantum numbers satisfy, we have

$$\begin{aligned} |m| &= 0, 1, 2, 3, \ldots \\ l &= |m|, |m| + 1, |m| + 2, \ldots \\ n &= l + 1, l + 2, l + 3, \ldots \end{aligned} \qquad (10\text{–}36)$$

These conditions are more conveniently expressed in the form

$$\begin{aligned} n &= 1, 2, 3, 4, \ldots \\ l &= 0, 1, 2, \ldots, n - 1 \\ m &= -l, -l + 1, \ldots, 0, \ldots, +l - 1, +l \end{aligned} \qquad (10\text{–}37)$$

It is easy to see that equations (10–36) and (10–37) are equivalent.†

Now let us combine equations (10–29), (10–30), and (10–32) to obtain an expression for the allowed values of total energy for the atom containing a single *bound* electron. This gives

$$E = -\frac{\hbar^2\beta^2}{2\mu} = -\frac{\hbar^2}{2\mu}\frac{\mu^2Z^2e^4}{\hbar^4\gamma^2} = -\frac{\hbar^2\mu^2Z^2e^4}{2\mu\hbar^4n^2}$$

which is

$$E_n = -\frac{\mu Z^2e^4}{2\hbar^2n^2}, \qquad n = 1, 2, 3, 4, \ldots \qquad (10\text{–}38)$$

Comparing this equation with (5–14), we see that the eigenvalues predicted by quantum mechanics are identical with those predicted by the Bohr theory.‡ Schroedinger's derivation of this equation (1926) constituted the first convincing verification of the validity of the theory of quantum mechanics.

It is important to note that, whereas the form of the eigenfunctions depends on the values of all three quantum numbers n, l, m, *the eigenvalues depend only on the quantum number* n. Since for a given value of n there are generally a number of different possible values of l and m, it is apparent

† According to equations (10–36), the minimum value of l is equal to $|m|$, and the minimum value of $|m|$ is 0. Thus the minimum value of l is 0 and the minimum value of n, which is equal to $l + 1$, is $0 + 1 = 1$. Since n increases by integers without limit, the possible values of n are $n = 1, 2, 3, 4, \ldots$. For a given n, the maximum value of l is the one satisfying the relation $n = l + 1$, that is, $l = n - 1$. Consequently the possible values of l are $l = 0, 1, 2, \ldots, n - 1$. Finally, for a given l, the largest value which $|m|$ can assume is $|m| = l$. Thus the maximum value of m is $+l$ and the minimum value is $-l$, and it can assume only the values $m = -l, -l + 1, \ldots, 0, \ldots, +l - 1, +l$.

‡ Remember that in (5–14) m will be replaced by μ when proper account is taken of the finite nuclear mass.

that there will be situations in which two or more completely different eigenfunctions are degenerate because they correspond to exactly the same eigenvalue. The degeneracy with respect to m arises whenever the potential depends only upon r, while the l degeneracy is a consequence of the particular form of the r dependence of the Coulomb potential $V = -Ze^2/r$ (cf. section 7, Chapter 5). From equations (10–37) it is easy to see how many degenerate eigenfunctions there are which correspond to a particular

TABLE (10–1)
Possible Values of l and m for $n = 1, 2, 3$

n	1	2		3		
l	0	0	1	0	1	2
m	0	0	$-1, 0, +1$	0	$-1, 0, +1$	$-2, -1, 0, +1, +2$
Number of degenerate eigenfunctions for each l	1	1	3	1	3	5
Number of degenerate eigenfunctions for each n	1	4		9		

eigenvalue E_n. The possible values of the quantum numbers for $n = 1, 2,$ and 3 are shown in table (10–1). From this table, or from equations (10–37), it is apparent that:

1. For each value of n, there are n possible values of l.
2. For each value of l, there are $(2l + 1)$ possible values of m.
3. For each value of n, there are a total of n^2 degenerate eigenfunctions.

5. Eigenfunctions and Probability Densities

The eigenfunctions are

$$\psi_{nlm}(r, \theta, \phi) = R_{nl}(r)\, \Theta_{lm}(\theta)\, \Phi_m(\phi)$$

and we know that

$$\Phi_m(\phi) = e^{im\phi}$$

$$\Theta_{lm}(\theta) = \text{(polynomial in } \cos\theta)$$

$$R_{nl}(r) = e^{-(\text{constant})r}\,\text{(polynomial in } r)$$

All the bound state eigenfunctions have basically the same structure, except that with increasing values of n and l the polynomials in r and in $\cos\theta$ become increasingly more complicated. In addition, there exist eigenfunctions for unbound states (e.g., an ionized hydrogen atom plus an unbound electron in which $E > 0$) which have a somewhat different structure, but we shall not be concerned at present with situations in which unbound state eigenfunctions are needed. We list below the bound state eigenfunctions for the first three values of n.

$$
\begin{aligned}
\psi_{100} &= \frac{1}{\sqrt{\pi}}\left(\frac{Z}{a_0}\right)^{3/2} e^{-Zr/a_0} \\
\psi_{200} &= \frac{1}{4\sqrt{2\pi}}\left(\frac{Z}{a_0}\right)^{3/2}\left(2 - \frac{Zr}{a_0}\right) e^{-Zr/2a_0} \\
\psi_{210} &= \frac{1}{4\sqrt{2\pi}}\left(\frac{Z}{a_0}\right)^{3/2} \frac{Zr}{a_0} e^{-Zr/2a_0} \cos\theta \\
\psi_{21\pm1} &= \frac{1}{8\sqrt{\pi}}\left(\frac{Z}{a_0}\right)^{3/2} \frac{Zr}{a_0} e^{-Zr/2a_0} \sin\theta\, e^{\pm i\phi} \\
\psi_{300} &= \frac{1}{81\sqrt{3\pi}}\left(\frac{Z}{a_0}\right)^{3/2}\left(27 - 18\frac{Zr}{a_0} + 2\frac{Z^2r^2}{a_0^2}\right) e^{-Zr/3a_0} \\
\psi_{310} &= \frac{\sqrt{2}}{81\sqrt{\pi}}\left(\frac{Z}{a_0}\right)^{3/2}\left(6 - \frac{Zr}{a_0}\right)\frac{Zr}{a_0} e^{-Zr/3a_0} \cos\theta \\
\psi_{31\pm1} &= \frac{1}{81\sqrt{\pi}}\left(\frac{Z}{a_0}\right)^{3/2}\left(6 - \frac{Zr}{a_0}\right)\frac{Zr}{a_0} e^{-Zr/3a_0} \sin\theta\, e^{\pm i\phi} \\
\psi_{320} &= \frac{1}{81\sqrt{6\pi}}\left(\frac{Z}{a_0}\right)^{3/2} \frac{Z^2r^2}{a_0^2} e^{-Zr/3a_0}(3\cos^2\theta - 1) \\
\psi_{32\pm1} &= \frac{1}{81\sqrt{\pi}}\left(\frac{Z}{a_0}\right)^{3/2} \frac{Z^2r^2}{a_0^2} e^{-Zr/3a_0} \sin\theta\cos\theta\, e^{\pm i\phi} \\
\psi_{32\pm2} &= \frac{1}{162\sqrt{\pi}}\left(\frac{Z}{a_0}\right)^{3/2} \frac{Z^2r^2}{a_0^2} e^{-Zr/3a_0} \sin^2\theta\, e^{\pm 2i\phi}
\end{aligned}
\tag{10–39}
$$

These are written in terms of the parameter

$$a_0 \equiv \frac{\hbar^2}{\mu e^2} = 0.529 \times 10^{-8}\ \text{cm} \tag{10–40}$$

which the reader will recognize as the radius of the smallest orbit of a Bohr hydrogen atom.

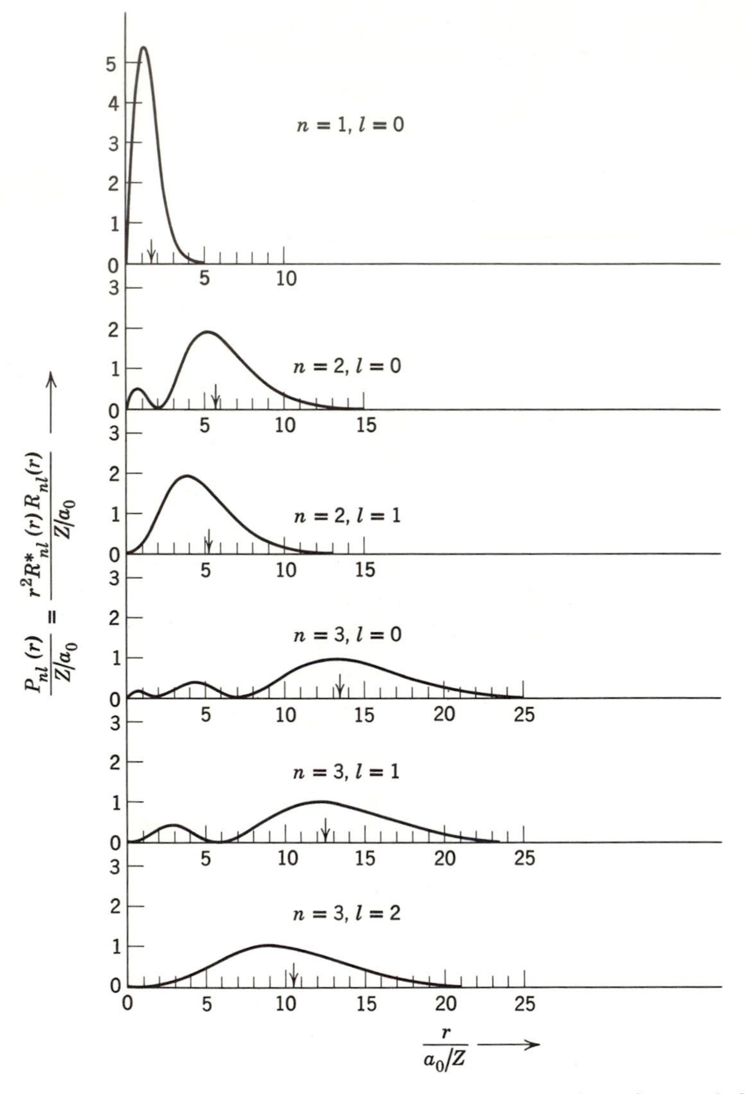

Figure 10–2. The radial probability densities of the one-electron atom for $n = 1, 2, 3$.

The eigenfunctions (10–39) are normalized. That is,

$$\int_0^\infty \int_0^\pi \int_0^{2\pi} \psi^*_{nlm}(r, \theta, \phi)\, \psi_{nlm}(r, \theta, \phi) r^2 \sin\theta\, dr\, d\theta\, d\phi = 1 \quad (10\text{–}41)$$

where $r^2 \sin\theta\, dr\, d\theta\, d\phi$ is the unit of volume in the system of spherical coordinates, and where the integral extends over all space. Also the eigenfunctions are orthogonal, so that

$$\int_0^\infty \int_0^\pi \int_0^{2\pi} \psi^*_{n'l'm'}(r, \theta, \phi)\, \psi_{nlm}(r, \theta, \phi) r^2 \sin\theta\, dr\, d\theta\, d\phi = 0 \quad (10\text{–}42)$$

unless $n' = n$, $l' = l$, and $m' = m$.

An interpretation of the eigenfunctions is best obtained by considering the form of the corresponding probability density functions $\psi^*_{nlm}\psi_{nlm} = R^*_{nl}\Theta^*_{lm}\Phi^*_m R_{nl}\Theta_{lm}\Phi_m$. As these are functions of three variables, we cannot plot them directly. However, an adequate understanding of their three dimensional behavior can be had by discussing separately their dependence on each variable. We consider first the r dependence in terms of the *radial probability density* $P_{nl}(r)$, defined by

$$P_{nl}(r)\, dr \equiv \int_0^\pi \int_0^{2\pi} \psi^*_{nlm}\psi_{nlm} r^2 \sin\theta\, dr\, d\theta\, d\phi$$
$$= r^2 R^*_{nl}(r)\, R_{nl}(r)\, dr \int_0^\pi \int_0^{2\pi} \Theta^*_{lm}\Theta_{lm}\Phi^*_m\Phi_m \sin\theta\, d\theta\, d\phi$$

The integrals over θ and ϕ are equal to unity because each of the functions R_{nl}, Θ_{lm}, and Φ_m is separately normalized. (They are also separately orthogonal.) Thus

$$P_{nl}(r)\, dr = r^2 R^*_{nl}(r)\, R_{nl}(r)\, dr \quad (10\text{–}43)$$

Now $\psi^*_{nlm}\psi_{nlm} r^2 \sin\theta\, dr\, d\theta\, d\phi$ is the probability of finding the electron in the volume element $r^2 \sin\theta\, dr\, d\theta\, d\phi$. Since $P_{nl}(r)\, dr$ is obtained from this quantity by integrating over the entire range of θ and ϕ, $P_{nl}(r)\, dr$ is the probability of finding the electron anywhere with a radial coordinate between r and $r + dr$. In figure (10–2), $P_{nl}(r)$ is plotted in units of Z/a_0 as a function of $r/(a_0/Z)$ for the cases $n = 1$, 2, and 3, and all the possible l values for these values of n. The radial probability density $P_{nl}(r)$ does not depend on the quantum number m.

We see that the radial probability densities for each eigenstate have appreciable values only in reasonably restricted ranges of the r axis. Thus, when the atom is in an eigenstate, there is a high probability that the radial coordinate of the electron will be found within a reasonably well-defined range; that is, the electron would quite probably be found within a certain *shell* limited by two concentric spherical surfaces centered on the nucleus. A study of these figures will demonstrate that the characteristic radii of these shells is determined primarily by the quantum number n, although there is a small l dependence. This can be seen in a more quantitative way by using the expectation value of the radial coordinate of the electron to characterize the radius of the shell. The expectation value is obtained by evaluating the integral

$$\overline{r_{nl}} = \int_0^\infty r P_{nl}(r)\, dr = \int_0^\infty \int_0^\pi \int_0^{2\pi} \psi^*_{nlm}\, r\, \psi_{nlm} r^2 \sin\theta \, dr\, d\theta\, d\phi$$

The result is

$$\overline{r_{nl}} = \frac{n^2 a_0}{Z}\left\{1 + \tfrac{1}{2}\left[1 - \frac{l(l+1)}{n^2}\right]\right\} \tag{10–44}$$

The values of $\overline{r_{nl}}$ are indicated in figure (10–2) with small arrows. It is apparent that $\overline{r_{nl}}$ depends primarily on n, since the l dependence is suppressed by the factor $\frac{1}{2}$ and the factor $1/n^2$. It is interesting to compare this expression with equation (5–12), which gives the radii of the circular orbits of a Bohr atom. The equation is

$$r_{\text{Bohr}} = \frac{n^2 a_0}{Z}$$

Quantum mechanics shows that the $\overline{r_{nl}}$ are of approximately the same size, and have approximately the same n dependence, as the radii of the circular Bohr orbits.

All electrons in eigenstates with common n values have roughly similar radial probability densities and approximately the same expectation value for the radial coordinate, independent of the values of l or m. According to the commonly used terminology, such electrons are said to be in the same shell.† Each shell also has the property that the associated eigenvalues all have exactly the same value E_n.

Some insight into the reason why the eigenvalues depend only upon n may be obtained by computing the expectation value of the potential energy of the atom. This is

$$\bar{V} = \int_0^\infty \int_0^\pi \int_0^{2\pi} \psi^*_{nlm}\left(\frac{-Ze^2}{r}\right)\psi_{nlm} r^2 \sin\theta\, dr\, d\theta\, d\phi = -Ze^2\overline{\left(\frac{1}{r}\right)}$$

Evaluating $\overline{(1/r)}$ leads to the result

$$\overline{\left(\frac{1}{r}\right)} = \frac{Z}{a_0 n^2} = \frac{\mu Z e^2}{\hbar^2 n^2}$$

† Since at present we are considering one-electron atoms, this terminology refers to two or more different atoms in each of which the electron is in the same shell. However, this terminology is also used for multi-electron atoms in which two or more electrons can literally be in the same shell.

Since this does not depend on l or m, the expectation value of the potential energy,

$$\bar{V} = -\frac{\mu Z^2 e^4}{\hbar^2 n^2}$$

does not either. This is reflected in the behavior of the total energy E, because it is a characteristic property of any potential of the form $V(r) \propto 1/r$ that the total energy of a particle moving in the potential is always equal to exactly one-half its average potential energy.

Finally, we call attention to a point which will later be seen to be of considerable importance. Inspection of the eigenfunctions (10–39) will verify that for values of r which are small compared to a_0/Z (where the exponential term is slowly varying), the radial dependence of the eigenfunctions is

$$\psi_{nlm} \propto r^l, \qquad r \to 0 \tag{10–45}$$

Therefore the radial dependence of the probability density for small r is

$$\psi^*_{nlm}\psi_{nlm} \propto r^{2l}, \qquad r \to 0 \tag{10–45$'$}$$

As a consequence, the value of $\psi^*_{nlm}\psi_{nlm}$ in a volume element near $r = 0$ is relatively large only for $l = 0$, and decreases very rapidly with increasing l.

Consider again the probability density,

$$\psi^*_{nlm}\psi_{nlm} = R^*_{nl}R_{nl}\Theta^*_{lm}\Theta_{lm}\Phi^*_m\Phi_m$$

From equation (10–21), we have

$$\Phi^*_m(\phi)\,\Phi_m(\phi) = e^{-im\phi}e^{im\phi} = 1$$

Thus the probability density does not depend on the coordinate ϕ. The three dimensional behavior of $\psi^*_{nlm}\psi_{nlm}$ is therefore completely specified by the product of the quantity $R^*_{nl}(r)\,R_{nl}(r) = P_{nl}(r)/r^2$ and the quantity $\Theta^*_{lm}(\theta)\,\Theta_{lm}(\theta)$, which plays the role of a directionally dependent modulation factor. The form of $\Theta^*_{lm}(\theta)\,\Theta_{lm}(\theta)$ is conveniently presented in terms of a polar diagram as shown in figure (10–3). The origin of the diagram is at the point $r = 0$ (the nucleus), and the z axis is taken along the direction from which the angle θ is measured. The distance from the origin to the curve, measured at an angle θ, is equal to the value of $\Theta^*_{lm}(\theta)\,\Theta_{lm}(\theta)$ for that angle. Such a diagram can also be thought of as representing the complete directional dependence of $\psi^*_{nlm}\psi_{nlm}$ by visualizing the three dimensional surface obtained by rotating the diagram about the z axis through the 360° range of the angle ϕ. The distance, measured in the direction specified by the angles θ and ϕ from the origin to a point on the surface, is equal to $\Theta^*_{lm}(\theta)\,\Theta_{lm}(\theta)\,\Phi^*_m(\phi)\,\Phi_m(\phi)$ for those values of θ and ϕ.

In figure (10–4) we illustrate a typical example of the dependence of the form of the function $\Theta^*_{lm}(\theta)\,\Theta_{lm}(\theta)$ on the quantum number m, by a set of polar diagrams for $l = 3$, and the seven possible values of m for this

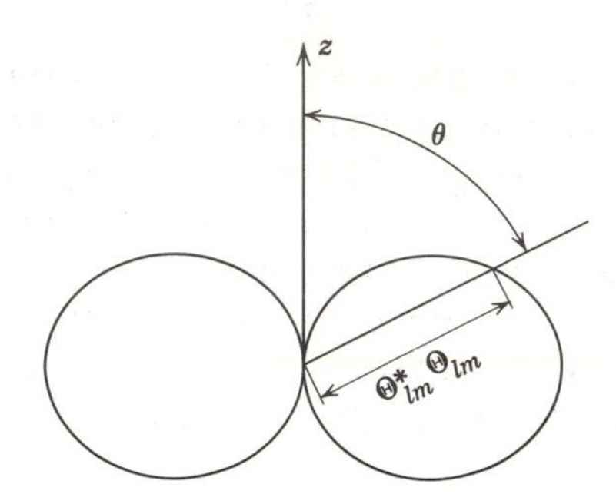

Figure 10–3. A polar diagram of the directionally dependent modulation factor of the probability density.

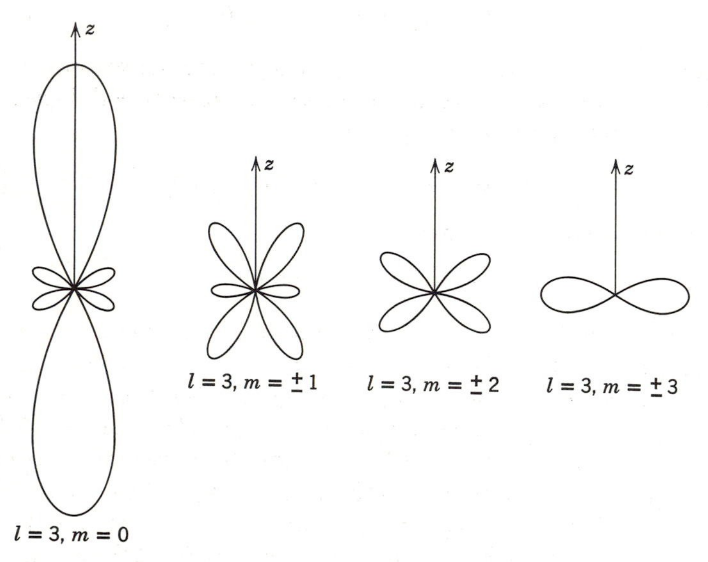

Figure 10–4. Polar diagrams of the directionally dependent modulation factors of the probability densities for $l = 3$; $m = 0, \pm 1, \pm 2, \pm 3$. From L. Pauling and E. B. Wilson, *Introduction to Quantum Mechanics*, McGraw-Hill Book Co., New York, 1935.

value of l, i.e., $|m| = 0, 1, 2, 3$. Note the way in which the region of concentration of $\Theta^*_{lm}\Theta_{lm}$ (and therefore of $\psi^*_{nlm}\psi_{nlm}$) shifts from the z axis to the plane perpendicular to the z axis as the absolute value of m increases. Some important features of the dependence of $\Theta^*_{lm}(\theta)\,\Theta_{lm}(\theta)$

on the quantum number l can be indicated by presenting a set of polar plots with $|m| = l$ and $l = 0, 1, 2, 3, 4$, as illustrated in figure (10–5). For the eigenstate with $n = 1$, $l = m = 0$, which is the ground state of the atom, $\psi^*_{nlm}\psi_{nlm}$ depends on neither θ nor ϕ and the probability density is spherically symmetric. For other states, the concentration of probability density in the plane perpendicular to the z axis, when $|m| = l$, becomes more and more pronounced with increasing l.

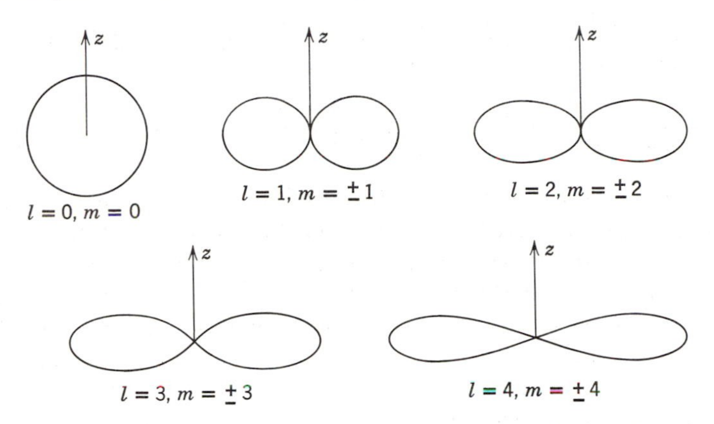

Figure 10–5. Polar diagrams of the directionally dependent modulation factors of the probability densities for $l = 0, 1, 2, 3, 4$; $|m| = l$. From L. Pauling and E. B. Wilson, *Introduction to Quantum Mechanics*, McGraw-Hill Book Co., New York, 1935.

The set of quantum numbers nlm determines the complete three dimensional behavior of $\psi^*_{nlm}\psi_{nlm}$, with the behavior as a function of the direction relative to the z axis depending only on l and m. For a typical set of quantum numbers, $R^*_{nl}(r)\, R_{nl}(r)$ has zeros for several values of r, and $\Theta^*_{lm}(\theta)\, \Theta_{lm}(\theta)$ has zeros for several values of θ. Thus the probability density for an electron bound in the Coulomb potential of a nucleus has, in general, a set of spherical and conical nodal surfaces on which its value is zero. These surfaces are analogous to the nodal points in the probability density of a particle bound in a one dimensional potential, and they are a consequence of the fact that the wave functions for a bound particle must be standing waves. Of course, it is not possible to perform a sequence of measurements on the location of an electron in a particular eigenstate with quantum numbers nlm, and thereby locate the nodal surfaces and determine the direction of the z axis. This would amount to finding a preferred direction in a space which should be spherically symmetric because the potential $V = -Ze^2/r$ is spherically symmetric. It cannot be done because any eigenstate, except the spherically symmetric eigenstate $n = 1$, $l = m = 0$, is degenerate with several other eigenstates with the same value of n and so cannot be investigated separately. All that can be measured experimentally is the average probability density for the whole set of states which are degenerate with each other. For instance, it would be possible to make measurements on the probability density for atoms whose energy is E_2. The results would be

$$\begin{aligned}\tfrac{1}{4}[\psi^*_{200}\psi_{200} + \psi^*_{21-1}\psi_{21-1} + \psi^*_{210}\psi_{210} + \psi^*_{211}\psi_{211}] \\ = \frac{1}{128\pi}\left(\frac{Z}{a_0}\right)^3 e^{-Zr/a_0}\left[\left(2 - \frac{Zr}{a_0}\right)^2 + \left(\frac{Zr}{a_0}\right)^2(\tfrac{1}{2}\sin^2\theta + \tfrac{1}{2}\sin^2\theta + \cos^2\theta)\right] \\ = \frac{1}{128\pi}\left(\frac{Z}{a_0}\right)^3 e^{-Zr/a_0}\left[\left(2 - \frac{Zr}{a_0}\right)^2 + \left(\frac{Zr}{a_0}\right)^2\right]\end{aligned} \tag{10-46}$$

This is a spherically symmetric distribution, and it cannot be used to determine the direction of the z axis. Thus there is no contradiction of the fact that this direction was initially chosen in a completely arbitrary way. On the other hand, consider a situation in which the orientation of the z axis is not arbitrary because there is a preferred direction defined by an external magnetic or electric field applied to the atom. In such a field, as we shall see later, the eigenstates are not degenerate and measurements of the probability density of a particular eigenstate could be performed. In fact, such measurements could be used to determine the direction of the external field.

We shall show shortly that the quantum numbers l and m are also related to the magnitude of the orbital angular momentum L of the electron, and to its z component L_z, by the equations

$$\begin{aligned} L &= \sqrt{l(l+1)}\,\hbar \\ L_z &= m\hbar \end{aligned} \tag{10-47}$$

We mention this now because it is an important clue to the interpretation of the dependence of $\psi^*_{nlm}\psi_{nlm}$ on l and m. Consider the case $m = l$. Then $L_z = l\hbar$, which is almost equal to $L = \sqrt{l(l+1)}\,\hbar$. In this case the angular momentum vector must point nearly in the direction of the z axis. For a Bohr atom this would mean that the orbit of the electron would lie nearly in the plane perpendicular to the z axis. With increasing values of l, $l\hbar \rightarrow \sqrt{l(l+1)}\,\hbar$ and $L_z \rightarrow L$, so the angle between the angular momentum vector and the z axis decreases. In terms of the Bohr picture, this demands that the orbit lie more nearly in the plane perpendicular to the z axis. An inspection of the polar diagrams will demonstrate the correspondence between these features of $\psi^*_{nlm}\psi_{nlm}$ and the picture of a Bohr orbit. For $m = 0$ we have $L_z = 0$, and the angular momentum vector must be perpendicular to the z axis. In a Bohr atom this would mean that the plane of the orbit contained the z axis. Some indication of this behavior can be seen in the polar diagram for $l = 3$, $m = 0$.

Although there are many points at which the quantum mechanical theory of the one-electron atom corresponds quite closely to the Bohr theory, there are certain striking differences. In both theories the ground state corresponds to the quantum number $n = 1$. But in the Bohr theory the orbital angular momentum for this state is $L = n\hbar = \hbar$, whereas in quantum mechanics $L = \sqrt{l(l+1)}\,\hbar = 0$ since $l = 0$ when $n = 1$. There is an overwhelming accumulation of experimental evidence which shows that the quantum mechanical prediction is the correct one. It would be very difficult to modify the Bohr theory in a way which would allow for zero angular momentum states, since the orbit for such a state would be a radial oscillation passing directly through the nucleus. In quantum mechanics the electron does not move in a well-defined orbit, and so no problem arises.

6. Angular Momentum Operators

In the next sections we shall discuss the orbital angular momentum of the eigenstates of a one-electron atom. We shall start by simply calculating the expectation values for this quantity, but we shall find that it is possible to obtain a very much more precise evaluation. In the course of this discussion a new and broader insight into the general properties of the theory of quantum mechanics will be obtained.

In classical mechanics the angular momentum of a particle, relative to the origin of a certain coordinate system, is a vector quantity $\mathbf{L}$ which is given by the equation

$$\mathbf{L} = \mathbf{r} \times \mathbf{p} \tag{10–48}$$

The vector $\mathbf{r}$ is the position of the particle relative to the origin, and the vector $\mathbf{p}$ is the linear momentum of the particle. These vectors are indicated in figure (10–6). The "cross product" $\mathbf{r} \times \mathbf{p}$ is defined to mean that the vector $\mathbf{L}$ is perpendicular to the plane formed by the vectors $\mathbf{r}$ and $\mathbf{p}$, and points in the direction so that $\mathbf{r}$, $\mathbf{p}$, and $\mathbf{L}$ have the same relative spatial orientation as do the x, y, z axes of a right-handed coordinate system. Furthermore, the cross product is defined so that the magnitude of the vector $\mathbf{L}$ is

$$L = rp \sin\theta \tag{10–49}$$

where r and p are the magnitudes of $\mathbf{r}$ and $\mathbf{p}$, and θ is the angle between these two vectors. From the definition of $\mathbf{r} \times \mathbf{p}$, it is easy to show that the three rectangular components of $\mathbf{L}$ are

$$\begin{aligned} L_x &= yp_z - zp_y \\ L_y &= zp_x - xp_z \\ L_z &= xp_y - yp_x \end{aligned} \tag{10–50}$$

where x, y, z are the components of $\mathbf{r}$, and p_x, p_y, p_z are the components of $\mathbf{p}$.

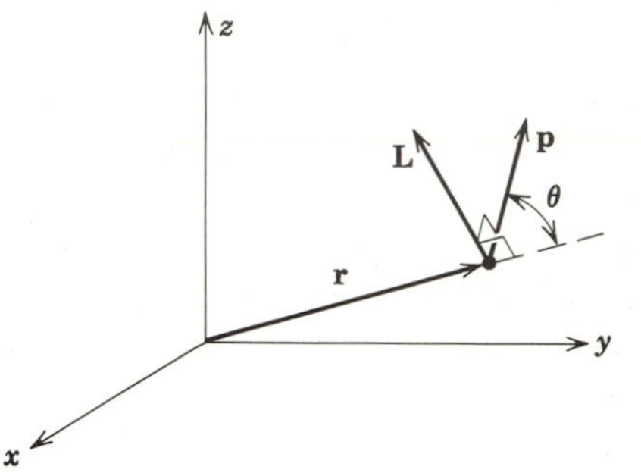

Figure 10–6. The position, linear momentum, and angular momentum vectors of a moving particle.

In order to discuss the dynamical quantity angular momentum in quantum mechanics, we construct the associated operators by replacing p_x, p_y, p_z with their quantum mechanical equivalents $-i\hbar\,\partial/\partial x$, $-i\hbar\,\partial/\partial y$, $-i\hbar\,\partial/\partial z$ (cf. 7–77). Thus the operators for the three components of angular momentum are

$$\begin{aligned} L_{x_{\text{op}}} &= -i\hbar\left(y\frac{\partial}{\partial z} - z\frac{\partial}{\partial y}\right) \\ L_{y_{\text{op}}} &= -i\hbar\left(z\frac{\partial}{\partial x} - x\frac{\partial}{\partial z}\right) \\ L_{z_{\text{op}}} &= -i\hbar\left(x\frac{\partial}{\partial y} - y\frac{\partial}{\partial x}\right) \end{aligned} \tag{10–51}$$

It is desirable to transform these expressions into the spherical coordinates r, θ, ϕ. When this is done, it is found that

$$\begin{aligned} L_{x_{\text{op}}} &= i\hbar\left(\sin\phi\frac{\partial}{\partial\theta} + \cot\theta\cos\phi\frac{\partial}{\partial\phi}\right) \\ L_{y_{\text{op}}} &= i\hbar\left(-\cos\phi\frac{\partial}{\partial\theta} + \cot\theta\sin\phi\frac{\partial}{\partial\phi}\right) \\ L_{z_{\text{op}}} &= -i\hbar\frac{\partial}{\partial\phi} \end{aligned} \tag{10–52}$$

We shall also be interested in the square of the magnitude of the vector **L**, which is

$$L^2 = L_x^2 + L_y^2 + L_z^2 \tag{10–53}$$

In spherical coordinates the associated operator is

$$L^2_{\text{op}} = -\hbar^2\left[\frac{1}{\sin\theta}\frac{\partial}{\partial\theta}\left(\sin\theta\frac{\partial}{\partial\theta}\right) + \frac{1}{\sin^2\theta}\frac{\partial^2}{d\phi^2}\right] \tag{10–54}$$

Let us calculate the expectation values of the components of L, and of the square of its magnitude, for an electron in the eigenstate nlm of a one-electron atom.† According to equation (7–80), extended to three dimensions, they are given by

$$\overline{L_x} = \iiint \Psi^* L_{x_{\text{op}}} \Psi r^2 \sin\theta\, dr\, d\theta\, d\phi$$

$$= \iiint e^{iE_nt/\hbar}\psi^*_{nlm} L_{x_{\text{op}}} e^{-iE_nt/\hbar}\psi_{nlm} r^2 \sin\theta\, dr\, d\theta\, d\phi$$

So

$$\overline{L_x} = \iiint \psi^*_{nlm} L_{x_{\text{op}}} \psi_{nlm} r^2 \sin\theta\, dr\, d\theta\, d\phi$$

Similarly,

$$\overline{L_y} = \iiint \psi^*_{nlm} L_{y_{\text{op}}} \psi_{nlm} r^2 \sin\theta\, dr\, d\theta\, d\phi$$

$$\overline{L_z} = \iiint \psi^*_{nlm} L_{z_{\text{op}}} \psi_{nlm} r^2 \sin\theta\, dr\, d\theta\, d\phi \tag{10–55}$$

$$\overline{L^2} = \iiint \psi^*_{nlm} L^2_{\text{op}} \psi_{nlm} r^2 \sin\theta\, dr\, d\theta\, d\phi$$

where the integrations are over the entire range of the variables. To evaluate these integrals, we must first evaluate the quantities

$$L_{x_{\text{op}}}\psi_{nlm}, \quad L_{y_{\text{op}}}\psi_{nlm}, \quad L_{z_{\text{op}}}\psi_{nlm}, \quad L^2_{\text{op}}\psi_{nlm}$$

In doing this, we shall find that the last two lead to an interesting result. Consider first the quantity

$$L_{z_{\text{op}}}\psi_{nlm} = -i\hbar\frac{\partial\psi_{nlm}}{\partial\phi}$$

† Strictly speaking, we shall calculate the components and squared magnitude of a vector **L** which is the orbital angular momentum of the electron plus the orbital angular momentum of the nucleus, relative to the center of mass of the atom. This is because the eigenfunctions ψ_{nlm} actually describe the motion of an electron of reduced mass μ, which is moving about a fixed nucleus in such a way that the orbital angular momentum of the reduced mass electron is equal to the total orbital angular momentum of the atom. Cf. section 2.

Since

$$\psi_{nlm} = R_{nl}(r)\,\Theta_{lm}(\theta)\,\Phi_m(\phi)$$

we have

$$-i\hbar\frac{\partial\psi_{nlm}}{\partial\phi} = R_{nl}(r)\,\Theta_{lm}(\theta)\left[-i\hbar\frac{d\Phi_m(\phi)}{d\phi}\right]$$

According to equation (10–21),

$$\Phi_m(\phi) = e^{im\phi}$$

So

$$\frac{d\Phi_m(\phi)}{d\phi} = ime^{im\phi} = im\Phi_m(\phi)$$

Thus

$$\begin{aligned}-i\hbar\frac{\partial\psi_{nlm}}{\partial\phi} &= R_{nl}(r)\,\Theta_{lm}(\theta)[-i\hbar im\Phi_m(\phi)]\\ &= m\hbar R_{nl}(r)\,\Theta_{lm}(\theta)\,\Phi_m(\phi)\end{aligned}$$

and

$$L_{z_{\mathrm{op}}}\psi_{nlm} = m\hbar\psi_{nlm} \tag{10–56}$$

Next evaluate

$$L^2_{\mathrm{op}}\psi_{nlm} = R_{nl}(r)(-\hbar^2)\left[\frac{\Phi_m(\phi)}{\sin\theta}\frac{d}{d\theta}\left(\sin\theta\frac{d\Theta_{lm}(\theta)}{d\theta}\right) + \frac{\Theta_{lm}(\theta)}{\sin^2\theta}\frac{d^2\Phi_m(\phi)}{d\phi^2}\right]$$

Differentiating equation (10–21), we have

$$\frac{d^2\Phi_m(\phi)}{d\phi^2} = -m^2\Phi_m(\phi)$$

So

$$L^2_{\mathrm{op}}\psi_{nlm} = R_{nl}(r)\,\Phi_m(\phi)(-\hbar^2)\left[\frac{1}{\sin\theta}\frac{d}{d\theta}\left(\sin\theta\frac{d\Theta_{lm}(\theta)}{d\theta}\right) - \frac{m^2}{\sin^2\theta}\Theta_{lm}(\theta)\right]$$

Now $\Theta_{lm}(\theta)$ is a function which is a solution to the differential equation (10–18). This is

$$\frac{1}{\sin\theta}\frac{d}{d\theta}\left(\sin\theta\frac{d\Theta_{lm}(\theta)}{d\theta}\right) - \frac{m^2}{\sin^2\theta}\Theta_{lm}(\theta) = -l(l+1)\Theta_{lm}(\theta)$$

where we have evaluated α from (10–24). Therefore

$$\begin{aligned}L^2_{\mathrm{op}}\psi_{nlm} &= R_{nl}(r)\,\Phi_m(\phi)(-\hbar^2)[-l(l+1)\Theta_{lm}(\theta)]\\ &= l(l+1)\hbar^2 R_{nl}(r)\,\Theta_{lm}(\theta)\,\Phi_m(\phi)\end{aligned}$$

and we have

$$L^2_{\mathrm{op}}\psi_{nlm} = l(l+1)\hbar^2\psi_{nlm} \tag{10–57}$$

In equation (10–56), we see that $L_{z_{\mathrm{op}}}\psi_{nlm}$ is equal to $m\hbar\psi_{nlm}$. Thus the result of operating on the one-electron eigenfunction ψ_{nlm} with the differential operator $L_{z_{\mathrm{op}}}$ is simply to multiply that eigenfunction by the constant $m\hbar$. Similarly, equation (10–57) shows that the result of operating on the one-electron eigenfunction ψ_{nlm} with the differential operator L^2_{op} is to multiply that eigenfunction by the constant $l(l+1)\hbar^2$. These results are certainly not typical of the behavior of differential operators. For instance, if we operate on the function $f(x) = x^2$ with the differential operator d/dx, we obtain a very different function $f'(x) = 2x$. As another example, the reader may verify that the result of operating on ψ_{nlm} with the operators $L_{x_{\text{op}}}$ or $L_{y_{\text{op}}}$ is to produce new functions of r, θ, ϕ in which these variables enter in a different manner from the way they enter in the function ψ_{nlm}. That is,

$$L_{x_{\text{op}}}\psi_{nlm} \neq (\text{constant})\psi_{nlm}$$
$$L_{y_{\text{op}}}\psi_{nlm} \neq (\text{constant})\psi_{nlm} \tag{10–58}$$

Using equations (10–56) and (10–57), it is trivial to evaluate $\overline{L_z}$ and $\overline{L^2}$ from equations (10–55). These values are

$$\overline{L_z} = \iiint \psi^*_{nlm} L_{z_{\text{op}}} \psi_{nlm} r^2 \sin\theta \, dr \, d\theta \, d\phi$$
$$\overline{L_z} = m\hbar \iiint \psi^*_{nlm} \psi_{nlm} r^2 \sin\theta \, dr \, d\theta \, d\phi$$
$$\overline{L_z} = m\hbar \tag{10–59}$$

and

$$\overline{L^2} = \iiint \psi^*_{nlm} L^2_{\text{op}} \psi_{nlm} r^2 \sin\theta \, dr \, d\theta \, d\phi$$
$$\overline{L^2} = l(l+1)\hbar^2 \iiint \psi^*_{nlm} \psi_{nlm} r^2 \sin\theta \, dr \, d\theta \, d\phi$$
$$\overline{L^2} = l(l+1)\hbar^2 \tag{10–60}$$

where we have employed the normalization condition (10–41). Let us compare these results with the relations (10–47) which were stated without proof in the previous section. These relations are

$$L_z = m\hbar \tag{10–61}$$

$$L^2 = l(l+1)\hbar^2 \tag{10–62}$$

It is apparent that equations (10–59) and (10–60) are consistent with (10–61) and (10–62) but do not prove their validity, as the latter pair make very much stronger statements about the values of L_z and L^2, for the eigenstate nlm, than do the former. In order to complete the proof of the validity of equations (10–61) and (10–62), it will be necessary to make a short digression into the general properties of the theory of quantum mechanics.

7. Eigenvalue Equations

Consider a particle moving under the influence of a time independent potential. For simplicity, consider the motion to be one dimensional. Let f be any dynamical quantity useful in describing the motion of the particle. Then the average value of f, obtained in a series of measurements made when the particle is in a state of motion described by the wave function Ψ, is the expectation value $\bar{f}$. According to equation (7–80), it has the value

$$\bar{f} = \int \Psi^* f_{\text{op}} \Psi \, dx$$

where the integral is taken over the entire range of the variable. The quantity $\bar{f}$ provides a partial specification of the behavior of f. To obtain a more complete specification, we must learn something about the *fluctuations* of f. This can be done by studying the quantity $\overline{\Delta f^2}$, which is the expectation value of the square of the difference between the results of a particular measurement of f and the average of these measurements $\bar{f}$. That is,

$$\overline{\Delta f^2} \equiv \overline{(f - \bar{f})^2} \tag{10–63}$$

It is clear that information relating to the fluctuations in f is provided by $\overline{\Delta f^2}$. For instance, consider a case in which f can have only a single value (e.g., the total energy of a particle in an eigenstate); then f will always be observed to be exactly equal to the value $\bar{f}$, and $\overline{(f - \bar{f})^2}$ will be zero. However, if f fluctuates about its average value $\bar{f}$, then $\overline{(f - \bar{f})^2}$ will be greater than zero. To evaluate $\overline{\Delta f^2}$, we apply equation (7–80) to (10–63) and obtain

$$\overline{\Delta f^2} = \int \Psi^* [(f - \bar{f})^2]_{\text{op}} \Psi \, dx$$

Since $\bar{f}$ is a constant, the operator $[(f - \bar{f})^2]_{\text{op}}$ is simply $(f_{\text{op}} - \bar{f})^2$, and we have

$$\begin{aligned}
\overline{\Delta f^2} &= \int \Psi^* (f_{\text{op}} - \bar{f})^2 \Psi \, dx \\
&= \int \Psi^* (f_{\text{op}}^2 - 2 f_{\text{op}} \bar{f} + \bar{f}^2) \Psi \, dx \\
&= \int \Psi^* f_{\text{op}}^2 \Psi \, dx - 2\bar{f} \int \Psi^* f_{\text{op}} \Psi \, dx + \bar{f}^2 \int \Psi^* \Psi \, dx
\end{aligned}$$

Using equation (7–80) on the first two terms of the right side, and using the normalization condition on the third term, we obtain

$$\overline{\Delta f^2} = \overline{f^2} - 2\bar{f}\bar{f} + \bar{f}^2$$

or

$$\overline{\Delta f^2} = \overline{f^2} - \bar{f}^2 \tag{10–64}$$

Thus the quantity $\overline{\Delta f^2}$ is equal to the difference between the expectation value of the square of f, and the square of the expectation value of f.

Now let us assume that the particle is in a certain eigenstate so that

$$\Psi(x, t) = e^{-iEt/\hbar}\,\psi(x)$$

Then we have

$$\bar{f} = \int e^{iEt/\hbar}\psi^* f_{\text{op}} e^{-iEt/\hbar}\psi\, dx$$

According to equation (7–80), f_{op} never contains derivatives with respect to time. Consequently the term $e^{-iEt/\hbar}$ can pass through f_{op} as if it were a constant, and we have

$$\bar{f} = \int e^{iEt/\hbar} e^{-iEt/\hbar}\psi^* f_{\text{op}}\psi\, dx$$

or

$$\bar{f} = \int \psi^* f_{\text{op}}\psi\, dx \tag{10–65}$$

Similarly we can show that

$$\overline{f^2} = \int \psi^* f_{\text{op}}^2\psi\, dx \tag{10–66}$$

when the particle is in an eigenstate.

For a given dynamical quantity f, and a given eigenfunction ψ, we can evaluate $\overline{\Delta f^2}$ from the equations above. In general, it is found that $\overline{\Delta f^2} > 0$. However, as mentioned previously, situations exist in which $\overline{\Delta f^2} = 0$. We shall now show that this happens when the relation between the dynamical quantity f and the eigenfunction ψ is such that

$$f_{\text{op}}\psi = F\psi \tag{10–67}$$

where F is a constant. If this is the case, then in equation (10–65) we have

$$\bar{f} = \int \psi^* f_{\text{op}}\psi\, dx = \int \psi^* F\psi\, dx = F\int \psi^*\psi\, dx = F$$

so

$$\bar{f}^2 = F^2$$

In equation (10–66) we have

$$\overline{f^2} = \int \psi^* f_{\text{op}}^2 \psi \, dx = \int \psi^* f_{\text{op}} f_{\text{op}} \psi \, dx$$

$$= \int \psi^* f_{\text{op}} F \psi \, dx = F \int \psi^* f_{\text{op}} \psi \, dx = FF$$

so

$$\overline{f^2} = F^2$$

Then, from equation (10–64), we find

$$\overline{\Delta f^2} = \overline{f^2} - \bar{f}^2 = F^2 - F^2 = 0 \qquad (10\text{–}68)$$

Thus, when the relation (10–67) is true, the quantity $\overline{\Delta f^2}$, which is a measure of the fluctuations in the dynamical quantity f, is equal to zero. In any normal case this is adequate to prove that f can have only the single value $f = \bar{f} = F$. In exceptional cases equation (10–68) alone does not constitute a proof of this statement, but it is also necessary to show that

$$\overline{f^k} - \bar{f}^k = 0, \qquad \text{for any integer } k \qquad (10\text{–}69)$$

However, it is easy to see that when equation (10–67) is true the steps immediately preceding equation (10–68) can be repeated k times to show that the requirement (10–69) is satisfied. Thus we conclude that *when the relation between f and ψ is*

$$f_{\text{op}} \psi = F \psi \qquad (10\text{–}70)$$

the dynamical quantity f can have only the definite, precise value

$$f = F \qquad (10\text{–}70')$$

In the mathematical literature an equation of the form (10–70) is called an *eigenvalue equation*, the function ψ is called an *eigenfunction of the operator* f_{op}, and the constant F is called an *eigenvalue of the operator*. We have come across most of this terminology before in connection with a particular eigenvalue equation. Consider the time independent Schroedinger equation (10–8), and rewrite it

$$\left\{ \sum_{j=1}^{N} - \frac{\hbar^2}{2m_j} \nabla_j^2 + V \right\} \psi = E \psi$$

Comparing the curly bracket with the classical definition of the total energy of the system (10–1), we see that this equation may be written

$$e_{\text{op}} \psi = E \psi \qquad (10\text{–}71)$$

where E is a constant numerically equal to the total energy of the system, and where the total energy operator e_{op} is obtained from the corresponding dynamical quantity by replacing p_{x_j} with $-i\hbar\,\partial/\partial x_j$, etc. This operator is often called the *Hamiltonian*. For one particle in one dimension it is simply

$$e_{\text{op}} = -\frac{\hbar^2}{2m}\frac{d^2}{dx^2} + V$$

In the process of studying the particular eigenvalue equation (10–71), we have learned much about mathematical properties which are actually common to all such equations. For instance, we know that, if the function ψ and its derivative are required to satisfy conditions of continuity, finiteness, and single-valuedness, there will be solutions to equation (10–71) only for a certain set of eigenvalues $E_1, E_2, E_3, \ldots E_n, \ldots$ and eigenfunctions $\psi_1, \psi_2, \psi_3, \ldots, \psi_n, \ldots$. The same is true of the general eigenvalue equation (10–70). In interpreting the significance of equation (10–71), we concluded that the set of eigenvalues comprise the only physically possible values of the dynamical quantity whose corresponding operator appears in the equation. (That is, the allowed values of the total energy are the eigenvalues $E_1, E_2, \ldots$.) If we extend this to the general eigenvalue equation, we obtain a statement which, in advanced treatments of the subject, is usually taken as the *fundamental postulate of the theory of quantum mechanics*:

A measurement of the value of a dynamical quantity f can yield only one of the eigenvalues F_n of the corresponding operator f_{op}, where f_{op} is obtained from f by replacing p_x with $-i\hbar\,\partial/\partial x$, etc., and where the relation between F_n and f_{op} is

$$f_{\text{op}}\psi_n = F_n\psi_n$$

in which the eigenfunction ψ_n and its derivative satisfy conditions of continuity, finiteness, and single-valuedness.

This one-sentence postulate, and the postulates which provide a physical interpretation of the eigenfunctions, contain the complete theory of quantum mechanics for time independent phenomena. The theory is extended to include time dependent phenomena by introducing the operator $i\hbar\,\partial/\partial t$.

We may illustrate the theory developed in this section with some simple examples. Consider first the traveling wave eigenfunction

$$\psi = e^{iKx} \tag{10–72}$$

We know that this is an eigenfunction of e_{op}, as the function (10–72) was obtained by solving (10–71) for the case $V = 0$, and that the eigenvalue associated with this eigenfunction is $E = \hbar^2K^2/2m$. The function (10–72) is also an eigenfunction of the momentum operator $p_{op} = -i\hbar\, \partial/\partial x$ since

$$p_{op}\psi = -i\hbar \frac{\partial}{\partial x} \psi = -i\hbar \frac{\partial}{\partial x} e^{iKx} = -i\hbar iKe^{iKx} = \hbar K\psi$$

or

$$p_{op}\psi = P\psi$$

where the eigenvalue is $P = \hbar K$. We see that this function is simultaneously an eigenfunction of two operators. Therefore, according to equation (10–70′), a particle whose motion is described by this function must have a definite value of total energy $E = \hbar^2K^2/2m$ and a definite value of momentum $P = \hbar K$. Similarly, we may show that the traveling wave eigenfunction $\psi = e^{-iKx}$ is an eigenfunction of e_{op} associated with the eigenvalue $E = \hbar^2K^2/2m$, and is also an eigenfunction of p_{op} associated with the eigenvalue $P = -\hbar K$. Therefore a particle whose motion is described by this function has the definite value $E = \hbar^2K^2/2m$ for its total energy, and the definite value $P = -\hbar K$ for its momentum.

Next consider the standing wave eigenfunction

$$\psi = \sin Kx \tag{10–73}$$

We know that this is also an eigenfunction of e_{op} associated with the eigenvalue $E = \hbar^2K^2/2m$. However, it is not an eigenfunction of p_{op} since

$$p_{op}\psi = -i\hbar \frac{\partial}{\partial x} \sin Kx = -i\hbar K \cos Kx \neq (\text{constant})\psi$$

Although the energy of a particle whose motion is described by this function has the definite value $E = \hbar^2K^2/2m$, the momentum has no definite value. According to the postulate stated above, a measurement of the momentum of the particle can yield only one of the eigenvalues of the momentum operator. For a given value K there are just two eigenvalues of p_{op}, namely $P = \hbar K$ and $P = -\hbar K$. A measurement of the momentum of the particle would yield either of these two values and, in a large number of measurements, they would occur with equal probability.† It is evident that these properties of the standing wave eigenfunctions and traveling wave eigenfunctions are in complete agreement with the interpretation presented in the first section of Chapter 8.

† This is a consequence of the fact that $\bar{p} = 0$ for this eigenfunction, as the reader may verify.

8. Angular Momentum of One-Electron-Atom Eigenfunctions

We return now to the functions ψ_{nlm} which arise in treating the problem of a one-electron atom. Being solutions to the time independent Schroedinger equation for this problem, the ψ_{nlm} are necessarily eigenfunctions of the total energy operator. According to equation (10–38), the associated eigenvalues are $E_n = \mu Z^2e^4/2\hbar^2n^2$. From equations (10–56) and (10–57), we see that these functions are also eigenfunctions of the operators for the z component of the orbital angular momentum $L_{z_{op}}$, and for its square L^2_{op}, with associated eigenvalues $m\hbar$ and $l(l+1)\hbar^2$. However, equations (10–58) state that they are not eigenfunctions of the operators for the x and y components of orbital angular momentum. Also, it is easy to show that they are not eigenfunctions of any of the linear momentum operators. Therefore, when an electron is in a state described by ψ_{nlm}, the total energy of the atom is definitely E_n, the magnitude of the total orbital angular momentum is definitely

$$L = \sqrt{l(l+1)}\,\hbar \tag{10–74}$$

and its z component is definitely

$$L_z = m\hbar \tag{10–75}$$

but there is no definite value of the x or y components of orbital angular momentum, or of any component of linear momentum.

The fact that ψ_{nlm} does not describe a state of definite linear momentum is certainly reasonable, since any physical picture of the motion of an electron in a state having a definite magnitude and z component of orbital angular momentum calls to mind some sort of rotation in which linear momentum could not be constant. However, the fact that ψ_{nlm} does not describe a state with a definite x and y component of orbital angular momentum is somewhat mysterious because, in classical mechanics, the orbital angular momentum vector of an isolated system moving under the influence of a spherically symmetrical potential $V(r)$ would be fixed in space, and all three components would have definite values since there would be no torques acting on such a system. The quantum mechanical result is a manifestation of an uncertainty principle relation which can be shown to exist between any two components of orbital angular momentum. Because the z component of orbital angular momentum has the precise value $m\hbar$, this relation requires that the values of the x and y components be indefinite. Upon evaluating $\overline{L_x}$ and $\overline{L_y}$, it is found that both of these quantities are equal to zero. Thus the orientation of the orbital angular momentum vector must be constantly changing in such a way that its x and y components rapidly fluctuate about an average value of zero.

Many of the properties of the orbital angular momentum can be conveniently represented in terms of vector diagrams. Consider the set of states having a common value of the quantum number l. For each of these states the length of the orbital angular momentum vector, in units of $\hbar$, is $L/\hbar = \sqrt{l(l+1)}$. In the same units, the z component of this vector is $L_z/\hbar = m$ which, according to equations (10–37), may assume any integral value from $L_z/\hbar = -l$ to $L_z/\hbar = +l$, depending on the value of m. The

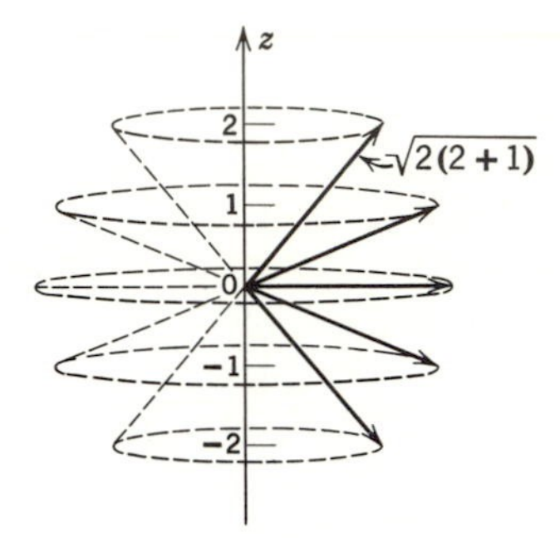

Figure 10–7. Illustrating the angular momentum vectors precessing about the z axis in the five possible states for $l = 2$.

case of $l = 2$ is illustrated in figure (10–7). The figure depicts the angular momentum vectors of the five states corresponding to the five possible values of m for this value of l. Since L_x and L_y fluctuate about their average values of zero, these vectors must precess randomly in the conical surfaces surrounding the z axis. The actual spatial orientation of the angular momentum vector is known with the greatest precision for the states with $m = \pm l$; but even for these states there is some uncertainty since the vector can be anywhere on a cone of half-angle $\cos^{-1}[l/\sqrt{l(l+1)}]$. In the classical limit $l \rightarrow \infty$, and this angle becomes vanishingly small. Thus, in this limit, the angular momentum vector for the states $m = \pm l$ is constrained to lie almost along the z axis and is therefore essentially fixed in space in agreement with the behavior predicted by the classical theory.

Since $L_z = m\hbar$ with m an integer, and since the direction of the z axis was chosen arbitrarily, it follows that quantum mechanics requires the component *in any direction* of the orbital angular momentum to be an integral multiple of $\hbar$. This is known as *spatial quantization*. If the reader will review the derivation of the eigenvalue equations (10–56) and (10–57), it will be evident that both spatial quantization and the quantization of the magnitude of the orbital angular momentum according to the rule $L = \sqrt{l(l+1)}\,\hbar$ occur not only for the Coulomb potential $V(r) = -Ze^2/r$, but *for any potential which depends only on the radial coordinate r.*

BIBLIOGRAPHY

Leighton, R. B., *Principles of Modern Physics*, McGraw-Hill Book Co., New York, 1959.

Pauling, L., and E. B. Wilson, *Introduction to Quantum Mechanics*, McGraw-Hill Book Co., New York, 1935.

Schiff, L. I., *Quantum Mechanics*, McGraw-Hill Book Co., New York, 1955.

EXERCISES

1. Write an expression for the probability flux *vector* for a system containing one particle in a three dimensional space.

2. Separate (10–10) into an equation in the rectangular coordinates of the center of mass of the atom, x, y, z, defined by (10–11), and an equation in the rectangular coordinates of the electron relative to the nucleus, X, Y, Z, defined by $X = x_2 - x_1$, $Y = y_2 - y_1$, $Z = z_2 - z_1$.

3. Transform the equation in X, Y, Z into an equation in the r, θ, ϕ of (10–12), and thereby verify (10–15). *Hint:* Consult some reference on intermediate calculus for an explanation of the use of Jacobians.

4. Solve equation (10–23) by the procedure indicated in the paragraph containing that equation.

5. Solve equation (10–31) by the procedure indicated in the paragraph containing that equation.

6. Consider a rigid rod of length R and negligible mass, with one end fixed at the origin in such a way that it can rotate freely and the other end carrying a particle of mass M. Adapt the time independent Schroedinger equation (10–15) to this system. Solve the equation and show that the eigenvalues are

$$E = \frac{\hbar^2 l(l+1)}{2I}, \qquad l = 0, 1, 2, 3, \ldots$$

where $I = MR^2$ is the moment of inertia about whatever axis the rod may rotate. Note that there is no zero point energy for this system because all the coordinates are free to assume all possible values. Compare these results with the results of exercise 3, Chapter 5.

7. Evaluate the expression analogous to (10–46) for $n = 3$, and show that it is spherically symmetric.

8. Verify equations (10–50).

9. Use some of the equations obtained in exercise 3 to verify equations (10–52) and (10–54).

10. Evaluate $\bar{p}$, $\overline{p^2}$, and $\overline{\Delta p^2}$ for the eigenfunction $\psi = \sin Kx$, and comment on their values. *Hint:* Use box normalization.

CHAPTER 11

Magnetic Moments, Spin, and Relativistic Effects

1. Orbital Magnetic Moments

In the first part of this chapter we shall discuss some experiments which provide very convincing evidence for spatial quantization. Actually the experiments do not directly measure the z component of the angular momentum vector $\mathbf{L}$, but instead measure the z component of a closely related quantity $\boldsymbol{\mu}$, the *magnetic moment* vector. To begin the discussion, we shall evaluate the *orbital magnetic moment* $\boldsymbol{\mu}_l$ of an atomic electron. This will be done by using a combination of simple electromagnetic theory, the Bohr theory, and quantum mechanics. A completely quantum mechanical calculation will not be presented because such a treatment would require a more advanced knowledge of electromagnetic theory than has been assumed in this book, and also because, owing to the conceptual simplicity of the description of the atom provided by the Bohr theory, there are advantages in using that theory when introducing something new. Of course, the Bohr theory does not always yield quantitatively correct results—but for the calculations we shall perform here it does.

Consider, then, an electron of mass m and charge $-e$ moving with velocity v in a circular Bohr orbit of radius r. This charge circulating in a loop constitutes a current of magnitude

$$i = \frac{e}{c}\frac{1}{\tau} = \frac{e}{c}\frac{v}{2\pi r} \tag{11–1}$$

where τ is the orbital period of the electron. It is well known from elementary electromagnetic theory that such a current loop produces a magnetic field which is the same at large distances as that of a magnetic dipole located at the center of the loop and oriented perpendicular to its plane. For a current i in a loop of area A, the magnitude of the magnetic moment of the equivalent dipole is

$$\mu_l = iA \tag{11–2}$$

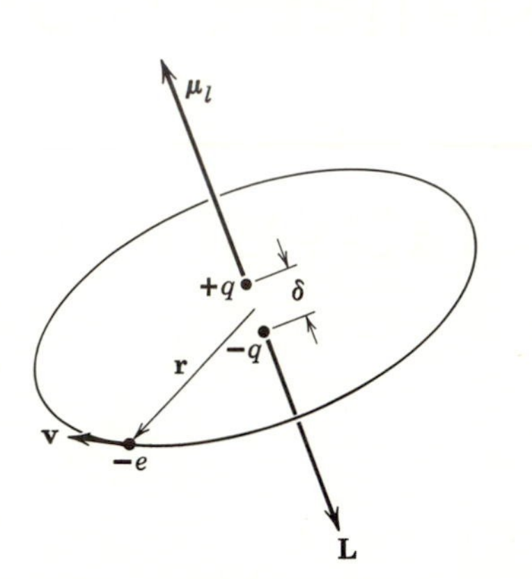

Figure 11–1. The quantities associated with an electron moving in a circular Bohr orbit

The magnetic moment is a measure of the strength and orientation of the dipole and is defined as

$$\boldsymbol{\mu}_l \equiv q\boldsymbol{\delta} \tag{11–3}$$

where q is the magnetic strength of each of the poles and $\boldsymbol{\delta}$ is a vector extending from the negative (south) pole to the positive (north) pole. For a current loop produced by a negatively charged particle, the direction of $\boldsymbol{\mu}_l$ is opposite the direction of the angular momentum vector $\mathbf{L}$, whose orientation and magnitude are given by the equation

$$\mathbf{L} = \mathbf{r} \times \mathbf{p} = m\mathbf{r} \times \mathbf{v} \tag{11–4}$$

These quantities are illustrated in figure (11–1). Evaluating i from equation (11–1) and A for the circular Bohr orbit, we find from (11–2) that the magnitude of the magnetic moment is

$$\mu_l = \frac{e}{c}\frac{v}{2\pi r}\pi r^2 = \frac{evr}{2c} \tag{11–5}$$

Since $\mathbf{r}$ and $\mathbf{v}$ are perpendicular in a circular orbit, the magnitude of the angular momentum is simply

$$L = mrv \tag{11–6}$$

From these two equations we see that the ratio of the magnitude of the magnetic moment to the magnitude of the angular momentum is a constant,

$$\frac{\mu_l}{L} = \frac{evr}{2cmrv} = \frac{e}{2mc} \tag{11–7}$$

This constant is usually written $g_l \mu_B / \hbar$, where

$$\mu_B = e\hbar/2mc = 0.927 \times 10^{-20} \text{ ergs/gauss} \tag{11–8}$$

and

$$g_l = 1 \tag{11–9}$$

The quantity μ_B forms a natural unit for the measurement of atomic magnetic moments and is called the *Bohr magneton.* The quantity g_l is called the *orbital g factor.* It is introduced, even though it appears redundant, because we shall later come across a g factor which is not unity. In terms of these quantities, we may rewrite equation (11–7) as a vector equation specifying both the magnitude of $\boldsymbol{\mu}_l$ and its orientation relative to $\mathbf{L}$:

$$\boldsymbol{\mu}_l = -\frac{g_l \mu_B}{\hbar} \mathbf{L} \tag{11–10}$$

The ratio of μ_l to L does not depend on the size of the orbit or on the orbital frequency. By making a calculation similar to the one above for an elliptical orbit, it can be shown that μ_l/L is independent of the shape of the orbit. That this ratio is completely independent of the details of the orbit suggests that its value might not depend on the details of the mechanical theory used to evaluate it, and this is actually the case. Upon evaluating μ_l quantum mechanically (which will not be done here), and dividing by the quantum mechanical expression $L = \sqrt{l(l+1)}\,\hbar$, the ratio of μ_l to L is found to have the same value that we obtained from the Bohr theory. Granting this, the reader will accept that the correct quantum mechanical expressions for the magnitude and the z component of the magnetic moment vector are

$$\mu_l = \frac{g_l \mu_B}{\hbar} L = \frac{g_l \mu_B}{\hbar} \sqrt{l(l+1)}\,\hbar = g_l \mu_B \sqrt{l(l+1)} \tag{11–11}$$

and

$$\mu_{l_z} = -\frac{g_l \mu_B}{\hbar} L_z = -\frac{g_l \mu_B}{\hbar} m\hbar = -g_l \mu_B m \tag{11–12}$$

(The minus sign follows from the minus sign in equation 11–10.)

2. Effects of an External Magnetic Field

Using the definition (11–3), let us evaluate the forces acting on a system with magnetic moment $\boldsymbol{\mu}_l$ placed in an external magnetic field $\mathbf{H}$, which is uniform in the region of the equivalent dipole. This is illustrated in figure (11–2). In this field there is a force parallel to the direction of $\mathbf{H}$ and of magnitude qH acting on the positive pole, and a force of the same

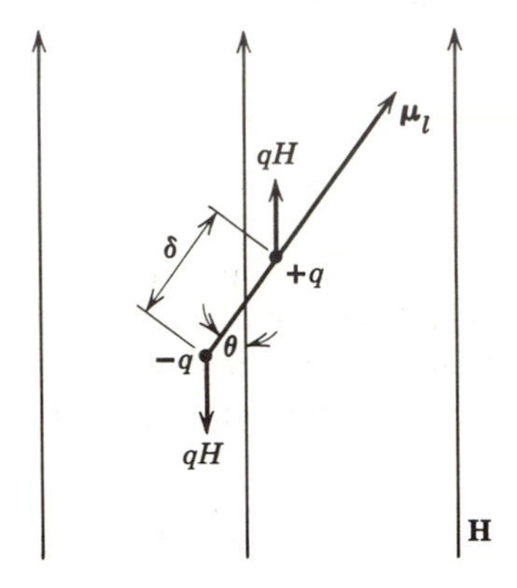

Figure 11–2. A magnetic moment in a uniform external magnetic field.

magnitude acting in the opposite direction on the negative pole. There is no net translational force, but there is a torque of magnitude

$$T = qH\frac{\delta}{2}\sin\theta + qH\frac{\delta}{2}\sin\theta = q\delta\,H\sin\theta = \mu_l H\sin\theta$$

where θ is the angle between $\boldsymbol{\mu}_l$ and $\mathbf{H}$. The torque vector is perpendicular to both $\boldsymbol{\mu}_l$ and $\mathbf{H}$, and it points in the direction which would tend to align $\boldsymbol{\mu}_l$ with $\mathbf{H}$. Its magnitude and direction can thus be expressed by the equation

$$\mathbf{T} = \boldsymbol{\mu}_l \times \mathbf{H} \tag{11–13}$$

In an external magnetic field $\mathbf{H}$, the system with magnetic moment $\boldsymbol{\mu}_l$ has an orientational potential energy ΔE which depends on the angle θ between the two vectors. The energy ΔE can be evaluated by summing the potential energy of each of the poles of the dipole, if we define the zero of potential energy to be at its center. For a magnetic field which is uniform in the region of the dipole, this is

$$\Delta E = -zqH - (-z)(-q)H = -2\frac{\delta}{2}\cos\theta\,qH = -\mu_l H\cos\theta \tag{11–14}$$

where z is the coordinate of the positive pole, and $-z$ is the coordinate of the negative pole, both measured along an axis parallel to **H**. This can be written in terms of the "dot product" notation as

$$\Delta E = -\boldsymbol{\mu}_l \cdot \mathbf{H} \tag{11–15}$$

Equation (11–14) can be taken as the definition of this notation. The orientational potential energy ΔE is minimized when $\boldsymbol{\mu}_l$ and **H** are aligned. However, in the absence of some dissipative mechanism, the vector $\boldsymbol{\mu}_l$ cannot obey the natural tendency to align itself with the vector **H**. Instead, $\boldsymbol{\mu}_l$ will precess around **H** in such a way that θ and ΔE remain constant.

This precessional motion is a consequence of the fact that the torque **T** is always perpendicular to the angular momentum vector **L** since, from equations (11–10) and (11–13), we have

$$\mathbf{T} = -\frac{g_l \mu_B}{\hbar} \mathbf{L} \times \mathbf{H}$$

Substituting this into Newton's law of motion for angular momentum,

$$\frac{d\mathbf{L}}{dt} = \mathbf{T}$$

we obtain the equation of motion of the vector **L**:

$$\frac{d\mathbf{L}}{dt} = -\frac{g_l \mu_B}{\hbar} \mathbf{L} \times \mathbf{H} \tag{11–16}$$

From the equation we see that the change in **L** in the time dt is

$$d\mathbf{L} = -\frac{g_l \mu_B}{\hbar} \mathbf{L} \times \mathbf{H}\, dt$$

These vectors are illustrated in figure (11–3).† Since $d\mathbf{L}$ is perpendicular to **L**, its magnitude L remains constant. In the time dt the tip of the vector **L** moves through an angle

$$d\phi = \frac{dL}{L \sin\theta} = \frac{g_l \mu_B L H \sin\theta\, dt}{\hbar L \sin\theta} = \frac{g_l \mu_B H\, dt}{\hbar}$$

Thus **L**, and also $\boldsymbol{\mu}_l$, precess about **H** with an angular frequency of magnitude

$$\omega_L = \frac{d\phi}{dt} = \frac{g_l \mu_B}{\hbar} H$$

† This figure, which shows **L** precessing about **H**, should not be confused with figure (11–1), which shows the electron circulating in its orbit.

As the sense of the precession is in the direction of **H**, we may write

$$\boldsymbol{\omega}_L = \frac{g_l \mu_B}{\hbar} \mathbf{H} \tag{11–17}$$

This phenomenon is known as the *Larmor precession*, and ω_L is called the *Larmor frequency*. It is quite analogous to the precession of the spin

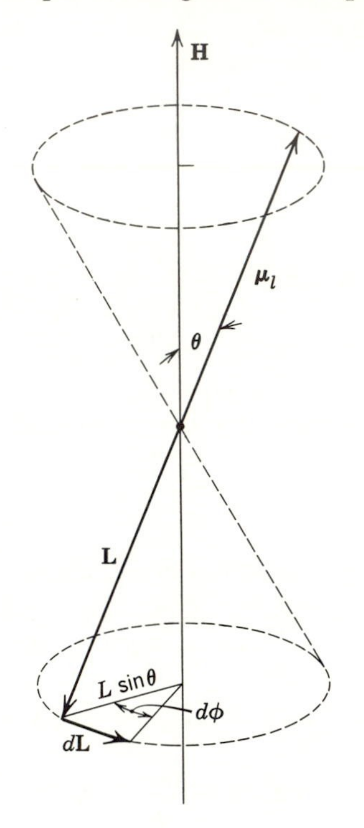

Figure 11–3. Illustrating a magnetic moment precessing about an external magnetic field.

angular momentum of a top in a gravitational field. In fact, equation (11–16) is identical in form with the equation of a top.

An experiment capable of detecting spatial quantization of the magnetic moment of an atom must be able to measure μ_{l_z}, or some quantity which depends on μ_{l_z}. Equation (11–15) suggests placing the atom in a magnetic field oriented along the direction of the z axis and measuring the potential energy $\Delta E = -\boldsymbol{\mu}_l \cdot \mathbf{H} = -\mu_{l_z} H$ by observing, for instance, the effect of the magnetic field on the optical spectrum of the atom. Such an experiment will be described in Chapter 13 under the heading *Zeeman effect*. Here we shall consider instead a different technique which involves sending a beam of atoms through a magnetic field that deflects each atom by an amount which depends on μ_{l_z}.

We have seen above that in a uniform magnetic field there is no net translational force acting on the magnetic moment. However, if the field is not uniform there will be a translational force. Consider a magnetic field which is non-uniform, but does not vary with y and is symmetric with respect to the yz plane.† In the yz plane the field then has a component only along the z axis. Assume that a magnetic dipole producing a magnetic moment $\boldsymbol{\mu}_l$ is straddling the yz plane with $\boldsymbol{\mu}_l$ at an angle θ

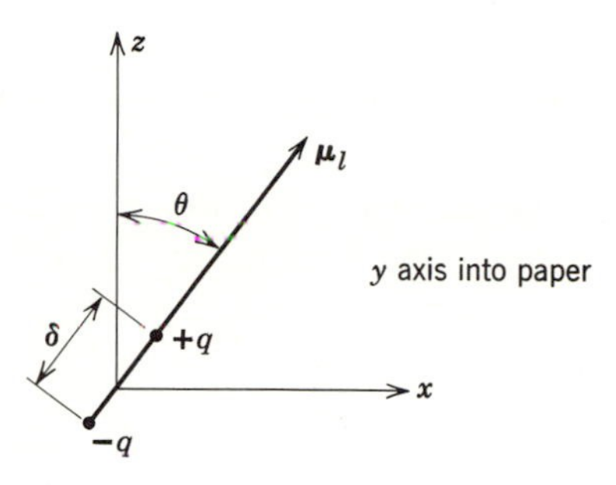

Figure 11–4. A magnetic moment in a non-uniform external magnetic field.

from the z axis, as indicated in figure (11–4). At the center of the dipole the field is H_0, which is entirely in the z direction. At the positive and negative poles the field in the z direction is approximately

$$(H_z)_{+q} = H_0 + \frac{\partial H_z}{\partial x}x + \frac{\partial H_z}{\partial z}z$$

$$(H_z)_{-q} = H_0 + \frac{\partial H_z}{\partial x}(-x) + \frac{\partial H_z}{\partial z}(-z)$$

where $x = (\delta/2)\sin\theta$ and $z = (\delta/2)\cos\theta$. Thus the net z force is

$$F_z = q(H_z)_{+q} - q(H_z)_{-q} = \frac{\partial H_z}{\partial x}q\delta\sin\theta + \frac{\partial H_z}{\partial z}q\delta\cos\theta$$

or

$$F_z = \frac{\partial H_z}{\partial x}\mu_{l_x} + \frac{\partial H_z}{\partial z}\mu_{l_z}$$

Since $\boldsymbol{\mu}_l$ is not fixed in space, we must evaluate the average of F_z. It is

$$\overline{F_z} = \frac{\partial H_z}{\partial x}\overline{\mu_{l_x}} + \frac{\partial H_z}{\partial z}\overline{\mu_{l_z}}$$

† Such a field is shown in figure (11–5).

In the classical theory $\overline{\mu_{l_x}} = 0$ because of the Larmor precession. In quantum mechanics $\overline{\mu_{l_x}} = -(g_l\mu_B/\hbar)\,\bar{L}_x = 0$ because $\bar{L}_x = 0$. In both theories $\overline{\mu_{l_z}} = \mu_{l_z}$ since it has a fixed value. Consequently we have

$$\overline{F_z} = \frac{\partial H_z}{\partial z}\mu_{l_z} \tag{11–18}$$

A similar calculation shows the average net x component of force to be $\overline{F_x} = (\partial H_x/\partial z)\,\mu_{l_z}$, but this vanishes because $\partial H_x/\partial z = 0$ near the yz plane owing to the symmetry of the field.

3. The Stern-Gerlach Experiment and Electron Spin

In 1922 Stern and Gerlach measured the possible values of μ_{l_z} for Ag atoms by sending a beam of these atoms through a non-uniform magnetic

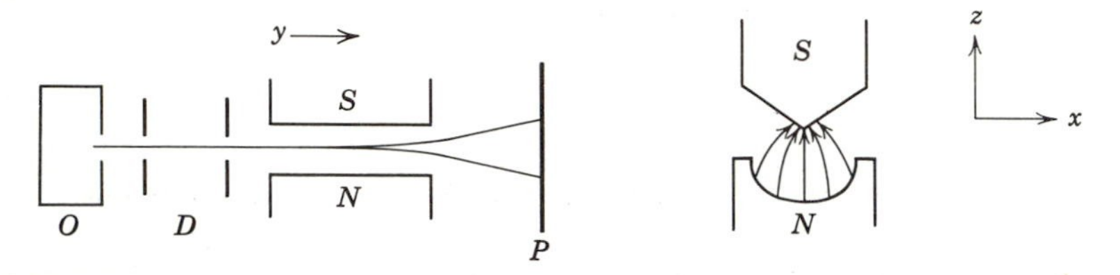

Figure 11–5. An apparatus used to measure the z component of the magnetic moment of an atom.

field. A schematic diagram of their apparatus is shown in figure (11–5). A beam of neutral atoms is formed by evaporating Ag in an oven O. The beam is collimated by the diaphragms D and enters a magnet N-S. The cross sectional view of the magnet shows that it produces a field of the type described in the last section. Since the atoms are neutral, the only force acting is the force $\overline{F_z}$ specified by equation (11–18).† The force acting on each atom of the beam therefore depends on the value of μ_{l_z} for that atom, and the beam is analyzed into components according to the various values of μ_{l_z}. The deflected atoms strike a metallic plate P, upon which they condense and leave a visible trace.

According to classical theory, μ_{l_z} can have *any* value from $-\mu_l$ to μ_l. The predictions of quantum mechanics, as summarized by equation (11–12), are that μ_{l_z} can have only the *discrete* values

$$\mu_{l_z} = -g_l\mu_B m \tag{11–19}$$

† The experiment cannot be performed with ionized atoms because they would experience a force $F = Hev/c$, acting directly on the charge, which is very much larger than the force acting on the magnetic moment.

where m is one of the integers

$$m = -l, -l + 1, \ldots, 0, \ldots, +l - 1, +l \qquad (11\text{–}19')$$

Thus classical theory predicts that the deflected beam would be spread into a continuous band, while quantum mechanics predicts that the deflected beam would be split up into several discrete components. Furthermore, quantum mechanics predicts that this should happen for all orientations of the magnet, i.e., for any choice of the direction of the z axis. Stern and Gerlach found that the beam of Ag atoms is split into two discrete components; one component is bent in the positive z direction, and the other component is bent in the negative z direction. The experiment was repeated using different orientations of the magnet and several other kinds of atoms, and it was always found that the deflected beam is split into two, or sometimes more, discrete components. These results are, qualitatively, *very direct experimental proof of the existence of spatial quantization.*

However, they are not in quantitative agreement with equation (11–19). According to that equation the number of possible values of μ_{l_z} is equal to the number of possible values of m, which is $(2l + 1)$. Since l is an integer, this is always an odd number. Also, for any value of l one of the possible values of m is zero. Thus the fact that the beam of Ag atoms is split into only two components, both of which are deflected, implies either that something is wrong with the Schroedinger theory of quantum mechanics, *or* that the theory is incomplete.

The theory is not wrong, but, as it stands, it is incomplete. This is shown most clearly by an experiment performed in 1927 by Phipps and Taylor, who used the Stern-Gerlach technique on a beam of hydrogen atoms. This experiment is particularly significant because the atoms contain a single electron and the theory we have developed makes the unambiguous prediction that, the atoms being in the ground state, the quantum number l will have the value $l = 0$. Then there is only one possible value of m, namely $m = 0$, and we expect that the beam will be unaffected by the magnetic field since μ_{l_z} will be equal to zero. However, Phipps and Taylor found that the beam is split into two symmetrically deflected components. Thus there is certainly some magnetic moment in the atom which we have not hitherto considered.

One possibility would be a magnetic moment due to a spinning motion of the charged nucleus about its own axis. The magnitude of such a magnetic moment would be of the order of $e\hbar/2Mc$, where M is the mass of the proton. But the magnetic moment measured experimentally (from the size of the splitting) is of the order of $\mu_B = e\hbar/2mc$, which is a thousand times larger. Therefore the nucleus cannot be responsible for the observed magnetic moment—its source must be the electron.

This leads us to the reasonable assumption, which is supported by other evidence to be discussed later, that *an electron has an intrinsic magnetic moment* $\boldsymbol{\mu}_s$ *due to the existence of an intrinsic angular momentum* **S** *called the spin.* We also assume that the magnitude S and z component S_z, of the spin angular momentum, are related to two quantum numbers s and m by the equations

$$S = \sqrt{s(s+1)}\,\hbar \tag{11–20}$$

$$S_z = m_s\hbar \tag{11–21}$$

which are identical with those for orbital angular momentum, and that the relation between the spin magnetic moment and the spin angular momentum is of the same form as the relation for the orbital case; i.e.,

$$\boldsymbol{\mu}_s = -\frac{g_s\mu_B}{\hbar}\mathbf{S} \tag{11–22}$$

$$\mu_{s_z} = -g_s\mu_B m_s \tag{11–23}$$

The quantity g_s is called the *spin g factor*. Now, from the experimental observation that the beam of hydrogen atoms is split into two symmetrically deflected components, it is apparent that μ_{s_z} can assume just two values which are equal in magnitude but opposite in sign. If we make the final assumption that the possible values of m_s differ by unity and range from $-s$ to $+s$, as is true of the quantum numbers for orbital angular momentum, then we conclude that the two possible values of m_s are

$$m_s = -\tfrac{1}{2}, +\tfrac{1}{2} \tag{11–24}$$

and that s has the single value

$$s = \tfrac{1}{2} \tag{11–25}$$

By measuring the splitting of the beam of hydrogen atoms, it is possible to evaluate the magnetic force $\overline{F_z}$ which the hydrogen atoms experience in traversing the magnetic field. From equations (11–18) and (11–23), this is

$$\overline{F_z} = -\frac{\partial H_z}{\partial z}\mu_B g_s m_s$$

Since μ_B is known and $\partial H_z/\partial z$ can be measured, this equation allows the evaluation of the quantity $g_s m_s$. Within the accuracy of the experiment it is found that

$$g_s m_s = \pm 1$$

Since we have concluded that $m_s = \pm\frac{1}{2}$, this implies

$$g_s = 2 \tag{11-26}$$

Thus the spin g factor for an electron is twice the value of its orbital g factor.† On the other hand, $\boldsymbol{\mu}_s$ and $\mathbf{S}$ are antiparallel, just as $\boldsymbol{\mu}_l$ and $\mathbf{L}$ are, because the relative orientation of either pair of vectors depends only on the fact that an electron is negatively charged.

The addition of electron spin to the quantum mechanical theory requires the addition of the quantum number m_s to the set n, l, m, in order to identify completely the mathematical form of an eigenfunction. However, it is not necessary to indicate specifically the value of the quantum number s since it always has the same value, $\frac{1}{2}$. Thus an eigenfunction must carry four subscripts, and it is written

$$\psi_{nlm_lm_s} \tag{11-27}$$

where we use the symbol m_l for the symbol we have previously written as m in order to have a symmetrical notation. Such an eigenfunction contains the information that the z component of the spin is $+\frac{1}{2}\hbar$ when $m_s = +\frac{1}{2}$, and is $-\frac{1}{2}\hbar$ when $m_s = -\frac{1}{2}$. The eigenfunctions $\psi_{nlm_lm_s}$ can be written as a product of a spatial eigenfunction ψ_{nlm_l} times a spin eigenfunction σ_{m_s}. However, the mathematical structure of the spin eigenfunctions is quite different from that of the spatial eigenfunctions. The ψ_{nlm_l} are the continuous functions of the continuous variables r, θ, ϕ, which we have studied; but the "spin variable" can only have two values corresponding to the two possible values of m_s, so the σ_{m_s} are not continuous functions. There are several ways of representing these functions, the most commonly used being the *Pauli matrix representation* in which they are written as two row, single column matrices. This representation is very closely related to the Heisenberg matrix formulation of quantum mechanics mentioned in section 1, Chapter 7, and is treated in advanced books on quantum mechanics. Since for our purposes it is possible to avoid situations in which it is necessary to know the explicit form of the spin eigenfunctions, they will not be treated in this book.

As is implied by its name, the spin angular momentum $\mathbf{S}$ can be thought of as analogous to the angular momentum of a particle spinning about an axis through its center. Using this in the Bohr theory, we can make an analogy between the motion of a spinning electron around its nucleus and the motion of a spinning planet around its sun. However, this analogy must not be taken too seriously as *electron spin is an essentially non-classical*

† Very accurate experiments by Lamb (1947), using different techniques, show that g_s is actually very slightly different from 2. Its precise value is $g_s = 2.00232$.

quantity. The reason is that the quantum number s which specifies the magnitude of the spin angular momentum has the fixed value $\frac{1}{2}$. Therefore we cannot take $\mathbf{S}$ to the classical limit by letting $s \rightarrow \infty$, as we did in the last chapter for $\mathbf{L}$ by letting $l \rightarrow \infty$. Another way of saying the same thing is that in the classical limit the magnitude of the spin $S = \sqrt{\frac{1}{2}(\frac{1}{2} + 1)}\,\hbar$ is completely negligible, and so spin must be essentially non-classical.

Although electron spin is completely compatible with the Schroedinger theory of quantum mechanics, the theory does not predict its existence and it must be introduced as a separate postulate. The reason for this is associated with the fact that the Schroedinger theory is an approximation which ignores relativistic effects (cf. section 2, Chapter 7). In 1928 Dirac developed a relativistic theory of quantum mechanics. Utilizing essentially the same postulates as the Schroedinger theory, plus the requirement that quantum mechanics conform with the theory of relativity, Dirac showed that an electron must have an inherent angular momentum and magnetic moment with exactly the properties described above. This put electron spin on a firm theoretical foundation. A quantitative treatment of the Dirac theory would not be appropriate in this book, and we shall continue discussing spin as an ad hoc addition to the Schroedinger theory. However, some of the more interesting features of the Dirac theory will be described qualitatively in the following chapters.

4. The Spin-Orbit Interaction

The existence of electron spin actually was postulated first by Goudsmit and Uhlenbeck in 1925. These gentlemen were trying to understand why certain lines of the optical spectra of hydrogen and the alkali elements are composed of a closely spaced pair of lines. This is the *fine structure*, which had been treated by Sommerfeld in the old quantum theory as due to a splitting of the atomic energy levels because of a small contribution to the total energy resulting from the relativistic variation of electron mass with velocity (cf. section 7, Chapter 5). The results of the Sommerfeld theory, as expressed by equation (5–22), were in good numerical agreement with the fine structure of the hydrogen spectrum. But they were also in good agreement with the fine structure of the alkali element spectra. This was perplexing because in the old quantum theory the spectroscopically active electron of an alkali element (an element with a single valence electron) would be moving in an orbit of large radius at relatively low velocity, so that relativistic corrections should be negligible. Consequently doubt arose concerning the validity of Sommerfeld's explanation of the origin of the fine structure. In considering other possibilities, Goudsmit and Uhlenbeck proposed that an electron has an inherent angular momentum and magnetic moment, whose z components are specified by a quantum number m_s which can assume the two values $-\frac{1}{2}$ and $+\frac{1}{2}$. The splitting of the atomic energy levels could then be understood as due to an orientational potential energy of the inherent electron magnetic moment in the atom's own magnetic field, which would be either positive or negative depending on the sign of m_s.

In this section we shall evaluate the orientational potential energy,

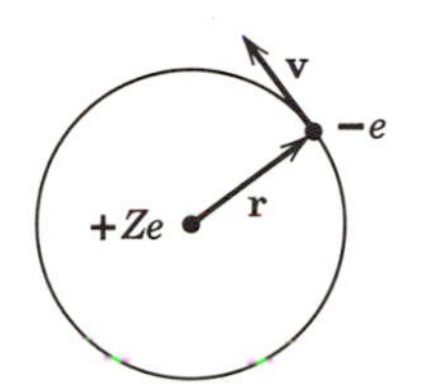

Figure 11–6. An electron moving in a circular Bohr orbit, as seen by the nucleus.

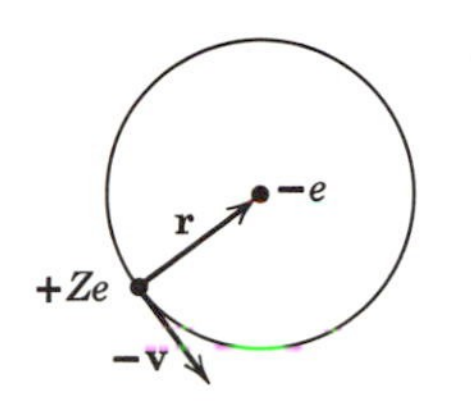

Figure 11–7. An electron moving in a circular Bohr orbit, as seen by the electron.

called the *spin-orbit energy*, by a calculation which involves a judicious mixture of the Bohr theory and the theory of relativity. The justification of this calculation is found in the fact that its results agree with those of a rigorous calculation using the Dirac theory.

Consider an electron of charge $-e$ moving in a Bohr orbit around a nucleus of charge $+Ze$, as shown in figure (11–6). Let the velocity of the electron with respect to the nucleus be $\mathbf{v}$, and its position relative to the nucleus be $\mathbf{r}$. From the point of view of the electron, the nucleus is moving around it with velocity $-\mathbf{v}$. This is shown in figure (11–7). The charged nucleus moving with velocity $-\mathbf{v}$ constitutes a current element

$$\mathbf{j} = -Ze\mathbf{v}/c$$

According to Ampere's law, this produces a magnetic field which, at the position of the electron, is

$$\mathbf{H} = \frac{\mathbf{j} \times \mathbf{r}}{r^3} = -\frac{Ze}{c}\frac{\mathbf{v} \times \mathbf{r}}{r^3}$$

It is convenient to express this in terms of the electric field $\mathbf{E}$ acting on the electron. According to Coulomb's law,

$$\mathbf{E} = +Ze\frac{\mathbf{r}}{r^3}$$

From the last two equations we have

$$\mathbf{H} = -\frac{1}{c}\mathbf{v} \times \mathbf{E} \tag{11–28}$$

From a study of the behavior of Maxwell's equations under the application of a Lorentz transformation, it can be shown that this equation is of general validity; any particle moving with velocity $\mathbf{v}$ through an arbitrary electric field $\mathbf{E}$ will experience the magnetic field $\mathbf{H}$ given by the equation.

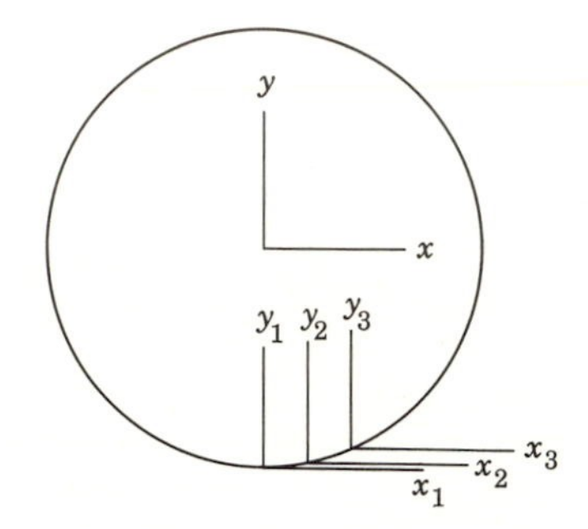

Figure 11–8. The frames of reference used in calculating the Thomas precession.

According to equation (11–17), the magnetic field will cause a Larmor precession of the spin magnetic moment with angular frequency

$$\boldsymbol{\omega}_L = \frac{g_s \mu_B}{\hbar}\mathbf{H} \tag{11–29}$$

where we have replaced g_l by g_s to obtain the correct value of the spin magnetic moment. The magnetic field will, according to equation (11–15), also produce an orientational potential energy

$$\Delta E = -\boldsymbol{\mu}_s \cdot \mathbf{H} = \frac{g_s \mu_B}{\hbar}\mathbf{S} \cdot \mathbf{H} \tag{11–30}$$

Both of these quantities have been evaluated in a frame of reference in which the electron is at rest, but what we are really interested in is their values in the normal frame of reference in which the nucleus is at rest. Although it is not obvious that transforming back to the normal frame of reference will change the apparent values of $\boldsymbol{\omega}_L$ or ΔE, this really happens due to a kinematical effect of the theory of relativity called the *Thomas precession.*

Consider the situation as seen by an observer in the frame of reference xy, in which the nucleus is at rest. Figure (11–8) illustrates this. The electron is momentarily at rest in the frame x_1y_1 at the instant t_1, and momentarily at rest in the frame x_2y_2 at the slightly later instant t_2. The axes of both xy and x_2y_2 have been constructed parallel to the axes of x_1y_1, as seen by an observer instantaneously moving with the electron in x_1y_1. Nevertheless, we shall show that the observer in xy sees the axes of x_2y_2 rotated slightly relative to his own axes. He sees the axes of the frame x_3y_3 rotated even more, etc. Thus he sees that the set of axes in which the electron is instantaneously at rest are precessing, relative to his own set of axes, as the electron goes around the nucleus—even though the observers instantaneously at rest relative to the electron contend that each set of axes $x_{n+1}y_{n+1}$ is parallel to the preceding set x_ny_n.

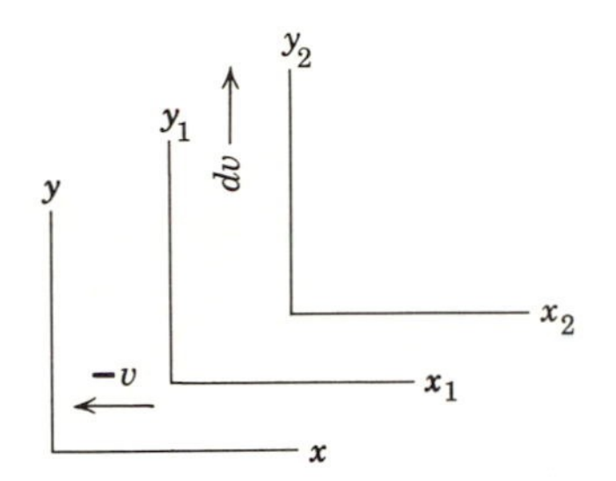

Figure 11–9. The frames of reference used in calculating the Thomas precession, as seen in the x_1y_1 frame.

Figure (11–9) shows xy, x_1y_1, and x_2y_2, from the point of view of the observer in x_1y_1. Since the electron is moving with velocity **v** relative to the nucleus, the axes xy are moving with velocity $-\mathbf{v}$ in the direction of the negative x_1 axis relative to x_1y_1. As seen in x_1y_1, the electron is accelerating toward the nucleus with acceleration **a** in the direction of the positive y_1 axis. If the time interval $t_2 - t_1$ is very small, the change in velocity of the electron in that interval is

$$d\mathbf{v} = \mathbf{a}(t_2 - t_1) = \mathbf{a}\, dt \tag{11–31}$$

and this will be the velocity of x_2y_2 as seen by x_1y_1. Now let us use equations (1–16) to evaluate the components of $\mathbf{V}_a$, the velocity of x_2y_2 as seen by xy. These give

$$V_{a_x} = \frac{dv_x - v_x}{1 - \dfrac{v_x\, dv_x}{c^2}} = \frac{0 + v}{1 - \dfrac{-v \cdot 0}{c^2}} = v$$

$$V_{a_y} = \frac{dv_y\sqrt{1 - \dfrac{v_x^2}{c^2}}}{1 - \dfrac{v_x\, dv_x}{c^2}} = dv\sqrt{1 - \frac{v^2}{c^2}}$$

Using equations (1–16) to evaluate the components of $\mathbf{V}_b$, the velocity of xy as seen by x_2y_2, we have

$$V_{b_x} = \frac{v_x\sqrt{1 - \dfrac{dv_y^2}{c^2}}}{1 - \dfrac{dv_y v_y}{c^2}} = \frac{-v\sqrt{1 - \dfrac{dv^2}{c^2}}}{1 - \dfrac{dv \cdot 0}{c^2}} = -v\sqrt{1 - \frac{dv^2}{c^2}}$$

$$V_{b_y} = \frac{v_y - dv_y}{1 - \dfrac{dv_y v_y}{c^2}} = -dv$$

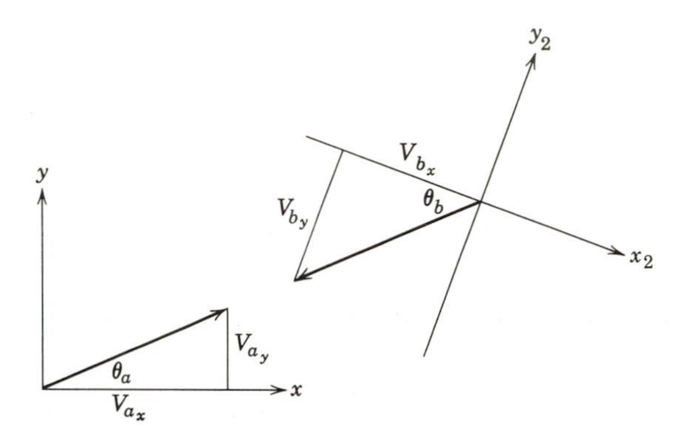

Figure 11–10. An exaggerated illustration of the Thomas precession.

The reader may verify that the magnitude of $\mathbf{V}_a$ is equal to the magnitude of $\mathbf{V}_b$, in conformity with Einstein's postulate of the equivalence of inertial frames. Now let us calculate the angle between the vector $\mathbf{V}_a$ and the x axis of the xy frame. It is

$$\theta_a = \frac{V_{a_y}}{V_{a_x}} = \frac{dv\sqrt{1 - \dfrac{v^2}{c^2}}}{v}$$

The angle between the vector $\mathbf{V}_b$ and the x axis of the x_2y_2 frame is

$$\theta_b = \frac{V_{b_y}}{V_{b_x}} = \frac{-dv}{-v\sqrt{1 - \dfrac{dv^2}{c^2}}}$$

Figure (11–10) shows the x_2y_2 and xy frames from the point of view of xy. According to the postulate of equivalence, $\mathbf{V}_a$ and $\mathbf{V}_b$ must be exactly opposite in direction. Since the angles between the x axes and the relative velocity vectors are not the same, the x_2y_2 frame appears to be rotated relative to the xy frame. The angle of rotation is

$$d\theta = \theta_b - \theta_a = \left[\frac{dv}{v\sqrt{1 - \dfrac{dv^2}{c^2}}} - \frac{dv}{v}\sqrt{1 - \frac{v^2}{c^2}}\right]$$

As dv is a differential, we may neglect dv^2/c^2. Then

$$d\theta = \left[\frac{dv}{v} - \frac{dv}{v}\sqrt{1 - \frac{v^2}{c^2}}\right]$$

Now v represents the velocity of an electron in an atom, so $v^2/c^2 \lesssim 10^{-4}$. Therefore we may obtain a very good approximation to $d\theta$ by taking only the first two terms in the binomial expansion of $\sqrt{1 - v^2/c^2}$. We have

$$d\theta = \frac{dv}{v}\left[1 - \sqrt{1 - \frac{v^2}{c^2}}\right] \simeq \frac{dv}{v}\left[1 - \left(1 - \frac{v^2}{2c^2}\right)\right]$$

$$= \frac{dv\, v^2}{2vc^2} = \frac{v\, dv}{2c^2} = \frac{va\, dt}{2c^2}$$

where we have evaluated dv from equation (11–31). The axes in which the electron is instantaneously at rest appear to precess, relative to the nucleus, with angular frequency

$$\omega_T = d\theta/dt = va/2c^2$$

This frequency is called the *Thomas frequency*. Inspection of the figures will verify that the sense of the precession is given by the vector equation

$$\boldsymbol{\omega}_T = -\frac{1}{2c^2}\mathbf{v} \times \mathbf{a} \tag{11–32}$$

Relative to frames in which the electron is at rest, its spin magnetic moment precesses with angular frequency $\boldsymbol{\omega}_L$. But these frames are themselves precessing with angular frequency $\boldsymbol{\omega}_T$ relative to the normal frame in which the nucleus is at rest. Consequently the spin magnetic moment is seen in the normal frame to precess with angular frequency

$$\boldsymbol{\omega} = \boldsymbol{\omega}_L + \boldsymbol{\omega}_T \tag{11–33}$$

According to equations (11–28) and (11–29),

$$\boldsymbol{\omega}_L = -\frac{g_s\mu_B}{c\hbar}\mathbf{v} \times \mathbf{E}$$

Evaluating μ_B and g_s from equations (11–8) and (11–26), we have

$$\boldsymbol{\omega}_L = -\frac{2e\hbar}{c\hbar 2mc}\mathbf{v} \times \mathbf{E} = -\frac{e}{mc^2}\mathbf{v} \times \mathbf{E} \tag{11–34}$$

To evaluate $\boldsymbol{\omega}_T$, we may use Newton's law to express the acceleration of the electron in terms of the electric field:

$$\mathbf{a} = -e\mathbf{E}/m$$

With this, equation (11–32) becomes

$$\boldsymbol{\omega}_T = \frac{e}{2mc^2}\mathbf{v} \times \mathbf{E} \tag{11–35}$$

Thus the precessional frequency in the normal frame of reference is

$$\boldsymbol{\omega} = -\frac{e}{mc^2}\mathbf{v} \times \mathbf{E} + \frac{e}{2mc^2}\mathbf{v} \times \mathbf{E} = -\frac{e}{2mc^2}\mathbf{v} \times \mathbf{E} \tag{11–36}$$

This equation was first derived in 1926 by Thomas.

Comparing equations (11–36) and (11–34), we see that transforming the spin precession frequency from the frame in which the electron is at rest to the normal frame reduces its magnitude by exactly a factor of 2. The same is true of the orientational potential energy since, as can be seen from equations (11–29) and (11–30), the magnitude of that quantity is proportional to the magnitude of the precession frequency. Therefore the value of the orientational potential energy in the normal frame of reference, which we shall designate by the symbol $\Delta E_{\mathbf{S}\cdot\mathbf{L}}$, is

$$\Delta E_{\mathbf{S}\cdot\mathbf{L}} = \frac{1}{2}\frac{g_s\mu_B}{\hbar}\mathbf{S}\cdot\mathbf{H} \tag{11–37}$$

It is convenient to express this in terms of the quantity $\mathbf{S}\cdot\mathbf{L}$. To this end we use equation (11–28) to write

$$\mathbf{H} = -\frac{1}{c}\mathbf{v} \times \mathbf{E}$$

and express $\mathbf{E}$ in terms of the force $\mathbf{F}$ acting on the electron, using

$$-e\mathbf{E} = \mathbf{F}$$

In the Coulomb potential of the nucleus or, more generally, in any spherically symmetric potential $V(r)$, the force $\mathbf{F}$ is

$$\mathbf{F} = -\frac{dV(r)}{dr}\frac{\mathbf{r}}{r}$$

Combining these three equations, we have

$$\mathbf{H} = -\frac{1}{ec}\frac{1}{r}\frac{dV(r)}{dr}\mathbf{v} \times \mathbf{r}$$

Multiplying the numerator and denominator by m, and using equation (11–4), we obtain†

$$\mathbf{H} = \frac{1}{emc}\frac{1}{r}\frac{dV(r)}{dr}\mathbf{L} \tag{11–38}$$

Thus equation (11–37) can be written

$$\Delta E_{\mathbf{S}\cdot\mathbf{L}} = \frac{g_s \mu_B}{2emc\hbar}\frac{1}{r}\frac{dV(r)}{dr}\mathbf{S}\cdot\mathbf{L}$$

Evaluating g_s and μ_B, we obtain

$$\Delta E_{\mathbf{S}\cdot\mathbf{L}} = \frac{1}{2m^2c^2}\frac{1}{r}\frac{dV(r)}{dr}\mathbf{S}\cdot\mathbf{L} \tag{11–39}$$

Although our calculation of the spin-orbit energy was essentially based on the Bohr theory, the results, as expressed by equation (11–39), are in exact agreement with the results of a relativistically correct quantum mechanical treatment.

Let us make an estimate of $\Delta E_{\mathbf{S}\cdot\mathbf{L}}$ for a typical state of a one-electron atom, to check whether it is of the same order of magnitude as the observed fine structure splitting of the corresponding energy level. Consider hydrogen in the $n = 2$, $l = 1$ state. The potential $V(r)$ is

$$V(r) = -e^2 r^{-1}$$

so

$$\frac{dV(r)}{dr} = e^2 r^{-2}$$

and

$$\Delta E_{\mathbf{S}\cdot\mathbf{L}} = \frac{e^2}{2m^2c^2}\frac{1}{r^3}\mathbf{S}\cdot\mathbf{L}$$

The magnitude of $\mathbf{S}\cdot\mathbf{L}$ is approximately $\hbar^2$. The average value of $1/r^3$ for the $n = 2$ state is approximately $1/2^3 a_0^3$, where $a_0 = \hbar^2/me^2$. So

$$|\Delta E_{\mathbf{S}\cdot\mathbf{L}}| \sim \frac{e^2}{2m^2c^2}\frac{m^3e^6}{2^3\hbar^6}\hbar^2 \sim 10^{-16}\text{ ergs} \sim 10^{-4}\text{ev}$$

Since $\mathbf{S}\cdot\mathbf{L}$ can be either positive or negative, depending on whether $m_s = +\frac{1}{2}$ or $-\frac{1}{2}$, the energy level is split by about 2×10^{-4} ev. Comparing

† From the definition of the cross product it is apparent that

$$\mathbf{r} \times \mathbf{v} = -\mathbf{v} \times \mathbf{r}$$

this with the energy of the $n = 2$, $l = 1$ level of hydrogen, $E_2 = -3.4$ ev, we see that the predicted splitting is about one part in 10^4—in good agreement with the splitting required to explain the fine structure of the lines of the hydrogen spectrum associated with this level.

It is also interesting to estimate the magnitude of the magnetic field H acting on the spin magnetic moment of an electron in the $n = 2$, $l = 1$ state of hydrogen. From equation (11–30) we have

$$|\Delta E_{\mathbf{S}\cdot\mathbf{L}}| \sim \mu_s H$$

where

$$\mu_s \sim \mu_B \sim 10^{-20} \text{ erg-gauss}^{-1}$$

Therefore

$$H \sim \frac{10^{-16} \text{ ergs}}{10^{-20} \text{ erg-gauss}^{-1}} = 10^4 \text{ gauss}$$

This is a strong magnetic field.

5. Total Angular Momentum

In the absence of the spin-orbit interaction, and of any external torques, the magnitudes and z components of the spin and the orbital angular momentum vectors **L** and **S** would have the fixed values specified by the quantum numbers l, m_l, m_s.† But in actuality there is a strong magnetic field, the orientation of which is determined by **L**, that produces a torque on the spin magnetic moment, the orientation of which is determined by **S**. We have seen in section 2 that such a torque will not change the magnitude of the vector **S**. Nor will the reaction torque acting on **L** change its magnitude. But this torque does provide a coupling between **L** and **S** which makes the orientation of each vector dependent on the orientation of the other. As a result, neither of the vectors will have a fixed z component because both will precess about the vector

$$\mathbf{J} = \mathbf{L} + \mathbf{S} \tag{11–40}$$

which is their vectorial sum. The vector **J** is called the *total angular momentum vector.*

According to classical physics, the magnitude and all three components of the total angular momentum vector would be fixed if there were no external torques acting on the atom. This is only partly true in quantum mechanics because, as the reader might guess, this theory predicts that

† It is not necessary to specify the value of the quantum number s since it can only have the value $s = \frac{1}{2}$.

the magnitude and z component of $\mathbf{J}$ are fixed, whereas its x and y components have no definite value. By using techniques closely related to those we used in studying the properties of the orbital angular momentum vector, it can be shown that the magnitude and z component of the total angular momentum vector are specified by two quantum numbers j and m_j according to the equations†

$$J = \sqrt{j(j+1)}\,\hbar$$
$$J_z = m_j \hbar \tag{11–41}$$

where

$$m_j = -j, -j+1, \ldots, j-1, j \tag{11–42}$$

Let us determine the possible values of the quantum number j. Taking the z component of equation (11–40), we have

$$J_z = L_z + S_z$$

which is

$$m_j\hbar = m_l\hbar + m_s\hbar$$

or

$$m_j = m_l + m_s \tag{11–43}$$

Since the maximum possible value of m_l is l, and the maximum possible value of m_s is $s = \frac{1}{2}$, the maximum possible value of m_j is

$$(m_j)_{\max} = l + \tfrac{1}{2}$$

According to equation (11–42), this is also equal to the maximum possible value of j. In common with the other quantum numbers we have been discussing, the possible values of j differ by integers. Therefore these values must be members of the series

$$j = l + \tfrac{1}{2}, l - \tfrac{1}{2}, l - \tfrac{3}{2}, l - \tfrac{5}{2}, \ldots$$

To determine the point at which this series terminates, we may use the vector inequality

$$|\mathbf{L} + \mathbf{S}| \geqslant \big||\mathbf{L}| - |\mathbf{S}|\big|$$

the validity of which may easily be demonstrated by constructing some simple vector diagrams. This inequality gives

$$|\mathbf{J}| \geqslant \big||\mathbf{L}| - |\mathbf{S}|\big|$$

or

$$\sqrt{j(j+1)}\,\hbar \geqslant \left|\sqrt{l(l+1)}\,\hbar - \sqrt{\tfrac{1}{2}(\tfrac{1}{2}+1)}\,\hbar\right|$$

† In the Heisenberg matrix formulation of quantum mechanics it is proved that *all* angular momentum vectors (orbital, spin, and all combinations) obey quantization equations of exactly the same form.

We leave it as an exercise for the reader to show that when $l \neq 0$ only the two members of the series

$$j = l + \tfrac{1}{2}, l - \tfrac{1}{2} \tag{11–44}$$

satisfy this inequality. When $l = 0$, $\mathbf{L} = 0$, and the only possible value of j is

$$j = \tfrac{1}{2}, \qquad l = 0 \tag{11–44′}$$

The content of equations (11–42), (11–44), and (11–44′) can be formally represented in terms of the rules of vector addition, if we construct a

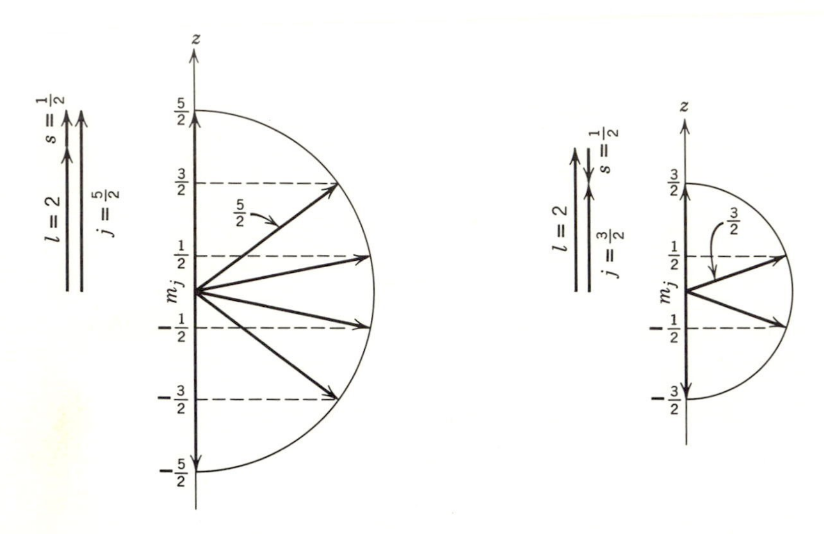

Figure 11–11. Vector addition diagrams for the quantum numbers $l = 2$, $s = \frac{1}{2}$.

set of vectors whose lengths are proportional to the values of the quantum numbers l, s, and j. As an example, consider $l = 2$. Then the two possible values of j are $\frac{5}{2}$ and $\frac{3}{2}$. For $j = \frac{5}{2}$: $m_j = -\frac{5}{2}, -\frac{3}{2}, -\frac{1}{2}, +\frac{1}{2}, +\frac{3}{2}, +\frac{5}{2}$. For $j = \frac{3}{2}$: $m_j = -\frac{3}{2}, -\frac{1}{2}, +\frac{1}{2}, +\frac{3}{2}$. Vector addition diagrams for the two values of j are shown in figure (11–11). The interpretation of these diagrams needs no discussion. The diagrams represent only the rules for adding the quantum numbers l and s to obtain the possible values of the quantum numbers j and m_j. *If* the relation between the magnitude of an angular momentum vector, such as $\mathbf{L}$, and its associated quantum number were $L = l\hbar$ instead of $L = \sqrt{l(l+1)}\,\hbar$, these diagrams would also represent the addition of $\mathbf{L}$ and $\mathbf{S}$ to obtain $\mathbf{J}$ and J_z. Although this is not true, such diagrams are often presented in elementary discussions of atomic structure as a simplified representation of the addition of the vectors. This is a useful representation to keep in mind, but only if it is also kept in mind that it is an approximation, and that a more accurate representation of the addition of these vectors would have an appearance similar to that shown in figure (11–12) for the case $l = 2, j = \frac{5}{2}, m_j = \frac{3}{2}$. The figure attempts to show the vectors **L** and **S** precessing about the vector **J**, which in turn is precessing about the z axis.

When the spin-orbit interaction is considered, we may no longer use the quantum numbers m_l and m_s to identify the function ψ because it is no

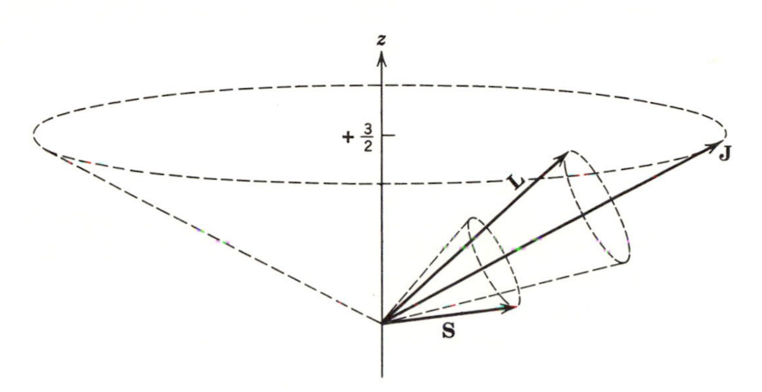

Figure 11–12. Illustrating the **L** and **S** vectors precessing about the **J** vector, and the **J** vector precessing about the z axis.

longer an eigenfunction of the operators associated with the dynamical quantities L_z and S_z. That is to say, L_z and S_z do not have definite values and so we cannot characterize ψ by definite values of m_l and m_s. But in losing m_l and m_s as good quantum numbers, we gain an appreciation of the fact that the quantum numbers j and m_j can be used instead. As J and J_z have the definite values $\sqrt{j(j+1)}\,\hbar$ and $m_j\hbar$, it is apparent that ψ must be an eigenfunction of the operators associated with these dynamical quantities. Therefore the eigenfunctions can be written

$$\psi_{nljm_j} \tag{11–45}$$

where the four quantum numbers identify the form of the function. Furthermore, these quantum numbers can be used and the functions (11–45) are proper solutions to the Schroedinger equation, even if there were no spin-orbit interaction, since we know that J and J_z would certainly still have definite values because there are no external torques acting on the atom.

This brings up the interesting point that there is a redundancy of quantum numbers if the spin-orbit interaction is neglected. That is, n, l, m_l, m_s, j, m_j would all be good quantum numbers, but, since there are only three spatial coordinates and one spin coordinate, only four quantum numbers are needed. In other words, if we ignore the spin-orbit interaction, we know from (11–27) that there are solutions to the Schroedinger equation, written

$$\psi_{nlm_lm_s} \tag{11–46}$$

which are eigenfunctions of the operators associated with: the total energy, the magnitude of the orbital angular momentum, the z component of the orbital angular momentum, and the z component of the spin angular momentum. However, there are also solutions to the Schroedinger equation, written as (11–45), which are eigenfunctions of the operators associated with: the total energy, the magnitude of the orbital angular momentum, the magnitude of the total angular momentum, and the z component of the total angular momentum. The relation between the $\psi_{nlm_lm_s}$ and the ψ_{nljm_j} can best be understood by considering the procedures involved in evaluating the spin-orbit energy $\Delta E_{\mathbf{S}\cdot\mathbf{L}}$ by the methods of perturbation theory.

In using these methods we first solve the Schroedinger equation, neglecting the spin-orbit interaction, and obtain a set of eigenfunctions, say the $\psi_{nlm_lm_s}$ which we found first. Then these unperturbed eigenfunctions can be used to evaluate $\Delta E_{\mathbf{S}\cdot\mathbf{L}}$. However, the $\psi_{nlm_lm_s}$ are degenerate since the total energy of the state specified by the quantum numbers nlm_lm_s depends only on the quantum number n. Therefore it is necessary to follow the procedures of degenerate perturbation theory, which require a preliminary investigation of the way that the perturbing spin-orbit interaction mixes those eigenfunctions that are degenerate with each other. From this we obtain a new set of eigenfunctions which consist of certain linear combinations of the set $\psi_{nlm_lm_s}$. In carrying this through, it is found that the new set are precisely the eigenfunctions ψ_{nljm_j}. We shall not bother to write down the ψ_{nljm_j} in terms of linear combinations of the $\psi_{nlm_lm_s}$, since they are quite complicated and we shall not need to know their explicit form. It is more important that we make sure that the reader understands the physical significance of these results. The point is that the set of degenerate unperturbed eigenfunctions $\psi_{nlm_lm_s}$ cannot be used directly in a perturbation calculation of the spin-orbit energy, because in the presence of this small perturbation L_z and S_z have no definite values. Thus the effect of the perturbation is to mix completely functions with different values of m_l and m_s, forming a new set of functions. This new set of functions are just the ψ_{nljm_j}, which describe the fact that J and J_z have definite values even in the presence of the spin-orbit interaction.

This discussion suggests that the evaluation of the spin-orbit energy might be considerably simplified by initially starting with the set of degenerate unperturbed eigenfunctions ψ_{nljm_j}. Since J and J_z have definite values whether or not the spin-orbit interaction is present, it is apparent that the application of this perturbation cannot change their values. Consequently the perturbation cannot produce a large mixing of the ψ_{nljm_j}, *even though they are degenerate*. In this case non-degenerate perturbation theory can be used directly (cf. section 4, Chapter 9). According to that theory (equation 9–13 and the following discussion extended to three dimensions), the approximate value of the shift in the energy of the state specified by the quantum numbers $nljm_j$, due to the presence of the spin-orbit interaction energy $\Delta E_{\mathbf{S}\cdot\mathbf{L}}$, is

$$\overline{\Delta E_{\mathbf{S}\cdot\mathbf{L}}} = \iiint \psi^*_{nljm_j} (\Delta E_{\mathbf{S}\cdot\mathbf{L}})_{op} \psi_{nljm_j} r^2 \sin\theta \, dr \, d\theta \, d\phi$$

which is equal to the expectation value of $\Delta E_{\mathbf{S}\cdot\mathbf{L}}$ in that state. Using (11–39), we have

$$\overline{\Delta E_{\mathbf{S}\cdot\mathbf{L}}} = \iiint \psi^*_{nljm_j} \left(\frac{1}{2m^2c^2} \frac{1}{r} \frac{dV(r)}{dr} \mathbf{S}\cdot\mathbf{L} \right)_{op} \psi_{nljm_j} r^2 \sin\theta \, dr \, d\theta \, d\phi$$

This is

$$\overline{\Delta E_{\mathbf{S}\cdot\mathbf{L}}} = \frac{1}{2m^2c^2} \iiint \psi^*_{nljm_j} \frac{1}{r} \frac{dV(r)}{dr} (\mathbf{S}\cdot\mathbf{L})_{op} \psi_{nljm_j} r^2 \sin\theta \, dr \, d\theta \, d\phi \tag{11–47}$$

For a Coulomb potential $V(r) = -Ze^2/r$, so $dV(r)/dr = Ze^2/r^2$, and this becomes

$$\overline{\Delta E_{\mathbf{S}\cdot\mathbf{L}}} = \frac{Ze^2}{2m^2c^2} \iiint \psi^*_{nljm_j} \frac{1}{r^3} (\mathbf{S}\cdot\mathbf{L})_{op} \psi_{nljm_j} r^2 \sin\theta \, dr \, d\theta \, d\phi$$

To evaluate $(\mathbf{S}\cdot\mathbf{L})_{op}$, consider the equation

$$\mathbf{J} = \mathbf{L} + \mathbf{S}$$

defining **J**. Taking the dot product of this equation times itself, and employing the fact that $\mathbf{L}\cdot\mathbf{S} = \mathbf{S}\cdot\mathbf{L}$, we have

$$\mathbf{J}\cdot\mathbf{J} = \mathbf{L}\cdot\mathbf{L} + \mathbf{S}\cdot\mathbf{S} + 2\mathbf{S}\cdot\mathbf{L}$$

So

$$\mathbf{S}\cdot\mathbf{L} = \tfrac{1}{2}(\mathbf{J}\cdot\mathbf{J} - \mathbf{L}\cdot\mathbf{L} - \mathbf{S}\cdot\mathbf{S})$$

This is

$$\mathbf{S}\cdot\mathbf{L} = \tfrac{1}{2}(J^2 - L^2 - S^2) \tag{11–48}$$

Therefore

$$(\mathbf{S}\cdot\mathbf{L})_{op} = \tfrac{1}{2}(J^2 - L^2 - S^2)_{op} = \tfrac{1}{2}(J^2_{op} - L^2_{op} - S^2_{op})$$

Then we have

$$\overline{\Delta E_{\mathbf{S}\cdot\mathbf{L}}} = \frac{Ze^2}{4m^2c^2}\iiint \psi^*_{nljm_j}\frac{1}{r^3}(J^2_{\text{op}} - L^2_{\text{op}} - S^2_{\text{op}})\psi_{nljm_j} r^2 \sin\theta\, dr\, d\theta\, d\phi$$

Now ψ_{nljm_j} is an eigenfunction of J^2_{op}, L^2_{op}, and S^2_{op}, corresponding to the eigenvalues $j(j+1)\hbar^2$, $l(l+1)\hbar^2$, and $\frac{1}{2}(\frac{1}{2}+1)\hbar^2$.† Therefore

$$(J^2_{\text{op}} - L^2_{\text{op}} - S^2_{\text{op}})\psi_{nljm_j} = [j(j+1) - l(l+1) - \tfrac{3}{4}]\hbar^2\psi_{nljm_j}$$

and we have

$$\overline{\Delta E_{\mathbf{S}\cdot\mathbf{L}}} = \frac{Ze^2\hbar^2}{4m^2c^2}[j(j+1) - l(l+1) - \tfrac{3}{4}] \times \iiint \psi^*_{nljm_j}\frac{1}{r^3}\,\psi_{nljm_j} r^2 \sin\theta\, dr\, d\theta\, d\phi \quad (11\text{–}49)$$

The integrals over θ and ϕ both give unity owing to the normalization properties of the ψ, Thus

$$\overline{\Delta E_{\mathbf{S}\cdot\mathbf{L}}} = \frac{Ze^2\hbar^2}{4m^2c^2}[j(j+1) - l(l+1) - \tfrac{3}{4}]\int R^*_{nl}(r)\frac{1}{r^3}R_{nl}(r)r^2\, dr$$

The integral over r can be evaluated. It gives

$$\int R^*_{nl}(r)\frac{1}{r^3}R_{nl}(r)r^2\, dr = \overline{r^{-3}} = \frac{Z^3}{a_0^3 n^3 l(l+\frac{1}{2})(l+1)}$$

Note that when $l = 0$ this equation says $\overline{r^{-3}} \to \infty$. For this value of l, $j = \frac{1}{2}$ and $[j(j+1) - l(l+1) - \frac{3}{4}] = 0$. Thus $\overline{\Delta E_{\mathbf{S}\cdot\mathbf{L}}}$ appears to be indeterminate for $l = 0$. However, this is merely the result of evaluating $\overline{r^{-3}}$ in perturbation theory. A more accurate evaluation of this quantity shows that it is very large, but finite, for $l = 0$. Since $[j(j+1) - l(l+1) - \frac{3}{4}]$ is strictly zero in this case, we have

$$\overline{\Delta E_{\mathbf{S}\cdot\mathbf{L}}} = 0, \qquad l = 0 \quad (11\text{–}50)$$

For other values of l it is

$$\overline{\Delta E_{\mathbf{S}\cdot\mathbf{L}}} = \frac{Z^4e^2\hbar^2}{4m^2c^2a_0^3}\,\frac{[j(j+1) - l(l+1) - \frac{3}{4}]}{n^3 l(l+\frac{1}{2})(l+1)}$$

In terms of the *fine structure constant* $\alpha \equiv e^2/\hbar c \simeq 1/137$, and the absolute value of the unperturbed energy of the state, $E_n = -mZ^2e^4/2\hbar^2n^2$, this

† All these functions are eigenfunctions of S^2_{op}, since S^2 can have only the single value $\frac{1}{2}(\frac{1}{2}+1)\hbar^2$.

can be written

$$\overline{\Delta E_{\mathbf{S}\cdot\mathbf{L}}} = \frac{Z^2|E_n|\alpha^2}{2n} \frac{[j(j+1) - l(l+1) - \frac{3}{4}]}{l(l+\frac{1}{2})(l+1)}$$

The value of the spin-orbit energy depends on the value of j. The two possibilities are $j = l + \frac{1}{2}$ or $j = l - \frac{1}{2}$, which correspond respectively to the situations in which $\mathbf{S}$ is roughly parallel to $\mathbf{L}$ or antiparallel to $\mathbf{L}$. Evaluating $\overline{\Delta E_{\mathbf{S}\cdot\mathbf{L}}}$ for these possibilities, we find

$$\overline{\Delta E_{\mathbf{S}\cdot\mathbf{L}}} = \begin{cases} +\dfrac{Z^2|E_n|\alpha^2}{n(2l+1)(l+1)}, & j = l + \frac{1}{2},\ l \neq 0 \\[2ex] -\dfrac{Z^2|E_n|\alpha^2}{nl(2l+1)}, & j = l - \frac{1}{2},\ l \neq 0 \end{cases} \tag{11–51}$$

6. Relativistic Corrections for One-Electron Atoms

In section 4 of this chapter we estimated the shift in the energy of a typical level of the hydrogen atom, due to the spin-orbit interaction, to be about 10^{-4} times the energy of the level. In section 7 of Chapter 5 we estimated that the shift in the energy of such a level, due to the relativistic dependence of mass on velocity, is also about 10^{-4} times the energy of the level. The energy shifts due to relativistic effects other than the spin are thus quite comparable to those due to the spin-orbit interaction for the hydrogen atom, and we cannot expect to be able to compare the predictions of equation (11–50) or (11–51) with the experimental data until we have taken these effects into account. A complete treatment of the relativistic effects can only be given in terms of the Dirac theory; but results which are almost complete can be obtained with the Schroedinger theory by treating the additional relativistic corrections as a perturbation.

The starting point of Schroedinger quantum mechanics consists in writing the classical expression for E, which is defined as the sum of the kinetic energy T and the potential energy V. That is,

$$E = T + V = \frac{p^2}{2m} + V \tag{11–52}$$

This equation is not relativistically correct since, according to equations (1–22′) and (1–25),

$$T_{\text{Rel}} = (c^2p^2 + m^2c^4)^{1/2} - mc^2$$

where we use m for the rest mass. Let us write the relativistically correct expression for E, and then evaluate it in the limit of small values of p.

This procedure gives

$$E_{\mathrm{Rel}} = T_{\mathrm{Rel}} + V = (c^2p^2 + m^2c^4)^{1/2} - mc^2 + V$$

$$E_{\mathrm{Rel}} = mc^2\left[\left(1 + \frac{p^2}{m^2c^2}\right)^{1/2} - 1\right] + V$$

$$E_{\mathrm{Rel}} \simeq mc^2\left[1 + \frac{p^2}{2m^2c^2} - \frac{p^4}{8m^4c^4} - 1\right] + V$$

$$E_{\mathrm{Rel}} \simeq \frac{p^2}{2m} - \frac{p^4}{8m^3c^2} + V$$

We have taken the first three terms in the binomial expansion. Comparing this with equation (11–52), we see that the error in the classical expression for E is approximately

$$\Delta E_{\mathrm{Rel}} = -\frac{p^4}{8m^3c^2} = -\frac{T^2}{2mc^2} = -\frac{(E - V)^2}{2mc^2} = -\frac{E^2 + V^2 - 2EV}{2mc^2}. \tag{11–53}$$

If we use the eigenfunctions ψ_{nljm_j}, we can use non-degenerate perturbation theory to evaluate the shift in the total energy E of the corresponding state due to the addition of the term ΔE_{Rel}, which may be thought of as an addition to V. According to the theory, the approximate value of the energy shift is

$$\overline{\Delta E_{\mathrm{Rel}}} = -\frac{1}{2mc^2}\iiint \psi^*_{nljm_j}(E^2 + V^2 - 2EV)_{\mathrm{op}}\psi_{nljm_j} r^2 \sin\theta \, dr \, d\theta \, d\phi$$

Now

$$V = -Ze^2/r \quad \text{and} \quad E = E_n, \quad \text{a constant}$$

So we obtain

$$\overline{\Delta E_{\mathrm{Rel}}} = -\frac{E_n^2}{2mc^2} - \frac{Z^2e^4}{2mc^2}\iiint \psi^*_{nljm_j}\frac{1}{r^2}\psi_{nljm_j} r^2 \sin\theta \, dr \, d\theta \, d\phi$$

$$-\frac{E_nZe^2}{mc^2}\iiint \psi^*_{nljm_j}\frac{1}{r}\psi_{nljm_j} r^2 \sin\theta \, dr \, d\theta \, d\phi$$

Evaluation of the integrals yields

$$\overline{\Delta E_{\mathrm{Rel}}} = -\frac{Z^2|E_n|\alpha^2}{n}\left(\frac{2}{2l + 1} - \frac{3}{4n}\right) \tag{11–54}$$

Comparison of equations (11–51) and (11–54) shows that the energy shift we have just calculated is, in fact, comparable in magnitude to the energy shift due to the spin-orbit interaction. Since both quantities have the same Z dependence, this is true for all one-electron atoms.

Adding the two energy shifts to the unperturbed energy, we obtain an expression for the total energy,

$$E = E_n + \overline{\Delta E_{\mathbf{S}\cdot\mathbf{L}}} + \overline{\Delta E_{\text{Rel}}}$$

of the state specified by the quantum numbers $nljm_j$. It is

$$E = -|E_n| \left\{1 + \frac{Z^2\alpha^2}{n}\left(-\frac{1}{(2l+1)(l+1)} + \frac{2}{2l+1} - \frac{3}{4n}\right)\right\}$$

$$= -|E_n| \left\{1 + \frac{Z^2\alpha^2}{n}\left(\frac{1}{l+1} - \frac{3}{4n}\right)\right\}, \qquad j = l + \tfrac{1}{2}, l \neq 0$$

and

$$E = -|E_n| \left\{1 + \frac{Z^2\alpha^2}{n}\left(\frac{1}{l(2l+1)} + \frac{2}{2l+1} - \frac{3}{4n}\right)\right\}$$

$$= -|E_n| \left\{1 + \frac{Z^2\alpha^2}{n}\left(\frac{1}{l} - \frac{3}{4n}\right)\right\}, \qquad j = l - \tfrac{1}{2}, l \neq 0$$

For both values of j this can be written

$$E = -\frac{\mu Z^2 e^4}{2\hbar^2 n^2}\left\{1 + \frac{Z^2\alpha^2}{n}\left(\frac{1}{j+\frac{1}{2}} - \frac{3}{4n}\right)\right\}, \qquad l \neq 0 \qquad (11\text{–}55)$$

For $l = 0$, $\overline{\Delta E_{\mathbf{S}\cdot\mathbf{L}}} = 0$ and $E = E_n + \overline{\Delta E_{\text{Rel}}}$, which is

$$E = -\frac{\mu Z^2 e^4}{2\hbar^2 n^2}\left\{1 + \frac{Z^2\alpha^2}{n}\left(2 - \frac{3}{4n}\right)\right\}, \qquad l = 0 \qquad (11\text{–}56)$$

where we have evaluated E_n, using the reduced electron mass μ.

Let us compare these results with the results of Sommerfeld's relativistic Bohr theory of the energy levels of a one-electron atom, given in equation (5–22). This is

$$E = -\frac{\mu Z^2 e^4}{2\hbar^2 n^2}\left\{1 + \frac{Z^2\alpha^2}{n}\left(\frac{1}{n_\theta} - \frac{3}{4n}\right)\right\} \qquad (11\text{–}57)$$

The quantum number n_θ appearing in this expression can assume the values $n_\theta = 1, 2, 3, \ldots, n$. Since the quantum number j can assume the values $j = \frac{1}{2}, \frac{3}{2}, \frac{5}{2}, \ldots, n - \frac{1}{2}$,† the quantity $j + \frac{1}{2}$ can assume the values $j + \frac{1}{2} = 1, 2, 3, \ldots, n$. Thus, for $l \neq 0$, the values of total energy predicted by our calculations are in exact agreement with the values predicted by Sommerfeld. For $l = 0$ there is a discrepancy. But this discrepancy is

† For $l = 0, j = \frac{1}{2}$; otherwise $j = l \pm \frac{1}{2}$. Since $l = 0, 1, 2, \ldots, n - 1$, the minimum value of j is $\frac{1}{2}$ and the maximum value is $n - 1 + \frac{1}{2} = n - \frac{1}{2}$.

removed by the Dirac theory, which shows that there is one more relativistic correction to the energy, whose existence we could not have guessed because it has no classical analog. This correction is zero for $l \neq 0$. For $l = 0$ its value is such that, when it is added to the right side of (11–56), the resulting equation has the same form as equation (11–55). Therefore the results of a complete relativistic treatment by the Dirac theory can be expressed by the single equation

$$E = -\frac{\mu Z^2 e^4}{2\hbar^2 n^2}\left\{1 + \frac{Z^2\alpha^2}{n}\left(\frac{1}{j+\frac{1}{2}} - \frac{3}{4n}\right)\right\} \tag{11–58}$$

These results are in exact agreement with the predictions of Sommerfeld's theory. Since the Sommerfeld theory was based on the Bohr theory, it is only a very rough approximation to physical reality. In contrast, the Dirac theory represents an extremely refined expression of our understanding of physical reality. From this point of view the agreement between equations (11–57) and (11–58) is one of the most amazing coincidences to be found in the study of physics.

The energy levels of the hydrogen atom, as predicted by the Bohr, Sommerfeld, and Dirac theories, are shown in figure (11–13) for the first three values of n. In order to make the energy level splittings visible, the displacements of the Sommerfeld and Dirac energy levels from those given by the Bohr theory have been exaggerated by a factor of $(137)^2 = 1.88 \times 10^4$. Thus the diagrams would be completely to scale if the value of the fine structure constant were 1 instead of 1/137. The values of the quantum number m_j, which specifies the spatial orientation of the total angular momentum vector, are not shown on the Dirac energy level diagram because the energy of the atom does not depend on this quantum number (unless the atom is in an external electric or magnetic field). Although we have not mentioned it before, a quantum number analogous to m_j is required in the Sommerfeld theory to specify the spatial orientation of the plane of the orbit. Since the energy of the atom does not depend on this quantum number (if the atom is not in an external field), its values are not indicated on the Sommerfeld diagram. Also not shown are the energy levels of hydrogen measured experimentally by optical spectroscopy, which are in extremely good agreement with the levels of the Sommerfeld and Dirac theories.

The only essential difference between the results of these two theories is that the Dirac theory predicts that for most levels there is a degeneracy, in addition to the trivial degeneracy with respect to spatial orientation mentioned above, because the energy depends on the quantum numbers n and j but not on the quantum number l. Since there are generally two values of l corresponding to the same value of j, the Dirac theory predicts that most levels are actually double. This prediction was verified experimentally in 1947 by Lamb, who showed that for $n = 2$ and $j = \frac{1}{2}$ there are two levels, which actually do not quite coincide†. The $l = 0$ level lies above the $l = 1$ level by about one-tenth the separation between that level and the $n = 2$, $l = 1$, $j = \frac{3}{2}$ level. This is called the *Lamb shift*. It can

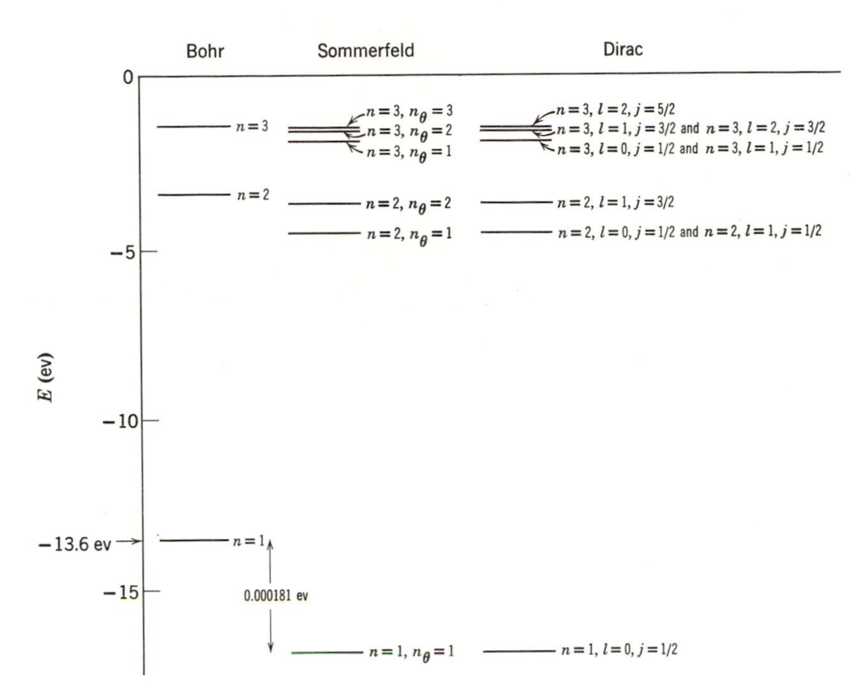

Figure 11–13. Energy levels of the hydrogen atom according to Bohr, Sommerfeld, and Dirac. The displacements from the Bohr energy levels are exaggerated by a factor of $(137)^2$.

be understood quantitatively in terms of the theory of *quantum electrodynamics*, as can the slight departure of g_s from 2 mentioned in the footnote on page 337, but we shall not be able to discuss this theory here.

We close our treatment of the one-electron atom by mentioning one final effect which influences the atomic energy levels. This is the *hyperfine splitting*, which is due to an interaction between the magnetic field produced by the moving electron and an inherent magnetic moment of the

† Lamb's experiment involved measuring the frequency of quanta absorbed in transitions between the $n = 2$, $l = 0$, $j = \frac{1}{2}$ and $n = 2$, $l = 1$, $j = \frac{1}{2}$ levels. The energy difference between these levels is so small that the frequency is in the microwave radio range. Since measurements of radio frequencies can be made very accurately, it was possible to measure the energy difference to five significant figures!

nucleus. As nuclear magnetic moments are smaller than atomic magnetic moments by three orders of magnitude, the hyperfine splitting is smaller than the spin-orbit splitting by the same factor. Nevertheless, we shall see in Chapter 13 that this effect can be described quantitatively in terms of Schroedinger quantum mechanics. In fact, every aspect of the behavior of a one-electron atom can be explained in detail by the theories of modern physics. As we have seen, these explanations can be quite complicated. Fortunately the degree of complication does not increase in proportion to the number of electrons in the atom.

BIBLIOGRAPHY

Leighton, R. B., *Principles of Modern Physics*, McGraw-Hill Book Co., New York, 1959.

Ruark, A. E., and H. C. Urey, *Atoms, Molecules, and Quanta*, McGraw-Hill Book Co., New York, 1930.

White, H. E., *An Introduction to Atomic Spectra*, McGraw-Hill Book Co., New York, 1934.

EXERCISES

1. Use Ampere's law to evaluate the magnetic field produced by a circular current loop at a point on the axis of symmetry far from the loop. Then evaluate the magnetic field produced at the same point by a magnetic dipole located at the center of the loop and lying along the axis of symmetry. Show that the fields are the same if the current in the loop and its area are related to the magnetic moment of the dipole by equation (11–2). Next extend the argument to show that if this relation is obeyed the fields are the same at all points far from the loop or dipole. Finally, extend the argument to show that this is true for a plane current loop of arbitrary shape.

2. Evaluate μ_l/L for an electron moving in the $n = 2$, $n_\theta = 1$ elliptical orbit of an atom of the type treated in the Sommerfeld theory. Compare with equation (11–7). *Hint:* Use some of the equations derived in exercise 3, Chapter 4, and exercise 4, Chapter 5.

3. The first footnote of section 3 states that a Stern-Gerlach experiment cannot be performed with ionized atoms. It might be thought, though, that the experiment can be done if the axis of the magnetic field is bent to follow the curvature produced by the Hev/c force, and if a beam is used which has such small transverse dimensions that the particles never get far from the axis, because then all the particles of the beam will feel essentially the same Hev/c force. However, the uncertainty principle shows that it is impossible to produce a beam in which the transverse dimensions remain sufficiently small. Prove this.

4. The Thomas precession can also be described in terms of a time dilation, between the reference frame in which the nucleus is at rest and the reference frames in which the electron is instantaneously at rest, which leads to a disagreement between an observer at the nucleus and the observers at the electron concerning the time required for each to make a complete revolution about the other. Work out the details of this description, and compare the results with equation (11–32).

5. Verify equation (11–44), and also verify the vector inequality which leads to that equation.

CHAPTER 12

Identical Particles

1. Quantum Mechanical Description of Identical Particles

One final topic of the theory of quantum mechanics remains to be discussed before we can study the structure of multi-electron atoms. It concerns the question of how to give a proper quantum mechanical description of a system containing two or more *identical particles.* The nature of the question can best be illustrated by a specific example.

Consider a box containing two electrons. These two identical particles move around in the box, bouncing from the walls and occasionally scattering from each other. In a classical description of this system, the electrons travel in sharply defined trajectories so that constant observation of the system would allow us, in principle, to distinguish between the two electrons even though they are identical particles. That is, if at some instant we label one of the electrons 1 and the other electron 2, we could constantly observe the motion of the electrons without disturbing the system and would be able to say which electron is 1 and which electron is 2 at any later instant. In a quantum mechanical description this could not be done because the uncertainty principle would not allow us to observe constantly the motion of the electrons without completely changing the behavior of the system. An equivalent statement is that in quantum mechanics the finite extent of the wave function associated with each particle will lead to an overlapping of these wave functions which would make it impossible to tell which wave function was associated with which particle. A good example is the two electrons in an He atom. When this system is in its ground state the wave functions of the two electrons overlap completely. Of course, there is also overlap of the wave functions associated with the electron and proton of an H atom, but this does not lead to any problems in distinguishing one particle from the other because the particles are not identical.

We see that there is a fundamental difference between the classical and quantum mechanical description of a system containing identical particles. A correct quantum mechanical theory of such systems must be so formulated as to take explicit account of the *indistinguishability* of identical particles. When this is done, the theory leads to the prediction of some very interesting effects. These effects have no classical analogy because indistinguishability is a purely quantum mechanical situation.

Let us look for a way of writing wave functions describing the two identical particles in the box which will contain a mathematical expression of the ideas developed above. To simplify the argument, we shall assume that there are no interactions between the two particles. Then the particles will bounce between the walls of the box and their wave functions will certainly overlap, but they will not scatter from each other. We shall show later that the results of the following discussion are of quite general validity despite this simplification. The time independent Schroedinger equation for the system is

$$-\frac{\hbar^2}{2m}\left(\frac{\partial^2\psi_T}{\partial x_1^2}+\frac{\partial^2\psi_T}{\partial y_1^2}+\frac{\partial^2\psi_T}{\partial z_1^2}\right)-\frac{\hbar^2}{2m}\left(\frac{\partial^2\psi_T}{\partial x_2^2}+\frac{\partial^2\psi_T}{\partial y_2^2}+\frac{\partial^2\psi_T}{\partial z_2^2}\right) + V_T\psi_T = E_T\psi_T \quad (12\text{–}1)$$

where

$$m = \text{the mass of either particle}$$
$$x_1, y_1, z_1 = \text{the coordinates of particle 1}$$
$$x_2, y_2, z_2 = \text{the coordinates of particle 2}$$

We have written this directly from equation (10–8) for the case $N = 2$ and $m_1 = m_2 = m$. Since this equation specifies the identity of the particles with the labels 1 and 2, in using it we stand a chance of violating the requirements of indistinguishability. This will be discussed more carefully later. In equation (12–1),

$$\psi_T(x_1, \ldots, z_2) = \text{the eigenfunction for the whole system}$$
$$V_T(x_1, \ldots, z_2) = \text{the potential energy for the whole system}$$
$$E_T = \text{the total energy for the whole system}$$

Since we have assumed that there is no interaction between the two particles, the potential energy of the whole system is simply the sum of the potential energies of each particle in its interaction with the walls of the box. Each potential energy will depend only on the coordinates of one particle and, since the particles are identical, the two potential functions are the same. Thus

$$V_T(x_1, \ldots, z_2) = V(x_1, y_1, z_1) + V(x_2, y_2, z_2) \quad (12\text{–}2)$$

We leave it as an exercise for the reader to show that for the potential (12–2) solutions exist to equation (12–1) of the form

$$\psi_T(x_1, \ldots, z_2) = \psi(x_1, y_1, z_1)\, \psi(x_2, y_2, z_2) \tag{12–3}$$

where $\psi(x_1, y_1, z_1)$ and $\psi(x_2, y_2, z_2)$ satisfy identical time independent Schroedinger equations. These equations have three spatial coordinates as independent variables. Consequently, each of the eigenfunctions requires three quantum numbers to specify the mathematical form of its dependence on the three spatial coordinates. In addition, one more quantum number is required to specify the orientation of the spin angular momentum vector of the particle.† However, we shall shorten the notation by using a single quantum number from the series $\alpha, \beta, \gamma, \delta, \ldots$ to designate a particular set of the four quantum numbers required to specify the state of a particle. Then a particular eigenfunction for particle 1 would be written

$$\psi_\alpha(x_1, y_1, z_1)$$

We further shorten the notation by writing this

$$\psi_\alpha(1)$$

This eigenfunction contains the information that particle 1 is in the state described by α. An eigenfunction indicating that particle 2 is in the state β would be written

$$\psi_\beta(2)$$

Then the total eigenfunction $\psi_T(x_1, \ldots, z_2)$ for the case in which particle 1 is in the state α, and particle 2 is in the state β, is

$$\psi_T(x_1, \ldots, z_2) = \psi_\alpha(1)\, \psi_\beta(2)$$

We write this

$$\psi_{\alpha\beta} \equiv \psi_\alpha(1)\, \psi_\beta(2) \tag{12–4}$$

An eigenfunction indicating that particle 1 is in state β and particle 2 is in state α would be

$$\psi_{\beta\alpha} \equiv \psi_\beta(1)\, \psi_\alpha(2) \tag{12–5}$$

It is interesting to evaluate the probability density functions for systems described by the two eigenfunctions (12–4) and (12–5). These are

$$\psi^*_{\alpha\beta}\psi_{\alpha\beta} = \psi^*_\alpha(1)\, \psi^*_\beta(2)\, \psi_\alpha(1)\, \psi_\beta(2) \tag{12–6}$$

and

$$\psi^*_{\beta\alpha}\psi_{\beta\alpha} = \psi^*_\beta(1)\, \psi^*_\alpha(2)\, \psi_\beta(1)\, \psi_\alpha(2) \tag{12–7}$$

Now, since the two identical particles are indistinguishable, we should be able to exchange their labels without changing a physically measurable

† Particles other than electrons also have spin.

quantity such as the probability density. Let us carry out this operation on $\psi_{\alpha\beta}^*\psi_{\alpha\beta}$. We find

$$\psi_{\alpha\beta}^*\psi_{\alpha\beta} = \psi_\alpha^*(1)\,\psi_\beta^*(2)\,\psi_\alpha(1)\,\psi_\beta(2) \underset{\substack{1\to 2\\ 2\to 1}}{\longrightarrow} \psi_\alpha^*(2)\,\psi_\beta^*(1)\,\psi_\alpha(2)\,\psi_\beta(1) = \psi_{\beta\alpha}^*\psi_{\beta\alpha}$$

From equations (12–6) and (12–7) it is apparent that $\psi_{\alpha\beta}^*\psi_{\alpha\beta} \neq \psi_{\beta\alpha}^*\psi_{\beta\alpha}$, and so a relabeling of the particles actually does change the probability density function calculated from the eigenfunction (12–4). The same is true for (12–5). Therefore we must conclude that these are not acceptable eigenfunctions for the description of a system containing two identical particles. The suspicion which we expressed after writing equation (12–1) has been justified.

2. Symmetric and Antisymmetric Eigenfunctions

It is possible to construct an eigenfunction which satisfies the time independent Schroedinger equation (12–1), and yet has the acceptable property that its probability density function is not changed by a relabeling of the particles. In fact, there are two ways of doing this. Consider the two linear combinations of $\psi_{\alpha\beta}$ and $\psi_{\beta\alpha}$,

$$\psi_S = \frac{1}{\sqrt{2}}\left[\psi_{\alpha\beta} + \psi_{\beta\alpha}\right] \tag{12–8}$$

and

$$\psi_A = \frac{1}{\sqrt{2}}\left[\psi_{\alpha\beta} - \psi_{\beta\alpha}\right] \tag{12–9}$$

If the reader has done the exercise suggested below equation (12–2), it will be apparent that $\psi_{\alpha\beta}$ and $\psi_{\beta\alpha}$ are degenerate. If not, it should still be apparent since the total energy of a system containing a particle in state α and a particle in state β will not depend on which particle is in which state, if the particles are identical. This is called *exchange degeneracy*. In the footnote on page 189 we have shown that any linear combination of degenerate eigenfunctions will be a solution to the time independent Schroedinger equation. Therefore the two new eigenfunctions ψ_S and ψ_A are both solutions to equation (12–1). The eigenfunctions ψ_S and ψ_A are, of course, also degenerate. The probability densities for these two eigenfunctions are

$$\psi_S^*\psi_S = \tfrac{1}{2}(\psi_{\alpha\beta}^*\psi_{\alpha\beta} + \psi_{\beta\alpha}^*\psi_{\beta\alpha}) + \tfrac{1}{2}(\psi_{\beta\alpha}^*\psi_{\alpha\beta} + \psi_{\alpha\beta}^*\psi_{\beta\alpha}) \tag{12–10}$$

and

$$\psi_A^*\psi_A = \tfrac{1}{2}(\psi_{\alpha\beta}^*\psi_{\alpha\beta} + \psi_{\beta\alpha}^*\psi_{\beta\alpha}) - \tfrac{1}{2}(\psi_{\beta\alpha}^*\psi_{\alpha\beta} + \psi_{\alpha\beta}^*\psi_{\beta\alpha}) \tag{12–11}$$

From these it is easy to show that ψ_S and ψ_A are normalized if $\psi_{\alpha\beta}$ and $\psi_{\beta\alpha}$ are normalized. Furthermore, since (12–4) and (12–5) show that $\psi_{\alpha\beta} \rightarrow \psi_{\beta\alpha}$ and $\psi_{\beta\alpha} \rightarrow \psi_{\alpha\beta}$ when we exchange the labels of the particles, we see by inspection that both $\psi_S^*\psi_S$ and $\psi_A^*\psi_A$ are unchanged by this operation. This is also true for other measurable quantities such as the probability flux. Thus the eigenfunctions ψ_S and ψ_A, as well as the wave functions associated with these eigenfunctions, provide an acceptable description of a system containing two identical particles. Although the

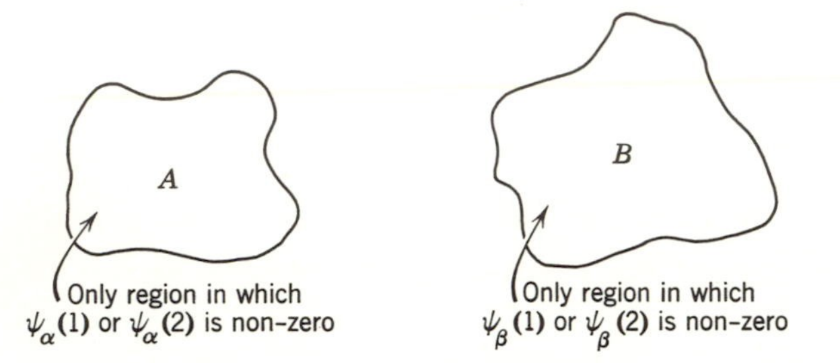

Figure 12–1. Illustrating the classical limit for a system of two indistinguishable particles.

numbers 1 and 2 do appear when we write in full the expressions for ψ_S and ψ_A, this labeling does not violate the requirements of indistinguishability because the value of any physically measurable quantity, which can be obtained from the eigenfunctions, is independent of the assignment of the labels. As far as the eigenfunctions themselves are concerned, the operation of exchanging the labels of the particles leaves one unchanged and multiplies the other by -1. By inspection we see that

$$\begin{aligned} \psi_S &\underset{\substack{1\to 2\\ 2\to 1}}{\longrightarrow} \psi_S \\ \psi_A &\longrightarrow -\psi_A \end{aligned} \tag{12–12}$$

The eigenfunction ψ_S is called a *symmetric eigenfunction*, and the eigenfunction ψ_A is called an *antisymmetric eigenfunction*. The behavior of ψ_A is not objectionable since an eigenfunction is not physically measurable.

An instructive interpretation of the role played by the first and second parentheses in the expressions (12-10) and (12-11) for $\psi_S^*\psi_S$ and $\psi_A^*\psi_A$ is obtained by considering these expressions in the classical limit. In this limit, the eigenfunctions for one particle can be sufficiently localized that two different one-particle eigenfunctions do not overlap. That is, there will exist two non-overlapping regions of space A and B, as indicated in figure (12–1), such that $\psi_\alpha(1)$ or $\psi_\alpha(2)$ is non-zero only when the coordinates of particle 1 or particle 2 are in A, and such that $\psi_\beta(1)$ or $\psi_\beta(2)$ is non-zero only when the coordinates of particle 1 or particle 2 is in B. Then the two-particle eigenfunction $\psi_{\alpha\beta} = \psi_\alpha(1)\,\psi_\beta(2)$ will vanish unless the coordinates of particle 1 are in A and the coordinates of particle 2 are in B. But, if this is so, $\psi_{\beta\alpha} = \psi_\beta(1)\,\psi_\alpha(2)$ must vanish. Thus, if there is no region in which both of the one-particle eigenfunctions have non-zero values, either one or the other of the two particle eigenfunctions $\psi_{\alpha\beta}$ or $\psi_{\beta\alpha}$ will always be zero. Then the second parentheses in the expressions for $\psi_S^*\psi_S$ and $\psi_A^*\psi_A$ will always vanish, and we have

$$\psi_S^*\psi_S = \psi_A^*\psi_A = \tfrac{1}{2}(\psi_{\alpha\beta}^*\psi_{\alpha\beta} + \psi_{\beta\alpha}^*\psi_{\beta\alpha}) \tag{12–13}$$

We see that in the classical limit the quantum mechanical expressions for the probability densities of the symmetric and antisymmetric eigenfunctions are equal. A moment's consideration should demonstrate to the reader that their common value (the first parentheses in the general expressions for $\psi_S^*\psi_S$ and $\psi_A^*\psi_A$) is completely analogous to the classical expression for the probability density of a system containing two identical particles, if one is in the state α and the other in the state β, but it is not known which particle is in which state. The quantity which is responsible for the difference between $\psi_S^*\psi_S$ and $\psi_A^*\psi_A$ when the one-particle eigenfunctions overlap (the second parentheses in their general expression) contains terms which correspond to some sort of interference between the two-particle eigenfunctions $\psi_{\alpha\beta}$ and $\psi_{\beta\alpha}$. The presence of this quantity is a result of the quantum mechanical requirements of indistinguishability.

3. The Exclusion Principle

We shall now prove that, if at some certain instant the wave function $\Psi(x_1, \ldots, z_N, t)$ describing a system of N identical particles is either symmetric or antisymmetric with respect to exchanging the labels of any two of the particles, then under any circumstances the wave function must always preserve the same symmetry. To do this we consider the Schroedinger equation as a differential equation governing the time dependence of $\Psi(x_1, \ldots, z_N, t)$, which can be written

$$e_{\text{op}}\Psi(x_1, \ldots, z_N, t) = i\hbar\,\frac{\partial\Psi(x_1, \ldots, z_N, t)}{\partial t}$$

This equation allows us to evaluate the wave function at time $t + dt$, if we know it at time t, because

$$\Psi(x_1, \ldots, z_N, t + dt) = \Psi(x_1, \ldots, z_N, t) + \frac{\partial\Psi(x_1, \ldots, z_N, t)}{\partial t}\,dt$$

$$= \Psi(x_1, \ldots, z_N, t) + \frac{e_{\text{op}}}{i\hbar}\Psi(x_1, \ldots, z_N, t)\,dt$$

Now the derivative terms of the total energy operator

$$e_{\text{op}} = \sum_{i=1}^{N} -\frac{\hbar^2}{2m}\left(\frac{\partial^2}{\partial x_i^2} + \frac{\partial^2}{\partial y_i^2} + \frac{\partial^2}{\partial z_i^2}\right) + V(x_1, \ldots, z_N, t)$$

are obviously symmetric with respect to exchange of the labels of any two particles since such an exchange will leave the summation unchanged. This is also true of the potential energy term since its value cannot depend on the labeling of the identical particles, even if there are interactions between the particles. Since e_{op} is a symmetric operator, the term $(e_{\text{op}}/i\hbar)\Psi(x_1, \ldots, z_N, t)$ has the same symmetry as $\Psi(x_1, \ldots, z_N, t)$. Consequently, if $\Psi(x_1, \ldots, z_N, t)$ is symmetric, $\Psi(x_1, \ldots, z_N, t + dt)$ will be symmetric and, if $\Psi(x_1, \ldots, z_N, t)$ is antisymmetric, $\Psi(x_1, \ldots, z_N, t + dt)$ will be antisymmetric. By continuing this integration process to $t + 2\,dt$, etc., we prove that the symmetry character of the wave function must be preserved. If we consider the particular case of an eigenstate wave function

$$\Psi(x_1, \ldots, z_N, t) = e^{-iEt/\hbar}\,\psi(x_1, \ldots, z_N)$$

it is clear that the symmetry character of the eigenfunctions is the same as the symmetry character of the wave functions,

For a system of two non-interacting identical particles, we know that the eigenfunction must be either symmetric or antisymmetric. We have just shown that, in the more general case of a system of several identical particles which are interacting with each other, the eigenfunction *may* be either symmetric or antisymmetric. That is, we have shown that it is consistent for the eigenfunction always to be symmetric or always to be antisymmetric. The question then arises: *Do* eigenfunctions describing real systems of several interacting identical particles have a definite symmetry, either symmetric or antisymmetric, and, if so, which one? The answer to this question is known. However, it is provided not by the theory of quantum mechanics, but by experiment.

As a result of an analysis of experimental information concerning the energy levels of certain atoms, in 1925 Pauli enunciated the *exclusion principle:*

In a multi-electron atom there can never be more than one electron in the same quantum state.

It was soon established from the analysis of other experimental data that the exclusion principle represents a property of electrons and not, specifically, of atoms; the exclusion principle operates in any system containing electrons. Now consider the two-particle antisymmetric eigenfunction

(12–9) for the case in which both particles 1 and 2 are in the same state α. It is

$$\psi_A = \frac{1}{\sqrt{2}}(\psi_{\alpha\alpha} - \psi_{\alpha\alpha}) \equiv 0 \tag{12–14}$$

The eigenfunction vanishes identically; if two particles are described by the antisymmetric eigenfunction, they cannot both be in the same quantum state. Thus the property of electrons expressed by the exclusion principle is exactly the property of the antisymmetric eigenfunctions, and we conclude that:

A system containing several electrons must be described by an antisymmetric eigenfunction.

We know the form of the antisymmetric eigenfunction for a system containing two non-interacting particles. Note that it can be written as a determinant

$$\psi_A = \frac{1}{\sqrt{2}}\begin{pmatrix} \psi_\alpha(1) & \psi_\alpha(2) \\ \psi_\beta(1) & \psi_\beta(2) \end{pmatrix} \tag{12–15}$$

since, upon evaluating the determinant, this is

$$\psi_A = \frac{1}{\sqrt{2}}[\psi_\alpha(1)\,\psi_\beta(2) - \psi_\beta(1)\,\psi_\alpha(2)] = \frac{1}{\sqrt{2}}[\psi_{\alpha\beta} - \psi_{\beta\alpha}]$$

For a system containing N non-interacting particles the antisymmetric eigenfunction can be written, in a completely analogous form, as

$$\psi_A = \frac{1}{\sqrt{N!}}\begin{pmatrix} \psi_\alpha(1) & \psi_\alpha(2) & \cdots & \psi_\alpha(N) \\ \psi_\beta(1) & \psi_\beta(2) & \cdots & \psi_\beta(N) \\ \cdot & \cdot & \cdots & \cdot \\ \cdot & \cdot & \cdots & \cdot \\ \cdot & \cdot & \cdots & \cdot \\ \psi_\nu(1) & \psi_\nu(2) & \cdots & \psi_\nu(N) \end{pmatrix} \tag{12–16}$$

This is called a *Slater determinant*. Since the value of this determinant consists of sums and differences of $N!$ terms comprised of products of one-particle eigenfunctions, such as $\psi_\alpha(1)\,\psi_\beta(2)\cdots\psi_\nu(N)$ and similar terms in which the particle labels are permuted, ψ_A is simply a linear combination of degenerate eigenfunctions. Therefore ψ_A will also be a solution to the time independent Schroedinger equation. The constant $1/\sqrt{N!}$ is such that ψ_A will be normalized if each of the products of one-particle eigenfunctions is normalized. Because a determinant vanishes if any two rows are identical, it is apparent that ψ_A has the required property of vanishing if more than one particle is in the same state (e.g., if $\beta = \alpha$).

It is possible to formulate most quantum mechanical calculations in such a way that the interactions between the identical particles are considered perturbations. Then, for unperturbed eigenfunctions, functions of the form (12–16) can be used. However, if the interactions between the particles cannot be ignored, the antisymmetric eigenfunctions cannot be written as (12–16), since $\psi_\alpha(1)\, \psi_\beta(2) \cdots \psi_\nu(N)$, and all the other terms, will not be solutions to the time independent Schroedinger equation. In this case antisymmetric eigenfunctions which vanish identically if two particles are in the same state can still be obtained by taking sums and differences of the $N!$ degenerate solutions to the time independent Schroedinger equation which will always exist for a system with N identical particles.

In this and the following chapters we shall study the multi-electron systems whose analysis led Pauli and others to the conclusion that such systems must be described by antisymmetric eigenfunctions. We shall see that this property of electrons plays an extremely important role in governing the behavior of systems containing these particles. Atoms, and therefore the entire universe, would be *radically different* if electrons were such as to require a description in terms of symmetric instead of antisymmetric eigenfunctions.

The symmetry character of particles other than electrons is also a question which can be settled only by experiment. It is found that systems of protons (the nuclei of H atoms), or neutrons (uncharged particles of approximately proton mass), or certain other particles, must also be described by antisymmetric eigenfunctions. Thus an eigenfunction describing a system of non-interacting protons must be of the form (12–16). On the other hand, it is found that systems of quanta, or certain other particles, must be described in terms of symmetric eigenfunctions. Also, systems containing particles which are tightly bound aggregates of an even number of antisymmetric particles (e.g., a system containing alpha particles which consist of two neutrons and two protons) must be described by eigenfunctions which are symmetric with respect to exchange of the labels of any two of the aggregate particles, because such an exchange is equivalent to an even number of exchanges of the labels of antisymmetric particles, and the net effect is to multiply the eigenfunction for the system by an even power of -1. A symmetric eigenfunction for a system of N non-interacting particles can be obtained by taking the sum of the $N!$ terms that can be formed from the product $\psi_\alpha(1)\, \psi_\beta(2) \cdots \psi_\nu(N)$ and the similar terms obtained by permuting the particle labels, and then dividing by the normalization constant $\sqrt{N!}$. If the particles are interacting, a symmetric eigenfunction can still be constructed by taking the sum of the $N!$ degenerate solutions to the time independent Schroedinger equation.

It is interesting to note that there must be some connection between the symmetry character of a particle and its intrinsic spin angular momentum, since all the antisymmetric particles also have half-integral spin just as the electron has, while all the others have zero or integral spin. However, the reason for this is not completely understood at present. Consequently, the symmetry character of a particle must be considered a basic property, like mass, charge, and spin, which can be determined only by experiment.

4. Additional Properties of Antisymmetric Eigenfunctions

As mentioned in section 3, Chapter 11, it is often convenient to write the complete eigenfunction containing both the spatial and spin coordinates of a particle as a product of a spatial eigenfunction and a spin eigenfunction. This is possible providing there are no interactions explicitly involving the spin (e.g., the spin-orbit interaction), so that its z component remains constant and m_s is a good quantum number. As a particular example, the one-electron-atom eigenfunctions $\psi_{nlm_lm_s}$ may always be written

$$\psi_{nlm_lm_s} = \psi_{nlm_l}\sigma_{m_s}$$

since, if the $\psi_{nlm_lm_s}$ are valid, m_s must be a good quantum number. In the general case in which this restriction is satisfied, we may write the eigenfunction for a particle carrying the label 1 as

$$\psi_\alpha(1) = \psi_a(1)\,\sigma_{m_{s_a}}(1) \tag{12–17}$$

where the single quantum number a stands for a set of three spatial quantum numbers. Now consider a system containing two particles in which there are no explicit interactions, spin or otherwise, between the particles. Then, according to (12–4) and (12–17), we may write the eigenfunctions describing these two particles as the product

$$\psi_{\alpha\beta} = \psi_\alpha(1)\,\psi_\beta(2) = \psi_a(1)\,\psi_b(2)\,\sigma_{m_{s_a}}(1)\,\sigma_{m_{s_b}}(2)$$

Using a notation analogous to that defined by (12–4), this can be written

$$\psi_{\alpha\beta} = \psi_{ab}\sigma_{m_{s_a}m_{s_b}} \tag{12–18}$$

This is not an acceptable eigenfunction for a system of two electrons (or other identical particles of an antisymmetric character) because it is not antisymmetric with respect to an exchange of the numbers 1 and 2 which label the electrons. However, it is possible to construct a properly antisymmetrized eigenfunction by taking a symmetric or antisymmetric linear combination of spatial eigenfunctions times a linear combination of spin eigenfunctions of the opposite symmetry, so that the product which forms the complete eigenfunction is antisymmetric. Consider the spin eigenfunctions. Since for an electron m_s can assume just two values $-\frac{1}{2}$ and $+\frac{1}{2}$, there are four possibilities

$$\sigma_{+\frac{1}{2}+\frac{1}{2}},\ \sigma_{+\frac{1}{2}-\frac{1}{2}},\ \sigma_{-\frac{1}{2}+\frac{1}{2}},\ \sigma_{-\frac{1}{2}-\frac{1}{2}}$$

From these we can construct four different spin eigenfunctions,

$$\begin{aligned} &\frac{1}{\sqrt{2}}(\sigma_{+\frac{1}{2}-\frac{1}{2}} - \sigma_{-\frac{1}{2}+\frac{1}{2}}) && \text{singlet} \\ &\left.\begin{aligned} &\sigma_{+\frac{1}{2}+\frac{1}{2}} \\ &\frac{1}{\sqrt{2}}(\sigma_{+\frac{1}{2}-\frac{1}{2}} + \sigma_{-\frac{1}{2}+\frac{1}{2}}) \\ &\sigma_{-\frac{1}{2}-\frac{1}{2}} \end{aligned}\right\} && \text{triplet} \end{aligned} \tag{12–19}$$

All four are normalized. The first one, which is called the *singlet* spin eigenfunction, is obviously antisymmetric with respect to an exchange of the particle labels. The last three form the so-called *triplet* of spin eigenfunctions. They are all obviously symmetric with respect to an exchange of the labels. To obtain complete eigenfunctions which are antisymmetric, we multiply the antisymmetric singlet spin eigenfunction by a symmetric spatial eigenfunction,

$$\frac{1}{\sqrt{2}}(\psi_{ab} + \psi_{ba}) \tag{12–20}$$

and multiply each of the symmetric triplet spin eigenfunctions by an antisymmetric spatial eigenfunction,

$$\frac{1}{\sqrt{2}}(\psi_{ab} - \psi_{ba}) \tag{12–21}$$

So there are the following four different antisymmetric complete eigenfunctions for a system containing two non-interacting electrons:

$$\psi_A = \begin{cases} \frac{1}{\sqrt{2}}(\psi_{ab} + \psi_{ba})\frac{1}{\sqrt{2}}(\sigma_{+\frac{1}{2}-\frac{1}{2}} - \sigma_{-\frac{1}{2}+\frac{1}{2}}) \\ \frac{1}{\sqrt{2}}(\psi_{ab} - \psi_{ba})\sigma_{+\frac{1}{2}+\frac{1}{2}} \\ \frac{1}{\sqrt{2}}(\psi_{ab} - \psi_{ba})\frac{1}{\sqrt{2}}(\sigma_{+\frac{1}{2}-\frac{1}{2}} + \sigma_{-\frac{1}{2}+\frac{1}{2}}) \\ \frac{1}{\sqrt{2}}(\psi_{ab} - \psi_{ba})\sigma_{-\frac{1}{2}-\frac{1}{2}} \end{cases} \tag{12–22}$$

The first describes the *singlet state*, and the others describe the three *triplet states*. All four are normalized, orthogonal, and degenerate.

A physical interpretation of the singlet and triplet states can be obtained by evaluating, for each state, the magnitude and z component of the *total spin angular momentum vector* $\mathbf{S}'$, which is

$$\mathbf{S}' = \mathbf{S}_1 + \mathbf{S}_2 \tag{12–23}$$

the sum of the spin angular momenta of the two electrons. We cannot do this in detail here because we have not studied the properties of the spin eigenfunctions; we shall merely state the results and give some

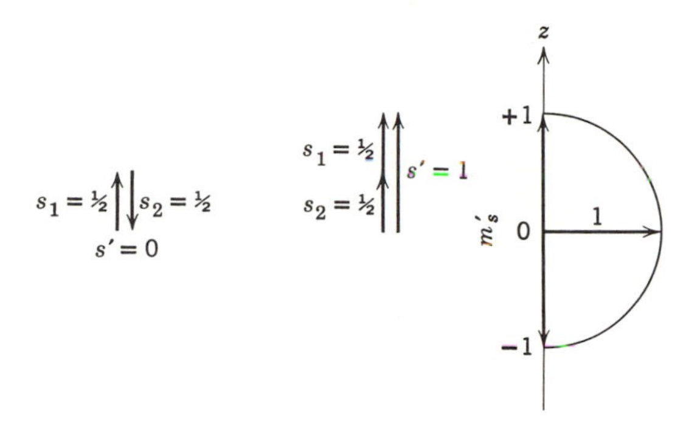

Figure 12–2. Vector addition diagrams for the quantum numbers $s_1 = \frac{1}{2}$, $s_2 = \frac{1}{2}$.

arguments for their validity. The reader should, however, have no difficulty in accepting the results, as they are completely analogous to those governing the addition of the spin and orbital angular momentum vectors to form the total angular momentum vector. It is found that the magnitude and z component of the total spin angular momentum vector are given by the equations

$$\begin{aligned} S' &= \sqrt{s'(s'+1)}\,\hbar \\ S'_z &= m'_s\hbar \end{aligned} \tag{12–24}$$

where

$$m'_s = -s', \ldots, +s'$$

and where

$$s' = 0, 1 \tag{12–25}$$

for the singlet and triplet states, respectively. Equation (12–25) can be proved by the same kind of arguments as were used in proving equation (11–44). These rules can formally be represented by diagrams in which two vectors of length $s = \frac{1}{2}$, which must not be interpreted as angular momentum vectors, are added to form a vector of length $s' = 0$ or 1. This is illustrated in figure (12–2). For the singlet state, $m'_s = 0$. The value of this quantum number for the three triplet states is shown in table (12–1). The properties of the states in which the spin eigenfunction is $\sigma_{+\frac{1}{2}+\frac{1}{2}}$ or $\sigma_{-\frac{1}{2}-\frac{1}{2}}$ are easy to interpret. In these states, the z components of the spin vectors of the two electrons are of the same sign. Thus the spins combine into a total spin with z component for $m_s' = \pm 1$, which is twice that of a single electron. The quantum number specifying the magnitude of the total spin vector for these two states would then necessarily be $s' = 1$. For the states in which the spin eigenfunction is $(1/\sqrt{2})(\sigma_{+\frac{1}{2}-\frac{1}{2}} - \sigma_{-\frac{1}{2}+\frac{1}{2}})$ or $(1/\sqrt{2})(\sigma_{+\frac{1}{2}-\frac{1}{2}} + \sigma_{-\frac{1}{2}+\frac{1}{2}})$, it is clear that

TABLE (12–1)
Quantum Numbers for the Singlet and Triplet States

Spin eigenfunction	Designation	s'	m_s'
$\frac{1}{\sqrt{2}}(\sigma_{+\frac{1}{2}-\frac{1}{2}} - \sigma_{-\frac{1}{2}+\frac{1}{2}})$	singlet	0	0
$\sigma_{+\frac{1}{2}+\frac{1}{2}}$	triplet	1	+1
$\frac{1}{\sqrt{2}}(\sigma_{+\frac{1}{2}-\frac{1}{2}} + \sigma_{-\frac{1}{2}+\frac{1}{2}})$	triplet	1	0
$\sigma_{-\frac{1}{2}-\frac{1}{2}}$	triplet	1	−1

we must have $m'_s = 0$. Although it is not so apparent why the quantum number specifying the magnitude of the total spin vector is $s' = 0$ for the antisymmetric state and $s' = 1$ for the symmetric state, these results follow directly from the theory of spin eigenfunctions.† The triplet and singlet states are often spoken of as the states in which the spin vectors of the two electrons are parallel and antiparallel, respectively. Although it is not literally true, this description certainly contains an essence of truth and provides a useful physical picture.

If the spin vectors of the two electrons are "parallel," the spin eigenfunction is symmetric and, to make the complete eigenfunction antisymmetric, the spatial eigenfunction must be antisymmetric. Let us consider such a situation for a case in which the spatial coordinates of the two electrons are very similar, i.e., $x_1 \simeq x_2$, $y_1 \simeq y_2$, $z_1 \simeq z_2$. Then $\psi_a(1) \simeq \psi_a(2)$ and $\psi_b(1) \simeq \psi_b(2)$. Consequently

$$\psi_{ab} = \psi_a(1)\,\psi_b(2) \simeq \psi_a(2)\,\psi_b(1) = \psi_b(1)\,\psi_a(2) = \psi_{ba}$$

† These results are certainly also reasonable from an intuitive point of view because they mean that all three symmetric spin eigenfunctions correspond to the same value of the quantum number $s' = 1$, i.e., to the same physical situation in which the electron spin vectors are "parallel."

In this case the value of the antisymmetric spatial eigenfunction is

$$\frac{1}{\sqrt{2}}(\psi_{ab} - \psi_{ba}) \simeq \frac{1}{\sqrt{2}}(\psi_{ab} - \psi_{ab}) = 0$$

As a result, the probability density will be very small when the two electrons are close together. Since there is only a very small chance of finding the two electrons close together, the two particles act as if they repel each other. This has nothing to do with a Coulomb repulsion because we assumed at the very beginning of this section that there is no explicit interaction between the two electrons. Furthermore, it is easy to show that, if the spatial eigenfunction describing the two electrons is symmetric, the probability density is twice as large as average when they are close together. Thus, if the spin vectors of the two electrons are "antiparallel," the spin eigenfunction is antisymmetric, the spatial eigenfunction must be symmetric, and the two particles act as if they attract each other because there is a large chance of finding them close together. We see that the requirement that the complete eigenfunction be antisymmetric with respect to an exchange of the labels of the two electrons leads to a coupling between their spin and spatial coordinates. They act as if they move under the influence of a force whose sign depends on the relative orientation of their spins. This is called an *exchange force*. It is a purely quantum mechanical effect and has no classical analog. However, it does lead to directly observable results, such as those described in the next section.

5. The Helium Atom

The simplest real system containing two identical particles is the neutral He atom. On the right side of the energy level diagram in figure (12–3), we show the energies of the ground state and the first four excited states of He. The energy of the ground state is obtained experimentally by measuring the energy required to ionize He to form He^{+}. The sum of this ionization energy and the ground state energy of the one-electron atom He^{+}, which is -54.1 ev, is just equal to the energy of the ground state of He.†
The energies of the excited states relative to the ground state are obtained from an analysis of the He spectrum.

To treat He with the theory of quantum mechanics, we assume first

† The magnitude of the energy of the ground state is thus defined as the energy required to ionize completely the He atom to form He^{2+}. In spectroscopy it is, however, conventional to define the energy of the ground state as the energy required to ionize singly the atom; a spectroscopist would put $E = 0$ at the value we label $E = -54.1$ ev.

that the two electrons are moving under the influence of the same nuclear Coulomb potential, but that they are non-interacting. The interaction between the two electrons is then taken into account by means of perturbation theory. Thus, we consider an unperturbed system whose energy

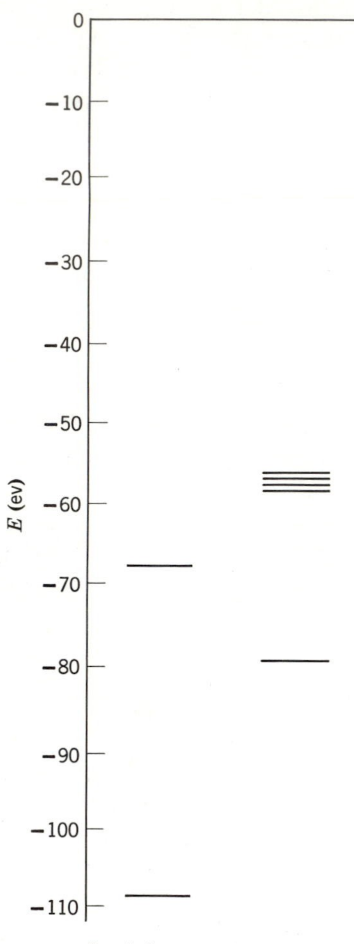

Figure 12–3. Right: An energy level diagram showing the ground state and the first four excited states of the helium atom. Left: The energy levels predicted by ignoring the electron-electron interaction.

states are given by the sum of two expressions of the form (10–38). That is, we take

$$E = -\mu Z^2 e^4/2\hbar^2 n_1^2 - \mu Z^2 e^4/2\hbar^2 n_2^2 \tag{12–26}$$

where $n_1 = 1, 2, 3, \ldots$; $n_2 = 1, 2, 3, \ldots$; $Z = 2$. In the ground state of this system, the quantum numbers n_1 and n_2 are both equal to 1. In the first excited state, one of the quantum numbers is equal to 1 and the other is equal to 2. The energies predicted for these states, in this approximation, are shown on the left side of the energy level diagram of figure (12–3). It is apparent that the unperturbed system provides a poor description of the energy levels of the actual system. This means that the interactions between the two electrons play an important role in the behavior of the He atom.

The interactions consist primarily of the Coulomb repulsion which the two negatively charged electrons exert on each other. It produces a positive potential energy which raises the total energy of the atom. In addition, there are magnetic interactions involving the spins of the electrons; but they are very small compared to the interaction of their charges and can be neglected. Thus we take for the perturbing potential the quantity

$$v = \frac{(-e)(-e)}{r_{12}} = \frac{e^2}{r_{12}} \tag{12–27}$$

where r_{12} is the distance between the two electrons. For unperturbed eigenfunctions we must use the antisymmetric eigenfunctions (12–22) and, for each energy state of the unperturbed system, there will, in general, be a number of degenerate unperturbed eigenfunctions. The degeneracy is due to exchange degeneracy, and also to the degeneracy of the energy states of the individual electrons with respect to the quantum numbers l and m_l.

For the ground state of the unperturbed system, there is, however, only a single eigenfunction. The lm_l degeneracy is absent because when $n_1 = n_2 = 1$ the other spatial quantum numbers can only have a single value $l_1 = m_{l_1} = l_2 = m_{l_2} = 0$. Furthermore, since all the spatial quantum numbers for both electrons are the same, we see from equation (12–22) that exchange degeneracy is absent as there is only one possible antisymmetric eigenfunction:

$$\frac{1}{\sqrt{2}}(\psi_{aa} + \psi_{aa})\frac{1}{\sqrt{2}}(\sigma_{+\frac{1}{2}-\frac{1}{2}} - \sigma_{-\frac{1}{2}+\frac{1}{2}})$$

The ground state of the unperturbed system can only be the singlet state because the spatial part of the antisymmetric eigenfunctions, the use of which is required by the exclusion principle, vanishes identically for the triplet states when all the spatial quantum numbers for the two electrons are the same. Of course, the fact that all the spatial quantum numbers are the same does not contradict the exclusion principle because the spin quantum numbers are different. We see that the properties of spin enter *implicitly*, even in the properties of the spatial part of the unperturbed eigenfunction. However, when we use the eigenfunction to calculate the energy shift resulting from the perturbation, it is not necessary to consider spin *explicitly*. Because the perturbation does not involve spin, the spin eigenfunctions enter only as a product in the term $\psi^* v \psi$ which must be evaluated in the calculation. Since the spin eigenfunctions are normalized, this product is equal to unity. Consequently in the present calculation we may simply ignore the spin part of the complete antisymmetric eigenfunction, and take for the properly normalized unperturbed eigenfunction

$$\psi_{aa} = \psi_{100}(1)\,\psi_{100}(2) = \frac{1}{\pi}\left(\frac{Z}{a_0}\right)^3 e^{-Zr_1/a_0} e^{-Zr_2/a_0}$$

where we have evaluated ψ_{100} from (10–39). Then, according to the three dimensional extension of equation (9–13), the energy shift due to the perturbation is

$$E' - E \simeq \int\!\!\int \psi_{100}^*(1)\,\psi_{100}^*(2)\,\frac{e^2}{r_{12}}\,\psi_{100}(1)\,\psi_{100}(2)\,d\tau_1\,d\tau_2$$

where

$$d\tau_1 = r_1^2 \sin\theta_1\,dr_1\,d\theta_1\,d\phi_1$$
$$d\tau_2 = r_2^2 \sin\theta_2\,dr_2\,d\theta_2\,d\phi_2$$

This integral has been evaluated. It gives

$$E' - E \simeq \frac{5}{4} Z \frac{\mu e^4}{2\hbar^2}$$

Combining this result with (12–26) for $n_1 = n_2 = 1$, we have

$$E' \simeq \left(-2Z^2 + \frac{5}{4} Z\right)\frac{\mu e^4}{2\hbar^2} \tag{12–28}$$

The value of E' calculated from this equation for $Z = 2$ agrees with the energy of the ground state of He to within 5 percent. Perturbation theory gives a reasonably accurate result, even though the perturbation is obviously not small. Much more accurate results can be obtained by using a modification of perturbation theory in which the error is of the order of the cube of a small quantity. Expression (12–28) also predicts the ground state energy of a two-electron ion with $Z > 2$, such as Li^+, Be^{2+}, B^{3+}. The agreement improves with increasing Z because the Coulomb interaction between the nucleus and each electron increases more rapidly than the Coulomb interaction between the two electrons, so the perturbation is smaller and the perturbation calculation more accurate.

Next consider the first excited state of the unperturbed system for He.

This is a highly degenerate state because there are four degenerate one-electron states for $n = 2$, and there are also exchange degeneracies. However, much of the degeneracy is removed when the interaction between the two electrons is applied as a perturbation. Before considering this quantitatively, it is worth while to describe what happens qualitatively.

The increase in the energy of the atom produced by the Coulomb repulsion between the two electrons depends on the average distance between the two electrons. This, in turn, depends on the quantum numbers of the spatial part of the eigenfunction, and also on its symmetry character. Let us consider first the dependence on the spatial quantum numbers, ignoring temporarily the symmetry properties and the requirements of indistinguishability. Then we may list the possible sets of quantum numbers for the two electrons in the first excited state of the unperturbed system as follows:

$$n_1 = 1, l_1 = 0, m_{l_1} = 0; \quad n_2 = 2, l_2 = 0, m_{l_2} = 0$$
$$n_1 = 1, l_1 = 0, m_{l_1} = 0; \quad n_2 = 2, l_2 = 1, m_{l_2} = -1$$
$$n_1 = 1, l_1 = 0, m_{l_1} = 0; \quad n_2 = 2, l_2 = 1, m_{l_2} = 0$$
$$n_1 = 1, l_1 = 0, m_{l_1} = 0; \quad n_2 = 2, l_2 = 1, m_{l_2} = 1$$

Since the probability density of electron 1 is spherically symmetrical, the average distance between the two electrons depends only on the radial probability densities for the two particles. These quantities, defined in equation (10–43), are specified completely by the quantum numbers n and l. By inspecting figure (10–2), the reader will see that the average distance between the two electrons is somewhat larger for the set of quantum numbers $n_1 = 1$, $l_1 = 0$; $n_2 = 2$, $l_2 = 0$ than it is for the set $n_1 = 1$, $l_1 = 0$; $n_2 = 2$, $l_2 = 1$. Since the perturbing potential is inversely proportional to the distance between the two electrons, the total energy of the atom will be increased more for the state specified by $n_1 = 1$, $l_1 = 0$; $n_2 = 2$, $l_2 = 1$ than for the state specified by $n_1 = 1$, $l_1 = 0$; $n_2 = 2$, $l_2 = 0$. Thus the degenerate energy level for the first excited state of the unperturbed system is split into two components by the perturbation. If we now take into account the requirements of indistinguishability, we shall see that each of these components is split again into two components. This is a result of the fact that each of the states, such as the one specified by the quantum numbers $n_1 = 1$, $l_1 = 0$; $n_2 = 2$, $l_2 = 0$, is comprised of four exchange degenerate states corresponding to the four antisymmetric complete eigenfunctions (12–22) which must be used to describe the system of two electrons. The splitting occurs because the average distance between the two electrons is smaller in the singlet state than it is in the triplet states, so the increase in the energy of the atom is larger for the singlet state than it is for the triplet states. The net effect of the perturbation is thus to increase the energy of the first excited state of the unperturbed system and split the energy level into four components. In the energy level diagram of figure (12–3), the reader will see that this conclusion is qualitatively in agreement with the experimentally observed energy levels of He.

Let us use perturbation theory to evaluate quantitatively the effect of the perturbation. This will not be so difficult as the reader might guess, for the following reasons:

1. As in the ground state calculation, we may neglect the spin part of the antisymmetric complete eigenfunctions and use for unperturbed eigenfunctions only the spatial part.

2. Since the total angular momentum of the atom (the sum of the orbital and spin angular momenta of the two electrons) cannot change if there are no external torques, and since the perturbation does not cause interactions between the spin angular momenta and the orbital angular momenta, it follows that neither the total spin angular momentum nor the total orbital angular momentum can change as a result of the application of the perturbation. Therefore the perturbation cannot induce appreciable mixing between the degenerate states specified by $n_1 = 1, l_1 = 0; n_2 = 2, l_2 = 0$, for which the magnitude of the total orbital angular momentum is 0, and the degenerate states specified by $n_1 = 1, l_1 = 0; n_2 = 2, l_2 = 1$, for which the magnitude of the total orbital angular momentum is $\sqrt{1(1+1)}\,\hbar$. Consequently it is possible to treat the degenerate states for these two sets of quantum numbers separately.

3. For simplicity, we shall actually carry through the calculation only for the degenerate states specified by $n_1 = 1$, $l_1 = 0$; $n_2 = 2$, $l_2 = 0$. For these states there is no spatial quantum number degeneracy since m_{l_1} and m_{l_2} can have only the single value $m_{l_1} = m_{l_2} = 0$. There is, however, an exchange degeneracy which arises because the spatial part of the singlet and triplet eigenfunctions,

$$\frac{1}{\sqrt{2}}[\psi_{ab} + \psi_{ba}] = \frac{1}{\sqrt{2}}[\psi_{100}(1)\,\psi_{200}(2) + \psi_{200}(1)\,\psi_{100}(2)] \quad (12\text{–}29)$$

and

$$\frac{1}{\sqrt{2}}[\psi_{ab} - \psi_{ba}] = \frac{1}{\sqrt{2}}[\psi_{100}(1)\,\psi_{200}(2) - \psi_{200}(1)\,\psi_{100}(2)] \quad (12\text{–}29')$$

correspond to the same total energy for the unperturbed system.

4. As stated in 2, the total spin angular momentum of the atom cannot change as a result of the application of the perturbation. For this reason the perturbation cannot induce appreciable mixing between the degenerate eigenfunctions (12–29) and (12–29′) because the first corresponds to the

singlet state in which the two spin vectors are "antiparallel" and the magnitude of the total spin angular momentum is 0, while the second corresponds to the triplet states in which the two spin vectors are "parallel" and the magnitude of the total spin angular momentum is $\sqrt{1(1+1)}\,\hbar$. Therefore it is possible to evaluate the energy shift due to the perturbation by applying ordinary perturbation theory to each of the eigenfunctions, even though they are degenerate, just as we did for the degenerate eigenfunctions ψ_{nljm_j} when calculating the spin-orbit energy shift.

Thus we use the three dimensional extension of (9–13) with the perturbation (12–27) and the unperturbed eigenfunctions (12–29) and (12–29′), and obtain for the energy shift due to the perturbation the expression

$$E'_{\pm} - E \simeq \frac{1}{2}\int\!\!\int [\psi^*_{100}(1)\,\psi^*_{200}(2) \pm \psi^*_{200}(1)\,\psi^*_{100}(2)]\,\frac{e^2}{r_{12}} \times [\psi_{100}(1)\,\psi_{200}(2) \pm \psi_{200}(1)\,\psi_{100}(2)]\,d\tau_1\,d\tau_2$$

where E'_+ corresponds to the singlet state described by (12–29) and E'_- corresponds to the triplet state described by (12–29′). This is

$$\begin{aligned} E'_{\pm} \simeq E &+ \frac{1}{2}\int\!\!\int \Big[\psi^*_{100}(1)\,\psi^*_{200}(2)\,\frac{e^2}{r_{12}}\,\psi_{100}(1)\,\psi_{200}(2) \\ &\qquad + \psi^*_{200}(1)\,\psi^*_{100}(2)\,\frac{e^2}{r_{12}}\,\psi_{200}(1)\,\psi_{100}(2)\Big]\,d\tau_1\,d\tau_2 \\ &\pm \frac{1}{2}\int\!\!\int \Big[\psi^*_{100}(1)\,\psi^*_{200}(2)\,\frac{e^2}{r_{12}}\,\psi_{200}(1)\,\psi_{100}(2) \\ &\qquad + \psi^*_{200}(1)\,\psi^*_{100}(2)\,\frac{e^2}{r_{12}}\,\psi_{100}(1)\,\psi_{200}(2)\Big]\,d\tau_1\,d\tau_2 \end{aligned} \qquad (12\text{–}30)$$

The reader should check these results by performing a routine degenerate perturbation theory calculation, as described in section 4, Chapter 9, for the case of two degenerate eigenfunctions $\psi_{100}(1)\,\psi_{200}(2)$ and $\psi_{200}(1)\,\psi_{100}(2)$. This is easy to do because the calculation carried out in that section can be immediately applied to the present problem. Equations (9–26) and (9–27) show that (12–29) and (12–29′) are exactly the right linear combinations of $\psi_{100}(1)\,\psi_{200}(2)$ and $\psi_{200}(1)\,\psi_{100}(2)$, and that (12–30) is the correct expression for the perturbed energy.

Let us compare the results of our calculation with equations (12–10) and (12–11), and with the interpretation which was given to these equations. This comparison will identify the first integral in equation (12–30) as an expression which, from classical considerations, would be written for the expectation value of the perturbation in a system containing two identical particles in the one-particle states $n = 1$, $l = 0$ and $n = 2$, $l = 0$. The same comparison identifies the second integral as a manifestation of the quantum mechanical requirements of indistinguishability. This integral takes into account the effective attraction or repulsion in the singlet or triplet states by increasing the energy shift for the former and decreasing it for the latter. The second integral is called the *exchange integral,* and

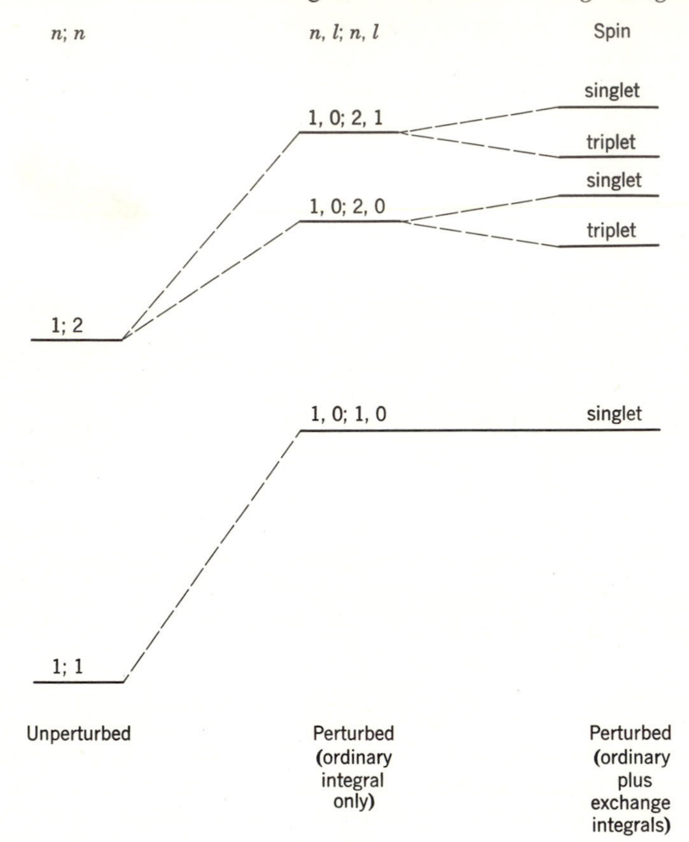

Figure 12–4. Illustrating the effects which contribute to the energies of the ground state and the first four excited states of the helium atom.

the first integral is called the *ordinary integral.* To evaluate these integrals we evaluate ψ_{100} and ψ_{200} from (10–39), which gives

$$\psi_{100} = \frac{1}{\sqrt{\pi}}\left(\frac{Z}{a_0}\right)^{3/2} e^{-Zr/a_0}, \qquad \psi_{200} = \frac{1}{4\sqrt{2\pi}}\left(\frac{Z}{a_0}\right)^{3/2}\left(2 - \frac{Zr}{a_0}\right)e^{-Zr/a_0}$$

By substituting these into (12–30), and carrying through the integration, values for E'_- and E'_+ are found. These values agree with the experimentally observed energies of the first two excited states of He to about 1 percent.

The perturbation calculation is more accurate for the excited states than for the ground state because the average distance between the electrons is larger and the perturbation is weaker.

A perturbation calculation for the unperturbed states, in which one set of quantum numbers is $n = 1$, $l = 0$ and the other set is $n = 2$, $l = 1$, leads to results which are similar to (12–30) except that the numerical value of the ordinary integral turns out to be somewhat larger. The important features and results of the theoretical description of the first few energy levels of He are represented schematically by the set of energy level diagrams shown in figure (12–4).

Not shown in the diagram is a fine structure splitting of the triplet energy levels, the existence of which is inferred from the spectrum of He. It is due to the very small interactions involving the spins of the electrons, and other relativistic effects, that split each of the triplet levels (except those in which the total orbital angular momentum is zero) into three levels with a separation about 10^{-4} times the separation between associated singlet and triplet levels. A theoretical treatment of this effect can be given in terms of the equations developed in the previous chapter. There is also a hyperfine splitting of the levels, due to extremely small interactions involving the spin of the nucleus, which produces a separation of about 10^{-3} times the separation produced by the fine structure splitting, and which can be treated by methods to be developed in Chapter 13. But we shall not treat either of these effects here.

Of particular interest is the absence of a triplet energy level associated with the singlet ground state level, which in our treatment we have shown to be a consequence of the exclusion principle. This is quite true, but historically the argument was made in the opposite order. The experimental fact that the He spectrum shows this triplet to be absent provided the primary evidence that led Pauli to the discovery of the exclusion principle.

6. The Fermi Gas

All particles which must be described by antisymmetric eigenfunctions are given the generic name *Fermi particles*, in honor of a man prominent in developing the theory of their properties. In the last section we presented a detailed discussion of a system containing two Fermi particles. Here we shall discuss, in somewhat less detail, a system containing a large number of these particles.

Consider a system in which a "gas" of freely moving, non-interacting, Fermi particles is confined to a certain region. Such a system, which is called a *Fermi gas*, can be used to describe some of the important features of the behavior of the electrons confined to a multi-electron atom. A Fermi gas can also be used to describe certain features of the motion of the protons and neutrons confined to a nucleus. Possibly the best physical example of a Fermi gas is the conduction electrons in a metal. These electrons, which are responsible for the electrical conductivity of the metal, move through it quite freely because the forces exerted by the various atoms tend to cancel, and the forces which the electrons exert on each other are small. However, the conduction electrons are confined to the metal by non-canceling attractive forces at the surfaces. In this section we shall develop some of the properties of a Fermi gas, such as its total energy and the energy distribution of the individual particles, and shall illustrate their significance for the conduction electrons. We shall see that these properties are very directly influenced by the exclusion principle.

Each particle of a Fermi gas is moving independently in a potential which is a constant (that we define as zero) inside a certain region, and which becomes larger than the total energy at the boundaries of the region. In most cases it is an adequate approximation to assume that the potential becomes infinitely large at the boundaries. Then the identical Schroedinger equations describing the motion of each of the identical particles assume a particularly simple form,

$$-\frac{\hbar^2}{2m}\left(\frac{\partial^2\psi}{\partial x^2}+\frac{\partial^2\psi}{\partial y^2}+\frac{\partial^2\psi}{\partial z^2}\right)=E\psi \qquad (12\text{–}31)$$

in which $\psi(x, y, z) = 0$ at the boundaries of the region. The reader may easily verify that solutions to this equation can be written

$$\psi(x, y, z) = X(x)\ Y(y)\ Z(z) \qquad (12\text{–}32)$$

in which

$$-\frac{\hbar^2}{2m}\frac{d^2X(x)}{dx^2} = E_xX(x) \qquad (12\text{–}32')$$

and similarly for $Y(y)$ and $Z(z)$, and where

$$E = E_x + E_y + E_z$$

It will not restrict the generality of our calculation to assume that the region containing the Fermi gas is a cube of side a. Then, defining the origin of the coordinates at the center of the cube and the axes parallel to its edges, the condition on $X(x)$ is

$$X(x) = 0, \qquad x \leqslant -a/2, \qquad x \geqslant a/2 \qquad (12\text{–}33)$$

and similarly for $Y(y)$ and $Z(z)$. Equation (12–32′) and condition (12–33) are identical with (8–89) and (8–90), which describe the problem of a one dimensional infinite square well potential. Consequently the allowed values of E_x can be written directly from (8–96), and we have

$$E_x = \frac{\pi^2 \hbar^2 n_x^2}{2ma^2} \tag{12–34}$$

where $n_x = 1, 2, 3, 4, \ldots$. Similar expressions obtain for E_y and E_z. Thus the allowed values of total energy of one of the identical particles is given by the equation

$$E = \frac{\pi^2 \hbar^2}{2ma^2} (n_x^2 + n_y^2 + n_z^2) \tag{12–35}$$

where $n_x = 1, 2, 3, 4, \ldots$; $n_y = 1, 2, 3, 4, \ldots$; $n_z = 1, 2, 3, 4, \ldots$. The three spatial quantum numbers n_x, n_y, n_z specify the quantum states of the particle.

It is useful to calculate from this equation the quantity $N(E)\,dE$, the number of quantum states for which the total energy of the particle is within the limits E to $E + dE$. The procedure is similar to that used in section 6, Chapter 2. We imagine a uniform cubic lattice constructed in one octant of a rectangular coordinate system in such a way that the three coordinates of each point of the lattice are equal to a possible set of quantum numbers n_x, n_y, n_z. Then the number of possible sets of quantum numbers is equal to $N(r)\,dr$, the number of points contained between concentric shells of radii r and $r + dr$, where

$$E = \frac{\pi^2 \hbar^2}{2ma^2} r^2 \tag{12–36}$$

By construction, $N(r)\,dr$ is just equal to the volume enclosed by the shells:

$$N(r)\,dr = \frac{1}{8} 4\pi r^2\,dr = \frac{\pi r^2\,dr}{2}$$

According to equation (12–36),

$$dE = \frac{\pi^2 \hbar^2}{ma^2} r\,dr$$

and

$$E^{1/2} = \frac{\pi \hbar}{2^{1/2} m^{1/2} a} r$$

so

$$N(r)\,dr = \frac{\pi}{2} \frac{2^{1/2} m^{1/2} a}{\pi \hbar} E^{1/2} \frac{ma^2}{\pi^2 \hbar^2} dE$$

Since $N(r)\,dr = N(E)\,dE$, we have

$$N(E)\,dE = \frac{m^{3/2} V}{2^{1/2} \pi^2 \hbar^3} E^{1/2}\,dE \tag{12–37}$$

where we write $V = a^3$ for the volume of the cubical region in which the particle of mass m and kinetic energy E is confined.† It can be shown that equation (12–37) is the correct expression for the number of quantum states in the energy range E to $E + dE$ for the general case of a particle confined to a region of volume V and any arbitrary shape.

The eigenfunction (12–32) describing the motion of one of the identical particles is a standing wave with nodes at the boundaries of the region. The eigenfunction describing the entire system of particles is composed of products of such standing wave eigenfunctions and, to satisfy the exclusion principle, it must be written in the antisymmetrized form (12–16). If we discuss only the energetics of a Fermi gas, it will not be necessary to deal

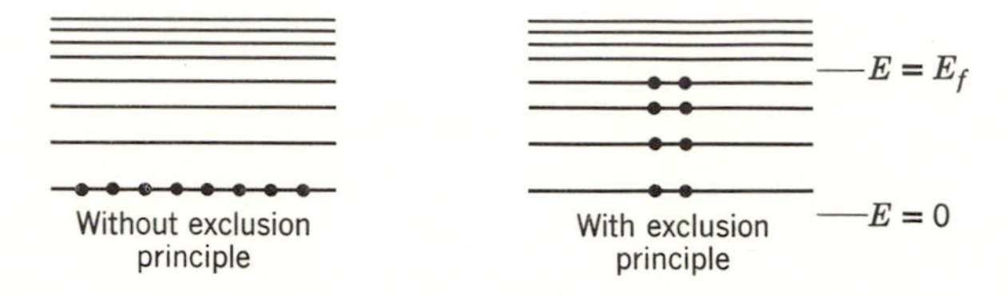

Figure 12–5. Illustrating the effect of the exclusion principle.

with this eigenfunction. However, it will be very necessary to take account of the exclusion principle. This can be done simply by imposing the auxiliary requirement that no more than two particles have the same spatial quantum numbers n_x, n_y, n_z. The exclusion principle will be satisfied because one particle can have the spin quantum number $m_s = -\frac{1}{2}$ and the other can have $m_s = +\frac{1}{2}$.

It is easy to see that the exclusion principle plays an important role in determining the minimum total energy of the Fermi gas. Consider the situation illustrated schematically in the two diagrams of figure (12–5). The diagram on the left represents the occupied quantum states when a system of eight identical particles, which do not obey the exclusion principle, has the minimum possible total energy. All the particles would be in the state of lowest energy. The other diagram represents the occupied quantum states for the minimum possible total energy of a system in which the particles do obey the exclusion principle. Only two particles can be in the same spatial state, and so the total energy of the system is considerably larger. The diagrams also illustrate the way in which the number of quantum states per unit energy increases according to $E^{1/2}$. For a system of N Fermi particles, the lowest total energy is achieved when there are two particles in each of the energy states between $E = 0$ and $E = E_f$.

† Since the potential energy is zero in the region, the total energy E is equal to the kinetic energy.

The particles fill the states up to the *Fermi energy* E_f. This energy may be evaluated by requiring that the integral of $2N(E)\,dE$ from 0 to E_f be equal to the number of particles in the system.† That is,

$$N = \int_0^{E_f} 2N(E)\,dE$$

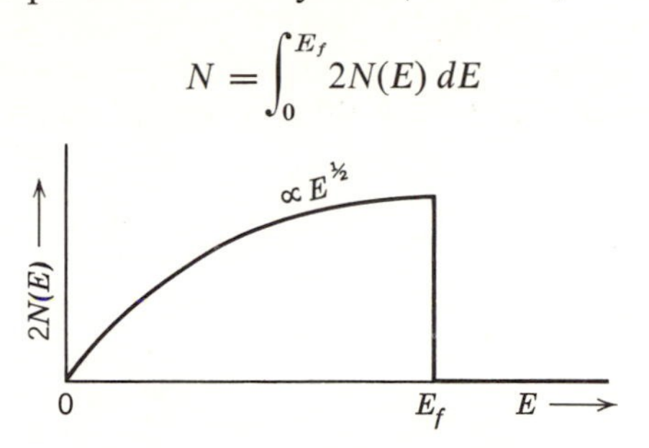

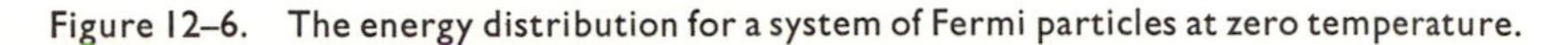

Figure 12–6. The energy distribution for a system of Fermi particles at zero temperature.

The factor of 2 enters because there are two particles per spatial quantum state. Evaluating $N(E)\,dE$ from (12–37), we have

$$N = \frac{2^{1/2}m^{3/2}V}{\pi^2\hbar^3}\int_0^{E_f} E^{1/2}\,dE = \frac{2^{1/2}m^{3/2}V}{\pi^2\hbar^3}\frac{2}{3}E_f^{3/2}$$

$$N = \frac{2^{3/2}m^{3/2}V}{3\pi^2\hbar^3}E_f^{3/2}$$

Thus

$$E_f = 3^{2/3}\pi^{4/3}\frac{\hbar^2}{2m}\left(\frac{N}{V}\right)^{2/3}$$

or

$$E_f = 3^{2/3}\pi^{4/3}\frac{\hbar^2}{2m}\rho^{2/3} \tag{12–38}$$

where ρ is the number of particles per unit volume. The number of particles with energies between E and $E + dE$ in a minimum energy Fermi gas is just $2N(E)\,dE$ for $E \leqslant E_f$. As shown in figure (12–6), there are no particles with $E > E_f$. We leave it as an exercise for the reader to prove that the average energy of the particles in this energy distribution is

$$\bar{E} = \tfrac{3}{5}E_f \tag{12–39}$$

Thus the total energy of the Fermi gas is

$$E_t = N\bar{E} = \tfrac{3}{5}NE_f \tag{12–40}$$

† The use of an integral is justified if we are treating a system containing a large number of particles. Then the integral is a good approximation to the sum over the discrete energy states which, strictly speaking, should be used.

It should be noted that, for Fermi particles of a given mass, E_f and $\bar{E}$ depend only on the density of the Fermi gas.

Let us evaluate some of these quantities for the Fermi gas of conduction electrons in a metal. Consider Cu, which has one conduction electron per atom. In this case ρ is equal to the density of atoms, which is Avogadro's number times the density of Cu divided by its atomic weight. Thus

$$\rho = 6 \times 10^{23} \times 9/64 \simeq 8 \times 10^{22}\ \text{cm}^{-3}$$

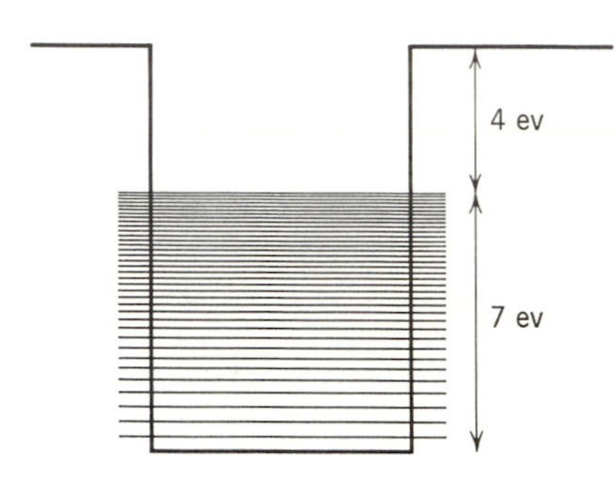

Figure 12–7. The potential experienced by the conduction electrons in copper, and the states which these electrons fill.

Using this value of ρ, and setting m equal to the mass of the electron, equation (12–38) gives

$$E_f \simeq 10 \times 10^{-12}\ \text{ergs} \simeq 7\ \text{ev}$$

This is the energy of the highest filled state relative to the bottom of the potential. The energy of this state relative to the potential energy outside the metal is just the quantity W, appearing in equation (3–8), that is measured in the photoelectric effect. For Cu it is

$$W \simeq 4\ \text{ev}$$

Knowing E_f and W, we may draw a diagram representing the potential energy in which the conduction electrons move, and also the energy states which they fill, as illustrated in figure (12–7). For a 1 cm^3 sample of Cu there are $\simeq 4 \times 10^{22}$ filled energy states! This fact justifies the approximation discussed in the footnote on page 385. The approximation of assuming that the potential is infinitely large at the boundaries is justified by the fact that the highest filled state is far below the top of the potential.

The conduction electrons of Cu will have a minimum total energy when the total energy content of the metal itself is a minimum, i.e., when the temperature is at absolute zero. But even at $T = 0$ the average energy of the conduction electrons is $\bar{E} = 3E_f/5 \simeq 4$ ev. This is a direct result of the exclusion principle. For a system composed of particles which do not obey this principle (e.g., a classical gas of distinguishable particles), all the particles will be in the ground state when the system is at $T = 0$, and their energy will be essentially zero. For such particles to have an average energy of 4 ev, the energy must be supplied by thermal agitation and the required temperature satisfies the equation

$$\tfrac{3}{2}kT = \tfrac{3}{5}E_f \simeq 4 \text{ ev}$$

where k is Boltzmann's constant. This equation gives the high temperature

$$T \simeq 5 \times 10^4 \text{ °K}$$

which can be taken as a measure of the temperature equivalent of the average energy of conduction electrons.

The fact that Fermi energy E_f is very much larger than thermal agitation energy at normal temperatures,

$$(\tfrac{3}{2}kT)_{300\text{°K}} \simeq 0.025 \text{ ev}$$

leads to some interesting consequences. For a Fermi gas at $T = 0$ all the states below the Fermi energy are completely filled. As the temperature of the metal containing the Fermi gas is raised, there is a tendency for the electrons to be excited to higher energy states by collisions with the thermally agitated atoms. But, according to the exclusion principle, it is only possible for an electron to make a transition to a state which is not filled. Since at normal temperatures the thermal energy which can be supplied to an electron is quite small compared to the Fermi energy, only the electrons whose energies are very near E_f can be given enough energy to allow a transition to an unfilled state at $E > E_f$. Electrons with energies appreciably less than E_f cannot be excited because there is not enough thermal energy available. Consequently, only a small fraction of the total number of conduction electrons can take part in the thermal excitation, and the contribution of the electrons to the heat capacity of the metal will be very small. For this reason the heat capacity of a metal at normal temperatures is observed experimentally to be just what would be calculated from the thermal motion of the atoms alone. When this explanation was first proposed, it answered a long standing puzzle because, in classical physics, the heat capacity arising from the thermal motion of the conduction electrons should be quite comparable to that arising from the thermal motion of the atoms.

Figure (12–6) shows the energy distribution for the particles of a Fermi gas at $T = 0$. Energy distributions for a Fermi gas at several other values of T are shown in figure (12–8). Note that at each temperature the drop to zero occurs in an energy range of width about equal to kT, because particles of energy much less than $E_f - kT$ cannot be excited so the energy distribution of these particles cannot be different from the $T = 0$ distribution. The curves in this figure are plotted from the equation

$$\mathcal{N}(E)\,dE = 2N(E)\,F(E)\,dE \tag{12–41}$$

The quantity $N(E)$ is the density of states for Fermi particles (12–37) and is proportional to $E^{1/2}$. The quantity $F(E)$ is the *Fermi probability distribution*

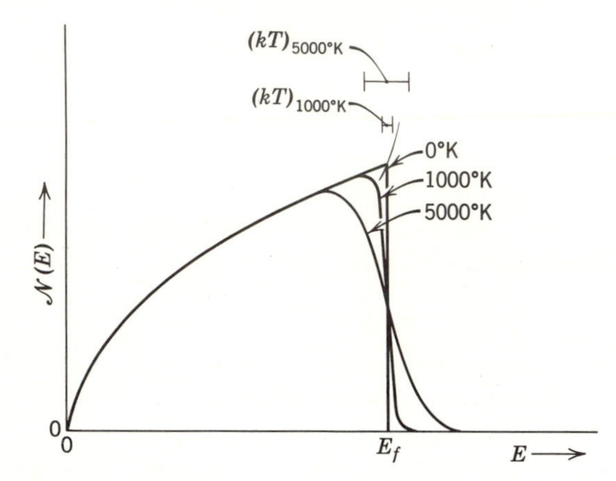

Figure 12–8. The energy distributions for a system of Fermi particles at several temperatures.

which specifies the probability that a Fermi particle, in a system in thermal equilibrium at temperature T, will be in a state with energy E. It is

$$F(E) = \frac{1}{e^{(E-E_0)/kT} + 1} \tag{12–42}$$

where E_0 is a constant which depends on T in such a way that $E_0 = E_f$ when $kT \ll E_f$.

The role of the Fermi probability distribution in the description of a system of Fermi particles is analogous to that of the Boltzmann probability distribution (2–23) in the description of a system of classical distinguishable particles. The latter probability distribution

$$P(E) = \frac{1}{e^{E/kT}} \tag{12–43}$$

specifies the probability that a particle of such a system, which is in thermal equilibrium at temperature T, will be in a state with energy E. The difference between the probability distributions (12–42) and (12–43) reflects the difference between the behavior of Fermi particles and classical particles and is primarily an effect of the exclusion principle. For a Fermi gas with $kT \gg E_f$, the thermal excitation is very large, all the particles are highly excited, and the states are very sparsely occupied. Then the chance that a Fermi particle would be prevented from making a transition to a state, because that state was already filled, is so small that the exclusion principle is no longer important. But in the limit $kT \gg E_f$ the constant E_0 in equation (12–42) is large and negative and the term $+1$ is negligible compared to the exponential. Then the probability distributions $F(E)$ and $P(E)$ are identical except for a multiplicative constant which only affects their normalization.

To complete this discussion, we mention that particles which must be described in quantum mechanics by symmetric eigenfunctions are given the generic name *Bose particles.* Even though they are indistinguishable, these particles do not obey the exclusion principle because a symmetric eigenfunction, such as (12–8), does not vanish if all the quantum numbers of two or more particles are the same. The probability that a particle in a *Bose gas* in thermal equilibrium at temperature T will be in a state with energy E is given by the *Bose probability distribution*,

$$B(E) = \frac{1}{e^{(E-E_0)/kT} - 1} \qquad (12\text{–}44)$$

In general, the constant E_0 depends on T in such a way that the Bose probability distribution becomes identical, within a multiplicative constant, with the Boltzmann probability distribution at high temperatures. However, the two probability distributions differ at low temperatures where a large number of particles occupy the same state since, even though they both describe particles which do not obey the exclusion principle, the particles are indistinguishable in the Bose case and distinguishable in the Boltzmann case. The He atom is an example of a Bose particle because it is a tightly bound aggregate containing an even number of antisymmetric particles. The extremely interesting properties of a system of He atoms at low temperatures (liquid He) are due primarily to the symmetry character of these atoms. The quantum is another example of a Bose particle. In this particular case the constant E_0 in equation (12–44) is identically zero, and the probability distribution is

$$B(E)_{\text{quanta}} = \frac{1}{e^{E/kT} - 1} \qquad (12\text{–}45)$$

By setting $E = h\nu$, this equation can be used to evaluate the average energy carried by quanta of frequency ν in an enclosure containing a large number of quanta in equilibrium at temperature T. The resulting expression is identical with Planck's equation (2–31) for the average energy contained in electromagnetic standing waves in such an enclosure. Thus the energy density of a black body cavity can be evaluated by thinking of it as containing a system of standing waves, or as containing a Bose gas of quanta.

In Chapter 2 we showed with a simple system how the Boltzmann probability distribution is developed from the theory of statistical mechanics. The Fermi and Bose probability distributions are developed from a variant of that theory, called *quantum statistical mechanics*, which differs from the classical theory only in the definition of an experimentally distinguishable division of the total energy of the system and, for Fermi particles, in the application of the exclusion principle. This is an important application of quantum mechanics which we shall not treat here only because it would divert us from its even more important application to the study of multi-electron atoms.

BIBLIOGRAPHY

Bohm, D., *Quantum Theory*, Prentice-Hall, Englewood Cliffs, N.J., 1951.
Schiff, L. I., *Quantum Mechanics*, McGraw-Hill Book Co., New York, 1955.
Sherwin, C. W., *Introduction to Quantum Mechanics*, Henry Holt and Co., New York, 1959.

EXERCISES

1. Show that equation (12–3) is a solution to (12–1).

2. Using the procedure suggested immediately below (12–30), show that the expression agrees with the results of a routine degenerate perturbation theory calculation.

3. Consider two electrons moving in a narrow impenetrable walled tube of length a. Write proper eigenfunctions for the ground and first excited states of this system, assuming that the electrons are non-interacting. Evaluate the energy of the ground state and first excited state of the system in this approximation, and discuss the degeneracy of these states. Now consider the perturbation due to the Coulomb interaction between the electrons. Write equations for the energy shifts of the ground state and first excited state, and evaluate these shifts. Explain the physical origin of the splitting of the first excited state by the application of the perturbation.

4. Verify equations (12–31) and (12–32).

5. Verify equation (12–39).

6. Repeat the calculation of exercise 3, Chapter 2 for a system of Fermi particles, and also for a system of Bose particles. Compare the first calculation with equation (12–42) and the second with (12–44), and compare both with (12–43).

CHAPTER 13

Multi-electron Atoms

1. Introduction

A multi-electron atom contains a nucleus of charge $+Ze$ surrounded by Z electrons of charge $-e$. Each electron moves under the influence of an attractive Coulomb force exerted by the nucleus and repulsive Coulomb forces exerted by all the other $Z - 1$ electrons, as well as certain other forces involving the spin angular momenta which are usually weaker than those involving the charges. To find an exact solution of the extremely complicated Schroedinger equation for such a system is essentially impossible. Nevertheless, there exists a completely adequate theoretical description of multi-electron atoms.

In section 5, Chapter 12, we treated a multi-electron atom with $Z = 2$ by first considering an unperturbed system describing a simplified atom in which the interaction between the electrons is ignored, and then improving the accuracy of the description by using perturbation theory to evaluate the effect of this interaction. The modus operandi for atoms with larger values of Z also consists in taking a simplified description of the atom as an unperturbed system, and then using perturbation theory to evaluate the corrections. However, if there are a number of electrons, their interactions cannot be completely ignored in the unperturbed system because the effect of these interactions is much too large to be treated as a perturbation. A useful unperturbed system must take into account the main effects of the Coulomb interactions between the electrons. On the other hand, the unperturbed system must not be so complicated that the Schroedinger equation describing the system is insolvable. In effect, the latter requirement means that in the unperturbed system the electrons must be non-interacting—only then can the Schroedinger equation for the unperturbed system be separated into a set of Z equations which can be solved without too much difficulty because each of them involves the coordinates of a single electron.

An unperturbed system with the required properties can be obtained by assuming that each electron moves *independently* in a spherically symmetrical *net potential* $V(r)$, where r is the radial coordinate of the electron with respect to the nucleus. This net potential is the sum of the spherically symmetrical attractive Coulomb potential due to the nucleus and a spherically symmetrical repulsive potential which represents the *average* effect of the repulsive Coulomb forces between the electron and its $Z - 1$ colleagues. The average repulsive potential depends on the average radial probability densities of the electrons. Of course, these probability densities are not known until after the equations governing the behavior of the unperturbed system have been solved. This might seem to be an intractable situation, but actually it is not. It can be handled by demanding that the average repulsive potential be *self-consistent*. That is, if we calculate the probability densities for the correct average repulsive potential, and then evaluate an average repulsive potential from these probability densities, we demand that the potential with which we end up must be identical with the potential with which we started. This condition is sufficient to determine the correct average repulsive potential.

Two different self-consistent treatments will be presented. In section 2, the electrons in the unperturbed system will be treated as a Fermi gas. Then it will not be necessary to solve the Schroedinger equations for the unperturbed system explicitly, because many quantum mechanical properties of a system of non-interacting electrons have already been included in the derivation of the equations specifying the properties of a Fermi gas. In section 3, we shall describe a more accurate treatment of the unperturbed system which involves solving the Schroedinger equations for the spherically symmetrical potentials. The results of this treatment will provide the basis for a discussion of the ground state of multi-electron atoms, and of the periodic table of the elements. Then we shall use perturbation theory to correct for the fact that the effect of the Coulomb interaction between the electrons is only approximately represented by the spherically symmetrical repulsive potential, and also to correct for other interactions such as those involving spin. This will allow us to give a description of the excited states of atoms. Finally, we shall use time dependent perturbation theory to evaluate the probability that an atom will make a transition from one excited state to another. From all these considerations, we shall be able to understand the optical and X-ray spectra of multi-electron atoms.

2. The Thomas-Fermi Theory

About 1928, Thomas and Fermi independently developed a simple theory of the ground state of multi-electron atoms. The theory is based on treating the electrons as a minimum energy Fermi gas confined to a region by a spherically symmetrical net potential energy $V(r)$. Then equation (12–38) can be used to relate the depth of the potential, at any value of r, to the density of the Fermi gas for that value of r. This relation is obtained by assuming that the depth of the potential is everywhere such that the energy levels are filled to the very top. That is,

$$E_f = -V(r) \tag{13–1}$$

This condition assures that the total energy of the atom is a minimum. Solving (12–38) for the density ρ we then have

$$\rho = \frac{2^{3/2}m^{3/2}}{3\hbar^3\pi^2}[-V(r)]^{3/2} \tag{13–2}$$

This procedure is justified if the distance in which the value of $V(r)$ does not change significantly is large enough, compared to the de Broglie wavelengths of the electrons, that a number of electrons can be localized in a volume in which $V(r)$ is essentially constant. The condition is satisfied for atoms of large Z, except very near the center and very far from the center. The theory does not apply for atoms of very small Z because the condition is not satisfied anywhere.

Self-consistency is obtained by requiring that the electron density be such as to reproduce the net potential $V(r)$. The potential is evaluated in terms of ρ by using Gauss' law. This law states that the integral over any closed surface of the normal component of the electric field E times the element of surface area dS is equal to 4π times the charge contained within the surface. Consider a spherical surface of radius r centered on the nucleus. Then we have a completely spherically symmetrical situation in which Gauss' law has the simple form

$$4\pi r^2 E = 4\pi\left[Ze + \int_0^r -e\rho(r')4\pi r'^2\,dr'\right] \tag{13–3}$$

The left side of the equation is the area of the surface times the electric field at the surface, which is constant in magnitude and everywhere normal to it. The first term in the bracket is the nuclear charge. The integral is the total charge due to the electrons contained within the surface; $-e\rho(r')$ is the charge per unit volume, and $4\pi r'^2\,dr'$ is the volume element.

Now the electric field is related to the force acting on an electron, and thus to the net potential $V(r)$ acting on the electron, by the equations

$$-eE = F = -\frac{dV(r)}{dr} \tag{13-4}$$

When we write this as

$$E = -\frac{1}{e}\frac{d[-V(r)]}{dr}$$

equation (13–3) becomes

$$r^2\frac{d[-V(r)]}{dr} = -Ze^2 + 4\pi e^2\int_0^r \rho(r')r'^2\,dr'$$

Differentiating both sides with respect to r gives

$$\frac{d}{dr}\left(r^2\frac{d[-V(r)]}{dr}\right) = 4\pi e^2\rho(r)r^2$$

Using equation (13–2) to express $\rho(r)$ in terms of $-V(r)$, we have

$$\frac{1}{r^2}\frac{d}{dr}\left(r^2\frac{d}{dr}[-V(r)]\right) = \frac{4e^2 2^{3/2}m^{3/2}}{3\pi\hbar^3}[-V(r)]^{3/2} \tag{13-5}$$

It is convenient to rewrite this in terms of the variable x, defined by the equation

$$r = \frac{1}{2}\left(\frac{3\pi}{4}\right)^{2/3}\frac{\hbar^2}{me^2Z^{1/3}}\,x = \frac{0.885a_0}{Z^{1/3}}\,x \tag{13-6}$$

and to make the substitution

$$-V = \frac{Ze^2}{r}\chi \tag{13-7}$$

In terms of these quantities, the differential equation (13–5) becomes

$$x^{1/2}\frac{d^2\chi(x)}{dx^2} = \chi(x)^{3/2} \tag{13-8}$$

To obtain an explicit solution to this differential equation, we must know the "boundary conditions" for the function $\chi(x)$. They are obtained from the following considerations. For $r \to 0$, it is clear that the net potential energy of an electron approaches the Coulomb potential energy due to the nucleus alone. That is,

$$V(r) \to -Ze^2/r, \qquad r \to 0$$

From equations (13–6) and (13–7), we see this means

$$\chi(x) \to 1, \qquad x \to 0 \tag{13-9}$$

For $r \to \infty$, an electron experiences a potential energy due to the nuclear charge $+Ze$ shielded by the charge $-(Z-1)e$ of the other electrons. Thus

$$V(r) \to -e^2/r, \qquad r \to \infty$$

or

$$rV(r) \to -e^2, \qquad r \to \infty$$

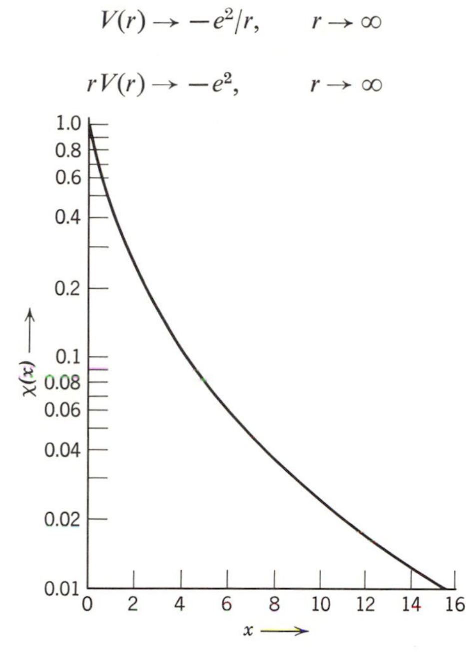

Figures 13–1. The plot of a function which arises in the Thomas-Fermi theory.

We make the approximation that this condition is

$$rV(r) \to 0, \qquad r \to \infty \tag{13–10}$$

which amounts to assuming that e^2 is very small compared to Ze^2. This approximation is another reason why the Thomas-Fermi theory is not accurate far from the center of an atom. Using equations (13–6) and (13–7) to write (13–10) in terms of $\chi(x)$, we have

$$\chi(x) \to 0, \qquad x \to \infty \tag{13–11}$$

The solution of equation (13–8), with the boundary conditions (13–9) and (13–11), has been carried through by several people. Numerical methods are required, and so the results must be displayed as a table or a plot. Such a plot is shown in figure (13–1).

The function $\chi(x)$ determines the net potential $V(r)$ and the electron density $\rho(r)$. In the next section these quantities will be plotted for a particular atom, and compared with the predictions of a more accurate theory. It will be found that, even for an atom of fairly small Z, the

Thomas-Fermi theory gives a good description of the average radial behavior of the net potential and the electron density. Here we shall only call attention to the fact that equation (13–6) says the factor connecting the radial coordinate r and the dimensionless variable x is proportional to $Z^{-1/3}$. Since the same dimensionless variable is used for all atoms, this means that for increasing Z the radial scale of $\rho(r)$ contracts according to $Z^{-1/3}$. The Thomas-Fermi theory predicts that the "size" of atoms decreases slowly with increasing Z.

3. The Hartree Theory

A more accurate theory of the unperturbed system for the ground state of a multi-electron atom was developed by Hartree in 1928 and the following years. It consists in solving the Schroedinger equation for a system of Z electrons moving independently in spherically symmetrical net potentials. The equation is

$$\frac{-\hbar^2}{2m}\sum_{i=1}^{Z}\nabla_i^2\psi_T + \sum_{i=1}^{Z}V_i(r_i)\psi_T = E_T\psi_T \tag{13–12}$$

where ∇_i^2 is the Laplacian operator in spherical coordinates for the ith electron (cf. equation 10–15), and where $V_i(r_i)$ is the potential energy of the ith electron. The eigenvalue E_T is the total energy of the atom, and the eigenfunction ψ_T is the eigenfunction describing the motion of all its Z electrons. That is,

$$\psi_T = \psi_T(r_1, \theta_1, \phi_1, \ldots, r_i, \theta_i, \phi_i, \ldots, r_Z, \theta_Z, \phi_Z) \tag{13–13}$$

Spin is ignored, except that the exclusion principle is satisfied in a way which will be described later. Since the electrons are non-interacting, we expect that there is a solution to equation (13–12) of the form

$$\psi_T = \psi(r_1, \theta_1, \phi_1)\,\psi(r_2, \theta_2, \phi_2)\cdots\psi(r_Z, \theta_Z, \phi_Z) \tag{13–14}$$

In fact there is such a solution. The equation splits into Z one-particle Schroedinger equations such as

$$\frac{-\hbar^2}{2m}\nabla_i^2\psi(r_i, \theta_i, \phi_i) + V_i(r_i)\,\psi(r_i, \theta_i\, \phi_i) = E_i\psi(r_i, \theta_i, \phi_i) \tag{13–15}$$

where

$$E_T = \sum_{i=1}^{Z}E_i \tag{13–16}$$

Initially the potentials $V_i(r_i)$ are not known. However, the problem can still be solved by the following self-consistent treatment.

1. A reasonable guess about the form of the $V_i(r_i)$ is obtained by setting

$$V_i(r_i) = V(r_i), \qquad \text{for all } i \tag{13–17}$$

where $V(r_i)$ is the net potential evaluated from the Thomas-Fermi theory for the atom with Z electrons. With these common values of $V_i(r_i)$, all the one-particle Schroedinger equations (13–15) are identical.

2. These equations are solved by numerical integration, and the one-particle eigenfunctions $\psi_\alpha(r_i, \theta_i, \phi_i)$, $\psi_\beta(r_i, \theta_i, \phi_i)$, . . . , $\psi_\iota(r_i, \theta_i, \phi_i)$, . . . are found. They are listed in order of increasing energy of the corresponding eigenvalues. Each of the quantum numbers α, β, . . . , ι, . . . stands for a complete set of spatial and spin quantum numbers for one electron.

3. To obtain the ground state of the atom, the one-particle states are filled in such a way as to minimize the total energy (13–16) and yet satisfy the exclusion principle. That is, the one-particle states are filled in order of increasing energy with one electron in each state. Then the one-particle eigenfunctions describing the Z electrons are

$$\psi_\alpha(r_1, \theta_1, \phi_1), \psi_\beta(r_2, \theta_2, \phi_2), \ldots, \psi_\iota(r_i, \theta_i, \phi_i), \ldots, \psi_\zeta(r_Z, \theta_Z, \phi_Z)$$

4. Next a more accurate estimate of the potentials $V_i(r_i)$ is calculated by integrating the equation obtained by combining Gauss's law (equation 13–3) and the relation (13–4) between the electric field and the potential energy. In calculating the potential acting on a particular electron, say the kth electron, the quantity $\rho(r)$ in equation (13–3) is set equal to the sum of the quantum mechanical probability densities for all the filled states, except the state κ occupied by the kth electron, averaged over a spherical surface of radius r. That is,

$$\rho(r) = \frac{1}{4\pi r^2} \int_0^\pi \int_0^{2\pi} \sum_{\substack{\iota=\alpha \\ \iota \neq \kappa}}^{\zeta} \psi_\iota^*(r, \theta, \phi)\, \psi_\iota(r, \theta, \phi) r^2 \sin\theta \, d\theta \, d\phi \tag{13–18}$$

Thus the potential acting on the kth electron is evaluated from a spherically symmetrical charge distribution due to the nucleus plus the average radial dependence of charge carried by all the electrons except the kth. The $V_i(r_i)$ which are obtained will, in general, be somewhat different from the initial estimates taken in step 1.

5. If they are different, the entire procedure is repeated, starting at step 2 and using the new set of $V_i(r_i)$. After several cycles ($2 \to 3 \to 4 \to 2 \to 3 \to 4 \to \cdots$) it is found that the $V_i(r_i)$ obtained at the end of a cycle are essentially the same as those used in the beginning. Then this set of $V_i(r_i)$ is the set of self-consistent potentials, and the eigenfunctions calculated from these potentials are the set of one-particle eigenfunctions which describe the motion of the electrons in the multi-electron atom.

There are several obvious approximations in the Hartree theory. One is that the charge distribution due to each electron is effectively treated as if it were spherically symmetrical. The justification of this will be discussed shortly. Another approximation is that antisymmetric eigenfunctions, such as (12–16), are not used to describe the system of Fermi particles. This is not a serious fault, because the exclusion principle is satisfied by means of the auxiliary requirement of step 3 that only one electron can be in each quantum state. A modification of the Hartree theory to include antisymmetric eigenfunctions has been given by Fock (1930). In comparing the results of the two theories, it is found that the error introduced by not using antisymmetric eigenfunctions is small. Consequently, most calculations have been made with the original Hartree theory because it is relatively easier to handle. Even so, the calculations are very lengthy and, as yet, results have been obtained only for a limited number of atoms. However, the essential features of the Hartree description of multi-electron atoms are quite well understood.

It is found that all the self-consistent potentials $V_i(r_i)$ are very similar so that, to a good approximation, we may speak of a single net potential $V(r)$ in which all the electrons of the multi-electron atom are moving independently. The one-particle eigenfunctions for these electrons in this spherically symmetrical potential are closely related to the one-electron-atom eigenfunctions discussed in Chapter 10. In fact, they can be written

$$\psi_{nlm_lm_s}(r, \theta, \phi) = R_{nl}(r)\, \Theta_{lm_l}(\theta)\, \Phi_{m_l}(\phi)\sigma_{m_s} \tag{13–19}$$

The eigenfunctions are labeled by the same set of quantum numbers n, l, m_l, m_s as are used for the one-electron-atom eigenfunctions, and these quantum numbers are related to each other just as specified by equations (10–37). The spin eigenfunction σ_{m_s} is exactly the same function as in the case of a one-electron atom. Furthermore, the functions $\Theta_{lm_l}(\theta)$ and $\Phi_{m_l}(\phi)$ are also exactly the same. The reason is the Schroedinger equation (13–15) for an electron in a *spherically symmetrical* net potential is just equation (10–15). This equation leads directly to equations (10–17) and (10–18), and to their solutions $\Theta_{lm_l}(\theta)$ and $\Phi_{m_l}(\phi)$. Consequently, all the discussion in Chapter 10 concerning the θ and ϕ dependence of the one-electron-atom eigenfunctions applies directly to the θ and ϕ dependence of the multi-electron atom one-particle eigenfunctions.

As an example, equation (10–46) shows that the sum of the probability densities for the one-electron-atom eigenfunctions with $n = 2$, $l = 1$, and all possible values of m_l, is spherically symmetrical. This statement is certainly also true for the case of $n = 2$, $l = 0$, and it can be shown to be true for any given n and l. From the discussion above, we conclude that the same statement also applies to the one-particle eigenfunctions for a multi-electron atom. Now, when a multi-electron atom is in its ground state, the lowest energy one-particle states are completely filled. This means that for almost all values of n and l there are electrons in states with all possible values of m_l. Since the sum of the probability densities for these electrons is spherically symmetrical, their total charge distribution

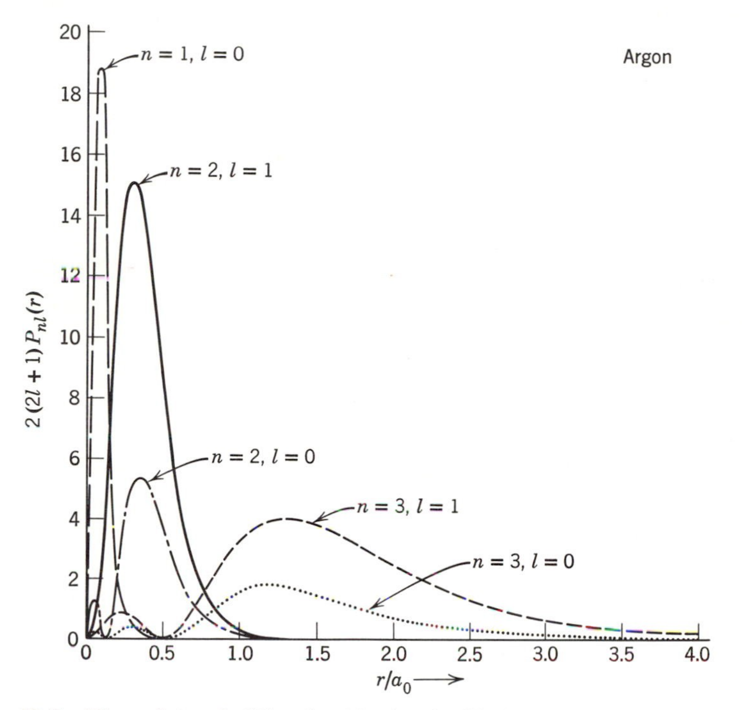

Figure 13–2. The radial probability densities for the filled quantum states of the argon atom according to the Hartree theory.

is symmetrical also. At most, only a few electrons in the highest energy states, for which the states with all possible values of m_l might not be filled, can contribute to any asymmetries in the charge distribution. This is the justification for treating the charge distributions as spherically symmetrical in the Hartree theory.

The functions $R_{nl}(r)$ determining the radial behavior of the eigenfunctions are not the same as for the case of a one-electron atom because the net potential $V(r)$, which enters in the differential equation for these functions, does not have the same r dependence as the Coulomb potential of a one-electron atom. Typical examples of the radial behavior of the one-particle eigenfunctions are shown in figure (13–2). In this figure we plot the results of a Hartree theory calculation for the A atom ($Z = 18$) in terms of the quantities $2(2l + 1)r^2R_{nl}^2(r) = 2(2l + 1)P_{nl}(r)$, where $P_{nl}(r)$ is the radial probability density defined by equation (10–43). The electrons of this atom completely fill the 18 possible one-particle quantum states for the n and l values shown in the figure. Since there are $2l + 1$ possible values of m_l for each l, and for each of these there are 2 possible values of m_s, there are 2 states with $n = 1$, $l = 0$; 2 with $n = 2$, $l = 0$; 6 with $n = 2$, $l = 1$; 2 with $n = 3$, $l = 0$; and 6 with $n = 3$, $l = 1$. These states are the ones which are filled in the ground state of the atom because, as we shall see later, they are the states of lowest energy. The quantity $2(2l + 1)P_{nl}(r)$ is, therefore, the radial probability density for the state with quantum numbers n and l, times the number of electrons in that state. It is the total radial probability density for the electrons in that state, and the sum of these quantities gives $P(r)$, the *total radial probability density* for the atom, which is the probability of finding *some* electron with a radial coordinate in the region of r. Thus

$$P(r) = \sum_{n,l} 2(2l + 1)P_{nl}(r) \tag{13–20}$$

This is plotted for the A atom in figure (13–3) as $P(r)_H$. In the same figure we plot $P(r)_{T-F}$, which is the total radial probability density for the A atom evaluated from the Thomas-Fermi theory. We see that the Thomas-Fermi theory gives a good description of the average behavior of the total radial probability density predicted by the Hartree theory, except that for large values of r it does not decrease rapidly enough. The Thomas-Fermi theory is also inaccurate for small r, but it is not apparent in a plot of $P(r)$ because this is dominated by the r^2 factor for small r. It is possible to obtain a fairly direct experimental measurement of the total radial probability density by scattering electrons from the atom. The angular distribution of the scattered electrons depends, in a manner which will be described in Chapter 15, on the total radial probability density of the atomic electrons.† It can also be measured by scattering X-rays from the atom, using a technique which will be explained in Chapter 14. Although neither of these measurements is accurate enough to resolve the oscillatory structure of $P(r)_H$, they do confirm its average behavior.

In figure (13–3) we also indicate the radial dependence of the net potential $V(r)$ in which, according to the Hartree theory, each electron

† The reader will recall that this is quite different from the information provided by scattering alpha particles from an atom. The reason is that in the electron experiments the mass of the scattered particle equals the mass of an atomic electron, whereas in the alpha particle experiments the mass of the scattered particle is so large compared to the mass of an atomic electron that it essentially cannot be scattered by these electrons.

of the A atom is moving. This is given by plotting the radial dependence of the *effective charge* $Z(r)$ defined by the equation

$$V(r) = -\frac{e^2 Z(r)}{r} \tag{13–21}$$

The behavior of $Z(r)$ is easy to understand. For $r \rightarrow 0$, the potential energy of an electron is dominated by the Coulomb attraction of the

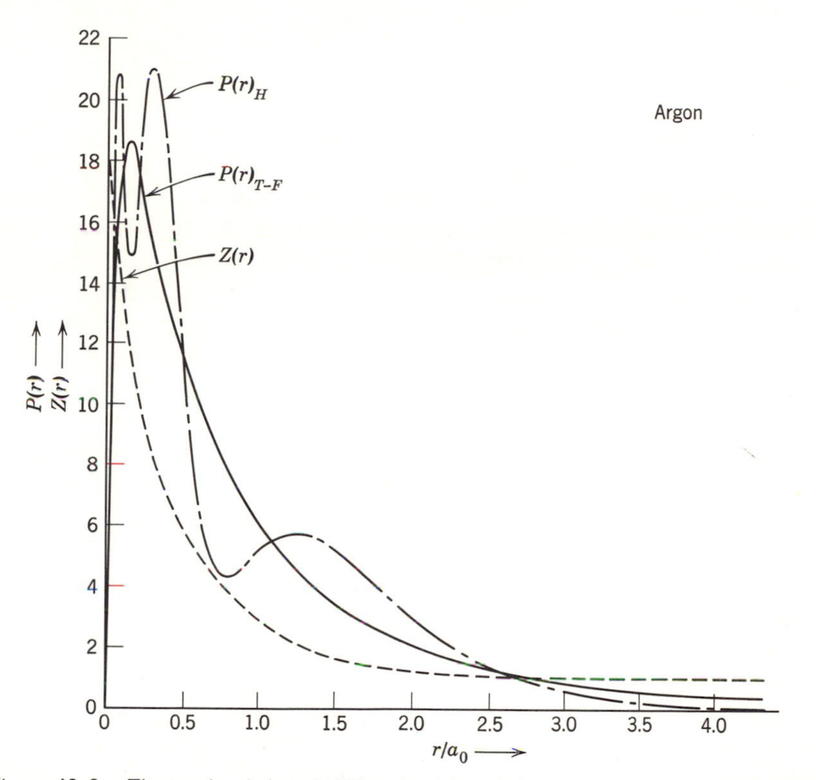

Figure 13–3. The total radial probability densities of the argon atom according to the Hartree and Thomas-Fermi theories, and the effective charge of that atom according to the Hartree theory.

nucleus of charge $+Ze$. So $V(r) \rightarrow -e^2 Z/r$, and $Z(r) \rightarrow Z = 18$. With increasing r, $Z(r)$ decreases because the magnitude of $V(r)$ decreases more rapidly than the Coulomb potential due to the nucleus, as the electrons with radial coordinates less than r shield part of the nuclear charge. For $r \rightarrow \infty$, $V(r) \rightarrow -e^2/r$, and $Z(r) \rightarrow 1$, because the nuclear charge $+Ze$ is shielded by the charge $-(Z - 1)e$ of the other electrons in the atom.

By inspecting the plots of $P_{nl}(r)$ we see that, for all the electrons in states with common values of the quantum number n, the probability densities are large only in essentially the same range of r. All these electrons are said to be in the same *shell*—terminology we have used before in connection with one-electron atoms. Furthermore, the range of r in which the probability densities are large (the "thickness" of each shell) is restricted enough that $Z(r)$ has a reasonably well-defined value in that range. These circumstances form the basis of a very approximate description of the motion of the electrons in a multi-electron atom, in which all the electrons in a particular shell are considered to be moving in a Coulomb potential,

$$V_n(r) = -\frac{e^2 Z_n}{r} \tag{13–22}$$

where Z_n is a constant equal to $Z(r)$ evaluated at the average value of r for the shell (the "radius" of the shell). As an example, we find from the figures (13–2) and (13–3) for the A atom: $Z_1 \simeq 16$, $Z_2 \simeq 8$, $Z_3 \simeq 2.5$. In the approximation (13–22), the one-electron atom equations specifying the expectation value of r, the total energy, etc., can be used if we replace Z by Z_n. This is quite accurate for the $n = 1$ shell, but it is only accurate to within a factor of 2 or 3 for the $n = 3$ shell of the A atom. Nevertheless, the approximation is useful because it provides a readily understandable picture of many of the basic results of the Hartree theory.

In the Hartree theory, shells for small n are of very small radii because for these shells there is little shielding, and the electrons feel the full Coulomb attraction of the nucleus. In fact, the radius of the $n = 1$ shell is smaller than that of the $n = 1$ shell of the H atom by about a factor of $1/Z_1$. This can be understood in our approximate description from equation (10–44). The electrons in these *inner shells* are in a region of large negative potential energy, so their total energies are correspondingly large and negative. The results of the Hartree theory show that the magnitude of the total energy of an electron in the $n = 1$ shell is larger than that of an electron in the $n = 1$ shell of the H atom by a factor of about Z_1^2. This can be understood by referring to equation (10–38). On the other hand, electrons in shells of large n are highly shielded from the nuclear Coulomb attraction by the electrons in the inner shells. The electrons in these *outer shells* feel only a small net attraction to the nucleus. The radius of the outermost shell of atoms of different atomic number can be estimated from equation (10–44), using the appropriate values of Z_n. This equation predicts an increase in radius, with increasing n value of the outermost shell for atoms of increasing atomic number, which is *somewhat* less rapid than n^2, because Z_n slowly becomes larger with increasing atomic number since the shielding of the nuclear charge is not perfect. Actually the Hartree theory does predict an increase in the radius of the outermost shell, but one which is *much* less rapid than n^2,† because the electrons in the outermost shell do have a small probability, which we have ignored in our approximate description, of being found in the region near the center of the atom. Although the probability is small, it is important because near the center the electrons feel a very strong attraction that tends to decrease the radius of the shell. This effect becomes more important with increasing atomic number and, along with the effect of the increase in Z_n, leads to the important consequence that according to the Hartree theory the radius of the outermost shell is roughly the same for all atoms. Finally, we can see from our approximate description why the Hartree theory predicts that the total energy of the electrons in the outermost shell of all atoms is comparable to that of an electron in the ground state of the H atom. It is because in all atoms Z_n for the outermost shell is always of the order of unity, so the potential energy is comparable to the potential energy acting on the electron in the H atom. Each of the predictions of the Hartree theory mentioned in this paragraph are borne out by the observed properties of atoms.

In a one-electron atom, all the electrons in a certain shell have exactly the same total energy, if the very small effects like spin-orbit interaction are ignored.‡ The reason is that in the one-electron atom the total energy depends only on the quantum number n, and so is the same for all electrons in the same shell, even though they may have different values of the quantum number l. In a multi-electron atom this is not true. This statement should not be surprising because we have mentioned on several previous occasions that the fact that the total energy of a one-electron atom does not depend on l is a consequence of the fact that the potential is exactly proportional to $1/r$. In a multi-electron atom the electrons are moving in a net potential $V(r)$ which is not exactly proportional to $1/r$, and the total energy E_{nl} of these electrons depends on l as well as on n. However, since we are ignoring spin-orbit interactions and certain other effects, the total energy does not depend on the quantum number m_s which determines the spatial orientation of the spin, nor on the quantum number m_l which determines the spatial orientation of the "orbit." The important features of the dependence of E_{nl} on n were described approximately in the previous paragraph in terms of the energy equation for a one-electron atom with Z replaced by Z_n. That is, we took

$$E_{nl} \sim -\mu Z_n^2 e^4/2\hbar^2 n^2 \tag{13–23}$$

† This statement does not contradict the Thomas-Fermi theory prediction of a gradual decrease in the overall radius with increasing atomic number.

‡ By this we mean that the total energies of the electrons in a number of different one-electron atoms are the same if the electrons have the same values of the quantum number n.

For small n, this becomes less negative with increasing n much more rapidly than for the case of a one-electron atom because of the dependence of Z_n on n. For large n, Z_n does not change rapidly, so E_{nl} has a less rapid n dependence similar to that of the total energy of a one-electron atom. Now, the results of the Hartree theory show that E_{nl} is actually more negative than is indicated by the approximation (13–23). Furthermore, the difference is largest for $l = 0$ and diminishes progressively with increasing l. This difference is quite appreciable. In fact, for large values of n, the dependence of E_{nl} on l can be stronger than it is on n.

The reason for the dependence of the total energy on l is easy to understand in terms of the behavior of $Z(r)$ and the probability density $\psi^*_{nlm_lm_s}\psi_{nlm_lm_s}$ in the region near $r = 0$. In (10–45′) we pointed out that

$$\psi^*_{nlm_lm_s}\psi_{nlm_lm_s} \propto r^{2l}, \quad r \to 0 \tag{13–24}$$

This was with reference to one-electron-atom eigenfunctions, but it is equally true for multi-electron-atom eigenfunctions because the term $-l(l + 1)/r^2$ in the differential equation (10–27), which governs the radial behavior of the eigenfunctions, completely dominates the term $(2\mu/\hbar^2)\,[E - V(r)]$ for small r.† Consequently, in the region near $r = 0$ the exact form of $V(r)$ is unimportant and all eigenfunctions for spherically symmetrical potentials have the r^l radial behavior of (10–45). Now consider two electrons in the same shell of a multi-electron atom, one with $l = 0$ and the other with $l = 1$. According to (13–24), there is some chance of finding the $l = 0$ electron in the region of small r, but only a very small chance of finding the $l = 1$ electron in that region, since $r^0 \simeq 1$ and $r^2 \simeq 0$ for small r. Although the $l = 0$ electron will spend only part of its time near $r = 0$, this makes an appreciable difference in the average value of its potential energy because $Z(r)$ becomes very large near $r = 0$ so the net potential energy $V(r)$ becomes very negative. Thus the average potential energy of the $l = 0$ electron will be more negative than the average potential energy of the $l = 1$ electron. Similarly, the average

† An instructive interpretation of the first of these terms can be obtained by making the substitution $u = rR$ in equation (10–27). This gives

$$\frac{d^2u}{dr^2} + \left\{-\frac{l(l+1)}{r^2} + \frac{2\mu}{\hbar^2}[E - V(r)]\right\}u = 0$$

or

$$-\frac{\hbar^2}{2\mu}\frac{d^2u}{dr^2} + \left\{\frac{l(l+1)\hbar^2}{2\mu r^2} + V(r)\right\}u = Eu$$

which is seen to be equivalent to the Schroedinger equation for motion in one dimension, with the term $l(l + 1)\hbar^2/2\mu r^2 = L^2/2\mu r^2$ added to the potential $V(r)$. This term is often called the *centrifugal potential*, for reasons which can be seen by considering the classical energy conservation equation for a particle of mass μ moving under the influence of a potential $V(r)$. As the particle will move in a plane containing the origin, it can be potential energy of the $l = 1$ electron will be more negative than that for an $l = 2$ electron in the same shell because $r^2 \gg r^4$ for small r. On the other hand, the average kinetic energy of the $l = 0$ electron is larger than that for the $l = 1$ electron, since it spends more time in the region of larger negative potential energy. It happens that these two effects exactly balance out for a potential in which $Z(r)$ is a constant; thus, for the exact $1/r$ Coulomb potential of a one-electron atom, the total energy is independent of l. But in all other cases the total energy depends on l as well as on n. In a multi-electron atom, the l dependence of the average potential energy dominates the l dependence of the average kinetic energy, and the total energy E_{nl}, which is the sum of the average potential and kinetic energies, is more negative for $l = 0$ than it is for $l = 1$, more negative for $l = 1$ than it is for $l = 2$, etc. For high Z multi-electron atoms, in which $Z(r)$ increases particularly rapidly with decreasing r in the region near $r = 0$, and for outer shells of large n in which the n dependence of E_{nl} is particularly weak, the l dependence of E_{nl} is actually stronger than the n dependence. We shall see that this plays an important role in the structure of multi-electron atoms.

All the electrons in a particular shell have probability densities which are of approximately the same form in the region where their values are large. However, the probability densities are significantly different in the region near $r = 0$. As we have seen, the result is that the total energies of the electrons in the same shell are not all the same. For this reason, it is convenient to speak of each shell as being composed of a number of *subshells*, one for each value of l. All the electrons in the same subshell have the same quantum numbers n and l. Therefore, they have exactly the same total energy (in the approximation in which we neglect spin-orbit interactions and other effects). Also they have exactly the same radial probability densities $P_{nl}(r)$.

---

described by the coordinates r, θ, and the equation is

$$E = \tfrac{1}{2}\mu\left(\frac{dr}{dt}\right)^2 + \tfrac{1}{2}\mu\left(r\frac{d\theta}{dt}\right)^2 + V(r)$$

Also the orbital angular momentum of the particle is a constant,

$$L = \mu r^2 \frac{d\theta}{dt}$$

so the equation can be written

$$E = \tfrac{1}{2}\mu\left(\frac{dr}{dt}\right)^2 + \left\{\frac{L^2}{2\mu r^2} + V(r)\right\}$$

which is seen to be the classical energy conservation equation for motion in one dimension, with the term $L^2/2\mu r^2$ added to the potential $V(r)$. This positive term acts like a repulsive potential, tending to keep the particle away from the origin. The higher the value of L, the stronger is this effect.

TABLE (13–1)
The Periodic Table of the Elements

	s^1	s^2
1s	1 H	
2s	3 Li	4 Be
3s	11 Na	12 Mg
4s	19 K	20 Ca
5s	37 Rb	38 Sr
6s	55 Cs	56 Ba
7s	87 Fr	88 Ra

	d^1	d^2	d^3	d^4	d^5	d^6	d^7	d^8	d^9	d^{10}
3d	21 Sc	22 Ti	23 V	24 Cr $4s^1 3d^5$	25 Mn	26 Fe	27 Co	28 Ni	29 Cu $4s^1 3d^{10}$	30 Zn
4d	39 Y	40 Zr	41 Nb $5s^1 4d^4$	42 Mo	43 Tc	44 Ru $5s^1 4d^7$	45 Rh $5s^1 4d^8$	46 Pd $5s^0 4d^{10}$	47 Ag $5s^1 4d^{10}$	48 Cd
5d	57 — La Lanthanides	72 Hf	73 Ta	74 W	75 Re	76 Os	77 Ir	78 Pt $6s^1 5d^9$	79 Au $6s^1 5d^{10}$	80 Hg
6d	89 — Ac Actinides									

	p^1	p^2	p^3	p^4	p^5	p^6
1s						2 He $1s^2$
2p	5 B	6 C	7 N	8 O	9 F	10 Ne
3p	13 Al	14 Si	15 P	16 S	17 Cl	18 A
4p	31 Ga	32 Ge	33 As	34 Se	35 Br	36 Kr
5p	49 In	50 Sn	51 Sb	52 Te	53 I	54 Xe
6p	81 Tl	82 Pb	83 Bi	84 Po	85 At	86 Rn
7p						

	f^1	f^2	f^3	f^4	f^5	f^6	f^7	f^8	f^9	f^{10}	f^{11}	f^{12}	f^{13}	f^{14}
Lanthanides 4f	58 Ce $5d^1 4f^1$	59 Pr $5d^0 4f^3$	60 Nd $5d^0 4f^4$	61 Pm ?	62 Sm $5d^0 4f^6$	63 Eu $5d^0 4f^7$	64 Gd $5d^1 4f^7$	65 Tb $5d^1 4f^8$	66 Dy ?	67 Ho ?	68 Er ?	69 Tm $5d^0 4f^{13}$	70 Yb $5d^0 4f^{14}$	71 Lu $5d^1 4f^{14}$
Actinides 5f	90 Th $6d^2 5f^0$	91 Pa ?	92 U $6d^1 5f^3$	93 Np ?	94 Pu ?	95 Am $6d^0 5f^7$	96 Cm ?	97 Bk ?	98 Cf ?	99 Es ?	100 Fm ?	101 Md ?	102 No ?	

4. The Periodic Table

Some of the most important properties of the elements are periodic functions of the atomic number Z which specifies the number of electrons in an atom of the element. This can be demonstrated by constructing a periodic table of the elements such as the one presented in table (13–1). Each element is represented in the table by its chemical symbol, and also by its atomic number. (The entries sometimes appearing below the chemical symbol specify the quantum numbers for the ground state of an atom of the element in a way which will be explained later.) Elements with similar chemical and physical properties are in the same column. For instance, all elements in the first column are alkalis and have a valence of $+1$; all elements in the last column are noble gases and have a valence of 0. We assume that the reader is at least partially familiar with the periodic properties of the elements from his study of elementary chemistry. Our task here is to interpret these properties in terms of the theory of multi-electron atoms.

This interpretation is based upon knowing the way in which the outer filled subshells of multi-electron atoms are ordered according to energy. In principle, this information could be obtained entirely from the Hartree theory calculations described in the last section. However, these calculations have been carried through for only a limited number of elements. Consequently, much of the information we need must be obtained from experiment. This can be done by studying the spectrum emitted when excited atoms of an element make transitions to their ground states. In such transitions an electron drops from some high energy one-particle state to the lowest energy state which is unoccupied. This state is then the highest energy filled one-particle state for the ground state of the atom. From an analysis of the spectral lines which are emitted, it is possible to identify the quantum numbers n and l of this one-particle state. This amounts to identifying an outer subshell which is the highest energy subshell filled in the ground state of the atom. By making such an analysis on a large number of elements, it is found that, except for minor deviations which will be described later, the ordering according to energy of the outer filled subshells is as shown in table (13–2). The first column identifies the subshell by the quantum numbers n and l. The second column does the same thing except that n and l are given by the *spectroscopic notation*. This notation is commonly used in discussing the spectrum and energy levels of atoms. The number gives the value of n, and the letter gives the value of l according to the scheme shown in table (13–3). The third column of table (13–2) is $2(2l + 1)$, which is, as mentioned in the last section, the number of possible quantum states for the value of l characteristic of the subshell. Thus this column gives the maximum number of electrons that can occupy the subshell without violating the exclusion principle.

TABLE (13–2)
The Energy Ordering of the Outer Filled Subshells

Quantum numbers n, l	Designation of subshell	Capacity of subshell $2(2l+1)$	
—	—	—	
—	—	—	
6, 2	$6d$	10	
5, 3	$5f$	14	
7, 0	$7s$	2	
6, 1	$6p$	6	
5, 2	$5d$	10	
4, 3	$4f$	14	
6, 0	$6s$	2	↑
5, 1	$5p$	6	Increasing energy
4, 2	$4d$	10	(less negative)
5, 0	$5s$	2	
4, 1	$4p$	6	
3, 2	$3d$	10	
4, 0	$4s$	2	
3, 1	$3p$	6	
3, 0	$3s$	2	
2, 1	$2p$	6	
2, 0	$2s$	2	
1, 0	$1s$	2	← Lowest energy (most negative)

TABLE (13–3)
The Spectroscopic Notation for l

l	0	1	2	3	4	5	6	...
Spectroscopic notation	s	p	d	f	g	h	i	...

Many of the features of this experimentally observed ordering of the outer filled subshells according to energy have also been explicitly obtained by Hartree theory calculations. For instance, calculations for the K atom show that the energy of an electron in the $4s$ subshell should, in fact, be lower than the energy of an electron in the $3d$ subshell of this atom. Furthermore, all the features which are observed experimentally can be understood at least qualitatively in terms of the description of multi-electron atoms provided by the Hartree theory. In our discussion in the last section we found that this theory predicts that the energy of the subshells becomes more negative with decreasing values of n and with decreasing values of l. We see this immediately in table (13–2). The $1s$ subshell, which is the only subshell in the $n = 1$ shell, has the lowest energy. The two subshells of the $n = 2$ shell are both of higher energy and, of these, the $2s$ subshell is of lower energy than the $2p$ subshell. In the $n = 3$ shell the subshells $3s$, $3p$, $3d$ are also ordered in energy according to the predictions of the Hartree theory. However, the energy of the $4s$ subshell is actually lower than the energy of the $3d$ subshell because, for reasons described in the last section, the l dependence of the energy E_{nl} of the subshells becomes more important than the n dependence for outer subshells with large values of n. Continuing up the list, we see that the ordering of the subshells always satisfies the rule that, for a given n, the subshell with the lowest l has the lowest energy and that, for a given l, the subshell with the lowest n has the lowest energy. Near the top of the list the l dependence of E_{nl} is so much stronger than the n dependence that the energy of the $7s$ subshell is lower than the energy of the $5f$ subshell. It should be emphasized that table (13–2) does not necessarily give the energy ordering of all the subshells in any particular atom, but only the energy ordering of the subshells which happen to be the outer subshells for that atom. For instance, the energy of the $4s$ subshell is lower than that of the $3d$ subshell for K atoms and the next few atoms of the periodic table, but for atoms near the end of the periodic table the $3d$ subshell is of lower energy than the $4s$ subshell because for these atoms they are inner subshells and the n dependence of E_{nl} is so strong that it dominates the l dependence.

The characteristics of an atom depend on the motion of its electrons. The motion of an electron is specified by the set of four quantum numbers which specify its quantum state. However, in the approximation represented by the Hartree theory, only the quantum numbers n and l are important. Therefore, in this approximation an atom can be characterized by specifying the n and l quantum numbers of all the electrons, i.e., by specifying the subshells occupied by the various electrons. This is called the *configuration*. The ordering according to energy of the outer filled subshells being known, it is trivial to determine the configuration of any atom in its ground state because in this state the electrons must fill all the subshells in such a way as to minimize the total energy of the atom and yet not exceed the capacity $2(2l + 1)$ of any subshell. The subshells will be filled in order of increasing energy, as listed in table (13–2). Consider first the H atom. The single electron occupies the $1s$ subshell. For the He atom both electrons are in the $1s$ subshell (as we saw in our discussion of the ground state of the He atom in section 5, Chapter 12). The configuration of H is written

$$^{1}\text{H}: \quad 1s^1$$

The configuration of He is written

$$^{2}\text{He}: \quad 1s^2$$

The superscript on the subshell designation specifies the number of electrons which it contains; the superscript on the chemical symbol specifies the value of Z for the atom. In the ^{3}Li atom one of the electrons must be in the $2s$ subshell because the capacity of the $1s$ subshell is only 2. The configuration of this atom is

$$^{3}\text{Li}: \quad 1s^22s^1$$

The ^{4}Be atom completes the $2s$ subshell and has the configuration

$$^{4}\text{Be}: \quad 1s^22s^2$$

In the six elements from ^{5}B to ^{10}Ne the additional electrons fill the $2p$ subshell. The configurations of ^{5}B and ^{10}Ne are

$$^{5}\text{B}: \quad 1s^22s^22p^1$$

$$^{10}\text{Ne}: \quad 1s^22s^22p^6$$

The reader will note that the periodic table of the elements presented in table (13–1) is divided vertically into a series of blocks with each row labeled by the subshell which, according to table (13–2), the elements of the row are filling. With this information, it is easy to write the configuration of any atom using a procedure that will become apparent by comparison with the configurations listed above. The configuration of certain atoms, for which the last few electrons are observed to be in different subshells than would be predicted by this scheme, are indicated by entries below the chemical symbol of the element. To illustrate the way this information is used to write the configuration of an atom, we present several typical examples:

$$^{19}\text{K}: \quad 1s^22s^22p^63s^23p^64s^1$$

$$^{23}\text{V}: \quad 1s^22s^22p^63s^23p^64s^23d^3$$

$$^{24}\text{Cr}: \quad 1s^22s^22p^63s^23p^64s^13d^5$$

$$^{43}\text{Tc}: \quad 1s^22s^22p^63s^23p^64s^23d^{10}4p^65s^24d^5$$

$$^{44}\text{Ru}: \quad 1s^22s^22p^63s^23p^64s^23d^{10}4p^65s^14d^7$$

$$^{46}\text{Pd}: \quad 1s^22s^22p^63s^23p^64s^23d^{10}4p^64d^{10}$$

$$^{57}\text{La}: \quad 1s^22s^22p^63s^23p^64s^23d^{10}4p^65s^24d^{10}5p^66s^25d^1$$

$$^{58}\text{Ce}: \quad 1s^22s^22p^63s^23p^64s^23d^{10}4p^65s^24d^{10}5p^66s^25d^14f^1$$

$$^{59}\text{Pr}: \quad 1s^22s^22p^63s^23p^64s^23d^{10}4p^65s^24d^{10}5p^66s^24f^3$$

We see that in certain cases the actual configurations observed for the elements do not strictly adhere to the predictions of table (13–2). For instance, this table says that the energy of the $3d$ subshell is greater than the energy of the $4s$ subshell when these subshells are filling. Yet in ^{24}Cr, and also in ^{29}Cu, one of the electrons that could be in the $4s$ subshell is actually in the $3d$ subshell. Similar situations are observed to occur for the $4d$ and $5s$ subshells. In ^{43}Tc the $5s$ subshell is filled in the normal manner. But in ^{45}Rb there is only one electron in the $5s$ subshell; in ^{46}Pd both electrons have left the $5s$ subshell and moved to the $4d$ subshell. The ^{78}Pt and ^{79}Au configurations show that the same kind of thing can happen for the $5d$ and $6s$ subshells. From these circumstances we conclude that the energy differences between the $3d$ and $4s$, the $4d$ and $5s$, and the $5d$ and $6s$ subshells must be so small while they are being filled that, although generally the ordering of these subshells is as shown in table (13–2), in certain cases the ordering can actually be reversed. Configurations which disagree with table (13–2) are also observed in ^{57}La and in the *lanthanides* ($Z = 58$ to 71)—more commonly called the *rare earths*. The table predicts that after the completion of the $6s$ subshell the $4f$ subshell should fill. But in ^{57}La, and several rare earths, there is one $5d$ electron. A similar situation occurs in the group of elements following ^{89}Ac, which are called the *actinides* ($Z = 90$ to 103). From the same argument as above we interpret these observations to mean that the energy differences between the $5d$ and $4f$ subshells, and between the $6d$ and $5f$ subshells, are very small while these subshells are being filled.

On the other hand, certain predictions of table (13–2) are always obeyed. Since none of the configurations is exceptional for elements in the first two and the last six columns of the periodic table (13–1), we conclude that every p subshell is always of higher energy than the preceding s or d subshell while these subshells are being filled, and that in these circumstances every s subshell is always of higher energy than the preceding p subshell. Therefore there must be large energy differences between the subshells concerned while they are being filled. In fact, the energy differences between every s subshell and the preceding p subshell are *particularly* large, and it is easy to understand why. Since for a given n the energy of a subshell becomes higher with increasing l, an s subshell is always the first subshell to be occupied in a new shell. Consequently, when an electron is added to a configuration with a completed p subshell and goes into the subshell of next highest energy, which according to table (13–2) is always an s subshell, the electron will be the first one in a new shell. Compared to the electrons in the preceding subshell, its average radial coordinate will be considerably larger, its average potential energy will be considerably less negative, and its total energy will be considerably higher—much higher than for the usual increase in total energy in going from one subshell to the next.

The fact that there is a particularly large energy difference between every *s* subshell and the preceding *p* subshell has some very important consequences. Consider atoms of the elements ^{10}Ne, ^{18}A, ^{36}Kr, ^{54}Xe, ^{86}Rn in which a *p* subshell is just completed. Because of the very large difference between the energy of an electron in the *p* subshell and the energy it would have if it were in the *s* subshell, the first excited state of these atoms is unusually far above the ground state. As a result, these atoms are particularly difficult to excite and are very inert. Furthermore, they produce no external electric field because they consist of sets of completely filled subshells and, therefore, have charge distributions which are spherically symmetrical.† Also we shall see later that atoms with completely filled subshells produce no external magnetic fields. Such atoms cannot interact with other atoms to produce chemical compounds, and they constitute the *noble gas* elements. In addition, ^{2}He is a noble gas because for this atom the first unfilled subshell is an *s* subshell (even though it does not contain a filled *p* subshell) so it has an unusually high first excited state, and also because the atom consists of completely filled subshells and so produces no external electric or magnetic fields. That ^{2}He is a noble gas is indicated by its being listed in the last column of the periodic table instead of the second column. An element such as ^{20}Ca is not a noble gas, even though it consists of completely filled subshells, because in its first excited state an electron goes to a 3*d* subshell so the excited state is not far above the ground state and the atom is not particularly inert.

Another aspect of the particular inertness of the noble gases can be obtained by plotting, for the various elements, the measured values of the magnitude of the total energy of an electron in the highest energy filled subshell. This is equal to the energy required to remove the electron from the atom, which is the ionization energy of the atom. Figure (13–4) shows such a plot. We see that the ionization energy oscillates about an average value which is essentially independent of Z, in agreement with our earlier conclusion that the total energy of the electrons in the outer shells is

† There is obviously no external monopole electric field (the field produced by a single charge) for any neutral atom. Also, there is no external dipole electric field (the field produced by two displaced charges of opposite sign) because of the symmetry of the charge distribution. However, there is an external quadrupole electric field (the field produced by two identical dipoles arranged along a line with one charge of the same sign from each dipole coinciding) unless the charge distribution of the atom is spherically symmetrical. Cf. equation (10–46).

roughly the same throughout the periodic table. However, the oscillations are quite pronounced, and it is apparent that the total energy of an electron in the highest energy filled subshell of a noble gas is considerably more negative than average. These electrons are very tightly bound, and the atoms are very difficult to ionize.

We also see that the ionization energy is particularly small for the elements ^{3}Li, ^{11}Na, ^{19}K, ^{37}Rb, and ^{55}Cs (^{87}Fr has not been measured).

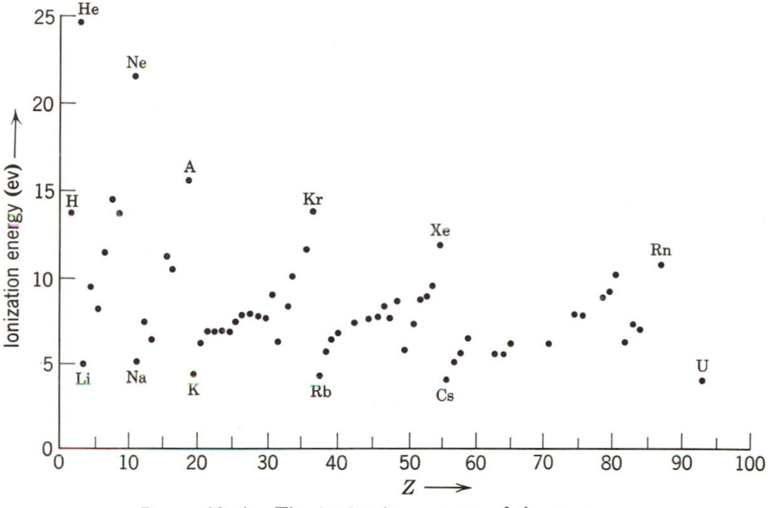

Figure 13–4. The ionization energy of the atoms.

These are the *alkalis*. They contain a single weakly bound electron in an *s* subshell. Alkali elements are very active chemically because they are quite eager to get rid of the weakly bound electron and revert to the more stable arrangement obtained with completely filled subshells. These elements have a valence of $+1$. At the other extreme are the *halogens*, ^{9}F, ^{17}Cl, ^{35}Br, ^{53}I, ^{85}At, which have one less electron than is required to fill their *p* subshell. These elements are very prone to capture an electron, and they have a valence of -1.

For the first three rows of the periodic table (13–1), the properties of the elements, such as valence and ionization energy, change uniformly from the alkali element with which the row begins to the noble gas with which it ends. In the fourth row of the periodic table this situation is no longer always true. The elements ^{21}Sc through ^{28}Ni, which are called the *first transition group*, have quite similar chemical properties and also almost the same ionization energies. These elements occur during the filling of the 3*d* subshell. The radius of this subshell is considerably less than that of the 4*s* subshell, which is completely filled for all the first transition group except ^{24}Cr. The filled 4*s* subshell tends to shield the 3*d* electrons from outside influences, and so the properties of these elements are all quite similar, independent of exactly how many 3*d* electrons they contain. The chemical properties of ^{29}Cu are somewhat different from those of the first transition group because it has only a single 4*s* electron in the outermost subshell. To a lesser extent this is also true for ^{24}Cr. The element ^{30}Zn consists of a set of completely filled subshells and so is somewhat more inert, as can be seen from its ionization energy. Similar transition groups arise in the filling of the 4*d* and 5*d* subshells.

An extreme example of the same situation is found in the rare earths, ^{58}Ce through ^{71}Lu. These are the elements in which the 4*f* subshell is filling. This subshell lies deep within the 6*s* subshell, which is completely filled in all the rare earths. The 4*f* electrons are so well shielded from the outside that the chemical properties of these elements are almost identical. The same thing happens in the actinides, ^{90}Th through ^{103}?. In this group the 5*f* subshell is filling inside the filled 7*s* subshell.

We finish our discussion of the ground state of multi-electron atoms and the periodic table by calling attention to two points. First, the study of the electronic structure of atoms gives no indication of why only 92 elements are found in nature. This implies that the instability of atoms with $Z > 92$ must be due to an instability of the atomic nuclei. We shall see in Chapter 16 that this is indeed the case. However, it is possible to produce such nuclei artificially in certain nuclear reactions. It is found that they have absolutely no difficulty in collecting electrons and forming atoms which are perfectly stable as long as the nuclei remain stable. The members of the actinide group with $Z > 92$, which are called the *transuranic elements*, are formed in just this way.

The second point we emphasize is the importance of the antisymmetric symmetry character of electrons—the property of electrons which causes them to obey the exclusion principle. If they did not obey this principle, all the electrons of a multi-electron atom would be in the 1*s* subshell because it is the subshell of lowest energy. If this were the case, all atoms would have very small radii, very high first excited states and ionization energies, and no external electric or magnetic fields. All elements would be extremely inert and there would be no chemical compounds, no life, and no readers or writers of books on modern physics!

5. Excited States of Atoms

In the previous sections we have described qualitatively the energies of electrons in the subshells that are being filled in various atoms. Now let us consider a quantitative example of the energies of electrons in all the subshells of a particular atom. Consider the typical atom ^{26}Fe, whose electron energies are shown in the energy level diagram of figure (13–5). The diagram is plotted on an energy scale which is linear from 0 to -10 ev, and logarithmic for more negative energies. The measured values of the total energy E_{nl} of an electron in the various occupied subshells of the ^{26}Fe atom in its ground state are plotted, and are also written to the right

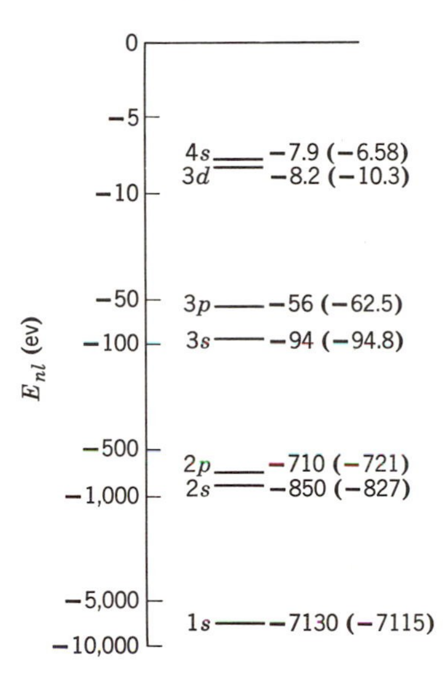

Figure 13–5. An energy level diagram for the filled subshells of the iron atom.

of the levels. These values are obtained from an analysis of the optical and X-ray spectra produced by the atom. The procedures involved in such analyses will become apparent later. Shown in parentheses are the values of E_{nl} obtained from a Hartree theory calculation for the ^{26}Fe atom. The reader will note that the agreement between theory and experiment is generally quite good. He will also note that these data confirm many of our previous statements regarding the energies of the various subshells. For instance, the energy difference between each p subshell and the s subshell of higher energy is seen to be particularly large. Also the energy difference between the $3d$ and $4s$ subshells is seen to be very small. It is also interesting to note that the energy of a $3d$ electron in the ^{26}Fe atom is actually slightly more negative than that of a $4s$ electron, despite the fact that the observed configurations of this atom and the others in the fourth row of the periodic table show that generally the $4s$ subshell fills before the $3d$ subshell. This is not at all paradoxical, and it is predicted by the Hartree theory. Because of repulsive effects, the presence of an electron in a particular subshell changes somewhat the electric field produced by the other electrons in that subshell. When the energy difference between two subshells is very small, this can lead to the situation found in the ^{26}Fe atom.

Figure (13–5) shows only the subshells—that is, the one-particle energy levels—which are occupied in the ground state of the atom. There are also many unoccupied one-particle energy levels of higher energy. Just as for the ^{1}H atom, there are an infinite number of discrete one-particle levels with $E_{nl} < 0$, and a continuum of levels with $E_{nl} > 0$. In an excited state of the atom, some of its electrons go to the higher energy one-particle levels.

Fortunately, the number of electrons in an atom which make such transitions is small, because of the nature of the processes responsible for the excitation of an atom. For instance, in a discharge tube atoms are excited by the bombardment of energetic electrons. One of the energetic electrons collides with an atom, knocking one or two atomic electrons into levels of higher energy. In such a process it is improbable that many atomic electrons are simultaneously excited. Another typical process involves the excitation of atoms by the absorption of quanta. At most, one quantum is absorbed by any particular atom and, particularly for higher energy quanta, the quantum tends to interact with only one or two electrons. Thus the excited states of an atom are not so complicated as they could be. The configuration of an atom in an excited state differs from its ground state configuration only in that one or a very few electrons are in higher energy subshells, and there are corresponding vacancies in the subshells from which they came. For simplicity, we often speak as if only one electron in an atom can be excited. Even so, it is apparent that there are still a very large number of different kinds of configurations for an atom in its excited states.

However, it is possible to divide the different kinds of excitations into two types. The first type corresponds to excitations in which an electron in one of the higher energy subshells occupied in the ground state of the atom is excited. Since the subshells of higher energy are also the outer subshells, this type corresponds to the excitation of one of the outer electrons of the atom. The second type corresponds to the excitation of an electron in one of the lower energy subshells, which is one of the inner subshells. Typical examples of these two types of excitations are illustrated in figure (13–6) for a ^{26}Fe atom. [These diagrams employ the same pseudo-logarithmic energy scale as figure (13–5).] In an excitation of the first type, the excited electron may make a transition to any of the discrete or continuum one-particle levels of energy higher than the energy of its original level. This is not true of an excitation of the second type. With the exception of the $3d$ level, all the levels through the $4s$ level contain their complete complement of electrons. Consequently the exclusion principle demands that the excited electron go to one of the one-particle levels above the $4s$ level, or possibly to the $3d$ level. This means that the energy required to produce an excitation of the second type is much higher

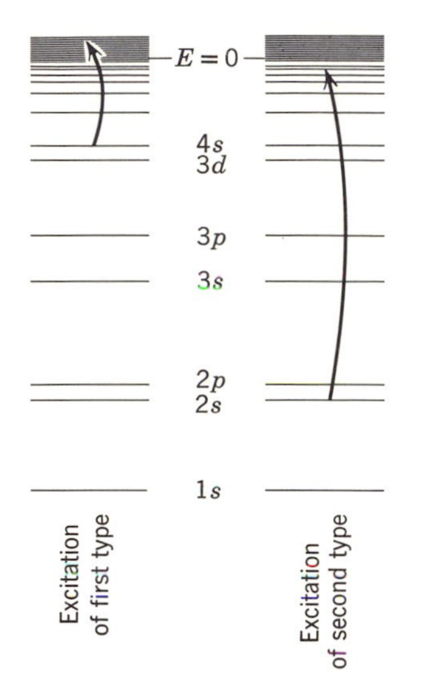

Figure 13–6. Illustrating excitations of the first type (optical excitations) and excitations of the second type (X-ray excitations).

than that required to produce an excitation of the first type. The latter energy is typically only a few electron volts, but an excitation of the second type for the case illustrated in the diagram requires more than 800 ev. After an excitation of the second type, the very highly excited atom will eventually return to its ground state. It does this by emitting quanta. These are the very high energy quanta which form the X-ray spectrum of the atom. The much lower energy quanta which are emitted when an atom returns to its ground state after an excitation of the first type form the optical spectrum of the atom (the spectrum in and near the visible range). X-rays and X-ray spectra constitute the subject of the next chapter. In this chapter we discuss only optical spectra. Consequently we consider here only excitations of the first type, which are often called *optical excitations*. These excitations are produced, for instance, by exposing atoms to low voltage electrical discharges or flames.

6. Alkali Atoms

In an optical excitation, usually it is only the electrons in a partially filled subshell that are excited, because the energy of these electrons is usually considerably less negative than the energy of the electrons of any completely filled subshell. There are exceptions to this rule. An example of an exception is ^{26}Fe, for which the energy of an electron in the completely filled 4s subshell is, as we have seen, actually slightly less negative than the energy of an electron in the partially filled 3d subshell. However, the rule is normally obeyed.

In an atom of an alkali element the rule is always obeyed. These atoms contain a set of completely filled subshells, the one of highest energy being a p subshell, plus a single additional electron in the next s subshell. As mentioned before, the energy of the electrons in a filled p subshell is particularly negative with respect to the energy of an electron in the next s subshell. Consequently these electrons are not excited in any of the low energy processes which lead to optical excitations. In essence, an alkali atom consists of an inert noble gas core plus a single electron moving in an external shell. The analysis of the optical spectrum of an alkali atom in terms of its excited states is relatively simple because the excited states can be described completely by describing the single *optically active* electron, and the core containing completely filled subshells can be ignored. The total energy of the core does not change, so the total energy of the atom is a constant plus the total energy of the optically active electron. It is convenient in discussing the excited states of an alkali atom to define the zero of total energy in such a way that the total energy of the atom is equal to that of the optically active electron. Using this definition, we present in figure (13–7) diagrams showing the energies of the ground state and the first few excited states of the alkali elements ^{3}Li and ^{11}Na, obtained from an analysis of the spectra of these elements, and also the energy levels of ^{1}H for $n = 2, 3, 4, 5, 6$. Each energy level is labeled by the quantum numbers n and l of the optically active electron, i.e., by its configuration. These diagrams do not show a fine structure splitting, which will be discussed later.

The energy level diagrams also do not show the results of Hartree calculations which have been made for these atoms, since the discrepancies between theory and experiment are generally too small to be indicated. The Hartree theory works particularly well in calculating the energy levels of the optically active electron of an alkali element because the net potential $V(r)$, due to the nucleus plus the electrons of the core, actually is spherically symmetrical, as assumed in the theory. In terms of the Hartree theory, it is easy to understand the structure of these energy level diagrams and their relation to the diagram for ^{1}H. The dependence of the energy of the optically active electron on its quantum numbers n and l is just as we have described on previous occasions. For a given n, the energy is

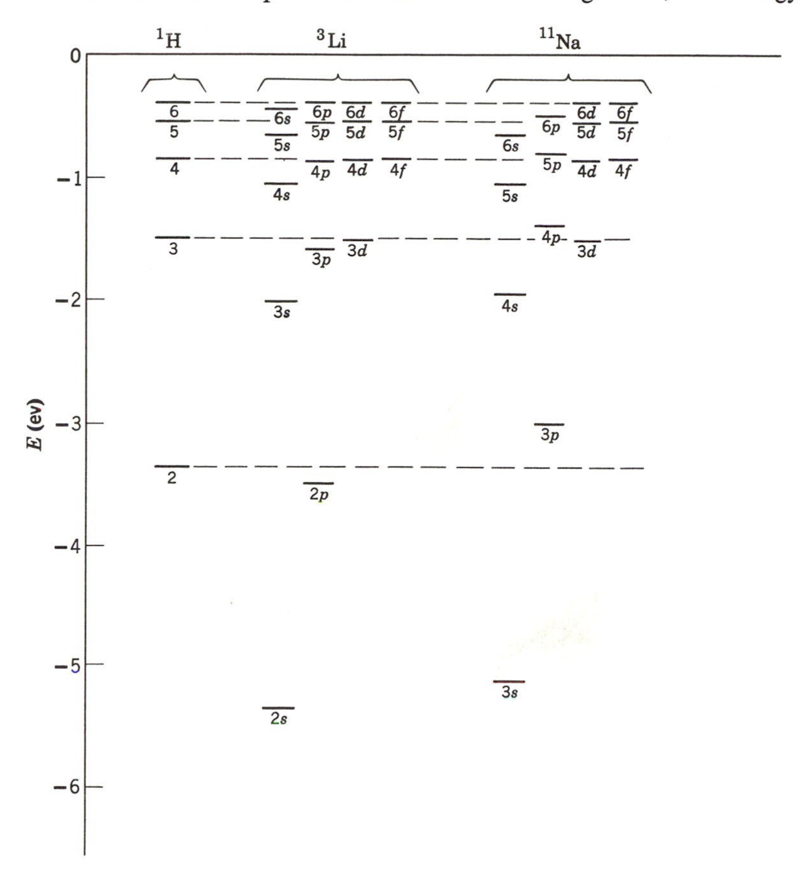

Figure 13–7. Some energy levels of hydrogen, lithium, and sodium atoms.

most negative for the smallest value of l because the electron spends more time near the center of the atom, where it feels the full nuclear charge. In the ground state of the ^{3}Li atom, the optically active electron is in the $2s$ subshell and its energy is about 2 ev more negative than an $n = 2$ electron in a ^{1}H atom. But, in the first excited state, the optically active electron is in the $2p$ subshell and its energy is only about 0.2 ev more negative than an $n = 2$ electron in ^{1}H. In ^{11}Na the l dependence makes the $4s$ level more negative than the $3d$ level. However, for the large radii subshells with large values of n, the l dependence becomes less important and the energy levels of the optically active electron become very close to the energy levels of an electron in a ^{1}H atom, because the shielding of the nuclear charge $+Ze$ by the charge $-(Z-1)e$ of the electrons in the core of the alkali atom becomes essentially complete for an electron in a subshell of radius large compared to the radius of the core. This occurs for smaller values of n in ^{3}Li than in ^{11}Na because the radius of the core is smaller for the former atom. Putting it another way, although $\psi^*\psi \propto r^{2l}$ near $r = 0$ for all n, (10–39) shows that the proportionality constant decreases with increasing n so that $\psi^*\psi$ becomes smaller near $r = 0$. The result is that the l dependence becomes less important when n becomes very large.

TABLE (13–4)

The Quantum Defects for the Sodium Atom

Subshell	s	p	d	f
μ	1.35	0.86	0.01	$\sim$0

Note that, in going from the inner filled subshells, through the outer filled subshells, and then through the unfilled subshells of a particular atom, the l dependence of E_{nl} first increases in relative importance, is of maximum relative importance in the outermost filled subshells, and then decreases in relative importance.

The energy levels of the ^{1}H atom are given by the equation

$$E = -Rch/n^2$$

where R is the Rydberg constant. For an alkali atom a similar equation can obviously be written

$$E = -Rch/n'^2 \tag{13–25}$$

if the n', which are not integers, are given the proper values. These quantities are related to the quantum numbers n by the equation

$$n' = n - \mu \tag{13–26}$$

The μ are called *quantum defects.* By fitting equation (13–25) to the observed energy levels of various alkali elements, it is found that for a given element the quantum defects are essentially the same for all subshells with the same quantum number l. The quantum defects for ^{11}Na are listed in table (13–4). These results explain the basis of Rydberg's series formula (5–5) for the alkali elements.

The lines of the optical spectra emitted by alkali elements show a fine structure splitting which indicates that all energy levels are double, except those for $l = 0$. This is due to a spin-orbit interaction acting on the optically active electron. The other relativistic effects, which are just as important as the spin-orbit interaction in the case of a one-electron atom, are generally quite negligible. We can see this by using equation (5–13) to estimate the average velocity of an optically active electron. In this equation we must replace Z by the quantity Z_n defined by (13–22). As Z_n is roughly equal to unity for the optically active electrons of all atoms, (5–13) shows that the average value of v/c is roughly equal to $1/n$ times the average value of this ratio for an electron in the ground state of the ^{1}H atom; that is, $v/c \sim (1/n) \times 10^{-2}$. Since the quantum number n for the optically active electron of an atom increases with increasing Z, relativistic effects other than the spin-orbit effect, which are of the order of $(v/c)^2$, are negligible for the optically active electrons of all atoms except those of very small Z.†

The shift of the energy levels of an alkali element due to a spin-orbit interaction acting on the optically active electron can be understood by considering equation (11–47). That equation leads immediately to an equation analogous to (11–49). This is

$$\overline{\Delta E_{\mathbf{S}\cdot\mathbf{L}}} = \frac{\hbar^2}{4m^2c^2}[j(j+1) - l(l+1) - s(s+1)]\overline{\left(\frac{1}{r}\frac{dV(r)}{dr}\right)} \qquad (13\text{–}27)$$

where

$$\overline{\left(\frac{1}{r}\frac{dV(r)}{dr}\right)} \equiv \int\int\int \psi^*_{nljm_j}\frac{1}{r}\frac{dV(r)}{dr}\psi_{nljm_j} r^2 \sin\theta \, dr\, d\theta\, d\phi \qquad (13\text{–}28)$$

Just as for a one-electron atom, when the spin-orbit interaction is included m_l and m_s are no longer good quantum numbers and the eigenfunctions describing the behavior of the optically active electron must be those which are specified by the quantum numbers n, l, j, m_j. These quantum numbers obey the same rules as before. Specifically,

$$\begin{aligned} &s = \tfrac{1}{2} \\ &j = l + \tfrac{1}{2} \text{ or } l - \tfrac{1}{2}, \qquad l \neq 0 \\ &j = \tfrac{1}{2}, \qquad l = 0 \end{aligned} \qquad (13\text{–}29)$$

For $l = 0$, equation (13–27) shows that $\overline{\Delta E_{\mathbf{S}\cdot\mathbf{L}}} = 0$. For other values of l, $\overline{\Delta E_{\mathbf{S}\cdot\mathbf{L}}}$ assumes two different values, one positive and the other negative,

† These relativistic effects are not so negligible for electrons in the inner subshells. For a $1s$ electron in the ^{92}U atom, the average value of v/c is about 0.5. However, this affects only the fixed value of the total energy of the core, and therefore in optical excitations it affects only the definition of the zero of total energy.

according to whether $j = l + \frac{1}{2}$ or $j = l - \frac{1}{2}$. Except for $l = 0$, each energy level is split into two components, one of slightly higher energy for the spin and orbital angular momenta "parallel" and one of slightly lower energy for these angular momenta "antiparallel." The magnitude of the splitting is proportional to the expectation value of $(1/r)(dV/dr)$. Since both $1/r$ and the derivative of the net potential $V(r)$ become large for small r, the expectation value is very sensitive to the behavior of ψ_{nljm_j} and $V(r)$ near $r = 0$. In fact, it is so sensitive that it is not possible to obtain accurate values of the splitting from Hartree theory calculations. Nevertheless, the Hartree theory certainly predicts the general features of the experimentally observed values of fine structure splitting.

TABLE (13–5)
Spin-Orbit Splittings in the Sodium Atom

Subshell	$3p$	$4p$	$5p$	$6p$
Spin-orbit splitting	21.3×10^{-4} ev	7.63×10^{-4}	3.05×10^{-4}	1.74×10^{-4}

For a given alkali element, the theory predicts that the splitting will decrease with increasing n because, with increasing values of this quantum number, the eigenfunction ψ_{nljm_j} becomes smaller in the region near $r = 0$ where $(1/r)(dV/dr)$ is large, so the expectation value of this quantity decreases. This behavior is seen, for instance, in the experimentally measured energy levels of ^{11}Na associated with an electron excited to the p subshells. The splittings of these energy levels are listed in table (13–5). In section 3 we showed that according to the Hartree theory the dependence of the net potential $V(r)$ on r, near $r = 0$, becomes more pronounced with increasing Z. Thus the Hartree theory predicts that dV/dr increases with increasing Z near $r = 0$, and therefore that the spin-orbit splitting will increase with increasing Z. This behavior can also be seen in the experimental data. In table (13–6) we list for various alkali elements the observed splittings of the energy levels for an electron excited to the first p subshell. Note that the spin-orbit splitting becomes quite large for large values of Z.

Figure (13–8) shows a few of the lowest energy levels of the ^{11}Na atom. Levels for which all the quantum numbers except n are the same are grouped in columns, and the value of n for each level is labeled by the configuration of the optically active electron. The quantum numbers l and j are given according to the standard spectroscopic notation by a symbol such as $^2P_{1/2}$, which is read "doublet P one-half." The letter specifies the value of l according to the scheme we have become familiar

TABLE (13–6)
Spin-Orbit Splittings in a Number of Alkali Atoms

Element	^{3}Li	^{11}Na	^{19}K	^{37}Rb	^{55}Cs
Subshell	$2p$	$3p$	$4p$	$5p$	$6p$
Spin-orbit splitting	0.42×10^{-4} ev	21×10^{-4}	72×10^{-4}	295×10^{-4}	687×10^{-4}

with, except that it is conventional to use capitals. The subscript gives the value of j. The superscript is equal to $(2s + 1) = (2 \cdot \frac{1}{2} + 1)$, and is also equal to the number of components into which the levels are split by the

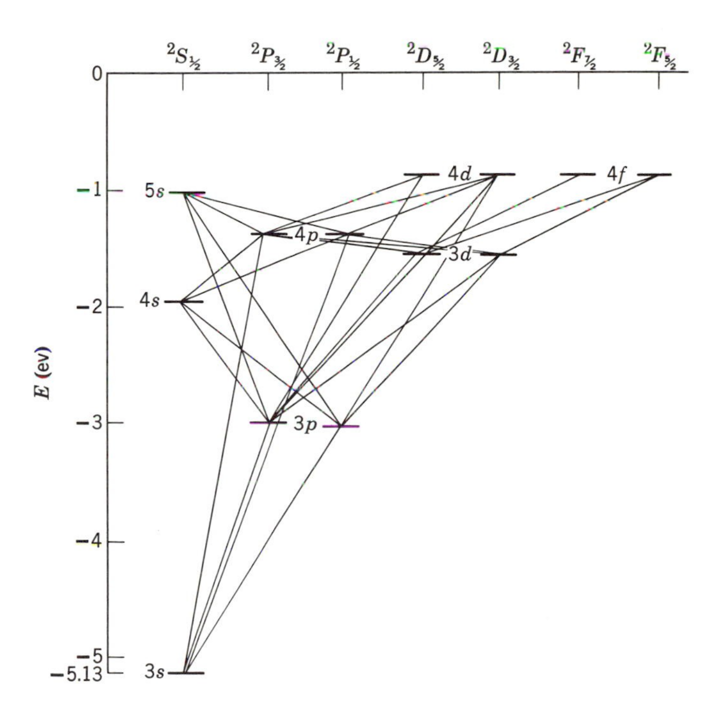

Figure 13–8. Some energy levels of the sodium atom, and the possible transitions between these levels.

spin-orbit interaction, except for $^2S_{1/2}$ levels that are not split. Since the spin-orbit energy is positive for $j = l + \frac{1}{2}$ and negative for $j = l - \frac{1}{2}$, each $^2P_{3/2}$ level should be slightly higher than the associated $^2P_{1/2}$ level, and similarly for the $^2D_{5/2}$ and $^2D_{3/2}$ and for the $^2F_{7/2}$ and $^2F_{5/2}$ levels. This splitting is, of course, much too small to be seen on the scale of the diagram; however, we have indicated it for the $3p\,{}^2P_{3/2}$ and $3p\,{}^2P_{1/2}$ levels.

We have also indicated in figure (13–8) the possible transitions between energy levels which produce the spectral lines of ^{11}Na that are associated with the few levels shown. If the reader visualizes the complete set of energy levels and transitions, he will certainly be impressed with its complexity, and with the tremendous amount of work expended by Rydberg, and others, in obtaining energy level diagrams from an analysis of the observed spectra. The analysis became possible only after Rydberg discovered that the spectral lines could be grouped into various series. For alkali elements there are four important types of series. These arise from the transitions indicated below.†

"Sharp" series:	$n\,{}^2S_{1/2} \to n_0\,{}^2P_{3/2,1/2}$	doublets of lines
"Principal" series:	$n\,{}^2P_{3/2,1/2} \to n_0\,{}^2S_{1/2}$	doublets of lines
"Diffuse" series:	$n\,{}^2D_{5/2,3/2} \to n_0\,{}^2P_{3/2,1/2}$	triplets of lines
"Fundamental" series:	$n\,{}^2F_{7/2,5/2} \to n_0\,{}^2D_{5/2,3/2}$	triplets of lines

In a particular series, n_0 is fixed and n is variable. The intense yellow doublet of the ^{11}Na spectrum is the first member of the principal series for $n_0 = 3$. That is, the two lines are due to the two transitions $3\,{}^2P_{3/2} \to 3\,{}^2S_{1/2}$ and $3\,{}^2P_{1/2} \to 3\,{}^2S_{1/2}$. All the lines of the sharp and principal series are doublets since they involve transitions from a single ${}^2S_{1/2}$ level to a closely spaced pair of ${}^2P_{3/2}$ and ${}^2P_{1/2}$ levels, or vice versa. The lines of the diffuse and fundamental series might be expected to be quartets since the initial and final energy levels are always both members of a pair of levels. However, it is observed that the transitions $n\,{}^2D_{5/2} \to n_0\,{}^2P_{1/2}$ and $n\,{}^2F_{7/2} \to n_0\,{}^2D_{3/2}$ do not occur. From an analysis of many spectra, it is found that these are particular examples of a general *selection rule* for the quantum number j. Transitions only occur for which

$$\Delta j = 0, \pm 1 \tag{13–30}$$

Because of this rule the lines of the diffuse and fundamental series are triplets. In figure (13–8) we can also see the operation of a selection rule on the quantum number l. Transitions are indicated in the diagram only when

$$\Delta l = \pm 1 \tag{13–31}$$

From a study of the spectra of the alkali elements, it is found that this

† The series were named in the early days of spectroscopy for somewhat obscure reasons. These names are the origin of the scheme by which the quantum numbers $l = 0, 1, 2, 3$, are represented by s, p, d, f.

selection rule on l is obeyed whenever there is a single optically active electron. Thus equations (13–30) and (13–31) are both also obeyed in the case of the ^{1}H atom. We shall discuss the origin of selection rules in section 12.

7. Atoms with Several Optically Active Electrons

An atom of a typical element contains a core of completely filled subshells surrounding the nucleus, plus several electrons in a partially filled subshell. Since any of the electrons in a partially filled subshell may participate in an optical excitation, all such electrons are optically active. It is apparent, then, that a description of the excited states that give rise to the optical spectrum of such an atom requires a description of the behavior of all the optically active electrons. Despite the fact that there are a number of different effects which determine the behavior of these electrons, it is possible to understand their influence on the energy of the excited states by treating them one at a time, in order of decreasing importance, with each new effect considered a perturbation on the effects already treated.

As a starting point of this perturbation treatment, we consider each of the optically active electrons to be moving independently in the Hartree net potential $V(r)$ due to the core plus the other optically active electrons. In this Hartree approximation the energy of each optically active electron is completely specified by its quantum numbers n and l, and the dependence of this energy E_{nl} on the two quantum numbers is very similar to that of a single optically active electron in an alkali atom with the same core.† The total energy of the atom is a constant plus the total energy of the optically active electrons. Consequently, the energy of the atom is determined completely in this approximation by the configuration of the optically active electrons, i.e., by the n and l quantum numbers of these electrons. Now, associated with each configuration of the optically active electrons are generally a number of different possible sets of quantum numbers m_l and m_s. Since the energy does not depend on these quantum numbers, there is usually a quantum number degeneracy for each configuration. Also, there is generally an exchange degeneracy because the energy does not depend on which of the identical particles has a particular set of quantum numbers. We see that in this approximation a number of degenerate energy levels are associated with each configuration. However, many of these degeneracies are removed when the perturbations are taken

† The reason is that $V(r)$ is very similar to the net potential due to the core alone.

into account. The situation is analogous to the case of an alkali atom in which some of the degeneracies associated with each configuration of the optically active electron are removed when the spin-orbit interaction is treated as a perturbation.

The perturbations which we must include correct for the effects that are ignored in the Hartree approximation. The two most important corrections are for:

1. the fact that the Hartree potential acting on each optically active electron describes only the *average* effect of the Coulomb interactions between that electron and all the other optically active electrons, and
2. the magnetic interactions between the spin and orbital angular momenta of each optically active electron.

There are also relativistic corrections, corrections for interactions between the spins of the optically active electrons because of magnetic interactions between the associated magnetic moments, and certain other corrections; but all of them are very small and can generally be neglected.

We have seen one manifestation of the residual Coulomb interaction effect (correction 1) in our treatment of the ^{2}He atom. In this atom the energy level for the $1s2s$ configuration is split into two levels, one corresponding to the triplet states for which the energy is lowered, and one corresponding to the singlet state for which the energy is raised. The same is true of the $1s2p$ configuration. The reader will recall that this happens because the requirement that the complete eigenfunction be antisymmetric introduces an interaction between the relative orientation of the spins of the electrons and their relative spatial coordinates. The average distance between the two electrons is larger in the triplet states when the spins are "parallel" than it is in the singlet state when the spins are "antiparallel." Consequently the positive Coulomb repulsion energy is smaller in the triplet states, for which the magnitude of the total spin has the constant value $S' = \sqrt{1(1+1)}\,\hbar$, than it is in the singlet state, for which the magnitude of the total spin has the constant value $S' = 0$. Of course, the magnitudes of the individual spin angular momentum vectors are not changed by this effect of the residual Coulomb interaction. It is shown by an analysis of the experimentally observed spectra, and can also be understood theoretically from calculations similar to those we made for ^{2}He, that this effect is important in all atoms with two or more optically active, electrons. That is, for such atoms there is a tendency for the spin angular momenta of the optically active electrons to couple in such a way that the magnitude of the total spin angular momentum S' is constant, and the energy is usually smallest for the state in which S' is largest.

The residual Coulomb interaction effect also produces a tendency for the orbital angular momenta of the optically active electrons to couple in such a way that the magnitude of the total orbital angular momentum L' is constant. We did not see an example of this in the excited states of the ^{2}He atom because in all the configurations we studied one electron is a $1s$ electron and, consequently, there is at most only one non-zero orbital angular momentum present. However, it is easy to understand the origin of the forces tending to couple the orbital angular momenta by considering the polar plots of the probability density for an electron moving in a spherically symmetrical potential [cf. figures (10–4) and (10–5)]. Choosing the z axis so that the orbital angular momentum has its maximum possible

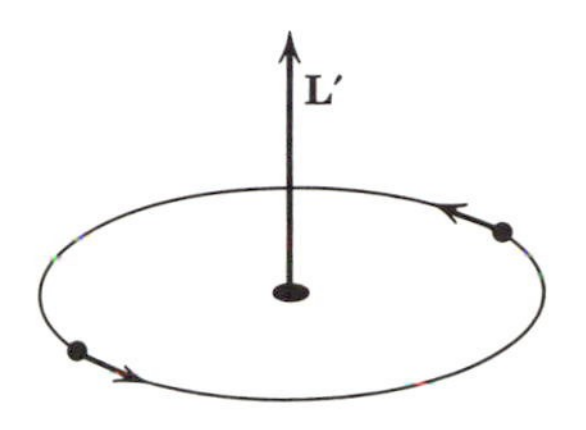

Figure 13–9. Illustrating why two optically active electrons tend to couple to give a maximum total orbital angular momentum vector.

component along that axis, and considering the corresponding polar plot, we see that the probability density of the electron is concentrated near the plane perpendicular to the orbital angular momentum vector. Thus the charge density produced by the electron is largest near this plane. Now, considering two or more electrons, it is apparent that the repulsions between their charge densities will have the effect of producing interactions between their orbital angular momentum vectors. These interactions are torques acting between the orbital angular momentum vectors which do not tend to change the magnitudes of the individual orbital angular momentum vectors but only to make them precess about the total orbital angular momentum vector in such a way that its magnitude L' remains constant. The question then arises: Which of the possible values of L' corresponds to the state of smallest energy? This is not so easy to answer because there are several opposing tendencies. However, the physical basis of the one which is usually most important can be understood by considering two electrons in a Bohr atom, as illustrated in figure (13–9). Because of the Coulomb repulsion between the electrons, the most stable arrangement is obtained when the electrons stay at the opposite ends of a diameter. In this state of minimum energy, the electrons rotate together with individual orbital angular momentum vectors parallel, and with the magnitude of the total orbital angular momentum vector L' a maximum. This conclusion is confirmed by an analysis of the spectra produced by atoms with several optically active electrons. That is, for such atoms there is a tendency for the orbital angular momenta of the optically active electrons to couple in such a way that the magnitude of the total orbital angular momentum L' is constant, and the energy is usually smallest for the state in which L' is largest.

In contrast to the tendencies produced by the residual Coulomb interaction, the spin-orbit interaction (correction 2) produces a tendency for the spin angular momentum vector of each optically active electron to couple with its *own* orbital angular momentum vector in such a way as to leave the magnitudes of these vectors constant while they precess about their resultant total angular momentum vector, which is of constant magnitude J. We know that this is due to torques arising from the interaction of the magnetic moment connected with the spin angular momentum and the magnetic field connected with the orbital angular momentum. We also know that the energy is smallest for the state in which J is smallest.

8. *LS* Coupling

Because the residual Coulomb and spin-orbit interactions tend to produce effects which are in opposition to each other, atoms in which both interactions are of comparable importance are very difficult to treat. Fortunately, such atoms are in the minority. For atoms of small and intermediate Z, the effects of the residual Coulomb interactions are usually quite large compared to those of the spin-orbit interactions. We can see evidence of this by comparing the energy difference ($\sim$1 ev) associated with the residual Coulomb interaction coupling of the two spin angular momenta in the first excited states of ^{2}He with the energy difference ($\sim 10^{-4}$ ev) associated with the spin-orbit interaction coupling of the spin and orbital angular momenta in the first excited states of ^{3}Li. Atoms of small and intermediate Z are said to be examples of *LS coupling*, which is sometimes also called *Russell-Saunders coupling* after two spectroscopists who first explained the structure of the energy levels of these atoms. In studying coupling of this type, the perturbation due to the residual Coulomb interactions is treated first since it is the most important, and the perturbation due to the spin-orbit interactions is temporarily ignored. Then the individual spin angular momenta $\mathbf{S}_i$ of the optically active electrons couple to form a total spin angular momentum $\mathbf{S}'$, where

$$\mathbf{S}' = \mathbf{S}_1 + \mathbf{S}_2 + \cdots + \mathbf{S}_i + \cdots \tag{13–32}$$

and where $\mathbf{S}'$ has a constant magnitude satisfying the quantization equation

$$S' = \sqrt{s'(s'+1)}\,\hbar \tag{13–33}$$

Also the individual orbital angular momenta $\mathbf{L}_i$ of the optically active electrons couple to form a total orbital angular momentum $\mathbf{L}'$, where

$$\mathbf{L}' = \mathbf{L}_1 + \mathbf{L}_2 + \cdots + \mathbf{L}_i + \cdots \tag{13–34}$$

and where $\mathbf{L}'$ has a constant magnitude satisfying the quantization equation

$$L' = \sqrt{l'(l'+1)}\,\hbar \tag{13–35}$$

These vectors couple in such a way that all their magnitudes S_i and L_i also remain constant. Because of the perturbation, the energy of the atom depends on S' and L'. Then the quantum states of the same configuration, but associated with different values of S' and L', no longer have the same energy. The state with the maximum possible values of S' and L' usually has the minimum energy.

The dominant residual Coulomb interactions having been taken into account as a first perturbation, the smaller spin-orbit interactions are included in *LS* coupling as an additional perturbation. This is done by considering a spin-orbit interaction between the angular momentum vectors $\mathbf{S}'$ and $\mathbf{L}'$. This interaction couples these two vectors in such a way that the magnitude J' of the total angular momentum

$$\mathbf{J}' = \mathbf{L}' + \mathbf{S}' \tag{13–36}$$

is constant, and S' and L' remain constant. The magnitude J' is also quantized according to the usual equation

$$J' = \sqrt{j'(j'+1)}\,\hbar \tag{13–37}$$

As a result of this additional perturbation, the energy of the atom depends also on J'. The state with the minimum possible value of J' has the minimum energy.

Figure (13–10) illustrates the way the various angular momentum vectors combine in *LS* coupling in the state which is normally the one of minimum energy for two optically active electrons with quantum numbers $l_1 = 1$, $s_1 = \frac{1}{2}$ and $l_2 = 2$, $s_2 = \frac{1}{2}$. The spin angular momenta $\mathbf{S}_1$ and $\mathbf{S}_2$ precess about their sum $\mathbf{S}'$, and this vector has its maximum possible magnitude (corresponding to $s' = 1$). The orbital angular momenta $\mathbf{L}_1$ and $\mathbf{L}_2$ precess about their sum $\mathbf{L}'$, which also has its maximum possible magnitude (corresponding to $l' = 3$). Also $\mathbf{S}'$ and $\mathbf{L}'$ precess about their sum $\mathbf{J}'$, but this vector has its minimum possible magnitude (corresponding to $j' = 2$). Finally, $\mathbf{J}'$ precesses about the z axis in such a way that its component J_z' along that axis is a constant which satisfies the quantization equation

$$J_z' = m_j'\hbar \tag{13–38}$$

where

$$m_j' = -j', -j' + 1, \ldots, +j' - 1, +j' \tag{13–39}$$

Figure (13–10) is drawn for $m_j' = j'$. The quantization of the magnitude of the total angular momentum J' and its z component J_z' is a necessary consequence of the absence of external torques acting on the atom

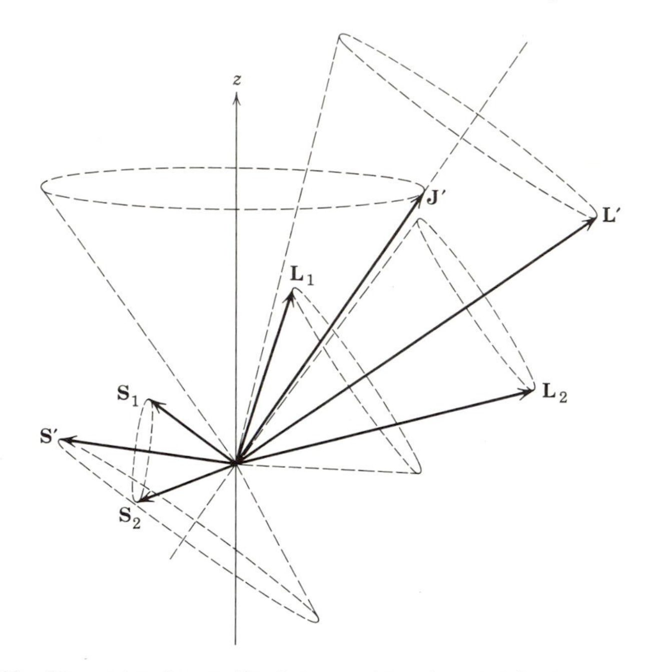

Figure 13–10. Illustrating the coupling between the various angular momentum vectors in a typical *LS* coupling state of minimum energy.

Figure (13–10) illustrates only one of the quantum states that can be formed in *LS* coupling by two optically active electrons with quantum numbers $l_1 = 1$, $s_1 = \frac{1}{2}$ and $l_2 = 2$, $s_2 = \frac{1}{2}$. In fact, there are twelve different sets of states, with different quantum numbers s', l', j', that can be formed by these two electrons. And each of these sets consists of $2j' + 1$ different quantum states corresponding to the $2j' + 1$ different possible values of m_j'. The rule specifying the possible values of m_j' is expressed by equation (13–39). The rules specifying the possible values of s', l', j' are conveniently expressed with reference to vector addition diagrams employing vectors whose lengths are proportional to the values of the quantum numbers. For the two electrons in question, these diagrams have the form shown in figure (13–11). The reader may verify that the possible values of s', l', j' shown in the vector diagrams agree with those obtained from the equations

$$\begin{aligned} s' &= |s_1 - s_2|, |s_1 - s_2| + 1, \ldots, s_1 + s_2 \\ l' &= |l_1 - l_2|, |l_1 - l_2| + 1, \ldots, l_1 + l_2 \\ j' &= |s' - l'|, |s' - l'| + 1, \ldots, s' + l' \end{aligned} \tag{13–40}$$

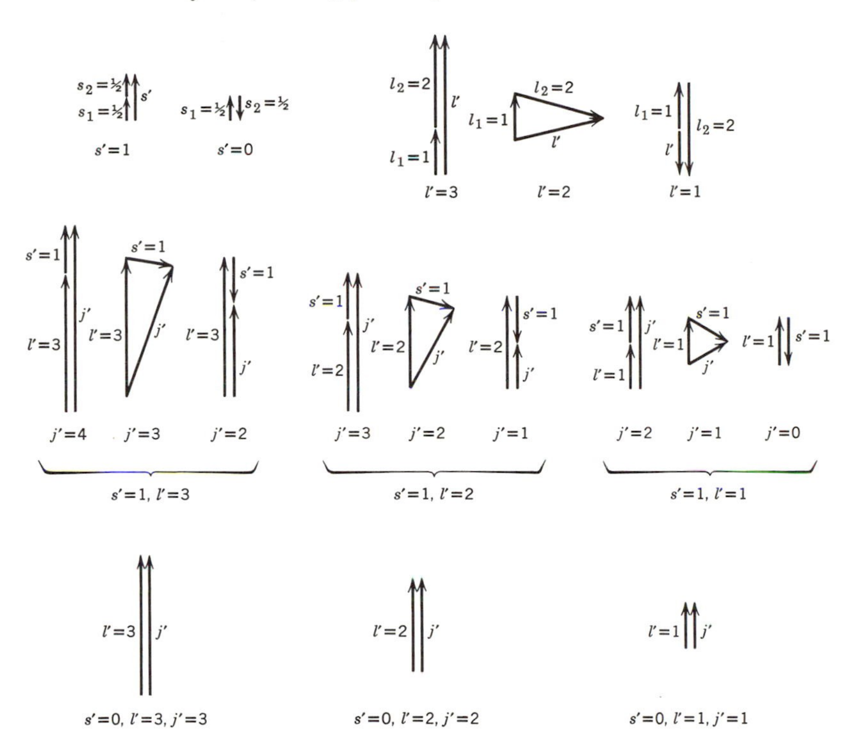

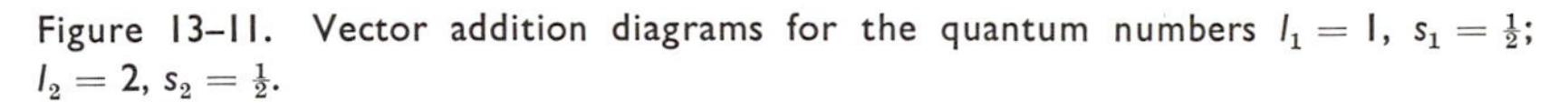

Figure 13–11. Vector addition diagrams for the quantum numbers $l_1 = 1$, $s_1 = \frac{1}{2}$; $l_2 = 2$, $s_2 = \frac{1}{2}$.

Since $s_1 = s_2 = \frac{1}{2}$, the first equation gives

$$s' = 0, 1 \tag{13–40′}$$

These equations can be proved by the same kind of arguments as were used in proving equation (11–44). The equation for s' is, of course, the same as (12–25).

Equations (13–40) are the rules specifying the possible values of s', l', j'

for the general case of LS coupling of any two electrons. For any N electrons, the rules can be written

$$s' = |\mathbf{s}_1 + \mathbf{s}_2 + \cdots + \mathbf{s}_N|_{\min}, |\mathbf{s}_1 + \mathbf{s}_2 + \cdots + \mathbf{s}_N|_{\min} + 1, \ldots, s_1 + s_2 + \cdots + s_N$$
$$l' = |\mathbf{l}_1 + \mathbf{l}_2 + \cdots + \mathbf{l}_N|_{\min}, |\mathbf{l}_1 + \mathbf{l}_2 + \cdots + \mathbf{l}_N|_{\min} + 1, \ldots, l_1 + l_2 + \cdots + l_N$$
$$j' = |s' - l'|, |s' - l'| + 1, \ldots, s' + l' \qquad (13\text{–}41)$$

which reduce to (13–40) for $N = 2$. Since the magnitudes of the vectors $\mathbf{s}_1, \mathbf{s}_2, \ldots, \mathbf{s}_N$ are all $\frac{1}{2}$, the first equation gives

$$s' = 0, 1, 2, 3, \ldots, N/2, \qquad \text{for } N \text{ even}$$
$$s' = \tfrac{1}{2}, \tfrac{3}{2}, \tfrac{5}{2}, \tfrac{7}{2}, \ldots, N/2, \qquad \text{for } N \text{ odd} \qquad (13\text{–}41')$$

As an example of the second equation, consider three electrons with quantum numbers $l_1 = 1, l_2 = 2, l_3 = 4$. For this case the minimum value of l' is $|\mathbf{l}_1 + \mathbf{l}_2 + \mathbf{l}_3|_{\min} = 1$, and the maximum value of l' is $l_1 + l_2 + l_3 = 7$. As another example, if $l_1 = 1$, $l_2 = 1$, $l_3 = 1$, then l' ranges from $|\mathbf{l}_1 + \mathbf{l}_2 + \mathbf{l}_3|_{\min} = 0$ to $l_1 + l_2 + l_3 = 3$. We shall soon see that, for configurations in which there are two or more electrons with the same n and l quantum numbers, the exclusion principle puts certain restrictions on the possible values of s', l', j' in addition to those indicated by equations (13–40) and (13–41).

In the Hartree approximation the eigenfunctions ψ_T describing the system of optically active electrons are labeled by the quantum numbers of n, l, m_l, m_s of each of the electrons. In this approximation, all the ψ_T associated with a particular configuration of these electrons are degenerate. In the next approximation, the residual Coulomb and spin-orbit interactions of the optically active electrons are treated in LS coupling as perturbations. These perturbations thoroughly mix the degenerate ψ_T to form an equal number of new eigenfunctions ψ'_T, which are certain linear combinations of the ψ_T.† Each of the ψ'_T describes a state in which the optically active electrons couple so as to yield a particular set of S', L', J', J'_z. Consequently the ψ'_T are labeled by specifying the configuration of the optically active electrons plus the quantum numbers s', l', j', m'_j. Since the energy of the atom depends in LS coupling on the values of S', L', J', as well as on the configuration, most of the degeneracy is removed. The eigenfunctions ψ'_T have only a trivial degeneracy with

† This is completely analogous to the mixing of degenerate one-electron eigenfunctions by a spin-orbit perturbation (cf. section 5, Chapter 11). As in that case, it will not be necessary for us to know the exact form of these linear combinations.

respect to the quantum number m'_j which specifies the spatial orientation of the total angular momentum vector. The energy levels of the atom are labeled in *LS* coupling by the configuration plus s', l', j', using the spectroscopic notation introduced previously. If we are not interested in the value of m'_j, the same notation can also be used to label the quantum states. For instance, the energy level and quantum states of an atom in which

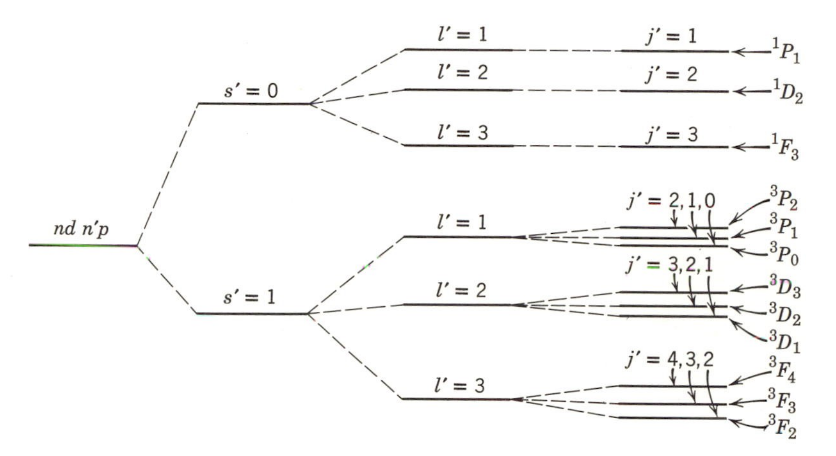

Figure 13–12. The splitting of the energy levels in a typical *LS* coupling configuration.

the two optically active electrons have the quantum numbers $n = 1$, $l = 0$ and $n = 2$, $l = 1$, and couple in such a way that $s' = 1$, $l' = 1$, $j' = 2$, are identified by the symbol $1s2p\ ^3P_2$. The first part of the symbol gives the configuration, and the last part gives the other quantum numbers according to the scheme described in section 6.

The splitting of the single degenerate energy level of a particular configuration, say the *nd n'p* configuration of an atom with two optically active electrons, due to the residual Coulomb and spin-orbit perturbations is illustrated in figure (13–12). For obvious reasons we cannot present explicit equations from which the energies of all these levels can be evaluated. However, we can write an equation which gives the j' dependence of the spin-orbit interaction energy. This is the effect which produces the *fine structure* splitting of the levels for $s' = 1$, and a given l', into triplets of levels. Consider equation (13–27), which is

$$\overline{\Delta E_{\mathbf{S}\cdot\mathbf{L}}} = \frac{\hbar^2}{4m^2c^2}\overline{\left(\frac{1}{r}\frac{dV(r)}{dr}\right)}[j(j+1) - l(l+1) - s(s+1)]$$

This equation was developed for an atom with one optically active electron, but the same equation written with primes gives the interaction energy of the total spin and orbital angular momentum vectors $\mathbf{S}'$ and $\mathbf{L}'$ of any atom in which these vectors can be defined, that is, any atom in which *LS* coupling is valid. Thus in *LS* coupling the total energy of the atom contains a term

$$\overline{\Delta E_{\mathbf{S}' \cdot \mathbf{L}'}} = K[j'(j'+1) - l'(l'+1) - s'(s'+1)] \tag{13–42}$$

where the quantity

$$K = \frac{\hbar^2}{4m^2c^2}\overline{\left(\frac{1}{r}\frac{dV(r)}{dr}\right)}$$

has the same value for all energy levels of a configuration with common values of s' and l'. Let us calculate from this equation the separation in

TABLE (13–7)

Fine Structure Splittings in the Calcium Atom

Configuration	Levels	Separation	Levels	Separation	Ratio Exp.	Ratio Theo.
$3d3d$	${}^3P_1, {}^3P_0$	16.7×10^{-4} ev	${}^3P_2, {}^3P_1$	33.3×10^{-4} ev	1.99	2/1
$4s4p$	${}^3P_1, {}^3P_0$	64.9×10^{-4} ev	${}^3P_2, {}^3P_1$	131.2×10^{-4} ev	2.02	2/1
$4s3d$	${}^3D_2, {}^3D_1$	16.9×10^{-4} ev	${}^3D_3, {}^3D_2$	26.9×10^{-4} ev	1.59	3/2
$3d4p$	${}^3D_2, {}^3D_1$	33.1×10^{-4} ev	${}^3D_3, {}^3D_2$	49.6×10^{-4} ev	1.50	3/2

energy between the adjacent levels of a configuration with common values of s' and l', which are called energy levels of the same *multiplet*. If the quantum number associated with the level of lower energy is j', the quantum number associated with the level of higher energy is $j' + 1$, and the separation is

$$K[(j'+1)(j'+2) - l'(l'+1) - s'(s'+1)] - K[j'(j'+1) - l'(l'+1) - s'(s'+1)] = K[(j'+1)(j'+2) - j'(j'+1)] = 2K(j'+1) \tag{13–43}$$

The separation in energy of adjacent levels of the same multiplet is proportional to the total angular momentum quantum number of the level of higher energy. This is called the *Lande interval rule*. In table (13–7) we present some energy separations between adjacent levels of multiplets which are observed in the ^{20}Ca atom, and compare the ratios of these separations with the ratios predicted from the Lande interval rule. Since we compare ratios, the quantity $2K$ cancels out. The excellent agreement between the experimentally observed and theoretically predicted ratios, which is typical for atoms of small and intermediate Z, provides convincing evidence of *LS* coupling in these atoms.

In discussing the coupling of the angular momenta of several optically active electrons, we have not yet considered the effect of the exclusion principle. But, for configurations in which two or more of these electrons have *the same n and l quantum numbers* (are in the same subshell), the exclusion principle cannot be ignored because it imposes important restrictions on the way in which their angular momenta can couple. To determine these restrictions, we consider the Hartree approximation in which n, l, m_l, m_s are good quantum numbers for the optically active electrons. Then the exclusion principle states simply that no two electrons can have the same set of four quantum numbers. We shall see that this limits the possible values of the quantum numbers s', m_s', l', m_l', j', m_j' which determine the magnitudes and z components of $\mathbf{S}'$, $\mathbf{L}'$, $\mathbf{J}'$, and which also are all good quantum numbers in the Hartree approximation. Now, although in LS coupling the z components of $\mathbf{S}'$ and $\mathbf{L}'$ are changed by the application of the residual Coulomb and spin-orbit perturbations, S', L', J', J_z' are not changed.† Consequently, any restrictions we find in the Hartree approximation concerning the quantum numbers s', l', j', m_j' will be equally applicable in LS coupling.

Consider first a configuration ns^2, in which two s electrons have the same quantum number n and so are in the same s subshell. Since for $l = 0$ it is only possible that $m_l = 0$, these two electrons actually have the same values of the three quantum numbers n, l, m_l. To satisfy the exclusion principle, their m_s quantum numbers must be different; one electron must have $m_s = +\frac{1}{2}$ and the other must have $m_s = -\frac{1}{2}$. According to (12–23), (12–24), and (11–21), m_s' is equal to the sum of the m_s quantum numbers of the two electrons. Thus the only possible value for this quantum number is $m_s' = +\frac{1}{2} - \frac{1}{2} = 0$. Similarly, m_l' is equal to the sum of the m_l quantum numbers of the two electrons, and so has only the single possible value $m_l' = 0 + 0 = 0$. According to (11–43), $m_j' = m_l' + m_s'$, so the only possible value of this quantum number is $m_j' = 0 + 0 = 0$. Since the three quantum numbers m_s', m_l', m_j' can have only the single value 0, it is apparent from equations such as (12–24) that the quantum numbers s', l', j' also can only have the single value 0. Thus the two electrons can couple to form only a single quantum state. This is the state $^1S_0$, in which the spin vectors of the two electrons are "antiparallel." If it were not for the exclusion principle the m_s quantum numbers of the two electrons could be equal, and there would be the three additional $^3S_1$ states in which the spin vectors are "parallel." These conclusions should be

† In the presence of the perturbations, S_z' and L_z' have no definite values and m_s' and m_l' are not good quantum numbers. However, since S', L', J', J_z' have definite values whether or not the perturbations are present, it is apparent that the application of the perturbations cannot change their values.

familiar to the reader because we have already used them in the case of the $1s^2$ ground state configuration of the ^{2}He atom.

Consider next the configuration np^2, in which two optically active electrons are in the same p subshell. This case is more complicated, but it can be handled by the same procedures used above. In table (13–8) we list all the possible sets of values of m_l and m_s for the two electrons which

TABLE (13–8)

Possible Quantum Numbers for an np^2 Configuration

Number	m_l	m_s	m_l	m_s	m'_s	m'_l	m'_j
1	+1	$+\frac{1}{2}$	+1	$-\frac{1}{2}$	0	+2	+2
2	+1	$+\frac{1}{2}$	0	$+\frac{1}{2}$	+1	+1	+2
3	+1	$+\frac{1}{2}$	0	$-\frac{1}{2}$	0	+1	+1
4	+1	$+\frac{1}{2}$	−1	$+\frac{1}{2}$	+1	0	+1
5	+1	$+\frac{1}{2}$	−1	$-\frac{1}{2}$	0	0	0
6	+1	$-\frac{1}{2}$	0	$-\frac{1}{2}$	−1	+1	0
7	+1	$-\frac{1}{2}$	−1	$+\frac{1}{2}$	0	0	0
8	+1	$-\frac{1}{2}$	−1	$-\frac{1}{2}$	−1	0	−1
9	0	$+\frac{1}{2}$	+1	$-\frac{1}{2}$	0	+1	+1
10	0	$+\frac{1}{2}$	0	$-\frac{1}{2}$	0	0	0
11	0	$+\frac{1}{2}$	−1	$+\frac{1}{2}$	+1	−1	0
12	0	$+\frac{1}{2}$	−1	$-\frac{1}{2}$	0	−1	−1
13	−1	$+\frac{1}{2}$	0	$-\frac{1}{2}$	0	−1	−1
14	−1	$+\frac{1}{2}$	−1	$-\frac{1}{2}$	0	−2	−2
15	−1	$-\frac{1}{2}$	0	$-\frac{1}{2}$	−1	−1	−2

do not violate the exclusion principle. For each set we also list the corresponding values of the quantum numbers m'_s, m'_l, m'_j. There are 15 different sets of m_l and m_s quantum numbers for the two electrons which satisfy the exclusion principle, and a number of others, such as $m_l = +1$, $m_s = +\frac{1}{2}$; $m_l = +1$, $m_s = +\frac{1}{2}$, which are ruled out because they violate this principle. The problem now is to identify the allowed quantum states, specified in Table (13–8) in terms of the quantum numbers m'_s, m'_l, m'_j, with the specification of these states in terms of the quantum numbers s', l', j'. First we use equations (13–40), with $l_1 = l_2 = 1$, and find that the various possible combinations of s', l', j' values are only those which are specified in the following list by the spectroscopic notation: $^1S_0$, $^1P_1$, $^1D_2$, $^3S_1$, $^3P_0$, $^3P_1$, $^3P_2$, $^3D_1$, $^3D_2$, $^3D_3$. The $^3D_3$ quantum states are immediately ruled out because, for these states with $j' = 3$, equation (13–39) shows that there would necessarily be m'_j values of $+3$ and -3, but there are none in the table. Since there are no $^3D_3$ states, there can be no $^3D_2$ or $^3D_1$ states. All these states correspond to $\mathbf{S}'$ and $\mathbf{L}'$ vectors of the same magnitude; they are states of the same multiplet, and they stand or fall together. Now, entry number 1 in table (13–8) says that there must be quantum states with $s' \geqslant 0$ and $l' \geqslant 2$, since $m_s' = -s', \ldots, s'$ and $m_l' = -l', \ldots, l'$. These requirements can be satisfied only by the quantum states ${}^1D_2$. There are five such states corresponding to the five values $m_j' = -2, -1, 0, 1, 2$. Entry number 2 says that there must be states with $s' \geqslant 1$ and $l' \geqslant 1$. This requires the presence of the states ${}^3P_0, {}^3P_1, {}^3P_2$. For ${}^3P_0$ there is one quantum state corresponding to $m_j' = 0$. For ${}^3P_1$ there are three states corresponding to $m_j' = -1, 0, 1$. For ${}^3P_2$ there are five corresponding to $m_j' = -2, -1, 0, 1, 2$, The number of states we have identified so far is $5 + 1 + 3 + 5 = 14$. Only a single state is left, and this must be a state with $m_j' = 0$ because all the other m_j' values of the table have been used. It is clear then that this must be the single quantum state ${}^1S_0$. Thus, in the Hartree approximation the possible quantum states for two electrons with the configuration np^2 are only those associated with the symbols ${}^1S_0$, ${}^1D_2$, ${}^3P_{0,1,2}$. Since these restrictions are expressed in terms of the quantum numbers s', l', j', they are equally valid in *LS* coupling.

In table (13–9) we list the quantum states which are allowed by the exclusion principle for configurations containing several electrons in the same subshell. Each symbol gives the values of the quantum numbers s' and l' of an allowed multiplet of quantum states. The possible values of j' and m_j' for the states of any multiplet can be determined in terms of s' and l' from equations (13–39) and (13–41). Entries are given for configurations ranging from no electrons in the subshell up to the maximum number of electrons consistent with the exclusion principle. For no electrons, $s' = l' = j' = 0$, which is described by the symbol ${}^1S_0$. For one electron in any subshell, $l' = l$ and, since $s' = \frac{1}{2}$, the allowed states comprise a doublet. The allowed states for the other configurations are determined by calculations similar to those we used in determining the allowed states of the ns^2 and np^2 configurations, or by more elegant calculations based on the mathematical theory of groups.

It is particularly interesting to note the symmetries in table (13–9) about the half-filled subshell configurations. The number of states is greatest for this configuration, and the states for a configuration in which a subshell is filled except for k electrons are exactly the same as the states for the configuration in which there are just k electrons in the subshell. This is a striking demonstration of the effect of the exclusion principle because, if it were not for this principle, the number of states would increase monotonically as the number of electrons in the subshell increased. There are several interesting consequences of this property. The most important is that when a subshell is completely filled the only allowed

TABLE (13–9)

Possible Quantum States for Configurations Containing Several Electrons in the Same Subshell

ns^0	^1S					
ns^1		^2S				
ns^2	^1S					
np^0	^1S					
np^1		^2P				
np^2	$^1S,\ ^1D$		^3P			
np^3		$^2P,\ ^2D$		^4S		
np^4	$^1S,\ ^1D$		^3P			
np^5		^2P				
np^6	^1S					
nd^0	^1S					
nd^1		^2D				
nd^2	$^1S,\ ^1D,\ ^1G$		$^3P,\ ^3F$			
nd^3		$^2D,\ ^2P,\ ^2D,\ ^2F,\ ^2G,\ ^2H$		$^4P,\ ^4F$		
nd^4	$^1S,\ ^1D,\ ^1G,\ ^1S,\ ^1D,\ ^1G,\ ^1H,\ ^1I$		$^3P,\ ^3F,\ ^3P,\ ^3D,\ ^3F,\ ^3G,\ ^3H$		^5D	
nd^5		$^2D,^2P,^2D,^2F,^2G,^2H,^2S,^2D,^2F,^2G,^2I$		$^4P,\ ^4F,\ ^4D,\ ^4G$		^6S
nd^6	$^1S,\ ^1D,\ ^1G,\ ^1S,\ ^1D,\ ^1G,\ ^1H,\ ^1I$		$^3P,\ ^3F,\ ^3P,\ ^3D,\ ^3F,\ ^3G,\ ^3H$		^5D	
nd^7		$^2D,\ ^2P,\ ^2D,\ ^2F,\ ^2G,\ ^2H$		$^4P,\ ^4F$		
nd^8	$^1S,\ ^1D,\ ^1G$		$^3P,\ ^3F$			
nd^9		^2D				
nd^{10}	^1S					

state is the same as when the subshell is completely empty. This is the quantum state $^1S_0$, in which S', L', and J' all vanish. We proved this for the s subshell, and it is easy to prove for any other subshell. Let us consider the np^6 configuration and construct table (13–10), which is

TABLE (13–10)

Possible Quantum Numbers for an np^6 Configuration

Number	m_l	m_s	m_l	m_s	m_l	m_s	m_l	m_s	m_l	m_s	m_l	m_s	m_s'	m_l'	m_j'
1	$+1$	$+\frac{1}{2}$	$+1$	$-\frac{1}{2}$	0	$+\frac{1}{2}$	0	$-\frac{1}{2}$	-1	$+\frac{1}{2}$	-1	$-\frac{1}{2}$	0	0	0

similar to the table we used in studying the np^2 configuration. The table has only a single entry, and this entry is obviously the single state $^1S_0$. The conclusion that the total spin and orbital angular momentum of a completely filled subshell are zero is confirmed by analysis of the optical spectra of noble gas atoms. Furthermore, if there is no net spin or orbital angular momentum, there can be no net magnetic moment, so this conclusion can also be confirmed by Stern-Gerlach experiments on noble gas atoms—or by the fact that the magnetic moment of ^{47}Ag, which in the ground state contains a single 5s electron outside a set of completely filled subshells, is measured in these experiments to be the same as the magnetic moment of ^{1}H, which in the ground state contains a single 1s electron.

We are now in a position to understand the various features of a diagram showing the optically excited energy levels of a typical atom of small or

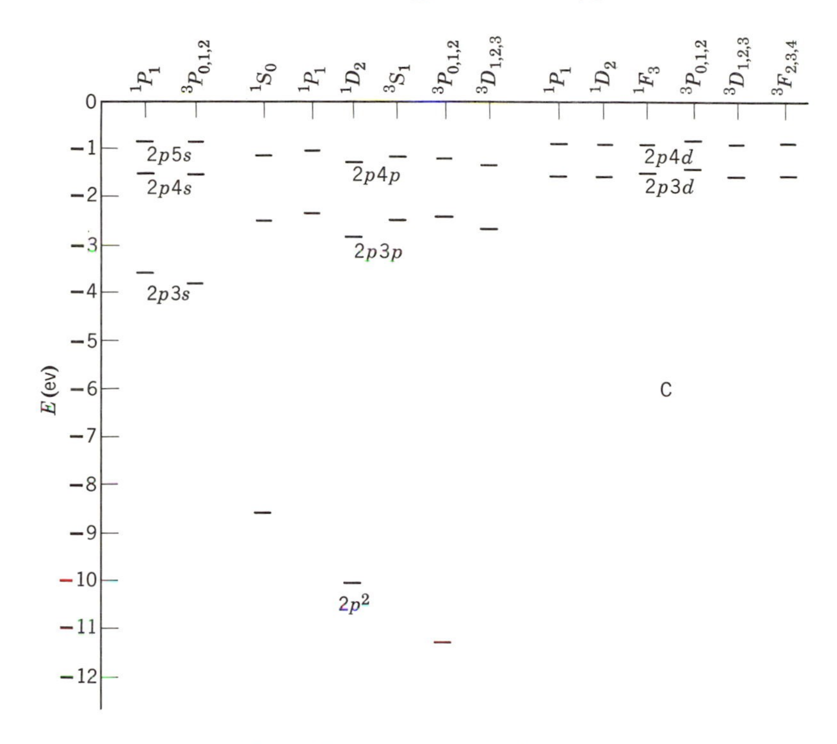

Figure 13–13. Some energy levels of the carbon atom.

intermediate Z. Consider the energy level diagram for ^{6}C presented in figure (13–13). The ground state of this atom has the configuration $1s^2 2s^2 2p^2$, so there are two p electrons which are optically active. The zero of the energy scale is so defined that the magnitude of the total energy of the atom in its ground state is equal to the energy required to ionize singly the atom. Consequently, the diagram is directly comparable with energy level diagrams for alkali atoms and ^{1}H, in which the zero of energy is defined in the same way. The energy levels are labeled by the configuration of the two optically active electrons, and by the spectroscopic symbol specifying s', l', j'. The fine structure splitting according to the value of j' is not shown because it is much too small to be seen on the scale of the diagram. Consider first the average energy of the levels of the various configurations. In the configuration of lowest average energy, $2p^2$, both electrons remain in the same subshell that they are in for the ground state of the atom. In the other configurations, one electron remains in that subshell and one is in a subshell of higher energy. Note that the average energies of the configurations depend on the n and l quantum numbers of the electron in the higher energy subshell in essentially the same way as if this electron were the single optically active electron in an alkali atom. The configuration of lowest average energy provides an example of an np^2 configuration for which the exclusion principle allows only the ${}^1S_0$, ${}^1D_2$, and ${}^3P_{0,1,2}$ states. The ${}^3P_{0,1,2}$ states are of lower energy than the ${}^1S_0$ and ${}^1D_2$ states because they correspond to a larger value of s', and the ${}^1D_2$ states are of lower energy than the ${}^1S_0$ state because they correspond to a larger value of l'. Note that the s' dependence is stronger than the l' dependence. It is almost always found that the energy associated with the residual Coulomb interaction coupling of the spin angular momenta is larger than the energy associated with the residual Coulomb interaction coupling of the orbital angular momenta. Of the three closely spaced energy levels for the ${}^3P_{0,1,2}$ states that would be seen on a larger diagram, the one for the ${}^3P_0$ state is of lowest energy because it corresponds to the smallest value of j'. Thus the ground state of the atom is the state $2p^2\,{}^3P_0$. The exclusion principle does not apply to the configurations other than the one of lowest average energy, because in these configurations the electrons are in different subshells. Consequently, all the states predicted by equation (13–40) are found. In the $2p3s$ configuration the energy level corresponding to maximum s' is lowest in energy, just as in the $2p^2$ configuration. Deviations from this rule, and from the rule that the maximum l' gives the lowest energy, are seen in the configurations of higher average energy, but there are no deviations from the rule that the minimum j' gives the minimum energy. Not shown in the diagram are a few energy levels which have been identified as belonging to the configuration $2s2p^3$.

The optical spectrum of the ^{6}C atom can be constructed from its energy level diagram by evaluating the energy and frequency of quanta emitted in all possible transitions which do not violate the following selection rules:

1. Transitions can occur only between configurations which differ in the n and l quantum numbers of a *single* electron. This means that two or more electrons cannot simultaneously make transitions between subshells.

2. Furthermore, the l quantum number which is different in the two configurations must differ by one unit:

$$\Delta l = \pm 1 \tag{13–44}$$

This is the same rule as (13–31).

3. Transitions can occur only between states in these configurations for which the differences in the s', l', j' quantum numbers are

$$\begin{aligned} \Delta s' &= 0 \\ \Delta l' &= 0, \pm 1 \\ \Delta j' &= 0, \pm 1 \quad (\text{but not } j' = 0 \rightarrow j' = 0) \end{aligned} \tag{13–45}$$

The first equation prohibits transitions between singlet and triplet states, and vice versa. Nevertheless, transitions are observed between the $2p^2\,{}^1D_2$ states and the $2p^2\,{}^3P_{0,1,2}$ states because all excitations of atoms to singlet states eventually lead to the population of the $2p^2\,{}^1D_2$ states and, when these states are highly populated, the total number of transitions per second to the $2p^2\,{}^3P_{0,1,2}$ states becomes appreciable even though the probability that any single atom will make this transition is extremely small because it violates the selection rule. The second of the equations above is similar to (13–31) but not quite so restrictive. The third equation is the same rule as (13–30). All these selection rules apply in any atom which is an example of *LS* coupling.

9. *JJ* Coupling

Atoms of small Z, such as ^{6}C, are known to be good examples of *LS* coupling because the energy levels are grouped in the characteristic *LS* multiplets with a j' dependence of the energy that is very small compared to

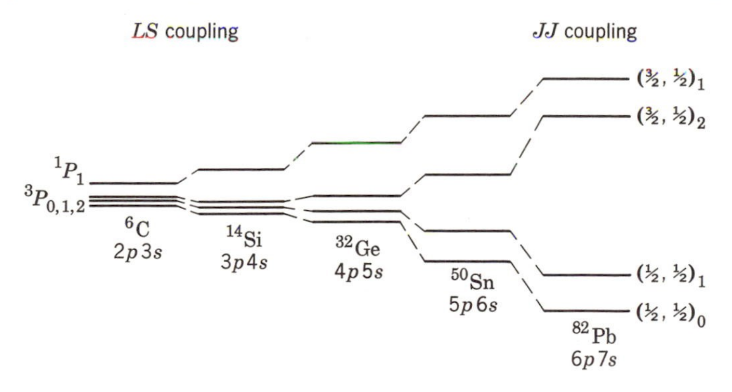

Figure 13–14. Illustrating the transition from *LS* to *JJ* coupling. From H. E. White, *Introduction to Atomic Spectra*, McGraw-Hill Book Co., New York, 1934.

the dependence on s' and l', and because the Lande interval rule (13–43) is accurately obeyed. However, with increasing Z the j' dependence rapidly becomes larger, and the Lande interval rule is obeyed less accurately. This is shown schematically in figure (13–14), which indicates the disposition of energy levels in the $2p3s$ configuration of ^{6}C, and of the levels in corresponding configurations of higher Z atoms that are in the same column of the periodic table. Significant departures from LS coupling are often seen in atoms of large Z because, as we saw in section 6, the spin-orbit interactions become large in these atoms. In fact, for some of the largest Z atoms the perturbation due to the spin-orbit interactions becomes appreciably larger than the perturbation due to the residual Coulomb interaction. These atoms are said to be examples of *JJ coupling*. In coupling of this type, we assume first that the dominant perturbation due to the spin-orbit interactions causes the spin and orbital angular momentum vectors of each optically active electron to couple, forming a total angular momentum vector for that electron. Thus, for the ith electron $\mathbf{L}_i$ and $\mathbf{S}_i$ couple to form

$$\mathbf{J}_i = \mathbf{L}_i + \mathbf{S}_i \tag{13–46}$$

The magnitude of this vector is quantized according to the usual equation

$$J_i = \sqrt{j_i(j_i + 1)}\,\hbar \tag{13–47}$$

The energy of the atom then depends on the quantum numbers j_i of all the electrons and is smallest when these quantum numbers have their smallest possible values. Next, the weaker perturbation due to the Coulomb interactions is taken into account. This perturbation couples the individual total angular momentum vectors to form the total angular momentum vector

$$\mathbf{J}' = \mathbf{J}_1 + \mathbf{J}_2 + \cdots + \mathbf{J}_i + \cdots \tag{13–48}$$

which obeys the same quantization equations (13–37), (13–38), and (13–39), as in LS coupling. The energy of the atom then also depends, but to a lesser extent, on the quantum number j'. However, it is not possible to make a general statement about the values of the j' which yield the smallest energy. The possible values of the j_i and of j' are given by obvious adaptations of equations (13–41). These quantum numbers must be used to identify the quantum states and energy levels.† In figure (13–14), the four levels of the $6p7s$ configuration of ^{82}Pb are labeled with the spectroscopic notation for these quantum numbers. This atom is a fairly good example of JJ coupling because the $(\frac{3}{2}, \frac{1}{2})_{1,2}$ levels, for which $j_1 = \frac{3}{2}$ and $j_2 = \frac{1}{2}$, are of nearly the same energy and are well separated from the $(\frac{1}{2}, \frac{1}{2})_{1,0}$ levels, for which $j_1 = j_2 = \frac{1}{2}$. The subscript gives the value of j'. Note that the smallest j' gives the smallest energy in one pair of levels and the largest energy in the other pair. The exclusion principle operates in

† Of course, s' and l' are not good quantum numbers in an atom which is an example of JJ coupling.

JJ coupling, for configurations in which two or more electrons are in the same subshell, in much the same way as in *LS* coupling. Selection rules for *JJ* coupling are the same as those for *LS* coupling except that the restrictions on $\Delta s'$ and $\Delta l'$ are replaced by the restriction

$$\Delta j_i = 0, \pm 1 \tag{13–49}$$

for the electron which makes a transition between subshells, and

$$\Delta j_i = 0 \tag{13–49'}$$

for all the other electrons.

We do not give a detailed discussion of *JJ* coupling here because examples of it are not very frequently found in the behavior of the optically active electrons in atoms. Most atoms are either good or fair examples of *LS* coupling However, examples of *JJ* coupling are very often found in the behavior of the protons and neutrons in nuclei. We shall study this in section 8, Chapter 16, which concerns the *nuclear shell model.* The model explains many properties of nuclei by assuming that the motion of the protons and neutrons is the same as the motion of the electrons in an atom, except that the net potential $V(r)$ has a different r dependence, and that these particles experience a very large spin-orbit interaction.

10. The Zeeman Effect

In 1896 it was observed by Zeeman that placing an atom in a magnetic field splits each of the spectral lines which it emits into several components. This is called the *Zeeman effect.* For fields up to a few thousand gauss, the splitting is proportional to the strength of the magnetic field and is small compared to the fine structure splitting of the spectral lines of the atom. This means that in the presence of a magnetic field the energy levels of the atom which are involved in the emission of the spectral lines must be split into several components. In certain special cases it was possible to understand the *Zeeman splitting* of the energy levels in terms of a classical theory developed by Lorentz. But, in the general case, even a qualitative explanation of the observed splittings could not be given until the development of quantum mechanics and the introduction of electron spin.

In terms of the modern theory the Zeeman effect is easy to understand. Except when it is in a $^1S_0$ state, an atom will have a net magnetic moment $\boldsymbol{\mu}$ due to the spin and orbital magnetic moments of its optically active electrons.† According to equation (11–15), when this magnetic moment is

† The other electrons are in completely filled subshells which do not contribute to $\boldsymbol{\mu}$.

in the external magnetic field $\mathbf{H}$ it will have an orientational potential energy

$$\Delta E = -\boldsymbol{\mu} \cdot \mathbf{H}$$

or

$$\Delta E = -\mu_{\mathbf{H}} H \tag{13–50}$$

where $\mu_{\mathbf{H}}$ is the component of $\boldsymbol{\mu}$ along the direction of $\mathbf{H}$. Each energy level will be split into several energy levels corresponding to the several possible values of $\mu_{\mathbf{H}}$. To calculate these values, we evaluate $\boldsymbol{\mu}$ by using (11–10) and (11–22) for each optically active electron. That is, we take

$$\boldsymbol{\mu} = \sum_i -\frac{g_l \mu_B}{\hbar} \mathbf{L}_i + \sum_i -\frac{g_s \mu_B}{\hbar} \mathbf{S}_i$$

$$\boldsymbol{\mu} = -\frac{\mu_B}{\hbar}\left[\sum_i \mathbf{L}_i + 2\sum_i S_i\right]$$

where the summation goes over all the optically active electrons, and where (11–9) and (11–26) are used to evaluate g_l and g_s. Now let us assume that the atom in question is an example of LS coupling. Then

$$\boldsymbol{\mu} = -\frac{\mu_B}{\hbar}[\mathbf{L}' + 2\mathbf{S}'] \tag{13–51}$$

Note that $\boldsymbol{\mu}$ is not antiparallel to

$$\mathbf{J}' = \mathbf{L}' + \mathbf{S}'$$

unless $\mathbf{L}' = 0$ or $\mathbf{S}' = 0$, because the orbital and spin g factors are different.

In LS coupling, $\mathbf{L}'$ and $\mathbf{S}'$ precess about $\mathbf{J}'$ because of the Larmor precession of $\mathbf{S}'$ in the atomic magnetic field associated with $\mathbf{L}'$, with a precessional frequency ω_L which is proportional to the strength of the atomic magnetic field (cf. 11–29). It is apparent from equation (13–51) that $\boldsymbol{\mu}$ must also precess about $\mathbf{J}'$ with the same precessional frequency. In the presence of the external magnetic field $\mathbf{H}$, there will also be a precession of $\boldsymbol{\mu}$ about the direction of this field with a precessional frequency which is proportional to its strength. Thus the motion of $\boldsymbol{\mu}$ is quite complicated. However, if the external field is weak compared to the atomic field, the precession of $\boldsymbol{\mu}$ about the external field will be slow compared to its precession about $\mathbf{J}'$, and it is not so complicated to evaluate the orientational potential energy ΔE. We have seen in section 4, Chapter 11, that the strength of an atomic magnetic field is typically of the order of 10^4 gauss.

Consider, then, the case of an external magnetic field H which is weak compared to $\sim 10^4$ gauss. Since $\boldsymbol{\mu}$ precesses much more rapidly about $\mathbf{J}'$

than about **H**, we may evaluate $\mu_{\mathbf{H}}$ by finding $\mu_{\mathbf{J}'}$, which is the average component of $\boldsymbol{\mu}$ in the direction of $\mathbf{J}'$, and then taking the component of $\mu_{\mathbf{J}'}$ in the direction of **H**. That is,

$$\mu_{\mathbf{J}'} = \mu \frac{\boldsymbol{\mu} \cdot \mathbf{J}'}{\mu J'} = -\frac{\mu_B}{\hbar} \frac{(\mathbf{L}' + 2\mathbf{S}') \cdot (\mathbf{L}' + \mathbf{S}')}{J'}$$

and

$$\mu_{\mathbf{H}} = \mu_{\mathbf{J}'} \frac{\mathbf{J}' \cdot \mathbf{H}}{J'H} = \mu_{\mathbf{J}'} \frac{J'_z}{J'} = -\frac{\mu_B}{\hbar} \frac{(\mathbf{L}' + 2\mathbf{S}') \cdot (\mathbf{L}' + \mathbf{S}')J'_z}{J'^2}$$

where we have chosen the z axis to be in the direction of H. Evaluating the dot product gives

$$\mu_{\mathbf{H}} = -\frac{\mu_B}{\hbar} (L'^2 + 2S'^2 + 3\mathbf{L}' \cdot \mathbf{S}') \frac{J'_z}{J'^2}$$

Writing equation (11–48) with primes, we have

$$3\mathbf{L}' \cdot \mathbf{S}' = \tfrac{3}{2}(J'^2 - L'^2 - S'^2)$$

So

$$\mu_{\mathbf{H}} = -\frac{\mu_B}{\hbar} (L'^2 + 2S'^2 + \tfrac{3}{2}J'^2 - \tfrac{3}{2}L'^2 - \tfrac{3}{2}S'^2) \frac{J'_z}{J'^2}$$

$$\mu_{\mathbf{H}} = -\frac{\mu_B}{\hbar} \frac{(3J'^2 + S'^2 - L'^2)}{2J'^2} J'_z$$

Then according to equation (13–50), the orientational potential energy is

$$\Delta E = -\mu_{\mathbf{H}} H = \frac{\mu_B H}{\hbar} \frac{(3J'^2 + S'^2 - L'^2)}{2J'^2} J'_z \qquad (13\text{–}52)$$

The value of this term is of the order of $\mu_B H$, which is small compared to all the other effects that contribute to the total energy of the atom. For instance, the spin-orbit interaction energy is large compared to this term because it is of the order of μ_B times the strength of the atomic magnetic field, which is large compared to H. Consequently, we may use perturbation theory to calculate $\overline{\Delta E}$, the contribution of ΔE to the total energy of the atom. Using the LS coupling eigenfunctions ψ'_T, which are labeled by the quantum numbers s', l', j', m'_j, we have†

$$\overline{\Delta E} = \int \psi'^*_T \, \Delta E_{\text{op}} \psi'_T \, d\tau \qquad (13\text{–}53)$$

† Although the eigenfunctions ψ'_T are degenerate with respect to the quantum number m'_j, we may use the ordinary perturbation theory expression (13–53) to evaluate $\overline{\Delta E}$ because J'_z has a definite value whether or not the perturbation is present, so the perturbation cannot produce a large mixing of the eigenfunctions.

where the integration is taken over all the coordinates of all the optically active electrons, and where ΔE_{op} is the operator formed from ΔE by replacing S'^2, L'^2, J'^2, J'_z with their associated operators. Since ψ'_T is an eigenfunction of all these operators with eigenvalues $s'(s'+1)\hbar^2$, $l'(l'+1)\hbar^2$, $j'(j'+1)\hbar^2$, $m'_j\hbar$, respectively, equations (13–52) and (13–53) immediately give

$$\overline{\Delta E} = \frac{\mu_B H}{\hbar} \frac{[3j'(j'+1)\hbar^2 + s'(s'+1)\hbar^2 - l'(l'+1)\hbar^2]}{2j'(j'+1)\hbar^2} m'_j\hbar$$

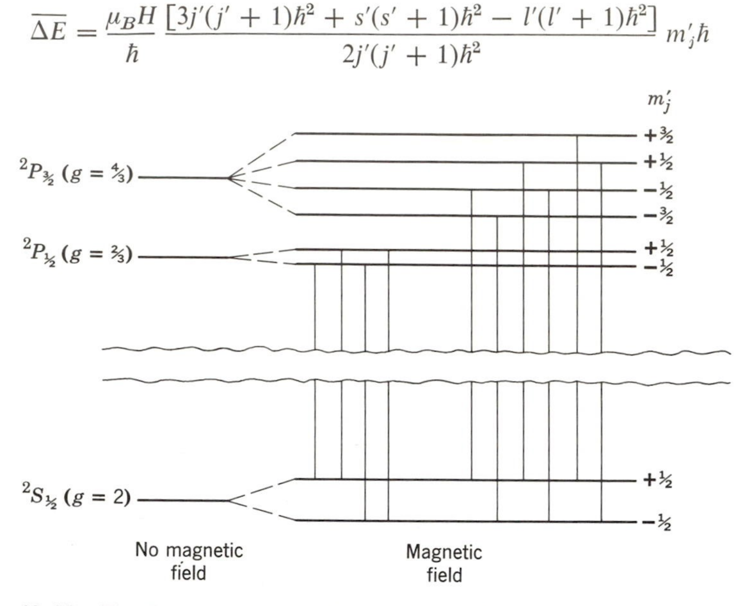

Figure 13–15. The Zeeman splitting of a typical pair of energy levels, and the possible transitions between these levels.

It is convenient to write this as

$$\overline{\Delta E} = \mu_B H g m'_j \tag{13–54}$$

where

$$g \equiv 1 + \frac{j'(j'+1) + s'(s'+1) - l'(l'+1)}{2j'(j'+1)} \tag{13–54'}$$

The quantity g is called the *Lande g factor*. Note that $g = 1 = g_l$ for $s' = 0$, and $g = 2 = g_s$ for $l' = 0$, as would be expected. From equation (13–54), we see that in the presence of an external field of strength H each energy level will be split into $2j' + 1$ components, one for each of the values of m'_j, and that the magnitude of the splitting will be different for levels with different Lande g factors.

We illustrate equation (13–54) by the energy level diagram of figure (13–15), which shows the splitting of $^2P_{1/2,3/2}$ and $^2S_{1/2}$ energy levels in a weak magnetic field. Also shown are the possible transitions between the $^2P_{1/2,3/2}$ and the $^2S_{1/2}$ states (which, of course, are in a different configuration). Transitions do not occur if they violate the selection rule

$$\Delta m_j' = 0, \pm 1 \quad (\text{but not } m_j' = 0 \rightarrow m_j' = 0 \text{ if } \Delta j' = 0) \qquad (13\text{–}55)$$

Even with this simplification, the Zeeman effect splits each of the spectral lines into a pattern of components which can be quite complicated for lines arising from transitions to or from states of large j'. However, all spectral lines arising from transitions between singlet states are split into a simple pattern of two components symmetrically disposed about a third component which has the same frequency as the single zero field line, because $s' = 0$ for singlet states so all the g factors have the same value, $g = 1$. By drawing a diagram similar to the one in figure (13–15) for transitions between singlet states, the reader should show for himself that all the transitions which satisfy (13–55) lead to only three spectral lines, even though there may be more than three possible transitions, because there are only three possible energy differences for these transitions. We mentioned before that in certain special cases it was possible to explain the Zeeman effect in terms of a classical theory. The reader can probably guess that these were the cases in which the spectral lines are split into three components, because these cases correspond to transitions between states in which the non-classical phenomena of spin does not play its usual role as $s' = 0$. These cases are called the *normal Zeeman effect*; the other, and more general, cases are called the *anomalous Zeeman effect*. This terminology was introduced long before the modern theory provided a complete understanding of the Zeeman effect and, from the point of view of this theory, it is not very appropriate.

An external magnetic field **H**, which is weak compared to the atomic magnetic fields that couple **S**′ and **L**′ to form **J**′, cannot disturb this coupling and only causes a relatively slow precession of **J**′ about the direction of **H**. However, if **H** is strong compared to the atomic magnetic fields, it overpowers these fields and destroys the coupling of **S**′ to **L**′. In this case **S**′ and **L**′ precess independently about the direction of **H**. This is the so-called *Paschen-Bach effect*, which is observed for external fields that are comparable to or larger than $\sim 10^4$ gauss. In this case,

$$\boldsymbol{\mu} = -\frac{\mu_B}{\hbar}(\mathbf{L}' + 2\mathbf{S}')$$

and

$$\mu_{\mathbf{H}} = -\frac{\mu_B}{\hbar}(L_z' + 2S_z')$$

where we have chosen the z axis in the direction of **H**. Then

$$\overline{\Delta E} = \frac{\mu_B H}{\hbar} (L'_z + 2S'_z)$$

From this we immediately obtain

$$\overline{\Delta E} = \mu_B H(m'_l + 2m'_s) \tag{13–56}$$

Of course, m'_l and m'_s are good quantum numbers for an atom in a strong external field because L'_z and S'_z have definite values. It is observed that the selection rules for these two quantum numbers are

$$\begin{aligned} \Delta m'_s &= 0 \\ \Delta m'_l &= 0, \pm 1 \end{aligned} \tag{13–57}$$

All spectral lines are split by the Paschen-Bach effect into three components, just as in the normal Zeeman effect.

The Zeeman effect plays an important role in experimental spectroscopy. By analyzing the Zeeman splitting of the spectral lines of an atom, the spectroscopist determines the Zeeman splitting of the energy levels of the atom. This can be used to identify the quantum number j' of each level because $2j' + 1$ is equal to the number of components into which the level is split. Furthermore, the magnitude of the splitting between any two components gives the value of $\mu_B H g$ and, μ_B and H being known, this gives the value of g for the energy level. Since the value of g depends on s', l', j', it can be used to confirm the assignment of these quantum numbers. The quantum number s' is assigned by counting the number of levels of the multiplet of which the level in question is a member. This number is $2s' + 1$ (except that when $s' > l'$ it is $2l' + 1$). The quantum number l' for the levels of a multiplet is assigned, after assigning s', by analyzing the separations between adjacent levels in terms of the Lande interval rule (13–43), and finding the value of l' which, in consideration of the restrictions imposed by the third of equations (13–41), leads to the values of j' that give the observed separations.† This procedure also evidently leads to the assignment of the quantum numbers j'. The quantum numbers l' for all the levels of a particular configuration being known (the levels can be grouped into configurations by the similarity of their energies), the l quantum numbers of that configuration are identified by using the second of equations (13–41). Identification of the n quantum numbers associated with the various l quantum numbers is not difficult, if the n quantum numbers of the ground state configuration are known,

† This is applicable only in the usual case in which the atom is an example of LS coupling.

by making use of the fact that the energy of the subshells with common values of l increases monotonically with increasing n. The identification of the n quantum numbers of the ground state configurations of the atoms is based on the same fact.

11. Hyperfine Structure

When an interferometer is used to measure the spectrum of a typical atom with extremely high resolution,† many of the spectral lines are found to consist of several components. The separation between these components is typically three orders of magnitude smaller than the separation characteristic of the fine structure of spectral lines. These very closely spaced components constitute the *hyperfine structure* of the spectral lines. They imply the existence of a *hyperfine splitting* of the energy levels of the atom.‡

Hyperfine structure was discovered in 1891 by Michelson. It was first explained in 1924 by Pauli as a result of the orientational potential energy of a magnetic dipole moment associated with the atomic nucleus, in the magnetic field associated with the motion of the atomic electrons. The *nuclear magnetic dipole moment* $\boldsymbol{\mu}_i$ is due to a *nuclear spin angular momentum* $\mathbf{I}$. The magnitude and z component of $\mathbf{I}$ are quantized according to the equations

$$I = \sqrt{i(i+1)}\,\hbar \qquad I_z = m_i\hbar \tag{13–58}$$

where

$$m_i = -i, -i+1, \ldots, +i-1, +i$$

These are the same as the quantization equations for the electron spin $\mathbf{S}$. However, the equations relating the nuclear magnetic dipole moment to the nuclear spin,

$$\boldsymbol{\mu}_i = \frac{g_i\mu_N}{\hbar}\mathbf{I} \qquad \mu_{i_z} = g_i\mu_N m_i \tag{13–59}$$

are not quite the same as (11–22) and (11–23), which relate the electron

† Interferometers are described in section 3, Chapter 1.

‡ We are considering only the spectrum emitted by a source in which all atoms have the same atomic weight, as well as the same atomic number. For a source containing atoms of the same atomic number but a variety of atomic weights (a variety of "isotopes"), there is an additional multiplicity of the spectral lines, but this is primarily a trivial effect arising from the slight dependence of the reduced electron mass μ on the atomic weight.

magnetic dipole moment to the electron spin. The factor g_i is of the order of unity, but it is positive for some nuclei and negative for others. The quantity μ_N, which is called the *nuclear magneton*, is defined as

$$\mu_N \equiv \frac{e\hbar}{2Mc} = \frac{1}{1836}\frac{e\hbar}{2mc} = \frac{1}{1836}\mu_B \tag{13–60}$$

where M is the mass of the nucleus of a ^{1}H atom, and m is the mass of an electron. Since the nuclear magneton μ_N is three orders of magnitude smaller than the Bohr magneton μ_B, nuclear magnetic moments are smaller than electron magnetic moments by the same factor. Consequently the orientational potential energy of the nuclear magnetic moment (hyperfine structure) is three orders of magnitude smaller than the orientational potential energy of the atomic magnetic moment (fine structure), because both magnetic moments experience comparable magnetic fields.

The interaction of the nuclear magnetic dipole moment with the atomic magnetic field couples the nuclear spin angular momentum vector **I** to the total angular momentum vector **J** of the electrons.† Because of this coupling, both vectors precess about the resultant *grand total angular momentum vector*

$$\mathbf{F} = \mathbf{I} + \mathbf{J} \tag{13–61}$$

The magnitudes of **I** and **J** remain constant, but their z components do not. Consequently, when the hyperfine structure interaction is considered, m_i and m_j are not good quantum numbers. However, the two quantum numbers which specify the magnitude and z component of **F** according to the equations

$$\begin{aligned} F &= \sqrt{f(f+1)}\,\hbar \\ F_z &= m_f\hbar \end{aligned} \tag{13–62}$$

where

$$m_f = -f, -f+1, \ldots, +f-1, +f$$

are good quantum numbers and can be used to identify the quantum states and the energy levels. For instance, the term in the total energy of the atom which gives rise to the hyperfine splitting of the energy levels can be written

$$\overline{\Delta E_{\mathbf{I}\cdot\mathbf{J}}} = K[f(f+1) - j(j+1) - i(i+1)] \tag{13–63}$$

This is completely analogous to equation (13–42); equation (13–63) is obtained by evaluating the orientational potential energy of the nuclear magnetic moment in the atomic magnetic field, which is proportional

† To simplify the notation, in this section we shall not put primes on **J** or on the associated quantum numbers.

to the expectation value of $\mathbf{I} \cdot \mathbf{J}$, and equation (13–42) is obtained by evaluating the orientational potential energy of the electron magnetic moment in the atomic magnetic field, which is proportional to the expectation value of $\mathbf{S} \cdot \mathbf{L}$.

The term $\overline{\Delta E_{\mathbf{I} \cdot \mathbf{J}}}$ assumes different values according to the value of the quantum number f. The possible values of f are determined by an equation analogous to the third of equations (13–41), and there are $2i + 1$ of these values (except that when $i > j$ there are $2j + 1$ values). Thus the energy levels will generally be composed of $2i + 1$ hyperfine structure components. Consequently, the quantum number specifying the magnitude of the spin of an atomic nucleus can be determined by simply measuring the hyperfine structure of the spectrum of the atom, analyzing the spectrum into an energy level diagram, and then counting the number of hyperfine structure components of the energy levels. It is found that the quantum number i is integral (0, 1, 2, . . .) for nuclei of atoms in which the atomic weight A is even, with $i = 0$ when the atomic number Z is also even; i is half-integral ($\frac{1}{2}, \frac{3}{2}, \frac{5}{2}, \ldots$) for nuclei of atoms in which the atomic weight A is odd.†

The quantity K in equation (13–63) depends on the magnetic field at the nucleus and on the nuclear magnetic moment. If the Hartree eigenfunctions describing the motion of the atomic electrons are available, the magnetic field can be calculated to an accuracy of about 10 percent. Then, within this accuracy, the measured value of $\overline{\Delta E_{\mathbf{I} \cdot \mathbf{J}}}$ can be used to evaluate the magnitude and sign of the nuclear magnetic dipole moment. As stated above, it is found that for all nuclei the magnitude of the dipole magnetic moment vector is of the order of one nuclear magneton (i.e., $|g_i| \sim 1$), but that the vector is parallel to the spin vector in some nuclei and antiparallel in others (i.e., g_i is + or −).

Measurements of hyperfine structure also give information about the *electric quadrupole moment* Q of the nucleus of the atom producing the spectrum. This quantity specifies the extent to which the charge distribution of the nucleus departs from spherical symmetry. In all cases, the charge distribution is that of an ellipsoid with an axis of symmetry along the direction about which the nuclear spin angular momentum vector precesses; for $Q > 0$ the ellipsoid is elongated in the direction of the axis, for $Q < 0$ it is flattened in the direction of the axis, and for $Q = 0$ the charge distribution has spherical symmetry.‡ For $Q \neq 0$, the nucleus will have an orientational potential energy in an atomic electric field which

† We are refering to the integer A which is closest to the atomic weight of a particular "isotope" of an atom. With this stipulation, the atomic weight is always very close to an integer.

‡ A precise definition of Q is given in section 9, Chapter 16.

is not spherically symmetrical. In a typical case, the magnitude of this energy is comparable to the magnitude of the orientational potential energy of the nuclear magnetic dipole moment in the atomic magnetic field. However, the dependence of the energy on the quantum numbers specifying the orientation of the nucleus in the atom is different for the electric quadrupole and magnetic dipole interactions, so it is possible to separate these two effects experimentally. When the electric quadrupole interaction is present, the hyperfine splitting of the energy levels is observed to depart from the predictions of equation (13–63), which treats only the magnetic dipole interaction. By analyzing these departures, and then calculating the atomic electric field at the nucleus from Hartree eigenfunctions, it has been possible to evaluate the sign and the approximate magnitude of the electric quadrupole moment of a number of nuclei. It is found that $Q = 0$ for all nuclei with $i = 0$ or $i = \frac{1}{2}$. This can be understood from symmetry considerations; for instance, when $i = 0$ there is nothing to define an axis of symmetry about which an electric quadrupole moment can be measured. For nuclei with $i \geqslant 1$, Q is usually measurably different from 0, and $Q > 0$ occurs most often. The nucleus of the ^{71}Lu atom has one of the largest known quadrupole moments. The charge distribution of this nucleus is elongated in the direction of the axis of symmetry, and the ratio of the semimajor axis of the ellipsoid to its semiminor axis is 1.20 ± 0.05.

Techniques for measuring the magnetic dipole moments of nuclei from an analysis of hyperfine structure are limited in their accuracy because they depend upon calculating the strength of the atomic magnetic field. In the years following 1933, Rabi and others developed several different techniques which are extremely accurate. All of them involve measuring the orientational potential energy of the dipole in an external magnetic field of known strength. For atoms in which the total magnetic dipole moment produced by the atomic electrons is zero (an atom in a $^1S_0$ state), it is possible to measure the nuclear magnetic dipole moment by sending a beam of atoms through an apparatus similar to that used by Stern and Gerlach. In other cases the energy of a nuclear magnetic moment in an external magnetic field can be investigated by using a field which is strong enough, compared to the atomic magnetic field, to destroy the coupling of **I** and **J**. In the most accurate of these techniques the magnetic moment is measured, not in terms of the deflection of an atomic beam, but in terms of the energy of the radiofrequency quanta which produce transitions between two energy levels that are split by hyperfine splitting in an external magnetic field of known strength. These measurements, and the quantum mechanics upon which they are based, form a very interesting subject which we reluctantly pass over in order to continue the discussion of atomic structure and spectra.

12. Transition Rates and Selection Rules

Consider an isolated atom in an excited state of energy E_l. The atom will eventually make a spontaneous transition from that state to a state of lower energy E_k, emitting a quantum of electromagnetic radiation. We would like to calculate a quantity called the *transition rate for spontaneous emission*, which is the probability per unit time that the atom will make such a transition. This calculation will enable us to understand the *selection rules* governing transitions between any two energy states, as well as the lifetimes of these states and the widths of the spectral lines emitted in such transitions.

The theory that has been developed so far in this book does not tell us how to treat spontaneous emission, but it does tell us how to treat the related process of absorption. In this process, an atom acted on by electromagnetic radiation makes a transition from the state of energy E_k to the state of higher energy E_l, absorbing a quatum. The electromagnetic radiation acts as a small perturbation on the atom, and time dependent perturbation theory can be used to calculate the probability per unit time that this perturbation will produce such a transition. This probability per unit time is the *transition rate for absorption.* We shall first calculate this transition rate and then establish a relation between it and the transition rate for spontaneous emission.

An atom in a field of electromagnetic radiation experiences interactions between its magnetic moments and the magnetic field, and also between its electric charges and the electric field. The magnetic interaction is usually completely negligible compared to the electric interaction, and we shall consider here only the latter. Although the electric field is described by a vector $\mathbf{E}$, it is possible to treat separately the interaction produced by each component of this vector. Each component, such as E_x, is an oscillatory function of time and also of position. However, the wavelengths which are characteristic of the absorption or emission optical spectrum of an atom ($\lambda \sim 10^3$ A) are so large, compared to the dimensions of the atom ($r \sim 1$ A), that at any instant E_x is essentially constant over any region of space occupied by the atom. Consequently, we can ignore the dependence on position and describe E_x at a given atom by the function

$$E_x = 2E_x^0(\nu) \cos 2\pi\nu t \tag{13–64}$$

where the amplitude $2E_x^0(\nu)$ depends, in general, on the frequency ν. It is convenient to write this as†

$$E_x = E_x^0(\nu)(e^{2\pi i\nu t} + e^{-2\pi i\nu t}) \tag{13–64′}$$

† The procedure for proving the equality of (13–64) and (13–64′) is indicated in the footnote on page 172.

The electric charges of the atom have a potential energy in the electric field E_x. Let us define the zero of potential energy and the origin of the x axis to be at the center of the atom, and calculate the total potential energy of the atom due to the x component of the electric field by summing the potential energy of each charge. This gives

$$v_x = -E_x \sum_j -ex_j \tag{13–65}$$

The summation is taken over each electron of charge $-e$ and coordinate x_j, but the charged nucleus does not contribute as its x coordinate is 0. Evaluating E_x, we have the perturbing potential

$$v_x = E_x^0(\nu)(e^{2\pi i\nu t} + e^{-2\pi i\nu t}) \sum_j ex_j \tag{13–66}$$

which is an oscillatory function of time.

We find the effect of this perturbation on an atom, which at $t = 0$ is in the quantum state of energy E_k, by solving equation (9–32), which is

$$\frac{da_m(t)}{dt} = -\frac{i}{\hbar}\sum_n a_n(t)e^{-i(E_n-E_m)t/\hbar}v_{x_{mn}} \tag{13–67}$$

with the initial conditions for $a_n(t)$ at $t = 0$:

$$a_n(0) = \begin{matrix} 0, & n \neq k \\ 1, & n = k \end{matrix} \tag{13–68}$$

The quantity $v_{x_{mn}}$ is the matrix element of the perturbation v_x between the states n and m. That is,

$$v_{x_{mn}} \equiv \int \psi_m^* v_x \psi_n \, d\tau$$

where the integration is taken over all the coordinates of the volume element $d\tau$. The quantities $a_n(t)$ are the coefficients in the expansion

$$\Psi'(x, t) = \sum_n a_n(t)\,\Psi(x, t)$$

of the perturbed wave function in terms of the unperturbed wave functions, and the combination $a_n^*(t)\,a_n(t)$ is equal to the probability of finding the atom in the quantum state of energy E_n at time t. Now, if we require that the perturbation be small enough that

$$a_n(t) \begin{matrix} \ll 1, & n \neq k, \\ \simeq 1, & n = k, \end{matrix} \qquad t > 0 \tag{13–69}$$

we may neglect all terms on the right side of equation (13–67) except for $n = k$. This gives

$$\frac{da_m(t)}{dt} \simeq -\frac{i}{\hbar} a_k(t)e^{-i(E_k-E_m)t/\hbar}v_{x_{mk}}$$

By setting $m = l$, we obtain the differential equation for the coefficient $a_l(t)$, in which we are particularly interested. In this equation we may make the approximation that $a_k(t) = 1$. Then it reads

$$\frac{da_l(t)}{dt} \simeq -\frac{i}{\hbar} e^{-i(E_k - E_l)t/\hbar} v_{x_{lk}}$$

Upon evaluating $v_{x_{lk}}$, and doing a little rearranging, this becomes

$$\frac{da_l(t)}{dt} \simeq -\frac{i}{\hbar} \int \psi_l^* \sum_j ex_j \psi_k \, d\tau [e^{i(E_l - E_k + h\nu)t/\hbar} + e^{i(E_l - E_k - h\nu)t/\hbar}] E_x^0(\nu)$$

or

$$\frac{da_l(t)}{dt} \simeq -\frac{i}{\hbar} \mu_{x_{lk}} [e^{i(E_l - E_k + h\nu)t/\hbar} + e^{i(E_l - E_k - h\nu)t/\hbar}] E_x^0(\nu)$$

where

$$\mu_{x_{lk}} \equiv \int \psi_l^* \sum_j ex_j \, \psi_k \, d\tau \qquad (13\text{–}70)$$

Integrating the differential equation from $t = 0$ to $t = t'$, with the initial conditions (13–68), we obtain

$$a_l(t') \simeq \mu_{x_{lk}} \left[\frac{1 - e^{i(E_l - E_k + h\nu)t'/\hbar}}{E_l - E_k + h\nu} + \frac{1 - e^{i(E_l - E_k - h\nu)t'/\hbar}}{E_l - E_k - h\nu} \right] E_x^0(\nu) \qquad (13\text{–}71)$$

Let us consider the behavior of this equation. Since the perturbation is small, the quantity $\mu_{x_{lk}} E_x^0(\nu)$ will be small. As the absolute values of the numerator of each of the two terms in the bracket are always in the range 0 to 2, the entire right side is small except for frequencies for which the denominator of one of the terms becomes small. Thus $a_l(t')$ is appreciable only when

$$E_l - E_k \pm h\nu \simeq 0 \qquad (13\text{–}72)$$

Since $E_l > E_k$, this can be satisfied only for the $-$ sign. Thus the probability of finding the atom in the state of energy E_l, which is $a_l^*(t')\, a_l(t')$, is appreciable only when the frequency of the perturbing electromagnetic radiation satisfies approximately the relation $E_l - E_k = h\nu$. This is precisely Einstein's relation between the frequency and the energy of the quantum absorbed in the transition.

From these arguments we see that the first term in the bracket of (13–71) may be neglected. Doing this, then multiplying the remaining equation into its complex conjugate, and finally using the relation discussed in the footnote on page 172, we find

$$a_l^*(t)\, a_l(t) = 4\mu_{x_{lk}}^* \mu_{x_{lk}} \frac{\sin^2 \left[\dfrac{(E_l - E_k - h\nu)}{2\hbar} t \right]}{(E_l - E_k - h\nu)^2} E_x^0(\nu)^2 \qquad (13\text{–}73)$$

where we have dropped the primes to simplify the notation. The reader should compare this with equation (9–41), which expresses the result of a perturbation that is "switched on" at $t = 0$ and remains constant afterwards. Both equations have a t^2 dependence for small t. In equation (9–41) this goes over to a t^1 dependence, which is what would be expected intuitively, when the equation is integrated. The same is true for equation (13–73). To correspond with physical reality, this equation must be integrated over ν because an atom is never subjected to electromagnetic radiation of perfectly sharply defined frequency, but instead to electromagnetic radiation which covers a band of frequencies. The integration can be taken from $+\infty$ to $-\infty$, since the integrand is small except for a restricted range of ν. The calculation is the same as that preceding equation (9–44), and the results are

$$P_{x_{lk}} \simeq \frac{1}{\hbar^2} \mu^*_{x_{lk}} \mu_{x_{lk}} E^0_x(\nu_{lk})^2 t \tag{13–74}$$

The quantity $P_{x_{lk}}$ is the probability that an atom initially in the state of energy E_k, and subjected to the x component of the electric field of a continuous spectrum of electromagnetic radiation, will be found in the state of energy E_l. The quantity $E^0_x(\nu_{lk})^2$ is the square of the value of $E^0_x(\nu)$ for a frequency satisfying the relation

$$E_l - E_k = h\nu_{lk} \tag{13–75}$$

The contribution of the x component of the electric field to the transition rate for absorption is thus

$$R_{x_{lk}} \equiv \frac{dP_{x_{lk}}}{dt} \simeq \frac{1}{\hbar^2} \mu^*_{x_{lk}} \mu_{x_{lk}} E^0_x(\nu_{lk})^2 \tag{13–76}$$

Similar expressions are obtained for the contributions from the y and z components. The complete transition rate for absorption is

$$R_{lk} \simeq \frac{1}{\hbar^2} [\mu^*_{x_{lk}} \mu_{x_{lk}} E^0_x(\nu_{lk})^2 + \mu^*_{y_{lk}} \mu_{y_{lk}} E^0_y(\nu_{lk})^2 + \mu^*_{z_{lk}} \mu_{z_{lk}} E^0_z(\nu_{lk})^2] \tag{13–77}$$

Let us assume that the radiation is unpolarized. Then the amplitudes of the three components of the electric field are all equal, and

$$R_{lk} \simeq \frac{1}{\hbar^2} E^0_x(\nu_{lk})^2 [\mu^*_{x_{lk}} \mu_{x_{lk}} + \mu^*_{y_{lk}} \mu_{y_{lk}} + \mu^*_{z_{lk}} \mu_{z_{lk}}] \tag{13–78}$$

It is convenient to express this result in terms of $\bar{\rho}$, the time average of the energy content per unit volume of the electromagnetic radiation. Taking the time average of both sides of equation (2–4), we obtain

$$\bar{\rho} = \frac{1}{4\pi} \overline{E^2} \tag{13–79}$$

where $\overline{E^2}$ is the time average of the electric field, and where we have deleted the subscript for transverse. Since the amplitudes of the three components of the electric field are assumed to be all equal, the same is true of the time averages of the squares of these components. That is,

$$\overline{E_x^2} = \overline{E_y^2} = \overline{E_z^2}$$

Then, since

$$\overline{E^2} = \overline{E_x^2} + \overline{E_y^2} + \overline{E_z^2}$$

we have

$$\overline{E^2} = 3\overline{E_x^2} \tag{13–80}$$

By squaring both sides of equation (13–64), and averaging over t, we obtain a relation between $\overline{E_x^2}$ and $E_x^0(\nu_{lk})^2$:

$$\overline{E_x^2} = 2E_x^0(\nu_{lk})^2 \tag{13–81}$$

Then, from equations (13–78) through (13–81), we obtain

$$R_{lk} \simeq \frac{2\pi}{3\hbar^2}\,\bar{\rho}(\nu_{lk})[\mu^*_{x_{lk}}\mu_{x_{lk}} + \mu^*_{y_{lk}}\mu_{y_{lk}} + \mu^*_{z_{lk}}\mu_{z_{lk}}] \tag{13–82}$$

where we have explicitly indicated that the energy density is to be evaluated at the frequency ν_{lk}. This equation gives the desired expression for the transition rate for absorption.

We have seen that there is an appreciable probability of a transition from the state of energy E_k to the state of higher energy E_l, with the absorption of a quantum, only when equation (13–72) is satisfied with the $-$ sign. However, the equation can also be satisfied with the $+$ sign for a transition from the state of energy E_l to the state of lower energy E_k, if a quantum is *emitted.* This phenomenon, in which the perturbation produced by an electromagnetic field induces a transition where a quantum is emitted, is called induced emission. The *transition rate for induced emission* R_{kl} can be evaluated by starting at equation (13–69) and interchanging k and l everywhere except in the condition $E_l > E_k$ and in the symbol ν_{lk}. In this case the first term in the bracket of equation (13–71) is the one which contributes; but otherwise everything is the same and we immediately obtain

$$R_{kl} \simeq \frac{2\pi}{3\hbar^2}\,\bar{\rho}(\nu_{lk})[\mu^*_{x_{kl}}\mu_{x_{kl}} + \mu^*_{y_{kl}}\mu_{y_{kl}} + \mu^*_{z_{kl}}\mu_{z_{kl}}] \tag{13–83}$$

It is easy to see from equation (13–70) that

$$\mu^*_{x_{kl}}\mu_{x_{kl}} = \mu^*_{x_{lk}}\mu_{x_{lk}}, \text{ etc.}$$

So, comparing (13–83) and (13–82), we see that

$$R_{lk} = R_{kl} \tag{13–84}$$

The transition rates for absorption and induced emission are equal. This is an example of the principle of *microscopic reversibility*, which is very important in the theory of statistical mechanics.

The process of induced emission must not be confused with the process of spontaneous emission. According to equation (13–83), the transition rate for induced emission is proportional to the energy density $\bar{\rho}$ of the electromagnetic radiation applied to the atom. If no electromagnetic radiation is applied, there will be no induced emission. However, there will be spontaneous emission, whether or not electromagnetic radiation is applied to the atom, because the transition rate for this process does not involve $\bar{\rho}$. The transition rate for spontaneous emission S_{kl} is an inherent characteristic of the atom and is not influenced by the environment in which the atom is placed. Let us return now to our original purpose—the evaluation of S_{kl}.

This can be done in two quite different ways. The more rigorous treatment is based on the theory of *quantum electrodynamics*.† We present here the other treatment, given by Einstein some years before quantum mechanics was developed, which leads to the same results and is much simpler. The idea is to establish a relation between the transition rates for absorption, induced emission, and spontaneous emission, from considerations of thermodynamic equilibrium. Imagine a cavity containing a large number of atoms which are emitting and absorbing electromagnetic radiation. When this system is in equilibrium at temperature T:

1. The electromagnetic radiation must have the black body spectral distribution characteristic of that temperature.
2. The total number of atoms in each energy state must be distributed according to the Boltzmann probability distribution for that temperature.
3. The total number of atoms in each energy state must, as a consequence of the second condition, remain constant.

Condition 3 requires that the total number of transitions per unit time from the state of lower energy E_k to the state of higher energy E_l equal the total number of transitions in the opposite direction. Now the total number of transitions per unit time from the lower energy state to the higher energy state is

$$\mathcal{N}_{lk} = N_k R_{lk} \tag{13–85}$$

where N_k is the total number of atoms in the lower energy state, and where

† Cf. footnote on page 462.

the transition rate for absorption R_{lk} gives the probability per unit time that one of these atoms will make a transition to the higher energy state. According to (13–82), R_{lk} can be written

$$R_{lk} = B_{lk}\bar{\rho}(\nu_{lk}) \tag{13–86}$$

This defines the proportionality constant B_{lk} between the transition rate for absorption and the energy density, which is called *Einstein's coefficient of absorption*.† Using this quantity, equation (13–85) becomes

$$\mathcal{N}_{lk} = N_k B_{lk}\bar{\rho}(\nu_{lk}) \tag{13–87}$$

The total number of transitions per unit time from the higher energy state to the lower energy state is

$$\mathcal{N}_{kl} = N_l[R_{kl} + S_{kl}] \tag{13–88}$$

where N_l is the total number of atoms in the higher energy state, R_{kl} is the transition rate for induced emission, and S_{kl} is the transition rate for spontaneous emission. From (13–83) we can write

$$R_{kl} = B_{kl}\bar{\rho}(\nu_{lk}) \tag{13–89}$$

which defines *Einstein's coefficient of induced emission* B_{kl}. We also write

$$S_{kl} = A_{kl} \tag{13–90}$$

which defines *Einstein's coefficient of spontaneous emission* A_{kl}. The coefficients B_{lk}, B_{kl}, and A_{kl} all depend only on the inherent characteristics of the atoms, and equation (13–90) emphasizes the fact that we assume the transition rate for spontaneous emission to be independent of the intensity of the electromagnetic radiation applied to the atoms. From the last two equations, we may write equation (13–88) as

$$\mathcal{N}_{kl} = N_l[B_{kl}\bar{\rho}(\nu_{lk}) + A_{kl}] \tag{13–91}$$

Then we satisfy condition 3 by equating $\mathcal{N}_{kl}$ and $\mathcal{N}_{lk}$, and obtain

$$\frac{N_k}{N_l} = \frac{B_{kl}\bar{\rho}(\nu_{lk}) + A_{kl}}{B_{lk}\bar{\rho}(\nu_{lk})} \tag{13–92}$$

The quantity N_k/N_l can be evaluated by applying condition 2. The Boltzmann distribution (2–23) says that at temperature T the probability that an atom will have energy E_k is proportional to $e^{-E_k/KT}$, and the probability that an atom will have energy E_l is proportional to $e^{-E_l/KT}$,

† At the time he developed this treatment, Einstein wrote equation (13–86) as a postulate because he did not know equation (13–82).

where K is Boltzmann's constant. Consequently, the ratio of the number of atoms with these two energies is

$$\frac{N_k}{N_l} = \frac{e^{-E_k/KT}}{e^{-E_l/KT}} = e^{(E_l - E_k)/KT} \tag{13–93}$$

According to equation (13–75), this is

$$\frac{N_k}{N_l} = e^{h\nu_{lk}/KT} \tag{13–94}$$

Combining this with equation (13–92) gives

$$e^{h\nu_{lk}/KT} = \frac{B_{kl}\bar{\rho}(\nu_{lk}) + A_{kl}}{B_{lk}\bar{\rho}(\nu_{lk})}$$

Solving for $\bar{\rho}(\nu_{lk})$, we obtain

$$\bar{\rho}(\nu_{lk}) = \frac{A_{kl}}{B_{lk}e^{h\nu_{lk}/KT} - B_{kl}} \tag{13–95}$$

Finally, we apply condition 1 by equating $\bar{\rho}(\nu_{lk})$ to the black body spectral distribution of energy density (2–31). In terms of the present notation this is

$$\bar{\rho}(\nu_{lk}) = \frac{8\pi h\nu_{lk}^3}{c^3}\frac{1}{e^{h\nu_{lk}/KT} - 1} \tag{13–96}$$

Then a comparison of equations (13–95) and (13–96) shows that the three Einstein coefficients must be related to each other as follows:

$$B_{lk} = B_{kl} \tag{13–97}$$

and

$$A_{kl} = \frac{8\pi h\nu_{lk}^3}{c^3} B_{kl} \tag{13–98}$$

Equation (13–97) shows that this treatment is consistent with equation (13–84). Equation (13–98) provides the desired connection between the transition rates for absorption and spontaneous emission. Using (13–83), (13–89), and (13–90), we obtain at last the transition rate for spontaneous emission:

$$S_{kl} \simeq \frac{32\pi^3\nu_{lk}^3}{3\hbar c^3}[\mu_{x_{kl}}^*\mu_{x_{kl}} + \mu_{y_{kl}}^*\mu_{y_{kl}} + \mu_{z_{kl}}^*\mu_{z_{kl}}] \tag{13–99}$$

The transition rate for spontaneous emission is seen to be proportional to the cube of the frequency (or cube of the energy) of the emitted quanta. In common with the transition rates for induced emission and absorption, it is also proportional to the sum of the products of quantities such as $\mu_{x_{kl}}$, and their complex conjugates. The quantity $\mu_{x_{kl}}$ is called the *matrix* *element of the x component of the electric dipole moment* of the atom. The quantities $\mu_{y_{kl}}$ and $\mu_{z_{kl}}$ are given corresponding names. This nomenclature is used because the term $\sum_j ex_j$ that occurs in the integrand of equation (13–70) defining $\mu_{x_{kl}}$ is the x component of the electric dipole moment of the atom. The matrix element of $\sum_j ex_j$ is its average value calculated with a weighting factor $\psi_k^*\psi_l$. This weighting factor is neither the probability density $\psi_l^*\psi_l$ of the initial quantum state, nor the probability density $\psi_k^*\psi_k$ of the final quantum state, but instead some sort of "mixed probability density" which involves the eigenfunctions of both the initial and the final states. The probability of an atom radiating a quantum depends, then, not only on what the atom does in the initial quantum state before it radiates, but also on what it is going to do in the final quantum state after it radiates. The reason is that when the atom radiates it is actually not in either the initial state or the final state, but in some linear combination of the two.

This point of view can be used to give a physical picture of what happens in a radiation process. Consider an induced transition of an atom from the initial state of energy E_l to the final state of lower energy E_k, accompanied by the emission of a quantum. According to (7–45), the wave function describing the atom can always be written

$$\Psi = \sum_n a_n e^{-iE_nt/\hbar}\psi_n \tag{13–100}$$

Before the application of the electromagnetic field which induces the transition, the atom is in the eigenstate of energy E_l. Only the coefficient a_l is non-zero, and

$$\Psi = a_l e^{-iE_lt/\hbar}\psi_l$$

The corresponding probability density is

$$\Psi^*\Psi = a_l^* a_l \psi_l^* \psi_l \tag{13–101}$$

This quantity specifies the charge distribution of the atom. It is constant in time. Consequently the electric dipole moment (as well as every other feature of the charge distribution) is constant in time. According to classical physics, the atom can radiate only if its electric dipole moment (or some other feature of the charge distribution) changes in time. Thus we can understand, in a sense even classically, why the atom cannot radiate as long as it is in an eigenstate. Now, when the electromagnetic field is applied, it produces a perturbation which mixes contributions from the wave function for the eigenstate of energy E_k into the wave function for the atom. The coefficient a_k increases with time from its initial value of zero, and the coefficient a_l decreases. Consider the wave function for the atom when both a_k and a_l are non-zero. It has the form

$$\Psi = a_l e^{-iE_l t/\hbar}\psi_l + a_k e^{-iE_k t/\hbar}\psi_k \tag{13–102}$$

The corresponding probability density is

$$\Psi^*\Psi = a_l^* a_l \psi_l^* \psi_l + a_k^* a_k \psi_k^* \psi_k + a_l^* a_k e^{i(E_l - E_k)t/\hbar}\psi_l^* \psi_k + a_k^* a_l e^{-i(E_l - E_k)t/\hbar}\psi_k^* \psi_l \tag{13–103}$$

The last two terms of $\Psi^*\Psi$ oscillate with the frequency

$$\nu = \frac{E_l - E_k}{h} \tag{13–104}$$

Eventually a_l is reduced to zero. Then the probability density is

$$\Psi^*\Psi = a_k^* a_k \psi_k^* \psi_k$$

With such a probability density the atom cannot radiate. But it is apparent that it can radiate when it is still in the state described by the wave function (13–102). In this state the probability density (13–103), and consequently the charge distribution, have components which oscillate with frequency (13–104) that is exactly equal to the frequency of the quantum of electromagnetic radiation which is emitted during the transition.

These arguments provide a good physical picture of the process of induced emission and absorption. It would appear that they do not apply to the process of spontaneous emission because in this process there is apparently no perturbing electromagnetic field to produce the mixing of the wave functions that leads to an oscillating probability density. However, the theory of quantum electrodynamics removes this difficulty. In this theory both the electromagnetic field and the atom are quantized.† Then it is found that there are interactions between these two quantized systems, even when the electromagnetic field is in the state of lowest energy, because some electromagnetic radiation must be present in this state as the electromagnetic field has a zero point energy like any other quantized system executing simple harmonic oscillations.

† The electromagnetic field was quantized in the very simple theory of Planck and Einstein, which states that it is comprised of quanta of energy $E = h\nu$. But a more complete theory is needed because it is necessary to know how an electromagnetic field in the quantum limit (containing few quanta) interacts with an external system of charges, just as it is necessary to know how an electromagnetic field in the classical limit (containing many quanta) interacts with an external system. The classical limit is described by Maxwell's theory; the quantum limit is described by the theory of quantum electrodynamics. In the latter theory, Schroedinger's quantum mechanics is applied to a system of quanta, but the quantization of this system is considerably more involved than the quantization of a system of particles since the number of quanta in a system can *change* when it interacts with an external system.

If the eigenfunctions for an atom are known, the electric dipole matrix elements between all pairs of excited states can be obtained by evaluating integrals such as (13–70). Then the transition rates for spontaneous emission between these states can be calculated from equation (13–99). These transition rates immediately lead to theoretical predictions of the relative intensities of the spectral lines emitted in transitions between the various states. It is found that the agreement between these predictions and the experimental data is quite good. The transition rate for spontaneous emission varies widely from one case to the next. For the transition of the ^1H atom from its first excited state to its ground state, this transition rate has the value $S_{kl} \simeq 10^8 \text{ sec}^{-1}$. This is typical of the orders of magnitude encountered, except that the transition rates between certain pairs of states are essentially zero. These are the transitions for which the spectral lines are observed to be absent or extremely weak. The transition rates are predicted to be zero in these cases because the integrals involved in the electric dipole matrix elements vanish identically. Thus *the selection rules are a set of conditions on the quantum numbers of the eigenfunctions of the initial and final states such that the electric dipole matrix elements are zero when calculated with a pair of eigenfunctions whose quantum numbers violate these conditions.*

As a simple example, we shall carry out part of the calculation of the electric dipole matrix elements for an atom with one optically active electron. In this case the core of filled subshells does not change in the transition, and it is necessary only to evaluate the matrix elements of e times the three rectangular coordinates of the optically active electron. Furthermore, in evaluating these matrix elements we need only to write the eigenfunctions for the optically active electron. All the other factors in the complete eigenfunction of the atom immediately integrate to unity because of the normalization conditions. The rectangular coordinates are related to the spherical coordinates, in terms of which the eigenfunctions are written, by the equations

$$\begin{aligned} x &= r \sin\theta \cos\phi \\ y &= r \sin\theta \sin\phi \\ z &= r \cos\theta \end{aligned} \qquad (13\text{–}105)$$

Using these coordinates, we have

$$\begin{aligned} \mu_{x_{n'l'm_l',nlm_l}} &= e\iiint \psi^*_{n'l'm_l'} r \sin\theta \cos\phi\, \psi_{nlm_l} r^2 \sin\theta \, dr\, d\theta\, d\phi \\ &= e\int_0^\infty R^*_{n'l'} R_{nl} r^3\, dr \int_0^\pi \Theta^*_{l'm_l'} \Theta_{lm_l} \sin^2\theta\, d\theta \int_0^{2\pi} \Phi^*_{m_l'} \Phi_{m_l} \cos\phi\, d\phi \end{aligned}$$

Consider the third integral,

$$\int_0^{2\pi} \Phi^*_{m_l'}\Phi_{m_l} \cos\phi\, d\phi = \int_0^{2\pi} e^{-im_l'\phi}e^{im_l\phi} \cos\phi\, d\phi$$
$$= \frac{1}{2}\int_0^{2\pi} [e^{-i(m_l'-m_l+1)\phi} + e^{-i(m_l'-m_l-1)\phi}]\, d\phi$$

We leave it as an exercise for the reader to show that this integral vanishes unless

$$m_l' - m_l = \pm 1 \tag{13–106}$$

If it does vanish, so does $\mu_{x_{n'l'm_l',nlm_l}}$. The same selection rule is obtained upon evaluating $\mu_{y_{n'l'm_l',nlm_l}}$. For $\mu_{z_{n'l'm_l',nlm_l}}$ we have

$$\mu_{z_{n'l'm_l',nlm_l}} = e\int_0^{\infty} R^*_{n'l'}R_{nl}r^3\, dr \int_0^{\pi} \Theta^*_{l'm_l'}\Theta_{lm_l} \sin\theta\cos\theta\, d\theta \int_0^{2\pi} \Phi^*_{m_l'}\Phi_{m_l}\, d\phi$$

Since the Φ_{m_l} are orthogonal, the third integral vanishes unless

$$m_l' - m_l = 0 \tag{13–107}$$

The transition rate $S_{n'l'm_l',nlm_l}$ will be non-zero only if either (13–106) or (13–107) is satisfied. Thus the complete selection rule for m_l is

$$m_l' - m_l = \Delta m_l = 0, \pm 1 \tag{13–108}$$

This is in agreement with the second of equations (13–57), which is valid for one or several optically active electrons. The spectral lines from transitions that satisfy equation (13–107) are linearly polarized in the z direction. The spectral lines from transitions that satisfy equation (13–106) have components, which are linearly polarized in both the x and y directions, that combine to form circularly polarized radiation. These results are verified by measurements of the polarization of the radiation emitted in the Paschen-Bach effect by the different spectral lines corresponding to different values of Δm_l.

The selection rules for l can be obtained by a straightforward evaluation of the integrals over $d\theta$. However, this is quite tedious, and part of the results can be obtained in a simpler way from considerations of the parities of the eigenfunctions. In section 4, Chapter 8, we defined the *parity* of a one dimensional eigenfunction. This is the quantity which describes the behavior of the eigenfunction when the sign of the coordinate is changed. The definition can be immediately extended to three dimensions. That is, eigenfunctions which satisfy the equation

$$\psi(-x, -y, -z) = +\psi(x, y, z) \tag{13–109}$$

are said to be of *even parity*, and eigenfunctions which satisfy the equation

$$\psi(-x, -y, -z) = -\psi(x, y, z) \tag{13–110}$$

are said to be of *odd parity*. It is not difficult to show that most eigenfunctions which are solutions to Schroedinger equations with potentials whose values are unchanged when the signs of the coordinates are changed will have a *definite parity* (either even or odd).†

It is convenient to express equations (13–109) and (13–110) in terms of spherical coordinates. By considering the diagrams in figure (13–16) we see that, when the signs of the rectangular coordinates are changed, the behavior of the spherical coordinates is

$$r \to r, \quad \theta \to \pi - \theta, \quad \phi \to \pi + \phi \tag{13–111}$$

† Write the time independent Schroedinger equation as

$$e_{\text{op}}(x, y, z)\psi(x, y, z) = E\psi(x, y, z)$$

Change the signs of the coordinates:

$$e_{\text{op}}(-x, -y, -z)\psi(-x, -y, -z) = E\psi(-x, -y, -z)$$

Since e_{op} contains only $V(x, y, z)$ plus terms like $\partial^2/\partial x^2$, $e_{\text{op}}(-x, -y, -z) = e_{\text{op}}(x, y, z)$ when $V(-x, -y, -z) = V(x, y, z)$. If this relation holds, the previous equation reads

$$e_{\text{op}}(x, y, z)\psi(-x, -y, -z) = E\psi(-x, -y, -z)$$

Since both $\psi(-x, -y, -z)$ and $\psi(x, y, z)$ satisfy the same differential equation, it must be that

$$\psi(-x, -y, -z) = K\psi(x, y, z)$$

where K is a constant. Now change the signs of the coordinates in this equation and obtain

$$\psi(x, y, z) = K\psi(-x, -y, -z)$$

Multiplying this equation into the previous equation, we obtain,

$$K^2 = 1$$

or

$$K = \pm 1$$

so

$$\psi(-x, -y, -z) = \pm\psi(x, y, z); \qquad \text{Q.E.D.}$$

This proof breaks down if there are two or more linearly independent eigenfunctions that satisfy the differential equation for the same value of E, i.e., if there is degeneracy. But even in this case the eigenfunctions will have definite parities if they are appropriately chosen linear combinations. The ψ_{nlm_l} are examples of degenerate eigenfunctions with definite parities, as shown by equation (13–112). An example of a degenerate eigenfunction without a definite parity is the traveling wave (15–64).

Now, if the reader will inspect the eigenfunctions for the hydrogen atom (10–39), he can verify that for those which are listed

$$\Phi_{m_l}(\pi + \phi) = (-1)^{|m_l|}\Phi_{m_l}(\phi)$$

and

$$\Theta_{lm_l}(\pi - \theta) = (-1)^{l+|m_l|}\Theta_{lm_l}(\theta)$$

Thus

$$R_{nl}(r)\,\Theta_{lm_l}(\pi - \theta)\Phi_{m_l}(\pi + \phi) = (-1)^{l+|m_l|}(-1)^{|m_l|}R_{nl}(r)\,\Theta_{lm_l}(\theta)\,\Phi_{m_l}(\phi)$$

or

$$\psi_{nlm_l}(r, \pi - \theta, \pi + \phi) = (-1)^l\psi_{nlm_l}(r, \theta, \phi) \qquad (13\text{–}112)$$

This is true for all the hydrogen atom eigenfunctions. In fact, since the transformation (13–111) does not affect the coordinate r, equation (13–112)

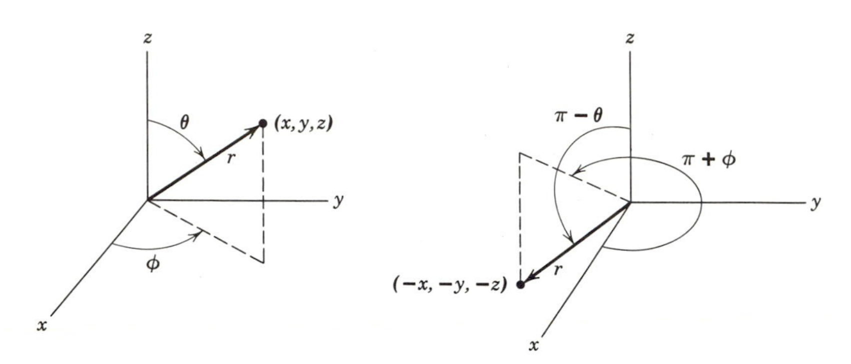

Figure 13–16. Illustrating the parity operation.

is true for all eigenfunctions for spherically symmetrical potentials because all such eigenfunctions can be written $R_{nl}(r)\,\Theta_{lm_l}(\theta)\,\Phi_{ml}(\phi)$ where $\Theta_{lm_l}(\theta)$ and $\Phi_{m_l}(\phi)$ are the same as for the hydrogen atom. Thus *the parity of all eigenfunctions for spherically symmetric potentials is determined by* $(-1)^l$; *the parity is even if l is even, and odd if l is odd.*

Now consider the matrix element of the x component of the electric dipole moment, which is

$$\mu_{x_{n'l'm_l',nlm_l}} = \int \psi^*_{n'l'm_l'} \sum_j ex_j\, \psi_{nlm_l}\, d\tau$$

The x component of the electric dipole moment is of odd parity since

$$\sum_j e(-x_j) = -\sum_j ex_j$$

This is also true of the y and z components. If the eigenfunctions $\psi_{n'l'm_l'}$ and ψ_{nlm_l} are of the same parity (both even or both odd), the entire integrands of $\mu_{x_{n'l'm_l',nlm_l}}$, $\mu_{y_{n'l'm_l',nlm_l}}$ and $\mu_{z_{n'l'm_l',nlm_l}}$ will be of odd parity. If this is the case, it is obvious that the integrals will vanish identically because the contribution from the point (r, θ, ϕ) will be canceled by the contribution from the point $(r, \pi - \theta, \pi + \phi)$, and the transition rates will be zero. Therefore there is the following selection rule for the relative parity of the initial and final eigenfunctions in an electric dipole transition:

$$\text{The parity must change.} \tag{13–113}$$

Equation (13–112) allows this to be expressed, in terms of the change in the quantum number l, as

$$\Delta l = \pm 1, \pm 3, \pm 5, \ldots \tag{13–113′}$$

This is part, but not all, of the selection rule (13–31). The rest of it can be justified by considering (13–108), which is $\Delta m_l = 0, \pm 1$. This selection rule makes it possible to understand why transitions are not allowed when $\Delta l = \pm 3, \pm 5, \ldots$ since in some of these transitions $\Delta m_l > 1$. As mentioned before, the selection rule (13–31) can also be obtained by a straightforward evaluation of the matrix elements.

To calculate the selection rules for the quantum number j, it is necessary to know the exact form of the eigenfunctions in the presence of the spin-orbit interaction. Therefore we can only assure the reader that calculations with these eigenfunctions do verify the selection rule (13–30). The same is true of the selection rules (13–45) for the quantum numbers of an atom with several optically active electrons that is an example of LS coupling. But one of these, $\Delta s' = 0$, has such a simple interpretation that we must mention it. The selection rule states that the coupling of the spin angular momentum vectors cannot change in the transition. The reason is simply that the interaction between electromagnetic radiation and spins takes place through the interaction of the magnetic field and the magnetic moments. These interactions are so very weak that the coupling of the spins is normally not affected.

However, faint spectral lines that violate the selection rule for s' can be seen in certain special circumstances (cf. the discussion at end of section 8). Thus this selection rule does not absolutely prohibit transitions that violate it, but only makes these transitions very unlikely. If the transition cannot take place by the normal means of an interaction involving the electric dipole moments and the electric fields, which is called an *electric dipole* interaction, there is a very small probability that it can take place through an interaction of the magnetic dipole moments and the magnetic fields that are also present in electromagnetic radiation. In fact, there is a very small probability of violating, by means of such *magnetic dipole* interactions, any of the selection rules quoted in this chapter. This can also be done by means of *electric quadrupole* interactions, which represent violations of our assumption that the wavelength of the radiation is so long compared to the dimensions of the atom that any variations of the electric fields over the atom can be neglected. Nevertheless, transitions going through electric dipole interactions are by far the most important for atoms.

13. Lifetimes and Line Widths

Consider a system of atoms in which $N(0)$ have been excited at time $t = 0$ by some process to a state of energy E_l. Let us calculate the number $N(t)$ remaining in that state at some later time t, assuming that the transition rates for spontaneous emission S_{kl} are known for all k for which $E_k < E_l$, and that these transition rates are very much larger than the transition rates for induced emission R_{kl} because the average density $\bar{\rho}$ of electromagnetic radiation is very small. Then the probability per second that one of the atoms in the state of energy E_l will make a transition out of that state into any of the states of lower energy E_k is the sum of the S_{kl}, for all k for which $E_k < E_l$. Call this probability λ; then

$$\lambda = \sum_k S_{kl} \tag{13–114}$$

In a time interval dt at t, the total number of such transitions will be, on the average,

$$N(t)\lambda \, dt$$

since during that interval there are approximately $N(t)$ atoms in the initial state and each one has a transition probability $\lambda \, dt$. This quantity is just the decrease in $N(t)$ during that interval. Thus

$$dN(t) = -N(t)\lambda \, dt$$

or

$$\frac{dN(t)}{N(t)} = -\lambda \, dt \tag{13–115}$$

Integrating from $t = 0$ to $t = t'$, we have

$$\int_0^{t'} \frac{dN(t)}{N(t)} = -\lambda \int_0^{t'} dt$$

which gives

$$\ln [N(t') - N(0)] = -\lambda t'$$

or

$$\ln \frac{N(t')}{N(0)} = -\lambda t'$$

So

$$\frac{N(t')}{N(0)} = e^{-\lambda t'}$$

or

$$N(t) = N(0)e^{-\lambda t} \tag{13–116}$$

where we have dropped the primes to simplify the notation. The number of atoms remaining in the initial state decreases exponentially from the original value $N(0)$ and, at the time $t = 1/\lambda$, the number remaining is reduced by a factor e^{-1}. The *lifetime* τ of this state is defined as the average time an atom spends in the state before making a transition. This is,

$$\tau = \frac{\int_0^\infty tN(t)\,dt}{\int_0^\infty N(t)\,dt} = \frac{\int_0^\infty N(0)te^{-\lambda t}\,dt}{\int_0^\infty N(0)e^{-\lambda t}\,dt}$$

Since this ratio of integrals has exactly the same form as (2–19), we may use the results of the calculation following that equation to write immediately

$$\tau = 1/\lambda \tag{13–117}$$

which is

$$\tau = \left[\sum_k S_{kl}\right]^{-1} \tag{13–118}$$

Because τ generally involves a sum over several S_{kl}, the variations in those quantities tend to average out and yield lifetimes which are of the order of $\tau \sim 10^{-9}$ sec for many states. However, there are a large number of exceptions.

There is a very interesting relation between the lifetime of an excited state of an atom and the width of the spectral lines arising from transitions involving that state. Spectral lines are not perfectly sharp; instead they have finite widths that can be measured, under favorable circumstances, by using equipment of the highest resolution. Usually the width is due primarily to a random Doppler shift of the frequencies of the emitted quanta, resulting from the random thermal motion of the atoms in the spectral source. However, this *Doppler broadening* can be reduced by using low temperature sources, and when this is done it is found that for lines originating from states of short lifetime a measurable width remains. This is a manifestation of the uncertainty principle relation (6–28), which says that a measurement of the energy of a system that is carried out in a time Δt must be uncertain by an amount ΔE, where

$$\Delta E \, \Delta t \sim \hbar$$

Consider a transition from some excited state to the ground state of the atom. A measurement of the frequency of the quantum emitted in this transition constitutes a measurement of the energy of the atom in the excited state. Furthermore, the measurement obviously must be carried out within a time comparable to τ, the lifetime of the excited state. Thus the uncertainty principle demands that the energy of the quantum can have any value within the range

$$\Delta E \sim \hbar/\tau \tag{13–119}$$

Then its frequency can have any value within the range

$$\Delta \nu = \frac{\Delta E}{h} \sim \frac{\hbar}{h\tau} = \frac{1}{2\pi\tau} \tag{13–119$'$}$$

and the associated spectral line will have a corresponding width. This *natural broadening* can be observed experimentally if the pressure in the spectral source is low. At high pressures the time between collisions of the emitting atom with other atoms becomes smaller than the lifetime of the excited state and, since it is no longer possible to carry out an undisturbed measurement for a time comparable to τ, the width of the line will be larger than its *natural width*. This is called *collision broadening*.

If a spectral line is emitted in a transition from some initial state to a final state of very much longer lifetime (e.g., the ground state), the natural width of the spectral line emitted in the transition is due entirely to the lifetime of the initial state. In these circumstances the natural width of the spectral line, expressed in energy units, is said to be equal to the *width* of the initial state. The idea that a state has a non-zero width if it has a finite lifetime comes into quantum theory directly through the uncertainty principle, as we have seen, and certainly is not in conflict with the statement made in Chapter 7 that the energy of a system of bound particles, such as an atom, can be equal only to one of a particular set of discrete (i.e., zero width) energies. We could make this statement because we ignored the possibility of transitions between states in Chapter 7 and, in that approximation, all states have infinite lifetimes or zero widths.

The idea of a non-zero width for a finite lifetime even comes into classical theory. Consider any resonant system, such as an electrical circuit containing inductance and capacitance. If it contains absolutely no resistance, the theory of electrical circuits predicts that the circuit has a perfectly sharp resonant frequency ν_0 and that, if driven by some source of variable frequency, it will accept power only when the source frequency is exactly equal to ν_0. But, of course, all circuits have some resistance. In this case, the theory predicts that the resonant frequency is not so well defined, and that the power P accepted from the source depends on the source frequency ν according to the relation

$$P(\nu) \propto \frac{1}{(\nu - \nu_0)^2 + (\Delta\nu/2)^2} \tag{13–120}$$

The quantity $P(\nu)$, which is plotted in figure (13–17), is the familiar resonance curve. The full width at half maximum $\Delta\nu$ is a simple function of

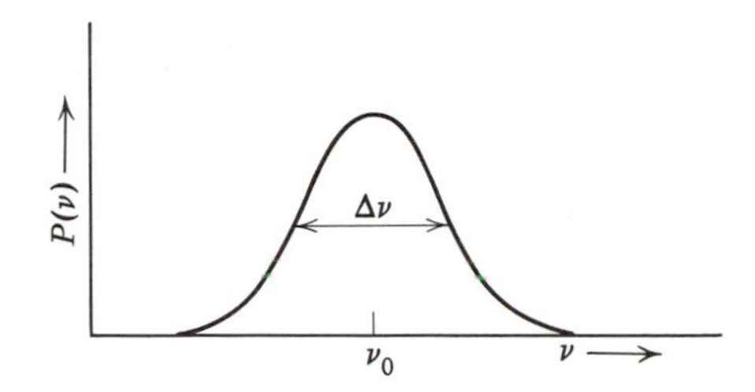

Figure 13–17. A resonance curve.

the resistance that increases with increasing resistance. Now, if the circuit accepts a certain amount of energy from the source and the source is then turned off, the energy content E of the circuit will decrease in time by dissipation in the resistance. The theory of electrical circuits shows that this decrease follows the law

$$E(t) = E(0)e^{-t/\tau} \tag{13–121}$$

where

$$\tau = \frac{1}{2\pi\,\Delta\nu} \tag{13–121'}$$

Of course, all these equations for electrical circuits have been verified experimentally.

Now consider an atom in its ground state, exposed to a source of electromagnetic radiation which is emitting quanta of all frequencies. The atom can absorb a quantum only if its energy is equal to the energy of one of its excited states but, since these states have non-zero widths, it is possible for a quantum to be absorbed in a transition to a particular state if its energy is within a range equal to the width of the state. In fact, quantum theory shows that the transition rate R for absorption to a particular state depends on the energy $h\nu$ of the quantum according to the relation

$$R(h\nu) \propto \frac{1}{(h\nu - h\nu_0)^2 + (h\,\Delta\nu/2)^2} \tag{13–122}$$

where $h\nu_0$ is the mean energy of the state and $h\,\Delta\nu$ is its full width at half maximum. Furthermore, the theory shows that the width $h\,\Delta\nu$ is related to the lifetime τ of the state by the equation

$$\tau = \frac{h}{2\pi h\,\Delta\nu} \tag{13–123}$$

The width $h\,\Delta\nu$ of a state is usually designated by the symbol Γ, and (13–122) is usually written in terms of the energy E of the quantum as

$$R(E) \propto \frac{1}{(E - E_0)^2 + (\Gamma/2)^2} \tag{13–124}$$

where E_0 is the energy of the state and Γ is its width. Then

$$\tau = \frac{\hbar}{\Gamma} \tag{13–125}$$

is the relation between the lifetime τ of a state and its width Γ. Since the transition rate for spontaneous emission is proportional to the transition rate for absorption, the former also has the functional dependence of (13–124). Therefore the natural width of a spectral line emitted in a transition from an initial state of width Γ to a final state of much narrower width is Γ, as has already been mentioned.

Note that equation (13–125) is in agreement with the limit set by the uncertainty principle relation (6–28). Also note that the results (13–122) and (13–123) of the quantum theory for resonant absorption of quanta by a state, which subsequently radiates, are essentially identical with the results (13–120) and (13–121′) of the classical theory for the resonant absorption of energy by a system, which subsequently dissipates. The natural spectral line shape predicted by (13–124) has been verified experimentally. Equation (13–125) cannot be tested by experiments measuring both τ and Γ because $\hbar$ is so very small that only one or the other is measurable. Nevertheless, it is assumed to be valid and is often used to predict the value of Γ from the measured value of τ, or vice versa.

BIBLIOGRAPHY

Herzberg, G., *Atomic Spectra and Atomic Structure*, Dover Publications, New York, 1944.

Leighton, R. B., *Principles of Modern Physics*, McGraw-Hill Book Co., New York, 1959.

Richtmyer, F. K., E. H. Kennard, and T. Lauritsen, *Introduction to Modern Physics*, McGraw-Hill Book Co., New York, 1955.

White, H. E., *An Introduction to Atomic Spectra*, McGraw-Hill Book Co., New York, 1934.

EXERCISES

1. Evaluate $V(r)$ and $\rho(r)$ in the Thomas-Fermi theory for the A atom ($Z = 18$), and compare with figure (13–3).

2. Use the approximation (13–22), the Z_n for the A atom, and the one-electron atom equation (10–44), to estimate the radii of the $n = 1, 2$, and 3 shells of the A atom. Compare with figure (13–3).

3. Verify that the vector diagrams of figure (13–11) are equivalent to equations (13–40). Also construct a similar set of vector diagrams for $l_1 = 2$, $s_1 = \frac{1}{2}$; $l_2 = 2$, $s_2 = \frac{1}{2}$.

4. Carry through a calculation as in table (13–8) for the configuration np^3, and thereby verify the results for that configuration listed in table (13–9).

5. Consider an atom which is an example of LS coupling, and an $nd\ n'd$ configuration of that atom in which both the ordering of the energy levels according to s', l', j', and the relative strengths of the s' and l' dependence of the energy, are normal. Draw a schematic energy level diagram for this configuration, using the same (exaggerated) scale for the fine structure splitting of all the levels within a given multiplet. Label each level with the spectroscopic notation.

6. Indicate which levels of the energy level diagram of exercise 5 would be absent if $n' = n$.

7. On the energy level diagram of exercise 6, draw to the same (highly exaggerated) scale the Zeeman effect splitting of each level into its various components under the influence of a weak magnetic field.

8. Count the total number of components, i.e., the total number of different quantum states in the configuration, obtained in exercise 7. Show that this equals the degeneracy of the configuration before the application of the perturbation, which is the product of degeneracy factors $2(2l + 1)$ for each of the two optically active electrons of the configuration.

9. In an atom which is an example of LS coupling, the relative separations between adjacent levels of a particular multiplet are 4:3:2:1. Use the procedure outlined at the end of section 10 to assign the quantum numbers s', l', j' for these levels. Repeat for multiplets with relative separations 7:5:3 and 9:7:5:3.

10. Verify equation (13–74).

11. Verify equation (13–112).

12. By a straightforward evaluation of the electric dipole matrix elements for the eigenfunctions (10–39), show that the selection rule (13–31) is valid for $n = 2 \rightarrow n = 1$ transitions in the H atom.

13. Evaluate a few of the electric dipole transition matrix elements for a one dimensional charged simple harmonic oscillator by using the eigenfunctions (8–118). Show that their values agree with the selection rule $\Delta n = \pm 1$. Compare with the discussion of section 8, Chapter 5.

14. Consider a gas of H_2 molecules at room temperature and pressure, in which the H atoms are making spontaneous transitions from the first excited state to the ground state. Use the fact that the transition rate for that transition is $\sim 10^8$ sec^{-1} to estimate the natural broadening of the resulting spectral line. Use the relation of elementary kinetic theory connecting the average distance between collisions to the density of the gas and the projected geometrical cross sectional area of the molecules, the perfect gas law, and the law of equipartition of energy to estimate the collision broadening. Use the law of equipartition of energy and the classical Doppler shift equation to estimate the Doppler broadening.

CHAPTER 14

X-rays

1. The Discovery of X-rays

Except for a brief description of the Compton effect, and a few other remarks, we have postponed the discussion of X-rays until the present chapter because it is particularly convenient to treat X-ray spectra after treating optical spectra. Although this ordering may have given the reader a distorted impression of the historical importance of X-rays, this impression will be corrected shortly as we describe the crucial role played by X-rays in the development of modern physics.

X-rays were discovered in 1895 by Roentgen while studying the phenomena of gaseous discharge. Using a cathode ray tube (cf. Chapter 3) with a high voltage of several tens of kilovolts, he noticed that salts of barium would fluoresce when brought near the tube, although nothing visible was emitted by the tube. This effect persisted when the tube was wrapped with a layer of black cardboard. Roentgen soon established that the agency responsible for the fluorescence originated at the point at which the stream of energetic electrons struck the glass wall of the tube. Because of its unknown nature, he gave this agency the name *X-rays*. He found that X-rays could manifest themselves by darkening wrapped photographic plates, discharging charged electroscopes, as well as by causing fluorescence in a number of different substances. He also found that X-rays can penetrate considerable thicknesses of materials of low atomic number, whereas substances of high atomic number are relatively opaque.† Roentgen took the first steps in identifying the nature of X-rays

† This property is the basis of the technique of X-ray photography of bones in living tissue, a technique which was used by surgeons only a few months after Roentgen's discovery.

by using a system of slits to show that (1) *they travel in straight lines*, and that (2) *they are uncharged*, because they are not deflected by electric or magnetic fields.

The discovery of X-rays aroused the interest of all physicists, and many joined in the investigation of their properties. In 1899 Haga and Wind performed a single slit diffraction experiment with X-rays which showed that (3) *X-rays are a wave motion phenomenon*, and, from the size of the diffraction pattern, their wavelength could be estimated to be $\lambda \sim 10^{-8}$ cm. In 1906 Barkla proved that (4) *the waves are transverse* by showing that they can be polarized by scattering from many materials.

There is, of course, no longer anything unknown about the nature of X-rays. They are electromagnetic radiation of exactly the same nature as visible light, except that their wavelength is several orders of magnitude shorter. This conclusion follows from comparing properties 1 through 4 with the similar properties of visible light, but it was actually postulated by Thomson several years before all these properties were known. Thomson argued that X-rays are electromagnetic radiation because such radiation would be expected to be emitted from the point at which the electrons strike the wall of a cathode ray tube. At this point, the electrons suffer very violent accelerations in coming to a stop and, according to classical electromagnetic theory, all accelerated charged particles emit electromagnetic radiations. We shall see later that this explanation of the production of X-rays is at least partially correct.

In common with other electromagnetic radiations, X-rays exhibit particle-like aspects as well as wave-like aspects. The reader will recall that the Compton effect (cf. Chapter 3), which is one of the most convincing demonstrations of the existence of quanta, was originally observed with electromagnetic radiation in the X-ray region of wavelengths.

2. Measurement of X-ray Spectra

The problem of measuring the wavelength spectrum of the radiation emitted from an X-ray tube seemed, at first, a very difficult one indeed. Roentgen had shown that the index of refraction of all substances for X-rays is very nearly unity. Consequently, a prism spectrometer is not a useful instrument for X-rays. An alternative is to use a diffraction grating. However, the wavelength of X-rays is so small compared to the line spacing of the best ruled gratings which can be produced that these devices would also appear to be useless. The problem was solved in 1912 when Laue suggested that a crystal might be used as a diffraction grating for X-rays because the atoms of a crystal are arranged in a regular lattice, with a separation of the lattice planes that is quite comparable to the wavelengths of X-rays. This suggestion was first put into practice by Friedrich and Knipping, and the technique was perfected by the Braggs.

Consider a crystal of NaCl. The macroscopic characteristics of such a crystal strongly suggest that on a microscopic scale the atoms are arranged in a regular cubic lattice. A two dimensional section of this lattice is indicated in figure (14–1). The figure also attempts to indicate the density distribution of the electrons within each atom. From our study of multi-electron atoms, we know that there is a region of very high

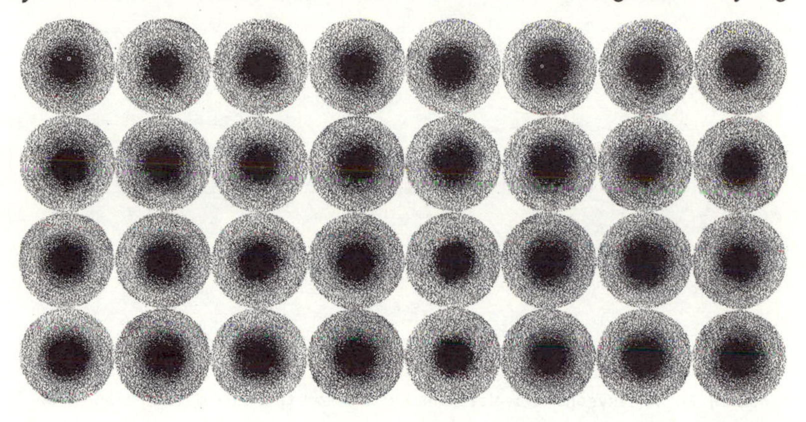

Figure 14–1. A two dimensional section of a sodium chloride crystal.

electron density quite near the center of the atom, and that the electrons in this region are very tightly bound to the atom. Now consider a beam of X-rays incident upon the crystal. The beam penetrates into the interior of the crystal; a fraction of it goes all the way through; but another fraction is scattered by the atoms of the crystal. There are several mechanisms which produce this scattering. All of these mechanisms will be discussed later, but at present we are interested only in the one in which the X-rays are scattered with no change in wavelength and no change in phase from the tightly bound electrons near the center of the atom. Each atom scatters X-rays by this mechanism into all angles, although the scattered intensity is largest in the direction of the incident beam.

In the region exterior to any atom, the intensity pattern of the X-rays which it scatters is the same as if all the scattering occurs exactly at its center. For this reason, it is possible to analyze the scattering by the crystal in terms of a model crystal consisting of a set of point scattering centers with the same relative locations as the centers of the atoms of the real crystal. These scattering centers consequently lie on a set of uniformly spaced lattice planes.

Imagine that a beam of monochromatic X-rays is incident on the model crystal at an angle θ measured from its surface. Part of the beam is scattered from the scattering centers of *each* lattice plane. Let us consider *one* such lattice plane. In the language of wave motion, the incident beam of X-rays can be described as a succession of wave fronts (surfaces of constant phase) moving over the scattering centers of the lattice plane. Each time a wave front crosses a scattering center the scattering center re-emits a spherical wave, with no change in phase, that moves out in all directions. Figure (14–2) shows several incident wave fronts and one set

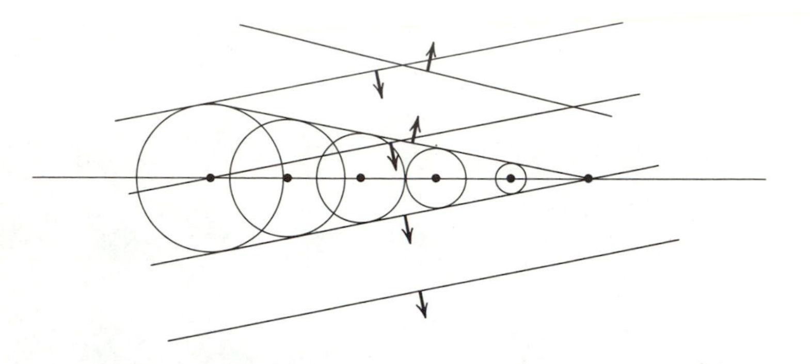

Figure 14–2. A Huygens' wavelet construction for the partial reflection of an X-ray beam from a lattice plane of a crystal.

of re-emitted waves. It is apparent from this Huygens' wavelet construction that the re-emitted waves combine to produce a set of reflected wave fronts, several of which are shown in the figure. These wave fronts constitute a beam reflected from the lattice plane. As can be seen from the figure, the angle of reflection is equal to the angle of incidence.

This result allows us to deal with an even simpler model crystal consisting of the set of lattice planes which are assumed to be partially reflecting surfaces. Figure (14–3) illustrates the incident beam passing through the crystal and being partially reflected from each lattice plane. In this figure the incident and reflected beams are represented as rays. Emerging from the face of the crystal is a set of reflected beams. These beams interfere with each other, and the interference is destructive unless each beam is in phase with all the others. For a given lattice plane spacing d, this condition is satisfied only for certain values of the angle of incidence and reflection θ. To determine these angles, consider a typical pair of reflected beams r and $r + 1$. Since there is no phase change on reflection, these beams will be in phase if the difference between the path length of the radiation following beam $r + 1$ and the path length of the radiation following beam r is equal to an integral number of wavelengths. That is they will be in phase if

$$\delta + \varepsilon = n\lambda, \qquad n = 1, 2, 3, \ldots$$

Since

$$d/\delta = \sin\theta$$

and

$$\varepsilon/\delta = \cos\phi = \cos(180° - 2\theta)$$

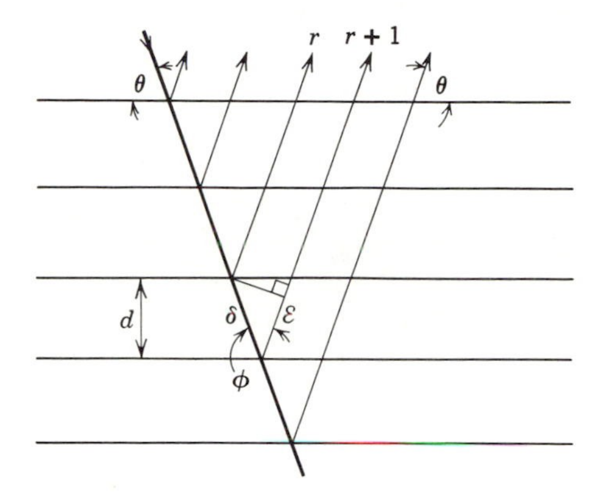

Figure 14–3. Illustrating the partial reflection of an X-ray beam from each lattice plane of a crystal.

this condition can be written

$$\frac{d}{\sin\theta} + \frac{d}{\sin\theta}\cos(180° - 2\theta) = n\lambda$$

This is

$$\frac{d}{\sin\theta}(1 - \cos 2\theta) = \frac{d}{\sin\theta}(1 - 1 + 2\sin^2\theta) = n\lambda$$

or

$$2d\sin\theta = n\lambda, \qquad n = 1, 2, 3, \ldots \tag{14–1}$$

This condition is called *Bragg's law.* For an incident beam of X-rays of wavelength λ, a reflected beam will be found only if θ is one of the solutions to this condition. The solution corresponding to $n = 1$ gives the so-called *first order reflection*; $n = 2$ gives the *second order reflection*, etc. The intensity of the first order reflected beam is greater than that of the second order, the second greater than that of the third, etc., because the intensity of the radiation scattered by atoms decreases with the increasing scattering angles corresponding to increasing values of n. For this reason it is often necessary to use the first order reflection although the higher orders give better spectral resolution.

Figure (14–4) shows an X-ray tube and crystal spectrometer. Following the design of Coolidge (1913), modern X-ray tubes are operated at high vacuum, and the electron beam is produced by a heated filament F. A high voltage applied between the filament and the anode accelerates the electrons into the anode A, which usually must be water cooled. For maximum X-ray yield, a heavy element such as tungsten is used for the

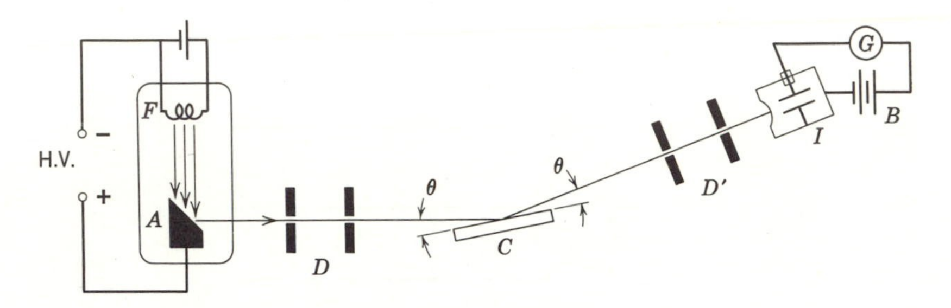

Figure 14–4. An apparatus used to measure X-ray spectra.

anode; however, it is possible to use almost any solid substance. X-rays emitted from the anode are defined into a beam by lead diaphragms D, and reflected from the crystal C. For a given angle θ, only those wavelengths are reflected for which Bragg's law is satisfied. The reflected beam is defined by a second set of diaphragms D' and is detected by an ionization chamber I. The X-rays enter through a thin window and pass between two plates, producing ionization of the gas contained in this region (ionization by X-rays will be discussed later). Ions are attracted to the insulated plate because of the electric field produced by the battery B, a current consequently flows through the galvanometer G, and the strength of the current is proportional to the intensity of the reflected X-ray beam. By measuring the galvanometer current as a function of the angle θ,† the spectrum of the X-rays produced by the anode is measured as a function of that parameter. If the lattice spacing d is known, this can be converted into a spectrum as a function of wavelength by using Bragg's law (14–1), since it is easy to identify the order of the reflection, n.

The lattice spacing d can be obtained for a simple cubic crystal like NaCl from a knowledge of Avogadro's number N, the density ρ of the crystal, and its molecular weight. If W is the molecular weight of NaCl,

† It is necessary to rotate both the ionization chamber and the crystal to keep both angles equal to θ.

then N molecules weigh W grams. They will occupy a volume W/ρ cm^3, and in that volume there will be $2N$ atoms. For NaCl, these atoms are arranged at the corners of a cube of length d. Thus

$$\frac{W}{\rho}\frac{1}{2N} = d^3$$

Substituting the known values of N, W, and ρ gives

$$d = 2.820 \times 10^{-8} \text{ cm} \quad \text{for NaCl}$$

This value has been verified by using Bragg's law to determine d in terms of the known wavelengths of certain X-rays. These wavelengths were actually obtained from measurements using a ruled diffraction grating of directly measurable line spacing. Although the line spacing of any ruled grating is very large compared to X-ray wavelengths, ruled gratings can still be used if the X-rays are incident at a grazing angle because, under these circumstances, the effective (i.e., projected) line spacing is very small. This technique is very difficult, and it is not appropriate for the routine measurement of X-ray spectra. However, it is valuable for calibration purposes, and it forms the basis of one of the most accurate experimental determinations of the value of Avogadro's number.

X-rays are extremely useful in studying the properties of crystals. By bombarding a crystal with a beam of X-rays of known wavelength composition, and determining the angles at which Bragg reflections occur, it is possible to determine the structure and dimensions of the crystal lattice. As an example, such measurements confirm our supposition that the atoms of NaCl are arranged in a regular cubic lattice. However, we shall not describe the results of such experiments because we are more interested in describing the results of the converse experiments, in which the wavelength composition of an X-ray beam is measured by using a spectrometer with a crystal of known properties. Typical results are indicated in figure (14–5), which represents the energy content per unit wavelength interval emitted by an X-ray tube with a ^{74}W anode. The wavelength spectrum is shown for two different values of accelerating voltage V, i.e., for two different values of the kinetic energy $E = eV$ of the electrons incident upon the anode. It is apparent that the observed spectrum can be decomposed into a set of sharp lines superimposed on a continuum. This decomposition is convenient because measurements using a variety of anode materials and accelerating voltages demonstrate that the properties of the continuum spectra are quite different from those of the line spectra. Some of these properties are described below.

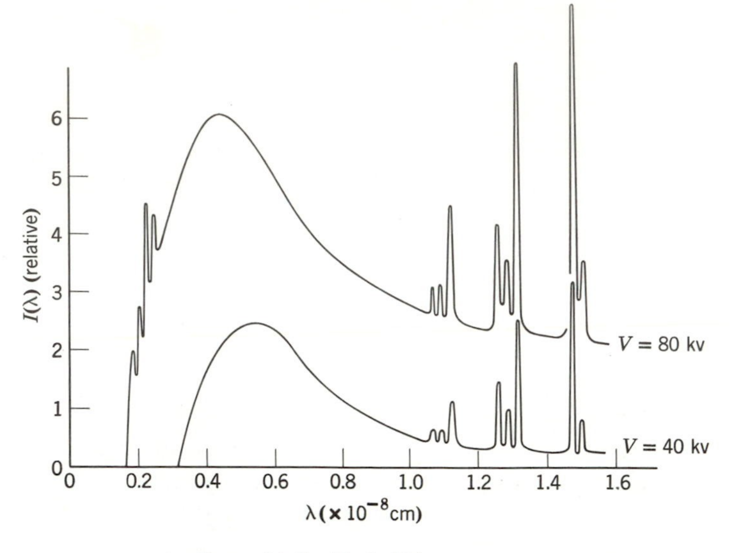

Figure 14–5. Typical X-ray spectra.

(I) Continuum Spectra

1. The shape of the continuum spectrum depends only on the energy of the electrons incident on the anode, and not on the nature of the anode.

2. In particular, the short wavelength cut off λ_0 is inversely proportional to the incident electron kinetic energy E. The cut-off frequency $\nu_0 = c/\lambda_0$ follows the empirical law

$$h\nu_0 = E \tag{14–2}$$

where h is in excellent numerical agreement with the value of Planck's constant obtained from the photoelectric effect, etc.

3. The total X-ray energy per incident electron, W, which is proportional to the integral over λ of the continuum part of the spectrum, obeys the empirical equation

$$W = kZE^2 \tag{14–3}$$

where

$$k \sim 0.7 \times 10^{-4} \quad \text{for } W \text{ and } E \text{ in Mev}$$

and

$$Z \text{ is the atomic number of the anode material}$$

The fraction of electron kinetic energy converted into X-ray energy is

$$W/E = kZE \tag{14–3′}$$

For the typical case $Z = 90$ and $E = 0.05$ Mev, W/E is only about 0.3 percent.

(II) Line Spectra

1. The wavelengths of the lines depend only on the nature of the anode (in fact, on its atomic number Z), and not on the energy of the incident electrons.

2. However, a line cannot appear in the spectrum unless the incident electron kinetic energy E satisfies the relation

$$E \geqslant E_T = h\nu = hc/\lambda \tag{14–4}$$

where ν is the frequency and λ the wavelength of the line in question.

3. The total X-ray energy emitted in a particular line increase with increasing incident electron kinetic energy according to the empirical equation

$$I \propto (E - E_T)^n \tag{14–5}$$

where

$$n \sim 1.5$$

The origins of the continuum and line spectra are as different as their properties. For this reason it is convenient to discuss them separately.

3. X-ray Line Spectra

The origin of the X-ray line spectra may be understood if we consider a particular one of the processes by which an energetic electron is slowed down as it passes into the anode of an X-ray tube. This process is the transfer of energy from the incident electron to the atoms of the anode by Coulomb interaction collisions between the incident electron and the electrons of the atoms.

An energetic electron incident upon an atom can transfer an appreciable fraction of its kinetic energy to an atomic electron only in a close collision with that electron. This statement may be justified and made quantitative, by considering the Coulomb interaction potential energy between two electrons:

$$V(r) = e^2/r$$

Evaluating from this equation the separation distance r at which the potential energy is $V(r) \sim 10^2$ ev, we find $r \sim 10^{-9}$ cm. Since the energy transfer from the incident to the atomic electron cannot be bigger than the maximum potential energy which they experience, we see that energy transfers greater than 100 ev occur only in collisions in which the distance of closest approach between the two electrons is less than 10^{-9} cm. This distance is about 10 percent of an atomic "diameter." Such close collisions are not very frequent because the atomic electrons in the outer shells can move out of the way of the incident electron and, although the atomic electrons in the inner shells cannot do this as they are tightly bound to the center of the atom, the probability that the incident electron will come near these electrons is small because the inner shells are small. Almost all the collisions are associated with an energy transfer of a few electron volts or, at most, a few tens of electron volts. In such collisions, it is energetically possible to excite only an atomic electron in one of the outer shells. The electron can be excited into a bound state, or into a continuum state leaving the atom ionized. In either case, the electron, or some other electron, will eventually drop back down through the various excited states to the ground state, and the atom will emit lines of its optical spectrum. In a small energy transfer collision there is not enough energy available to excite an atomic electron from an inner shell. These electrons are in highly negative energy levels† and, if they are to be excited, they must be excited all the way up to one of the slightly negative bound energy levels, or to one of the positive continuum energy levels, because all the intervening energy levels already contain their complete complement of electrons (cf. section 5, Chapter 13).

However, there are occasional collisions in which the energy transfer is a big fraction of the energy of the incident electron. Providing this energy is large enough (cf. equation 14–4), it is possible to excite an atomic electron of an inner shell from its highly negative energy level to one of the slightly negative or positive energy levels. This leaves the atom in a very highly excited state. The atom will eventually return to its ground state, and in this process it emits a set of very high energy quanta which are members of its X-ray line spectrum. As an example, let us assume that an electron is initially removed from the $n = 1$ shell. In the first step of the de-excitation process an electron from one of the shells of higher n drops into the hole in the $n = 1$ shell; for instance, an $n = 2$ shell electron could drop into the hole. This would leave a hole in the $n = 2$ shell, but the excitation energy of the atom would be considerably reduced because an electron in the $n = 2$ shell is much less tightly bound than an electron in the $n = 1$ shell. Energy is conserved by the emission of a quantum which has an energy equal to the decrease in the excitation energy of the atom. In the next step of the de-excitation process an electron from a shell of $n > 2$ drops into the hole in the $n = 2$ shell, a hole is produced in the shell from which it came, and a quantum of the appropriate energy is

† For example, the energy of an electron in the innermost shell of ^{74}W is -7×10^4 ev.

emitted. Note that the net effect of each of these steps is that a hole jumps to a shell of higher n. These steps continue until the hole has worked its way to the outermost shell, where it is filled by the electron initially ejected from the $n = 1$ shell, or some other electron.

The energy levels of an atom which are involved in the emission of its X-ray spectrum can be conveniently represented in terms of an energy level diagram. We present in figure (14–6) a diagram for the ^{92}U atom showing all its X-ray energy levels through $n = 4$. In this diagram the

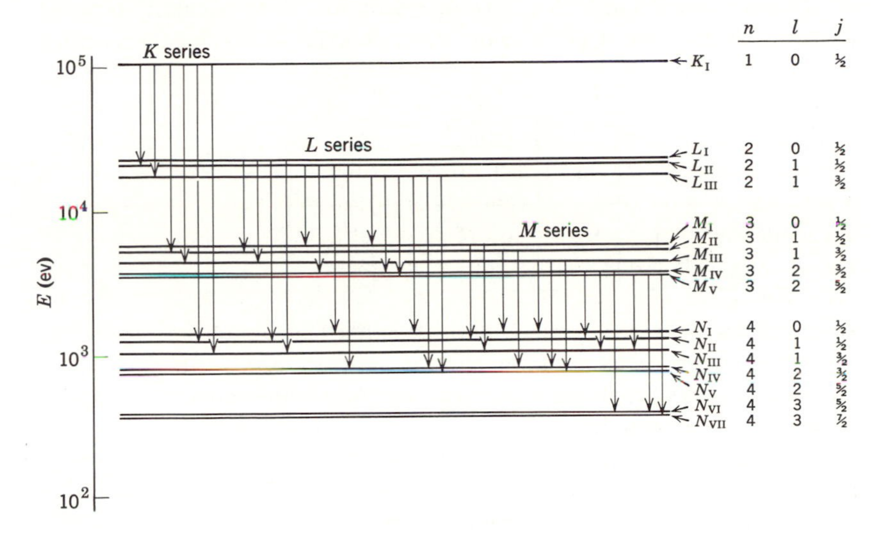

Figure 14–6. The higher energy X-ray levels for the uranium atom, and the possible transitions between these levels.

energy of the atom in its ground state is defined as zero. Each level gives, on a logarithmic scale, the energy of the atom when one electron of the indicated quantum numbers n, l, j is missing.† This can be thought of, then, as an energy level diagram for a hole with quantum numbers n, l, j. Since an electron of negative energy is missing, the energies of all these levels are positive. The energy levels are also identified by a notation commonly used in X-ray spectroscopy. In this notation the value of

† Note that the quantum numbers n, l, j, which completely specify the energy of an atom with one electron in a subshell, also completely specify the energy of an atom with one hole in the subshell. Because of the coupling of the individual spin and orbital angular momenta required by the exclusion principle for the electrons in a subshell, the magnitude of the total, orbital, and spin angular momenta of a subshell which has one hole is the same as if the subshell contained only the single electron corresponding to the hole. In either case the quantum number specifying the magnitude of the spin angular momentum is not written as it can only be $\frac{1}{2}$. Cf. discussion of table (13–9).

the quantum number n is given according to the scheme shown in table (14–1). The Roman numeral subscripts in figure (14–6) label the levels according to decreasing energy, but do not specify the values of l and j.

Because an X-ray energy level diagram gives the energy levels of a hole, the energy ordering of the levels with respect to the quantum numbers n, l, j is the reverse of that found in an optical energy level diagram giving the energy levels of an electron. This is true because, if the energy of an atom with one electron of quantum numbers n, l, j is particularly negative, the energy of an atom with a corresponding hole is particularly positive, as more energy must be given to the atom in order to remove the electron.

TABLE (14–1)
The Spectroscopic Notation for n

n	1	2	3	4	5	. . .
Spectroscopic notation	K	L	M	N	O	. . .

Keeping this difference in mind, all the familiar features of the dependence of atomic energy levels on the quantum numbers n, l, j can be found in an X-ray energy level diagram. In fact, they are more pronounced, particularly the splitting of the energy levels according to the value of j. In the L shell of ^{92}U this splitting is more than 2000 ev, and it is larger than the splitting according to the value of l. In such a case, it is hardly appropriate to describe the j dependence of the energy levels by the optical energy level terminology "fine structure splitting." This strong j dependence, which is characteristic of the inner shells of all atoms except those of very low Z, is partly due to the usual increase in spin-orbit splitting with decreasing n, but primarily due to other relativistic effects which become very large for the high velocity electrons that populate these shells (cf. section 6, Chapter 13).

As we have indicated, it is convenient to think of the production of X-ray line spectra in terms of the creation of a hole in one of its higher energy levels, and the subsequent jumping of the hole through its lower energy levels, with the emission at each jump of an X-ray quantum carrying the excess energy. The frequency ν of the quantum bears the usual relation to the energy E which it carries; $E = h\nu$. However, not all transitions are possible. There are the following set of selection rules for the change in quantum numbers of the hole:

$$\begin{aligned} \Delta l &= \pm 1 \\ \Delta j &= 0, \pm 1 \end{aligned} \qquad (14\text{–}6)$$

Note that these are the same as the selection rules (13–30) and (13–31) which govern the one optically active electron in an alkali atom.† In the X-ray energy level diagram for ^{92}U, we have shown the transitions that obey (14–6). The totality of X-rays which are emitted in these transitions (plus a few which are observed to be emitted very infrequently in violation of these rules) constitute the X-ray line spectrum of the atom. It is conventional to group those lines produced by transitions from the K shell into the so-called K series, from the L shell into the L series, etc. X-ray energy level diagrams are obtained experimentally from an analysis of measured X-ray spectra, and the analysis is greatly facilitated by this grouping.

One striking difference between X-ray and optical spectra is that X-ray spectra vary smoothly from element to element; there are none of the abrupt changes from one element to the next which occur in optical spectra. The reason is that the characteristics of X-ray spectra depend essentially on the binding energies of the electrons in inner shells. With increasing atomic number Z, these binding energies simply increase uniformly, owing to the higher nuclear charge, and are not affected by the periodic changes in the number of electrons in the outer shell. The regularity of X-ray spectra was first observed by Moseley. In 1913, shortly after the invention of the crystal spectrometer, he made a survey of X-ray spectra and obtained data for a number of elements on the wavelengths of two prominent lines, K_α and L_α. The first line is emitted in transitions from the K shell to the L shell, and the second is emitted in transitions from the L shell to the M shell. (Actually, K_α and L_α are each comprised of two lines, but Moseley could not resolve this structure.) Moseley found that these wavelengths could be fitted to within the experimental accuracy by the simple empirical formulas

For K_α:
$$1/\lambda = C_{K\alpha}(Z - 1)^2 \tag{14–7}$$

For L_α:
$$1/\lambda = C_{L\alpha}(Z - 7.4)^2 \tag{14–7'}$$

where $C_{K\alpha}$ and $C_{L\alpha}$ are constants having values of the order of the Rydberg constant for an infinitely massive nucleus, R_∞, defined by (5–15). Moseley interpreted these results on the basis of the Bohr theory of the atom, which had been proposed in the same year as his experiments. According to equation (5–14), the Bohr theory predicts that the energy of an electron in the shell of quantum number n is

$$E_n = -R_\infty hc\,\frac{Z^2}{n^2} \tag{14–8}$$

† This is a manifestation of the symmetry, described in the previous footnote, between the properties of a subshell with one electron and a subshell with a corresponding hole.

where we have introduced R_∞. As the excitation energy of an atom with a electron missing from the shell of quantum number n is equal to $|E_n|$, the energy E of the X-ray quantum emitted when the hole jumps from the n_i shell to the n_f shell is

$$E = R_\infty hcZ^2(1/n_i^2 - 1/n_f^2)$$

This energy is related to the frequency ν of the quantum by the equation

$$E = h\nu = hc/\lambda$$

so

$$hc/\lambda = R_\infty hcZ^2(1/n_i^2 - 1/n_f^2)$$

or

$$1/\lambda = R_\infty Z^2(1/n_i^2 - 1/n_f^2) \tag{14-9}$$

For K_α, $n_i = 1$ and $n_f = 2$; for L_α, $n_i = 2$ and $n_f = 3$. Therefore the Bohr theory predicts

For K_α: $$1/\lambda = [R_\infty(1/1^2 - 1/2^2)]Z^2 \tag{14-10}$$

For L_α: $$1/\lambda = [R_\infty(1/2^2 - 1/3^2)]Z^2 \tag{14-10'}$$

In each case the value of the constant in square brackets agrees well with the constant C in equations (14–7) and (14–7′). That these empirical equations contain $(Z - 1)$ and $(Z - 7.4)$, instead of Z, was explained by Moseley as due to shielding of the nuclear charge by atomic electrons (which was not accounted for in Bohr's derivation of equation 14–8). According to Moseley, the one electron remaining in the K shell during the transition of the hole from the K to L shells shields the nucleus so that the effective Z is reduced to $(Z - 1)$ for the emission of the K_α line. Similarly, the nine electrons remaining in the K and L shells shield the nucleus during the transition of a hole from the L to M shells. However, this shielding is not perfect, and the effective Z for the emission of the L_α line is reduced only to $(Z - 7.4)$. Moseley's description of shielding is closely related to the description connected with equation (13–22), and both are only approximations to the more accurate description provided by the Hartree theory. Furthermore, later measurements have shown that the empirical formulas (14–7) and (14–7′) are also only approximations.

Nevertheless, Moseley's work was an important step in the development of modern physics. His simple and successful application of the Bohr theory to X-ray line spectra provided one of the earliest confirmations of that theory. By using the empirical formulas (14–7) and (14–7′) to determine Z, Moseley established unambiguously the correlation of the nuclear charge of an atom with its ordering in the periodic table of the elements. He found, for instance, that the atomic number of ^{27}Co is less than that of ^{28}Ni even though its atomic weight is greater. He also showed that there were gaps in the periodic table, as it was then known, at $Z = 43$, 61, 72, and 75. Elements of these atomic numbers have all subsequently been discovered.

4. X-ray Continuum Spectra

The transfer of energy from an incident electron to the electrons in the atoms of an X-ray tube anode is the process which is primarily responsible for the slowing down of the incident electron, and it also leads (in the infrequent case of a large energy transfer) to the emission of the X-ray line spectra. But this process is not the only one which the incident electrons undergo. Another is the Coulomb interaction scattering of these electrons in close collisions with nuclei of the atoms in the anode. In a single collision of this type an incident electron suffers large accelerations and can be scattered through a large angle. Although the electron does not lose energy to the nucleus because nuclei are so massive that they do not recoil appreciably, the electron sometimes does lose energy by electromagnetic radiation. This radiation constitutes the continuum X-ray spectrum.†

Both classical theory and quantum theory predict the emission of electromagnetic radiation by a scattered electron. Classically, small amounts of radiation are emitted at all times in every scattering because Maxwell's equations lead to the conclusion that charged particles radiate whenever they accelerate. In the quantum theory this is not what happens. Instead, large amounts of radiation (quanta) are emitted occasionally in a scattering process. The situation can be compared to that which exists for the atom—the classical theory predicts the constant emission of radiation by the accelerated electron circling the nucleus, while the quantum theory predicts the occasional emission of radiation when the electron changes quantum states. In the case of the *bound* atomic electron, we know that conclusions drawn from the quantum theory are always correct and those drawn from the classical theory are generally incorrect. In the case of the *unbound* scattered electron, it is also true that the quantum theory always leads to the correct conclusions. However, in this case the classical theory does not lead to conclusions which are generally incorrect. In fact, the classical theory is capable of describing quite well certain features of the production of continuum spectrum X-rays. The contrast between

† X-rays of the continuum spectrum are often called *Bremsstrahlen*. This German word means braking radiation (i.e., deceleration radiation) and is a name derived from the theoretical explanation of their origin.

the two cases reflects the fact that the differences between a quantized and a non-quantized theory are not so fundamental in the case of an unbound particle as they are in the case of a bound particle.

According to equation (2–7) of the classical theory, a particle of charge ze, suffering an acceleration a, will emit energy in the form of electromagnetic radiation at the rate

$$R = \frac{2}{3}\frac{(ze)^2a^2}{c^3} \tag{14–11}$$

When this particle is scattered by a nucleus of charge Ze, it experiences a Coulomb force which is proportional to zZe^2 and, if the mass of the particle is m, its acceleration is proportional to zZe^2/m. Consequently the rate of emisssion of energy in electromagnetic radiation can be written

$$R \propto z^2a^2 \propto z^4Z^2/m^2 \tag{14–12}$$

The factor $1/m^2$ predicts that the emission of electromagnetic radiation is very much more important for electrons than it is for protons or other heavy particles. This is true; the energy emitted in continuum spectrum X-rays by a heavy charged particle in traversing matter is almost immeasurably small. The factor Z^2 predicts that anodes of high atomic number are much more efficient producers of these X-rays.† This is also verified by experiment. By evaluating the proportionality constant in (14–12), and then integrating over the (average) trajectory of the incident electron in coming to a stop in the anode, it is possible to estimate the total continuum spectrum X-ray energy W that is radiated. The results are in reasonable agreement with the empirical equation (14–3).

The classical theory is also capable of giving a good account of the polarization of the continuum spectrum X-rays. This theory predicts that the X-rays, emitted in the direction **r** during an acceleration **a**, are polarized with the electric vector perpendicular to **r** in the plane defined by **r** and **a**. Now the higher energy X-rays are all emitted in "head-on" collisions in which **a** is always approximately antiparallel to **i**, the direction of the incident beam of electrons. Consequently, such X-rays have a large net polarization. In the collisions which lead to the emission of lower energy X-rays, the direction of **a** is not so constrained and, as a result, the net polarization is small.

Classical electromagnetic theory even gives an accurate prediction of the angular distribution of the X-ray energy emitted in the continuum

† The Z^2 dependence also shows that the radiation emitted in scattering from an atomic electron (i.e., for $Z = 1$) is negligible.

spectrum. For incident electrons of non-relativistic energies, the angular distribution predicted by equation (2–6) is

$$I(\theta) \propto \sin^2 \theta \tag{14–13}$$

where θ is the angle between $\mathbf{r}$ and the average direction of acceleration, which is $-\mathbf{i}$.† For incident electrons of relativistic energies, certain relativistic transformation effects shift the angular distribution in the direction of $\mathbf{i}$.

The one aspect of the emission of continuum spectrum X-rays that cannot be explained by classical theory is the shape of the spectrum; this can be explained only by quantum theory. Consider the cut-off frequency (equation 14–2), which is certainly the most striking feature of the spectrum.

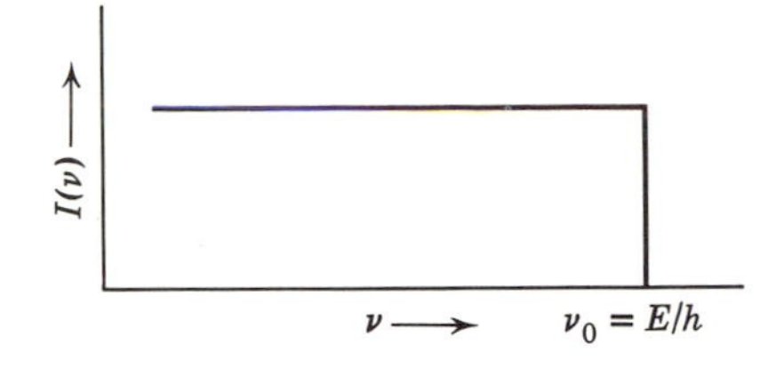

Figure 14–7. A thin anode X-ray continuum frequency spectrum.

This cut-off exists because the highest energy quantum that can be emitted by an electron of kinetic energy E has the energy $h\nu_0 = E$, where ν_0 is its frequency. In fact, this was one of the earliest pieces of evidence in support of Einstein's theory of quanta. Unfortunately, an explanation of the other features of the spectrum cannot be given by such elementary quantum theory, but only in terms of quantum electrodynamics. This subject is far beyond the level of this book, and we can only describe the results of the theory. For electrons of energy E incident upon an anode which is so thin that their energy remains essentially constant in passing through it, the spectrum is predicted by the theory to have the simple shape shown in figure (14–7). The quantity $I(\nu)$ is the X-ray energy per unit frequency interval. For comparison with the data presented in section 2, we convert this to $I(\lambda)$, the X-ray energy per unit wavelength interval. The relation between two quantities such as these has been derived before (cf. equation 2–27). It is

$$I(\lambda) = \frac{c}{\lambda^2} I(\nu)$$

† This angular distribution has the same form as the angular distribution of the energy emitted in radiofrequency radiation by electrons accelerated back and forth along a linear antenna. From the point of view of the classical theory, this process is essentially the same as that involved in X-ray emission.

Now, since

$$I(\nu) \simeq \text{constant}, \qquad \nu < \nu_0 = E/h$$

$I(\lambda)$ can be written

$$I(\lambda) \simeq \text{constant}/\lambda^2, \qquad \lambda > \lambda_0 = c/\nu_0 = ch/E$$

This spectrum is plotted in figure (14–8). The predicted wavelength spectrum for a "thin" anode is in good agreement with measurements

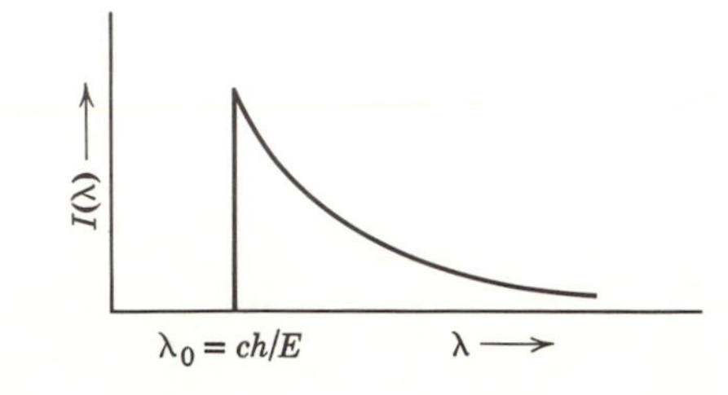

Figure 14–8. A thin anode X-ray continuum wavelength spectrum.

that have been made for thin anodes. It is also in agreement with the more typical measurements using "thick" anodes in which the incident electrons are brought to a stop. We show this in figure (14–9) by generating a thick anode spectrum from the sum of several thin anode spectra. Four

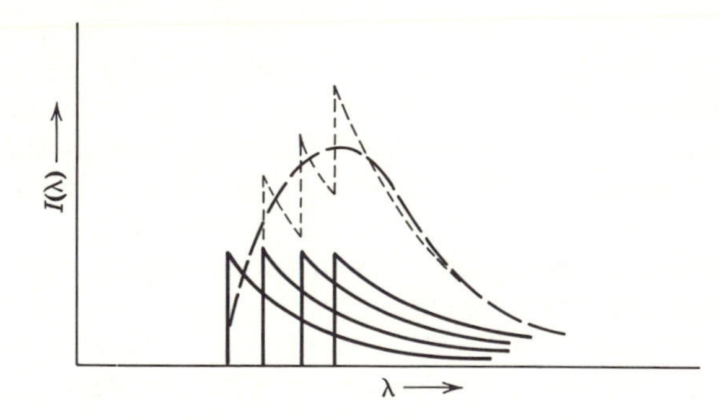

Figure 14–9. Illustrating the composition of a thick anode X-ray continuum wavelength spectrum.

values of E are used to represent the range of electron energies present in the anode. The dotted line shows the results of this approximate calculation. The dashed line indicates the results of an accurate calculation involving integration over the electron energy. Agreement with experimental data, such as presented in section 2, is excellent.

5. The Scattering of X-rays

Having discussed the nature and production of X-rays, we now turn our attention to their interaction with matter. Consider a beam of X-rays passing through a foil and into an X-ray detector, as indicated schematically in figure (14–10). A quantitative measurement of the interaction of the X-rays with the material contained in the foil is obtained by measuring the intensity of the attenuated beam striking the detector, and then, by removing the foil, measuring the intensity of the incident beam. The results are expressed in terms of the *attenuation* of the foil for these X-rays, which is defined as the fractional decrease in the intensity of the beam.

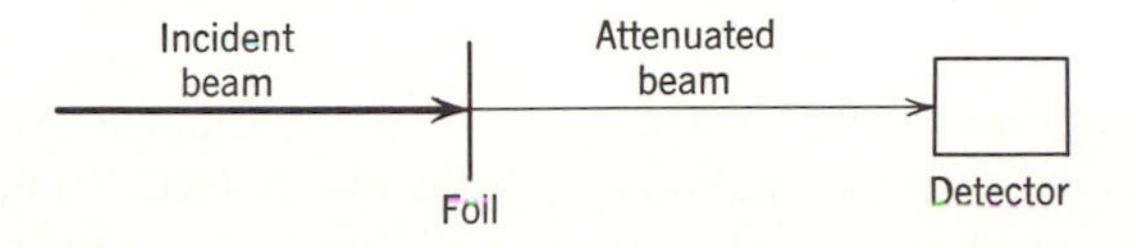

Figure 14–10. An apparatus used to measure the scattering and absorption of a beam of X-rays.

In passing through the foil, X-rays can be completely absorbed from the beam, and they can also be scattered. If the angle subtended at the foil by the detector is very small, then essentially all the scattered X-rays will miss the detector and the measurement will give the attenuation due to both absorption and scattering.† In such measurements it is found that the attenuation increases with the thickness of the foil and, generally, with the atomic number of the material which it contains. The attenuation also depends on the energy of the X-rays, but the dependence is not so easy to describe. Several very interesting phenomena are involved in the absorption and scattering processes which produce the attenuation. In this section we shall discuss those involved in the scattering process.

Consider the interaction of a beam of X-rays with the electrons in the atoms of the foil. We make the following assumptions.

1. The wavelength of the X-rays is small compared to the characteristic dimensions of the atoms:

$$\lambda \ll 10^{-8} \text{ cm} \tag{14–14}$$

† By varying this angle it is possible to separate experimentally the attenuation due to absorption from the attenuation due to scattering, and measure these quantities independently. The attenuation due to scattering can also be measured separately by an experiment in which the scattered X-rays are detected.

2. The energy of a quantum of the X-rays is large compared to the binding energy of the atomic electrons:

$$h\nu \gg \text{B.E.} \tag{14–15}$$

3. The energy of a quantum of the X-rays is small compared to the rest mass energy of an electron:

$$h\nu \ll mc^2 \tag{14–16}$$

The validity of these assumptions will be discussed later. For the moment let us explore their consequences. The first assumption implies that the phase relations between the X-rays scattered by the electrons in different atoms, or the same atom, will be essentially random, or incoherent. It rules out interference effects between different atoms (such as those discussed in section 2), as well as interference effects within the same atom. If this assumption is valid, the interaction of any electron with the X-rays is independent of the interactions of all the other electrons in the foil, and we therefore need to consider only one of these interactions. The second assumption allows us to ignore the binding of the electrons in the atoms and consider them to be free. The third assumption allows us to ignore the quantum aspects of X-rays which manifest themselves in Compton scattering.† Making these assumptions, we give a classical electromagnetic treatment of the scattering process due to Thomson (ca. 1900). In this treatment the X-rays are considered to be a beam of electromagnetic waves incident upon a single free electron. The interaction of the X-rays with the electron is then due to the interaction of its charge with the sinusoidally oscillating electric field incident upon it. (The interaction with the magnetic field is negligible.) We describe the instantaneous strength and direction of the electric field at the location of the electron by the vector **E**. This electric field acting on the electron produces a force $\mathbf{F} = -e\mathbf{E}$, and an acceleration

$$\mathbf{a} = \mathbf{F}/m = -e\mathbf{E}/m \tag{14–17}$$

where **E**, **F**, and **a** all oscillate sinusoidally with the frequency ν of the incident electromagnetic waves. Under the influence of this acceleration, the position of the electron will oscillate sinusoidally with frequency ν, and the electron will emit electromagnetic waves of the same frequency as, and in phase with, the incident waves. The energy emitted in these waves is supplied by the incident electromagnetic waves. Thus the electron absorbs energy from the beam of X-rays and scatters it in all directions.

The rate of emission of energy by the accelerated electron, that is, the

† Equation (3–17) shows that if (14–16) is valid the change of wavelength in the scattering of X-rays from a free electron is negligible.

rate at which the electron scatters energy, is given by equation (14–11) with $z = -1$:

$$S = \tfrac{2}{3}e^2a^2/c^3$$

where here we write S for scattered. Using (14–17), this becomes

$$S = \tfrac{2}{3}e^4E^2/m^2c^3 \tag{14–18}$$

As E changes constantly in time, S does also, and it is most convenient to deal with its time average,

$$\bar{S} = \overline{\tfrac{2}{3}e^4E^2/m^2c^3} = \tfrac{2}{3}e^4\overline{E^2}/m^2c^3 \tag{14–19}$$

The quantity $\bar{S}$ is the average energy emitted in all directions per second by the electron in the form of electromagnetic radiation, and we define this quantity to be the intensity of the scattered radiation.

Next let us evaluate $\bar{I}$, the intensity of the incident radiation. We define this quantity to be the average energy carried by the beam per second across a 1 cm² area oriented normal to the direction of the beam.† Equation (2–5) shows that the relation between $\bar{I}$ and the average energy density $\bar{\rho}$ is

$$\bar{I} = \bar{\rho}c \tag{14–20}$$

According to equation (2–4), $\bar{\rho}$ is

$$\bar{\rho} = \overline{E^2}/4\pi$$

where we delete the subscript for transverse. Thus $\bar{I}$ may be written

$$\bar{I} = c\overline{E^2}/4\pi \tag{14–21}$$

Now comparing equations (14–19) and (14–21), we see that both $\bar{S}$ and $\bar{I}$ are proportional to $\overline{E^2}$. Therefore, $\bar{S}$ and $\bar{I}$ are proportional to each other. That is

$$\bar{S} = \sigma_T\bar{I} \tag{14–22}$$

where σ_T is the proportionality constant. Evaluating σ_T from equations (14–19) and (14–21), we have

$$\sigma_T = \frac{8\pi}{3}(e^2/mc^2)^2 = 6.66 \times 10^{-25}\ \text{cm}^2 \tag{14–23}$$

This quantity, called the *Thomson cross section of the electron*, describes a characteristic property of a free electron. Through equation (14–22), σ_T tells what fraction of the intensity of a beam of X-rays is scattered into all directions by a free electron due to the *Thomson scattering* process.

† Note that the dimensions of $\bar{S}$ are ergs-sec^{-1}, while the dimensions of $\bar{I}$ are ergs-cm^{-2}-sec^{-1}.

It is reasonable to call σ_T a "cross section" because it has the dimensions of cm^2. But the name has a deeper significance that can be understood if we imagine the electron at the center of an imaginary disk oriented normal to the beam, and of area equal to σ_T cm^2. Then, if all the radiation in the beam hitting the disk were scattered, the ratio of the scattered intensity $\bar{S}$ to the incident intensity $\bar{I}$ would be just σ_T, in agreement with equation (14–22). This is easy to see by referring to figure (14–11) (in which the

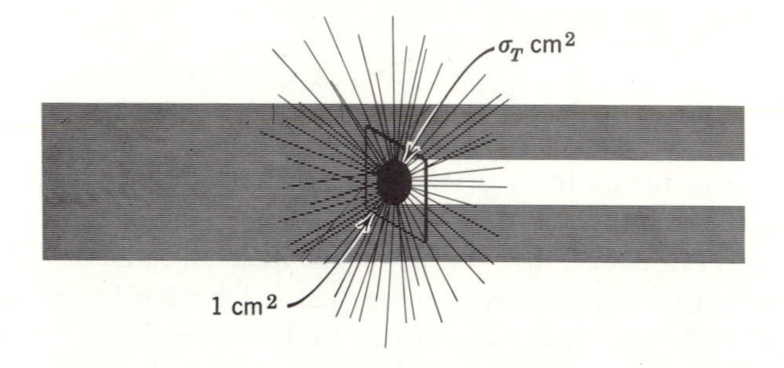

Figure 14–11. The geometrical interpretation of a cross section.

area of σ_T is exaggerated by about a factor of 10^{23}!). The average energy incident per second on the 1 cm^2 area is equal, by definition, to the incident intensity $\bar{I}$. The fraction of this energy that hits the imaginary disk is precisely the ratio $\sigma_T\ \text{cm}^2/1\ \text{cm}^2 = \sigma_T$. If all this radiation were scattered, the energy scattered per second, which is the scattered intensity $\bar{S}$, would be $\sigma_T \bar{I}$. This proves our point. The geometrical interpretation of a cross section is convenient for visualization, and even calculation, but it should

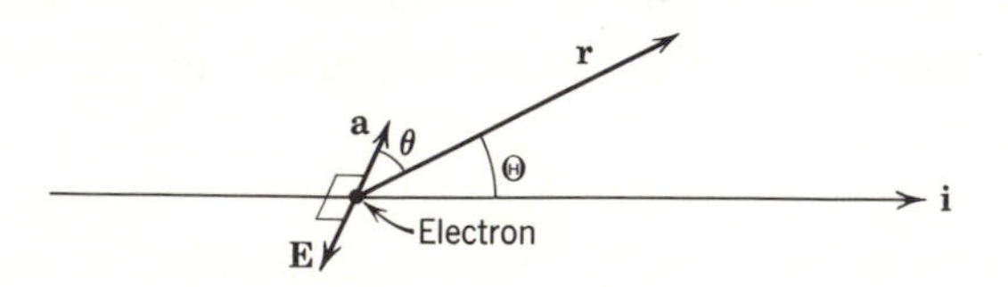

Figure 14–12. Illustrating the relation between the vectors which determine the polarization of scattered X-rays.

not be taken so literally as to lead to the conclusion that there actually exists a well-defined distance r, satisfying the equation $2\pi r^2 = \sigma_T$, such that all X-rays passing within this distance of the electron are scattered, whereas all other X-rays are not scattered.

Classical electromagnetic theory also makes predictions concerning the angular distribution of the X-rays scattered by an electron in the Thomson scattering process. Consider figure (14–12). The direction of the incident beam of X-rays is the vector **i**. The vector **E**, which is perpendicular to **i**, is the direction of the electric vector of a typical polarization component of the beam, and the acceleration **a** which this component gives to the electron lies in the opposite direction. The angular distribution of the radiation emitted by the electron suffering this acceleration is given by (14–13), where θ is the angle between the direction of acceleration **a** and the direction of emission of the radiation **r**. This angular distribution may

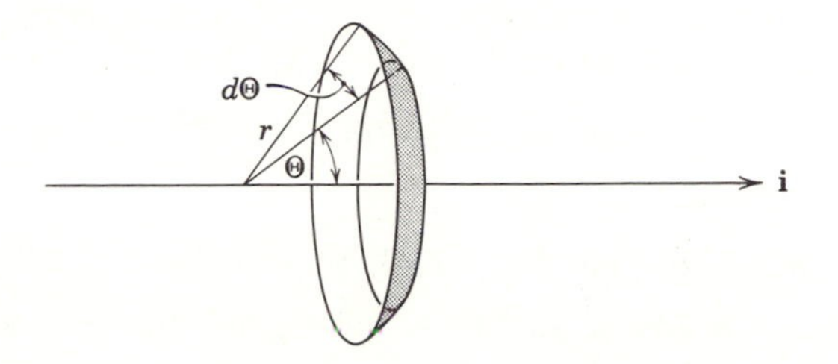

Figure 14–13. The element of solid angle between Θ and $\Theta + d\Theta$.

also be written in terms of the angle Θ between **r** and **i**, and the angle (not indicated on the figure) specifying the azimuthal orientation of **E**. Doing this, and then averaging over all possible azimuthal orientations to treat the case of an unpolarized incident beam, the angular distribution is obtained as a function of the angle Θ. It is

$$\bar{S}(\Theta) \propto (2 - \sin^2 \Theta) \tag{14–24}$$

where $\bar{S}(\Theta)$ is the average energy emitted per second per unit solid angle at the angle Θ.† The integral of $\bar{S}(\Theta)$ over all solid angles is equal to the average total energy emitted per second, or the intensity of the emitted radiation $\bar{S}$. Let us write (14–24) as

$$\bar{S}(\Theta) = K(2 - \sin^2 \Theta) \tag{14–24$'$}$$

and evaluate the integral:

$$\bar{S} = \int \bar{S}(\Theta)\, d\Omega$$

The element of solid angle $d\Omega$ is equal to the area intercepted on a sphere by cones of half-angles Θ and $\Theta + d\Theta$, divided by the square of the radius of the sphere. This is illustrated in figure (14–13). The area is $2\pi r^2 \sin \Theta\, d\Theta$, so

$$d\Omega = \frac{2\pi r^2 \sin \Theta\, d\Theta}{r^2} = 2\pi \sin \Theta\, d\Theta \tag{14–25}$$

† Since solid angle is dimensionless, the dimensions of $\bar{S}(\Theta)$ are ergs-sec^{-1}.

Consequently

$$\bar{S} = 2\pi K \int_0^\pi (2 - \sin^2 \Theta) \sin \Theta \, d\Theta \tag{14–26}$$

According to equations (14–22) and (14–23),

$$\bar{S} = \frac{8\pi}{3} (e^2/mc^2)^2 \bar{I} \tag{14–27}$$

By carrying out the integration of (14–26), and equating the resulting expression to (14–27), we evaluate K. With this value, (14–24′) becomes

$$\bar{S}(\Theta) = \tfrac{1}{2}(e^2/mc^2)^2(2 - \sin^2 \Theta)\bar{I}_0 \tag{14–28}$$

Defining the symbol

$$\begin{aligned} d\sigma_T/d\Omega &= \tfrac{1}{2}(e^2/mc^2)^2(2 - \sin^2 \Theta) \\ &= 3.98 \times 10^{-26}(2 - \sin^2 \Theta) \text{ cm}^2/\text{unit solid angle} \end{aligned} \tag{14–29}$$

equation (14–28) can be written

$$\bar{S}(\Theta) = \frac{d\sigma_T}{d\Omega} \bar{I} \tag{14–30}$$

The quantity $d\sigma_T/d\Omega$, which also has the dimensions cm^2, is called the *differential Thomson cross section of the electron.* It tells, through equation (14–30), what fraction of the intensity of a beam of X-rays is scattered in a Thomson scattering process by a free electron into a unit solid angle at the angle Θ—just as σ_T specifies, through equation (14–22), what fraction of the intensity is scattered in that process by a free electron into all solid angles. It is possible to give a geometrical interpretation to a differential cross section similar to that which we gave above for a cross section. However, we again caution the reader against taking these interpretations too literally.

Having completed our discussion of the Thomson scattering of X-rays by a free electron, let us return to the more realistic problem of the scattering of X-rays by an atom. If assumptions (14–14) and (14–15) are satisfied, this problem is trivial because each of the Z electrons of the atom scatters X-rays independently, and the total intensity scattered by the atom is Z times the intensity scattered by a single electron. Evaluating this intensity from equation (14–22), we obtain for the total $\bar{S}$

$$\bar{S} = Z\sigma_T \bar{I}$$

Defining

$$\sigma_{S_a} = Z\sigma_T \tag{14–31}$$

this can be written

$$\bar{S} = \sigma_{S_a} \bar{I} \tag{14–32}$$

The quantity σ_{S_a} is the *scattering cross section per atom*. It has the value given by (14–31) only if the assumptions mentioned above are satisfied. If they are, the *differential scattering cross section per atom* is also simply

$$\frac{d\sigma_{S_a}}{d\Omega} = Z\frac{d\sigma_T}{d\Omega} \tag{14–33}$$

This is related to the total $\bar{S}(\Theta)$ by the equation

$$\bar{S}(\Theta) = \frac{d\sigma_{S_a}}{d\Omega}\bar{I} \tag{14–34}$$

From these scattering cross sections per atom it is easy to calculate the predicted attenuation, due to scattering, of a foil containing a known number of atoms per unit volume.† In comparing this prediction with the results of the experiments described at the beginning of this section, agreement to within 10 or 20 percent is found for atoms with $Z \lesssim 15$ and for X-rays of wavelength near $\lambda = 0.2 \times 10^{-8}$ cm. This formed the basis of experiments performed by Barkla in 1909, which constituted the first measurements of Z, the number of electrons in an atom.

The quantum energy of X-rays of wavelength $\lambda = 0.2 \times 10^{-8}$ cm is $h\nu = 6 \times 10^4$ ev. As this wavelength and energy satisfy both assumptions (14–14) and (14–15) for atoms of small Z (and also satisfy assumption 14–16), we can understand why the cross sections per atom measured with these X-rays agree, approximately, with the predicted values. We can also understand why they do not agree for measurements made with X-rays of longer wavelength and atoms of larger Z. For these cases, the X-ray wavelength is comparable to, or larger than, the characteristic dimensions of the atoms, and also the electrons do not act as if they were free. For both these reasons the electrons of the atom do not interact independently with the incident beam of X-rays. In the limit of very long wavelengths and very low quantum energies, the interactions are the very opposite of being independent because, under these circumstances, the bound electrons of the atom are all oscillating under the influence of the incident electromagnetic waves with exactly the same phase, and the electromagnetic waves emitted by any one electron are completely in phase with those emitted by all the others. Because of this phase coherence the scattered intensity will be very much larger. It is easy to evaluate this by thinking of all the Z electrons oscillating together as one "big electron" of charge Ze and mass Zm. According to equation (14–17), the acceleration of this object will be the same as that of a single electron, but according to (14–11) the energy which it radiates will be larger than that radiated by a single electron by a factor of Z^2. Thus, in this limit the actual scattering

† Such a calculation will be carried out at the end of section 7.

cross section per atom, σ_{S_a}, will be Z^2 times the Thomson cross section of the electron, σ_T, and Z times larger than the prediction of equation (14–31). The actual differential scattering cross section per atom, $d\sigma_{S_a}/d\Omega$, will also be Z^2 times larger than the differential Thomson cross section of the electron, $d\sigma_T/d\Omega$, and Z times larger than the prediction of (14–33).

For X-rays whose wavelengths and quantum energies are such that the scattering from the various atomic electrons is neither completely coherent nor completely incoherent, the situation is much more complicated, but it can still be treated by classical electrodynamics. Assuming that the distribution of electrons in the atom is spherically symmetrical, the amplitude of the electromagnetic wave scattered by each volume element is evaluated, and then the contributions from all the volume elements are summed by adding amplitudes with relative phase factors that depend on the location of the volume element, the angle of scattering, and the wavelength. By squaring the net amplitude, the total intensity scattered at the angle in question is found. Dividing this by the incident intensity, the differential scattering cross section per atom is obtained. It can be written

$$\frac{d\sigma_{S_a}}{d\Omega} = \frac{d\sigma_T}{d\Omega}|F(\chi)|^2 \qquad (14\text{–}35)$$

where

$$F(\chi) = \int_0^\infty \rho(r)(\sin \chi r/\chi r) 4\pi r^2 \, dr \qquad (14\text{–}35')$$

with

$$\chi = 2K \sin(\Theta/2)$$

and

$$K = 2\pi/\lambda$$

In these expressions λ is the X-ray wavelength, Θ is the scattering angle, and $\rho(r)$ is the electron density of the atom in units of the electronic charge. Since

$$\int_0^\infty \rho(r) 4\pi r^2 \, dr = Z$$

we see that the function $F(\chi)$, which is called the *atomic form factor*, takes on its maximum value, the atomic number Z, when $\sin \chi r/\chi r = 1$. This condition is satisfied when $\chi \to 0$, which happens for all Θ when $K \to 0$, and which also happens for $\Theta \to 0$.

The behavior of $F(\chi)$ for $K \to 0$, or $\lambda \to \infty$, confirms the conclusions of the next to last paragraph. Its behavior for $\Theta \to 0$ shows the reason for the small angle peaking of the scattered intensity mentioned in Section 2. Figure (14–14) compares the differential scattering cross section per atom, measured with X-rays of wavelength $\lambda = 0.71 \times 10^{-8}$ cm incident upon ^{6}C atoms, with the prediction of equation (14–33). The discrepancy at small angles between the measured values of $d\sigma_{S_a}/d\Omega$ and the predictions of equation (14–33) provide convincing evidence of coherent effects in the scattering process. To compare the measured values $d\sigma_{S_a}/d\Omega$ with the predictions of equation (14–35), which takes these coherent effects into account, it is necessary to know the electron density that enters in the form factor (14–35′). If the electron density obtained from the Hartree theory of the atom is used, quite satisfactory agreement is obtained. Alternatively, equations (14–35) and (14–35′) can be used to

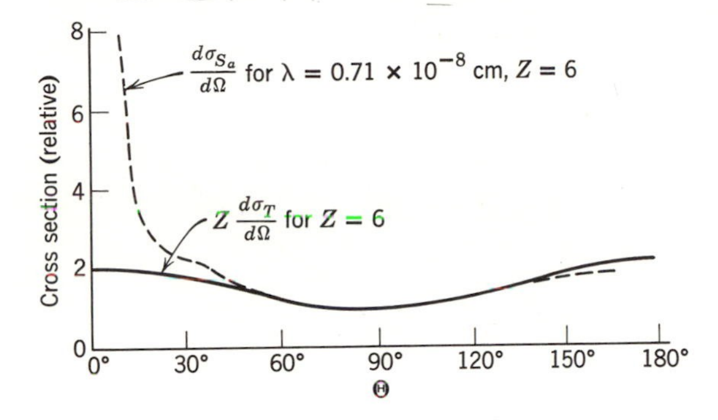

Figure 14–14. The differential scattering cross section per atom for $\lambda = 0.71$ A X-rays incident upon carbon. From A. H. Compton and S. K. Allison, *X-rays in Theory and Experiment*, D. Van Nostrand Co., Princeton, N.J., 1935.

extract the electron density of an atom, $\rho(r)$, from the measurements of its differential scattering cross section $d\sigma_{S_a}/d\Omega$. X-rays allow us to "see" not only the invisible structure of bones in living tissue, but also the invisible structure of atoms.

By integrating $d\sigma_{S_a}/d\Omega$ from equation (14–35) over all solid angles, the predicted value of the scattering cross section per atom, σ_{S_a}, is obtained as an expression involving the electron density $\rho(r)$. For cases in which this expression can be evaluated because $\rho(r)$ is known, the results are in reasonably good agreement with experiment, as long as λ is greater than about 0.2×10^{-8} cm. Experimental data are shown in figure (14–15). The ratio of the observed scattering cross section per atom to the prediction of equation (14–31) is plotted as a function of the X-ray wavelength. The curves are labeled with the atomic number of the atom. Note that these data provide examples of the fact, mentioned above, that the observed scattering cross sections per atom agree with equation (14–31) for $\lambda \simeq 0.2 \times 10^{-8}$ cm but become larger than this prediction with increasing λ owing to the effect of coherences.

Also note that the observed scattering cross sections per atom become smaller than the predictions of classical electromagnetic theory (with or without coherences) when λ becomes smaller than $\sim 0.2 \times 10^{-8}$ cm. For X-rays of these wavelengths, the quantum energy $h\nu$ is no longer very small compared to the rest mass energy of an electron, $mc^2 = 5.1 \times 10^5$ ev.

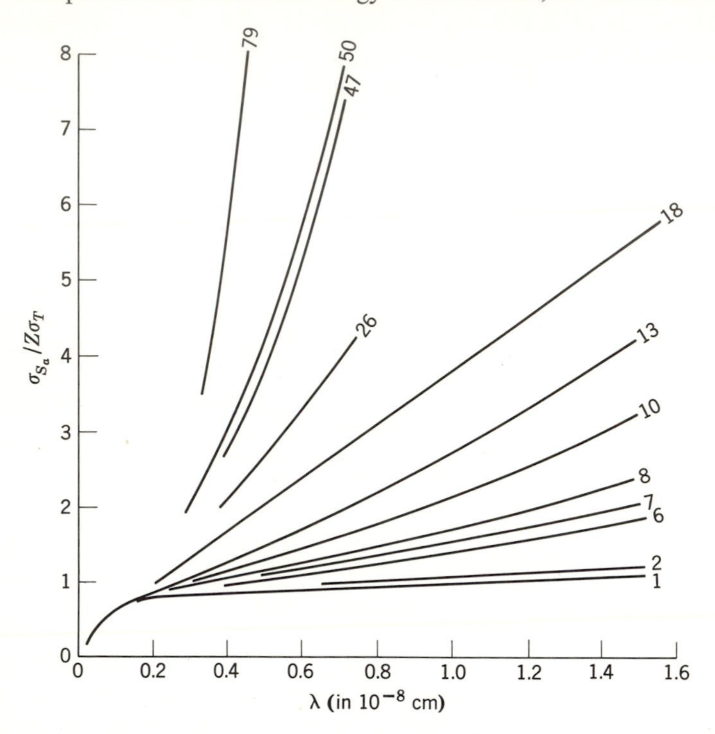

Figure 14–15. The scattering cross sections per atom for several atoms and a range of wavelengths. Each curve is labeled by the atomic number of the atom. From A. H. Compton and S. K. Allison, *X-rays in Theory and Experiment*, D. Van Nostrand Co., Princeton, N.J., 1935.

This violates assumption (14–16), and so the scattering can no longer be treated by classical theory; instead, quantum theory must be used. We have already discussed the energetics of the quantum theory for scattering of X-rays by free electrons. The reader will find it under the heading *The Compton Effect* in section 8, Chapter 3, and it would be appropriate at this point for him to review the discussion there.

The wavelength spectra presented in Chapter 3 are for the case of scattering from ^{6}C. For this small Z atom, the data show that the probability of scattering with an increase in the X-ray wavelength (Compton scattering) is quite comparable to the probability of scattering with no change in wavelength (Thomson scattering), even for incident wavelengths as large as 0.7×10^{-8} cm that correspond to quantum energies as low as 2×10^4 ev. This is not very typical of the situation for such low X-ray energies. For atoms of intermediate or large Z, the cross section for Thomson scattering is very large compared to its value for ^{6}C because of the coherences, but the cross section for Compton scattering is much the same as it is for ^{6}C since Compton scattering takes place only on *essentially free* electrons. For X-rays of low energy, this means that Compton scattering can involve only the electrons of relatively small binding energy in outer shells of the atom. As the energy of the X-rays increases, more and more electrons of the atom become available for Compton scattering. However, for intermediate energies, Thomson scattering can still occur from the very tightly bound electrons in inner shells—even from the quantum point of view—because these electrons are inseparably bound to a very massive nucleus and cannot absorb energy by recoiling, so the quantum is scattered with no change in energy or wavelength. For X-rays of energy large compared to the binding energy of the electrons in the innermost shells of the atom ($\sim 10^5$ ev for ^{92}U), all the electrons of the atom are essentially free in interactions with the X-rays. In this limit only Compton scattering can take place.

It is apparent that the dependence of the scattering cross section for a typical atom on the X-ray energy would be very difficult to calculate in the energy range we are now discussing.† In fact, calculations of the cross section for Compton scattering have been carried through only for a single completely free electron. These calculations were performed in 1928 by Klein and Nishina, who used Dirac's relativistic theory of quantum mechanics, and even the results of the calculations are complicated. They are

$$\frac{d\sigma_S}{d\Omega} = \frac{d\sigma_T}{d\Omega} \frac{1}{1 + 2\alpha \sin^2 \dfrac{\Theta}{2}} \left\{ 1 + \frac{4\alpha^2 \sin^4 \dfrac{\Theta}{2}}{(1 + \cos^2 \Theta)\left(1 + 2\alpha \sin^2 \dfrac{\Theta}{2}\right)} \right\} \tag{14–36}$$

$$\sigma_S = \sigma_T \left\{ \frac{3(1 + \alpha)}{4\alpha^2} \left[\frac{2 + 2\alpha}{1 + 2\alpha} - \frac{\ln(1 + 2\alpha)}{\alpha} \right] - \frac{3}{4} \left[\frac{1 + 3\alpha}{(1 + 2\alpha)^2} - \frac{\ln(1 + 2\alpha)}{2\alpha} \right] \right\} \tag{14–37}$$

where

$$\alpha \equiv \frac{h\nu}{mc^2} = \frac{h\nu}{0.51 \text{ Mev}} = \frac{\text{quantum energy}}{\text{electron rest mass energy}}$$

† However, the scattering cross section can be measured. Some data will be presented in section 7.

The cross sections for Compton scattering from a free electron are plotted in figures (14–16) and (14–17) from these two equations. For Compton scattering of X-rays of quantum energy large compared to the binding

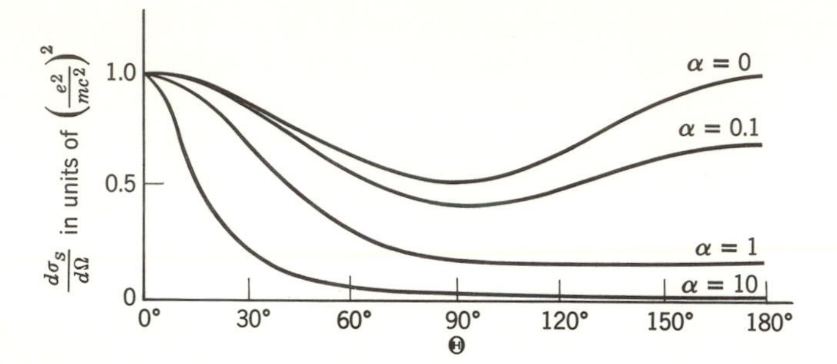

Figure 14–16. The differential scattering cross section for Compton scattering from a free electron.

energy of electrons in the innermost shell of the atom, an approximate estimate of the scattering cross sections per atom can be obtained by setting

$$\frac{d\sigma_{S_a}}{d\Omega} = Z\frac{d\sigma_S}{d\Omega} \tag{14–38}$$

and

$$\sigma_{S_a} = Z\sigma_S \tag{14–39}$$

where $d\sigma_S/d\Omega$ and σ_S are given by equations (14–36) and (14–37).

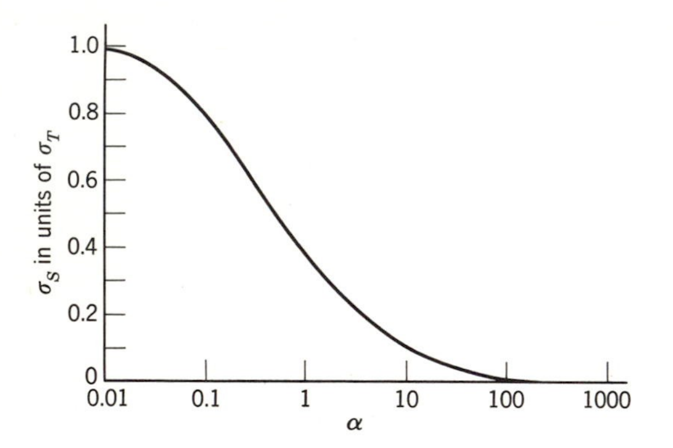

Figure 14–17. The scattering cross section for Compton scattering from a free electron.

As Compton scattering is described in terms of quanta, it is convenient in considering this process to express the intensities of the incident and scattered X-rays as fluxes of quanta. Thus we redefine $\bar{I}$ to be the average number of quanta carried by the incident beam per second across a 1 cm^2 area oriented normal to the beam, redefine $\bar{S}$ to be the average number of quanta scattered per second in all directions, and redefine $\bar{S}(\Theta)$ to be the average number of quanta scattered per second per unit solid angle at the angle Θ measured from the direction of the incident beam. These definitions differ from the previous ones only by a factor equal to $h\nu$, the energy carried per quantum. Since for Thomson scattering the energy of the scattered quantum is equal to the energy of the incident quantum, all the equations which we have written for Thomson scattering may be transformed into equations involving the new definitions simply by dividing both sides by the factor $h\nu$. Of course, this does not change the form of the equations, and we still have

$$\bar{S} = \sigma_{S_a}\bar{I} \tag{14–40}$$

$$\bar{S}(\Theta) = \frac{d\sigma_{S_a}}{d\Omega}\bar{I} \tag{14–41}$$

where for Thomson scattering the cross sections have exactly the same values as before. For Compton scattering the cross sections of the previous paragraph also satisfy equations (14–40) and (14–41) when the new definitions are used. Consequently, these new definitions allow all scattering processes (and also all absorption processes) to be discussed in the same language and, for this reason, we shall henceforth use these definitions, or equivalent ones, exclusively.

We close this section by pointing out that the Klein-Nishina equations, i.e., (14–36) and (14–37), show the free electron Compton cross sections reduce to the free electron Thomson cross sections in the limit of very low X-ray quantum energies. Furthermore, equation (3–17) shows the wavelength increase in Compton scattering from a free electron becomes negligibly small for very long wavelength X-rays because the wavelength change cannot be larger than 0.048×10^{-8} cm. Thus X-rays of very low quantum energies do not lose energy in Compton scattering. From both points of view, Thomson scattering acts as the low energy limit of Compton scattering for X-rays incident upon a free electron.

6. The Photoelectric Effect and Pair Production

In section 7, Chapter 3, we presented an elementary discussion of the *photoelectric effect* for electromagnetic radiation in the optical frequency range. The reader should review that discussion because the optical photoelectric effect is the limit of the X-ray photoelectric effect as the quantum energy becomes very small compared to the electron rest mass energy, i.e., in the limit $h\nu \ll mc^2$.

In this limit $h\nu$ is comparable to the binding energy of atomic electrons in outer shells, but it is very much smaller than the binding energy of electrons in inner shells. Thus the optical photoelectric effect can lead only to the ejection of optically active electrons. In contrast, the X-ray photoelectric effect for very large $h\nu$ can lead to the ejection of any electron.

The fundamental difference between the photoelectric effect and the Compton effect is that in the former process the incident quantum is completely absorbed by the atom, whereas in the latter process the quantum scatters from an essentially free electron and goes off suffering only a reduction in energy. We leave it as an exercise for the reader to demonstrate that in the Compton effect it is not possible to conserve *both* total relativistic energy and momentum if an essentially free electron absorbs all the energy of a quantum. On the other hand, the complete absorption of the incident quantum in the photoelectric effect does not violate energy-momentum conservation because in this process the atomic electron concerned is not essentially free, but instead is bound by electrostatic forces to the massive nucleus. This allows the nucleus to participate in the process, and the role it plays is to recoil in such a way as to conserve momentum. Since the nuclear mass is very large, and since $p = \sqrt{2ME}$, the recoiling nucleus can take up the required momentum without absorbing enough kinetic energy to affect significantly the energy balance between the quantum and the electron.

From this argument we can see that the probability of a photoelectric absorption on a particular atomic electron should increase with decreasing quantum energy, since for low energy quanta the electron is relatively more tightly bound to the nucleus and, therefore, it is easier for the nucleus to absorb the required momentum. However, when the quantum energy becomes smaller than the binding energy of the electron, the probability of photoelectric absorption on that electron abruptly drops to zero because there is no longer enough energy available to eject the electron from the atom. This behavior can be seen in the measured cross section for photoelectric absorption by ^{82}Pb atoms, which is presented in figure (14–18). The probability for photoelectric absorption is expressed in terms of a *photoelectric cross section per atom*, σ_{PE_a}, defined such that $\bar{A}$, the average number of quanta absorbed by this process per second per atom in a beam of $\bar{I}$ quanta per second per square centimeter, is

$$\bar{A} = \sigma_{PE_a}\bar{I} \tag{14–42}$$

This is an example of the use of a cross section to describe absorption processes. For the reasons mentioned above, the photoelectric absorption cross section increases rapidly with decreasing quantum energy but shows sharp breaks at quantum energies equal to the binding energies of the electrons in each of the subshells of the atom. These energies are labeled with the X-ray subshell designations introduced in section 3. Data extending to smaller quantum energies than those presented in figure (14–18) show that the photoelectric absorption cross section continues its general

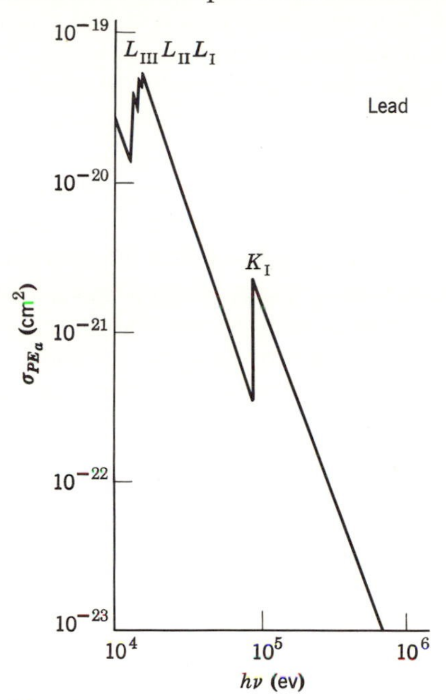

Figure 14–18. The photoelectric absorption cross section per atom for lead.

rise with decreasing energy, except for breaks of diminishing severity at the energies of the M and N subshells. Such data provide a direct measure of the binding energies of electrons in the various subshells of an atom and can be used to confirm or supplement the binding energy values obtained from an analysis of the X-ray line spectrum of the atom.

We shall not be able to calculate the cross section for photoelectric absorption. Such a calculation requires the theory of quantum electrodynamics, and we can only present some results. The predicted cross section per atom for photoelectric absorption on a *K shell electron* is

$$\sigma_{PE_a} = \sigma_T 4\sqrt{2}(1/137)^4 Z^5(mc^2/h\nu)^{7/2} \tag{14–43}$$

for $h\nu >$ binding energy of K electron, but $h\nu < mc^2$. Experiment shows that, in the energy region indicated, K shell absorption accounts for almost 90 percent of all photoelectric absorptions, and that equation (14–43) is in quite good agreement with the data. The very strong Z dependence of σ_{PE_a} reflects the fact that a K shell electron becomes much more tightly coupled to the nucleus with increasing Z, and so the nucleus finds it much easier to absorb the momentum which it must absorb in a photoelectric process.

For X-rays of quantum energies in excess of two times the electron rest mass energy, there is a mechanism in addition to the photoelectric effect through which the X-rays can be completely absorbed in interaction with matter. This is the process of *pair production*, which consists of the complete disappearance of a quantum and the subsequent *materialization* of part of its energy $h\nu$ into the rest mass energy of a pair of electrons. If $h\nu = 2mc^2$, the materialization of the two electrons uses up all the available energy. In the general case $h\nu > 2mc^2$, and the excess energy $h\nu - 2mc^2$ appears in the form of kinetic energy of the two electrons. One of these electrons is of the usual variety with mass m and charge $-e$. The other has the same mass m but has the positive charge $+e$ and is called a *positron*. Thus charge is conserved in the pair production process because the quantum is uncharged and the total charge of the electron and positron is $-e + e = 0$. Pair production cannot take place in free space because, as the reader can easily show, the disappearance of a quantum, followed by the appearance of an electron and positron, cannot conserve *both* total relativistic energy and momentum. However, the process can take place in the Coulomb field of an atom because there can be interactions between the pair of charged particles and the nucleus. These interactions allow the massive nucleus to absorb whatever momentum is required to conserve momentum, without affecting the energy balance. In the next section we shall describe more of the details, and some of the theory, of this extremely interesting process. But it seems best to complete our discussion of the attenuation of X-rays by matter before going into these subjects.

The cross section per atom for the absorption of X-rays by the pair production process can be calculated only with the aid of the Dirac theory. The results are

$$\sigma_{PR_a} = \sigma_T \frac{3}{8\pi} \frac{Z^2}{137} f(\alpha) \tag{14–44}$$

where $\alpha \equiv h\nu/mc^2$. The Z^2 dependence of the *pair production cross section per atom*, σ_{PR_a}, is a manifestation of the fact that the required momentum transferring interaction between the nucleus and the pair of charged particles is more readily achieved for atoms of large Z. The function $f(\alpha)$ has much the same form for all atoms. We present in figure (14–19) a graph of $f(\alpha)$ for the ^{82}Pb atom. This shows that σ_{PR_a} rises slowly from a value of zero at the *threshold energy* $h\nu = 2mc^2 = 1.02$ Mev and, for very large $h\nu$, approaches a constant value as $f(\alpha)$ approaches

$$f(\infty) = \frac{28}{9} \ln \frac{183}{Z^{1/3}} - \frac{2}{27} \tag{14–45}$$

We cannot write an equation for $f(\alpha)$ itself, as this function is obtained from a numerical integration. In an incident beam of $\bar{I}$ quanta per second

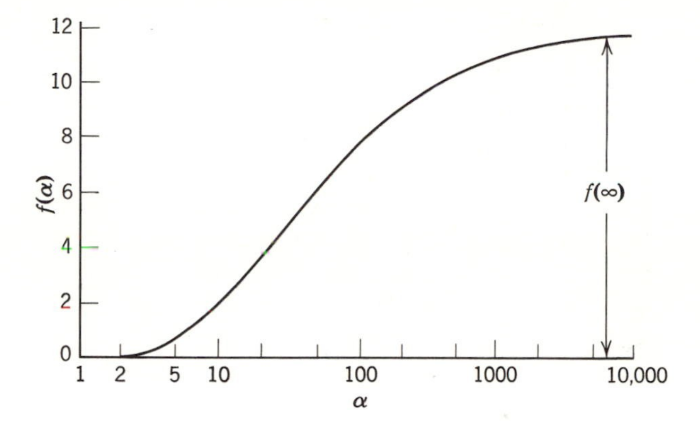

Figure 14–19. The plot of a function which arises in the theory of pair production. From W. Heitler, *The Quantum Theory of Radiation*, Oxford University Press, London, 1944.

per square centimeter, the average number $\bar{A}$ of quanta absorbed per second per atom by the pair production process is related to the cross section σ_{PR_a} by the usual relation

$$\bar{A} = \sigma_{PR_a} \bar{I} \tag{14–46}$$

7. Total Cross Sections and Attenuation Coefficients

We have now discussed all the scattering and absorption processes which contribute in a significant way to the attenuation of a beam of X-rays passing through matter. Let us next see how these processes combine to produce the total attenuation observed in an experiment, such as that described at the beginning of section 5. The average number $\bar{A}$ of quanta which are either scattered or absorbed per second per atom, in a beam of $\bar{I}$ incident quanta per second per square centimeter, can be evaluated by adding equations (14–40), (14–42), and (14–46):

$$\bar{A} = \sigma_{S_a} \bar{I} + \sigma_{PE_a} \bar{I} + \sigma_{PR_a} \bar{I} = \sigma_a \bar{I}$$

where

$$\sigma_a \equiv \sigma_{S_a} + \sigma_{PE_a} + \sigma_{PR_a} \tag{14–47}$$

The quantity σ_a is called the *total cross section per atom.* It specifies what fraction of the incident beam of X-rays undergoes any of the beam attenuating interactions with a given atom. Figure (14–20) shows the cross sections σ_{S_a}, σ_{PE_a}, σ_{PR_a}, and σ_a for the ^{82}Pb atom. For ^{82}Pb, the energy

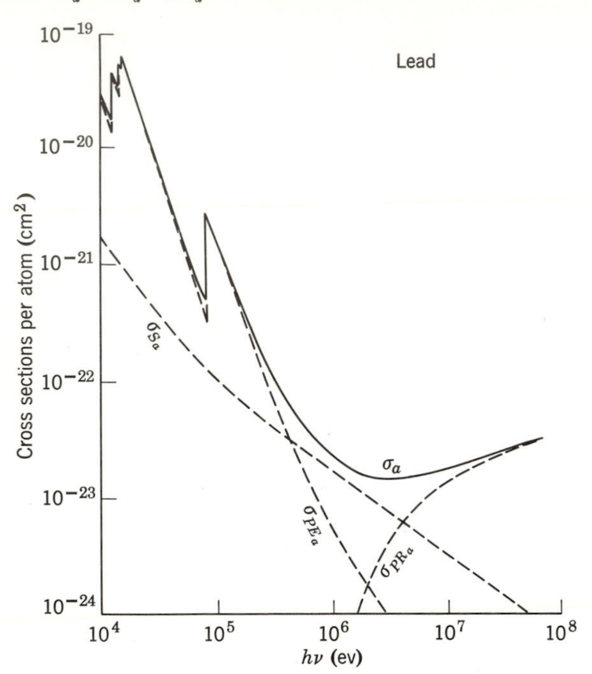

Figure 14–20. The scattering, photoelectric, pair, and total cross sections per atom for lead.

ranges in which each of the three processes makes the most important contribution to σ_a are approximately

$$\text{Photoelectric effect:}\quad h\nu < 5 \times 10^5 \text{ ev}$$

$$\text{Scattering:}\quad 5 \times 10^5 \text{ ev} < h\nu < 5 \times 10^6 \text{ ev}$$

$$\text{Pair production:}\quad 5 \times 10^6 \text{ ev} < h\nu$$

Because of the different Z dependencies of the cross sections for these processes, the energy ranges in which they dominate σ_a are quite different for atoms of low atomic number. For ^{13}Al these energy ranges are approximately

$$\text{Photoelectric effect:}\quad h\nu < 5 \times 10^4 \text{ ev}$$

$$\text{Scattering:}\quad 5 \times 10^4 \text{ ev} < h\nu < 1 \times 10^7 \text{ ev}$$

$$\text{Pair production:}\quad 1 \times 10^7 \text{ ev} < h\nu$$

Using the total cross section per atom σ_a, let us calculate the total attenuation of a beam of X-rays in passing through a foil of thickness t cm, containing N atoms per cm³. Consider the situation shown in figure (14–21). The flux (or intensity) of the beam, which is the average number of quanta crossing 1 cm² per sec, decreases as the beam penetrates the foil. Call the flux after penetrating x cm $\bar{I}(x)$; then $\bar{I}(0)$ is the incident flux and $\bar{I}(t)$ is the attenuated flux emerging from the foil. Consider a

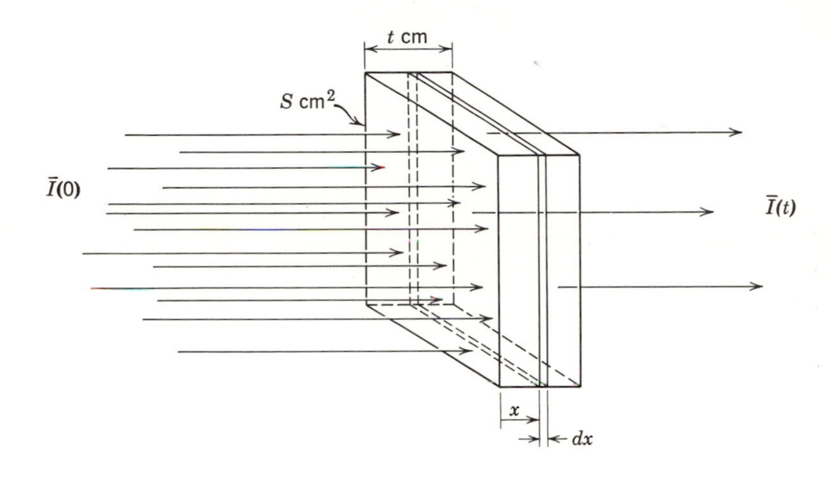

Figure 14–21. A beam of X-rays passing through a foil.

lamina of thickness dx at x. The number of atoms in the lamina is N times the volume of the lamina; it is $NS\,dx$, where S is the area of the foil in cm². The probability of a quantum being removed from the beam by *one* of these atoms is equal to the probability of the quantum striking an imaginary disk of area σ_a cm². This is equal to σ_a cm²/S cm², the ratio of the area of the disk to the area of the lamina. The net probability of a quantum being removed from the beam by *any* of the atoms in the lamina is equal to this ratio times the number of atoms; i.e., $(\sigma_a/S)NS\,dx = \sigma_a N\,dx$. As an average of $\bar{I}(x)$ quanta are incident upon 1 cm² of the lamina in 1 sec, the average number of quanta removed per cm² per sec from the beam by the lamina is $\bar{I}(x)\sigma_a N\,dx$. The number of quanta per cm² per sec emerging from the lamina, $\bar{I}(x + dx)$, is equal to the number incident minus the number removed; so

$$\bar{I}(x + dx) = \bar{I}(x) - \bar{I}(x)\sigma_a N\,dx$$

or

$$d\bar{I}(x) \equiv \bar{I}(x + dx) - \bar{I}(x) = -\bar{I}(x)\sigma_a N\,dx \qquad (14\text{–}48)$$

We find $\bar{I}(t)$ by integrating this expression over x:

$$\frac{d\bar{I}(x)}{\bar{I}(x)} = -\sigma_a N\, dx$$

$$\int_0^t \frac{d\bar{I}(x)}{\bar{I}(x)} = -\sigma_a N \int_0^t dx$$

$$\Big[\ln \bar{I}(x)\Big]_0^t = -\sigma_a N t$$

$$\ln \frac{\bar{I}(t)}{\bar{I}(0)} = -\sigma_a N t$$

$$\frac{\bar{I}(t)}{\bar{I}(0)} = e^{-\sigma_a N t}$$

$$\bar{I}(t) = \bar{I}(0) e^{-\sigma_a N t} \tag{14–49}$$

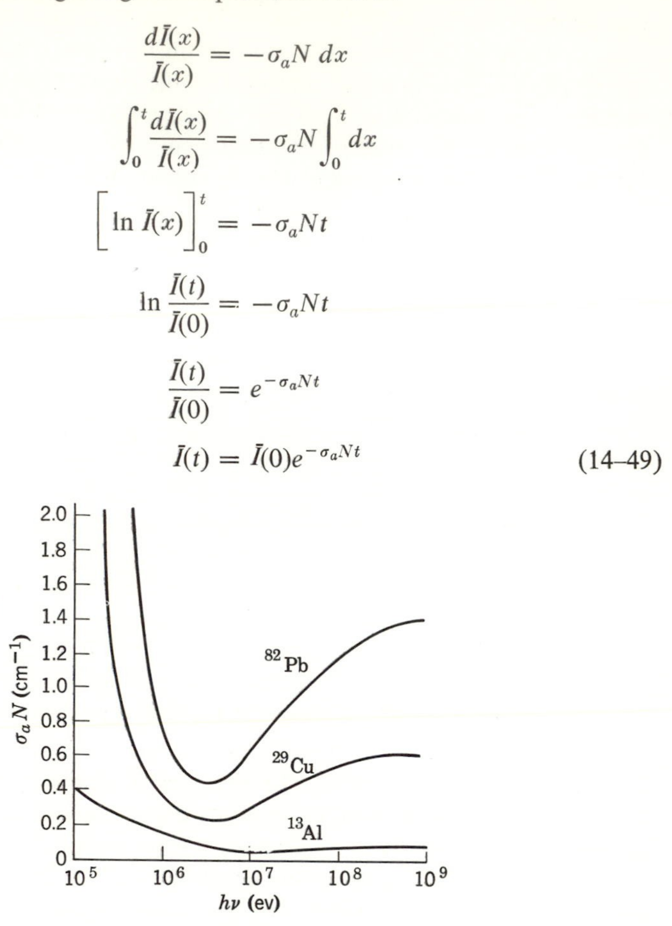

Figure 14–22. The attenuation coefficients for several atoms and a range of energies.

The flux of the attenuated beam decreases exponentially with increasing t. The quantity $\sigma_a N$, which is called the *attenuation coefficient*, has the dimensions cm^{-1} and is the reciprocal of the thickness of foil required to attenuate the beam by a factor e. This thickness is called the *attenuation length* Λ. That is

$$\Lambda = 1/\sigma_a N \tag{14–50}$$

Of course, the attenuation coefficient has the same energy dependence as the total cross section. Figure (14–22) shows the attenuation coefficients of ^{82}Pb, ^{50}Sn, and ^{13}Al for X-rays of relatively high energy.

8. Positrons and Other Antiparticles

The first experimental evidence for the pair production process, and the existence of positrons, was obtained in 1933 by Anderson during an investigation of the cosmic radiation. This radiation consists of a flux of very high energy X-rays and charged particles incident upon the earth from extra-terrestrial sources. Anderson was using a cloud chamber (a device filled with supersaturated vapor that condenses on the ions formed by charged particles traversing it and makes the paths of these particles visible) containing a thin lead plate, with the entire apparatus in a magnetic field. Upon exposing this apparatus to the cosmic radiation, it was found that very infrequently a pair of charged particles were ejected from some point in the lead plate. These events were assumed to be the result of the interaction of an X-ray quantum in the lead because no charged particle was seen to strike the point of ejection, whereas a quantum, being uncharged, could strike the point of ejection without being seen. The two charged particles ejected in these events were bent in opposite directions by the magnetic field. Therefore their charges were of the opposite sign. From other considerations it could be shown that the magnitudes of these charges were equal to one electronic charge and that the masses of the particles were at least approximately equal to one electronic mass. Since the discovery of these first few *electron-positron pairs*, countless others have been observed in many different experiments. In these experiments it has been established with great accuracy that the rest mass, charge, spin, symmetry character, and magnetic moment of the positron and electron are equal, except that the signs of the charge and magnetic moment of the positron are opposite those of the electron.

The discovery of the pair production process explained the origin of a discrepancy between the then current theory of X-ray attenuation and the measured attenuation coefficients of several materials for 2.6 Mev X-rays (gamma rays obtained from a radioactive source). As the theory originally did not include pair production, the predicted attenuation coefficients were too small; with the inclusion of the pair production process, good agreement is now obtained between experiment and theory. However, the real importance of Anderson's discovery was in the beautiful confirmation which it provided for Dirac's relativistic theory of quantum mechanics.

The Dirac theory leads to the prediction that the allowed values of total relativistic energy E for a free electron are

$$E = \pm\sqrt{c^2p^2 + (mc^2)^2} \tag{14–51}$$

where m is the electron rest mass. These are simply the solutions for E

of equation (1–25). But the solution with the minus sign corresponds to a negative total relativistic energy—a concept just as foreign to relativistic mechanics as a negative total energy is to classical mechanics. Instead of just throwing away the negative part of equation (14–51) on the grounds that it is not physically realistic, Dirac pursued the consequences of the entire equation. In doing this he was led to some very interesting conclusions. Consider figure (14–23), which is an energy level diagram

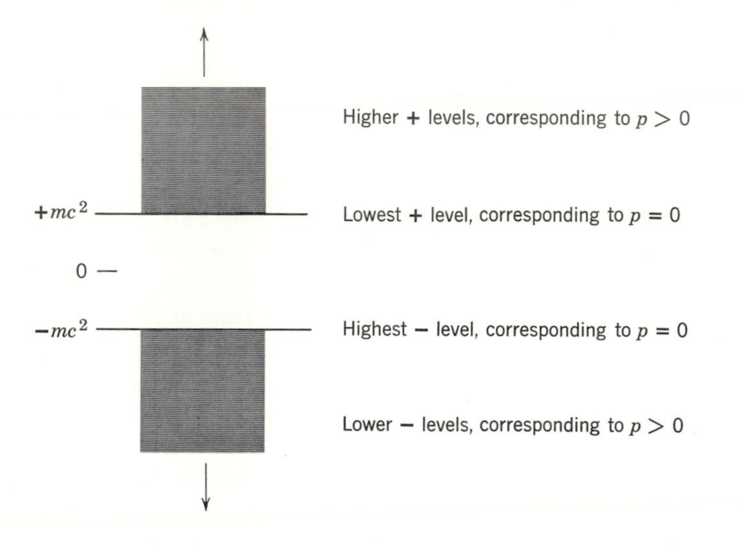

Figure 14–23. The energy levels of a free electron according to Dirac.

representing equation (14–51). If the indicated continuum of negative energy levels exists, all free electrons of positive energy should be able to make transitions into these levels, accompanied by the emission of quanta of the appropriate energies. This obviously disagrees with experiment because free electrons are not generally observed to emit spontaneously quanta of energy $h\nu \geqslant 2mc^2$. However, Dirac pointed out that this difficulty can be removed by assuming that all the negative energy levels are normally occupied at all points in space. Then the exclusion principle would prevent a free electron from dropping into any of these levels. According to this assumption, *a vacuum consists of a sea of electrons in negative energy levels.* This does not disagree with experiment. For instance, the negative charge could not be detected, as it is assumed to be uniformly distributed and therefore exerts no force on a charged body. Similar considerations will demonstrate to the reader that all the "usual" properties of a sea of negative energy electrons are such that its presence would not be apparent in any of the "usual" experiments. However,

Dirac's theory of the vacuum is not completely vacuous because it predicts certain new properties which can be tested by experiment.

The energy level diagram for a free electron suggests the possibility of exciting an electron in a negative energy level by the absorption of a quantum. Since all the negative energy levels are assumed to be occupied, the electron must be excited to one of the unoccupied positive energy levels. The minimum quantum energy required for this process is obviously $h\nu = 2mc^2$, and the process results in the production of an electron in a positive energy level plus a hole in a negative energy level. We demonstrate below that a hole in a negative energy level has the properties of a positron of positive energy. Consequently this is the pair production process observed experimentally by Anderson three years after its theoretical prediction by Dirac.†

Consider the excitation of an electron from a level of total relativistic energy $-E_1$ to a level of total relativistic energy $+E_2$. The energy of the quantum required is

$$h\nu = +E_2 - (-E_1) = E_2 + E_1 \tag{14–52}$$

If we use the symbol e^- to designate the electron left in the level $+E_2$, its total relativistic energy is

$$E_{e^-} = E_2$$

and, according to equations (1–22′) and (1–24), its kinetic energy is

$$T_{e^-} = E_{e^-} - mc^2 \tag{14–53}$$

where m is its rest mass. These relations are illustrated in the energy level diagram of figure (14–24). Now, there will be a positive total relativistic energy, which we call E_{e^+}, associated with the absence of an electron in the level of total relativistic energy $-E_1$. The relation between these two quantities is

$$E_{e^+} = -(-E_1) = E_1$$

From figure (14–24) we see that $-E_1$ can be decomposed as follows:

$$-E_1 = -mc^2 - T_{e^+}$$

which may be taken as a definition of the positive energy T_{e^+}. Thus

$$E_{e^+} = -(-mc^2 - T_{e^+}) = mc^2 + T_{e^+}$$

or

$$T_{e^+} = E_{e^+} - mc^2 \tag{14–54}$$

† In section 6 we mentioned that momentum conservation requires pair production to take place in the presence of a nucleus. However, the massive nucleus does not affect the energy balance, so an energy level diagram for a free electron can be used to describe the process.

By comparing this equation with (14–53), we identify T_{e^+} as a kinetic energy, and mc^2 as a positive rest mass energy. There are, then, a positive total relativistic energy E_{e^+}, a positive kinetic energy T_{e^+}, and a positive rest mass m associated with the hole in a negative energy level. Consequently the hole has all the mechanical properties (rest mass, inertia, etc.) of a particle of rest mass equal to one electronic mass. It is apparent that there will be a positive charge $+e$ associated with the absence of an

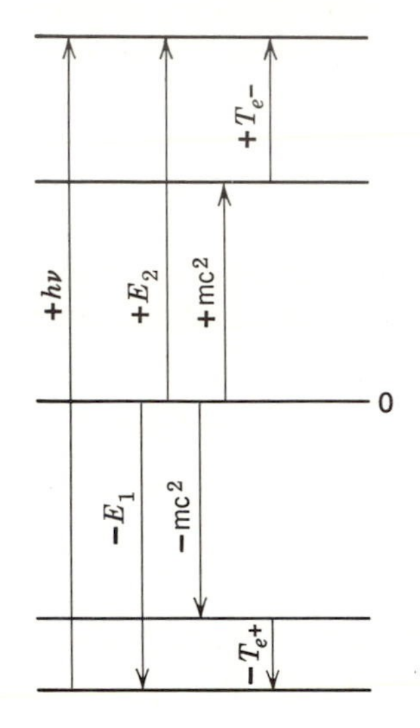

Figure 14–24. An energy level diagram illustrating pair production.

electron of charge $-e$. There are also a spin angular momentum and a magnetic moment associated with the absence of an electron in a previously filled energy level. The magnitude of the spin angular momentum is the same as for a level containing only a single electron (cf. footnote on page 485). The magnetic moment is of the same magnitude as for an electron, but the sign is opposite because the charge associated with the hole is opposite that of an electron. We see that in all respects the properties associated with a hole in a negative energy level are identical with the experimentally observed properties of a positron, which we have designated by the symbol e^+.

A positron is a completely stable particle as long as it is kept away from electrons. However, a positron passing through matter will eventually collide with an electron and the two particles will *annihilate*. In terms of the Dirac theory, this can be thought of as a transition in which the electron drops into the hole that constitutes the positron, with the liberated energy emitted in the form of quanta. Annihilation normally occurs when the positron is moving very slowly since then the probability of colliding with an electron is higher. If annihilation occurs on a free electron, two quanta must be emitted in opposite directions in order to conserve momentum. If the electron is bound to a nucleus, one quantum annihilation is possible. The average time required for a slow positron to annihilate in ^{82}Pb is $\sim 10^{-10}$ sec.

The electron and positron are but one example of a *particle* and its *antiparticle*. The existence of a number of other examples is predicted by theory and has been verified by experiment. For instance, the *antiproton*, a particle of charge $-e$ and mass equal to the proton mass M, was observed in 1955 by Segre and his co-workers. Since the proton rest mass energy $Mc^2 = 931$ Mev is very large, antiprotons can be generated only in the cosmic radiation or in high energy particle accelerators. The Segre experiment produced antiprotons in collisions between stationary protons and protons of 6200 Mev kinetic energy obtained from a large particle accelerator.

BIBLIOGRAPHY

Compton, A. H., and S. K. Allison, *X-rays in Theory and Experiment*, D. Van Nostrand Co., Princeton, N.J., 1935.

Evans, R. D., *The Atomic Nucleus*, McGraw-Hill Book Co., New York, 1955.

Richtmyer, F. K., E. H. Kennard, and T. Lauritsen, *Introduction to Modern Physics*, McGraw-Hill Book Co., New York, 1955.

EXERCISES

1. Verify equation (14–24).

2. Verify equation (14–28).

3. Prove that a free electron cannot absorb all the energy of a quantum, as stated in section 6.

4. Prove that pair production cannot take place in free space, as stated in section 6.

5. Using the data of figure (14–20), calculate the thickness of Pb required to attenuate a beam of 10^4 ev X-rays by a factor of 100.

6. Show that the attenuation length Λ is just equal to the average distance a quantum will travel before being scattered or absorbed.

7. A slow positron moving through matter can be captured by a free electron, forming an atomic-like system of transient existence called *positronium*. Evaluate the spectrum of electromagnetic radiation emitted in processes of this type.

CHAPTER 15

Collision Theory

1. Introduction

We have been able to go through a fairly complete discussion of the atom in terms of the quantum mechanical theory of the motion of particles bound in a potential, because most of the understanding of atoms has come from interpreting experimental measurements of the spectra which they emit in terms of the energy levels of their bound atomic electrons. For the nucleus this is not the case. From the very first the understanding of its properties has come largely from interpreting the results of experiments in which unbound particles collide with nuclei. As an example, the reader will recall that the very discovery of the nucleus came from the alpha particle scattering experiments. In preparation for our discussion of the nucleus, we must now extend to three dimensions the quantum mechanical theory of the collision of an unbound particle with a potential.†

In the previous chapter we described the processes which occur in the collision of electromagnetic quanta with atoms. This description was probably not very satisfying to the reader because we were generally able to state only results. A rigorous treatment of the scattering and absorption of *quanta* is not possible at the level of this book as it requires quantum electrodynamics and relativistic quantum mechanics. In this chapter we shall, however, consider only the scattering and absorption of *particles* and shall ignore spin and other relativistic effects. This will be adequate for most of our discussion of the nucleus and will allow us to treat the problem rigorously with the "old familiar" Schroedinger theory. At first we shall

† Collisions in one dimension were treated in the first sections of Chapter 8.

consider scattering. After certain preliminaries, we shall give an approximate treatment of this process, and then an exact treatment. Finally we shall treat absorption.

2. Laboratory-Center of Mass Transformations

In working with the Schroedinger equation for a system of two particles of mass m_1 and m_2 interacting with each other through a mutual potential energy $V(r)$, we have found that the problem can be reduced from one involving six spatial coordinates to one involving three by transforming to a frame of reference in which the center of mass of the system is at rest. In this *center of mass* (CM) frame the system can be described in terms of the motion of a single fictitious particle of reduced mass

$$\mu = \frac{m_1 m_2}{m_1 + m_2} \tag{15-1}$$

interacting with a center of force fixed at the origin. The distance r from this particle to the origin is equal to the separation of the two particles of mass m_1 and m_2, and the potential energy $V(r)$ of the fictitious particle is equal to the mutual potential energy of the two particles. Furthermore, the total (and kinetic) energy of the fictitious particle in the CM frame is equal to the total (and kinetic) energy in that frame of the system of two particles.† In the CM frame the system obtained by replacing one of the particles by a particle of mass μ, replacing the other by a particle of such large mass that it must remain fixed at the origin, but otherwise making no changes in the system or the equations governing its behavior, is completely equivalent to the original system. The equivalent system is much easier to work with because one of its particles remains fixed and, consequently, can be ignored except as a source of the potential $V(r)$.

Now, in evaluating the energy levels of such a system, we have never had to bother with the fact that the CM frame is generally moving, since this motion only adds a constant and non-quantized translational kinetic energy to the system which does not affect the energy differences that are observed in measurements of spectra. But, in treating scattering problems, we cannot ignore the motion of the CM frame because it affects the observed scattering angles and cross sections. In almost all experiments these quantities are measured in a frame of reference, called the *laboratory* (LAB) frame, which is moving with respect to the CM frame. The reason is that in almost all scattering experiments one of the particles, and not

† Cf. section 2, Chapter 10, and section 4, Chapter 5.

the center of mass, is initially stationary in the laboratory. Let us develop the relations between the description of a scattering process in these two frames. In doing this we shall deal with momentum vectors of precise magnitudes and directions; however, since the locations of the particles need not be specified, this will not violate the uncertainty principle.

Assume that the particle of mass m_2 is initially stationary in the LAB frame, and that the initial velocity of m_1 with respect to m_2 is $\mathbf{v}_{1L}$, as shown in figure (15–1). After the collision, m_1 moves off with velocity $\mathbf{v}'_{1L}$ at the

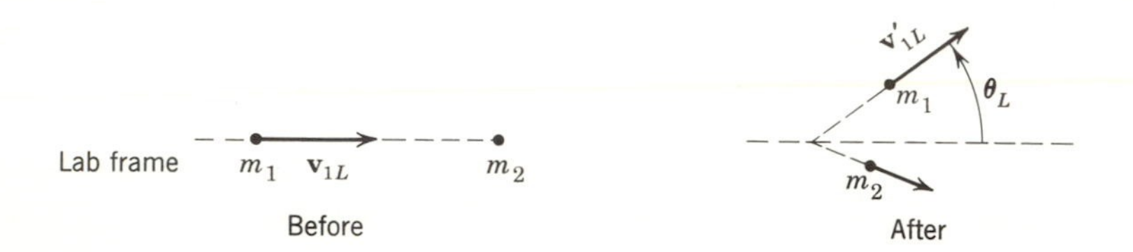

Figure 15–1. A collision between two particles, as seen in the LAB frame.

LAB scattering angle θ_L measured from its initial direction of motion, and m_2 recoils in some direction with some velocity. In the CM frame this process has the different appearance shown in figure (15–2), because the CM frame is moving with velocity $\mathbf{V}$ with respect to the LAB frame. By differentiating equations (5–16) with respect to time, and multiplying each of the resulting velocities by the mass of the associated particle, it is

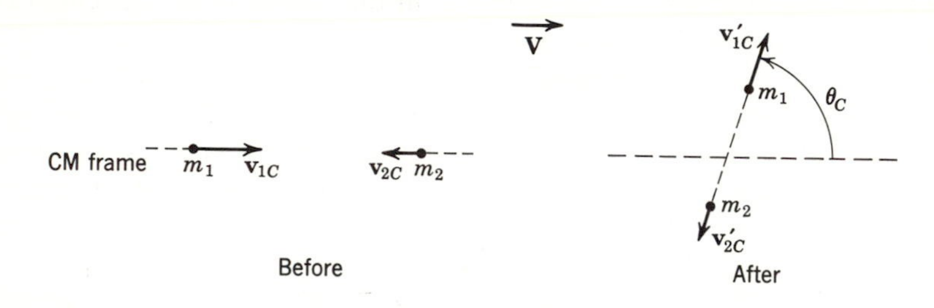

Figure 15–2. A collision between two particles, as seen in the CM frame.

easy to see that in the CM frame the total momentum must vanish.† The momentum vectors of m_1 and m_2 must therefore always be equal and opposite in this frame. Furthermore, the magnitudes of the momenta cannot change because energy is conserved. Therefore the result of a collision is simply to rotate the line along which these vectors are directed through the CM scattering angle θ_C, without changing their magnitudes. Thus

$$v'_{1C} = v_{1C}, \qquad v'_{2C} = v_{2C} \tag{15–2}$$

† This is also true relativistically. Cf. exercise 8, Chapter 1.

An equivalent description of the process in the CM frame is obtained by replacing m_1 by μ, with the initial velocity of μ equal to the relative velocity of the two particles (because the distance from μ to the origin is always equal to the distance between m_1 and m_2), and by letting $m_2 \to \infty$. This is indicated in figure (15–3). We have

$$v_{\mu C} = v_{1L} \tag{15–3}$$

and it is apparent that

$$v'_{\mu C} = v_{\mu C} \tag{15–4}$$

Figure 15–3. An equivalent description of a collision between two particles, as seen in the CM frame.

To determine the value of V, the velocity of the CM frame with respect to the LAB frame, we use the classical velocity transformation equations

$$v_{1C} = v_{1L} - V, \qquad v_{2C} = V \tag{15–5}$$

Their use, and the use of velocity independent masses, is justified because we are not treating relativistic problems. Now the condition that the total momentum vanish in the CM frame is

$$m_1 v_{1C} = m_2 v_{2C} \tag{15–6}$$

This is

$$m_1(v_{1L} - V) = m_2 V$$

$$m_1 v_{1L} = (m_1 + m_2)V$$

or

$$V = \frac{m_1}{m_1 + m_2}\, v_{1L} \tag{15–7}$$

To find the relation between θ_C and θ_L, we transform the velocity of m_1 after the collision from the CM frame to the LAB frame. The vectorial form of the classical velocity transformation equation which we use is

$$\mathbf{v}'_{1L} = \mathbf{V} + \mathbf{v}'_{1C} \tag{15–8}$$

This is illustrated in figure (15–4). Taking transverse components, we have

$$v'_{1L} \sin \theta_L = v'_{1C} \sin \theta_C \tag{15–9}$$

Longitudinal components give

$$v'_{1L} \cos \theta_L = V + v'_{1C} \cos \theta_C \tag{15–10}$$

Dividing equation (15–9) by (15–10), we find

$$\tan\theta_L = \frac{v'_{1C}\sin\theta_C}{V + v'_{1C}\cos\theta_C} = \frac{\sin\theta_C}{V/v'_{1C} + \cos\theta_C}$$

But

$$V = v_{2C} = \frac{m_1}{m_2}\,v_{1C} = \frac{m_1}{m_2}\,v'_{1C}$$

or

$$V/v'_{1C} = m_1/m_2$$

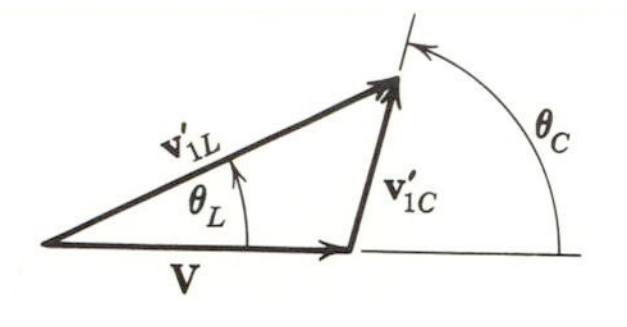

Figure 15–4. Illustrating the transformation of a velocity vector from the CM frame to the LAB frame.

Therefore

$$\tan\theta_L = \frac{\sin\theta_C}{m_1/m_2 + \cos\theta_C} \tag{15–11}$$

Now let us find the relations between the scattering cross sections and differential scattering cross sections for the two frames of reference. The definitions of the cross sections, and of the incident and scattered fluxes, may be directly adapted from equations (14–40) and (14–41) and the definitions immediately preceding those equations, if we rewrite them in terms of a single incident particle instead of many incident quanta. That is, we define the incident flux I to be the probability per second that the incident particle will be carried by the beam across a 1 cm^2 area normal to the beam direction, and define the scattered fluxes S and $S(\theta)$ to be, respectively, the probability per second that the incident particle will be scattered into all angles, and the probability per second that it will be scattered into a unit solid angle at an angle θ from the direction of the incident beam. These quantities are related by the scattering cross section σ and the differential scattering cross section $d\sigma/d\Omega$ as follows:

$$S = \sigma I \tag{15–12}$$

$$S(\theta) = \frac{d\sigma}{d\Omega}\,I \tag{15–13}$$

In the present context we simplify the notation by dropping the bar denoting average, and the subscript S for scattering. We also drop the subscript a denoting cross sections per atom, as it is no longer necessarily appropriate. However, we must add a subscript C for center of mass, or L for laboratory, to all these quantities in order to designate the frame of reference in which they are defined. We do not need to use some other symbol to distinguish between the quantities defined in the two alternative CM frame descriptions of the scattering process, since henceforth we shall be concerned only with the description of a particle of mass μ scattering from the potential $V(r)$ associated with the center of force fixed at the origin. From the Schroedinger theory we shall calculate the scattering cross sections for this description in the CM frame. These cross sections can then be transformed to the LAB frame, for comparison with experiment, by noting that the probability per second for the incident particle to be scattered must be the same whether measured in the CM frame or the LAB frame. That is,

$$S_L = S_C$$

Using equation (15–12), this becomes

$$\sigma_L I_L = \sigma_C I_C$$

But

$$I_L = I_C \tag{15–14}$$

because the motion of the particle μ relative to the origin before scattering (which determines I_C) is identical with the motion of the particle m_1 relative to the particle m_2 before scattering (which determines I_L). Therefore

$$\sigma_L = \sigma_C \tag{15–15}$$

Similarly, the probability per second for the incident particle to be scattered into the solid angle $d\Omega_L$ measured in the LAB frame must be the same as the probability per second that it will be scattered into the solid angle $d\Omega_C$ measured in the CM frame, if the same physical range of directions is described by both solid angles. That is,

$$S_L(\theta_L)\, d\Omega_L = S_C(\theta_C)\, d\Omega_C$$

so

$$\frac{d\sigma_L}{d\Omega_L} I_L\, d\Omega_L = \frac{d\sigma_C}{d\Omega_C} I_C\, d\Omega_C$$

and

$$\frac{d\sigma_L}{d\Omega_L} = \frac{d\sigma_C}{d\Omega_C}\frac{d\Omega_C}{d\Omega_L}$$

This can also be obtained by formally differentiating equation (15–15). The elements of solid angle have been evaluated in equation (14–25). They are

$$d\Omega_C = 2\pi \sin \theta_C \, d\theta_C$$

and

$$d\Omega_L = 2\pi \sin \theta_L \, d\theta_L$$

so

$$\frac{d\Omega_C}{d\Omega_L} = \frac{\sin \theta_C}{\sin \theta_L} \frac{d\theta_C}{d\theta_L} \tag{15–16}$$

As an exercise, the reader should use (15–11) to evaluate (15–16), and obtain

$$\frac{d\sigma_L}{d\Omega_L} = \frac{d\sigma_C}{d\Omega_C} \frac{[1 + (m_1/m_2)^2 + 2(m_1/m_2)\cos\theta_C]^{3/2}}{1 + (m_1/m_2)\cos\theta_C} \tag{15–17}$$

3. The Born Approximation

In this section we shall develop a method, due to Born, for calculating the approximate values of $d\sigma/d\Omega$ and σ. (The subscript C is to be understood because here, and throughout the remainder of this chapter, all cross sections are defined in the CM frame.) The first step is to give a quantum mechanical description for the motion of a particle of the incident beam. Equation (8–4), and the subsequent discussion, show that in one dimension an eigenfunction describing a free particle of mass μ traveling with velocity v in the positive direction along an axis, which we call here the z axis, is

$$\psi(z) = Ae^{iKz} \tag{15–18}$$

where

$$K = 2\pi/\lambda = 2\pi p/h = \mu v/\hbar \tag{15–18$'$}$$

and where A is a constant. The reader may show by substitution that the traveling wave eigenfunction (15–18) is also a solution to the three dimensional time independent Schroedinger equation for a free particle,

$$-\frac{\hbar^2}{2\mu}\left[\frac{\partial^2\psi}{\partial x^2} + \frac{\partial^2\psi}{\partial y^2} + \frac{\partial^2\psi}{\partial z^2}\right] = E\psi \tag{15–19}$$

where

$$E = p^2/2\mu = \hbar^2 K^2/2\mu \tag{15–19$'$}$$

In three dimensions, equation (15–18) describes a particle which is definitely known to be moving parallel to the z axis with velocity v, whose x and y coordinates are entirely unknown since $\psi^*(z)\,\psi(z)$ is obviously independent of x and y, and whose z coordinate is also entirely unknown since

$$\psi^*(z)\,\psi(z) = A^*e^{-iKz}Ae^{iKz} = A^*A \tag{15–20}$$

Thus the particle is moving somewhere in a beam, parallel to the z axis, of infinite transverse and longitudinal dimensions. Of course, this is not physically realistic since all beams are always limited in their transverse dimensions by diaphragms of finite aperture and in their longitudinal dimensions by the finite length of the apparatus. On the other hand, the dimensions of real beams are extremely large compared to the characteristic atomic or nuclear dimensions. Therefore equation (15–18) provides an accurate description of the incident particle in the region of importance where the atomic or nuclear potential which produces the scattering has any appreciable value.

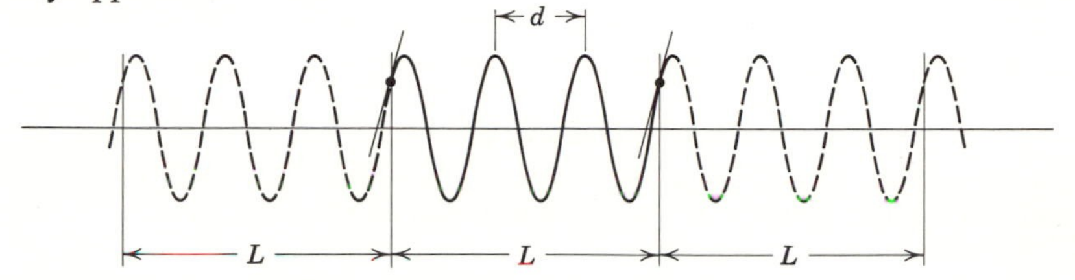

Figure 15–5. The spatial dependence of the real part of an eigenfunction in box normalization with periodic boundary conditions.

The unrealistic aspects of equation (15–18) are, however, the origin of certain problems concerning the normalization of the eigenfunction. In section 1, Chapter 8, we showed that these problems can always be handled and can usually be ignored. The present calculation provides an example of a case in which they cannot be ignored; we must use a three dimensional extension of the technique of box normalization with periodic boundary conditions. We set

$$A = L^{-3/2} \tag{15–21}$$

where L is the edge length of a very large cubical box surrounding the region of the scattering potential, and we restrict the range of the spatial variables to lie within the box. Then the eigenfunction is normalized because

$$\psi^*\psi = A^*A = A^2 = L^{-3} \tag{15–22}$$

and

$$\int \psi^*\psi \, d\tau = L^{-3}\int d\tau = L^{-3}L^3 = 1$$

where $d\tau$ is the volume element, and where the integration is now taken only over the volume of the box. We furthermore demand that the eigenfunction and its spatial derivative in the direction normal to the wall have the same values at corresponding points of the opposing walls of the box. The real (or imaginary) part of ψ would then typically have the behavior plotted in figure (15–5) as a function of one of the spatial variables, holding the other two constant. Using periodic boundary conditions, the eigenfunction will be completely periodic, with period L, in all three directions. Its behavior repeats indefinitely in the adjacent boxes (just as a scene observed within a cube with mirror walls repeats indefinitely), and we are justified in considering what happens only within a single box—that is, in restricting the ranges of the variables to the box.†

In most cases of physical interest the scattering potential is a spherically symmetrical function $V(r)$. We assume this to be true, although it is not a necessary restriction. It is then obviously convenient to describe the incident particle in terms of the spherical coordinates r, θ, ϕ instead of the rectangular coordinates x, y, z. Define the origin to be at the center of the potential and the polar axis to be along the z axis. This means

$$z = r \cos \theta$$

and ψ for a free particle of the incident beam can be written

$$\psi = L^{-3/2} e^{iKz} = L^{-3/2} e^{iKr \cos \theta} = L^{-3/2} e^{i\mathbf{K}\cdot\mathbf{r}} \tag{15–23}$$

where $\mathbf{K}$ is a vector of magnitude K directed along the beam direction, which is the direction of the z axis, and where $\mathbf{r}$ is a vector from the origin to the point (r, θ, ϕ). In this form the normalized eigenfunction for a free particle traveling in some other direction can be written

$$\psi = L^{-3/2} e^{i\mathbf{K}'\cdot\mathbf{r}} \tag{15–24}$$

where $\mathbf{K}'$ is a vector in the direction in question of magnitude equal to the value of K' appropriate to the reduced mass and velocity of the particle. The validity of equation (15–24) can be verified by the same arguments as were used for (15–23).

Now consider a particle in the incident beam impinging upon the potential $V(r)$. We want to calculate the probability per second that the particle will be scattered in some direction. If $V(r)$ is not too strong, we can treat this as a perturbation problem: What is the rate at which a constant (in time) perturbation $V(r)$ induces transitions from the initial eigenstate associated with the free particle eigenfunction (15–23) to a final eigenstate associated with a free particle eigenfunction (15–24)? Since the final eigenstate is in an essentially continuous range of final eigenstates because the possible eigenvalues $E' = \hbar^2 K'^2/2\mu$ are almost

† Another consideration, which serves to justify box normalization with periodic boundary conditions, is that the set of eigenfunctions satisfying these conditions forms a complete set of orthogonal functions in terms of which almost any function can be expanded. Cf. section 1, Chapter 8.

continuously distributed even with box normalization, the answer is given, approximately, by the three dimensional extension of Golden Rule No. 2 (equation 9–45). This is

$$R_{\mathbf{K}} \simeq \frac{2\pi}{\hbar} v^*_{\mathbf{K}'\mathbf{K}} v_{\mathbf{K}'\mathbf{K}} \rho_{\mathbf{K}'} \tag{15–25}$$

where we have used the vectors $\mathbf{K}$ and $\mathbf{K}'$, instead of quantum numbers, to label the initial and final states, and where $v_{\mathbf{K}'\mathbf{K}}$ is the matrix element of the potential taken between these states. That is,

$$v_{\mathbf{K}'\mathbf{K}} \equiv \int (L^{-3/2} e^{i\mathbf{K}'\cdot\mathbf{r}})^* V(r) L^{-3/2} e^{i\mathbf{K}\cdot\mathbf{r}}\, d\tau = L^{-3} \int V(r) e^{i(\mathbf{K}-\mathbf{K}')\cdot\mathbf{r}}\, d\tau = L^{-3} V_{\mathbf{K}'\mathbf{K}} \tag{15–25′}$$

with

$$V_{\mathbf{K}'\mathbf{K}} \equiv \int V(r) e^{i(\mathbf{K}-\mathbf{K}')\cdot\mathbf{r}}\, d\tau$$

The quantity $\rho_{\mathbf{K}'}$ of (15–25) is the number of possible eigenstates per unit energy interval for the particle associated with the final eigenfunction. As we have employed box normalization with periodic boundary conditions (required because the eigenfunctions appearing in $v_{\mathbf{K}'\mathbf{K}}$ must be normalized), the density of final states $\rho_{\mathbf{K}'}$ will have some finite value since the boundary conditions impose restrictions on the possible de Broglie wavelengths. As an example, consider $\mathbf{K}'$ parallel to one edge of the box. Then the real (or imaginary) part of ψ would typically have the appearance shown in figure (15–5), with the distance d equal to the de Broglie wavelength $\lambda' = 2\pi/K'$. The periodic boundary conditions can be satisfied for propagation parallel to one edge of the box only if L contains exactly an integral number of wavelengths of the traveling waves. Compare this with the case of free particle standing wave eigenfunctions in a box with impenetrable walls, which we treated in section 6, Chapter 12. In that case, the boundary conditions demand that ψ have nodes at the walls of the box. For the propagation direction parallel to one edge of the box, the condition can be satisfied if L contains either an integral number of wavelengths or a half-integral number of wavelengths. Consequently, in every wavelength or energy interval there are two times as many allowed wavelengths in the standing wave case as there are in the traveling wave case. However, for each possible wavelength there are two separate traveling waves, one propagating in one direction and another propagating in the opposite direction. The factors of two cancel out, not only for propagation in directions parallel to the edges of the box but also for propagation in all directions, and the number of possible eigenstates per unit energy interval is therefore the same in both cases. Thus we may use equation (12–37) which, in our present notation, is

$$\rho_{K'}\, dE' = \frac{\mu^{3/2} L^3 E'^{1/2}\, dE'}{2^{1/2}\pi^2 \hbar^3} \tag{15–26}$$

where

$$L^3 = V, \quad \text{the volume of the box}$$

and

$$E' = \hbar^2 K'^2 / 2\mu$$

Therefore

$$\rho_{K'} = \frac{\mu^{3/2} L^3}{2^{1/2}\pi^2 \hbar^3} \frac{\hbar K'}{2^{1/2}\mu^{1/2}} = \frac{\mu L^3 K'}{2\pi^2 \hbar^2} \tag{15–27}$$

This is not quite what we want because it is the density of all states associated with K', whereas we want $\rho_{\mathbf{K}'}$, the density of states associated with $\mathbf{K}'$ when that vector lies within some certain range of directions. Now it is clear that for a spherically symmetrical potential $V(r)$ the scattering angular distribution will not depend on the azimuthal angle ϕ. Consequently, it is appropriate to consider together all final states associated with vectors $\mathbf{K}'$ whose directions lie anywhere within the angular range θ to $\theta + d\theta$. The density $\rho_{\mathbf{K}'}$ of these states is smaller than $\rho_{K'}$ by a factor equal to the ratio of the solid angle $d\Omega = 2\pi \sin\theta\, d\theta$ contained within the range θ to $\theta + d\theta$ to the total solid angle 4π contained within the entire range of θ. That is,

$$\rho_{\mathbf{K}'} = \frac{d\Omega}{4\pi} \rho_{K'}$$

so

$$\rho_{\mathbf{K}'} = \frac{\mu L^3 K'}{8\pi^3 \hbar^2} d\Omega \tag{15–28}$$

Using this in (15–25), we have

$$R_{\mathbf{K}} \simeq \frac{2\pi}{\hbar} L^{-6} V^*_{\mathbf{K}'\mathbf{K}} V_{\mathbf{K}'\mathbf{K}} \frac{\mu L^3 K'\, d\Omega}{8\pi^3 \hbar^2} \tag{15–29}$$

Now let us calculate the probability per second that the particle in the initial eigenstate associated with the vector $\mathbf{K}$ will cross a 1 cm^2 area normal to the direction of $\mathbf{K}$. This is the incident flux I, and it can be evaluated as in the steps preceding equation (8–7) by using the three dimensional extension of the definition (7–31) of probability flux. Alternatively, we can use (8–8) and equate I to the product of the velocity of the particle and the probability density. That is,

$$I = v\psi^*\psi \tag{15–30}$$

With equations (15–18′) and (15–22), this becomes

$$I = \frac{K\hbar}{\mu} L^{-3} \tag{15–31}$$

Next, divide $R_{\mathbf{K}}$ by the element of solid angle $d\Omega$ to obtain the rate of transitions per unit solid angle into the final states associated with the vector $\mathbf{K}'$. Then we have the probability per second of scattering into a unit solid angle at the angle θ, which is $S(\theta)$, the scattered flux. Thus

$$S(\theta) \simeq \frac{\mu L^{-3} K'}{4\pi^2\hbar^3} V^*_{\mathbf{K}'\mathbf{K}} V_{\mathbf{K}'\mathbf{K}} \tag{15–32}$$

According to equation (15–13), the differential scattering cross section is

$$\frac{d\sigma}{d\Omega} = \frac{S(\theta)}{I} \simeq \frac{\mu}{K\hbar L^{-3}} \frac{\mu L^{-3} K'}{4\pi^2\hbar^3} V^*_{\mathbf{K}'\mathbf{K}} V_{\mathbf{K}'\mathbf{K}}$$

But $K' = \mu v'/\hbar = \mu v/\hbar = K$, because equation (15–4) shows $v' = v$. Therefore

$$\frac{d\sigma}{d\Omega} \simeq \left(\frac{\mu}{2\pi\hbar^2}\right)^2 V^*_{\mathbf{K}'\mathbf{K}} V_{\mathbf{K}'\mathbf{K}} \tag{15–33}$$

where

$$V_{\mathbf{K}'\mathbf{K}} \equiv \int V(r) e^{i(\mathbf{K}-\mathbf{K}')\cdot\mathbf{r}}\, d\tau \tag{15–33′}$$

with the integration taken over a very large box surrounding the scattering potential. This is the *Born approximation* for $d\sigma/d\Omega$. Note that the size of the box has dropped out since L does not appear in (15–33), and since contributions to the integral in (15–33′) will come only from the small region in which $V(r)$ has any appreciable value and therefore the value of the integral is independent of its limits.†

It is possible to carry out part of the integration of (15–33′) immediately. Set

$$\boldsymbol{\chi} = \mathbf{K} - \mathbf{K}' \tag{15–34}$$

and define a set of spherical coordinates r, Θ, Φ with an origin at the center of the potential and polar axis along the direction of $\boldsymbol{\chi}$. (They should not be confused with the spherical coordinates r, θ, ϕ whose polar axis lies along the direction of $\mathbf{K}$.) Then

$$(\mathbf{K} - \mathbf{K}')\cdot\mathbf{r} = \boldsymbol{\chi}\cdot\mathbf{r} = \chi r \cos\Theta$$

and

$$d\tau = r^2 \sin\Theta\, dr\, d\Theta\, d\Phi$$

† We use this fact in equation (15–35).

so

$$V_{\mathbf{K}'\mathbf{K}} = \int_0^\infty \int_0^\pi \int_0^{2\pi} V(r) e^{i\chi r \cos \Theta} r^2 \sin \Theta \, dr \, d\Theta \, d\Phi \tag{15–35}$$

$$V_{\mathbf{K}'\mathbf{K}} = \int_0^\infty \int_0^\pi V(r) e^{i\chi r \cos \Theta} 2\pi r^2 \sin \Theta \, dr \, d\Theta$$

The Θ integral can be evaluated by making the change of variable $Z = i\chi r \cos \Theta$. The result is

$$V_{\mathbf{K}'\mathbf{K}} = \int_0^\infty V(r) \left[\frac{e^{i\chi r} - e^{-i\chi r}}{i\chi r} \right] 2\pi r^2 \, dr \tag{15–36}$$

which is

$$V_{\mathbf{K}'\mathbf{K}} = \int_0^\infty V(r) \frac{\sin \chi r}{\chi r} 4\pi r^2 \, dr \tag{15–37}$$

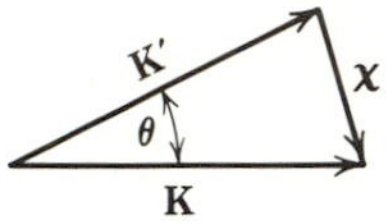

Figure 15–6. Illustrating the relation between the vectors which enter in the Born approximation.

Finally, let us express χ in terms of the scattering angle θ. Consider the vector diagram of figure (15–6), which illustrates the relation (15–34). From this figure it is apparent that

$$\chi = 2K \sin (\theta/2) \tag{15–38}$$

Now compare the results of our calculation, (15–33), (15–37), and (15–38), with the equation (14–35) for the differential cross section for coherent scattering of X-rays by an atom. The similarity is no accident because there is an alternative derivation of the Born approximation which leads to exactly the same results as we have obtained, but which is completely analogous to the derivation of (14–35). This derivation involves calculating the amplitude of the wave scattered in a given direction by every volume element in the region of the potential. Taking the relative phases into account, all these contributions to the total amplitude are added, and the result is squared to obtain the total scattering intensity.

4. Some Applications of the Born Approximation

Let us evaluate $d\sigma/d\Omega$ in the Born approximation for two illustrative cases. Consider an attractive Gaussian potential

$$V(r) = -V_0 e^{-(r/R)^2} \tag{15–39}$$

which is illustrated in figure (15–7). We have

$$V_{\mathbf{K}'\mathbf{K}} = -V_0 \int_0^\infty e^{-(r/R)^2} \frac{\sin \chi r}{\chi r} 4\pi r^2 \, dr$$

Upon integration this gives

$$V_{\mathbf{K}'\mathbf{K}} = -(2\pi)^{3/2} V_0 R^3 e^{-\frac{1}{2}\chi^2 R^2}$$

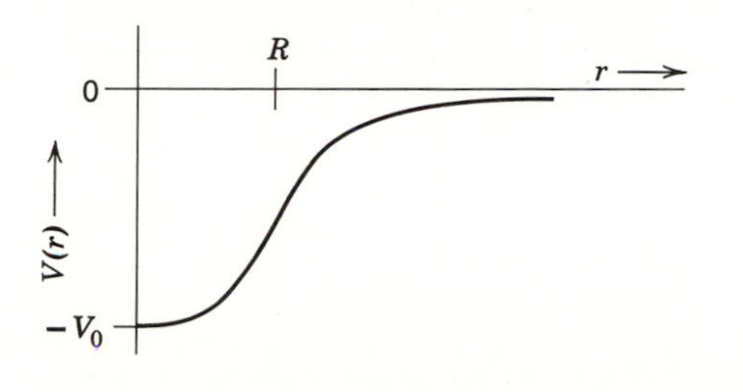

Figure 15–7. An attractive Gaussian potential.

Substituting into equation (15–33), we obtain

$$\frac{d\sigma}{d\Omega} \simeq \left(\frac{\mu}{2\pi\hbar^2}\right)^2 (2\pi)^3 V_0^2 R^6 e^{-\chi^2 R^2}$$

or

$$\frac{d\sigma}{d\Omega} \simeq \frac{2\pi\mu^2}{\hbar^4} V_0^2 R^6 e^{-4K^2 R^2 \sin^2(\theta/2)} \tag{15–40}$$

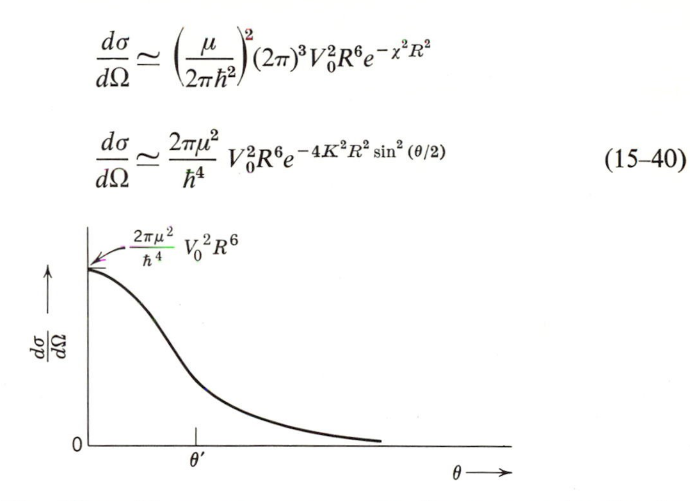

Figure 15–8. The differential scattering cross section for an attractive Gaussian potential.

The differential scattering cross section has the form shown in figure (15–8). This is a Gaussian-like function with a maximum at $\theta = 0$. It falls to e^{-1} times its maximum value at an angle θ' satisfying the relation

$$4K^2 R^2 \sin^2(\theta'/2) = 1$$

For energies high enough that $KR = (\sqrt{2\mu E}/\hbar)R \gg 1$, θ' will be small and we have

$$4K^2R^2(\theta'/2)^2 \simeq 1$$

or

$$\theta' \simeq \frac{1}{KR} \tag{15–41}$$

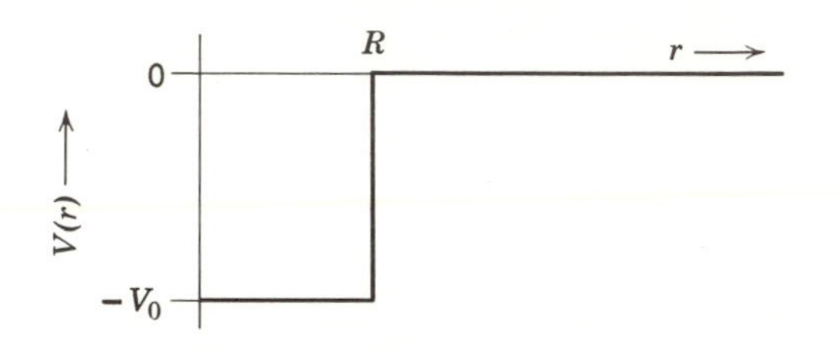

Figure 15–9. An attractive square well potential.

Consider next an attractive square well potential,

$$V(r) = \begin{matrix} -V_0, & r < R \\ 0, & r > R \end{matrix} \tag{15–42}$$

illustrated in figure (15–9). Here

$$V_{\mathbf{K}'\mathbf{K}} = -V_0 \int_0^R \frac{\sin \chi r}{\chi r} 4\pi r^2\, dr$$

and we obtain upon integration

$$V_{\mathbf{K}'\mathbf{K}} = -4\pi V_0 R^3 \frac{[\sin \chi R - \chi R \cos \chi R]}{(\chi R)^3}$$

So

$$\frac{d\sigma}{d\Omega} \simeq \left(\frac{\mu}{2\pi\hbar^2}\right)^2 16\pi^2 V_0^2 R^6 \frac{[\sin \chi R - \chi R \cos \chi R]^2}{(\chi R)^6}$$

or

$$\frac{d\sigma}{d\Omega} \simeq \frac{4\mu^2}{\hbar^4} V_0^2 R^6 \times \frac{\{\sin [2KR \sin (\theta/2)] - 2KR \sin (\theta/2) \cos [2KR \sin (\theta/2)]\}^2}{[2KR \sin (\theta/2)]^6} \tag{15–43}$$

The form of this differential scattering cross section is indicated in figure (15–10). At $\theta = 0$, $\chi R = 0$ but $[\sin \chi R - \chi R \cos \chi R]^2/(\chi R)^6 = 1/9$.

Consequently $d\sigma/d\Omega$ has a finite maximum at $\theta = 0$. It drops with increasing angle, reaching its first zero when $\sin \chi R - \chi R \cos \chi R = 0$ has its first non-zero root. This is

$$\chi R = 4.49$$

or

$$2KR \sin (\theta'/2) = 4.49$$

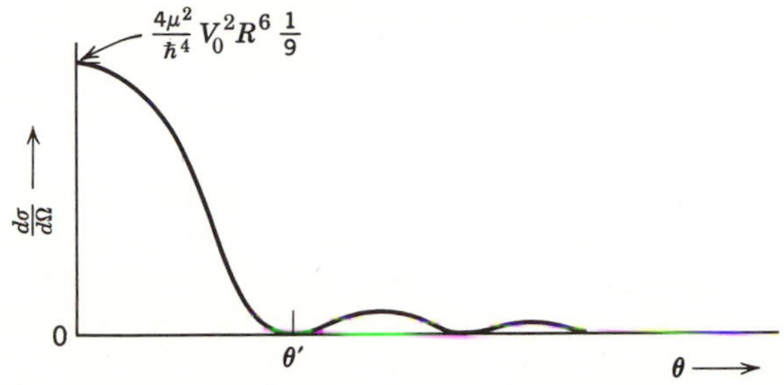

Figure 15–10. The differential scattering cross section for an attractive square well potential.

For high energies, $KR \gg 1$, $\theta' \ll 1$, and the value of θ at the first zero of $d\sigma/d\Omega$ is

$$\theta' \simeq \frac{4.49}{KR} \tag{15–44}$$

For a sharp edge potential, such as (15–42), $d\sigma/d\Omega$ has the characteristic behavior of an optical diffraction pattern. For a potential with a smooth edge, such as (15–39), the similarity between $d\sigma/d\Omega$ and an optical diffraction pattern is, perhaps, not so obvious because the secondary maxima are not present (or at least are weak). Nevertheless, the scattering of a wave function, in passing through a small region in which the potential changes in any way, is completely analogous to the scattering of an electromagnetic wave, in passing through a small region in which the index of refraction changes in an equivalent way (cf. equation 8–56). In both cases an angular distribution is obtained with most of the scattered intensity concentrated at angles less than

$$\theta' \sim \frac{1}{KR} = \frac{1}{2\pi}\frac{\lambda}{R}$$

This angle decreases with increasing K (increasing energy of the particle or increasing frequency of the electromagnetic quanta), and the angular distribution becomes more strongly peaked forward.

The scattering cross sections for the two potentials we have considered can be evaluated from their differential scattering cross sections by calculating

$$\sigma = \int \frac{d\sigma}{d\Omega} \, d\Omega \tag{15–45}$$

where the integral is taken over all solid angle. (This obvious equality follows from the definitions of the quantities involved.) We shall not actually carry this out because the important characteristics of the scattering cross sections are easy to see qualitatively. For both potentials, σ decreases with increasing K because the angular region in which $d\sigma/d\Omega$ has an appreciable value becomes smaller.

In closing, we must discuss the range of applicability of the Born approximation. The condition of validity of the perturbation theory underlying the approximation is (9–36), which states that the amplitude of the wave function for the scattered particle is small compared to the amplitude of the wave function for the incident particle. It is also necessary that the free particle wave function for (15–23) be a reasonable representation of the incident wave function and that the free particle wave function for (15–24) be a reasonable representation of the scattered wave function, in the region of the scattering potential where $V_{\mathbf{K}'\mathbf{K}}$ is evaluated. These conditions will *generally* be true if the energy E of the incident particle is large compared to the magnitude of the scattering potential, that is, if

$$E \gg |V(r)|, \qquad \text{for all } r \tag{15–46}$$

because then the scattering potential is a small perturbation which can usually produce only a small effect. However, (15–46) is neither necessary nor sufficient. For instance, the Born approximation turns out to be fairly accurate for the scattering of low energy electrons from atoms for reasons to be discussed later, and the Born approximation is not accurate in certain cases for the scattering of high energy particles. It is possible to give criteria more useful than (15–46), but this is complicated and will not be done here.

5. Partial Wave Analysis

In the previous section we developed an approximation method useful mainly for scattering at high energies. In this section we shall develop a method which is exact at all energies, but useful mainly for scattering at low energies. This is the method of *partial wave analysis*, originally applied in the nineteenth century by Rayleigh to the scattering of sound waves, and adapted in 1927 by Faxen and Holtsmark to the scattering of quantum mechanical waves.

Consider the scattering, by a potential $V(r)$, of the wave function associated with a particle in a beam which is incident upon the potential. The incident wave passes over the potential, and scattered from it is an outward moving wave. In the immediate vicinity of the potential both the incident and scattered waves generally are strongly distorted by the

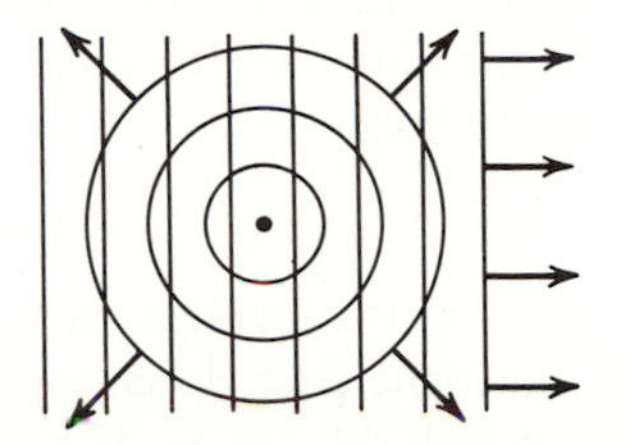

Figure 15–11. Illustrating incident and scattered waves as seen on a scale large compared to the size of the scattering potential.

potential. In fact, the waves may become so tangled that it is difficult to make a distinction between the incident and scattered waves. However, at large distances from the region in which the potential has any appreciable value these complications are no longer present, and the scattering has the simple appearance represented in figure (15–11) in terms of a set of nodal surfaces. If we do not look too closely, we see a plane wave incident upon the potential and a spherical wave scattered outward from it. An eigenfunction for this can be written

$$\psi = e^{iKz} + f(\theta)\frac{1}{r}e^{iKr}, \qquad r \to \infty \tag{15–47}$$

where we have introduced spherical coordinates r, θ, ϕ with origin at the center of $V(r)$, and z axis in the propagation direction of the incident wave. The first term is the eigenfunction (15–18) for a plane wave moving in the direction of increasing z, which represents the incident particle. (Here it will not be necessary to normalize the eigenfunction.) The second term is an eigenfunction for a spherical wave moving in the direction of increasing r, which represents the possible ways in which the scattered particle can move. The directional dependence of its amplitude is given by $f(\theta)$; there can be no ϕ dependence because of the symmetry of the scattering potential. The $1/r$ provides for the necessary inverse square law decrease in the intensity of the scattered wave. Later we shall verify that the second term is a solution to the time independent Schroedinger equation in the region $r \to \infty$ where $V(r) = 0$. For the present we shall assume this and evaluate the incident flux, the scattered flux, and the differential scattering cross section for the eigenfunction (15–47).

In evaluating the fluxes, we may treat the two terms of (15–47) separately since there are no interferences between the incident and scattered waves in the regions where the fluxes are measured. This is true because the incident flux is measured at large negative values of z where the second term is negligible owing to its $1/r$ dependence, and because the scattered flux is measured outside the transverse limits of the incident beam (which are always finite in a real scattering experiment). Thus the incident flux I is, in analogy to equation (15–30),

$$I = v(e^{iKz})^*(e^{iKz}) = ve^{-iKz}e^{iKz} = v \tag{15–48}$$

where $v = \hbar K/\mu$ is the velocity of the incident particle. This is the probability per second that the incident particle crosses a 1 cm^2 area normal to its direction of motion. Similarly, the probability per second that the scattered particle crosses a 1 cm^2 area normal to its direction of motion is

$$v\left[f(\theta)\frac{e^{iKr}}{r}\right]^*\left[f(\theta)\frac{e^{iKr}}{r}\right] = vf^*(\theta)f(\theta)\frac{1}{r^2}$$

The probability per second that it crosses dA cm^2 is

$$vf^*(\theta)f(\theta)\frac{dA}{r^2}$$

But dA/r^2 is equal, by definition, to the element of solid angle $d\Omega$, so the probability per second for scattering into the solid angle $d\Omega$ is

$$vf^*(\theta)f(\theta)\,d\Omega$$

Dividing by $d\Omega$, we have the probability per second for scattering into unit solid angle. This is $S(\theta)$, the scattered flux. Thus

$$S(\theta) = vf^*(\theta)f(\theta) \tag{15–49}$$

The scattered flux can also be evaluated from the three dimensional extension of the definition (7–31) of probability flux, but the method we have used is simpler. Now

$$d\sigma/d\Omega = S(\theta)/I$$

so we have

$$\frac{d\sigma}{d\Omega} = \frac{vf^*(\theta)f(\theta)}{v} = f^*(\theta)f(\theta) \tag{15–50}$$

and the calculation of $d\sigma/d\Omega$ is seen to be a matter of calculating $f(\theta)$.

This is done by matching the incident wave with the scattered wave in such a way as to obtain a solution to the Schroedinger equation, with the scattering potential $V(r)$, that behaves like (15–47) for $r \to \infty$. As a preliminary step, we must first find out how to decompose the plane wave described by

$$e^{iKz} = e^{iKr\cos\theta} \tag{15–51}$$

and associated with the linear momentum $K\hbar$, into a set of *partial waves*, each associated with the orbital angular momentum $\sqrt{l(l+1)}\,\hbar$. We must do this because the Schroedinger equation for a spherically symmetrical potential yields most easily a set of solutions of definite orbital angular momentum. Now the free particle eigenfunction (15–51) is a solution to the time independent Schroedinger equation (10–15) for the case $V(r) = 0$. We have seen that solutions to this equation have the general form

$$\psi(r, \theta, \phi) = R(r)\,\Theta(\theta)\,\Phi(\phi) \tag{15–52}$$

But, since (15–51) has no ϕ dependence, it suffices here to consider only solutions which are independent of ϕ. Therefore we take the solution to equation (10–17) for $m = 0$, which is $\Phi_0(\phi) = 1$. Then $\Theta(\theta)$ is an acceptable (never divergent) solution to equation (10–23), with the condition (10–24), for $m = 0$. These are the functions $\Theta_{l0}(\theta)$. They are called *Legendre polynomials* and are written $P_l(\cos\theta)$. The first few are

$$\begin{aligned} P_0(\cos\theta) &= 1 \\ P_1(\cos\theta) &= \cos\theta \\ P_2(\cos\theta) &= \frac{3\cos^2\theta - 1}{2} \\ P_3(\cos\theta) &= \frac{5\cos^3\theta - 3\cos\theta}{2} \end{aligned} \tag{15–53}$$

The reader should compare these functions with the θ dependence of (10–39) for $m = 0$. The Legendre polynomials are normalized to $2/(2l + 1)$. This is represented by the first of the equations

$$\int_{\cos\theta=-1}^{\cos\theta=1} P_{l'}(\cos\theta)\,P_l(\cos\theta)\,d(\cos\theta) = \begin{cases} \dfrac{2}{2l+1}, & l' = l \\ 0, & l' \neq l \end{cases} \tag{15–54}$$

The second equation represents the orthogonality of these functions. The function $R(r)$ is an acceptable solution to equation (10–27) with $V(r) = 0$. That equation is

$$\frac{1}{r^2}\frac{d}{dr}\left(r^2\frac{dR}{dr}\right) + \left\{-\frac{l(l+1)}{r^2} + \frac{2\mu}{\hbar^2}E\right\}R = 0$$

In the present case $E > 0$, so E is not quantized and the quantum number n does not arise. These solutions are written $j_l(Kr)$ and are called *spherical Bessel functions*. The first few are listed below.

$$
\begin{aligned}
j_0(Kr) &= \frac{1}{Kr}\sin Kr \\
j_1(Kr) &= \frac{1}{(Kr)^2}\sin Kr - \frac{1}{Kr}\cos Kr \\
j_2(Kr) &= \left[\frac{3}{(Kr)^3} - \frac{1}{Kr}\right]\sin Kr - \frac{3}{(Kr)^2}\cos Kr \\
j_3(Kr) &= \left[\frac{15}{(Kr)^4} - \frac{6}{(Kr)^2}\right]\sin Kr - \left[\frac{15}{(Kr)^3} - \frac{1}{Kr}\right]\cos Kr
\end{aligned}
\tag{15–55}
$$

The reader can verify that for $r \rightarrow 0$ they have the same r^l dependence as (10–45), in agreement with the remarks following (13–24). We shall verify their behavior for $r \rightarrow \infty$ in the following paragraph. Finally, we see that for a free particle a particular one of the solutions (15–52) can be written

$$j_l(Kr)\, P_l(\cos\theta) \tag{15–56}$$

This solution is associated with a definite orbital angular momentum about the origin of coordinates, of magnitude $\sqrt{l(l+1)}\,\hbar$.† A general solution in the form (15–52) for this case is

$$\sum_{l=0}^{\infty} a_l j_l(Kr)\, P_l(\cos\theta) \tag{15–57}$$

where the a_l are arbitrary constants. Now the particle described by the free particle eigenfunction (15–51) has the definite linear momentum $K\hbar$. But it can have any orbital angular momentum about the origin since its impact parameter with respect to the origin is completely unknown. Therefore we certainly cannot describe the particle by a single one of the eigenfunctions (15–56). However, we can describe it by the linear combination (15–57), since this contains all possible orbital angular momenta, and since it is a general solution to the Schroedinger equation for the case of a free particle. Thus we can write

$$e^{iKr\cos\theta} = \sum_{l=0}^{\infty} a_l j_l(Kr)\, P_l(\cos\theta) \tag{15–58}$$

† The reason is that it satisfies the L_{op}^2 eigenvalue equation (10–57).

To determine the a_l, we multiply both sides of this equation by $P_{l'}(\cos\theta)\,d(\cos\theta)$ and integrate over the entire range of $\cos\theta$:

$$\int_{-1}^{1} P_{l'}(\cos\theta)e^{iKr\cos\theta}\,d(\cos\theta) = \sum_{l=0}^{\infty} a_l j_l(Kr)\int_{-1}^{1} P_{l'}(\cos\theta)\,P_l(\cos\theta)\,d(\cos\theta)$$

According to (15–54), this picks out of the sum the single term for $l = l'$, yielding, upon transposition,

$$\frac{2}{2l'+1}\,a_{l'}j_{l'}(Kr) = \int_{-1}^{1} \underbrace{P_{l'}(\cos\theta)}_{u}\underbrace{e^{iKr\cos\theta}\,d(\cos\theta)}_{dv}$$

Integrating by parts, we have

$$\frac{2}{2l'+1}\,a_{l'}j_{l'}(Kr) = \left[\underbrace{P_{l'}(\cos\theta)}_{u}\underbrace{\frac{e^{iKr\cos\theta}}{iKr}}_{v}\right]_{-1}^{1} - \int_{-1}^{1}\underbrace{\frac{e^{iKr\cos\theta}}{iKr}}_{v}\underbrace{\frac{dP_{l'}(\cos\theta)}{d(\cos\theta)}\,d(\cos\theta)}_{du}$$

Integrating by parts again, we obtain

$$\frac{2}{2l'+1}\,a_{l'}j_{l'}(Kr) = \left[P_{l'}(\cos\theta)\frac{e^{iKr\cos\theta}}{iKr}\right]_{-1}^{1} - \left[\frac{dP_{l'}(\cos\theta)}{d(\cos\theta)}\frac{e^{iKr\cos\theta}}{(iKr)^2}\right]_{-1}^{1} + \int_{-1}^{1}\frac{e^{iKr\cos\theta}}{(iKr)^2}\frac{d^2P_{l'}(\cos\theta)}{d(\cos\theta)^2}\,d(\cos\theta)$$

In this form the determination of $a_{l'}$ can easily be effected by taking $r \to \infty$. Since the a_l are constants independent of r, the values obtained will be valid everywhere. For $r \to \infty$, the last two terms of the equation above will be negligibly small compared to the first term, since they are both proportional to r^{-2} while the first term is proportional to r^{-1}. Neglecting these terms, we have

$$\frac{2}{2l'+1}\,a_{l'}j_{l'}(Kr) = \frac{1}{iKr}\Big[P_{l'}(\cos\theta)e^{iKr\cos\theta}\Big]_{\cos\theta=-1}^{\cos\theta=1}, \qquad r\to\infty \tag{15–59}$$

To evaluate this expression, we use the relations

$$P_{l'}(\cos\theta) = \begin{matrix} 1, & \cos\theta = 1 \\ (-1)^{l'}, & \cos\theta = -1 \end{matrix}$$

which can be verified by comparison with equations (15–53). Since†

$$(-1)^{l'} = e^{il'\pi}$$

† This relation depends upon the equality $e^{iz} = \cos z + i\sin z$.

equation (15–59) gives

$$\frac{2}{2l'+1}\,a_{l'}j_{l'}(Kr) = \frac{1}{iKr}\left[e^{iKr} - e^{il'\pi}e^{-iKr}\right], \qquad r \to \infty$$

$$= \frac{e^{il'\pi/2}}{iKr}\left[e^{i(Kr-l'\pi/2)} - e^{-i(Kr-l'\pi/2)}\right], \qquad r \to \infty$$

But†

$$e^{il'\pi/2} = i^{l'} \tag{15–60}$$

and†

$$\frac{e^{i(Kr-l'\pi/2)} - e^{-i(Kr-l'\pi/2)}}{2i} = \sin\left(Kr - \frac{l'\pi}{2}\right) \tag{15–61}$$

Therefore this can be written

$$\frac{2}{2l'+1}\,a_{l'}j_{l'}(Kr) = 2i^{l'}\,\frac{\sin(Kr - l'\pi/2)}{Kr}, \qquad r \to \infty \tag{15–62}$$

Note that equation (15–62) is consistent with the behavior as $r \to \infty$ of the $j_{l'}(Kr)$ listed in (15–55) since, for all these functions,

$$j_{l'}(Kr) = \frac{\sin(Kr - l'\pi/2)}{Kr}, \qquad r \to \infty \tag{15–63}$$

as can be verified by neglecting the terms of (15–55) which vanish more rapidly than r^{-1}, and writing $\sin(Kr - l'\pi/2) = \sin(Kr)\cos(l'\pi/2) - \cos(Kr)\sin(l'\pi/2)$. Finally, we substitute (15–63) into (15–62) and solve for $a_{l'}$. We obtain

$$a_{l'} = (2l'+1)i^{l'}$$

This is true for all r. It is also true for all values of the quantum number. Consequently we may use it to evaluate the constants in equation (15–58) and find

$$e^{iKr\cos\theta} = \sum_{l=0}^{\infty}(2l+1)i^l j_l(Kr)\,P_l(\cos\theta) \tag{15–64}$$

This is the desired decomposition of the plane wave associated with the linear momentum Kr into a set of partial waves, each associated with the orbital angular momentum $\sqrt{l(l+1)}\,\hbar$. The terms $j_l(Kr)\,P_l(\cos\theta)$ specify the radial and angular dependence of the lth partial wave. The plane wave $e^{ikr\cos\theta}$ is obtained by mixing together the totality of partial waves, giving each an amplitude $(2l+1)$ and a phase factor i^l.

An instructive interpretation of the amplitude term $(2l+1)$ can be obtained from the following classical argument. Consider an imaginary

† These relations depend upon the equality $e^{iz} = \cos z + i\sin z$.

surface, perpendicular to the direction of propagation of the plane wave, and containing a set of circles of radii

$$d_l = l/K \tag{15–65}$$

centered on the intercept of the z axis and the surface. This is illustrated in figure (15–12). Classically, the orbital angular momentum about the origin of coordinates of a particle moving parallel to the z axis is equal to its

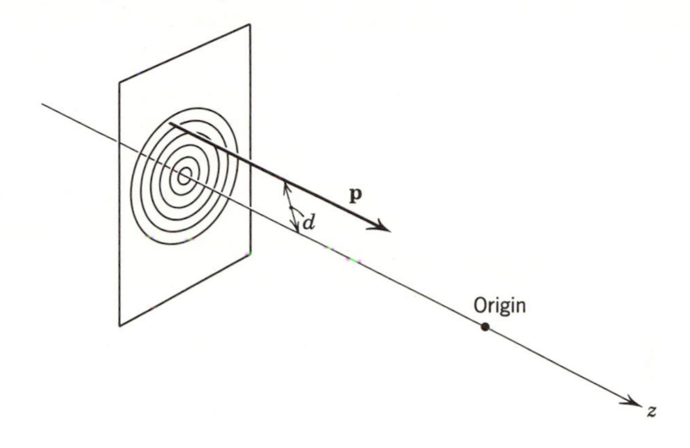

Figure 15–12. Illustrating the classical interpretation of the amplitude term which arises in the decomposition of a plane wave into partial waves.

impact parameter d times its linear momentum $p = K\hbar$. Thus all particles passing through a ring of inner radius d_l and outer radius d_{l+1} will have angular momentum between $(l/K)K\hbar = l\hbar$ and $[(l+1)/K]K\hbar = (l+1)\hbar$. In the classical limit, l is large compared to 1, so all particles passing through the ring can be characterized as having the angular momentum $l\hbar$. Now a particle which is a member of a uniform beam (one which would be described by $e^{ikr\cos\theta}$) is equally likely to have any impact parameter. Therefore the probability that it passes through the ring is just proportional to its area A. This is

$$A = \pi[d_{l+1}^2 - d_l^2] = \frac{\pi}{K^2}[(l+1)^2 - l^2]$$

$$A = \frac{\pi}{K^2}[l^2 + 2l + 1 - l^2] = \frac{\pi}{K^2}[2l+1] \tag{15–66}$$

or

$$A \propto (2l+1)$$

So $(2l+1)$ is the relative probability that a particle moving in a uniform beam will, in the classical limit, have an orbital angular momentum about the origin of $l\hbar$, which is the orbital angular momentum associated with the partial wave of quantum number l in the classical limit where $\sqrt{l(l+1)}\,\hbar \rightarrow l\hbar$.

Let us continue with the problem of determining the function $f(\theta)$ in the expression

$$\psi(r, \theta) = e^{iKr\cos\theta} + f(\theta)\frac{1}{r}e^{iKr}, \qquad r \rightarrow \infty \tag{15-67}$$

which, according to (15–47), describes the eigenfunction for the scattering of a particle by the potential $V(r)$, at positions far from the region where $V(r)$ has any appreciable value. Now, at *all* positions, $\psi(r, \theta)$ is a solution, having no ϕ dependence, to the time independent Schroedinger equation for the potential $V(r)$. From the discussion above, it is evident that its form is similar to the form of the solution (15–57) for $V(r) = 0$. That is,

$$\psi(r, \theta) = \sum_{l=0}^{\infty} b_l R_l(Kr)\, P_l(\cos\theta) \tag{15-68}$$

The angular dependence is exactly the same because the differential equation (10–23) which governs it never contains the potential. But the set of constants is different, and the radial dependence is different too because the $R_l(Kr)$ satisfy (10–27), i.e.,

$$\frac{1}{r^2}\frac{d}{dr}\left(r^2\frac{dR}{dr}\right) + \left\{-\frac{l(l+1)}{r^2} + \frac{2\mu}{\hbar^2}[E - V(r)]\right\}R = 0 \tag{15-69}$$

instead of this equation with $V(r) = 0$ which is satisfied by the $j_l(Kr)$. However, the $R_l(Kr)$ are *essentially* different from the $j_l(Kr)$ only for small r, where $V(r) \neq 0$. For large r, $V(r) = 0$, and both $R_l(Kr)$ and $j_l(Kr)$ satisfy the same differential equation. Thus, for large r these functions are essentially the same and, in particular, for $r \rightarrow \infty$ the $R_l(Kr)$ have the form

$$R_l(Kr) = \frac{\sin(Kr - l\pi/2 + \delta_l)}{Kr}, \qquad r \rightarrow \infty \tag{15-70}$$

For $r \rightarrow \infty$, each of the $R_l(Kr)$ can at most differ from the corresponding $j_l(Kr)$ of equation (15–63) by a *phase shift* δ_l, which arises from the different r dependence of the two functions in the region where $V(r) \neq 0$. It is not difficult to prove this by showing that the functions (15–70) are solutions to (15–69) for $r \rightarrow \infty$, *providing only that $V(r)$ approaches zero with increasing r faster than* $1/r$, but we shall not do this here. Instead, shortly we shall study some graphs of $j_l(Kr)$ and $R_l(Kr)$, for typical potentials $V(r)$, which will make it apparent that (15–70) is correct, and will also show the significance of the phase shifts.

By equating the right sides of equations (15–67) and (15–68), and writing $e^{ikr\cos\theta}$ in terms of (15–64), we obtain

$$\sum_{l=0}^{\infty}(2l+1)i^l j_l(Kr)\,P_l(\cos\theta) + f(\theta)\frac{e^{iKr}}{r} = \sum_{l=0}^{\infty} b_l R_l(Kr)\,P_l(\cos\theta), \qquad r\to\infty$$

Evaluating $j_l(Kr)$ for $r \to \infty$ from (15–63), and $R_l(Kr)$ in the same limit from (15–70), gives

$$\sum_{l=0}^{\infty}(2l+1)i^l \frac{\sin(Kr - l\pi/2)}{Kr} P_l(\cos\theta) + f(\theta)\frac{Ke^{iKr}}{Kr}$$
$$= \sum_{l=0}^{\infty} b_l \frac{\sin(Kr - l\pi/2 + \delta_l)}{Kr} P_l(\cos\theta), \qquad r\to\infty$$

This equation determines both $f(\theta)$, and the entire set of b_l, in terms of the phase shifts. To see this, we use (15–61), and a similar equation, to write the sine functions as complex exponentials. Then, after multiplying through by Kr, we have

$$\sum_{l=0}^{\infty}(2l+1)i^l\left[\frac{e^{i(Kr-l\pi/2)} - e^{-i(Kr-l\pi/2)}}{2i}\right]P_l(\cos\theta) + f(\theta)Ke^{iKr}$$
$$= \sum_{l=0}^{\infty} b_l\left[\frac{e^{i(Kr-l\pi/2+\delta_l)} - e^{-i(Kr-l\pi/2+\delta_l)}}{2i}\right]P_l(\cos\theta), \qquad r\to\infty$$

Grouping together the coefficients of e^{iKr} and e^{-iKr} gives

$$\left\{\sum_{l=0}^{\infty}\left[\frac{(2l+1)i^l}{2i}e^{-il\pi/2} - \frac{b_l}{2i}e^{-il\pi/2}e^{i\delta_l}\right]P_l(\cos\theta) + f(\theta)K\right\}e^{iKr}$$
$$+\left\{\sum_{l=0}^{\infty}\left[-\frac{(2l+1)i^l}{2i}e^{il\pi/2} + \frac{b_l}{2i}e^{il\pi/2}e^{-i\delta_l}\right]P_l(\cos\theta)\right\}e^{-iKr} = 0, \qquad r\to\infty$$

This can be satisfied for arbitrary r only if the coefficients of both e^{iKr} and e^{-iKr} vanish separately. The latter condition gives

$$\sum_{l=0}^{\infty}\left[-\frac{(2l+1)}{2i}i^l e^{il\pi/2} + \frac{b_l}{2i}e^{il\pi/2}e^{-i\delta_l}\right]P_l(\cos\theta) = 0$$

Since this must be true independent of the value of θ, the term multiplying each $P_l(\cos\theta)$ must be zero. For a typical l this requires

$$\frac{(2l+1)}{2i}i^l e^{il\pi/2} = \frac{b_l}{2i}e^{il\pi/2}e^{-i\delta_l}$$

or

$$b_l = (2l+1)i^l e^{i\delta_l} \tag{15–71}$$

Substituting this into the coefficient of e^{iKr}, and demanding that the coefficient vanish, we obtain

$$\sum_{l=0}^{\infty}\left[\frac{(2l+1)i^l}{2i}e^{-il\pi/2}-\frac{(2l+1)i^l}{2i}e^{i\delta_l}e^{-il\pi/2}e^{i\delta_l}\right]P_l(\cos\theta)=-f(\theta)K$$

or

$$f(\theta)=\frac{1}{2iK}\sum_{l=0}^{\infty}(2l+1)(e^{2i\delta_l}-1)P_l(\cos\theta) \tag{15–72}$$

which is

$$f(\theta)=\frac{1}{K}\sum_{l=0}^{\infty}(2l+1)e^{i\delta_l}\sin\delta_l P_l(\cos\theta) \tag{15–72'}$$

where we have used (15–60) and an equation similar to (15–61). Since the b_l are constants, they have the values specified by (15–71) everywhere, even though that equation was obtained at $r \to \infty$. A similar statement is not necessary for (15–72′) as the function $f(\theta)$ is defined only at $r \to \infty$.

Having the expression (15–72′) for $f(\theta)$, the scattering cross sections can be obtained immediately. According to (15–50), $d\sigma/d\Omega$ is

$$\begin{aligned}\frac{d\sigma}{d\Omega}&=f^*(\theta)\,f(\theta)\\&=\frac{1}{K^2}\left\{\sum_{l'=0}^{\infty}(2l'+1)e^{-i\delta_{l'}}\sin\delta_{l'}\,P_{l'}(\cos\theta)\right\}\left\{\sum_{l=0}^{\infty}(2l+1)e^{i\delta_l}\sin\delta_l\,P_l(\cos\theta)\right\}\end{aligned} \tag{15–73}$$

where we have used l' and l to indicate specifically that the two series are to be summed independently. According to (15–45), σ is

$$\sigma=\int\frac{d\sigma}{d\Omega}\,d\Omega=\int\frac{d\sigma}{d\Omega}\,2\pi\sin\theta\,d\theta=2\pi\int_{-1}^{1}\frac{d\sigma}{d\Omega}\,d(\cos\theta)$$

This is evaluated by multiplying the two series of (15–73) and integrating each of the resulting terms over $\cos\theta$. By applying (15–54), the reader may verify that all terms for $l' \neq l$ vanish, and that the result is

$$\sigma=\frac{2\pi}{K^2}\sum_{l=0}^{\infty}(2l+1)^2\sin^2\delta_l\,\frac{2}{2l+1}=\frac{4\pi}{K^2}\sum_{l=0}^{\infty}(2l+1)\sin^2\delta_l \tag{15–74}$$

We see that *the scattering cross sections $d\sigma/d\Omega$ and σ are completely determined by the value of K and the values of the phase shifts δ_l.*

The phase shifts are evaluated by solving equation (15–69) for each l and comparing the phase of $R_l(Kr)$, at some large value of r, with the phase of $j_l(Kr)$ at the same value of r. This is illustrated in figure (15–13) for a typical l and three different forms of the scattering potential $V(r)$. In these curves the function

$$u_l(Kr)=rR_l(Kr) \tag{15–75}$$

is plotted versus r. This is convenient because the amplitude of $u_l(Kr)$

behaves like $rr^l = r^{l+1}$, as $r \to 0$, so that it goes to zero at $r = 0$ for all l, including $l = 0$. The reason is $R_l(Kr) \propto r^l$, as $r \to 0$, in common with the radial dependence of the eigenfunctions for any spherically symmetrical potential [cf. (10–45) and the remarks following (13–24)]. In addition, the function $u_l(Kr)$ satisfies a differential equation

$$\frac{d^2u}{dr^2} + \left\{-\frac{l(l+1)}{r^2} + \frac{2\mu}{\hbar^2}[E - V(r)]\right\}u = 0 \tag{15–76}$$

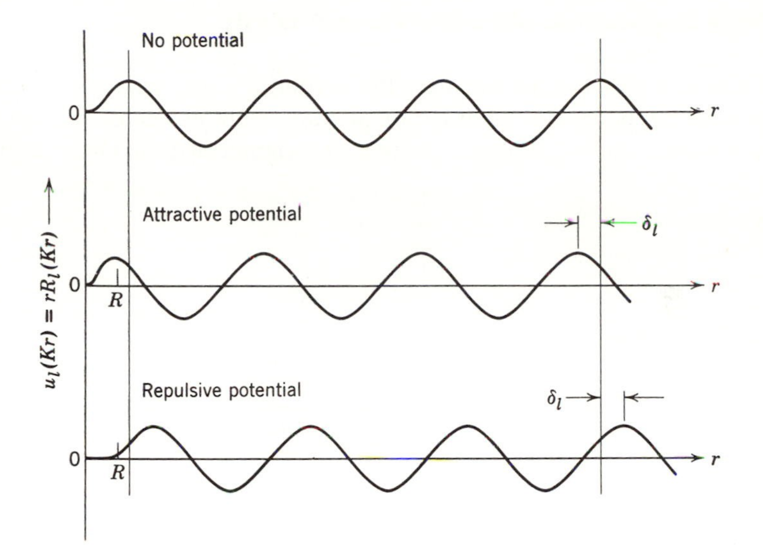

Figure 15–13. The radial dependence of an eigenfunction for three different forms of scattering potential.

which has a simpler form than the equation (15–69) for $R_l(Kr)$. [The equivalence of (15–69) and (15–76) can be verified by substituting (15–75) into the latter.] In figure (15–13) the top curve shows $u_l(Kr)$ for the case of no potential: $V(r) = 0$, all r. Here $R_l(Kr) = j_l(Kr)$, there is no phase shift for any l, and the phase of $u_l(Kr)$ at large r provides the zero of phase with respect to which the phase shifts that arise when $V(r) \neq 0$ can be measured. The middle curve shows $u_l(Kr)$ for the case of a potential: $V(r) < 0$, $r < R$ and $V(r) = 0$, $r > R$. With such an attractive potential, $[E - V(r)] > E$ in the region of the potential and, according to equation (15–76), the magnitude of $d^2u_l(Kr)/dr^2$ will be larger in that region than it is in the case of zero potential. Consequently, $u_l(Kr)$ oscillates more rapidly for $r < R$. For $r > R$, its behavior is unchanged except that its phase is shifted. We see from the curve that: With an *attractive potential* $u_l(Kr)$ is "*pulled in*" by its behavior at $r < R$, its *phase is advanced*, and the *phase shift is positive.*† The bottom curve shows $u_l(Kr)$ for the case of a repulsive potential: $V(r) > 0$, $r < R$ and $V(r) = 0$, $r > R$. Here $[E - V(r)] < E$ in the region of the potential, and the magnitude of $d^2u_l(Kr)/dr^2$ is decreased in this region. The result is that: With a *repulsive potential* $u_l(Kr)$ is "*pushed out*," its *phase is retarded*, and the *phase shift is negative.*†

6. Some Applications of Partial Wave Analysis

In this section we shall calculate the scattering cross sections for two quite different potentials in order to provide some examples of partial wave analysis, and to illustrate further the significance of phase shifts.

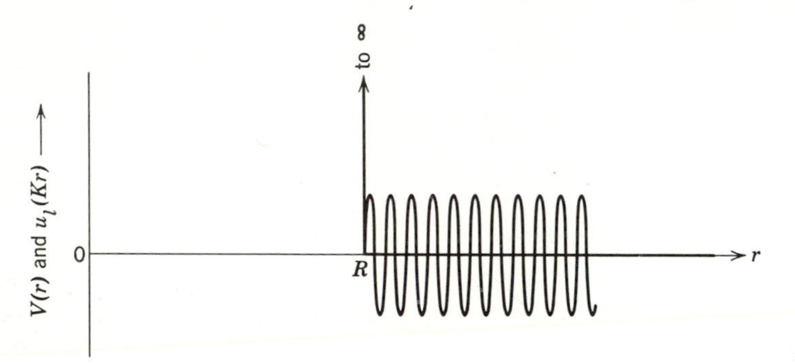

Figure 15–14. An eigenfunction and the scattering potential for the scattering of hard spheres.

Consider first the scattering, in the classical limit, of a small hard sphere by a large hard sphere (the scattering of a ball bearing by a billiard ball). The potential $V(r)$ can be represented as an infinite repulsion commencing at $r = R$, where the interaction radius R is the sum of the radii of the two spheres. The boundary condition for the $u_l(Kr)$ at $r = R$ is the same as at the edge of an infinite square well potential (cf. section 5, Chapter 8), and it demands that each $u_l(Kr)$ vanish at $r = R$. The potential $V(r)$, and a typical $u_l(Kr)$, are plotted in figure (15–14). In the classical limit, the interaction radius R is extremely large compared to the de Broglie wavelength $\lambda = 2\pi/K$, so

$$R \gg 2\pi/K$$

† The phase shift is positive for an attractive potential, and negative for a repulsive potential, only if the potential is not so strong as to make the magnitude of the phase shift greater than 180°. Cf. section 6.

or

$$KR \gg 1 \tag{15–77}$$

and $u_l(Kr)$ has been pushed out by a large number of wavelengths. It is clear, then, that the resulting phase shift δ_l of $u_l(Kr)$ might well have any value in its range of definition $-180°$ to $+180°$, and that for *similar* $u_l(Kr)$ the values of δ_l will be randomly distributed over this range. However, the statement is not true for all $u_l(Kr)$. For values of l greater than a certain l_{max}, $u_l(Kr)$ is not significantly changed at any r by the presence of the potential $V(r)$, and the phase shift δ_l is essentially zero. The reason is $u_l(Kr) = rj_l(Kr)$ in the absence of the potential, and it is a property of the spherical Bessel functions $j_l(Kr)$ that their values are very small for r less than $r_{max} \simeq l/k$, the location of their first maximum. [This property can be verified for the $j_l(Kr)$ of (15–55).] For sufficiently large l, $r_{max} > R$, so $u_l(Kr) = rj_l(Kr)$ is negligible in the region of the potential, and its presence cannot produce any significant change. A good estimate of l_{max} can be obtained by setting

$$l_{max}/K \simeq r_{max} \simeq R$$

or

$$l_{max} \simeq KR \tag{15–78}$$

The same estimate can be obtained from the classical argument connected with equation (15–65) by saying that a particle of angular momentum greater than $l_{max}\hbar$ will have an impact parameter greater than l_{max}/K, and thus, if $l_{max}/K = R$, its impact parameter will be greater than the radius of the potential. Classically, such a particle would not hit the potential and its motion could not be affected by its presence. Quantum mechanically, the potential also has no appreciable effect and the phase shift is negligible for the function $u_l(Kr)$ of corresponding angular momentum.

For the problem at hand, it happens to be particularly easy to calculate an accurate approximation to the scattering cross section,

$$\sigma = \frac{4\pi}{K^2} \sum_{l=0}^{\infty} (2l + 1) \sin^2 \delta_l$$

We set $\delta_l = 0$ for $l > l_{max}$, and let δ_l be random for $l < l_{max}$. Then the sum cuts off at $l = l_{max}$, and we may replace $\sin^2 \delta_l$ by its average value $\frac{1}{2}$. This gives

$$\sigma \simeq \frac{4\pi}{K^2} \frac{1}{2} \sum_{l=0}^{l_{max}} (2l + 1)$$

Furthermore, since l_{max} is large many terms contribute to the sum, and we may approximate it by an integral over dl. Thus

$$\sigma \simeq \frac{2\pi}{K^2} \int_0^{l_{max}} (2l + 1)\, dl \simeq \frac{2\pi}{K^2} \int_0^{l_{max}} 2l\, dl = \frac{2\pi (l_{max})^2}{K^2}$$

Using (15–78), this becomes

$$\sigma \simeq 2\pi K^2 R^2 / K^2$$

or

$$\sigma \simeq 2\pi R^2 \tag{15–79}$$

The differential scattering cross section is, however, not so easy to calculate as it is necessary to determine the exact behavior of the δ_l for values of l in the neighborhood of $l_{\max}$. This is because $d\sigma/d\Omega$ involves

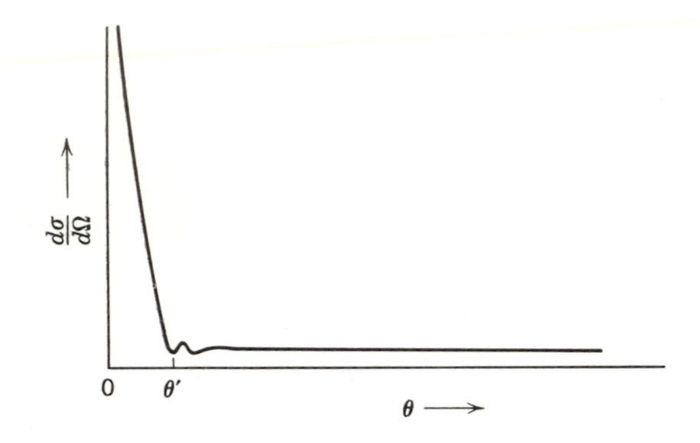

Figure 15–15. The differential scattering cross section for the scattering of hard spheres.

interferences between terms of different l (cf. equation 15–73) and is very sensitive to the values of δ_l. We therefore present only the results of the calculation in figure (15–15). The differential scattering cross section contains a strong peak at very small angles, with a width $\theta' \sim 1/KR$. At larger angles $d\sigma/d\Omega$ is isotropic. The integral $\int (d\sigma/d\Omega)\, d\Omega$ over the region of the peak has the value πR^2, and the integral over the isotropic region has the same value. The peak represents a diffraction effect which the theory predicts to be present, even in the classical limit. It cannot be observed experimentally in this limit because θ' is immeasurably small; a classical measurement of the scattering cross section yields the projected geometrical cross sectional area πR^2, and not $2\pi R^2$, because the diffraction scattering is missed. However, the diffraction scattering can be observed in experiments involving the scattering of nuclear particles by nuclei, for which θ' is not so small. Another example is the scattering of electromagnetic radiation from a reflecting sphere of radius R large compared to the wavelength, but not so large that $\theta' \sim \lambda/R$ is too small. The differential scattering cross section is found to have the form shown in figure (15–15) and the scattering cross section is found to be $\sigma = 2\pi R^2$. Half of this arises from the reflection of radiation hitting the sphere, and half arises from the diffraction of radiation, passing near the sphere, into the shadow on its dark side.

Partial wave analysis provides an exact procedure for solving scattering problems at all energies. However, its application at high energies is usually a very formidable task because K is large, so $KR \gg 1$, where R is the value of r beyond which the scattering potential $V(r)$ is negligible. This means that the expression (15–73) for $d\sigma/d\Omega$ will contain contributions from many terms, since for all l through $l_{max} \simeq KR$ there will be an appreciable phase shift. At high energies it is generally much easier to use the Born approximation, and this approximation is often of sufficient accuracy. The real utility of partial wave analysis is found at low energies where KR is not large. In this region the Born approximation normally cannot be applied because it is very inaccurate, but partial wave analysis is both exact and easy to apply. We shall illustrate this point by calculating the scattering cross sections for a low energy particle, incident upon a simple potential, in the limit $KR \ll 1$.

Equation (15–78) tells us that only the phase shift for $l = 0$ will be appreciable in the limit $KR \ll 1$. In terminology adopted from atomic spectroscopy, this is called *S-wave scattering*. For this case, the expressions (15–73) and (15–74) assume the simple forms

$$\frac{d\sigma}{d\Omega} = \frac{1}{K^2}\{e^{-i\delta_0} \sin \delta_0 \, P_0(\cos\theta)\}\{e^{i\delta_0} \sin \delta_0 \, P_0(\cos\theta)\}$$

or

$$\frac{d\sigma}{d\Omega} = \frac{\sin^2 \delta_0}{K^2}, \qquad KR \ll 1 \tag{15–80}$$

and

$$\sigma = \frac{4\pi \sin^2 \delta_0}{K^2}, \qquad KR \ll 1 \tag{15–81}$$

because $\sin \delta_l$ is essentially zero for $l = 1, 2, 3, \ldots$, and because $P_0(\cos\theta) = 1$. We see that $d\sigma/d\Omega$ is isotropic, and σ is the total solid angle 4π times $d\sigma/d\Omega$, in the limit $KR \ll 1$.

Let us calculate, in this limit, σ for the attractive square well potential,

$$V(r) = \begin{matrix} -V_0, & r < R \\ 0, & r > R \end{matrix} \tag{15–82}$$

which is plotted in figure (15–16). The phase shift δ_0 that this potential produces is evaluated by finding a solution to equation (15–76) for $l = 0$ which goes to zero at $r = 0$.† The equation is

$$\frac{d^2u}{dr^2} + \frac{2\mu}{\hbar^2}[E - V(r)]u = 0 \tag{15–83}$$

† Cf. equation (15–75).

In the region $r < R$, it can be written

$$\frac{d^2u}{dr^2} + K_0 u = 0 \tag{15–84}$$

where

$$K_0 = \sqrt{2\mu(E + V_0)}/\hbar \tag{15–84'}$$

The general solution to this familiar equation is

$$u = A \sin K_0 r + B \cos K_0 r, \qquad r < R$$

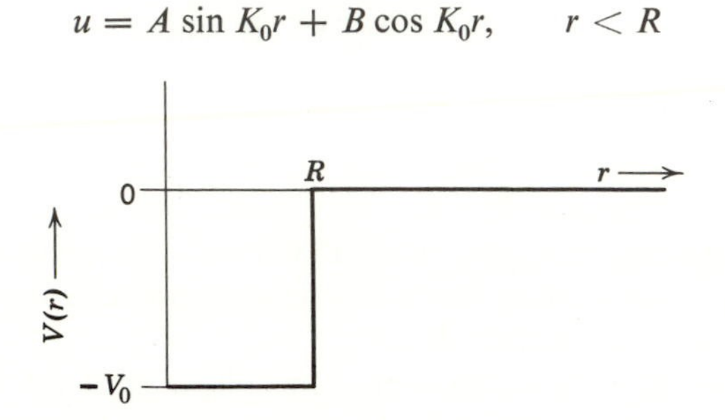

Figure 15–16. An attractive square well potential.

To obtain the required behavior at $r = 0$, we set $B = 0$ and take

$$u = A \sin K_0 r, \qquad r < R \tag{15–85}$$

In the region $r > R$, equation (15–83) can be written

$$\frac{d^2u}{dr^2} + K^2 u = 0 \tag{15–86}$$

where

$$K = \sqrt{2\mu E}/\hbar \tag{15–86'}$$

The general solution to this is

$$u = C \sin Kr + D \cos Kr, \qquad r > R \tag{15–87}$$

Both terms are needed in order to satisfy the requirement that u and du/dr be continuous at $r = R$. Before applying these requirements, it is convenient to rewrite (15–87) as

$$u = F \sin (Kr + \delta_0), \qquad r > R \tag{15–88}$$

The equivalence of (15–87) and (15–88) can be seen by expanding the latter to obtain $u = F \sin Kr \cos \delta_0 + F \cos Kr \sin \delta_0$, and setting $F \cos \delta_0 = C$, $F \sin \delta_0 = D$. Furthermore, by considering the function $R_0(Kr) = u_0(Kr)/Kr$, where $u_0(Kr)$ is given by (15–88), and comparing this at $r \to \infty$ with (15–70) for $l = 0$, it is seen that the quantity δ_0 of (15–88) is precisely the S-wave phase shift. Now apply the requirement that

(15–85) and (15–88) give the same value of u at $r = R$. This yields

$$A \sin K_0R = F \sin (KR + \delta_0) \tag{15–89}$$

The requirement that (15–85) and (15–88) give the same value of du/dr at $r = R$ yields

$$AK_0 \cos K_0R = FK \cos (KR + \delta_0) \tag{15–90}$$

Dividing (15–89) by (15–90), we obtain

$$\frac{K}{K_0} \tan K_0R = \tan (KR + \delta_0)$$

or

$$\tan^{-1}\left[\frac{K}{K_0} \tan K_0R\right] = KR + \delta_0$$

so

$$\delta_0 = \tan^{-1}\left[\frac{K}{K_0} \tan K_0R\right] - KR \tag{15–91}$$

Knowing the phase shift, we can calculate the scattering cross section from equation (15–81). The calculation is simple if we use the conditions $KR \ll 1$ and $K/K_0 \ll 1$, in the limit of very small E, and assume that K_0R does not happen to have such a value that $\tan K_0R \gg 1$. Then (15–91) becomes

$$\delta_0 \simeq \frac{K}{K_0} \tan K_0R - KR \simeq \sin \delta_0$$

or

$$\sin \delta_0 \simeq KR\left(\frac{\tan K_0R}{K_0R} - 1\right)$$

and (15–81) gives

$$\sigma \simeq 4\pi R^2\left(\frac{\tan K_0R}{K_0R} - 1\right)^2 \tag{15–92}$$

This approximate equation predicts $\sigma \rightarrow \infty$ for $K_0R = \pi/2, 3\pi/2, \ldots$ where $\tan K_0R \rightarrow \infty$. Of course, the equation cannot be trusted for these K_0R since they violate the assumption from which it was derived. The exact equation (15–91) shows that for these K_0R the phase shifts $\delta_0 = K_0R$ and, from (15–81), we see that for these δ_0 the scattering cross sections have the very large but finite values

$$\sigma = 4\pi/K^2 \tag{15–93}$$

Note that this can be written $\sigma = [4/(KR)^2]\pi R^2$ so that, since $KR \ll 1$, σ is very much larger than πR^2, the projected geometrical cross sectional area of the scattering potential. When σ takes on its maximum possible value for S-wave scattering (15–93), the scattering cross section is said to be at an *S-wave resonance*. Resonances for other partial waves occur if E is high enough that there are large phase shifts for $l > 0$. For instance, when $\delta_1 = \pi/2, 3\pi/2, \ldots$, the scattering cross section takes on a particularly large value and is said to be at a *P-wave resonance*.

The approximate equation (15–92) does correctly predict that $\sigma \rightarrow 0$ for K_0R which are solutions to the equation $\tan K_0R = K_0R$. For these K_0R, the S-wave phase shift is $\delta_0 = \pi, 2\pi, \ldots$,† and we find directly from (15–81) that $\sigma = 0$. This behavior of the scattering cross section can be seen in the *Ramsauer effect*, which is the term used to describe the extremely

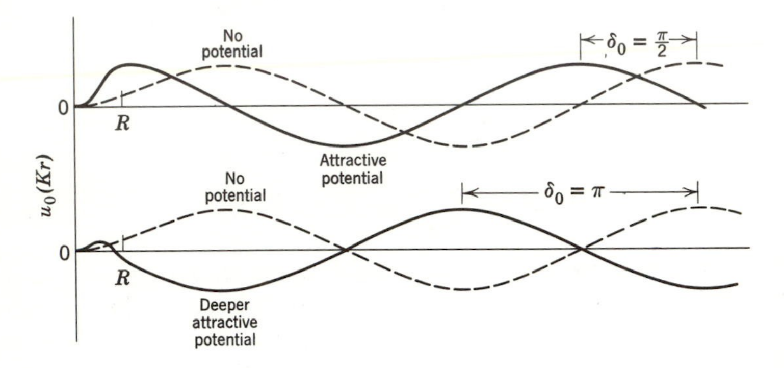

Figure 15–17. The radial dependence of an eigenfunction for two attractive square well potentials.

small value of σ measured in the scattering of electrons from atoms of a noble gas at an energy E of about 1 ev. At low energies only the S-wave is affected by the quite sharply defined attractive potential that a noble gas atom exerts on a passing electron, and at about 1 ev the S-wave is pulled in by just such an amount that $\delta_0 \simeq \pi$ and $\sigma \simeq 0$.

Figure (15–17) shows plots of the function $u_0(Kr)$ for an attractive square well potential of radius R which produces a phase shift $\delta_0 = \pi/2$, and for a deeper attractive square well potential of the same radius which produces a phase shift $\delta_0 = \pi$. Note that this figure, and figure (15–13), show that the amplitude of $u_l(Kr)$ is generally smaller in the region $r < R$ than in the region $r > R$, but that the amplitude is equal in the two regions when the partial wave is in resonance. It would be instructive for the reader to compare this, and the other properties of three dimensional scattering potentials, with the properties of one dimensional scattering potentials discussed in sections 3 and 4 of Chapter 8. It would also be instructive to compare the condition $K_0R = \pi/2$, for the lowest energy S-wave resonance in the potential (15–82), with the condition that a new $l = 0$ state is just bound in that potential. This can be done by using the

† $\delta_0 = 0$ is not of physical interest because it corresponds to no scattering potential.

solution to the Schroedinger equation for the potential (8–61), since it has exactly the same form as the Schroedinger equation (15–83) for a bound $l = 0$ state in the potential (15–82). The results of this comparison will show that when the potential binds an $l = 0$ state near zero total energy it also has an S-wave resonance near that energy. This is also true for higher l values and for potentials of any shape. The reason for this relation is easy to understand. At a low energy resonance the eigenfunction has a particularly large value in the region of the potential, and this is exactly the condition that characterizes a bound state. Furthermore, since for a high energy resonance the eigenfunction also has a particularly large value in the region of the potential, the eigenfunction for such a case has a certain analogy to the eigenfunction for a bound state. For this reason it is often said that a potential has *virtual states*, or *virtual levels*, at all energies for which it exhibits resonances, even though these energies are all positive and correspond to situations in which the particle is actually *unbound*.

For $KR \ll 1$, only the S-wave is affected by the scattering potential, and $d\sigma/d\Omega$ is isotropic. With increasing KR, partial waves of higher angular momentum are affected; for $KR = 1$ the P-wave will be significantly affected and the D-wave will be slightly affected. When these partial waves begin to contribute to the scattering, equation (15–73) shows that $d\sigma/d\Omega$ will no longer be isotropic. As more and more partial waves contribute to $d\sigma/d\Omega$, its structure becomes more and more pronounced. This can be seen by comparing the isotropic $d\sigma/d\Omega$, which we have just obtained for an attractive square well in the limit $KR \ll 1$, with the highly non-isotropic $d\sigma/d\Omega$ for such a potential in the high energy limit $KR \gg 1$, which we obtained by the Born approximation in section 4. The differential scattering cross section $d\sigma/d\Omega$ not only exhibits more structure with increasing KR, but it also becomes more sensitive to the exact form of the scattering potential $V(r)$. As an example, compare the complete independence of the structure of the isotropic $d\sigma/d\Omega$ on the form of $V(r)$ in the limit $KR \ll 1$ with the strong dependence shown in the limit $KR \gg 1$ for the two potentials treated in section 4.

This situation simply reflects the fact that the details of the form of the scattering potential can be important only if the de Broglie wavelength of the scattered particle, $\lambda = 2\pi/K$, is small enough, in comparison with the size of the potential, R, to allow the particle to resolve these details. These conclusions are of great practical importance in nuclear physics where the form of $V(r)$ is not known a priori but must be inferred from analyzing measurements of $d\sigma/d\Omega$. To study the details of $V(r)$, it is necessary to measure $d\sigma/d\Omega$ for high energy particles. The analysis of the measurements is usually quite difficult (unless the energy is so high that the Born approximation is sufficiently accurate), because many partial waves enter into equation (15–73). However, it is not impossible—thanks to high speed electronic computing machines.

We end this discussion by mentioning a very important scattering potential which cannot directly be treated by the method we have developed. This is the Coulomb potential $V(r) \propto 1/r$. As this potential does not approach zero with increasing r faster than $1/r$, the expression (15–70) is not valid. It is possible to modify the method in such a way that the potential can be treated, but because this is complicated we shall only present the results. It is found that for a particle of reduced mass μ and initial velocity v, scattering from a Coulomb potential $V(r) = zZe^2/r$, the differential scattering cross section is

$$\frac{d\sigma}{d\Omega} = \left(\frac{zZe^2}{2\mu v^2}\right)^2 \frac{1}{\sin^4(\theta/2)} \qquad (15\text{–}94)$$

Comparison with equation (4–18), and a moment's consideration, will demonstrate that these results of the quantum mechanical theory are identical with the results of Rutherford's classical theory of the scattering by a Coulomb potential. This remarkable situation is true only for a potential corresponding to an inverse square law of force, and it arises in the following way. From dimensional analysis, it can be shown that if the force exerted on a particle varies according to r^n, then the scattering cross section must vary according to h^{4+2n}. For the inverse square law $n = -2$, the scattering cross section is independent of the value of Planck's constant h, and this requires correspondence between the results of the classical and quantum mechanical calculations of the scattering cross section. In fact, for the Coulomb potential even the Born approximation yields a $d\sigma/d\Omega$ which is identical with (15–94), as the reader may verify with little difficulty. This is why the Born approximation is fairly accurate for the scattering of electrons from atoms at energies as low as $\sim$100 ev. The approximation is good in the outer shells of the atom because there the energy of the scattering electron is large compared to the non-Coulombic potential; it is good in the inner shells, although the energy is not large compared to the potential, because the potential is essentially Coulombic.

7. Absorption

The theory we have developed up to this point treats only the scattering produced by the potential $V(r)$ representing the interaction between the incident particle and the target particle. But there are also processes in which the incident particle is absorbed by the target particle. An example in atomic physics would be the capture of an incident electron by a target atom, leading to the formation of a stable negative ion or the re-emission of an electron of reduced energy. In nuclear physics there are many examples of the capture of an incident nuclear particle by a target nucleus, leading to some nuclear reaction. In this section we shall outline the procedures used for treating absorption processes in the Schroedinger theory. It is to be understood that we use the word "absorption" to mean all processes except scattering with no energy loss in the CM frame (elastic scattering); in particular, absorption includes scattering with energy loss in the CM frame (inelastic scattering). The word "scattering" is used here, as throughout this chapter and the remainder of the book, to mean only scattering with no energy loss in the CM frame.†

We have seen in equation (15–47) that the amplitude of the radially outward traveling waves, scattered at the angle θ, is proportional to the quantity $f(\theta)$. According to (15–72), this is

$$f(\theta) = \frac{1}{2iK}\sum_{l=0}^{\infty}(2l+1)(\eta_l - 1)P_l(\cos\theta) \tag{15–95}$$

where we write

$$\eta_l = e^{2i\delta_l} \tag{15–96}$$

Equations (15–68) and (15–70) show that the total eigenfunction, describing the incident plane waves plus the scattered waves, can be written as radial standing waves. These radial standing waves can be decomposed into radially inward and radially outward traveling waves. In doing this, it is found that the η_l of the relation (15–96) specify the amplitudes and phases of the radially outward traveling waves relative to the radially inward traveling waves. For instance, when $\eta_l = 1$ or $\delta_l = 0$ for all l, the inward and outward traveling waves have the same amplitudes and phases, and they combine to give a total eigenfunction describing incident plane waves plus *no* scattered waves. If $\eta_l = e^{2i\delta_l} \neq 1$ for some l, because $\delta_l \neq 0$, the inward and outward traveling waves have the same amplitudes but different phases, and they combine to give a total eigenfunction describing incident plane waves plus *some* scattered waves.

To allow for absorption as well as scattering, the relation (15–96) is broadened to read

$$\eta_l = A_l e^{2i\delta_l}, \qquad A_l \leqslant 1 \tag{15–97}$$

where the A_l are real. If $A_l < 1$ for some l, the amplitude of the associated outward traveling wave, which is proportional to A_l, will be smaller than

† Note that in the LAB frame there is always at least a small energy loss in "scattering" due to the recoil of the struck particle.

the amplitude of the associated inward traveling wave, which is proportional to 1. Then there is a net inward probability flux, proportional to the difference between the inward and outward intensity, i.e., proportional to $1 - A_l^2 = 1 - \eta_l^*\eta_l$. This implies absorption. The *absorption cross section* σ_A can be evaluated in terms of the η_l by using the three dimensional extension of the definition of probability flux (7–31) to calculate the net inward flux integrated over all angles. The *scattering cross section* σ_S can be evaluated just as we have done before. The results are

$$\sigma_A = \frac{\pi}{K^2}\sum_{l=0}^{\infty}(2l+1)(1-\eta_l^*\eta_l) \tag{15–98}$$

$$\sigma_S = \frac{\pi}{K^2}\sum_{l=0}^{\infty}(2l+1)(1-\eta_l^*)(1-\eta_l) \tag{15–99}$$

It is easy to verify that (15–99) reduces to (15–74), if $A_l = 1$ for all l. The *total cross section* σ_T is defined as

$$\sigma_T = \sigma_S + \sigma_A \tag{15–100}$$

Using (15–98), (15–99), and a little algebra, we find

$$\sigma_T = \frac{2\pi}{K^2}\sum_{l=0}^{\infty}(2l+1)\left(1 - \frac{\eta_l^* + \eta_l}{2}\right) \tag{15–101}$$

The *differential scattering cross section* $d\sigma_S/d\Omega$ is obtained immediately from (15–50) and (15–95). It is

$$\frac{d\sigma_S}{d\Omega} = f^*(\theta)\,f(\theta) = \frac{1}{4K^2}\left\{\sum_{l'=0}^{\infty}(2l'+1)(\eta_{l'}^* - 1)P_{l'}(\cos\theta)\right\} \times \left\{\sum_{l=0}^{\infty}(2l+1)(\eta_l - 1)P_l(\cos\theta)\right\} \tag{15–102}$$

Differential absorption or total cross sections are not defined.

These equations show that the absorption and scattering cross sections are not independent. If there is absorption there must be scattering because, in order to have no scattering, it is necessary that $\eta_l^* = \eta_l = 1$ for all l, and if so there can be no absorption. However, the converse is not true. If there is scattering there is not necessarily absorption because it is possible to have $\eta_l^*\eta_l = 1$, and still have $\eta_l^* \neq 1$, $\eta_l \neq 1$.

As a very simple example, consider the interaction in the classical limit of a small incident sphere with a large target sphere, and assume that the incident sphere is absorbed whenever $r \leqslant R$, where the interaction radius R is the sum of the radii of the two spheres. This problem of the perfectly absorbing sphere can be treated similarly to the problem of the perfectly reflecting sphere given in the previous section. In the present case, the total eigenfunction can contain only inward traveling waves for $l \leqslant l_{\text{max}} = KR$, and will contain the normal mixture of inward and outward traveling waves for $l > l_{\text{max}}$. Thus

$$\eta_l = 0, \qquad l \leqslant l_{\text{max}}$$
$$\eta_l = 1, \qquad l > l_{\text{max}}$$
$$l_{\text{max}} = KR$$

In both equations (15–98) and (15–99) the sums have contributions only for $l \leqslant l_{\text{max}}$, and

$$\sigma_A = \sigma_S = \frac{\pi}{K^2} \sum_{l=0}^{l_{\text{max}}} (2l + 1)$$

As in the work preceding (15–79), we may use an integral to approximate the sum. The result is

$$\sigma_A = \sigma_S = \frac{\pi(l_{\text{max}})^2}{K^2} = \frac{\pi K^2 R^2}{K^2} = \pi R^2 \tag{15–103}$$

and we also have

$$\sigma_T = \sigma_A + \sigma_S = 2\pi R^2 \tag{15–104}$$

The absorption cross section is the projected geometrical cross sectional area πR^2, as would be expected. But there is also a scattering cross section of equal magnitude. This is the diffraction scattering of incident particles passing near the absorbing sphere into the shadow on its far side, and it is exactly the same phenomenon that is found in the case of the reflecting sphere. As we have said before, the diffraction scattering cannot be observed in the classical domain because the angle $\theta' \sim 1/KR$, within which it occurs, is much too small. However, diffraction scattering is very observable and very important in the nuclear domain.

In more typical examples of problems involving absorption, it is not possible to evaluate the η_l by qualitative arguments, and these quantities must come instead from the solution of an appropriate Schroedinger equation. It actually is possible to write a Schroedinger equation for a situation in which quantum mechanical probability is absorbed, even though we have so far always found it to be conserved. This is done by using, as an interaction potential, the *complex function*

$$V(r) + iW(r) \tag{15–105}$$

where $V(r)$ and $W(r)$ are both real, and for absorption $W(r) \leqslant 0$ everywhere [for emission $W(r) > 0$]. There are several ways of demonstrating that the presence of an imaginary term in the interaction potential leads to absorption [or emission]. For instance, by starting at equation (7–26), and carrying through the subsequent calculation with a complex potential

$V(x) + iW(x)$ and a small region x_1 to x_2, the reader may show that the quantum mechanical probability absorbed per unit time in this region is $-(2W/\hbar)\Psi^*\Psi(x_2 - x_1)$, where W and $\Psi^*\Psi$ are evaluated at some typical point in the region. The rate of absorption of probability per unit length is then

$$R = -\frac{2W}{\hbar}\Psi^*\Psi \tag{15–106}$$

In three dimensions the same term gives the rate of absorption of probability per unit volume. Within recent years complex potentials have been widely used in describing the interactions between nuclear particles and nuclei.

BIBLIOGRAPHY

Bohm, D., *Quantum Theory*, Prentice-Hall, Englewood Cliffs, N.J., 1951
Schiff, L. I., *Quantum Mechanics*, McGraw-Hill Book Co., New York, 1955.

EXERCISES

1. Carry through the calculation leading to equation (15–17).

2. Verify that (15–18) is a solution to equation (15–19).

3. Use the Born approximation to calculate $d\sigma/d\Omega$ for the *screened Coulomb potential* $V(r) = (zZe^2/r)e^{-r/d}$, which provides a useful approximation to the potential between a charged particle and a neutral atom if d is set equal to the radius of the atom. Then let $d \to \infty$, and show that $d\sigma/d\Omega$ approaches the normal Coulomb differential scattering cross section (15–94) when $V(r)$ approaches the normal Coulomb potential.

4. For the Bessel functions listed in (15–56), verify that, as stated in sections 5 and 6: (a), $j_l(Kr) \propto r^l$, $r \to 0$; (b), $j_l(Kr) = \sin(Kr - l\pi/2)/Kr$, $r \to \infty$; (c), the first maximum of $j_l(Kr)$ is located at $r_{\max} \simeq l/K$.

5. Verify equation (15–74).

6. By using the procedure indicated in section 6, prove that the condition for the lowest energy S-wave resonance is the same as the condition that the potential just bind an $l = 0$ state near zero total energy.

7. Use partial wave analysis to calculate $d\sigma/d\Omega$ and σ for the scattering of hard spheres with $KR = 0.1$, considering only S-waves, and also with $KR = 0.5$, considering only S- and P-waves. Plot the $d\sigma/d\Omega$ obtained in both calculations. *Hint:* For $r > R$ it is necessary to use the general solution to the differential equation in r. This is $A_l j_l(Kr) + B_l n_l(Kr)$, where each $n_l(Kr)$ can be generated from the corresponding $j_l(Kr)$ by everywhere replacing cos by sin and replacing sin by $-$cos.

8. Derive equation (15–106).

CHAPTER **16**

The Nucleus

1. Introduction

It is fair to say that the properties of the atom are completely understood. The nature of the forces acting between its various parts is known, and quantum mechanics provides a complete technique for evaluating the effects of the forces. For the nucleus this is not the case. Despite great progress in recent years, the nature of the forces acting between its various parts is not completely known. Furthermore, it is not even certain that quantum mechanics will provide a completely adequate technique for evaluating the effects of these forces when they finally are known. There does not yet exist a well-established *theory* of the nucleus through which all its properties can be understood. However, there do exist a number of *models*, or rudimentary theories of restricted validity, each of which can explain a certain limited range of its properties. At the present stage, the study of the nucleus is largely the study of these models. An unavoidable consequence of this situation is that a discussion of the nucleus must lack much of the coherence which characterizes a discussion of the atom. On the other hand, it is just the fact that everything about the nucleus is not yet understood that makes the subject particularly interesting.

Let us introduce our study of the nucleus by recapitulating the pertinent information which we have learned incidental to our study of the atom, and by filling in certain other information.

1. We have briefly described (Chapter 4) Becquerel's discovery of the radioactivity of heavy atoms, and the identification of the alpha particles as doubly ionized He atoms. The beta particles were identified as fast electrons from a measurement of their charge to mass ratio. The gamma rays were assumed to be electromagnetic radiation, and this was confirmed

by showing that they can be diffracted from a crystal like other forms of that radiation. After Rutherford's discovery of the nucleus from an analysis of alpha particle scattering (Chapter 4), it became apparent that radioactivity involves the unstable nuclei of heavy atoms. Later in this chapter we shall study alpha, beta, and gamma emission by such nuclei, but first we shall concentrate on the properties of stable nuclei in their ground states.

2. We have learned (Chapter 4) that the mass of a nucleus is only slightly less than the mass of the corresponding atom. Thus the nuclear mass is approximately equal to the integer A times the mass of an H atom, or approximately equal to A times the mass of a *proton*, the nucleus of an H atom. The integer A is the integer closest to the atomic weight of the atom.† We have also learned that the charge of a nucleus is exactly equal to the atomic number Z times the negative of the charge of an electron, or exactly Z times the charge of a proton. The periodic table of the elements shows that A is roughly equal to $2Z$, except for the proton for which $A = Z = 1$.

3. Analysis of alpha particle scattering from nuclei of low A (Chapter 4) showed that the radii of such nuclei are of the order of 10^{-12} cm, where the radius is defined as the distance from the center of the nucleus at which the potential acting on the alpha particle first deviates from a Coulomb potential. Analysis of the rate of alpha emission from radioactive nuclei of high A (Chapter 8) indicated that the radii of these nuclei, defined in the same way, are $\simeq 9 \times 10^{-13}$ cm. Note that the ratio of the density of a nucleus to the density of bulk matter in the solid state is

$$\frac{\text{density of nucleus}}{\text{density of solid matter}} \sim \frac{[\text{volume of nucleus}]^{-1}}{[\text{volume of atom}]^{-1}} \sim \frac{[(10^{-12})^3]^{-1}}{[(10^{-8})^3]^{-1}} = 10^{12}$$

So the density of a nucleus is $\sim 10^{12}$ gm-cm^{-3}!

4. The word "radius" implies that the nucleus has a spherical shape. We have seen that the study of atomic hyperfine structure (Chapter 13) shows the charge distribution of nuclei is generally not spherical, but instead forms an ellipsoid of revolution with an axis of symmetry along the direction about which the nuclear spin angular momentum vector precesses. As a consequence, nuclei have a static electric quadrupole moment Q. We have learned that for nuclei of spin angular momentum quantum number $i \geqslant 1$, there are cases of $Q > 0$ (ellipsoid elongated in the direction of the symmetry axis) as well as cases of $Q < 0$ (ellipsoid flattened in the direction of the symmetry axis), but for $i = 0$ or $\frac{1}{2}$, $Q = 0$.

† For any particular isotope of an atom, the atomic weight is always very close to an integer. The term "isotope" will be explained in section 4.

However, the departures from sphericity are never severe; in extreme cases the ratio of the semimajor axis to the semiminor axis is $\simeq 1.2$. Thus it is often possible to ignore these departures and assume a spherical shape. Nuclei do not have static electric dipole moments, for this would correspond to a constant displacement of the center of the charge distribution from the center of the mass distribution, which cannot happen.

5. Nuclei do have static magnetic dipole moments. This is also shown in the study of atomic hyperfine structure (Chapter 13). We have seen that the magnitudes of these magnetic dipole moments are $\sim \mu_N = e\hbar/2Mc \sim 10^{-3}\mu_B$, where μ_N is the nuclear magneton, M is the proton mass, and μ_B is the Bohr magneton. Hyperfine structure also shows that the signs of the magnetic dipole moments (the relative orientation of the magnetic dipole moment vector and the nuclear spin angular momentum vector) are positive in some cases and negative in others.

6. The spin angular momentum quantum number i is directly measurable from atomic hyperfine structure. As we have learned (Chapter 13), i is integral when A is even, with $i = 0$ when Z is also even, and i is half-integral when A is odd. The magnitude I of the nuclear spin angular momentum is given in terms of i by the usual relation $I = \sqrt{i(i+1)}\,\hbar$.

7. Closely related to the question of nuclear spin is the question of the symmetry character of the eigenfunction for a system containing two or more nuclei of the same species (Chapter 12). This is studied by analyzing the spectra of diatomic molecules containing two identical nuclei and, as we have stated, it is found that nuclei of integral spin quantum number (even A) are of the symmetrical type and nuclei of half-integral spin quantum number (odd A) are of the antisymmetrical type.

2. The Composition of Nuclei

With one exception, $A \sim 2Z$ for all nuclei. (The exception is the proton for which $A = Z = 1$.) From this it is immediately apparent that a nucleus with a given value of A cannot be composed of A protons alone, because its Z would be $Z = A$ instead of $Z \sim A/2$. Before 1932 it was generally assumed that a nucleus is composed of A protons and $(A - Z)$ electrons. This gives the correct charge and also allows the mass to be correct. However, it was realized that the assumption that electrons are contained in nuclei presents serious difficulties.

Although the assumption made it easy to understand how electrons could be ejected from nuclei in the beta emission process, there were problems with the magnitude of the zero point energy necessary to confine a particle of mass as small as an electron in a region of space as small as the nucleus. By directly applying the uncertainty principle, and evaluating momentum from the equation $E = cp$ which obtains for highly relativistic particles, the reader may show that the kinetic energy of an electron confined to a region of dimensions 10^{-12} cm must be $\gg 100$ Mev. It was difficult to reconcile this with the fact that the typical kinetic energy of an electron ejected in the beta emission process is $\sim$1 Mev.

It was also difficult to reconcile the observation that all nuclear magnetic dipole moments are three orders of magnitude smaller than the magnetic dipole moment of an electron, under the assumption that electrons are contained in nuclei.

The insurmountable difficulty with the assumption came from a consideration of nuclei in which A is even and Z is odd, e.g., the nucleus of the N atom with $A = 14$, $Z = 7$. In conformity with the usual rules, the spin angular momentum quantum number of this nucleus is observed to be integral (specifically, $i = 1$), and the symmetry character is observed to be symmetrical. But, if it contains 14 protons and 7 electrons, its spin quantum number would necessarily be half-integral and its symmetry character would necessarily be antisymmetrical, because both the proton and the electron have spin quantum number $i = \frac{1}{2}$ and are of antisymmetrical symmetry character, and because $14 + 7 = 21$, an odd number. It is easy to see from the rules for the addition of spin and orbital angular momentum, presented in Chapter 13, that a system containing an odd number of particles with half-integral spin quantum number can only have total angular momentum (the nuclear spin angular momentum) corresponding to a half-integral quantum number. It is also easy to see, from the discussion in Chapter 12, that the symmetry character of a system containing an odd number of particles of antisymmetrical symmetry character must be antisymmetrical.

Some years before its discovery in 1932, Rutherford suggested the existence of a particle having the same mass, symmetry character, and spin quantum number as the proton, but zero charge. This is the particle we now call the *neutron*. One motivation for this suggestion was that the problems outlined above would not arise if nuclei were assumed to be composed of protons and neutrons. In particular, the spin and symmetry character of the N nucleus would agree with observation if it contained 7 protons and 7 neutrons. This would also, of course, be in agreement with the mass and charge required for $A = 14$ and $Z = 7$. A number of people tried to devise experiments which would detect the neutron. But this was difficult because, being uncharged, the neutron cannot directly produce ionization, and most devices for detecting particles depend upon ionization.

We shall describe the sequence of experiments and interpretations leading to the discovery of the neutron, as it is of great historical importance and also provides good examples of some of the early techniques of nuclear investigation. In 1919 Rutherford bombarded a target of N nuclei with 7.7-Mev alpha particles from a radioactive source and found that charged particles of long range were emitted from the target. The experimental apparatus, employing a ZnS scintillation screen (cf. section 3, Chapter 4) as a charged particle detector, is indicated schematically in figure (16–1). The *range* of a charged particle is the thickness of some standard absorbing material which it can traverse before losing all its kinetic energy by ionizing the atoms of the absorbing material, and coming

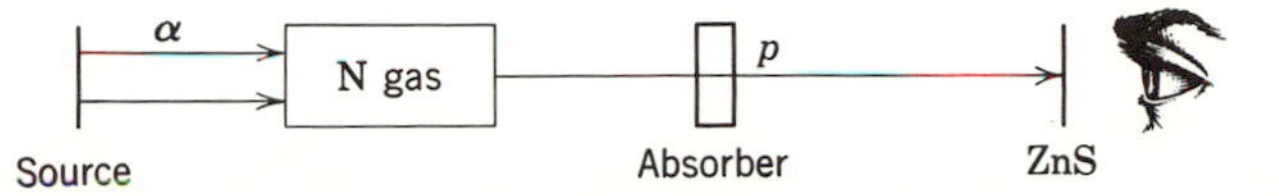

Figure 16–1. An apparatus used in the first artificially produced nuclear reaction.

to a stop. This quantity increases monotonically with the velocity of the particle and decreases monotonically with its charge. The range of the particles emitted from the N nuclei of the target was so long that they could unambiguously be identified as protons, since these particles have much longer ranges than alpha particles of comparable energies because they have higher velocities and smaller charges. Rutherford's experiment constituted the first artificially produced nuclear reaction

$$^{2}\text{He}^{4} + {}^{7}\text{N}^{14} \rightarrow {}^{8}\text{O}^{17} + {}^{1}\text{H}^{1} \tag{16–1}$$

The equation represents the reaction and introduces some useful notation: $^{2}\text{He}^{4}$ is the *bombarding particle*, an alpha particle which is the He nucleus with $Z = 2$, $A = 4$; $^{7}\text{N}^{14}$ is the *target nucleus*, an N nucleus with $Z = 7$, $A = 14$; $^{8}\text{O}^{17}$ is the *residual nucleus*, an O nucleus with $Z = 8$, $A = 17$; and $^{1}\text{H}^{1}$ is the *product particle*, a proton which is the H nucleus with $Z = 1$, $A = 1$.

Several other reactions of essentially the same type were investigated in succeeding years, but in 1930 a reaction of a different type was discovered by Bothe and Becker. They used an arrangement similar to the one shown in figure (16–1), except that the ZnS scintillation screen was replaced by a Geiger counter. This is a chamber filled with gas and containing a fine wire held at a high positive voltage with respect to the walls. Any ionization processes occurring within the chamber liberate one or more electrons, which are accelerated so strongly in the direction of the wire that they produce additional ionization by collisions with gas atoms. This process repeats, and an avalanche of electrons rapidly builds up.

Enough charge is collected by the wire, as the result of a *single* initial ionization, to produce an easily detectable electric pulse. Using alpha particles of kinetic energy 5.3 Mev, Bothe and Becker found that in the bombardment of Be or B nuclei uncharged particles, capable of traversing very thick absorbers with little attenuation, were emitted. They incorrectly assumed these to be very high energy quanta (gamma rays). A Geiger counter can detect quanta since they can produce electrons in the filling gas by the photoelectric effect, the Compton effect, or pair production.

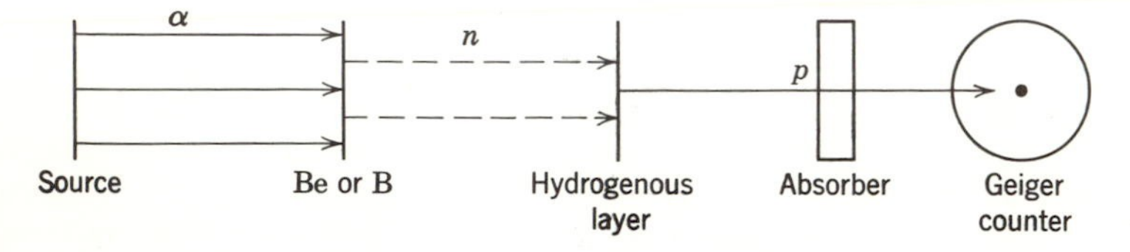

Figure 16–2. An apparatus used in the discovery of the neutron.

Further work on this reaction was done in 1932 by I. Curie and Joliot. They discovered, quite accidentally, that placing a layer of hydrogenous material in front of the Geiger counter caused the counting rate to increase! It was soon realized that the effect was due to the ejection of protons from the hydrogenous layer as a result of its bombardment by the uncharged particles. Protons, being charged, are counted much more efficiently by a Geiger counter than quanta or other uncharged particles. The apparatus used by Curie and Joliot is indicated in figure (16–2). By measuring the range in the absorber of protons ejected from the hydrogenous layer, the velocities or energies of these protons could be determined from the known range-velocity relation. Curie and Joliot found the proton energy $E_p \simeq 4.5$ Mev for Be target nuclei. Following Bothe and Becker, they assumed that quanta incident upon the hydrogenous layer ejected protons by a process analogous to Compton scattering. They also evaluated E_γ, the minimum energy of the quanta required to eject a 4.5-Mev proton. This gave $E_\gamma \simeq 50$ Mev, as the reader may verify by using the equations of section 8, Chapter 3 (with the electron mass replaced by the proton mass).

Shortly after this experimental development, Chadwick demonstrated that its interpretation was untenable, and that the uncharged particles emitted from the Be or B nuclei under alpha particle bombardment were not quanta but the long-sought neutrons. He showed first that the number of protons ejected from the hydrogenous layer was several orders of magnitude smaller than would be predicted by the Klein-Nishina equation (14–36 with the electron mass replaced by the proton mass), if the incident uncharged particles were 50-Mev quanta. He showed next that it is not energetically possible for a 50-Mev quantum to be liberated in a reaction in which a 5.3-Mev alpha particle is captured by a Be nucleus to form a C nucleus. This calculation is based upon a knowledge of the mass of the alpha particle, the mass of the Be nucleus, the mass of the C nucleus and Einstein's relation (1–23) between mass and energy. We shall do several such calculations in section 4. For the present, suffice it to say that a similar calculation showed that all the experimental results were consistent if it was assumed that the reaction involved in the case of a Be target nucleus is

$$^{2}\mathrm{He}^{4} + {}^{4}\mathrm{Be}^{9} \rightarrow {}^{6}\mathrm{C}^{12} + {}^{0}n^{1} \tag{16–2}$$

where ${}^{0}n^{1}$ is the neutron with $Z = 0$, $A = 1$. A similar reaction would explain the experiments performed with a B target nucleus. With this assumption, the ejection of protons from the hydrogenous layer could be understood as a simple collision between an incident particle, the neutron, and a stationary free particle of equal mass, the proton.† Then the minimum neutron energy required to eject a 4.5-Mev proton is $E_n = 4.5$ Mev, and this is in good agreement with the restrictions imposed by the masses of the entities appearing in (16–2) plus the mass-energy relation.

Since the discovery of the neutron, there has never been any doubt that a nucleus is composed of Z protons and $(A - Z)$ neutrons. Thus a nucleus contains $Z + (A - Z) = A$ *nucleons*, a word used for both protons and neutrons.

3. Nuclear Sizes and the Optical Model

Some of the first evidence for the sizes of nuclei, other than that which came from alpha particle scattering and emission, was obtained from the differences in the total binding energies of *mirror nuclei*. Two nuclei are said to form a set of mirror nuclei if they have common values of A, and if $Z = (A - Z) + 1$ for one and $(A - Z) = Z + 1$ for the other. Some examples are:

$$^{1}\mathrm{H}^{3} \text{ and } {}^{2}\mathrm{He}^{3}, \quad {}^{3}\mathrm{Li}^{7} \text{ and } {}^{4}\mathrm{Be}^{7}, \quad {}^{5}\mathrm{B}^{11} \text{ and } {}^{6}\mathrm{C}^{11},$$

$$^{7}\mathrm{N}^{15} \text{ and } {}^{8}\mathrm{O}^{15}, \quad {}^{11}\mathrm{Na}^{23} \text{ and } {}^{12}\mathrm{Mg}^{23}, \quad {}^{17}\mathrm{Cl}^{35} \text{ and } {}^{18}\mathrm{A}^{35}$$

In each set, one nucleus is unstable and eventually decays by the emission of an electron or a positron, forming the stable member of the set. The differences in the total binding energies of the two nuclei of each set can be

† The kinetic and binding energies of a proton in a hydrogenous compound are completely negligible compared to 4.5 Mev.

obtained directly from measurements of the energies of the electrons or positrons which are emitted in the decay, and can also be obtained indirectly, in some cases, from measurements of the masses of the nuclei and the mass-energy relation.

Now it is clear that, in addition to the repulsive Coulomb forces acting between each pair of protons in a nucleus, there must be attractive forces of some other type acting between all the nucleons. If it were not for these so-called *nuclear forces*, neutrons would not be bound in nuclei and, in addition, all nuclei for $Z \geqslant 2$ would break up by Coulomb repulsion.† It was assumed at an early date, and has subsequently been verified by several different kinds of evidence to be presented later, that the nuclear forces acting between pairs of nucleons are independent of whether the nucleons are protons or neutrons. With this assumption, the total binding energies of two mirror nuclei will differ only by the additional Coulomb repulsion energy ΔE_C of the nucleus with the extra proton. If it is also assumed that all the charge due to the protons, including that due to the extra proton, is uniformly distributed over a sphere of radius R, the additional Coulomb repulsion energy can be evaluated from classical electrostatics. The result is

$$\Delta E_C = \frac{6}{5}\frac{Ze^2}{R} \tag{16–3}$$

where Z is the atomic number of the nucleus without the extra proton. Using this equation to analyze the observed differences in total binding energies of a number of sets of mirror nuclei with A ranging from 3 to 38, it was found that values of R so determined could be represented by the equation

$$R = r_0 A^{1/3} \tag{16–4}$$

where r_0 is a constant. The analysis gave $r_0 \simeq 1.4 \times 10^{-13}$ cm. However, recent work on this problem has shown that quantum mechanical corrections to this simple classical calculation reduce r_0 somewhat. The value currently quoted is

$$r_0 \simeq 1.2 \times 10^{-13} \text{ cm} \tag{16–4$'$}$$

The quantum mechanical calculation uses a more realistic charge distribution for the extra proton and takes into account the exchange integral (cf. 12–30). Both factors tend to lower ΔE_C, thus requiring a compensating decrease in r_0.

† The nuclear forces are definitely not gravitational in nature. From the data presented in section 1 on the charge and mass of nuclei, the reader may easily show that gravitational forces are too weak to overcome the Coulomb forces by about 35 orders of magnitude.

There are several other methods of measuring the size of the nuclear charge distribution. We shall consider the one which is the most straightforward and accurate. It is the scattering of electrons from nuclei at energies of several hundred Mev. As nuclear forces do not act between nucleons and electrons, the scattering of an electron from a nucleus is due to the Coulomb interaction between the electron and the nuclear charge distribution. Therefore measurements of electron scattering should provide information about the nuclear charge distribution. The method can be thought of as the use of an "electron microscope" of extremely high resolution to "look at" the charge distribution. The resolution is extremely

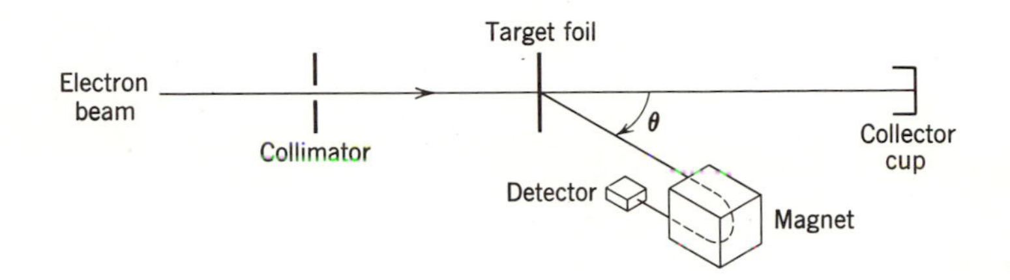

Figure 16–3. An apparatus used to measure the scattering of high energy electrons by nuclei.

high because the de Broglie wavelength λ of the electrons is extremely short. Using the expression $p = E/c$ to evaluate the momentum of the electrons, which are highly relativistic because their energy is very large compared to their rest mass energy $m_0c^2 = 0.51$ Mev, we find for 500-Mev electrons $\lambda = 2.4 \times 10^{-13}$ cm. Taking $R \sim 10^{-12}$ cm as an estimate of the size of the charge distribution, we see that for such electrons $\lambda \sim R/4$ and $KR = (2\pi/\lambda)R \sim 20$. According to the conclusions drawn in the preceding chapter, the differential cross section for the scattering of such electrons from the nuclear charge distribution should be very sensitive to its size, and even to the details of its shape.

We shall not attempt to describe the machines which are used to accelerate electrons (or other particles) to the high energies required in these experiments since, in the space available, we could not do justice to the subject.† However, it is worth while to describe the other aspects of the electron scattering experiments. The apparatus used in recent work by Hofstadter and colleagues is indicated in figure (16–3). The collimated electron beam passes through a thin foil, the atoms of which contain the nuclei to be investigated,‡ and it is stopped in an insulated metallic cup.

† Descriptions can be found in a number of elementary textbooks.

‡ An electron of several hundred Mev has such a high relativistic mass that the scattering produced by the atomic electrons is completely negligible.

The total charge collected by the cup is a measure of the incident flux. Electrons scattered into a certain angular range at θ enter a magnet, which bends them in a semicircle in the plane perpendicular to the paper. Upon leaving the magnet, they enter a scintillation detector similar to the old ZnS + human eye detector, but employing a more efficient crystal and a

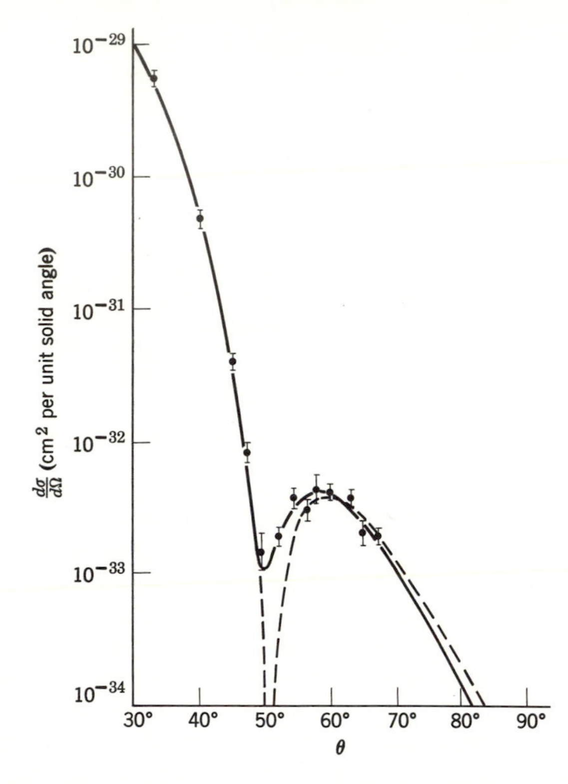

Figure 16–4. The differential scattering cross section for 420-Mev electrons incident upon carbon. From R. Hofstadter, *Annual Review of Nuclear Science*, Vol. 7, Annual Reviews, Stanford, 1957.

photoelectric cell. The radius of curvature of the electron path in the known magnetic field is a measure of the momentum or energy of the scattered electrons.

It is found that by far the most likely process is scattering with no energy loss in the CM frame. The differential scattering cross section measured for 420-Mev electrons, incident on the typical small A nucleus $^6C^{12}$, is shown in figure (16–4).† The points with accuracy estimates are

† This is actually $d\sigma/d\Omega$ in the LAB frame, but for electron scattering the difference between the LAB frame and the CM frame is quite negligible.

the experimental measurements of $d\sigma/d\Omega$. They are evaluated from the measured incident flux, the measured number of full energy electrons scattered per unit solid angle at each angle, and the number of nuclei per cm^2 in the target foil.

That absorption processes are observed to be improbable verifies our idea that the scattering should be explicable on the basis of a real scattering potential representing the Coulomb interaction between the electron and the nuclear charge distribution. That the electron energies are high suggests the additional idea that the Born approximation should be useful in the analysis of these experiments. It actually is found that for nuclei of small A (and therefore small nuclear charge and scattering potential) the Born approximation is capable of giving a reasonably good account of the measured $d\sigma/d\Omega$—if cognizance is taken of the fact that the electrons are highly relativistic. This is done by using free particle eigenfunctions obtained from the Dirac equation, instead of the Schroedinger equation, to evaluate the matrix elements entering in the Born approximation. The results of this relativistic Born approximation are

$$\frac{d\sigma}{d\Omega} = \left(\frac{Ze^2}{2E}\right)^2 \frac{\cos^2(\theta/2)}{\sin^4(\theta/2)} \frac{|F(\chi)|^2}{Z^2} \tag{16–5}$$

where E is the electron energy. The form factor $F(\chi)$ is

$$F(\chi) = \int_0^\infty \rho(r) \frac{\sin \chi r}{\chi r} 4\pi r^2 \, dr \tag{16–5'}$$

with

$$\chi = 2K \sin(\theta/2), \qquad K = \frac{p}{\hbar} \simeq \frac{E}{c\hbar}$$

where $\rho(r)$ is the nuclear charge density in units of proton charges. Just as in the case of the form factor that arises in the discussion of X-ray scattering (cf. equation 14–35), at small angles $F(\chi) \simeq Z$. Also, $\cos^2(\theta/2)$ is essentially unity at small scattering angles. Keeping these points in mind, writing $2E \simeq 2mc^2$ for the highly relativistic electrons, and then comparing equation (16–5) with equation (15–94), we see that at small scattering angles $d\sigma/d\Omega$ of the former equation is equal to the Coulomb differential scattering cross section for a point charge nucleus. At large scattering angles, $d\sigma/d\Omega$ is reduced below this value because $|F(\chi)|^2 < Z^2$, and because $\cos^2(\theta/2) < 1$. The first term describes the effect of the finite extension of the nuclear charge; the second term describes a relativistic effect which may be attributed to the electron spin.

The behavior of $d\sigma/d\Omega$, as calculated from equation (16–5), depends in a quite sensitive manner on the assumed form of the nuclear charge density

$\rho(r)$. Thus a comparison of the calculated and measured values of $d\sigma/d\Omega$ allows a determination of this quantity. The dashed curve in figure (16–4) shows the best fit obtained for $d\sigma/d\Omega$ calculated in the Born approximation. The solid curve in the figure shows $d\sigma/d\Omega$ calculated from a phase shift analysis using the same $\rho(r)$ as was used in the Born approximation calculation. Of course, the phase shift analysis is based upon the Dirac equation, and not upon the Schroedinger equation.

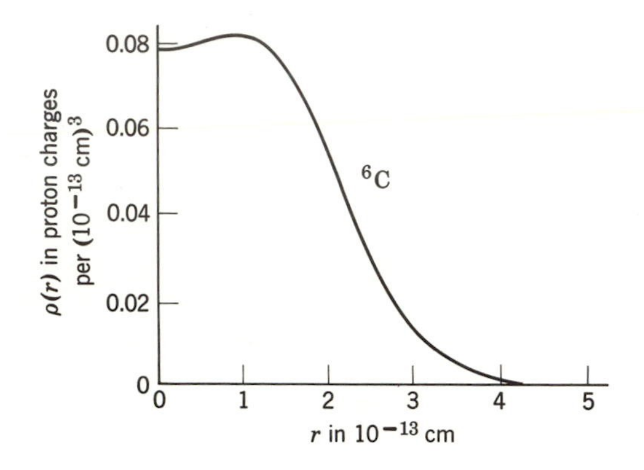

Figure 16–5. The nuclear charge density of carbon. From R. Hofstadter, *Annual Review of Nuclear Science*, Vol. 7, Annual Reviews, Stanford, 1957.

In figure (16–5) we plot the $\rho(r)$ used in these calculations. As essentially the same $\rho(r)$ is found in the analysis of $d\sigma/d\Omega$ measured in the scattering of electrons of several different energies from the C nucleus, and as the analyses are sensitive enough to distinguish variations in, for example, $\rho(0)$ of perhaps 10 percent, there is considerable confidence that this $\rho(r)$ must be quite close to the actual charge density of the nucleus.

For a given electron energy, the differential scattering cross sections measured for nuclei of larger A are found to be similar to those measured for C, except for a decrease with increasing A in the angle at which the first minimum is found, and a development of subsidiary maxima at larger angles. The experience gained in the previous chapter tells us that this means the size of the charge distribution increases with increasing A. This is seen in figure (16–6), which plots charge densities obtained by a partial wave analysis of the differential scattering cross sections measured for electrons of several energies and for a number of different nuclei. The forms of $\rho(r)$ given in the figure are not completely unique. For instance, the charge density in the central region for Ca, and heavier nuclei, is known to be constant only to within 10 or 20 percent. However, it is assumed that the central density is constant, and that the form of $\rho(r)$

can be described by the empirical equation

$$\rho(r) = \frac{\rho(0)}{1 + e^{(r-a)/b}} \tag{16–6}$$

With this assumption, the parameters a and b are accurately determined by the analysis of the scattering measurements to be

$$\begin{aligned} a &= 1.07A^{1/3} \times 10^{-13}\ \text{cm} \\ b &= 0.55 \times 10^{-13}\ \text{cm} \end{aligned} \tag{16–7}$$

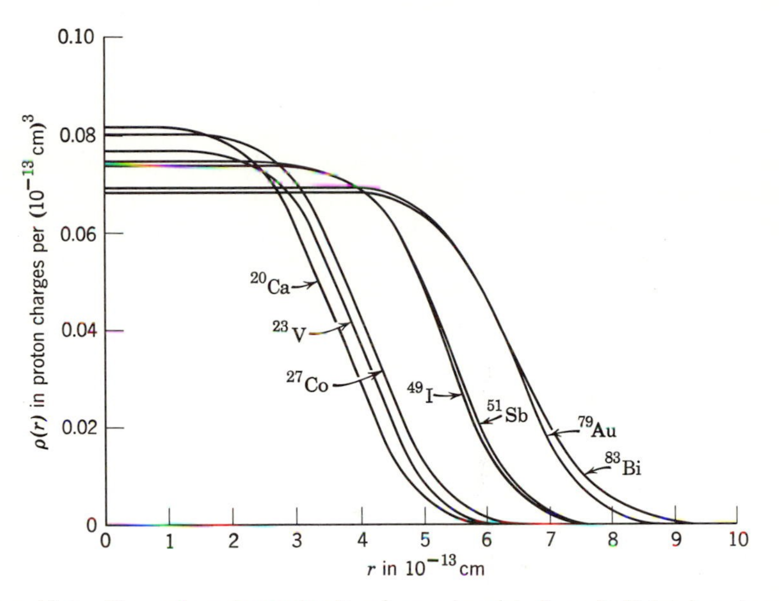

Figure 16–6. The nuclear charge density of several nuclei. From R. Hofstadter, *Annual Review of Nuclear Science*, Vol. 7, Annual Reviews, Stanford, 1957.

The parameter $\rho(0)$ is not determined very accurately by this analysis, but it can be simply by demanding that the integral of the charge density over the nuclear volume equal Z. Figure (16–6) consists of plots of equation (16–6) using the values of the parameters so determined.

We draw the following conclusions:

1. The charge density of nuclei, which must be essentially the distribution of the protons in the nuclei, is approximately constant in the interior region and falls fairly rapidly to zero at the surface.
2. The half-density radius a increases with increasing A according to an $A^{1/3}$ law.
3. The fall-off distance b is approximately the same for all nuclei.

4. The interior charge density $\rho(0)$ decreases slowly with increasing A.

5. If we assume that the distribution of protons in nuclei is approximately the same as the distribution of neutrons (there is good evidence for this assumption), the charge density $\rho(r)$ of a nucleus is related to its mass density $\rho_M(r)$ by the equation

$$\rho(r) \propto \frac{Z}{A} \rho_M(r) \qquad (16\text{–}8)$$

If so, the slow decrease of $\rho(0)$ with increasing A is completely consistent with the decrease in Z/A with increasing A ($Z/A \simeq 1/2$ for $A \simeq 40$; $Z/A \simeq 1/2.5$ for $A \simeq 240$), and with the assumption that the interior mass density $\rho_M(0)$ is approximately the same for all nuclei.

6. Although the coefficient of $A^{1/3}$ in the expression for a is 1.07, and not 1.2, equations (16–6) and (16–7) are actually consistent with (16–4) and (16–4′) because, within the accuracy of the analyses, both lead to the same root-mean-square radius, which can be shown to be the proper criterion for comparison. Consequently, this provides support for the assumption of the equality of the forces between various nucleons, which is the basis of the mirror nuclei analysis that leads to (16–4) and (16–4′).

Let us turn now to measurements of nuclear sizes from experiments which use nucleons as probes instead of electrons. Certain contrasts between these experiments and the electron experiments immediately arise. One is that a nucleon interacts with a nucleus by means of nuclear forces, and not by means of Coulomb forces (although there is also a Coulomb interaction if the nucleon is a proton). Since the interactions of nucleons with nuclei involve a different set of forces, these interactions measure a different set of attributes of the nuclei, and it should not be surprising if the measurements yield nuclear sizes which are not the same as those measured in the electron experiments. Another point is that the interaction between a nucleon and a nucleus must be described by a *complex* interaction potential. A real potential is not adequate because there is appreciable absorption (cf. section 7, Chapter 15) since the experiments show that, when a nucleon is incident upon a nucleus, the probability for absorption, including scattering with energy loss, is generally quite comparable to the probability for scattering with no energy loss.

In fact, the interpretation of the earliest experiments of this type was based on the assumption that, in its interaction with a nucleon, a nucleus behaves like a sphere of radius R which is completely absorbing. The experiments were measurements of the total cross sections σ_T of nuclei for neutrons of energies near 20 Mev. For a neutron of this energy, the de Broglie wavelength is $\lambda \simeq 6 \times 10^{-13}$ cm. Using $R \simeq 5 \times 10^{-13}$ cm as a reasonable estimate of the radius of a nucleus of intermediate A, we find $KR = (2\pi/\lambda)R \simeq 5$. We have then the case of a sphere which is assumed to be completely absorbing, with large KR. Consequently it should be possible to apply equation (15–104), which is

$$\sigma_T = 2\pi R^2 \tag{16–9}$$

The total cross section σ_T is measured by essentially the same beam attenuation technique that is used for X-rays (cf. section 5, Chapter 14), with a neutron detector consisting of a mixture of some hydrogenous compound and some compound that scintillates. Neutrons entering the

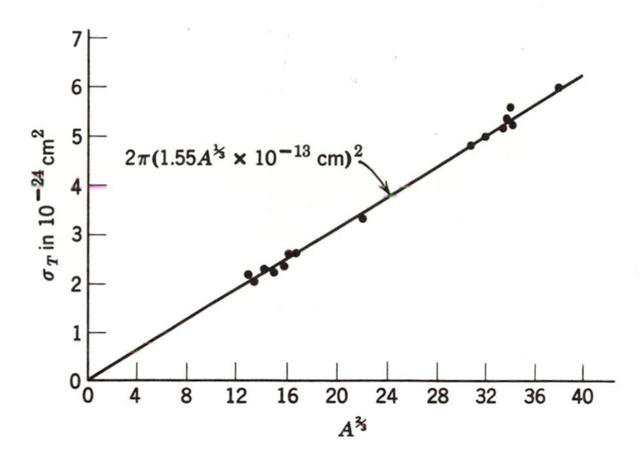

Figure 16–7. The total cross sections of several nuclei for 26-Mev neutrons.

detector collide with protons of the hydrogenous compound, transfer energy to the protons, and these charged particles subsequently produce scintillations. A number of different measurements have been made; figure (16–7) shows some data for 26-Mev neutrons. The solid curve is equation (16–9), using $R = 1.55A^{1/3} \times 10^{-13}$ cm. However, this is no longer accepted as a reliable estimate of the nuclear size. More sensitive measurements of the interaction between nucleons and nuclei show that the analysis based upon a completely absorbing sharp-edged sphere with $KR \to \infty$ is much too crude to yield accurate results.

These more sensitive measurements are of the differential scattering cross sections of nuclei for nucleons. In the past few years many accurate measurements have been made for both neutrons and protons, and for a variety of energies and nuclei. The experimental arrangements used are basically the same as that used in the electron scattering experiments. For protons, the energy sensitive detector can be a magnet plus scintillation counter, or simply a scintillation counter using a crystal such as NaI whose response is a linear function of the proton energy. Some data for

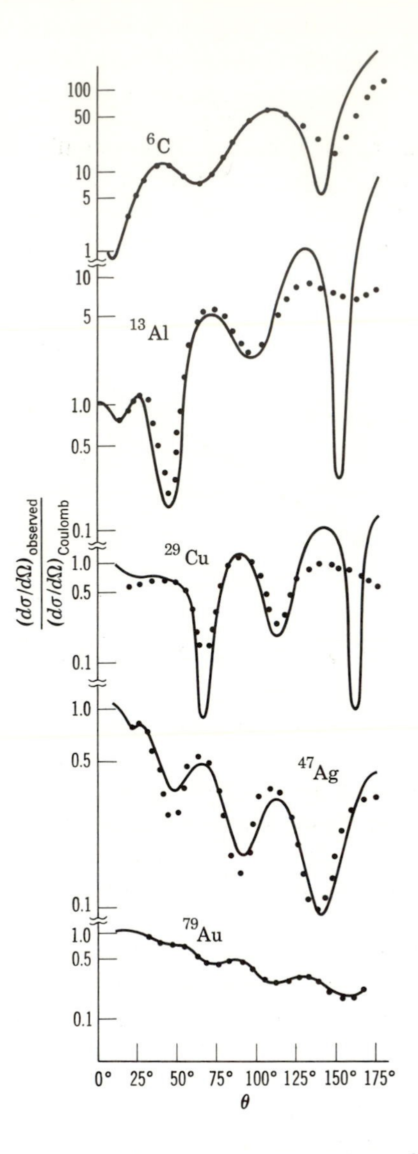

Figure 16–8. The differential scattering cross sections of several nuclei for 17-Mev protons. The points are experimental and the curves are theoretical fits. From A. E. Glassgold, *Revs. Mod. Phys.*, **30,** 419, (1958).

17-Mev protons are shown by the points in figure (16–8). Following convention, this figure plots the ratio in the CM frame of the observed differential scattering cross section (scattering with no energy loss in the CM frame) to the Coulomb differential scattering cross section for a point charge nucleus (equation 15–94). It should be noted, then, that the observed $d\sigma/d\Omega$ actually decreases rapidly with increasing θ owing to the $\sin^{-4}\theta/2$ dependence of the Coulomb $d\sigma/d\Omega$.

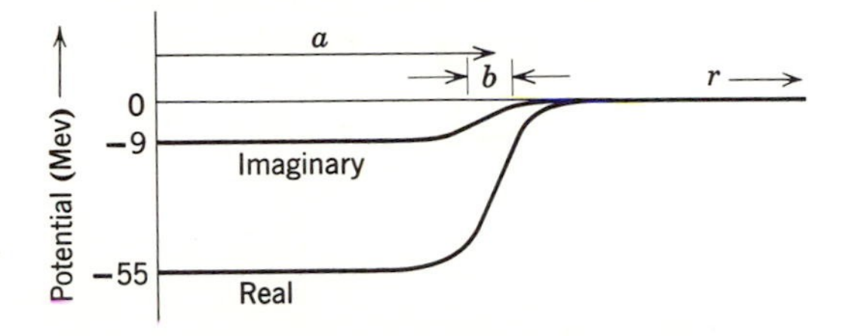

Figure 16–9. The real and imaginary parts of the optical model potential for the scattering of 17-Mev protons.

The solid curves are the best fits to the data obtained from a partial wave analysis using equation (15–102). In this analysis the interaction potential is taken to be of the form

$$\frac{V_0 + iW_0}{1 + e^{(r-a)/b}} + V_C(r) \tag{16–10}$$

The first term is a complex interaction potential representing the refracting and absorbing properties of the nucleus for the incident proton. The second term is the real Coulomb potential acting upon the proton and is evaluated from the nuclear charge density determined by electron scattering. For the analysis of the 17-Mev proton data, the following parameters were used in the first term:

$$\begin{aligned} V_0 &= -55 \text{ Mev} \\ W_0 &= -9 \text{ Mev} \\ a &= 1.25A^{1/3} \times 10^{-13} \text{ cm} \\ b &= 0.65 \times 10^{-13} \text{ cm} \end{aligned} \tag{16–10$'$}$$

The real and imaginary parts of the interaction potential are indicated in figure (16–9).

It is evident that a negative V_0 for an *unbound* proton moving through the nucleus is reasonable in light of the fact that the real part of the potential must, by definition, be attractive for a *bound* proton moving in the nucleus. It is even more evident that a negative W_0 is reasonable since, as indicated in (15–106), to have absorption the imaginary part of the potential must be negative. But the very form of the interaction potential makes it evident also that it is only an approximation to the true potential acting on the proton moving through the nucleus. This is because the true potential must certainly be a function of all the distances between the proton and the nucleons of the nucleus (or at least a number of these distances), and not just a function of the single distance between the proton and the center of the nucleus. Thus the approximation replaces the very complicated true potential by a much simpler potential which,

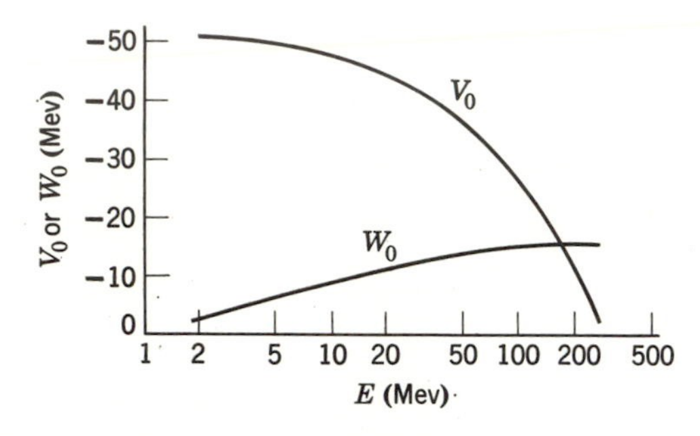

Figure 16–10. The approximate energy dependences of the depths of the real and imaginary parts of the optical model potential.

presumably, represents the *average* effect of the true potential—a procedure quite similar to that used in the Hartree treatment of the atom. This approximation is called the *optical model of the nucleus.* It was originally used around 1930, with a purely real potential. This was not adequate, and so the optical model fell into disfavor. However, with a complex potential it has experienced a very successful revival at the hands of Feshbach, Porter, Weisskopf and Saxon in the years after 1952.

As is true for any model, the justification of the optical model is that it provides a simple picture of certain processes and enables a large number of experiments involving these processes to be explained in terms of a small number of adjustable parameters. For instance, the set of parameters which provide an acceptable fit to the 17-Mev proton differential scattering cross sections also provides an acceptable fit to neutron total cross sections at a similar energy. In fact, the single set of parameters a and b quoted in (16–10′) is found to give an acceptable fit to the scattering and total cross sections for both protons and neutrons over a wide range of nuclei and energies. At a given energy, a single set of parameters V_0 and W_0 also is adequate to explain both cross sections for both particles and all nuclei. However V_0 and W_0 do exhibit a dependence on the energy E of the proton or neutron, as indicated in figure (16–10). There are still certain ambiguities in the values of V_0 and W_0 since, in some cases, different values of these parameters can give equally acceptable fits to the data. But the general behavior of their energy dependence is fairly well established, as are their values at low values of E. For instance, it is known that when E is 2 or 3 Mev, $-W_0$ can only be 1 or 2 Mev, because very pronounced resonances are observed in the scattering and total cross sections at these values of E. Resonances (cf. section 6, Chapter 15) involve constructive interferences between the radially inward traveling and radially outward traveling components of some partial wave. Therefore they can occur only if the outward traveling component is not absorbed and, as shown by equation (15–106), small absorption means small $-W_0$.

There is an interesting relation between W_0 and the rate of absorption of particles in the complex optical potential, which is of a form different from (15–106). Consider a particle of mass M moving with total energy E along the x axis through a medium of constant optical potential $-(V_0 + iW_0)$. The reader may verify by substitution that a solution to the time independent Schroedinger equation for this situation is

$$\psi(x) = e^{iKx}e^{-x/2\Lambda} \tag{16–11}$$

where

$$\left(\frac{1}{\Lambda}\right)^2 = \frac{4M}{\hbar^2}(E + |V_0|)\left[\sqrt{\left(\frac{|W_0|}{E + |V_0|}\right)^2 + 1} - 1\right] \tag{16–11$'$}$$

and

$$K^2 = \frac{M}{\hbar^2}(E + |V_0|)\left[\sqrt{\left(\frac{|W_0|}{E + |V_0|}\right)^2 + 1} + 1\right] \tag{16–11$''$}$$

Note that in the limit where $|W_0|/(E + |V_0|)$ is negligible compared to 1, equation (16–11$''$) reduces to the usual expression for K. When this quantity is only small compared to 1, equation (16–11$'$) reduces to

$$\Lambda \simeq \frac{\hbar}{2|W_0|}\left[\frac{2(E + |V_0|)}{M}\right]^{1/2} \tag{16–12}$$

To identify Λ, calculate the probability density:

$$\psi^*(x)\,\psi(x) = e^{-iKx}e^{-x/2\Lambda}e^{iKx}e^{-x/2\Lambda} = e^{-x/\Lambda}$$

We see that the probability density dies out exponentially with an attenuation length equal to Λ. Now let us evaluate Λ for $E = 20$ Mev, using the values of V_0 and W_0 taken from figure (16–10). We have: $E = 20$ Mev, $|V_0| = 45$ Mev, $|W_0| = 10$ Mev, $M =$ one proton mass. For these values (16–12) gives

$$\Lambda = 3.3 \times 10^{-13}\text{ cm}, \qquad E = 20\text{ Mev} \tag{16–13}$$

This is about one-half the average distance across a nucleus of average radius. Thus the probability density associated with a particle of this energy following such a path through a nucleus is attenuated by about a factor of e^2, and the particle has only about a 10 percent chance of not being absorbed. Nuclei are quite opaque to nucleons of energies in the range 20 Mev (and higher).† The situation is quite different at $E = 2$ Mev. At this energy $|V_0| = 50$ Mev, $|W_0| = 1.5$ Mev, and we obtain

$$\Lambda = 20 \times 10^{-13} \text{ cm}, \qquad E = 2 \text{ Mev} \tag{16–14}$$

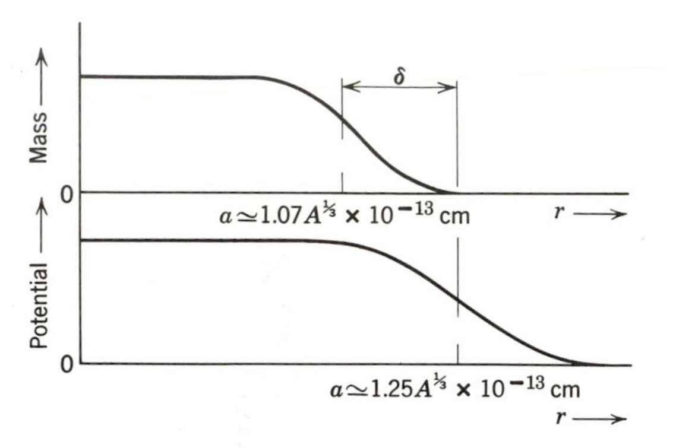

Figure 16–11. A comparison between the nuclear mass distribution and the nuclear interaction potential.

This is about three times the average distance across a nucleus of average radius. The probability density for a particle of this energy following such a path is attenuated only by a factor of about $e^{1/3}$, and the particle has about a 75 percent chance of not being absorbed. Nuclei are fairly transparent to nucleons of energies in the range 2 Mev (and lower).

Let us compare the results of the experiments that measure the charge or mass distribution with the results of the experiments that measure the distribution of the real or imaginary interaction potentials. This is done, for a typical nucleus, in figure (16–11). The figure plots the mass distribution, and the absolute value of the real interaction potential. It is extremely interesting to note that the spatial distribution of the nuclear interaction potential exerted by a nucleus on a passing nucleon extends

† This is part of the explanation for the large values of R found by analyzing the 20-Mev neutron total cross sections with equation (16–9); the opacity is high enough that there is appreciable probability of absorption even when the neutron passes through the outer nuclear surface. The rest of the explanation is that (16–9) assumes that the de Broglie wavelength of the neutron is negligibly small. Actually it is not, and, since the neutron can be absorbed if it passes within a de Broglie wavelength of the nucleus, this also increases the absorption probability and, therefore, the value of R calculated from (16–9).

beyond the spatial distribution of the nucleons in the nucleus by a distance δ which, for a typical nucleus, is only about 1×10^{-13} cm.† This means that *nuclear forces must be of very short range*. Thus nuclear forces are certainly very different from the long range gravitational and electric forces of common experience. An electron passing 2 or 3×10^{-8} cm from the surface of a nucleus feels an appreciable Coulomb force; a neutron passing 2 or 3×10^{-12} cm from the surface of the same nucleus feels essentially no force at all.

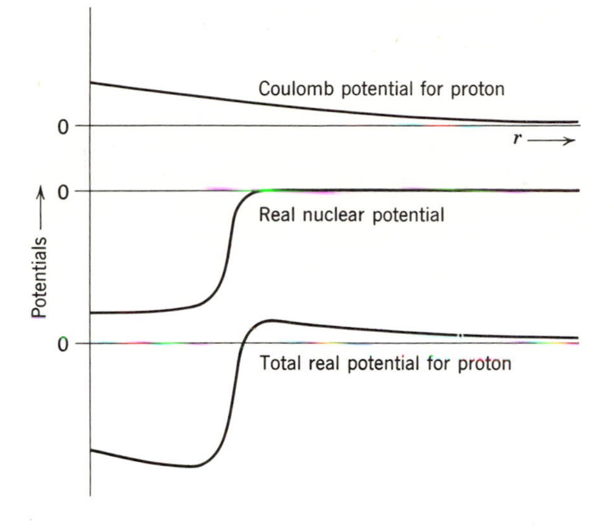

Figure 16–12. Illustrating the composition of the total real potential acting between a typical nucleus and a proton.

Of course, a proton passing 2 or 3×10^{-12} cm from the surface of the nucleus does feel a Coulomb force. This force is strong, but not so strong as the nuclear force the proton would feel on passing a little closer. An instructive comparison of the relative strengths of nuclear forces and Coulomb forces can be obtained by plotting the real nuclear potential, and the Coulomb potential acting on a proton, for a typical nucleus and a typical proton energy. This is done in figure (16–12). Because of the finite extension of the nuclear charge distribution, the Coulomb potential does not continue its Ze^2/r behavior for r less than the nuclear radius. For 20-Mev protons incident upon ^{92}U nuclei, the maximum positive

† The figure implies $\delta \propto A^{1/3}$; but more probably $\delta \simeq 1 \times 10^{-13}$ cm $\propto A^0$. This is not inconsistent since the values of a for the mass and (particularly) for the potential distributions are not known with too much certainty, so that δ is the relatively small difference between two large uncertain values.

value of the total real potential is about 15 Mev and the minimum negative value is about −45 Mev. Even for this extreme example of a large Z nucleus, we see that when r is less than the nuclear radius *the magnitude of the real nuclear potential is somewhat larger than the magnitude of the Coulomb potential.*

4. Nuclear Masses and Abundances

The detailed determinations of nuclear density distributions obtained from the recent electron scattering experiments show that the density in the interior of a nucleus is approximately the same for all nuclei. However, this was known at an early date because all the analyses of nuclear sizes, based on the assumption that the nucleus is a sharp-edged sphere of radius R and uniform interior density, lead to the result $R \propto A^{1/3}$ for which the density ρ = mass/volume $\propto A/(A^{1/3})^3$ is independent of A. In this section we shall study, primarily, the measurements of nuclear masses. Among other things, these measurements show that over a very broad range the binding energy per nucleon is approximately the same for all nuclei. This was also known at an early date and, along with the approximate equality of the nuclear density, it led to the assumption that all nuclei contain essentially the same material in essentially the same state, and differ from each other only in the amount of this material they contain. The assumption formed the basis of the very successful liquid drop model of the nucleus, which we shall discuss in the next section.

We know that the mass of a nucleus is almost equal to the mass of the corresponding atom. The masses of atoms of a particular Z, but possibly a mixture of A, can be obtained to an accuracy of several significant figures by the chemical techniques and a knowledge of Avogadro's number. But, for the more accurate determinations needed in the study of nuclei, it is necessary to use the physical techniques of *mass spectrometry* or *energy balance in nuclear reactions*. Both give information about the masses of atoms of a particular Z and A. From these masses, the masses of the corresponding nuclei can be evaluated by subtracting Z times the electron mass. In doing this, the mass equivalent of the electron binding energy is neglected. However, even in extreme cases, the error thereby introduced is less than 20 percent of the experimental uncertainties in the mass measurements.†

The first mass spectrometer was built in 1911 by Thomson. It used the same configuration of crossed electric and magnetic fields as the electron

† For $^{92}U^{238}$ the total electron binding energy is 2×10^5 ev $\simeq 0.0002$ mass unit. The mass of the atom is quoted as 238.1234 ± 0.0010 mass units.

charge to mass ratio measurements (cf. section 2, Chapter 3). A beam of atoms, ionized by electron bombardment, was passed through these fields which analyzed it into components according to the different charge to mass ratios present in the beam. It was usually possible to determine the degree of ionization of a component and thus determine its mass to within the somewhat low accuracy of the apparatus. With this apparatus Thomson discovered the existence of isotopes. Using a source containing

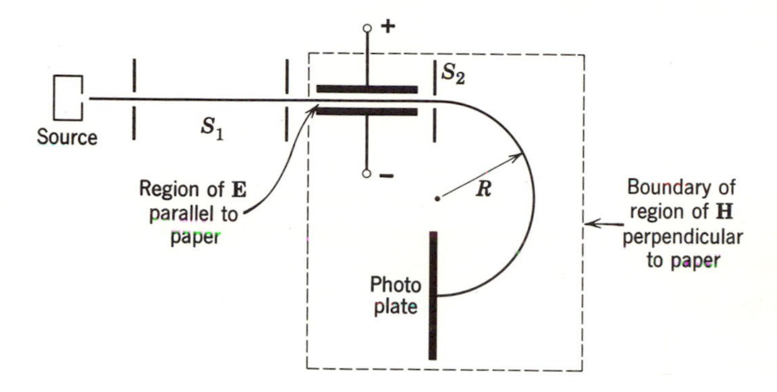

Figure 16–13. An apparatus used to measure atomic masses.

a mixture of noble gases, he found a component with mass corresponding to $A = 20$, and an associated weaker component with $A = 22$. A number of tests proved that these were both due to a noble gas, and this could only be Ne, which has a chemical atomic weight of 20.18. Thomson interpreted his results to mean that there are two chemically indistinguishable species of Ne atoms, called *isotopes*, one with $A = 20$ and relative abundance of about 91 percent, and one with $A = 22$ and relative abundance of about 9 percent. They are chemically indistinguishable because they have exactly the same structure of atomic electrons since their nuclei have the same Z, but they are physically distinguishable because they have different masses since their nuclei have different A. The nuclei of the Ne isotopes are: ${}^{10}\mathrm{Ne}^{20}$, ${}^{10}\mathrm{Ne}^{21}$, ${}^{10}\mathrm{Ne}^{22}$; the second occurs with relative abundance of about 0.3 percent and could not be detected with Thomson's apparatus.

A typical example of a more recent mass spectrometer is the type designed by Bainbridge (1933) and illustrated in figure (16–13). The source produces ionized atoms with charge $+Ze$, mass M, and a spectrum of velocities. These atoms pass through a region of crossed electric and magnetic fields which act as a velocity filter, passing only those with velocity v for which

$$ZeE = HZev/c$$

The terms on the left and right are the electric and magnetic forces, respectively. Atoms of velocity $v = Ec/H$ then enter a region of uniform magnetic field, are bent in a semicircle of radius R, and fall on a photographic plate where they produce an image. The distance from S_2 to the image is $2R$, where R satisfies the equation

$$HZev/c = Mv^2/R$$

The term on the right is the mass times the centrifugal acceleration. Solving for M, we have

$$M = R\frac{Ze}{c}\frac{H}{v} = \frac{RZeH^2}{Ec^2}$$

The mass can be determined from absolute measurements of the quantities on the right side of the last equation. In addition, much use is made of hydrocarbon molecules to calibrate the apparatus over a wide range of masses in terms of the carefully standardized masses of ${}^1H^1$ and ${}^6C^{12}$. With these techniques, extremely accurate measurements can be made. As an example, the mass of ${}^{18}A^{40}$ is quoted as

$$M_{{}^{18}A^{40}} = 39.975022 \pm 0.000029 \text{ amu}$$

An amu is one *atomic mass unit*. It is defined such that

$$M_{{}^{8}O^{16}} \equiv 16.000000 \text{ amu} \tag{16–15}$$

(Note that this differs slightly from the chemical mass scale which uses the normally occurring mixture of ${}^8O$ isotopes as a standard.)

Mass spectrometers, using detectors which are more linear than photographic plates, can provide accurate determinations of the relative abundances of the various isotopes. As an example, the abundances of the normally occurring mixture of ${}^8O$ isotopes are known to be

$$\begin{aligned} {}^8O^{16} &= 99.759\,\% \\ {}^8O^{17} &= 0.037\,\% \\ {}^8O^{18} &= 0.204\,\% \end{aligned}$$

Another technique of mass determination, which provides an accurate supplement and check for the technique of mass spectrometry, is the study of the energy balance in nuclear reactions. Consider a nuclear reaction such as (16–1),

$${}^2He^4 + {}^7N^{14} \rightarrow {}^8O^{17} + {}^1H^1$$

the one first observed by Rutherford. The general case may be written

$$a + A \rightarrow B + b \tag{16–16}$$

where a is the bombarding particle, A is the target nucleus, B is the residual nucleus, and b is the product particle. As mentioned in section 9, Chapter 1, it is found that in these reactions mass and kinetic energy are not separately conserved, but instead in any frame of reference there is conservation of total relativistic energy, $E = T + mc^2$, where T is kinetic energy and m is rest mass. For the reaction (16–16), the conservation of total relativistic energy in the LAB frame of reference reads

$$(T_a + m_a c^2) + m_A c^2 = (T_B + m_B c^2) + (T_b + m_b c^2) \qquad (16\text{–}17)$$

where T_a and m_a are the kinetic energy and rest mass of a, and so forth, and where $T_A = 0$ since A is stationary in the LAB frame. Because there

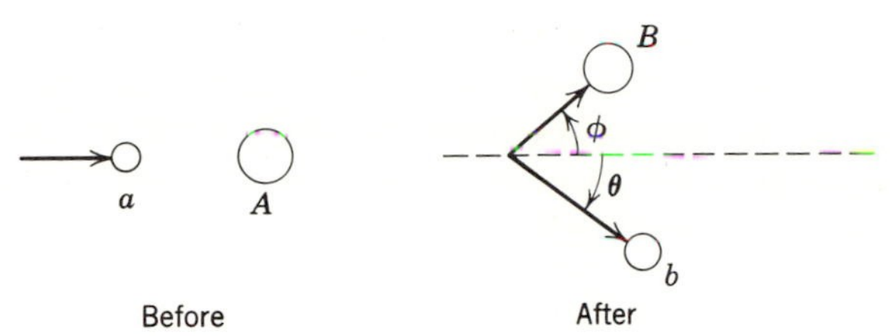

Figure 16–14. Illustrating the kinematics of a nuclear reaction.

can be an exchange of energy between kinetic energy and rest mass energy in these reactions, it is possible for the final kinetic energy $T_B + T_b$ to be greater, or less, than the initial kinetic energy T_a. The difference $T_B + T_b - T_a$ is called the Q *of the reaction*. That is,

$$Q = T_B + T_b - T_a \qquad (16\text{–}18)$$

From (16–17) this can also be written

$$Q = (m_a + m_A - m_B - m_b)c^2 \qquad (16\text{–}19)$$

We see that a measurement of the Q of a reaction gives information about the rest masses of the entities involved in the reaction. The Q can be measured by measuring T_a, T_b, and T_B. However, for T_B this is usually difficult and, fortunately, can be avoided by using a relation that comes from the conservation of momentum. Consider figure (16–14), which represents the kinematics of the reaction as seen in the LAB frame. By applying the conservation of linear momentum, an equation is obtained which makes it possible to eliminate T_B from (16–17). This is quite easy in the limit

$$T_a/m_a c^2 \ll 1, \qquad T_b/m_b c^2 \ll 1, \qquad T_B/m_B c^2 \ll 1$$

where the classical expressions such as $T_a = \frac{1}{2}m_a v_a^2$ and $p_a = m_a v_a$ can be used. Consequently, we shall leave it as an exercise for the reader to show that in this classical limit

$$Q = T_b\left(1 + \frac{m_b}{m_B}\right) - T_a\left(1 - \frac{m_a}{m_B}\right) - \frac{2}{m_B}(T_aT_bm_am_b)^{1/2}\cos\theta \quad (16\text{–}20)$$

This is of sufficient accuracy for the analysis of nuclear reactions at the energies which have been used in most experiments. Equation (16–20) shows that the Q can be determined by a measurement of T_a and T_b. [Even if the reaction is being studied in order to find the value of m_B in terms of the other masses, in practice this quantity would already be known with sufficient accuracy for use in (16–20)—particularly since $m_b/m_B \ll 1$ and $m_a/m_B \ll 1$ in typical cases.] The measurements of T_a and T_b use the techniques employed in the scattering experiments.

In equation (16–19), the masses refer to the rest masses of the nuclei A and B, and to the rest masses of the completely ionized nuclear particles a and b. However, to the accuracy of the approximation in which the mass equivalent of the electron binding energy is ignored, this equation can also be considered to read

$$Q = (M_a + M_A - M_B - M_b)c^2 \quad (16\text{–}21)$$

where the large M refer to the masses of the neutral atoms. This is obtained from equation (16–19) by adding $(Z_a + Z_A)mc^2$ to the first two terms and subtracting $(Z_B + Z_b)mc^2$ from the last two, where mc^2 is the rest mass energy of an electron. This procedure is valid since the relation

$$Z_a + Z_A = Z_B + Z_b \quad (16\text{–}22)$$

must be true in any nuclear reaction in order to have conservation of charge. Rutherford's reaction (16–1) provides a good example of the use of (16–20) and (16–21) to supply a connection between the atomic masses involved in a reaction. The measurements show that for this reaction $Q = -1.18$ Mev. This energy can be converted to its rest mass equivalent, -0.00126 amu, by dividing by c^2 and converting to the proper units, or by using the relation

$$1 \text{ amu} \leftrightarrow 931.14 \text{ Mev} \quad (16\text{–}23)$$

which comes from evaluating the rest mass energy of a particle of rest mass 1 amu. Then we have

$$-0.00126 \text{ amu} = M_{{}_2\mathrm{He}^4} + M_{{}_7\mathrm{N}^{14}} - M_{{}_8\mathrm{O}^{17}} - M_{{}_1\mathrm{H}^1}$$

This allows one of the atomic masses, say $M_{{}_8\mathrm{O}^{17}}$, to be determined in terms of the other three. The analysis of the energy balance in a large number of reactions has provided measurements of masses which accurately check the measurements by mass spectrometry. Furthermore, the agreement between these two methods provides excellent confirmation of the relativistic theory of mass and energy, upon which the energy balance analysis is based.

In table (16–1) are listed a few of the many atomic masses that have been measured, and also the mass of the neutron. Note that $M_{Z,A}$, the

TABLE (16–1)
Atomic Masses and Binding Energies

	Z	A	Mass (amu)	Binding Energy (Mev)	
				Total (ΔE)	Per Nucleon ($\Delta E/A$)
${}^{0}n^{1}$	0	1	1.008982 ($\pm$3)	—	—
${}^{1}H^{1}$	1	1	1.008142 ($\pm$3)	—	—
${}^{1}H^{2}$	1	2	2.014732 ($\pm$4)	2.22	1.11
${}^{2}He^{3}$	2	3	3.016977 ($\pm$11)	7.72	2.57
${}^{2}He^{4}$	2	4	4.003860 ($\pm$12)	28.3	7.07
${}^{4}Be^{9}$	4	9	9.01494 ($\pm$16)	58.0	6.45
${}^{8}O^{16}$	8	16	16.00000 ($\pm$0)	127.5	7.97
${}^{29}Cu^{63}$	29	63	62.94826 ($\pm$20)	552.1	8.75
${}^{50}Sn^{120}$	50	120	119.94012 ($\pm$72)	1020	8.50
${}^{74}W^{184}$	74	184	184.0052 ($\pm$11)	1476	8.02
${}^{92}U^{238}$	92	238	238.1234 ($\pm$10)	1803	7.58

mass of an atom of given Z and A, is always quite close to A amu. This is a consequence of defining the amu such that $M_{8,16} = 16.00000$. If the amu were defined such that $M_{1,1} = 1.00000$, which might at first seem more natural, then $M_{Z,A}$ would not be close to A amu for atoms of large A. The reader can demonstrate this by doing a little arithmetic.

The same procedure will also demonstrate the much more fundamental point that the mass of an atom is less than the mass of its constituent parts. For example, $M_{2,4} = 4.003860$ amu, whereas the mass of its constituent parts is the mass of two ${}^{1}H^{1}$ atoms plus the mass of two neutrons, i.e., $2M_{1,1} + 2M_{0,1} = 4.034248$ amu. This mass deficiency of the atom is due to a *mass deficiency* of its nucleus† resulting directly from the equivalence between energy and mass. We can see this by considering

† In the comparison just made, two electron rest masses are included in both $M_{2,4}$ and $2M_{1,1} + 2M_{0,1}$, and the mass equivalent of the electron binding energy is negligible. Thus the mass deficiency can only be due to the nucleus.

any one of the four nucleons in the $^2He^4$ nucleus. Since the nucleon is stably bound to the nucleus, it must be moving in some sort of an attractive real potential representing the net attraction of the other three nucleons. Furthermore, to be bound it must have a negative energy $E < 0$. The situation is depicted in figure (16–15). The minimum energy required to remove the nucleon from the nucleus, leaving it a free nucleon at $r \to \infty$, is $|E|$. Conversely, if a free nucleon comes in from $r \to \infty$ and combines with the other nucleons to form the nucleus, its energy must decrease by

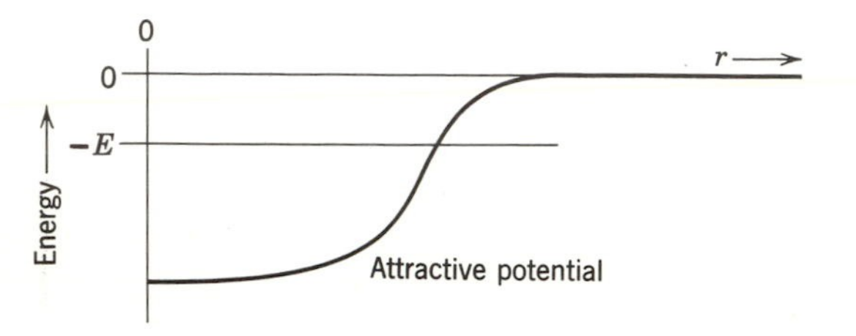

Figure 16–15. A schematic representation of the potential and total energies of a nucleon in a helium nucleus.

the amount $|E|$. (The excess energy will be carried off by a gamma ray, or by some other mechanism.) Since the same situation obtains for the other nucleons of the nucleus, we see that when a dispersed system of free nucleons combines to form a nucleus the total energy of the system must decrease by an amount ΔE, the *binding energy of the nucleus.* According to equation (1–24), the decrease ΔE in the total energy of the system must be accompanied by a decrease ΔM in its mass, where

$$\Delta M = \Delta E/c^2 \tag{16–24}$$

For $^2He^4$, the mass deficiency ΔM is $4.034248 - 4.003860 = 0.030388$ amu. Therefore the binding energy is $\Delta E = \Delta Mc^2 = 28.3$ Mev, where we have used (16–23). This figure is listed in the next to last column of table (16–1). The last column lists $\Delta E/A$, the binding energy of the nucleus divided by the number of nucleons it contains. For $^2He^4$, this *average binding energy per nucleon* is $28.3/4 = 7.07$ Mev.

One of the most important features of a nucleus is its average binding energy per nucleon. This quantity is plotted as a function of A in figure (16–16). The points are the data obtained from the measured masses in the manner just described. The smooth curve is obtained from an equation to be described later. Note that $\Delta E/A$ at first rises rapidly with increasing A, and then becomes reasonably constant at a value

$$\Delta E/A \sim 8 \text{ Mev} \tag{16–25}$$

However, it is not completely constant at this value; it maximizes at about 8.7 Mev for $A \simeq 60$ and then drops slowly to about 7.5 Mev for $A \simeq 240$. One consequence of this slow drop in $\Delta E/A$ is the famous phenomenon of *nuclear fission*, in which a large A nucleus, such as ${}^{92}U^{238}$, splits into two intermediate A nuclei. This happens because the final state is more stable than the initial state, since the average binding energy

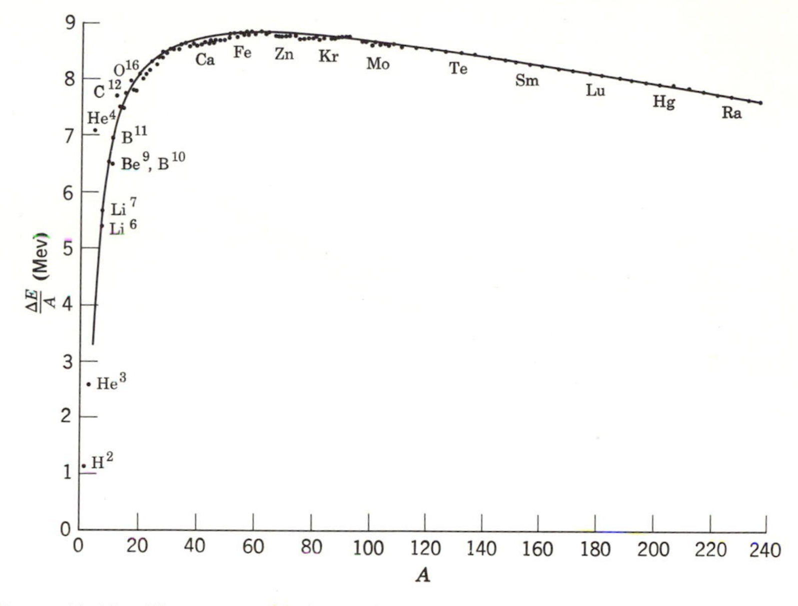

Figure 16–16. The average binding energy per nucleon for a number of nuclei. The smooth curve is from the semi-empirical mass formula. From R. B. Leighton, *Principles of Modern Physics*, McGraw-Hill Book Co., New York, 1959.

per nucleon increases from the value $\simeq 7.5$ Mev characteristic of the large A nucleus to the value $\simeq 8.5$ Mev characteristic of the two intermediate A nuclei. The energy liberated is about 1 Mev per nucleon, or about 1 Mev per nucleon $\times$ 200 nucleons = 200 Mev in total.† Alpha particle emission is a special case of fission in which one of the final nuclei is ${}^{2}He^{4}$, and it takes place for the same reason. The phenomenon of *nuclear fusion* consists of the combination of two or more nuclei of very small A to form a larger nucleus which has a higher average binding energy per nucleon, and therefore is more stable, because its value of A is nearer the value

† 200 Mev $= 3 \times 10^{-14}$ Btu, which does not seem like much energy in terms of this macroscopic unit. However, the mass, $200M_{1,1} = 200 \times 1.7 \times 10^{-24}$ gm $= 7 \times 10^{-25}$ lb, of the system emitting this energy is not very large either. The energy emitted per pound, $3 \times 10^{-14}/7 \times 10^{-25} = 4 \times 10^{10}$ Btu/lb, is phenomenal; it is about 10^{6} times larger than the energy emitted per pound from burning coal.

$A = 60$ for which $\Delta E/A$ maximizes. It might seem from this discussion that only a few nuclei near $A = 60$ would be stable. This is not true because there are other factors which tend to inhibit fission and fusion. Except for $A = 5$ and $A = 8$, there are nuclei for all values of A from 1 to beyond 200, which are either stable or have such immeasurably long lifetimes that they may be considered stable.

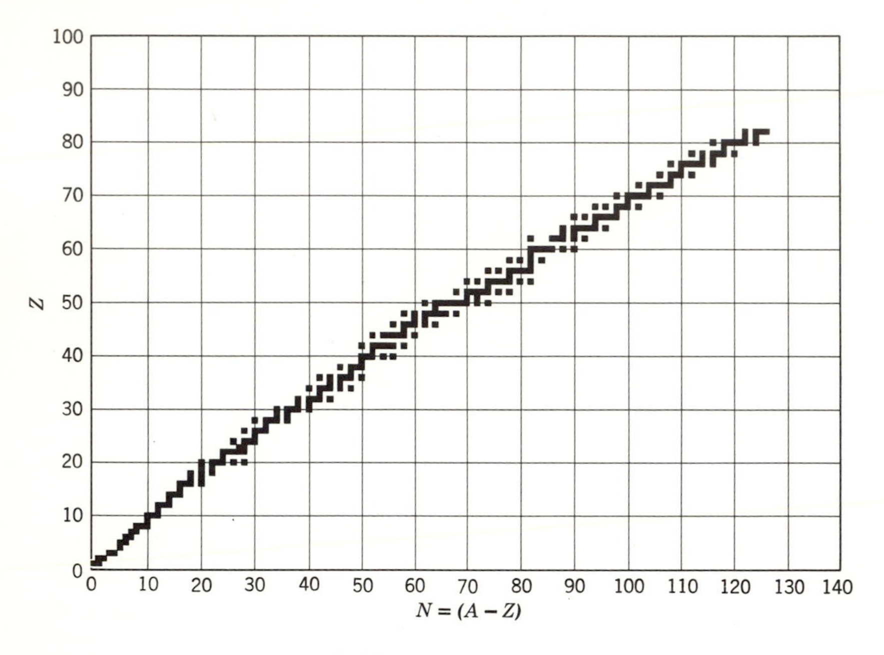

Figure 16–17. The distribution of stable nuclei.

It is interesting to consider the distribution of the Z and $(A - Z)$ values of these stable nuclei. We have seen that this information can be obtained from the mass spectrometer measurements. For instance, these measurements show that for $Z = 8$ there are stable nuclei for $A = 16, 17, 18$ (the isotopes of ^{8}O). The data are plotted in figure (16–17). Each stable nucleus is indicated by a square whose abscissa is $(A - Z) = N$, the number of neutrons in the nucleus, and whose ordinate is Z, the number of protons in the nucleus. Note that:

1. For each Z there are generally several values of N. For instance, for $Z = 10$ there are stable nuclei for $N = 10$, 11, and 12; these are the isotopes $^{10}Ne^{20}$, $^{10}Ne^{21}$, $^{10}Ne^{22}$.

2a. For small Z there is a tendency for stable nuclei to have $Z = N$.

2b. For large Z stable nuclei have $Z < N$. This is due to Coulomb repulsions between protons which produce a positive energy proportional to Z^2. If only this Coulomb effect were operating, the most stable nucleus with a given A would be obtained for $Z = 0$, $N = A$. However, the data show that for small Z, where the Coulomb effect is small, the tendency for $Z = N$ dominates and stability is obtained for $Z \simeq N$. For large Z the Coulomb effect becomes important and stability is obtained for $Z < N$. At $Z = 82$, $Z/N = 0.65$.

3. There is also a tendency for stable nuclei to have even Z and even N. This can be seen from table (16–2), which lists the number of stable nuclei

TABLE (16–2)
The Distribution of Stable Nuclei

A	N	Z	Number of Stable Nuclei
Even	Even	Even	166
	Odd	Odd	8
Odd	Even	Odd	57
	Odd	Even	53

of various types. The stable nuclei of odd Z and odd N are ${}^1H^2$, ${}^3Li^6$, ${}^5B^{10}$, ${}^7N^{14}$, ${}^{19}K^{40}$, ${}^{23}V^{50}$, ${}^{57}La^{138}$, and ${}^{71}Lu^{176}$. The reasons for the tendency to have even Z and even N, and also for the tendency to have $Z = N$, will be explained in section 7.

The preceding discussion has been written as if a nucleus of given A is free to make any adjustments in Z and N, consistent with the condition $Z + N = A$, which are required to achieve the condition of stability. This is not poetic license. Such adjustments can occur by the emission of electrons or positrons from the nucleus (beta emission), or by the capture of atomic electrons in the nucleus. These processes, which will be discussed in section 11, allow for changes in Z and N such that $Z + N$ remains constant; that is, they convert protons to neutrons or vice versa.

5. The Liquid Drop Model and the Semi-empirical Mass Formula

We shall now use the *liquid drop model of the nucleus*, and the information obtained from the study of the distribution of Z and N values for stable nuclei, to obtain a formula for the masses of these nuclei. This model is based upon the facts that for all nuclei, except those of very small A, the interior densities are approximately the same and the binding energies are approximately proportional to the masses of the nuclei ($\Delta E/A \sim$ constant, so $\Delta E \propto A$). Both facts can be compared with certain others concerning macroscopic drops composed of some incompressible liquid. For all such drops the interior densities are the same and the heats of vaporization are proportional to the masses of the drops. The latter comparison is significant because the heat of vaporization is the energy required to disperse the drop into its constituent molecules, and therefore it is analogous to the binding energy of the nucleus. In developing the mass formula, we shall use the model to suggest other analogies between a nucleus and a liquid drop.

The formula contains the sum of six terms,

$$M_{Z,A} = \sum_{i=0}^{5} f_i(Z, A) \tag{16–26}$$

The first term is the mass of the constituent parts of the atom,

$$f_0(Z, A) = 1.008142\,Z + 1.008982\,(A - Z) \tag{16–27}$$

where the coefficient of Z is the mass of the $^1H^1$ atom in amu and the coefficient of $(A - Z)$ is the mass of the neutron $^0n^1$ in amu. The remaining terms correct for the mass equivalents of several effects which contribute to the total nuclear binding energy. Of most importance is the term

$$f_1(Z, A) = -a_1 A \tag{16–28}$$

which accounts for a binding energy essentially proportional to the mass, or volume, of the nucleus. A similar term would be present for a liquid drop. The next term is

$$f_2(Z, A) = +a_2 A^{2/3} \tag{16–29}$$

a positive correction proportional to the surface area of the nucleus. In a liquid drop this term would represent the effect of the surface tension energy; it arises from the fact that a molecule at the surface of the drop experiences attractive forces only from one side, so its binding energy is less than the binding energy of a molecule in the interior. Therefore simply setting the total binding energy proportional to the volume of the drop overestimates the binding energy of the surface molecules, and a correction must be made which is proportional to the number of such molecules, or to the surface area. The term

$$f_3(Z, A) = +a_3 \frac{Z^2}{A^{1/3}} \tag{16–30}$$

accounts for the positive Coulomb energy of the charged nucleus, which is assumed to be a sphere of radius proportional to $A^{1/3}$. A similar term would be present for a charged liquid drop. The last two terms bring in properties specific to a nucleus. For one thing, there is the tendency to have $Z = N$. This can be accounted for by the term

$$f_4(Z, A) = +a_4 \frac{(Z - A/2)^2}{A} \tag{16–31}$$

which is zero for $Z = N = (A - Z)$, or $2Z = A$, but which is otherwise positive and increases with increasing departures from that condition. The form used in equation (16–31) is about the simplest one having these properties, but there is also some theoretical justification that will be presented in section 7. The tendency to have even Z and even N is accounted for by the term

$$f_5(Z, A) = \begin{cases} -f(A), & Z \text{ even}, \ (A - Z) = N \text{ even} \\ 0, & \begin{cases} Z \text{ even}, \ (A - Z) = N \text{ odd} \\ Z \text{ odd}, \ (A - Z) = N \text{ even} \end{cases} \\ +f(A), & Z \text{ odd}, \ (A - Z) = N \text{ odd} \end{cases} \tag{16–32}$$

which decreases the mass if both Z and N are even, and increases it if both Z and N are odd. The form of the function $f(A)$ is determined by fitting the data. It is found that the best fit for a simple power law is obtained with

$$f(A) = a_5 A^{-1/2} \tag{16–32$'$}$$

Gathering together equations (16–26) through (16–32′), we have

$$M_{Z,A} = 1.008142\, Z + 1.008982\, (A - Z) - a_1 A + a_2 A^{2/3} + a_3 Z^2 A^{-1/3} + a_4 (Z - A/2)^2 A^{-1} + \begin{pmatrix} -1 \\ 0 \\ +1 \end{pmatrix} a_5 A^{-1/2} \quad \text{(amu)} \tag{16–33}$$

This is called the *semi-empirical mass formula* because the parameters a_1 through a_5 are obtained by empirically fitting the measured masses. A formula of this type was first developed by Weizsacker (1935). Determinations of the parameters have since been made on several occasions. Green (1954) found

$$\begin{aligned} a_1 &= 0.01692 \\ a_2 &= 0.01912 \\ a_3 &= 0.000763 \qquad \text{(all in amu)} \\ a_4 &= 0.10178 \\ a_5 &= 0.012 \end{aligned} \tag{16–33$'$}$$

Using these parameters, quite a good fit is obtained for the measured masses of *all* stable nuclei except those of very small A; the discrepancy is generally less than several thousandths of an amu. A comparison is shown in figure (16–16). In that figure the smooth curve shows the average binding energy per nucleon evaluated from equation (16–33), and the points show the same quantity evaluated directly from the measured masses. The semi-empirical mass formula is extremely useful because it describes fairly accurately the masses of several hundred stable nuclei, and many more unstable nuclei, in terms of only five parameters.

6. Magic Numbers

The liquid drop model gives a good account of the average behavior of nuclei in regard to mass, and therefore in regard to stability. However, nuclei with certain values of Z and/or N show significant departures from this average behavior by being unusually stable. These values of Z and/or N are the *magic numbers*

$$2, 8, 20, 28, 50, 82, 126 \qquad (16\text{–}34)$$

The situation is analogous to the unusual stability of the electronic systems of the noble gas atoms containing a magic number 2, 10, 18, 36, 54, 86 of electrons. But in the nuclear case the effects indicating the unusual stability are not so pronounced as in the atomic case, and it is necessary to consider all of them in order to demonstrate conclusively the "magic" character of the numbers (16–34). These considerations, largely due to Mayer (1948), run as follows:

1. There is a tendency for nuclei to prefer magic Z and/or N. This can be seen from consulting figure (16–17). For example, there are 6 stable isotopes for $Z = 20$, whereas the average number of stable isotopes in that region of the periodic table is about 2. For $Z = 50$ there are 10 stable isotopes, although the average number in that region is about 4. There are 7 stable nuclei having $N = 82$, even though the average number in that region of N is only about 3. A number of similar examples can be seen.

2. Figure (16–16) shows that the average binding energy per nucleon is significantly higher, for nuclei with Z and/or N equal to 2 and 8, than it is for neighboring nuclei. The outstanding example is $^2He^4$, for which $Z = N = 2$. These effects are even more pronounced if a measure of stability more sensitive than the average binding energy per nucleon is considered. This is E_b, the *binding energy of the "last" neutron or proton* in the nucleus, which is the minimum energy required to separate a

neutron or a proton from the nucleus. As an example, the binding energy of the last neutron in $^{2}He^{4}$ (the energy required to produce the reaction $^{2}He^{4} \rightarrow {}^{2}He^{3} + {}^{0}n^{1}$) is 20.6 Mev. The binding energy of the last proton in $^{2}He^{4}$ is 19.8 Mev. These are abnormally high. Figure (16–18) shows a plot of the binding energy of the last neutron for a number of heavier nuclei. The abscissa is the number of neutrons in the nucleus; the ordinate is the difference between the actual value of the binding energy of the last neutron and the value predicted by the semi-empirical mass formula. The predicted value is a smooth function of N (except for the

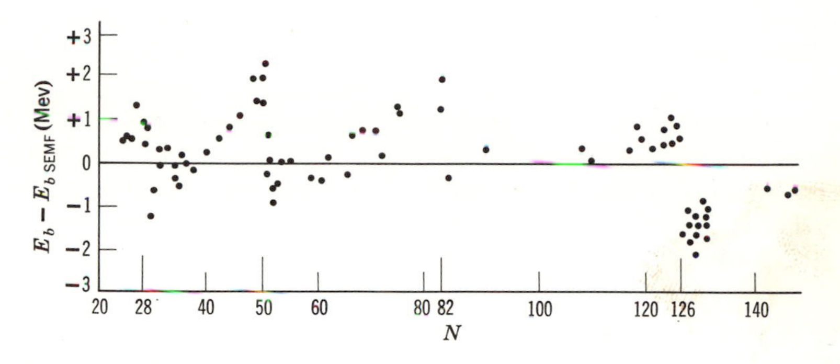

Figure 16–18. The binding energy of the last neutron for a number of nuclei. From J. A. Harvey, *Phys. Rev.*, **81**, 353 (1951).

effects of term 16–32), and it decreases slowly from around 7 Mev for intermediate N to around 5 Mev for large N. The unusual stability of nuclei with $N = 28$, 50, 82, 126 is demonstrated by the exceptionally large energy required to remove their last neutron.

3. There are a number of other somewhat less convincing pieces of evidence, such as the fact that for three of the four known spontaneous neutron emitters, $^{8}O^{17}$, $^{36}Kr^{87}$, and $^{54}Xe^{137}$, N equals a magic number plus 1. This implies an unusually small affinity for the extra neutron.

The analogy between nuclear and atomic magic numbers prompted many people to look for an explanation of the nuclear phenomenon analogous to the explanation of the atomic phenomenon. The reader will recall that the key point in that explanation is the formation of closed shells by the electrons moving *independently* in the atomic potential. But, when the nuclear magic numbers were first being discussed seriously (1948), it seemed very difficult to understand how any explanation based upon independent particle motion could be valid for the nucleus. The reason was that the very successful liquid drop model had been dominant for a number of years, and it seemed basic to this model that a nucleon in a nucleus interacts very strongly with its neighbors. With strong interactions, the nucleon would be constantly scattered in traveling through the nucleus and would follow an erratic path, resembling Brownian motion much more closely than the motion of an electron moving independently through its orbit in a closed shell.

From our present point of view there is, however, no problem in understanding the independent particle motion of nucleons in a nucleus. In fact, it is a very reasonable extrapolation of the optical model (1952). The calculations recorded in equations (16–13) and (16–14) show that the attenuation length Λ of a nucleon moving through a nucleus increases by about a factor of 6 as the total energy of the nucleon decreases from $E = +20$ Mev to $E = +2$ Mev, and that Λ is several times nuclear dimensions at the lower energy. An extrapolation of this behavior, through $E = 0$ to $E < 0$, predicts that, for the bound nucleons corresponding to $E < 0$, the attenuation length Λ is very long and, therefore, that the bound nucleons can move quite independently through the nucleus. Putting this in other words, a nucleon moving through a nucleus with $E > 0$ must have a certain probability of suffering collisions, which involve energy loss in the frame in which the CM of the nucleus is at rest, because it is moving in a complex potential. However, the energy dependence for $E > 0$ of the real and imaginary parts of this potential, indicated by figure (16–10), makes it easy to believe that for $E < 0$ there is no imaginary part to the potential, so that there can be no collisions with energy loss in the CM frame of the nucleus. But, since essentially all collisions between a nucleon and another nucleon involve energy loss in this frame, it follows that if the potential is purely real for the bound nucleons there can be no such collisions between these nucleons, and that they must therefore move independently through the nucleus.

7. The Fermi Gas Model

It is comforting to realize that the optical model analysis, of the experimental data concerning the motion through the nucleus of nucleons with $E > 0$, strongly suggests that nucleons with $E < 0$ should move independently through the nucleus. But it is even more comforting to realize that there is a simple theoretical explanation of why this should happen. The explanation is based on the *Fermi gas model of the nucleus*, which had been in existence for a number of years, but which had not been held in very high regard during the period of dominance of the liquid drop model. This model is essentially the same as the Fermi gas model of the conduction electrons in a metal considered in section 6, Chapter 12. It assumes that the neutrons and protons move in an attractive potential of nuclear dimensions, and that in the ground state these nucleons fill the energy levels in such a way as to minimize the total energy but not violate the exclusion principle. (Remember that both neutrons and protons are Fermi particles with spin quantum numbers $i = \frac{1}{2}$, just like electrons.) Figure (16–19) indicates the energy levels filled by the neutrons in the ground state of a nucleus. Since a proton is distinguishable from a neutron, the exclusion principle operates independently on the two types of nucleons, and we must imagine a separate and independent diagram representing the energy levels filled by the protons in the ground state of the nucleus.

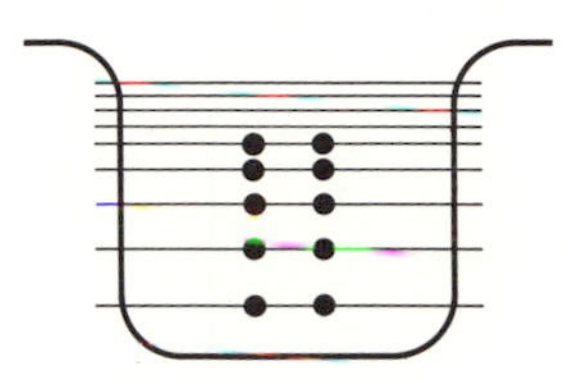

Figure 16–19. A schematic representation of the energy levels filled by the neutrons in the ground state of a nucleus.

It is immediately apparent from these diagrams why there can be essentially no collisions between the nucleons when the nucleus is in its ground state. The point is that all the levels which are energetically accessible are already filled, and so there can be no collisions except those in which two nucleons of the same type exactly exchange energy. But the net effect of such an exchange of two indistinguishable particles is the same as if there had been no collision at all. We can understand, then, why for the ground state of the nucleus the potential is real, and the nucleons are moving independently. We can also understand how the exclusion principle can be effective even in inhibiting collisions between a nucleon of low positive energy and the bound nucleons of a nucleus. In fact, calculations based on the Fermi gas model have shown that the decrease in the imaginary part of the optical model potential with decreasing energy (the energy dependence of W_0 in figure 16–10) is due primarily to the exclusion principle.

Furthermore, the Fermi gas model provides a good estimate of the depth of the real potential for the ground state of a nucleus. This is obtained by evaluating the Fermi energy (12–38), which is

$$E_f = \frac{3^{2/3}\pi^{4/3}\hbar^2}{2M}\,\rho^{2/3}$$

in terms of the nucleon mass M and the nucleon density ρ. Let us consider the Fermi gas of neutrons in a uniform spherical nucleus of radius

$$R = r_0 A^{1/3}$$

For a typical nucleus the number of neutrons is

$$N = 0.6A$$

Thus

$$\rho = \frac{N}{\frac{4}{3}\pi r_0^3 A}$$

gives

$$\rho = \frac{0.6A}{1.33\pi r_0^3 A} = \frac{0.45}{\pi r_0^3}$$

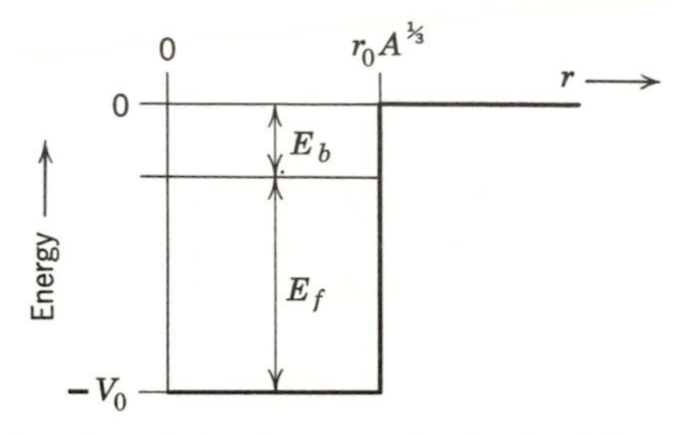

Figure 16–20. Illustrating the relation between the depth of the potential, the Fermi energy, and the binding energy of the last neutron.

and the Fermi energy is

$$E_f = \frac{3^{2/3}\pi^{4/3}\hbar^2(0.45)^{2/3}}{2M\pi^{2/3}r_0^2} \tag{16–35}$$

Using a radius constant $r_0 = 1.25 \times 10^{-13}$ cm consistent with the optical model potential radius of (16–10′), and evaluating the other parameters, we obtain

$$E_f \simeq 34 \text{ Mev}$$

The relations between the depth of the potential V_0, the Fermi energy E_f, and the binding energy E_b of the last neutron are shown in figure (16–20). As mentioned above, E_b is measured to be roughly 6 Mev for a wide range of nuclei. Thus the Fermi gas model predicts

$$|V_0| = E_f + E_b \simeq 34 + 6 = 40 \text{ Mev}$$

in satisfactory agreement with the value of about 50 Mev that is obtained by extrapolating to $E < 0$ the values obtained from the optical model (extrapolating the V_0 of figure 16–10).

The tendency for nuclei to have $Z = N$ finds a simple explanation in terms of the Fermi gas model. Consider a nucleus of very small Z, where this effect dominates because the Coulomb potential is very small. In this nucleus, there are two independent Fermi gases, the neutrons and the protons, moving in the same nuclear potential. The energy levels of these systems are indicated in figure (16–21). Except for the effect of a very small positive Coulomb potential acting on the proton system, the energy levels of the two systems are identical because for both systems the nuclear potential is the same. It is clear, then, that for a given A the total energy of the nucleus is minimized if the levels are filled with $Z = N$ since, if this condition were violated, nucleons would occupy levels which are of higher

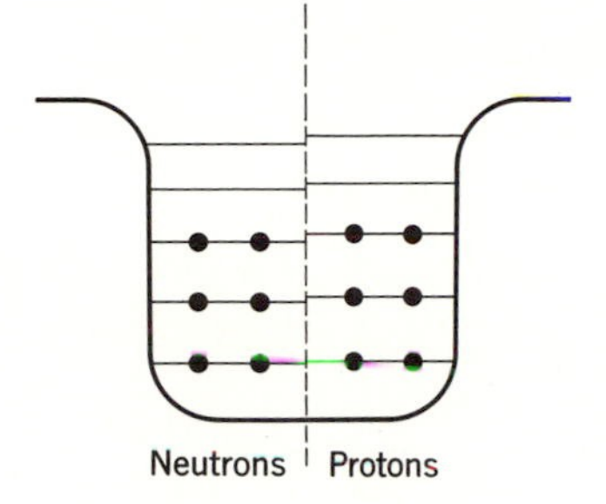

Figure 16–21. A schematic representation of independent Fermi gases of neutrons and protons in a nucleus.

energy than necessary. A quantitative treatment of this argument leads to the term (16–31) used in the semi-empirical mass formula to express the tendency for $Z = N$. According to the equations preceding (16–35), the Fermi energy for the neutron and proton systems can be written

$$E_{f_n} = C(N/A)^{2/3}, \qquad E_{f_p} = C(Z/A)^{2/3}$$

where C is a constant. Now use (12–40) to evaluate the total energy of the neutron and proton systems, as measured from the bottom of the potential, in terms of E_{f_n} and E_{f_p}. This gives

$$E_{t_n} = C' \frac{N^{5/3}}{A^{2/3}}, \qquad E_{t_p} = C' \frac{Z^{5/3}}{A^{2/3}}$$

where $C' = 3C/5$. The energy of the nucleus is the sum of these two terms, which is

$$E(Z, A) = C'A^{-2/3}[N^{5/3} + Z^{5/3}] \tag{16–36}$$

This is a function only of Z and A since

$$N + Z = A \tag{16–37}$$

As we already know, the minimum of the function (16–36), subject to the condition (16–37), is found for $N = Z = A/2$. In departing from this minimum the energy of the nucleus increases, and the corresponding correction in the semi-empirical mass formula will be proportional to this increase. Thus

$$f_4(Z, A) \propto E(Z, A) - E(Z, A)_{\min} = C'A^{-2/3}\{N^{5/3} + Z^{5/3} - 2[A/2]^{5/3}\}$$

Now let

$$D \equiv \frac{N - Z}{2} = N - \frac{A}{2} = \frac{A}{2} - Z$$

Then

$$f_4(Z, A) \propto A^{-2/3}\{[A/2 + D]^{5/3} + [A/2 + (-D)]^{5/3} - 2[A/2]^{5/3}\}$$

Use Taylor's series to expand the first two quantities in the curly bracket, keeping terms through D^2. This gives

$$\left[\frac{A}{2} + D\right]^{5/3} \simeq \left[\frac{A}{2}\right]^{5/3} + \frac{5}{3} D\left[\frac{A}{2}\right]^{2/3} + \frac{5}{3}\frac{2}{3}\frac{D^2}{2}\left[\frac{A}{2}\right]^{-1/3}$$

$$\left[\frac{A}{2} + (-D)\right]^{5/3} \simeq \left[\frac{A}{2}\right]^{5/3} + \frac{5}{3}(-D)\left[\frac{A}{2}\right]^{2/3} + \frac{5}{3}\frac{2}{3}\frac{(-D)^2}{2}\left[\frac{A}{2}\right]^{-1/3}$$

so

$$f_4(Z, A) \propto A^{-2/3}\left\{2\left[\frac{A}{2}\right]^{5/3} + \frac{5}{3}\frac{2}{3} D^2\left[\frac{A}{2}\right]^{-1/3} - 2\left[\frac{A}{2}\right]^{5/3}\right\}$$

$$\propto A^{-2/3}\frac{5}{3}\frac{2}{3} D^2\left[\frac{A}{2}\right]^{-1/3}$$

$$\propto A^{-1}D^2 = \frac{(A/2 - Z)^2}{A}$$

in agreement with (16–31).

The Fermi gas model also gives a partial explanation for the tendency of nuclei to have even Z and even N. Consider the even Z, even N nucleus represented in figure (16–21), and imagine constructing the nucleus of next higher A by adding one nucleon. To minimize the total energy, the nucleon will go into the lowest available energy level. This might be either a neutron level or a proton level, but let us assume that it is a neutron level, as shown in the figure. Then this nucleus will contain one additional neutron. Now construct the next nucleus in the series by adding yet another nucleon. The nucleon will also be a neutron, since there is still one vacancy in the neutron level and this level is still of lower energy than the unfilled proton level. We see that for only one of the three nuclei involved in this dicussion is N or Z odd, and for none of them are both N and Z odd. This explanation is indicative of what is happening, but it is not the whole story. The mass measurements, which demand the term (16–32) in the semi-empirical mass formula, show that there is also a *pairing energy* that increases the binding energy of a nucleus when N or Z becomes even. The origin of the pairing energy lies in the fact that the force between two nucleons is attractive. When the second nucleon is added to an energy level, the attractive force which it exerts on the nucleon that was already there produces a negative potential that lowers the energy of both these nucleons, i.e., lowers the energy level in question.

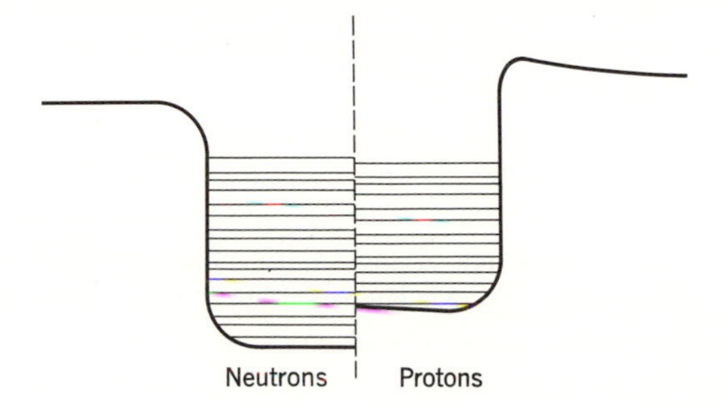

Figure 16–22. Illustrating why there are more neutrons than protons in a nucleus of large Z.

Before leaving the Fermi gas model, we present figure (16–22), which indicates the form of the potentials acting on the protons and neutrons in a nucleus of intermediate or large Z. Because of the Coulomb potential, the total potential acting on the protons is elevated. However, the highest filled level must be at essentially the same energy for both the neutrons and the protons if the nucleus is to be stable. This provides a very graphic explanation of the reason why there are more neutrons than protons in a nucleus of large Z.

8. The Shell Model

The Fermi gas model establishes the validity of treating the motion of the bound nucleons in a nucleus in terms of the independent motion of each nucleon in a common net potential $V(r)$, which represents the average attraction of all the other nucleons. In this model, certain quantum mechanical properties of a system of non-interacting Fermi particles are brought in by the use of equation (12–38), and the Schroedinger equation for $V(r)$ is not solved explicitly. This is adequate for the description of certain average properties of the nucleus but, for a more detailed description capable of accounting for the magic numbers, it is necessary to bring in more of the important quantum mechanical properties of such a system by actually solving the Schroedinger equation for $V(r)$. The treatment of the nucleus employing this procedure is called the *shell model of the nucleus*. The reader will note that the shell model of the nucleus is related to the Fermi gas model of the nucleus in much the same way that the Hartree theory of the atom is related to the Thomas-Fermi theory of the atom. He will also note that these nuclear models are much cruder than the corresponding atomic theories. In the self-consistent atomic theories the net potential $V(r)$ is completely determined by the theory; in the nuclear models the radius and exact form of the net potential $V(r)$ must be inserted ad hoc.

The procedure of the shell model is to find the neutron and proton energy levels for the potential $V(r)$ of a particular nucleus by solving the Schroedinger equation for this potential, and then to construct the nucleus by filling these levels in order of increasing energy with N neutrons and Z protons. Just as in the Hartree theory, the energies of the levels depend on the quantum numbers n and l, and each level has a capacity $2(2l + 1)$ corresponding to the 2 possible values of m_s and the $(2l + 1)$ possible values of m_l.† It was hoped that a form for the potentials $V(r)$ of the various nuclei could be found in which the ordering and spacing of the energy levels would be such that an unusually tightly bound level would completely fill in those nuclei having magic N or Z. However, it was found that there is no reasonable form for the $V(r)$ which leads even to the proper ordering of the energy levels.

The solution was provided by Mayer, and independently by Haxel, Jensen, and Suess (1949). These workers proposed that, in addition to the potential $V(r)$, there is a *strong inverted spin-orbit interaction*, proportional to $\mathbf{S} \cdot \mathbf{L}$, acting on each nucleon in a nucleus. Strong means that this interaction is much (about 20 times) larger than would be predicted by using equation (13–27), equating $V(r)$ to the net nuclear potential and m to the nucleon mass. Inverted means that the energy of the nucleon is decreased when $\mathbf{S} \cdot \mathbf{L}$ is positive and increased when it is negative, so that the sign of the spin-orbit interaction is inverted from the sign of the magnetic spin-orbit interaction experienced by an electron in an atom. However, as the magnitude of the spin-orbit interaction is proportional to $\mathbf{S} \cdot \mathbf{L}$ just as it is for an electron, the magnitude of the spin-orbit splitting will be approximately proportional to the value of the quantum number l, just as it is for an electron.‡

† The quantum numbers n, l, m_l, m_s, as well as the other quantum numbers, angular momentum vectors, and associated terminology, have the same significance as in the theory of the electronic structure of atoms.

‡ This is the Lande interval rule (13–43), which depends only on the assumption that the potential is proportional to $\mathbf{S} \cdot \mathbf{L}$, plus the statement that since $j = l \pm \frac{1}{2}$ the quantity $j + 1$ is approximately proportional to l.

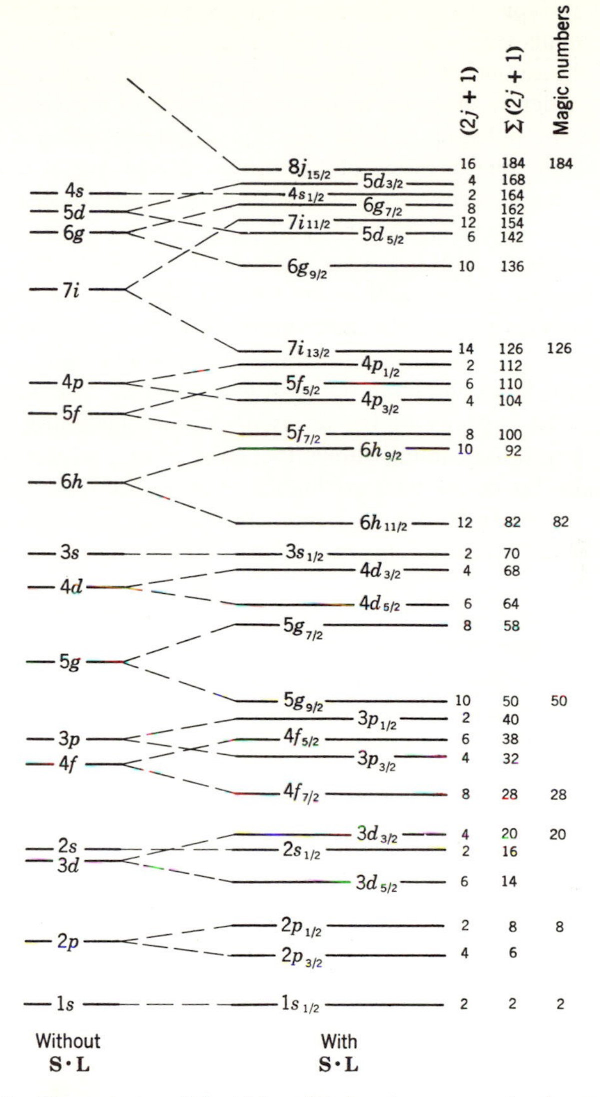

Figure 16–23. The ordering of the highest filled nucleon energy levels according to the shell model.

The net effect of the proposed spin-orbit interaction is indicated in figure (16–23). The part on the left shows the ordering and approximate spacing of the energy levels which nucleons are filling in nuclei with potentials $V(r)$ in the form of square wells with rounded edges, such as are used in the optical model; the part on the right shows the way these levels are split by the spin-orbit interaction. This is not an energy level diagram for any particular nucleus; it is a diagram giving the order in which levels appear as the radius of the potential increases with increasing A. Thus it gives the order in which the highest energy levels of the nucleus are filled. The diagram is analogous to table (13–2) giving the ordering of the highest filled electronic energy levels of the atoms, except that it also indicates the relative magnitudes of the separation between adjacent levels. In the presence of the spin-orbit interaction, m_l and m_s are no longer good quantum numbers, and n, l, j, m_j must be used. Also, the energies of the levels depend on j as well as on n and l, the larger j corresponding to the smaller energy since the interaction is inverted. Each of these levels has a capacity $(2j + 1)$ equal to the number of possible values of m_j. This is shown in the first column on the right in the figure. The second column shows the total capacity of the levels up to and including the level in question. The third column shows the same thing for each level which lies unusually far below the next higher level. Since these are the levels which will be unusually tightly bound, we see that the shell model with strong inverted spin-orbit interaction predicts precisely the magic numbers (16–34).†

Within the past few years, completely independent verification of the existence of a spin-orbit interaction has been obtained from experiments involving the scattering of low energy nucleons by nuclei. It is found that a scattered nucleon can have its spin angular momentum vector partially polarized as a result of interacting with the nucleus, and this can be explained only by the presence of a spin-orbit term in the potential describing the interaction. Furthermore, the sign of the polarization shows that the sign of this term is the same as that required by the shell model, and the amount of polarization shows that its strength (the spin-orbit potential is several Mev deep) agrees with the strength required for the shell model. It should be emphasized that the spin-orbit interaction for nucleons is not electromagnetic in origin. Instead it is an attribute of nuclear forces, whose origin is not well understood.

The spin-orbit shell model can do much more than explain the magic numbers and their consequences. For instance, it can also explain the spin angular momentum of the ground states of almost all the nuclei.

† It should be mentioned that the eigenfunctions obtained in solving the Schroedinger equation for the potential $V(r)$ show that the probability densities for nucleons in the eigenstates of this potential *do not* have the property of being large only in a fairly restricted range of r. Thus in the nuclear shell model there are no "shells" of fairly restricted thickness in which the nucleons are moving. The contrast in this point between the behavior of the nucleus and of the atom is due to the difference in the form of $V(r)$ for the two systems.

Consider nuclei for which both N and Z are magic, such as ${}^{8}O^{16}$, ${}^{20}Ca^{40}$, ${}^{82}Pb^{208}$. According to the model, they will contain only completely filled subshells of neutrons and protons, and the exclusion principle therefore demands that for both the neutron and the proton systems the total spin and orbital angular momenta couple to zero,† in agreement with the observation that $i = 0$ for these nuclei. For nuclei such as ${}^{7}N^{15}$, ${}^{8}O^{17}$, ${}^{19}K^{39}$, ${}^{82}Pb^{207}$, and ${}^{83}Bi^{209}$, which contain a magic number of nucleons of one type and a magic number plus or minus one of nucleons of the other type, the exclusion principle demands that the total angular momentum of the system be the total angular momentum of the extra nucleon or hole.† Thus figure (16–23) shows that these nuclei should have $i = \frac{1}{2}, \frac{5}{2}, \frac{3}{2}, \frac{1}{2}$, and $\frac{9}{2}$, respectively, also in agreement with observation. Now consider nuclei for which N and/or Z are further away from the magic numbers. These nuclei will contain subshells with more than one nucleon or hole, and the problem of how the spin and orbital angular momenta of the nucleons in these subshells couple is analogous to the same problem for the electronic structure of an atom. However, there are important differences. Most atoms seem to be examples of LS coupling, but most nuclei seem to be examples of JJ coupling. Of course, this is simply a consequence of the strong spin-orbit interaction in nuclei. It appears that another consequence is that the coupling in nuclei is much simpler than in atoms. Consider any nucleus with even N and even Z. As mentioned in section 1, all such nuclei are found to have $i = 0$. It is clear that this means that, whenever there are an even number of nucleons in a subshell, the total angular momenta of these nucleons simply couple to zero. If it is now assumed that the JJ coupling is strong enough that the addition of one more nucleon to the subshell does not destroy the coupling of the nucleons that were already there, the total angular momentum of the whole subshell will be that of the "odd" nucleon. With this *assumption, the entire angular momentum of an odd A nucleus will be due to the total angular momentum of the single odd nucleon in the highest occupied energy level, and i will be equal to the value of j for that level.* With only one or two exceptions, this assumption allows the observed values of i for all odd A nuclei to be explained in terms of figure (16–23). In doing this it is, however, necessary to allow for occasional interchanges of the ordering of certain closely spaced energy levels,‡ because of the effect of the pairing energy.

The spin-orbit shell model, plus the additional assumption just mentioned, provides a somewhat less satisfactory explanation of the magnetic moments of odd A nuclei. The prediction is that the magnetic moment of the nucleus is due to the magnetic moment of the single odd nucleon

† Cf. discussion associated with table (13–9).

‡ Recall that this also happens for the atomic energy levels.

because the contributions of the other nucleons cancel one another. Now it is easy to calculate the magnetic moment of a single nucleon of given j in terms of its orbital g factor (which is 1 for the proton and 0 for the neutron), and in terms of its spin g factor (which is such as to give z components of intrinsic magnetic moment $+2.79\mu_N$ for the proton and $-1.91\mu_N$ for the neutron, where the sign gives the orientation relative to the spin). For protons, two values are found corresponding to the two possibilities $j = l + \frac{1}{2}$ and $j = l - \frac{1}{2}$; for neutrons, two other values are found. A plot of these two values of magnetic moment as a function of j forms the so-called *Schmidt lines*. When the proton Schmidt lines are compared with the observed magnetic moments of odd Z, even N nuclei, and the neutron Schmidt lines are compared with those of even Z, odd N nuclei, it is seen that the predictions agree with the general trend of the observed values, but that the agreement is far from perfect. This shows that it is only an approximation to assume that an even number of nucleons always couple to give zero total angular momentum. The approximation is good enough to lead to the prediction of correct values of the nuclear spin angular momentum because this quantity is quantized; so, if there is some probability that the even number of nucleons occasionally have a non-zero total angular momentum, there must also be a corresponding probability that the odd nucleon have exactly the right total angular momentum to compensate and keep the total angular momentum of the system constant. But, if this happens, the magnetic moment of the system, which is not quantized, can be appreciably different from the predicted value. Apparently it does happen.

9. The Collective Model

The shell model is based upon the idea that the constituent parts of a nucleus move independently. The liquid drop model implies just the opposite, since in a drop of incompressible liquid the motion of any constituent part depends on the motion of all the neighboring parts. The conflict between these ideas is indicative of the current state of the theoretical understanding of the nucleus. It also emphasizes that a model provides a unified description of only a *limited* set of phenomena, completely without regard to the existence of conflicting models used for the description of other sets. Contrast this with a theory, such as relativity theory or quantum theory. A theory provides a description of a very *large* set of phenomena and, at the border lines between this set and other sets of phenomena, it fuses without conflict into the theories used for the description of the other sets.

As in every scientific field, the theoretical development of nuclear physics proceeds on two levels. On one level, people develop *models*, which correlate experimental results and provide simple descriptions of the phenomena involved. Then they constantly attempt to broaden the range of applicability of each model, to remove any apparent conflicts between the various models, and finally to fuse these models into a single unified model. On another level, people attempt to develop *theories*, which explain the properties of the models from the fundamental laws of nature. As an example, a theory of the nucleus would be able to explain the numerical values of the parameters used in the semi-empirical mass formula, in terms of the laws of force between nucleons.

The parallel development of models and theories was not very evident in our discussion of atomic physics, because that is a field in which there are now acceptable theories. At this stage, the models lose much of their importance and so are used in a discussion only if they are of particular historical interest, or are of pedagogical value, or provide convenient calculational tools. An example is Bohr's description of atomic structure, a model which we discussed at length for precisely these three reasons.

In later sections we shall describe some of the attempts which have been made to develop theories of the nucleus and theories of nuclear forces. In the present section we shall describe an example of one of the attempts to fuse two apparently contradictory models into a unified model. This is the *collective model* of A. Bohr and Mottelson (1953), which has been quite successful in combining certain features of the shell model and the liquid drop model. The model assumes that the nucleons in a nucleus move independently in a real potential. However, the potential is not the static spherically symmetrical potential $V(r)$ used in the shell model; instead it is a potential capable of undergoing deformations in shape. These deformations represent the collective motion of the nucleons in the nucleus—the type of motion associated with a liquid drop. As in the shell model, the nucleons fill the energy levels of the potential, which are split by the same spin-orbit interaction, and form a *core* containing an even number of protons and an even number of neutrons, plus (for odd A nuclei) a single odd nucleon. As is also true in the shell model, the odd nucleon plays an important role in determining the properties of the entire nucleus. However, in the collective model the odd nucleon is not entirely responsible for these properties because the core is not inert and can have angular momentum, etc. The instantaneous shape of the core is described by the instantaneous shape of the potential. Its departure from spherical symmetry affects the motion of the odd nucleon and, in turn, the motion of this nucleon affects the shape of the core. The latter coupling is due to what can be described as a centrifugal reaction exerted by the odd nucleon on the wall of the potential. The net effect is that a "tidal wave" circulates around the surface of the core, following the motion of the odd nucleon. As the reader might guess, calculations based on this model are very complicated. Consequently we shall only describe the results.

It is found that the collective model preserves all the features of the shell model which agree with experiment, and also provides new features which allow this agreement to be considerably extended. For instance, in this model part of the total angular momentum of the nucleus is carried in the form of orbital angular momentum by the tidal wave circulating around the surface of the core. This moving deformation constitutes a current and gives rise to a magnetic moment proportional to the associated orbital angular momentum. The proportionality constant is, however, smaller than the usual expression (11–7 with the sign changed and the electron mass replaced by the proton mass) by a factor Z/A, which is the ratio of the number of charged nucleons to the total number of nucleons. Consequently, that part of the spin angular momentum carried by the core does not contribute the same magnetic moment as it otherwise would, and the change is exactly what is required to explain the discrepancy between the Schmidt lines and the measured nuclear magnetic moments.

Another feature which can be explained quite nicely in terms of the collective model is the nuclear electric quadrupole moment Q. We have given a qualitative definition of this quantity on previous occasions (cf. section 1). Quantitatively, the electric quadrupole moment about the z axis is defined by the equation†

$$Q = \int \rho(x, y, z)[3z^2 - (x^2 + y^2 + z^2)]\, d\tau = Z[3\overline{z^2} - (\overline{x^2} + \overline{y^2} + \overline{z^2})] \tag{16–38}$$

where $\rho(x, y, z)$ is the nuclear charge density in units of proton charges, and where the integral is taken over the nuclear volume. The value of Q is Z, the number of protons in the nucleus, times the average over their charge distribution of the quantity $[3z^2 - (x^2 + y^2 + z^2)]$. It is clear then that $Q = 0$ if $\rho(x, y, z)$ is spherically symmetrical, since in that case $\overline{x^2} = \overline{y^2} = \overline{z^2}$. If $\rho(x, y, z)$ is not spherically symmetrical, it must at least have an axis of symmetry along the direction about which the nuclear spin angular momentum vector precesses. In (16–38) this is taken as the z axis, and the equation immediately shows that $Q > 0$ if $\rho(x, y, z)$ is elongated in the z direction so that $\overline{z^2} > \overline{x^2} = \overline{y^2}$, and that

† See also the footnote on page 646.

$Q < 0$ if $\rho(x, y, z)$ is flattened in the z direction so that $\overline{z^2} < \overline{x^2} = \overline{y^2}$. Figure (16–24) shows a plot of all the presently known values of Q, for stable odd A nuclei, measured under circumstances in which Q is

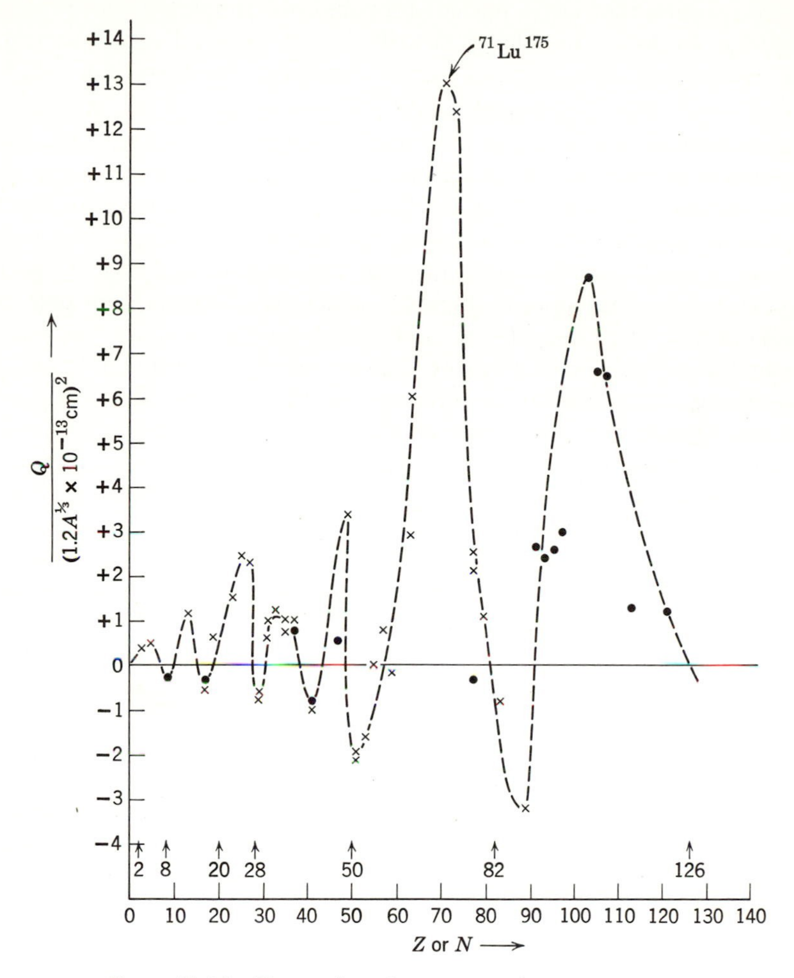

Figure 16–24. The quadrupole moments of stable odd A nuclei.

maximized because the nuclear spin angular momentum vector has its maximum possible component along the z axis ($m_i = i$). The abscissa is Z for odd Z, even N nuclei, and the values of Q for these nuclei are indicated with crosses; the abscissa is N for even Z, odd N nuclei, and the values of Q for these nuclei are indicated with dots. The ordinate is the ratio of Q to the square of the nuclear charge distribution radius.

Some of the features of this plot agree qualitatively with what would be expected from the shell model. For example, this model would predict $Q < 0$ for an odd Z, even N nucleus with Z equal to a magic number plus 1. The reason is that such a nucleus contains only completely filled proton subshells, for which the charge distribution is spherically symmetrical, plus one proton moving in an "orbit" near the x-y plane, for which $\overline{z^2} < \overline{x^2} = \overline{y^2}$. The shell model would also predict $Q > 0$ for an odd Z, even N nucleus with Z equal to a magic number minus one, since the quadrupole moment would be due entirely to a proton hole moving in an orbit near the x-y plane. Thus the fact that Q goes through zero from positive to negative values each time the number of odd protons goes through a magic number, and also the concomitant fact that Q goes through zero in the opposite direction somewhere between each pair of magic proton numbers, can be understood on the basis of the shell model.† However, the magnitudes of Q cannot be understood in terms of this model. According to the shell model, the quadrupole moment of *any* odd Z, even N nucleus is entirely due to the odd proton since the core containing an even number of nucleons has zero total angular momentum and must, therefore, have a charge distribution which appears to be spherically symmetric because there is no way to define a symmetry axis. Now it is apparent from equation (16–38) that the maximum possible value of Q due to a single odd proton is about $Q = R^2$, where R is the charge distribution radius. But figure (16–24) shows that Q is observed to be very much larger than this for many odd Z, even N nuclei. Also, in the shell model the quadrupole moment of all even Z, odd N nuclei would be very small because it could arise only from the very small displacement of the center of the charged core from the center of the entire nucleus. However, the figure shows that the value of Q for odd A nuclei is observed not to depend significantly on whether they are odd Z, even N or even Z, odd N. The collective model provides an explanation for all these features. It leads to large enough values of Q between the magic numbers simply because the core can be deformed so that many protons contribute to the total quadrupole moment. As the deformation can be due to either an odd proton or an odd neutron, the collective model explains why the values of Q are similar in both cases.

† Note that Q does not behave this way at 20. This, and other recent evidence, has been interpreted to mean that 20 is somewhat less magic than the other magic numbers. Also note that at 40 (corresponding to the filling of the $3p_{1/2}$ level) and 14 or 16 (corresponding to the filling of the $3d_{5/2}$ or $3s_{1/2}$ levels) Q behaves as it does at magic numbers. However, 40 and 14 or 16 are not considered magic because none of the other usual effects is observed.

10. Alpha Decay and Fission

Up to this point we have confined our attention to nuclei which are stable—at least in the sense that they have immeasurably long lifetimes. The qualification is necessary because, as we saw in section 4, all nuclei with A much larger than the value $A \simeq 60$, at which the average binding energy per nucleon $\Delta E/A$ maximizes, are energetically unstable to fission. Let us now discuss this interesting decay process, starting with *alpha decay*. In this particularly important special case of fission, a nucleus spontaneously emits an alpha particle $^{2}He^{4}$ because its initial mass $m_{Z,A}$ is greater than the sum of its final mass $m_{Z-2,A-4}$ and the alpha particle mass $m_{2,4}$. The energy equivalent of the mass difference is carried away as kinetic energy of the alpha particle. Ignoring the mass equivalents of the electron binding energies, this *alpha decay energy* E can be written in terms of atomic masses as†

$$E = [M_{Z,A} - (M_{Z-2,A-4} + M_{2,4})]c^2 \tag{16–39}$$

If the semi-empirical mass formula is used to estimate the masses in this equation, it predicts that E increases slowly with A, reaching values of 5 or 6 Mev for the heaviest nuclei. However, this prediction is not accurate enough to be useful because it ignores magic number effects, which are very important for nuclei of particular interest in alpha decay. Figure (16–25) shows the values of E for the nuclei in the alpha emitting range, obtained from direct measurements of the kinetic energy of the alpha particle with range or momentum analysis techniques, and/or from the use of equation (16–39) with measured masses. We see that the alpha decay energy E rises to a peak about 4 Mev above the general trend, and that the general trend slowly increases with A as predicted by the semi-empirical mass formula. The peak occurs just at $^{84}Po^{212}$, and it is easy to understand why. In emitting an alpha particle, this *parent* nucleus decays to the *daughter* nucleus $^{82}Pb^{208}$, which has magic $Z = 82$ and magic $N = 126$. The increase in alpha decay energy arises because the binding energy of the daughter is about 4 Mev higher than average. [Figure (16–18) shows that the increase is about 2 Mev because Z is magic and about 2 Mev because N is magic.] We also see from figure (16–25) that the energies of the known alpha emitters range from 8.9 Mev for $^{84}Po^{212}$ to 4.1 Mev for $^{90}Th^{232}$. Note, then, that all these energies are far below the maximum energy of the Coulomb barrier acting on the alpha particle, which is about 30 Mev. Therefore the alpha particle can escape

† There is no spontaneous emission of $^{0}n^{1}$, $^{1}H^{1}$, $^{1}H^{2}$, or $^{2}He^{3}$ from the naturally occurring nuclei because the values of E, calculated from equations similar to (16–39), are negative for all these processes.

the nucleus only by the process of barrier penetration discussed in section 3, Chapter 8.

In equation (8–58) of that section, we obtained the expression

$$T = e^{-2\int\sqrt{(2M/\hbar^2)[V(r)-E]}\,dr} \tag{16–40}$$

for the probability that an alpha particle of mass M will penetrate the potential barrier $V(r)$, if its kinetic energy far from the nucleus is E. This

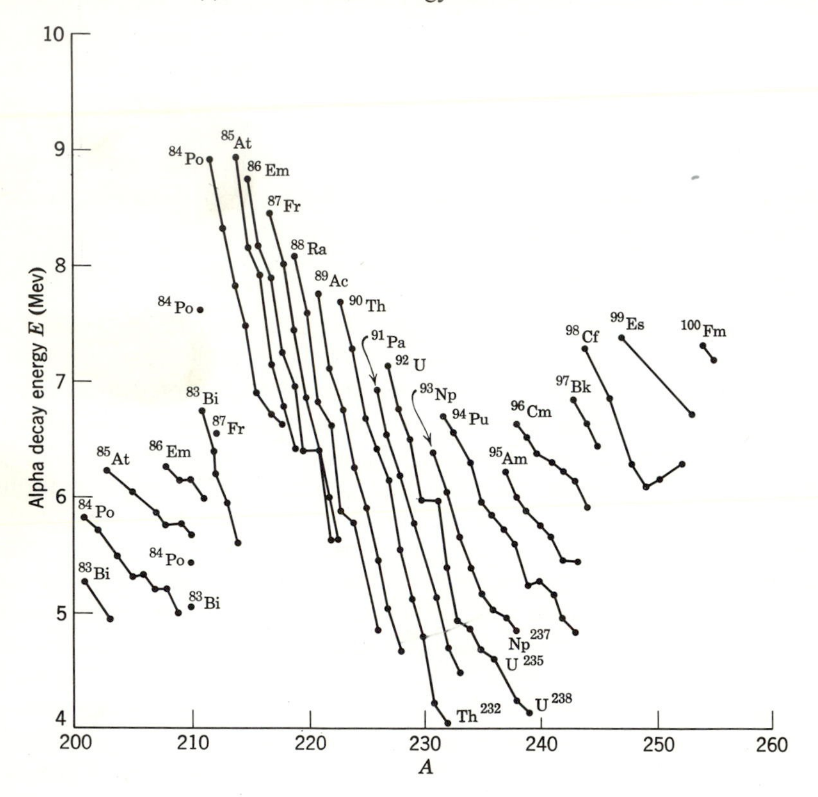

Figure 16–25. The alpha decay energy for the nuclei in the alpha emitting range. From I. Perlman, A. Ghiorso, and G. T. Seaborg, *Phys. Rev.*, **77**, 26 (1950).

equation is actually valid only in the case of emission of alpha particles of zero orbital angular momentum because only for that case is the time independent Schroedinger equation (15–76), in the radial coordinate r, the same as the time independent Schroedinger equation, in the rectangular coordinate x, which was used in the derivation of (16–40). For the emission of alpha particles of non-zero orbital angular momentum, the coefficient of u in equation (15–76) is more negative; this has the same effect as making the potential barrier $V(r)$ higher, and therefore the probability of penetration T is reduced. However, the reduction is negligible if the angular momentum is not too large, and usually it can be ignored.

Let us evaluate equation (16–40) by using the slightly simplified form for $V(r)$ indicated in figure (16–26). In the alpha decay of a parent nucleus of atomic number Z, the potential is

$$V(r) = \frac{2(Z-2)e^2}{r}$$

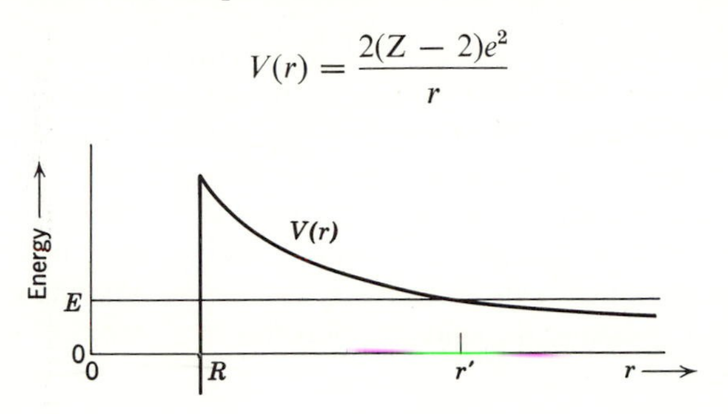

Figure 16–26. A slightly simplified alpha particle-nucleus potential.

The integral of (16–40) is taken from the nuclear potential radius R to the radius r' satisfying the equation

$$E = \frac{2(Z-2)e^2}{r'}$$

Thus

$$T = e^{-\frac{2\sqrt{2M}}{\hbar}\int_R^{r'}\sqrt{\frac{2(Z-2)e^2}{r} - E}\, dr}$$

Changing to the variable ρ defined by

$$\cos^2 \rho = \frac{E}{2(Z-2)e^2}\, r = \frac{r}{r'}$$

the integrand becomes $\sin^2 \rho$ and the integral immediately gives

$$T = e^{-\frac{2\sqrt{2ME}}{\hbar} r'\left[\cos^{-1}\left(\frac{R}{r'}\right) - \left(\frac{R}{r'}\right)\left(1-\frac{R}{r'}\right)^{1/2}\right]}$$

Since R/r' is fairly small compared to 1, the terms in the square bracket may be expanded. The result is

$$T \simeq e^{-\frac{2\sqrt{2ME}}{\hbar} r'\left[\frac{\pi}{2} - 2\left(\frac{R}{r'}\right)^{1/2}\right]}$$

Evaluating r', this becomes

$$T \simeq e^{\left[\frac{8M^{1/2}e}{\hbar}(Z-2)^{1/2}R^{1/2} - \frac{2\pi(2M)^{1/2}e^2}{\hbar}(Z-2)E^{-1/2}\right]}$$

which gives

$$T \simeq e^{2.97(Z-2)^{1/2}R^{1/2} - 3.95(Z-2)E^{-1/2}} \tag{16–41}$$

when E is written in units of Mev and R is written in units of 10^{-13} cm.

As in (8–59), we shall now assume that there is one alpha particle with energy E, which makes $v/2R$ attempts at the barrier per second, where v is its velocity inside the nucleus. There is little on which to base this assumption because little is known about what an alpha particle is doing in a nucleus before it is emitted. However, it does allow us to write for the *decay rate* λ, which is the probability per second that the nucleus will decay by alpha emission, the simple expression

$$\lambda = \frac{v}{2R} T \tag{16–42}$$

Substituting for T from (16–41), and taking logarithms, we have

$$\ln \lambda \simeq \ln \frac{v}{2R} + 2.97(Z-2)^{1/2}R^{1/2} - 3.95(Z-2)E^{-1/2} \tag{16–43}$$

This makes it apparent that λ is actually insensitive to the exact form of the coefficient of T in (16–42). In figure (16–27), the points show the measured values of $\ln \lambda$ for all the naturally occurring alpha emitters, and the curve shows the values of this quantity predicted by equation (16–43), using $R = 8 \times 10^{-13}$ cm and $Mv^2/2 = E$. Equation (16–43) can be plotted, approximately, as a function of $(Z-2)E^{-1/2}$ alone because there is essentially no change in $\ln (v/2R)$, and very little change in $2.97(Z-2)^{1/2}R^{1/2}$, over the restricted range of nuclei involved. The only change comes from $-3.95(Z-2)E^{-1/2}$, which varies because E varies from about 4 to about 9. We see that this leads to a variation of about 55 in $\ln \lambda$, corresponding to a variation in λ of about a factor of $e^{55} \simeq 10^{25}$! The correlation between the extremely rapid variation of λ and the slow variation of E was first pointed out by Geiger and Nuttall (1911) from a compilation of the experimental data. They devised an empirical formula giving the essential features of the energy dependence of the theoretical equation (16–43), which is due to Gamow, Condon, and Gurney (1928).

Now consider a system containing $N(0)$ unstable nuclei of the same species at time $t = 0$. If the probability per second that any one of the nuclei will decay is the decay rate λ, the number of nuclei remaining at time t is

$$N(t) = N(0)e^{-\lambda t} \tag{16–44}$$

and the *lifetime* τ, or average time required for one of the nuclei to decay, is

$$\tau = 1/\lambda \tag{16–45}$$

These relations are (13–116) and (13–117), which were obtained for the decay of a system of excited atoms by the emission of quanta. But a

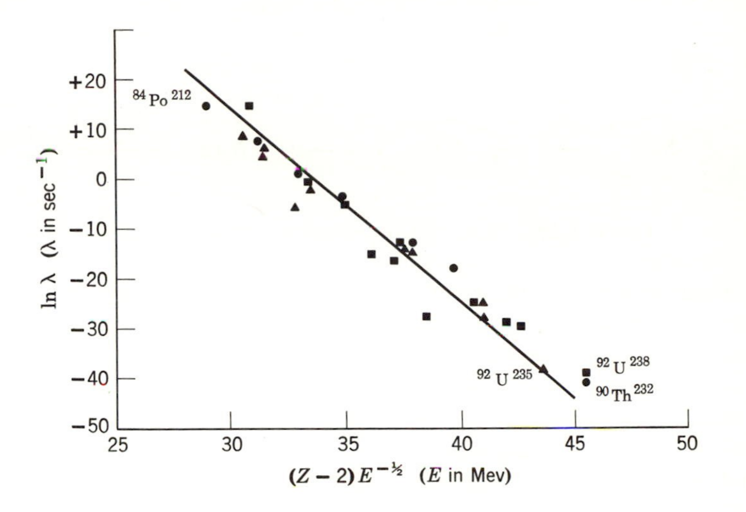

Figure 16–27. The relation between the alpha decay rates and alpha decay energies. ● = $4n$ series; ■ = $4n + 2$ series; ▲ = $4n + 3$ series. From R. B. Leighton, *Principles of Modern Physics*, McGraw-Hill Book Co., New York, 1959.

review of their derivation will show that they apply equally well to any system where a constant probability per second of decay can be defined.

In a more typical system, there are several different nuclear species which decay successively into each other. For instance, $^{92}U^{234}$ decays by alpha emission into $^{90}Th^{230}$; this nucleus decays by alpha emission into $^{88}Ra^{226}$; this nucleus, in turn, decays by alpha emission into $^{86}Em^{222}$; and this nucleus decays by alpha emission into $^{84}Po^{218}$. Consequently, a system, initially filled with $^{92}U^{234}$, will eventually contain a mixture of all these nuclei. The decay of such a *radioactive series* is indicated schematically in figure (16–28). The symbol $N_0(t)$ is the number of *parent* nuclei in the system at time t, and the symbol λ_0 is the probability per second for decay of these nuclei. Similar symbols are used for the numbers and decay rates of the *first daughter* nuclei, the *second daughter* nuclei, and so forth.

In this system, the relation (16–44) is valid for the parent only. However, analogous relations for the daughters can be obtained from the following set of equations:

$$\frac{dN_0(t)}{dt} = -N_0(t)\lambda_0$$

$$\frac{dN_1(t)}{dt} = N_0(t)\lambda_0 - N_1(t)\lambda_1 \qquad (16\text{–}46)$$

$$\frac{dN_2(t)}{dt} = N_1(t)\lambda_1 - N_2(t)\lambda_2$$

$$\vdots$$

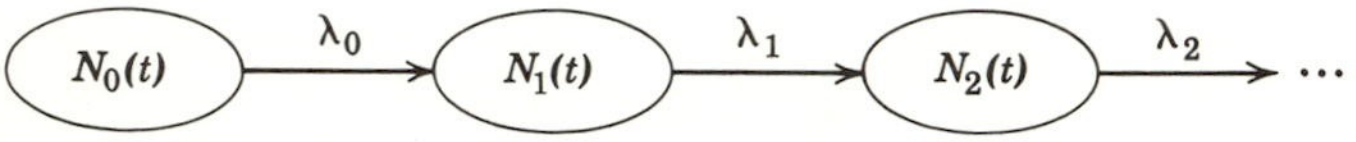

Figure 16–28. Illustrating the decay of a radioactive series.

The negative term on the right of each equation represents the loss of nuclei per second due to decay into the next nuclei of the series; the positive term represents the gain of nuclei per second due to decay from the previous nuclei of the series. A number of interesting solutions exist for this set of coupled differential equations, depending on the initial conditions and the constants λ.

One solution of particular importance is for the case when λ_0 is very small compared to all the other λ. Then the parent of the series decays very much more slowly than all its daughters and, after sufficient time, a condition of *equilibrium* will obtain. In equilibrium, all the $N(t)$ are constant, so the left sides of all the equations (16–46) go to zero. (This can be exactly true for all the equations except the first. For the first it can obviously be only an approximation; but it is a very good approximation if λ_0 is very small as assumed.) Applying this condition, we have

$$0 = N_0(t)\lambda_0 - N_1(t)\lambda_1$$

$$0 = N_1(t)\lambda_1 - N_2(t)\lambda_2$$

$$\vdots$$

which gives

$$N_0(t)\lambda_0 = N_1(t)\lambda_1 = N_2(t)\lambda_2 = \cdots \qquad (16\text{–}47)$$

The equilibrium obtains after a time such that $\lambda t \gg 1$ for all λt except $\lambda_0 t$, independent of the initial conditions. This case is important because the dependence of E, and therefore of λ, on A is usually such as to satisfy the requirement that λ_0 be very small compared to all the other λ [cf. the part of figure (16–25) between $A = 240$ and $A = 210$]. Note that the relation (16–47) shows that in equilibrium the number of nuclei of a particular species is inversely proportional to the decay rate of that species. This relation is often used in determining the values of the λ from measurements of the $N(t)$ and one known λ.

We are now in a position to understand how alpha emitting nuclei with very short lifetimes can be found in nature. For example, $^{84}Po^{212}$, with lifetime of about 10^{-6} sec, can be extracted from certain naturally occurring minerals which presumably have been in existence for millions of years. Of course, the reason is simply that the short lifetime alpha emitters are members of radioactive series with parents of long lifetime, and are in equilibrium in these series. There are three such series which occur naturally: the $4n$ *series* whose parent is $^{90}Th^{232}$ with lifetime $\tau = 2.01 \times 10^{10}$ years, the $4n + 2$ *series* whose parent is $^{92}U^{238}$ with lifetime $\tau = 6.52 \times 10^9$ years, and the $4n + 3$ *series* whose parent is $^{92}U^{235}$ with lifetime $\tau = 1.02 \times 10^9$ years. The names characterize the values of A for the nuclei which are members of the series. For instance, the parent of the $4n + 3$ series has A equal to 4 times an integer plus 3, where the integer is 58. Since each alpha decay reduces A by 4, all the daughters of this series will also have A equal to 4 times some smaller integer plus 3.

In addition to the $4n$, $4n + 2$, and $4n + 3$ series, there is evidently also room for a $4n + 1$ series. Actually there is such a series, and its parent is $^{93}Np^{237}$ with lifetime $\tau = 3.25 \times 10^6$ years. The series can be produced artificially by using a nuclear reaction to make the parent, but it is not found in nature since the lifetime of the parent is very short compared to the age of the earth (estimated from certain geological and cosmological evidence to be of the order of 10^9 years), and therefore any parent nuclei initially present have decayed away. In this connection note that figure (16–25) shows that the alpha decay energies of the parents of the three naturally occurring series happen to be particularly low. If they were less than 1 Mev higher, these parents too would have lifetimes short compared to the age of the earth, all the alpha emitters would have long since decayed away, and the naturally occurring elements would stop at $Z = 82$ instead of $Z = 92$. The same figure makes it clear why the naturally occurring elements do stop at $Z = 92$: it is because the alpha decay energies of all nuclei with $Z \geqslant 93$ are large enough to lead to lifetimes short compared to the age of the earth. Finally, a mental extrapolation of figure (16–25) to $Z \leqslant 82$, or an actual calculation of the alpha decay energies from the measured masses, shows that the corresponding elements are apparently stable to alpha decay because the alpha decay energies are so small that the lifetimes are immeasurably long. Strictly speaking, a number of nuclei with $Z \leqslant 82$ are unstable, but their lifetimes are so long that decay cannot be observed.

Alpha decay occurs essentially because the total Coulomb repulsion energy of the protons in the nucleus is decreased by the emission of a fragment containing two of these protons (the alpha particle). But an even larger decrease in this energy is obtained by the emission of a fragment containing a larger fraction of the protons. If the nucleus splits into two

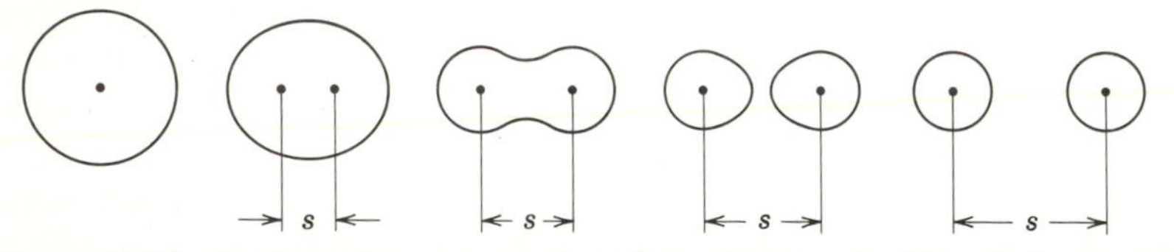

Figure 16–29. A schematic representation of the steps involved in a fission process.

parts, the total Coulomb repulsion energy of the system is obviously minimized when the parts contain equal numbers of protons. The calculation performed in section 4 shows that, when a nucleus of large Z undergoes such a *fission* process, the decrease in the energy equivalent of the total mass of the system is about 200 Mev. This is almost entirely due

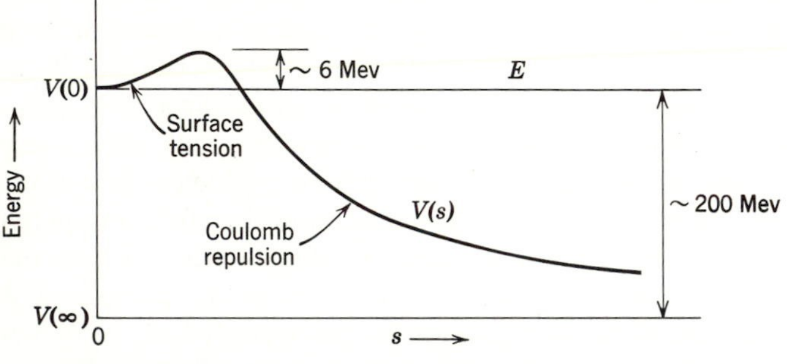

Figure 16–30. An energy diagram for a fissionable nucleus.

to the decrease in the total Coulomb repulsion energy of the system. Most of this energy is carried away by the two *fission fragments* as kinetic energy.

The steps which are involved in a fission process are indicated schematically by the set of drawings in figure (16–29). These drawings define (somewhat unprecisely) a parameter s characterizing the progress of the process. It is convenient to use this parameter and, in particular, to plot that part of the energy of the system which depends on s. This is shown in figure (16–30) by the function $V(s)$. Starting at small s, there is relatively little change in the Coulomb repulsion energy with increasing s, but the surface area of the nucleus increases rapidly. From the clearly appropriate point of view of the liquid drop model, the increase in surface area produces an increase in the surface tension energy. Thus $V(s)$ increases with increasing s for small s. As s continues to increase, the surface tension effect causes the nucleus to "neck down" and eventually to split. After it splits, the surface tension energy no longer changes with s, and $V(s)$ decreases with increasing s, following the decrease in the Coulomb repulsion energy of the two fragments. Since $V(s)$ first goes up and later comes down, it necessarily must pass through a maximum. Calculations based on the liquid drop model show that for a typical nucleus of large Z this maximum is about 6 Mev above $V(0)$. We already know that $V(0)$ is about 200 Mev above $V(\infty)$.

Figure (16–30) shows that nuclei are normally stable to decay by fission since they are sitting, with total energy $E = V(0)$, at the bottom of the depression in the potential $V(s)$. The process can take place by barrier penetration but, because the mass M entering in the exponent of expression (16–40) for the barrier penetrability is very large, the probability of barrier penetration is *extremely* small. For instance, if $^{92}U^{238}$ decayed only by this *spontaneous fission* process, its lifetime would be $\sim 10^{16}$ years. Nevertheless, spontaneous fission can actually be observed, because so much energy is liberated in the fission process that it is very easy to detect. One technique often used involves collecting the large number of ions produced by the fission fragments as they come to a stop in the gas filling of a device called an ionization chamber.

A process of much more importance is *induced fission*. Usually this is brought about by the nucleus capturing a low energy neutron. As the binding energy E_b of the last neutron in a nucleus of large Z is around 5 or 6 Mev, in favorable cases the capturing nucleus receives enough energy to put it over the top of the fission barrier. Very often this energy actually does go into vibrations of the type which lead to fission. Induced fission is perhaps the best example of the collective motions that are implied by the liquid drop model. The process is illustrated in terms of an energy diagram in figure (16–31). For $^{92}U^{233}$ and $^{92}U^{235}$, the binding energy E_b is enough to make the total energy E higher than the top of the barrier. For $^{92}U^{238}$ it is not quite enough, and the neutron must also carry in about 1 Mev of kinetic energy. The difference between the two cases reflects the fact that E_b, for the nuclei formed when $^{92}U^{233}$ and $^{92}U^{235}$ capture a neutron, is about 1 Mev greater than E_b for the nucleus formed when $^{92}U^{238}$ captures a neutron. The reason is that the pairing energy (16–32) is negative in the first case and zero in the second case. Fission can also be induced by the capture of a gamma ray of sufficient energy, or by the capture of a high energy proton (a low energy proton will not be captured because it is repelled from the nucleus by a Coulomb barrier about 15 Mev high).

The possibility of using fission to produce power in a *chain reaction* arises from the circumstance that 2 or 3 neutrons are emitted, on the average, in each fission process. This can be understood by considering figure (16–32). The figure plots the Z and N values of the nuclei which

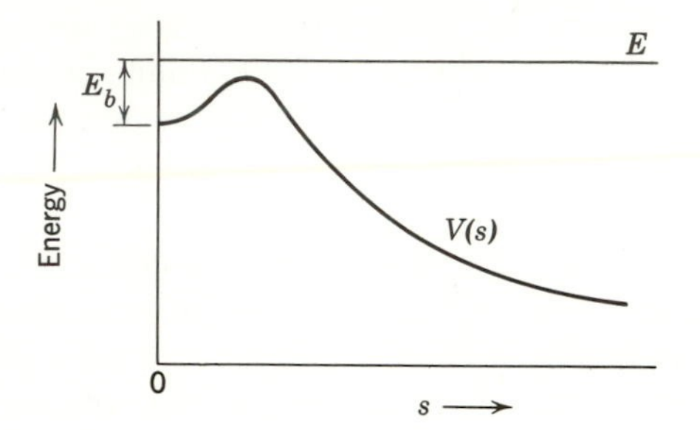

Figure 16–31. An energy diagram illustrating induced fission.

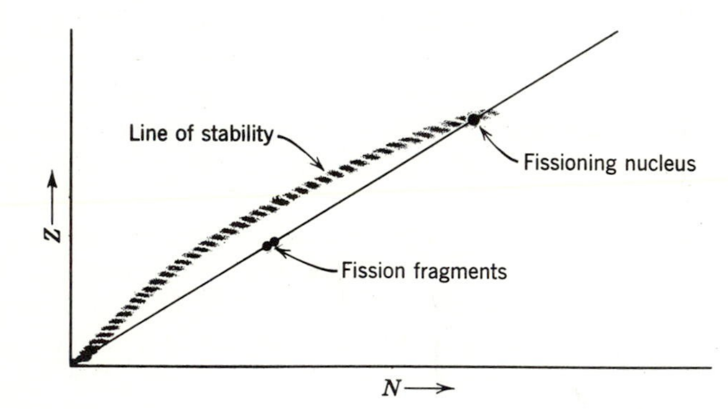

Figure 16–32. Illustrating why neutron emission occurs in fission.

are the most stable for each value of A, as in figure (16–17). These nuclei are represented by the shaded *line of stability*. The large dot indicates the fissioning nucleus. The two small dots indicate the two fragments which result from the fission. These fragments have the same Z/N ratio as the fissioning nucleus but, since their values of A are smaller by a factor of 2, their Z/N ratios are smaller than those of stable nuclei with these values of A—the fission fragments have too many neutrons. Most of the necessary adjustment in the Z/N ratio takes place by the beta decay process to be discussed in the next section, but part of it is achieved simply by emitting a few neutrons. These neutrons typically have kinetic energies of several Mev. However, they rapidly lose energy by processes involving scattering with energy loss in the material containing the fissioning nuclei and are no longer able to induce fission in ${}^{92}U^{238}$. This is why that isotope is usually not directly useful in a chain reaction. The slowing down of the neutrons emitted in fission is, however, not undesirable since, when their energy becomes so low that their de Broglie wavelength is larger than the radius of the ${}^{92}U$ nuclei, the cross section for capture by these nuclei increases with increasing wavelength or decreasing energy.† For a chain reaction

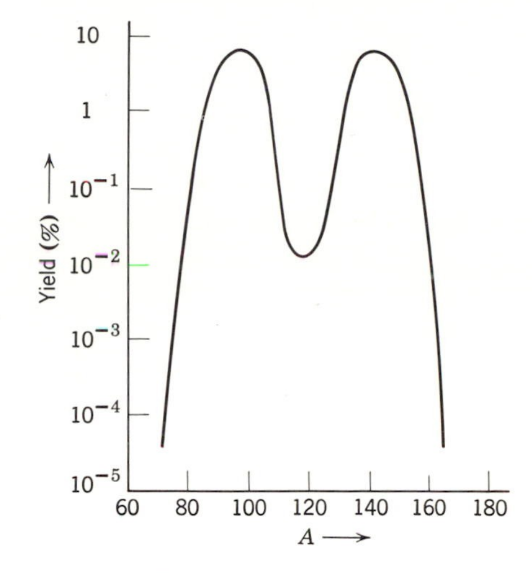

Figure 16–33. The distribution of the masses of fragments emitted in fission by thermal neutrons.

it is important to capture as many of the neutrons as possible; the reaction is self-sustaining only if at least one of the neutrons emitted in each fission is later captured. In *reactors*, advantage is taken of the energy dependence of the capture cross section by scattering the neutrons from some material of low atomic weight so that they lose energy by recoil and eventually come into thermal equilibrium at the very low kinetic energy $\frac{3}{2}kT \simeq 0.025$ ev. This is not done in a *bomb* because it takes too long ($\sim 10^{-6}$ sec). In *breeder reactors* a certain amount of ${}^{92}U^{238}$ can be used indirectly to produce fissions. These nuclei capture neutrons and, instead of fissioning, the resulting unstable nuclei ${}^{92}U^{239}$ undergo two successive beta decays, turning into the stable nuclei ${}^{94}Pu^{239}$ which have the same fission properties as ${}^{92}U^{233}$ and ${}^{92}U^{235}$.

We have simplified our discussion of fission by speaking as if the process always leads to two equal fragments. Actually this happens only very infrequently. Figure (16–33) shows the distribution in the A values of

† This should be apparent from the discussion of the previous chapter.

the fragments from fission of ${}^{92}U^{235}$ by neutrons of thermal energy. This distribution is typical. Since one peak occurs near A values corresponding to $N = 50$, while the other occurs near A values corresponding to $N = 82$, it is thought that these bimodal distributions arise from the influence of the magic numbers. However, the details of their origin are not well understood.

11. Beta Decay

We have also simplified our discussion of the radioactive series by ignoring the processes of *beta decay*, which are, in fact, an important feature of these series. This is illustrated by the processes occurring in the $4n$ series, plotted in figure (16–34). In addition to alpha decay processes,

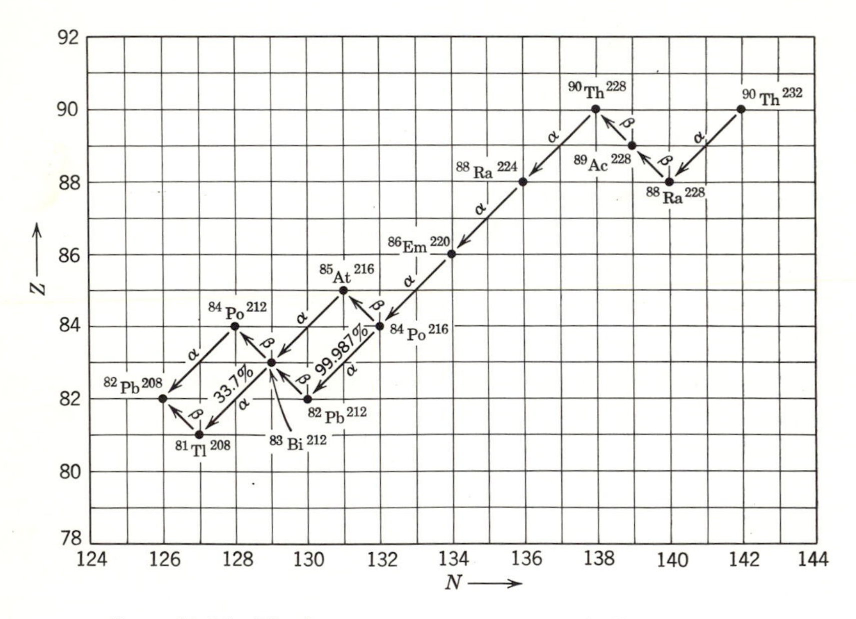

Figure 16–34. The decay processes occurring in the $4n$ series.

there are beta decay processes in which a nucleus Z, A emits a negatively charged electron and is transformed into the nucleus $Z + 1$, A. Thus, in these processes Z increases by 1, N decreases by 1, and the nucleus becomes the next nucleus lying on a line of slope 135° in the Z versus N plot. (There are also other types of beta decay processes that will be described later.) In alpha decay processes, Z decreases by 2, N decreases by 2, and the nucleus becomes the next but one nucleus lying on a line of slope 45° on the Z versus N plot.

The reason why these beta decay processes occur in the $4n$ series, as well as in the other series, can be understood by superimposing figure (16–34) on a plot of the Z and N values of the nuclei that are the most stable for each value of A, as shown in figure (16–35). As the slope of the line of stability is appreciably less than 45°, beta decay processes must occur in

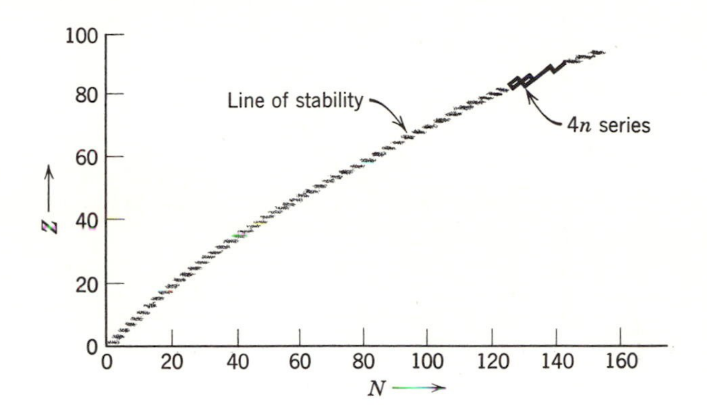

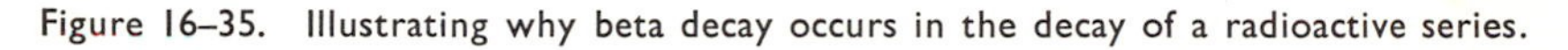
Figure 16–35. Illustrating why beta decay occurs in the decay of a radioactive series.

order that the average slopes of the radioactive series be essentially the same as the slope of the line of stability. The necessity of this simply reflects the requirement that each member of the radioactive series be "as stable as possible."

The decay of a radioactive series involves an interplay between the two competing processes of alpha decay and beta decay. Each member of the series will decay by the process having the largest decay rate. (The last statement is essentially correct if one decay rate is very much larger than the other. This is usually the case, but sometimes the decay rates are comparable, both processes occur, and the series *branches*. Cf. ${}^{84}Po^{216}$ and ${}^{83}Bi^{212}$ in the $4n$ series.) We have seen that the decay rate in alpha decay depends strongly on the energy available for the decay. We shall see that this is also true in beta decay.

To discuss the energy available in beta decay, it is convenient to plot $M_{Z,A}$, the mass of atoms of a given Z and A, as a function of Z for fixed A. Typical plots are indicated in figure (16–36) for odd A, and in figure (16–37) for even A. Except near magic N and/or magic Z, the masses are described quite well by the semi-empirical mass formula (16–33). For odd A, the values of $M_{Z,A}$ lie on a slightly skew parabola. The parabolic shape arises from the Z dependence of the term (16–31); the skewness arises from the Z dependence of the terms (16–27) and (16–30). For even A, there are two parabolas corresponding to the two possible values of the term (16–32). The elevated parabola corresponds to odd Z, odd N; the depressed

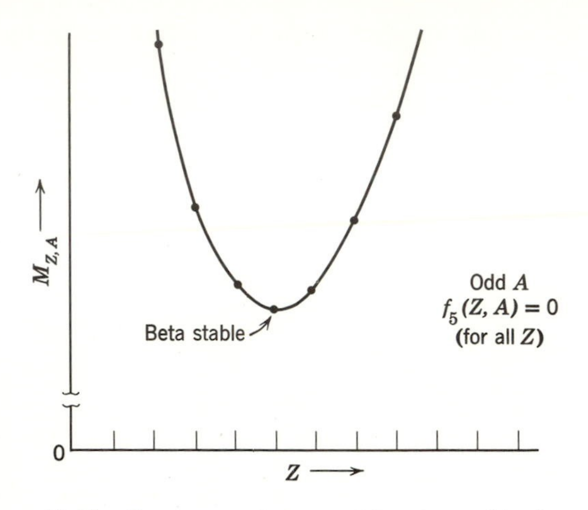

Figure 16–36. The masses of atoms with a given odd value of A.

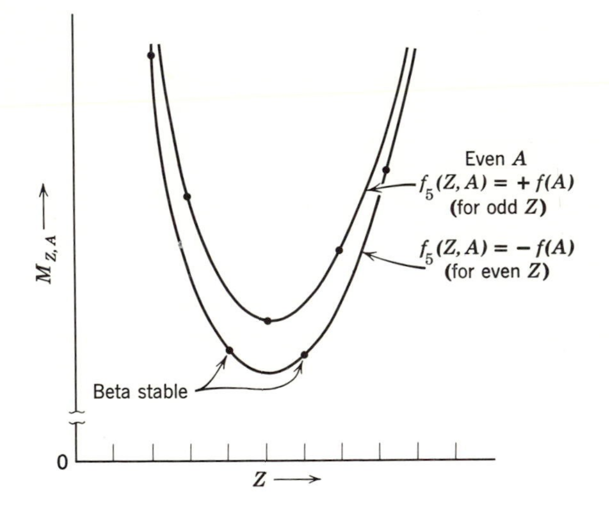

Figure 16–37. The masses of atoms with a given even value of A.

parabola corresponds to even Z, even N. From the plot for odd A, we see that for a given odd value of A there is only one nucleus which is stable to beta decay—the nucleus of the atom of smallest mass. (Very rarely an apparent exception to this rule is found when the difference between the masses for two adjacent values of Z at the bottom of the odd A parabola happens to be extremely small.) From the even A plot, we see that there are generally two nuclei with a given even value of A which are stable to beta decay, and that these two stable nuclei have even values of Z which differ by 2. Occasionally there are three beta stable nuclei for even A. Figure (16–38) indicates a plot of $M_{Z,A}$ representing this situation. An example is found for $A = 96$.

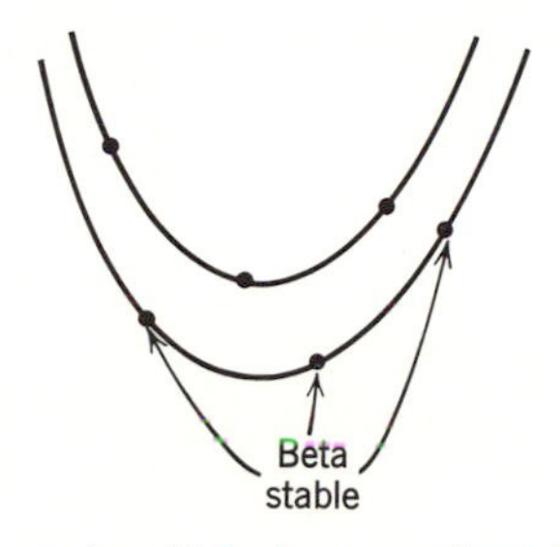

Figure 16–38. Illustrating a case in which there are three beta stable nuclei for a given even value of A.

Nuclei of given values of A, whose values of Z are such that they are not stable, can change their Z values to attain stability by three different beta decay processes. One is the process of *electron emission* that occurs in the radioactive series. As stated above, in this process a negatively charged electron is emitted by the nucleus, so that Z increases by 1 and N decreases by 1. The other two processes are *electron capture* and *positron emission*. In the former the nucleus captures a negatively charged atomic electron, and in the latter it emits a positively charged positron. In both, Z decreases by 1 and N increases by 1. The energy available for electron emission by the nucleus Z, A is

$$E = [M_{Z,A} - M_{Z+1,A}]c^2 \tag{16–48}$$

The energy available for electron capture by that nucleus is

$$E = [M_{Z,A} - M_{Z-1,A}]c^2 \tag{16–49}$$

The energy available for positron emission by the nucleus is

$$E = [M_{Z,A} - M_{Z-1,A} - 2m]c^2 \tag{16–50}$$

where m is the electron rest mass. The energy E ranges from a small fraction of 1 Mev to more than 10 Mev and is typically of the order of 1 Mev. Of course, in all these processes the available energy arises from the differences in the masses of the *nuclei* involved. These three equations are written in terms of the masses of the *atoms*, such as $M_{Z,A}$, simply because it is convenient and conventional to do so. A moment's reflection will demonstrate to the reader that, although these equations employ the usual approximation of neglecting the mass equivalents of the binding energies of the atomic electrons, the rest mass energies of all the electrons involved in the processes are properly taken into account. Note that there is a range in which the difference in masses between two atoms is such that electron capture is possible but positron emission is energetically forbidden. However, if $M_{Z,A}$ is greater than $M_{Z-1,A}$ by more than $2m$, both electron capture and positron emission can take place. In these circumstances positron emission is usually the more important process

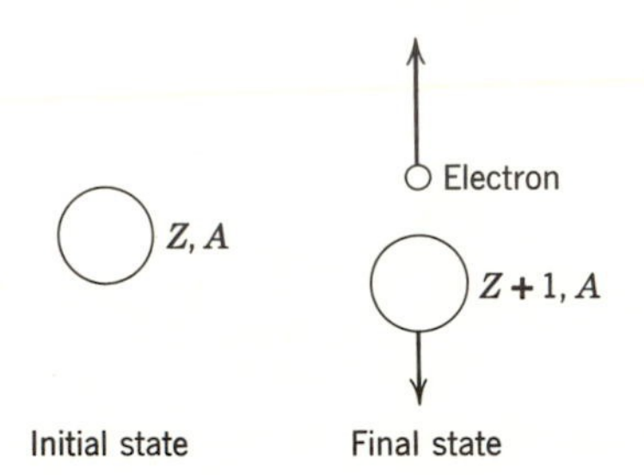

Figure 16–39. The beta decay process under the assumption that there are two particles in the final state.

for nuclei of small Z. For nuclei of intermediate and large Z, the wave functions of the atomic electrons are pulled in to the positively charged nucleus, and the wave function of a positron is repelled. Both of these effects tend to make electron capture relatively more probable and, particularly for the case of a K shell electron, electron capture usually dominates.

Now let us take up the interesting question of what happens to the available energy in beta decay processes. Consider the most common of these processes, electron emission. A nucleus Z,A, which we assume to be stationary in the initial state, emits an electron and recoils, as indicated in figure (16–39). Since there are just two particles in the final state, there should be only one way in which the total available energy E can be shared. In fact, since the nucleus is so massive, its recoil velocity is very low and it carries little kinetic energy. Thus the electron should carry away essentially all the energy E in the form of kinetic energy. However, measurements made at an early stage of the study of radioactivity, using momentum analyzing magnets and other techniques, showed that the electrons are emitted with a spectrum of kinetic energies T_e, as indicated in figure (16–40).†

For a number of years, the fact that electrons are emitted in the beta

† In some cases there are also a few groups of monoenergetic electrons in the measured spectra. They come from a process, called internal conversion, which has nothing to do with beta decay, and which will be explained in the next section.

decay process with a spectrum of energies was very mysterious and very disturbing. Electrons emitted at the end point $T_e^{\max}$ of the spectrum carry away all the available energy E, because $T_e^{\max}$ is observed to be equal to E within experimental accuracy. But the more typical electrons carry away much less energy than the energy E which, the mass differences show, must be released in the process, and it would appear that some of this energy has vanished! A number of suggestions aimed at resolving this dilemma were made. For instance, it was suggested that the electrons emitted from the nucleus could lose energy by collisions with the atomic

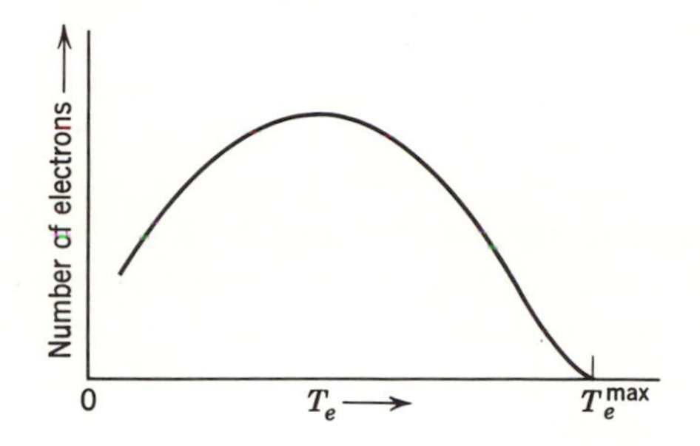

Figure 16–40. A typical beta decay electron kinetic energy spectrum.

electrons in emerging from the atom. If this were the case, and if the electron emitting atoms were enclosed in an isolated system, the lost energy would eventually be degraded into the form of heat, as would the energy actually carried away by the emitted electrons, so that the entire energy E would eventually be deposited in the system. But experiments performed by Ellis and Wooster (1927), using a calorimeter with very thick lead walls, showed that the energy deposited in the calorimeter per electron emission is equal to the average energy of the spectrum, which is less than half of E.

Another suggestion was that the electrons of kinetic energy lower than $T_e^{\max}$ are emitted in processes in which the final nucleus $Z + 1,A$ is left with excess energy in one of its excited states, and that these excited states are so close together that the spectrum of emitted electrons appears continuous. The excited nucleus would subsequently re-emit this energy in the form of a gamma ray. However, this was not tenable because there are a number of cases in which electrons are emitted with a spectrum of energies but no gamma rays are emitted, and also because other evidence shows that the excited states are not so closely spaced as would be required. The situation was grave enough that people were beginning seriously to consider abandoning the law of conservation of energy, until Pauli (1931) proposed a less repugnant alternative.

Pauli postulated that there is a particle called the *neutrino* which is emitted along with the electron in each electron emission process, but which is not observed because:

1. Neutrinos interact extremely weakly with matter.

This property allows the neutrinos, and the energy they carry, to pass through the thick lead walls of the Ellis and Wooster calorimeter. Pauli also postulated that:

2. Neutrinos are uncharged.
3. Neutrinos have spin angular momentum quantum number $s = \frac{1}{2}$.
4. Neutrinos have zero rest mass.

The second property is required in order that charge be conserved in the electron emission process. The third property allows total angular momentum to be conserved in this process. Consider a nucleus Z,A emitting an electron to become the nucleus $Z + 1,A$ and assume, for example, that A is even. Then the spin angular momentum quantum numbers i of *both* the initial and the final nuclei are integral (cf. item 6 of section 1). If only the electron with $s = \frac{1}{2}$ is emitted, it would be impossible for the final total angular momentum of the system to be equal to the initial integral total angular momentum, because the sum of a half-integral angular momentum (the electron) plus an integral angular momentum (the final nucleus) can only be half-integral. However, if a neutrino with $s = \frac{1}{2}$ is also emitted, there are two half-integral angular momenta which can combine with the integral angular momentum to form an integral total angular momentum equal to the initial value. The fourth property of neutrinos was postulated by Pauli to agree with the observation that the end point $T_e^{\max}$ of the electron kinetic energy spectrum is equal to the available energy E. When an electron happens to be emitted at the end point, it carries away all the available energy, and none is left for kinetic or rest mass energy of the neutrino.

In general, however, the electron will be emitted with kinetic energy T_e less than the available energy E, and the neutrino will be emitted with non-zero kinetic energy T_ν, such that

$$T_e + T_\nu = E = T_e^{\max} \tag{16–51}$$

With Pauli's postulate, there are three particles in the final state of the emission process, very little of the available energy is taken by the recoil of the massive nucleus, and therefore essentially all of it is divided between the electron and neutrino. As there are an infinite number of ways in which this division can be made, it is evident that the postulate allows the distribution of electron kinetic energies T_e to form a continuous spectrum.

The postulate also forms the basis of the theory of the shape of this spectrum. In its most elementary form due to Fermi (1934), the theory is quite simple. Consider a process in which a system makes a transition from an initial state, consisting of the nucleus Z,A, to a final state, consisting of the nucleus $Z + 1,A$ plus an electron moving off with momentum $\mathbf{p}_e$ and a neutrino moving off with momentum $\mathbf{p}_\nu$. We know that electrons do not exist inside nuclei, and there is no reason to believe that neutrinos do either. Therefore we must imagine the electron and the neutrino to be created in the transition—just as a quantum is created in an optical transition of an atom. Then the eigenfunction for the initial state is

$$\psi_i = \psi_{Z,A} \tag{16–52}$$

the eigenfunction describing the nucleus Z,A, and the eigenfunction for the final state is

$$\psi_f = \psi_{Z+1,A}\psi_e\psi_\nu \tag{16–53}$$

where we assume that in the final state there are no interactions between the nucleus, the electron, and the neutrino, so that the eigenfunction for the system may be written as a product of the eigenfunction $\psi_{Z+1,A}$ for the nucleus $Z + 1,A$, the eigenfunction ψ_e for the electron, and the eigenfunction ψ_ν for the neutrino. If there are no interactions, the eigenfunctions for the electron and the neutrino are free particle plane waves of the form (15–23). That is,

$$\psi_e = L^{-3/2}e^{i\mathbf{K}_e\cdot\mathbf{r}} \tag{16–54}$$

where

$$\mathbf{K}_e = \mathbf{p}_e/\hbar \tag{16–54$'$}$$

and

$$\psi_\nu = L^{-3/2}e^{i\mathbf{K}_\nu\cdot\mathbf{r}} \tag{16–55}$$

where

$$\mathbf{K}_\nu = \mathbf{p}_\nu/\hbar \tag{16–55$'$}$$

and where $\mathbf{r}$ is the position vector from the origin of coordinates to the center of the nuclei Z,A and $Z + 1,A$.† These eigenfunctions are normalized, using periodic boundary conditions, in a box of edge length L. We next use the three dimensional extension of Golden Rule No. 2 (equation 9–45) to calculate the rate at which the system will make transitions from the initial state to the final states. This says that the transition rate is approximately

$$\frac{2\pi}{\hbar} v_{fi}^* v_{fi} \rho_f \tag{16–56}$$

† We assume the nucleus Z,A to be stationary, and ignore the recoil of the nucleus $Z + 1,A$.

The quantity ρ_f is the energy density of final states of the system, and the quantity v_{fi} is the matrix element of the interaction potential V, the perturbation which causes the transitions, taken between the initial and final states. That is,

$$v_{fi} \equiv \int \psi_f^* V \psi_i \, d\tau \tag{16–56$'$}$$

As we know nothing about this interaction potential, we are at liberty to make the simplest possible assumption concerning its form. Following Fermi, we assume that V is a universal constant g, which is called the *beta decay coupling constant*. Then

$$v_{fi} = g \int \psi_f^* \psi_i \, d\tau = gL^{-3} \int \psi_{Z+1,A}^* e^{-i\mathbf{K}_e \cdot \mathbf{r}} e^{-i\mathbf{K}_\nu \cdot \mathbf{r}} \psi_{Z,A} \, d\tau \tag{16–57}$$

There will be contributions to the integral only from values of r of the order of the nuclear radii, since for larger values both $\psi_{Z+1,A}$ and $\psi_{Z,A}$ obviously vanish. But for electron emission processes, except those of unusually high energy, a quick calculation will show that K_e and K_ν are both always small enough that $\mathbf{K}_e \cdot \mathbf{r} \lessapprox 0.1$ and $\mathbf{K}_\nu \cdot \mathbf{r} \lessapprox 0.1$, for r of the order of or smaller than the nuclear radii. To a fairly good approximation, we may therefore set both exponentials in equation (16–57) equal to unity and obtain

$$v_{fi} \simeq gL^{-3} \int \psi_{Z+1,A}^* \psi_{Z,A} \, d\tau$$

or

$$v_{fi} \simeq gL^{-3} M \tag{16–58}$$

where the quantity

$$M \equiv \int \psi_{Z+1,A}^* \psi_{Z,A} \, d\tau \tag{16–58$'$}$$

is called the *nuclear matrix element*. Considering (16–56), we see that, since M is independent of the electron and neutrino momenta, these quantities enter into the transition rate, or into the shape of the electron spectrum, only through the density of final states ρ_f.

As the transitions are assumed to lead to final states of the system in which the stationary nucleus $Z + 1,A$ is in a single state (usually the ground state), ρ_f contains only the density of states of the electron and the neutrino. In fact, ρ_f is essentially the product of the densities of states of these particles. According to our discussion of section 3, Chapter 15, both particles do have finite densities of states because of the restrictions imposed by the periodic boundary conditions used to normalize the eigenfunctions (16–54) and (16–55). As in that discussion, we may immediately take over the results (12–37) of the calculation of the density of states for an electron of kinetic energy T_e in a box of length L. In our present notation, this is

$$N(T_e)\,dT_e = \frac{m^{3/2}L^3T_e^{1/2}\,dT_e}{2^{1/2}\pi^2\hbar^3} \tag{16–59}$$

Let us write (16–59) in terms of the electron momentum p_e, by using the equation

$$\frac{p_e^2}{2m} = T_e \tag{16–60}$$

Differentiation yields

$$\frac{p_e\,dp_e}{m} = dT_e$$

Multiplying this by the square root of (16–60), we obtain

$$\frac{p_e^2\,dp_e}{2^{1/2}m^{3/2}} = T_e^{1/2}\,dT_e$$

which allows us to write (16–59) as

$$N(T_e)\,dT_e = \frac{L^3p_e^2\,dp_e}{2\pi^2\hbar^3} \tag{16–61}$$

It can be shown that (16–61), which involves only the momentum p_e, is a relativistically correct expression, even though (16–59) comes from the non-relativistic Schroedinger equation and (16–60), which is non-relativistic.† Consequently, equation (16–61) may be applied to the relativistic electrons, and also to the relativistic neutrinos, emitted in the transitions we are dealing with. For the latter particles the equation reads

$$N(T_\nu)\,dT_\nu = \frac{L^3p_\nu^2\,dp_\nu}{2\pi^2\hbar^3} \tag{16–62}$$

Now equation (16–51) shows that in a given transition T_ν is determined by the value of T_e. In fact, by differentiating that equation we obtain

$$dT_\nu = -dT_e \tag{16–63}$$

Therefore T_e suffices to specify the final states of the transition. The number of final states, in which this electron kinetic energy lies between T_e and $T_e + dT_e$, is

$$d^2N = -N(T_\nu)\,dT_\nu\,N(T_e)\,dT_e = -\frac{L^6p_\nu^2\,dp_\nu\,p_e^2\,dp_e}{4\pi^4\hbar^6}$$

† The reason is that the periodic boundary condition restrictions which lead to the finite density of states are essentially restrictions on the allowed values of the de Broglie wavelength or, through the relativistically as well as classically valid expression $p = h/\lambda$, are restrictions on the allowed values of the momentum.

where the minus sign compensates for the fact that dT_ν and dT_e are intrinsically of the opposite sign. The number of these states per unit electron kinetic energy is

$$\frac{d^2N}{dT_e} = -\frac{L^6 p_\nu^2\, dp_\nu\, p_e^2\, dp_e}{4\pi^4\hbar^6\, dT_e} \tag{16-64}$$

To simplify this we note that for the zero rest mass neutrino the kinetic energy is the same as the total relativistic energy and, as in (3–10),

$$cp_\nu = T_\nu \tag{16-65}$$

Differentiating, we obtain

$$c\, dp_\nu = dT_\nu = -dT_e \tag{16-66}$$

where we have used equation (16–63). Substituting this in (16–64), we have

$$\frac{d^2N}{dT_e} = \frac{L^6 p_\nu^2 p_e^2}{4\pi^4\hbar^6 c}\, dp_e \tag{16-67}$$

The quantity p_ν^2 may be eliminated by using equations (16–65) and (16–51), which give

$$p_\nu^2 = \frac{T_\nu^2}{c^2} = \frac{(T_e^{\max} - T_e)^2}{c^2}$$

Thus equation (16–67) may be written

$$\frac{d^2N}{dT_e} = \frac{L^6(T_e^{\max} - T_e)^2}{4\pi^4\hbar^6 c^3}\, p_e^2\, dp_e \tag{16-68}$$

Taking (16–68) as the density of final states for which the electron momentum is between p_e and $p_e + dp_e$, and using (16–58), we have an expression for $R(p_e)\, dp_e$, the rate of transitions into these states. This is

$$R(p_e)\, dp_e \simeq \frac{g^2 M^* M}{2\pi^3\hbar^7 c^3}(T_e^{\max} - T_e)^2 p_e^2\, dp_e \tag{16-69}$$

The function $R(p_e)$ is the momentum spectrum of the electrons emitted in the transitions. These results appear in a convenient form since most of the experiments yield the momentum spectrum directly. They predict that a plot of the quantity $[R(p_e)/p_e^2]^{1/2}$ versus $(T_e^{\max} - T_e)$, or simply versus T_e, should yield a straight line. Figure (16–41) shows such a *Kurie plot* for the simplest of all electron emission processes,

$$^0n^1 \rightarrow {}^1\mathrm{H}^1 + e^- + \nu \tag{16-70}$$

the decay of the neutron $^0n^1$ into a proton $^1\mathrm{H}^1$ plus an electron e^- and a neutrino ν. The neutron decays this way because $[M_{0,1} - M_{1,1}]c^2 = +0.78$ Mev. The lifetime τ of the decay is about 1100 sec.

These data provide a fairly typical example of the kind of agreement obtained between the simple Fermi theory and experiment for the beta decay of low Z nuclei. The deviations at the low energy end of the spectrum are often found to be a result of experimental difficulties such as self-absorption of the low energy electrons in the source. However, for high Z nuclei there are definite deviations between experiment and the simple theory at the low energy end of the spectrum which are more pronounced,

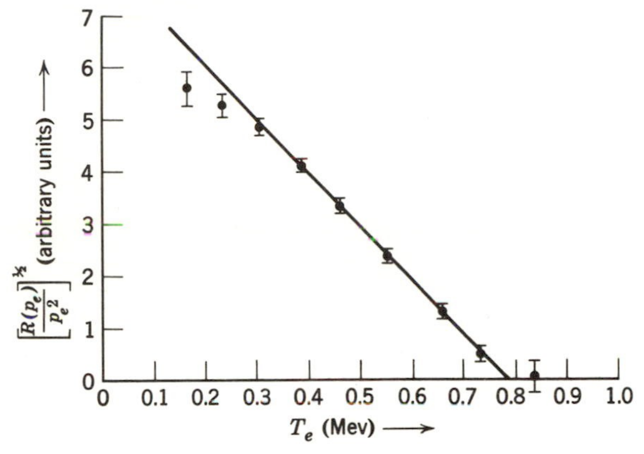

Figure 16–41. A Kurie plot for the beta decay of the neutron. From J. M. Robson, *Phys. Rev.*, **83,** 349 (1951).

and are of the opposite sense, compared to those of figure (16–41). These deviations for high Z nuclei arise from the use of plane waves (equation 16–54) for the electron eigenfunctions. This is an excellent approximation for the neutrino eigenfunctions because neutrinos interact so weakly that they really do behave like free particles. But electrons interact rather strongly with the Coulomb field of the final nucleus. This attractive interaction tends to make ψ_e larger than a plane wave in the region near the nucleus and thus increases the transition rate. As the Coulomb attraction is more important for small T_e, the effect is to enhance the low energy part of the spectrum. The simple Fermi theory also applies directly to positron emission, but the Coulomb effect has a different sign because the interaction is repulsive. This makes the positron eigenfunction smaller than a plane wave near the nucleus and leads to a depletion of the low energy part of the spectrum. When corrections for these Coulomb effects are included in the theory, its predictions are usually in good agreement with accurate experiments, even for electrons or positrons of quite low energy.

Of course, the question of the shape of the spectrum does not arise in electron capture. In this process there are only two particles in the final state, the nucleus $Z - 1, A$ and the neutrino. The latter particle always carries away essentially all the available energy E since the nucleus is so massive. Although only a neutrino is emitted in the primary process of electron capture, the process can be observed experimentally by detecting the X-rays which are subsequently emitted in the filling of the hole left in the electronic system of the atom.

By integrating (16–69) over the entire spectrum, the transition rate into all final states may be calculated. As in equation (13–118), this total probability per second that the initial nucleus will decay is equal to the reciprocal of its lifetime τ. Thus

$$\frac{1}{\tau} \simeq \frac{g^2 M^* M}{2\pi^3 \hbar^7 c^3} \int_0^{p_e^{\max}} (T_e^{\max} - T_e)^2 p_e^2 \, dp_e \tag{16–71}$$

To convert the integrand completely to a function of p_e, we write

$$\begin{aligned}(T_e^{\max} - T_e)^2 &= [T_e^{\max} + mc^2 - (T_e + mc^2)]^2 \\ &= (E_e^{\max} - E_e)^2\end{aligned}$$

where mc^2 is the rest mass energy and E_e is the total relativistic energy of the electron, and where we have used equations (1–22′) and (1–24). But, according to (1–25),

$$E_e = \sqrt{c^2 p_e^2 + m^2 c^4}$$

So

$$\frac{1}{\tau} \simeq \frac{g^2 M^* M}{2\pi^3 \hbar^7 c^3} \int_0^{p_e^{\max}} \left(\sqrt{c^2 p_e^{\max 2} + m^2 c^4} - \sqrt{c^2 p_e^2 + m^2 c^4}\right)^2 p_e^2 \, dp_e^2$$

Integrating this gives

$$\frac{1}{\tau} \simeq \frac{g^2 M^* M m^5 c^4}{2\pi^3 \hbar^7} F(\alpha) \tag{16–72}$$

where

$$F(\alpha) = -\tfrac{1}{4}\alpha - \tfrac{1}{12}\alpha^3 + \tfrac{1}{30}\alpha^5 + \tfrac{1}{4}\sqrt{1 + \alpha^2} \sinh^{-1} \sqrt{1 + \alpha^2} \tag{16–72′}$$

with

$$\alpha \equiv p_e^{\max}/mc$$

The function $F(\alpha)$ has the behavior

$$F(\alpha) \simeq \begin{matrix} \tfrac{1}{30}\alpha^5, & \alpha \gg 5 \\ \tfrac{2}{105}\alpha^7, & \alpha \ll 0.5 \end{matrix} \tag{16–72″}$$

and an intermediate behavior between these limits.

When the Coulomb corrections are included, $F(\alpha)$ is replaced by a function $F(\alpha, Z)$. An analytic expression for $F(\alpha, Z)$ cannot be given, but graphs and tables can be found in several references. For small Z, $F(\alpha, Z) \simeq F(\alpha)$. For $Z = 100$, $F(\alpha, Z) \simeq 100F(\alpha)$ in the case of electron emission, and $F(\alpha, Z) \simeq 0.1F(\alpha)$ in the case of positron emission, when α has a value corresponding to the typical available beta decay energy $E = 1$ Mev.

Equation (16–72) shows that the lifetime τ of a beta decaying nucleus depends strongly on $p_e^{\max}$, and therefore strongly on the available energy E. The observed lifetimes for the naturally occurring beta unstable nuclei range from values of the order of 1 sec for an available energy of several Mev to values of the order of 10^7 sec for available energies of several hundredths of an Mev. (Note, however, that the lifetime of an alpha decaying nucleus is an even stronger function of the energy available in the alpha decay.)

Equation (16–72) also shows that the quantity

$$F(\alpha, Z)\tau \simeq \frac{2\pi^3\hbar^7}{g^2m^5c^4}\frac{1}{M^*M} \tag{16–73}$$

depends only on a collection of universal constants, and on the value of the nuclear matrix element

$$M \equiv \int \psi^*_{Z\pm1,A}\psi_{Z,A}\, d\tau \tag{16–73$'$}$$

where we now indicate that the theory applies equally well to electron or positron emission. The product $F\tau$ is inversely proportional to M^*M and is smallest when it takes on its maximum possible value $M^*M = 1$. This occurs when the eigenfunction $\psi_{Z\pm1,A}$ is identical with the eigenfunction $\psi_{Z,A}$ because, in these circumstances, the normalization condition for eigenfunctions requires that (16–73′) yield $M = 1$. If the eigenfunctions $\psi_{Z\pm1,A}$ and $\psi_{Z,A}$ are not identical, $M^*M < 1$, and this quantity becomes smaller as the eigenfunctions become less similar. In fact, M, and therefore M^*M, are exactly zero if $\psi_{Z\pm1,A}$ and $\psi_{Z,A}$ are so dissimilar as to correspond to different values of nuclear spin i, or to different nuclear parities. The first restriction arises because the eigenfunctions $\psi_{Z\pm1,A}$ and $\psi_{Z,A}$, like any other eigenfunctions, are orthogonal with respect to all their quantum numbers, including the quantum number i. The second restriction arises because, if the parities of these two eigenfunctions differ, the entire integrand will be of odd parity,† and the contribution to the integral (16–73′) from each point (x, y, z) will be canceled by the contribution

† That is, (even)(odd) = (odd), and (odd)(even) = (odd).

from the point $(-x, -y, -z)$. These two restrictions constitute the *Fermi selection rules*:

$$\Delta i = 0$$

$$\text{The nuclear parity must not change.} \qquad (16\text{–}74)$$

that give conditions on the relative spins and parities of the states of the nuclei Z, A and $Z \pm 1, A$ which, if violated, cause the nuclear matrix element M to vanish identically. Transitions which satisfy the selection rules are called *allowed transitions.*

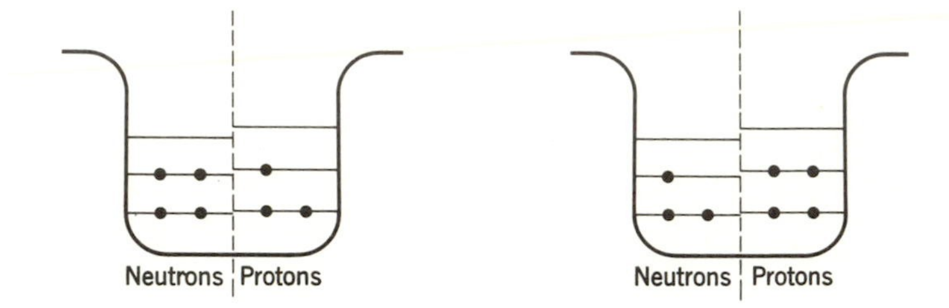

Figure 16–42. A schematic representation of the ground states of a typical set of mirror nuclei.

It immediately follows from the shell model that beta transitions between the ground states of any set of mirror nuclei are always allowed transitions. As mentioned in section 3, the two nuclei of a set of mirror nuclei consist of an odd number of protons and the same odd number of neutrons, plus an extra proton for one of the nuclei and an extra neutron for the other. The neutron and proton eigenstates filled by the nucleons of a typical set of mirror nuclei are indicated in figure (16–42). In the shell model, the nuclear spin is equal to the total angular momentum of the single odd nucleon and, since this nucleon is in the same eigenstate in each nucleus of the set, the spins of these two nuclei will be the same. Also, in the shell model the complete eigenfunctions $\psi_{Z\pm1,A}$ and $\psi_{Z,A}$ can be written as products of the eigenfunctions for each of the A independently moving nucleons, and the nuclear parity will be the product of the parities of each of these eigenfunctions. Since there are two nucleons in each eigenstate, except for the eigenstate containing the single odd nucleon, and since the product of the parities of two eigenfunctions of the same parity is always even,† the nuclear parity is equal to the parity of the eigenfunction of the single odd nucleon. As this nucleon is in the same eigenstate in each nucleus of the set, it follows that the parities of these two nuclei are the same. These arguments lead us not only to the conclusion that the

† That is, (even)(even) = (even), and (odd)(odd) = (even).

complete eigenfunctions $\psi_{Z\pm1,A}$ and $\psi_{Z,A}$ for a set of mirror nuclei are similar enough to satisfy the selection rules (16–74), but also to the much stronger conclusion that these two complete eigenfunctions should be essentially identical. They differ only in that the eigenfunction for one nucleon (the single odd nucleon) is a proton eigenfunction for one of the nuclei and a neutron eigenfunction for the other and, since these eigenfunctions are for the same eigenstate, they are identical except for Coulomb effects. But we have seen that, if $\psi_{Z\pm1,A} \simeq \psi_{Z,A}$, then $M^*M \simeq 1$. Thus we conclude that for all beta transitions between the ground states of sets of mirror nuclei the product $F\tau$ should assume its smallest possible value. This conclusion is verified by the experiments, which show that for all these transitions the $F\tau$ values are clustered around $F\tau \simeq 10^{3.5}$ sec, and that the $F\tau$ values for these transitions are the smallest which are ever observed. By using these experimental values of $F\tau$ in (16–73), and making the assumption that $M^*M = 1$ for these cases, the beta decay coupling constant can be evaluated. The result is

$$g \simeq 1.4 \times 10^{-49} \text{ erg-cm}^3 \tag{16–75}$$

Of course, there are many allowed transitions which do not take place between the ground states of mirror nuclei, since there are many other possibilities of satisfying the selection rules (16–74). Although M^*M is greater than zero for all these transitions, it is less than unity because the differences between $\psi_{Z\pm1,A}$ and $\psi_{Z,A}$ lead to cancellations which reduce the value of the integral (16–73′). With increasing A, these differences should become greater because the nuclei become larger and more complicated. This is seen in the experiments since the measured $F\tau$ values for these allowed transitions increase with increasing A. The largest $F\tau$ values for allowed transitions are around $F\tau \simeq 10^{5.5}$ sec.

If either part of the selection rules (16–74) is violated, $M^*M \equiv 0$. However, for such *forbidden transitions* it is observed that $F\tau$ does not become infinitely large, as predicted by equation (16–73), but increases only by about a factor of 10^3. This simply reflects the fact that (16–73) is not exact because it is based upon several approximations. One of these is the setting of $e^{-i\mathbf{K}_e\cdot\mathbf{r}}e^{-i\mathbf{K}_\nu\cdot\mathbf{r}} \equiv e^{-i\mathbf{K}\cdot\mathbf{r}}$ equal to unity in (16–57). A correction to this approximation is found by using the more accurate expression of (16–57) obtained by expanding $e^{-i\mathbf{K}\cdot\mathbf{r}}$ in a power series and taking the first two terms. This gives for the beta decay interaction potential matrix element

$$v_{fi} \simeq gL^{-3}\left[\int \psi^*_{Z\pm1,A}\psi_{Z,A}\,d\tau - i\mathbf{K}\cdot\int \psi^*_{Z\pm1,A}\mathbf{r}\psi_{Z,A}\,d\tau\right] \tag{16–76}$$

Situations can obviously arise in which the first term vanishes but the second does not. As $\mathbf{K} \cdot \int \psi^*_{Z\pm1,A} \mathbf{r} \psi_{Z,A}\, d\tau \leqslant \mathbf{K} \cdot \mathbf{r} \lessapprox 10^{-1}$ for beta decays of normal energy, this second term makes it possible to have forbidden transitions as long as the $F\tau$ values are larger than those for allowed transitions by a factor $\gtrsim (10^{-1})^{-2} = 10^2$. A similar possibility arises from a term of order $v/c \lessapprox 10^{-1}$, which should actually appear in the beta decay interaction potential matrix element because of relativistic effects.

The situation is further complicated by the experimental observation that there are certain beta transitions that violate the Fermi selection rules (16–74) but nevertheless have $F\tau$ values as small as for any of the transitions that are allowed by these rules. This implies that in these transitions the beta decay interaction potential V in (16–56′) is not a constant, as Fermi initially assumed, but is of a more complicated type that leads to an interaction potential matrix element which gives different selection rules. Theoretical investigations show that there are several types of matrix elements which should be possible, and that certain of them lead to the Fermi selection rules, while others lead to the *Gamow-Teller selection rules:*

$$\Delta i = 0, \pm 1 \qquad (\text{but not } i = 0 \rightarrow i = 0)$$
$$\text{The nuclear parity must not change} \tag{16–77}$$

By evaluating the consequences of the various possible types of matrix elements, and comparing these consequences with experiment, it is found that the beta decay interaction potential matrix element actually existing in nature contains a mixture of approximately equal amounts of the types leading to the Fermi selection rules and the types leading to the Gamow-Teller selection rules. These comparisons are not based on the shape of the beta decay spectrum, because this is always dominated by the same density of states factor for all types of matrix elements. We shall not go into the theory underlying these comparisons because these points are really intelligible only in the language of Dirac's relativistic theory of quantum mechanics. Similar motivation leads us to avoid making a distinction between *neutrinos* and *anti-neutrinos*.

However, we shall consider one recently discovered property of the beta decay interaction potential matrix element that is extremely interesting and not difficult to describe. It is found that this matrix element is not a *scalar* but instead is at least partly a *pseudo-scalar*. The distinction is that a scalar does not change when the parity operation is performed—that is, when the coordinates of each point (x, y, z) are changed to $(-x, -y, -z)$ —whereas a pseudo-scalar changes sign in this operation. An example of a scalar is the Coulomb interaction potential, or any other interaction potential that can be written $V(r)$. As the argument r is the magnitude of a vector $\mathbf{r}$, $V(r)$ does not change in the parity operation of reflecting $\mathbf{r}$

through the origin, even though $\mathbf{r} \to -\mathbf{r}$ in this operation. Since the matrix element of $V(r)$ is a certain average of this quantity, the matrix element is a scalar. In fact, with the exception of the beta decay interaction potential matrix element, the matrix elements of all the known interaction potentials of classical, atomic, and nuclear physics are scalars. For instance, consider the spin-orbit interaction potential matrix element, which is proportional to $\mathbf{S} \cdot \mathbf{L}$. We investigate what happens to $\mathbf{S} \cdot \mathbf{L}$ in the parity operation by writing $\mathbf{L} = \mathbf{r} \times \mathbf{p}$ and realizing that in this operation both $\mathbf{r}$ and $\mathbf{p}$ change sign.† Thus $\mathbf{L}$ does not change sign. To bring out the distinction between their behavior in the parity operation, vectors which change sign, such as $\mathbf{r}$ and $\mathbf{p}$, are called *polar vectors*, and vectors which do not change sign, such as $\mathbf{L}$, are called *axial vectors*. All angular momentum vectors, including $\mathbf{S}$, are axial vectors. Therefore we see that the spin-orbit interaction potential matrix element is a scalar because $\mathbf{S} \cdot \mathbf{L}$ does not change sign in the parity operation. An example of a pseudo-scalar is the quantity $\mathbf{P} \cdot \mathbf{A}$, where $\mathbf{P}$ is a polar vector and $\mathbf{A}$ is an axial vector. This is a pseudo-scalar since it changes sign in the parity operation because $\mathbf{P}$ changes sign and $\mathbf{A}$ does not. The beta decay interaction potential matrix element contains quantities of this type. As the parity operation is equivalent to leaving the point (x, y, z) fixed and reflecting the coordinate axes through the origin, the operation amounts to changing from a right-handed to a left-handed coordinate system and, as at least part of the beta decay interaction potential matrix element changes sign in this operation, the description of the beta decay processes which this matrix element produces will depend on whether a right-handed or a left-handed coordinate system is used. When this was first discovered, it was extremely surprising because it had been a fundamental assumption that the description of *all* physical processes must be independent of whether a right-handed or left-handed coordinate system is used.

This assumption was proved incorrect for beta decay in experiments performed by Wu and collaborators (1957), at the suggestion of Lee and Yang (1956). In these experiments, which we describe first in a right-handed coordinate system, the spin angular momentum vectors $\mathbf{I}$ of electron emitting ${}^{27}Co^{60}$ nuclei were aligned with respect to an external magnetic field by using the hyperfine structure interaction at temperatures so low that the thermal equilibrium energy kT is comparable to the hyperfine splitting energy. In these circumstances, a majority of atoms will be in the lowest energy state, and in this state $\mathbf{I}$ is essentially aligned with respect to the applied magnetic field. The angle between the direction of $\mathbf{I}$ and the direction of the momentum vector $\mathbf{p}$ was then measured for

† $\mathbf{p}$ changes sign since $m\,dx/dt \to m\,d(-x)/dt = -m\,dx/dt$, and similarly for the other momentum components.

each emitted electron, and the distribution of these angles was plotted. It was found that this distribution is asymmetric, with about a 30 percent preference for the emission of electrons whose momentum vectors $\mathbf{p}$ are in the opposite hemisphere from the spin vector $\mathbf{I}$. So in the average emission process $\mathbf{p} \cdot \mathbf{I} < 0$. But $\mathbf{p} \cdot \mathbf{I}$ is a pseudo-scalar since $\mathbf{p}$ is a polar vector and $\mathbf{I}$ is an axial vector. Thus in any left-handed coordinate system the sign of this pseudo-scalar will change, in the average emission process $\mathbf{p} \cdot \mathbf{I} > 0$, and the description of this process therefore depends on whether a right-handed or a left-handed coordinate system is used. This proves the beta decay interaction potential matrix element must be a pseudo-scalar, or at least contain a pseudo-scalar part. If it were purely scalar, the electron distribution would necessarily be symmetric so that in the average emission process $\mathbf{p} \cdot \mathbf{I} = 0$, and the description of this process would be independent of whether a right-handed or a left-handed coordinate system is used, even though the pseudo-scalar $\mathbf{p} \cdot \mathbf{I}$ changes sign, because $-0 = +0$.

Since the beta decay interaction potential matrix element is not a scalar, it is not true that for the interaction potential $V(-x, -y, -z) = V(x, y, z)$. Thus the eigenfunctions for the entire system do not necessarily have a definite parity (cf. footnote on page 465), and it is not necessary that the entire system conserve parity. Experiments indicate that it does not. The parity of the initial state consisting of the nucleus Z, A is not the same as the parity of the final state consisting of the nucleus $Z \pm 1$, A plus the emitted neutrino and electron or positron.

We close this section with brief descriptions of some other recent experiments involving beta decay and neutrinos that are of particular interest. In the period since 1950 a number of measurements have been made by Allen, Sherwin, and others on the recoil velocity of the nucleus remaining in the final state of a beta decay process. As the nucleus is very massive compared to the electron and neutrino, the recoil velocity is low enough that it can be measured by measuring electronically the time (several microseconds) required for the nucleus to travel a given distance (several centimeters) in a given direction through a field free vacuum. This measurement determines the momentum vector of the recoiling nucleus. By also measuring the momentum vector of the emitted electron, it is possible to show that linear momentum can be conserved only if a neutrino is emitted in the process. Furthermore, the required neutrino momentum agrees within the experimental accuracy (10 percent) with the neutrino energy required to conserve energy, if the neutrino rest mass is zero as assumed by Pauli.

Even after these recoil experiments had been performed, there was still a strong feeling in some quarters that the neutrino is not a particle but only a myth cleverly designed to provide for the conservation of energy, angular momentum, and linear momentum. There was a certain merit in this point of view as long as the neutrino remained undetectable. However, experiments by Cowan and Reines in 1956 succeeded in detecting the neutrino, and thereby gave it the status of any other particle. In these extremely difficult experiments they observed the reaction

$$\nu + {}^{1}\mathrm{H}^{1} \rightarrow {}^{0}n^{1} + e^{+} \qquad (16\text{–}78)$$

the capture of a neutrino ν by a proton ${}^{1}\mathrm{H}^{1}$ leading to a neutron ${}^{0}n^{1}$ and a positron e^{+}. This is the inverse of the reaction

$${}^{0}n^{1} + e^{+} \rightarrow {}^{1}\mathrm{H}^{1} + \nu \qquad (16\text{–}79)$$

which is an alternative form of the neutron decay process (16–70). A point which will be of interest shortly is that the reaction (16–79) should have about the same lifetime, $\tau \sim 10^{3}$ sec, as the reaction (16–70) because the matrix elements and densities of states governing the lifetimes contain essentially the same factors, albeit in a different order. The Cowan-Reines reaction (16–78) took place on the protons contained in a very large tank of hydrogenous scintillator exposed to the enormous flux of neutrinos emitted from the fission induced beta decays in a nuclear reactor, and the positrons were detected by the scintillations they produced in the same tank. Elaborate methods were used to minimize background scintillations. This was necessary because less than one reaction per minute was expected, despite the magnitude of the neutrino flux and the size of the hydrogenous target, since the interaction of neutrinos with matter is extremely weak. To show this, we make a rough estimate of the cross section for the reaction (16–78) as follows. Consider a neutrino of the typical energy 1 Mev incident upon a proton, and assume that there is a possibility of interaction between these two particles whenever they are close enough to allow appreciable overlap of their wave functions. This will occur when their relative separation is comparable to the de Broglie wavelength of the neutrino $\sim 10^{-10}$ cm. As their relative velocity is $\sim 10^{10}$ cm-sec^{-1}, the maximum time during which there can be overlap is $\sim 10^{-20}$ sec. Now assume that the lifetime for the reaction (16–78) is about the same as the lifetime $\sim 10^{3}$ sec for its inverse reaction (16–79); this assumption is justified by the principle of microscopic reversibility mentioned after equation (13–84). Then the probability that the reaction (16–78) will occur in the time $\sim 10^{-20}$ sec is $\sim 10^{-20}/10^{3} = 10^{-23}$, and the cross section σ for this interaction is the cross section $\sim(10^{-10}\ \mathrm{cm})^{2} = 10^{-20}\ \mathrm{cm}^{2}$ for collisions with overlapping wave functions times the probability just evaluated. This gives $\sigma \sim 10^{-20}\ \mathrm{cm}^{2} \times 10^{-23} = 10^{-43}\ \mathrm{cm}^{2}$. An accurate calculation, based on the method we used in evaluating the lifetimes of the processes leading to neutrino emission, gives

$$\sigma \simeq \frac{2g^2M^*M}{\pi\hbar^4 c}\frac{p_e^2}{v_e} \qquad (16\text{–}80)$$

For a positron of momentum p_e and velocity v_e corresponding to the typical kinetic energy 1 Mev, (16–80) predicts an interaction cross section $\sigma \sim 10^{-44}$ cm^2. The measurements of Cowan and Reines agree with this value, within the large experimental uncertainty.

Note that equation (16–80) shows clearly that the extreme smallness of the neutrino interaction cross section is a result of the extreme smallness of the beta decay coupling constant $g \sim 10^{-49}$ erg-cm^3. The cross section is so small that a neutrino emitted in a nuclear reaction in the interior of the sun can traverse its entire bulk with little probability of being absorbed. Beta decay interactions are very weak compared to nuclear interactions.

A numerical comparison can be made by dividing the beta decay coupling constant, which has the dimensions of energy times volume, by the volume of a typical nucleus. This gives a characteristic energy $\sim 10^{-49}$ erg-cm^3/$(5 \times 10^{-13})^3$ cm^3 $\sim 10^{-12}$ ergs $\sim$ 1 ev, which is smaller by a factor of $\sim 10^{-7}$ than the energy $\sim$10 Mev that characterizes nuclear interactions. Since the square of the beta decay coupling constant enters into measurable quantities, such as transition rates and cross sections, it is appropriate to say that *beta decay interactions are weaker than nuclear interactions by a factor of* $\sim 10^{-14}$. The reader will recall that the comparison of figure (16–12) indicates that electromagnetic interactions are only somewhat weaker than nuclear interactions; in section 16 a quantitative comparison will show that *electromagnetic interactions are weaker than nuclear interactions by a factor of* $\sim 10^{-3}$. This conclusion, plus the conclusion drawn in the first footnote of section 3, shows that *gravitational interactions are weaker than nuclear interactions by a factor of* $\sim 10^{-38}$.

12. Gamma Decay and Internal Conversion

The nucleus is a system of nucleons bound to a certain region of space. Since this system seems to obey the laws of quantum mechanics, it should have a set of quantized energy states. Heretofore we have considered only the lowest energy ground state, but in this section we turn to the higher energy excited states. Much of what has been learned about the excited states has come from a study of the high energy quanta, called gamma rays, which are emitted in transitions between some initial excited state of energy E_i and a final state of energy $E_f < E_i$. These spontaneous *gamma decay* transitions are completely analogous to the optical transitions that occur between the quantized energy states of an atom. In both cases, the transitions take place through the interaction of the charge moments and magnetic moments of the system with electromagnetic fields, and energy is conserved because the energy of the emitted quantum is

$$h\nu = E_i - E_f \tag{16–81}$$

There are a number of processes through which the nucleus can be excited to one of its higher energy states. For instance, it can happen that an alpha decay or beta decay process leaves the final nucleus in an excited state instead of its ground state. The possibilities are indicated schematically, for beta decay, in the energy level diagram of figure (16–43). The energy change in the beta decay is, of course, the available beta decay energy E, equal to the total energy carried away by the neutrino and the electron or positron. Usually only a few states of the final nucleus are excited because beta decays to states of certain spins and parities are forbidden by the beta decay selection rules, and because the strong dependence of the beta decay transition rate on E makes the transition rate negligible to states higher than several Mev above the ground state. In excitation following alpha decay, the even stronger dependence of the alpha decay transition rate on the alpha decay energy usually limits the excitation of the final nucleus to states somewhat less than 1 Mev above the ground state.† States of energies up to 6 or 8 Mev above the ground state can be excited when this binding energy is liberated by the capture of a low energy nucleon in a nucleus. States of even higher energy can be excited by the capture of a higher energy nucleon, processes in which a high energy nucleon is inelastically scattered from a nucleus, and by a number of other processes involving nuclear reactions.

After the excitation of a nucleus, it can decay by the emission of gamma rays. If the excitation energy is less than the binding energy of the last nucleon in the nucleus, it must decay by gamma emission, or by the related process of *internal conversion* that will be explained later in this section.

† Note that, when alpha decay between two nuclei leads to excited states of the final nucleus, the alpha particle energy spectrum will contain a *fine structure* of components of lower energy. It is also possible for the alpha decay to originate from excited states of the initial nucleus, and in this case the spectrum will contain *long range* components of higher energy. The similar situations which occur in beta decay result in several values of the available beta decay energy E, and electron or positron energy spectra which are a superposition of spectra corresponding to each value of E. Note, then, that information about the energies of the excited states can be obtained directly from a study of the energies of the particles emitted in alpha or beta decay. For these processes, and particularly the latter, a study of the lifetimes can also give information about the spins and parities of the excited states.

For excitation energies higher than the binding energy of the last nucleon, decay by nucleon emission is possible and, for reasons to be explained in the next section, with increasing excitation energy nucleon emission rapidly becomes the dominant decay mode. As a consequence, most gamma rays are emitted with energies less than about 6 or 8 Mev. In fact, their energies are generally much less than this because the de-excitation often proceeds in a *cascade* through a succession of low energy states, instead of by the emission of a single gamma ray which carries away all

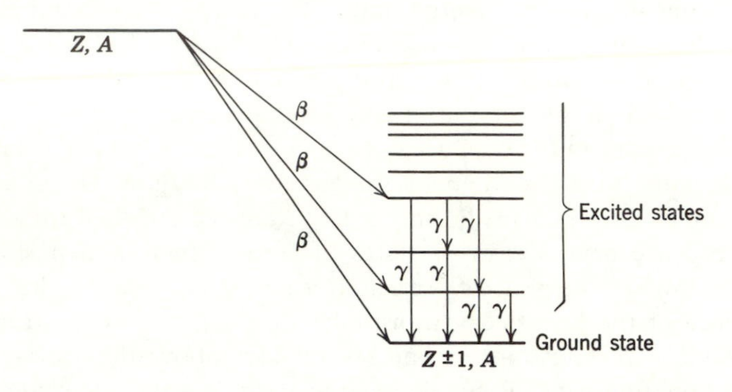

Figure 16–43. The excitation of a nucleus as a result of beta decay.

the excitation energy. The lower limit for the energy of these gamma rays is set by the minimum energy difference between the low-lying nuclear energy states. It is of the order of 10^3 ev.

Most of the methods for measuring the energies of gamma rays, or quanta in the energy range 10^3 ev to 6 or 8 Mev and higher, involve transferring their energy content to electrons by means of the photoelectric and Compton processes described in Chapter 14, or to electrons and positrons by means of the pair production process described in the same chapter. The energies of these charged particles can then be measured by standard techniques. In perhaps the most widely used method, the energy is transferred to electrons of the scintillating crystal NaI, and the electron energies are measured in the same crystal by measuring the intensity of the light pulse which it emits, due to scintillations produced by the electrons. For $h\nu \lesssim 0.3$ Mev, it is most probable that all the energy of a gamma ray will be transferred to a single electron of the I atoms by the photoelectric process. For 0.3 Mev $\lesssim h\nu \lesssim 7$ Mev, the Compton effect is the most probable, and only a fraction of the gamma ray energy is transferred to an electron in a single collision. In this range, it is necessary to use a large NaI crystal in order to absorb all the energy. For $h\nu \gtrsim 7$ Mev, the pair production process dominates, and the gamma ray energy is divided between the total energy of an electron and a positron. At even larger $h\nu$, the ranges of these particles are too long to allow their energies to be measured in an NaI crystal of convenient size and, instead, a device called a pair spectrometer is used. In this device the pair is produced in a foil of some high Z material, and the energies of the emitted electron and positron are measured by analyzing their momenta in a magnetic field. The methods just described have gamma ray energy resolutions of several percent at best, but other methods of different types and much better resolutions are also available. The one with the best resolution involves diffracting the gamma rays from a crystal lattice of known lattice spacing d. Although the gamma ray wavelengths $\lambda = c/\nu$ are as much as several orders of magnitude smaller than d, diffraction can be used by making the distances between the source, crystal lattice, and detector very large. This gives a large linear displacement even though the angular displacement $\theta \sim \lambda/d$ is small. However, as all the solid angles involved are very small, the diffraction method has low efficiency and can be used only in very favorable circumstances. The other high resolution methods also suffer from low efficiency.

The measured energy spectrum of gamma rays, emitted in transitions between the energy states of a nucleus, is used to determine the energies of these states—the connection being made by the same trial and error procedures that were used in the very early days of the study of atomic spectra and energy states. Of course, the energies of the nuclear states give important information about nuclear structure.

Another source of important information is the lifetimes of these states. If the lifetime $\tau \gtrsim 10^{-10}$ sec, it can be measured by electronically timing the delay between the instant the state was excited (e.g., by detecting the beta decay which led to the excitation) and the instant it decayed (by detecting the emitted gamma ray). In extreme cases, lifetimes as long as $\tau \sim 10^{8}$ sec are found. All gamma ray transitions with directly measurable lifetimes are called *isomeric transitions* to distinguish them from the more typical transitions, which have lifetimes $\tau \lesssim 10^{-10}$ sec that are not directly measurable. However, if $\tau \lesssim 10^{-16}$ sec, it can be measured indirectly by measuring the natural width of the gamma ray spectral line emitted in the transition, and then employing the relation (13–125). This technique can be used for nuclear transitions only with the aid of an ingenious method, called resonance absorption, that employs absorption in a target, containing some species of nuclei, as a spectrometer of extremely high resolution to measure the width of the spectral lines emitted by a source, containing nuclei of the same species. Gamma decay lifetimes as short as $\tau \sim 10^{-17}$ sec have been observed.

The tremendous variation which is found in the lifetimes for gamma decay is due, in part, to the dependence of τ on the energy of the emitted gamma ray. But an equally important part of this variation is due to the influence of the relative spins and parities of the initial and final nuclear states. Therefore analysis of the experimentally measured values of τ gives information about the spins and parities of the states. These quantities are just as important in characterizing the states as their energies. The analysis is based on the theory of gamma decay transitions in nuclei. As the theory is basically the same as the theory of optical transitions in atoms which we treated in section 12, Chapter 13, it would be appropriate for the reader to review that section.

In our treatment of optical transitions in atoms we considered only the *electric dipole* transitions, which arise from periodic oscillations in the electric dipole moment of the charge distribution, and we found that the electric dipole transition rate for spontaneous emission is proportional to the "square of the matrix element of the interaction potential for the associated moment,"† taken between the states involved in the transition. A completely rigorous treatment would also consider the *electric quadrupole*, *electric octupole*,‡ etc., transitions arising from periodic oscillations in the associated moments of the charge distribution, as well as the *magnetic dipole*, *magnetic quadrupole*, *magnetic octupole*,‡ etc., transitions arising from periodic oscillations in the associated moments of the current distribution (or of the magnetic moments produced by electron spin). In each case it would be found that the transition rate for spontaneous emission by a particular moment is proportional to the square of the matrix element of the interaction potential for that moment. We did not give the completely rigorous treatment because it is really not necessary for optical transitions.

The justification for neglecting all the magnetic optical transitions is that the current distribution is $\rho v/c$, where v is the characteristic velocity of the moving charges, and this is smaller than the charge distribution ρ by the factor v/c, which is typically $\sim 10^{-3}$. As a result, the magnetic moments are all smaller than the corresponding electric moments by this factor. (The magnetic moments produced by electron spin are of the same order as those produced by the current distribution.) Thus the square of the matrix element of the interaction potential, and therefore the

† We use this somewhat loose terminology as an abbreviation. The transition rate for spontaneous emission is proportional to the transition rate for absorption, and (13–77) shows that the latter quantity is actually proportional to a sum of terms each of which is the matrix element of the interaction (perturbation) potential for a component of the electric dipole moment times its own complex conjugate.

‡ Except for a special process involving the entire atom, there are no electric monopole transitions because nuclear charge conservation requires the associated moment to be constant. There are no magnetic monopole transitions because the associated moment is zero.

transition rate, is smaller for each of the magnetic moments than for the corresponding electric moment by the factor $(v/c)^2 \sim 10^{-6}$.

The justification for neglecting electric quadrupole, electric octupole, etc., optical transitions is that the matrix element of the interaction potential for each successively higher moment of the charge distribution is smaller than the matrix element for the preceding moment by the factor $2\pi R/\lambda$, where R is the radius of the charge distribution and λ is the wavelength of the radiation. This factor is also typically $\sim 10^{-3}$. The matrix element of the interaction potential for each successively higher moment of the current distribution decreases by the same factor. The factor arises because the interaction potential of a dipole moment in a spatially varying field involves products of the field times distances of the order of R, the interaction potential of a quadrupole moment involves products of first spatial derivatives of the field times distances of the order of R^2, the interaction potential of an octupole moment involves products of second spatial derivatives of the field times distances of the order of R^3, etc.† But,

† Equation (13–65) shows that this is true for the electric dipole moment interaction potential, and it is easy to verify for the higher electric moments. Let the external electric potential be $V(x, y, z)$ and the charge distribution be $\sum_j q_j$, where (x_j, y_j, z_j) are the coordinates of charge q_j, and where the sum is taken over all the charges. Then the total energy of the distribution is $\sum_j q_j V(x_j, y_j, z_j)$. Make a Taylor expansion of $V(x, y, z)$ about the origin (0, 0, 0) of the charge distribution. This gives

$$
\begin{aligned}
V(x, y, z) = [V_0] &+ \left[\left(\frac{\partial V}{\partial x}\right)_0 x + \left(\frac{\partial V}{\partial y}\right)_0 y + \left(\frac{\partial V}{\partial z}\right)_0 z\right] \\
&+ \left[\frac{1}{2}\left(\frac{\partial^2 V}{\partial x^2}\right)_0 x^2 + \frac{1}{2}\left(\frac{\partial^2 V}{\partial y^2}\right)_0 y^2 + \frac{1}{2}\left(\frac{\partial^2 V}{\partial z^2}\right)_0 z^2\right. \\
&+ \left.\left(\frac{\partial^2 V}{\partial x\,\partial y}\right)_0 xy + \left(\frac{\partial^2 V}{\partial x\,\partial z}\right)_0 xz + \left(\frac{\partial^2 V}{\partial y\,\partial z}\right)_0 yz\right] + \cdots
\end{aligned}
$$

Use this to evaluate the total energy. It is

$$
\begin{aligned}
\sum_j q_j V(x_j, y_j, z_j) = \left[V_0 \sum_j q_j\right] &+ \left[(-E_x)_0 \sum_j q_j x_j + (-E_y)_0 \sum_j q_j y_j + (-E_z)_0 \sum_j q_j z_j\right] \\
&+ \left[\frac{1}{2}\left(-\frac{\partial E_x}{\partial x}\right)_0 \sum_j q_j x_j^2 + \frac{1}{2}\left(-\frac{\partial E_y}{\partial y}\right)_0 \sum_j q_j y_j^2 + \frac{1}{2}\left(-\frac{\partial E_z}{\partial z}\right)_0 \sum_j q_j z_j^2\right. \\
&+ \left.\left(-\frac{\partial E_y}{\partial x}\right)_0 \sum_j q_j y_j x_j + \left(-\frac{\partial E_z}{\partial x}\right)_0 \sum_j q_j z_j x_j + \left(-\frac{\partial E_z}{\partial y}\right)_0 \sum_j q_j z_j y_j\right] + \cdots
\end{aligned}
$$

where we employ the relations $E_x = -\partial V/\partial x$, $E_y = -\partial V/\partial y$, $E_z = -\partial V/\partial z$ to introduce the components of the electric field. The first square bracket is the electric monopole interaction potential, and the sum it contains is the total charge; the second square bracket is the electric dipole interaction potential, and the three sums it contains are in a field which is a sinusoidal function of position with wavelength λ, such as $\sin(2\pi x/\lambda)$, each spatial derivative brings in a factor of $2\pi/\lambda$. Thus the interaction potential of each successively higher moment contains an extra factor of $2\pi R/\lambda$. The square of the matrix element, and thus the transition rate, is smaller for electric quadrupole transitions than for electric dipole transitions by the factor $(2\pi R/\lambda)^2 \sim 10^{-6}$, and the transition rates for the higher moments are smaller by successively higher even powers of this factor. All optical transitions in atoms, other than electric dipole transitions, have very small transition rates, or very long lifetimes. These transitions can be neglected because in the long lifetimes the atom will almost always be de-excited by collisions, or some other non-radiative process, before it gets a chance to radiate.

These simplifications do not obtain for gamma decay transitions in nuclei. For one thing, $(v/c)^2 \sim 10^{-2}$ and $(2\pi R/\lambda)^2 \sim 10^{-4}$, in the typical case $A = 200$ and $h\nu = 1$ Mev. Therefore the factor suppressing the rates for magnetic transitions is not so small, and neither is the factor suppressing the rates for transitions associated with the higher moments. Even more

---

the components of the electric dipole moment; the third square bracket is the electric quadrupole interaction potential, and the six sums it contains are the components of the electric quadrupole moment. This proves our point.

It is also interesting to develop the relation between the components of the electric quadrupole moment and the quantity Q defined by equation (16–38). In the case to which that definition applies, there is an axis of symmetry along the z axis, each of the last three components vanishes, and $\sum_j q_j x_j^2 = \sum_j q_j y_j^2$. Then the electric quadrupole interaction potential is

$$\frac{1}{2}\left[\left(\frac{\partial^2 V}{\partial x^2}\right)_0 + \left(\frac{\partial^2 V}{\partial y^2}\right)_0\right]\sum_j q_j x_j^2 + \frac{1}{2}\left(\frac{\partial^2 V}{\partial z^2}\right)_0 \sum_j q_j z_j^2$$

where we have changed back to the external electric potential V. But, for an external potential, Laplace's theorem of electrostatics demands

$$\frac{\partial^2 V}{\partial x^2} + \frac{\partial^2 V}{\partial y^2} + \frac{\partial^2 V}{\partial z^2} = 0$$

So the electric quadrupole interaction potential can be written

$$\frac{1}{2}\left(\frac{\partial^2 V}{\partial z^2}\right)_0 \left\{\sum_j q_j z_j^2 - \sum_j q_j x_j^2\right\} = \frac{1}{4}\left(\frac{\partial^2 V}{\partial z^2}\right)_0 Q$$

where

$$Q \equiv \sum_j q_j[2z_j^2 - 2x_j^2] = \sum_j q_j[2z_j^2 - x_j^2 - y_j^2]$$

$$= \sum_j q_j[3z_j^2 - (x_j^2 + y_j^2 + z_j^2)]$$

Equation (16–38) is obtained by letting the sum go to an integral and expressing the q_j in units of proton charges.

important is the fact that there are no collisions between nuclei, or other non-radiative processes (except internal conversion) if the excitation energy is less than the nucleon binding energy. Thus all states with excitation energy less than the nucleon binding energy must eventually decay by gamma ray emission (or internal conversion), even if the lifetime for decay is extremely long.

Although it is a task, which we shall avoid, even to write expressions for the matrix elements of the higher electric and magnetic moments, the real difficulty in evaluating these matrix elements arises from the lack of information concerning the nuclear eigenfunctions which they contain. This is the same difficulty that arises in evaluating the matrix elements for beta decay transitions. However, if some nuclear model is assumed, the eigenfunctions for that model can be found, the matrix elements can be evaluated, and predictions for the gamma decay transition rates for spontaneous emission can be obtained. The accuracy of these predictions will, of course, be only as good as the accuracy of the model. Using the shell model with a nucleus of radius R, Blatt and Weisskopf have obtained the following expressions for approximate values of the transition rates S for spontaneous emission in electric and magnetic transitions.

For electric transitions:

$$S \simeq S_0 \frac{4.4(L+1)}{L[1 \cdot 3 \cdot 5 \cdots (2L+1)]^2} \left(\frac{3}{L+3}\right)^2 \left(\frac{h\nu}{E_0}\right)^{2L+1} \left(\frac{R}{R_0}\right)^{2L} \tag{16–82}$$

For magnetic transitions:

$$S \simeq S_0 \frac{1.9(L+1)}{L[1 \cdot 3 \cdot 5 \cdots (2L+1)]^2} \left(\frac{3}{L+3}\right)^2 \left(\frac{h\nu}{E_0}\right)^{2L+1} \left(\frac{R}{R_0}\right)^{2L-2} \tag{16–82′}$$

where $L = 1$ for dipole transitions, $L = 2$ for quadrupole transitions, $L = 3$ for octupole transitions, etc., and where

$$S_0 = 10^{21}\ \text{sec}^{-1}$$
$$R_0 = 10^{-13}\ \text{cm}$$
$$E_0 = 197\ \text{Mev}$$

with E_0 the energy of a gamma ray for which $2\pi R_0/\lambda = 1$. Table (16–3) gives values of $\tau = 1/S$, in seconds, calculated from these equations for several values of $h\nu$, and for $R = 7 \times 10^{-13}$ cm corresponding to $R = 1.2A^{1/3} \times 10^{-13}$ cm with $A = 200$. If a state can decay only by the single gamma ray transition indicated, its lifetime is equal to this τ. Note that the relative magnitudes of S for electric and magnetic transitions with any common value of L, as well as the relative magnitudes of S for electric transitions with various values of L and the relative magnitudes of S for magnetic transitions with various values of L, agree with the implications of the previous paragraph. Also note that for each value of L the dependence of S on the energy $h\nu$ of the emitted gamma ray agrees with the implications of that paragraph since $\lambda = c/\nu$. We have seen this energy dependence before in the case of electric dipole transitions in atoms (the ν^3 factor in equation 13–99).

TABLE (16–3)

The Reciprocal of the Shell Model Transition Rates for Gamma Decay

Transition	L	10 Mev	1 Mev	0.1 Mev	0.01 Mev
Elec. dipole	1	4×10^{-19}	6×10^{-16}	4×10^{-13}	4×10^{-10}
Mag. dipole	1	4×10^{-17}	6×10^{-14}	4×10^{-11}	4×10^{-8}
Elec. quadrupole	2	1×10^{-16}	1×10^{-11}	1×10^{-6}	2×10^{-2}
Mag. quadrupole	2	1×10^{-14}	1×10^{-9}	1×10^{-4}	2×10^{0}
Elec. octupole	3	1×10^{-13}	1×10^{-6}	1×10^{1}	1×10^{8}
Mag. octupole	3	1×10^{-11}	1×10^{-4}	1×10^{3}	1×10^{10}
Elec. sixteenpole	4	1×10^{-10}	1×10^{-1}	1×10^{8}	1×10^{17}
Mag. sixteenpole	4	1×10^{-8}	1×10^{1}	1×10^{10}	1×10^{19}

Since the electric dipole transition has the largest transition rate for a given value of $h\nu$, the question immediately arises: Why do not all gamma ray decays proceed by an electric dipole transition? The answer is that in a gamma decay between initial and final states of given spin quantum numbers i_i and i_f, and given parities, only certain types of transitions are allowed by the following *selection rules*.

For electric transitions:

$$|i_i - i_f| \equiv \Delta i \leqslant L \leqslant i_i + i_f \qquad (\text{but not } i_i = 0 \rightarrow i_f = 0) \tag{16–83}$$

The nuclear parity must change if L is odd and must not change if L is even.

For magnetic transitions:

$$|i_i - i_f| \equiv \Delta i \leqslant L \leqslant i_i + i_f \qquad (\text{but not } i_i = 0 \rightarrow i_f = 0) \tag{16–83'}$$

The nuclear parity must change if L is even and must not change if L is odd.

The gamma decay will always proceed by the allowed transition having the largest transition rate. *Because of the strong L dependence of S, this means a transition with $L = \Delta i$.* If this value of L is odd, it will be an electric transition if the initial and final states are of the opposite parity, and a magnetic transition if these states are of the same parity. If this value of L is even, it will be an electric transition if these states are of the same parity, and a magnetic transition if they are of the opposite parity.

As is always the case, the selection rules for the various transitions arise from the form of the matrix elements of the interaction potentials which induce these transitions. The part of the rules relating the possible values of L to the values of i_i and i_f can be understood most easily if we realize that all these matrix elements must be such as to conserve angular momentum in the transition, and that the gamma ray must therefore carry away angular momentum as well as energy. In advanced treatments of this subject it is shown that, in the electric or magnetic transitions associated with a certain value of L, the quantum number specifying the magnitude of the angular momentum carried by the gamma ray is L, and also that a gamma ray cannot carry zero angular momentum. Granting these statements, and writing the first part of the selection rules as

$$L = |i_i - i_f|, |i_i - i_f| + 1, \ldots, i_i + i_f \qquad \text{(but not } i_i = i_f = 0\text{)}$$

we see, by comparison with (13–40), that this part simply represents the requirements of conservation of angular momentum.

To understand the second part of the selection rules, we must actually write down the matrix elements for some of the interaction potentials. We do this for the simplest cases—the electric and magnetic dipole interaction potential matrix elements, ignoring the motion of the CM of the nucleus. These are, respectively,

$$\mathbf{E} \cdot \int \psi_f^* \sum_j a_j \mathbf{r}_j \psi_i \, d\tau \tag{16–84}$$

$$\mathbf{H} \cdot \int \psi_f^* \sum_j (b_j \mathbf{L}_j + c_j \mathbf{S}_j) \psi_i \, d\tau \tag{16–84$'$}$$

The parameter a_j, which is $+e$ for protons and 0 for neutrons, relates the positions $\mathbf{r}_j$ of the nucleons to their electric dipole moments; the parameters b_j and c_j, which are the orbital and spin g factors for protons and neutrons, relate the orbital and spin angular momenta $\mathbf{L}_j$ and $\mathbf{S}_j$ of the nucleons to their magnetic dipole moments; the sums, taken over all the nucleons of the nucleus, give the total nuclear electric and magnetic dipole moments.† The matrix elements of these dipole interaction potentials initially contain inside the integral signs the dot product of the moments with the electric and magnetic fields **E** and **H**. However, both fields have a negligible spatial dependence over the nucleus as far as the dipole moments are

† The effect of the CM motion is to increase the contribution of neutrons to these matrix elements.

concerned, and can be factored out. Now, as we mentioned in the previous section, $\mathbf{r}_j$ is a polar vector and $\mathbf{L}_j$ and $\mathbf{S}_j$ are axial vectors. By inspecting Coulomb's law and Ampere's law, the reader can immediately verify that $\mathbf{E}$ is a polar vector and $\mathbf{H}$ is an axial vector. Thus the electric dipole interaction potential matrix element is a dot product of two polar vectors, the magnetic dipole interaction potential matrix element is a dot product of two axial vectors, and so we see that both are scalars, in agreement with a statement made in the previous section. We also see that (16–84) vanishes identically, unless ψ_i and ψ_f are of the opposite parity, because the polar vector $a_j\mathbf{r}_j$ is of odd parity and, if the entire integrand is of odd parity, the contribution to the integral from each point (x, y, z) will be canceled by the contribution from the point $(-x, -y, -z)$. We further see that (16–84′) vanishes identically, unless ψ_i and ψ_f are of the same parity, as the axial vector $b_j\mathbf{L}_j + c_j\mathbf{S}_j$ is of even parity. Both of these conclusions are in agreement with the second part of the selection rules for $L = 1$. Now imagine evaluating the dot product in the electric dipole interaction potential matrix element (16–84). This gives the sum of an integral containing x_j in the integrand, an integral containing y_j in the integrand, and an integral containing z_j in the integrand.† As we know, all the terms x_j, y_j, z_j are of odd parity. The electric quadrupole interaction potential matrix element consists of a similar set of integrals, each containing a term such as x_j^2, x_jy_j, y_j^2, etc., in the integrand.‡ As each of these terms is of even parity, it follows that all the integrals vanish identically unless ψ_i and ψ_f are of the same parity. This is in agreement with the selection rule quoted in (16–83) for $L = 2$. For $L = 3$ the integrals contain terms of odd parity such as x_j^3, $x_jy_jz_j$, etc., and this leads to the selection rules quoted in (16–83) for $L = 3$. Similar things happen in the integrals that comprise the magnetic transition matrix elements, and these lead to the selection rules quoted in (16–83′) for the higher values of L.

In many gamma decay transitions, particularly those in which L is large and $h\nu$ is small, several groups of monoenergetic electrons are emitted along with the gamma rays. The energies E of the groups are found empirically to be related to the energy $h\nu$ of the gamma rays by the equation

$$E = h\nu - W \tag{16–85}$$

where the value of W for the most prominent group is equal to the binding energy of a K shell electron of the gamma decaying atom, and the values

† In this form the interaction potential matrix element (16–84) times its own complex conjugate can easily be identified with the expression (13–77) that we derived for the electric dipole transition rate of atoms.

‡ Cf. footnote on page 645.

of W for the other groups are equal to the binding energies of the electrons in the L, M, N, etc., shells. This makes it quite evident that the various groups of electrons are emitted from the various shells of the atom, so that the energy of one of these electrons upon leaving the atom equals the energy of the gamma rays less its binding energy. The similarity between (16–85) and the photoelectric effect equation (3–8) suggested to its discoverers, Ellis and Meitner (1921), that the electrons are ejected through a process of *internal conversion*, in which an emitted gamma ray delivers all its energy to an atomic electron by the photoelectric effect before leaving the atom.

However, calculations of Taylor and Mott (1933) show that the probability for internal conversion by the photoelectric effect is very small compared to the probability for internal conversion by the direct transfer of energy from the nucleus to the electron through the Coulomb interaction between the two. In this process the nucleus is de-excited and the electron is ejected. Except for the differences in the interaction potential and in the initial and final eigenfunctions for the system, the calculations are quite similar to those involved in beta decay theory. In recent years Rose has used this picture of direct energy transfer to make a number of calculations of the ratio of the probability for emitting an electron from a given shell to the probability for emitting a gamma ray. This ratio is called the *internal conversion coefficient* α. The calculations of α should be very accurate because the factors involving the initial and final nuclear eigenfunctions cancel each other, leaving only factors involving the accurately known Coulomb interaction potential and the accurately known eigenfunctions for the initial bound electron and the final free electron. Figure (16–44) shows α_K, the value of α for the emission of a K shell electron, as calculated for the ^{40}Zr atom. Because the probability of internal conversion increases as the value at the nucleus of the bound electron eigenfunction increases, α_K rapidly becomes larger as Z becomes larger. For the same reason, at a given Z and $h\nu$, α_K is usually larger than α_L, or α_M, or α_N, etc. Furthermore, at a given Z and $h\nu$, the quantity α_K/α_L depends quite sensitively on the character of the gamma ray transition, that is, on the value of L and on whether the transition is electric or magnetic. Thus accurate measurements of α_K/α_L, which are relatively easy to make, provide a very good method of determining the character of a gamma ray transition and, therefore, the relative spins and parities of the nuclear states involved in the transition.

Internal conversion does not compete with gamma ray emission in the sense that one process inhibits the other. The processes are independent, and the probabilities for the two processes are additive. Thus the total rate S_t for transitions between the initial and final nuclear states is the sum

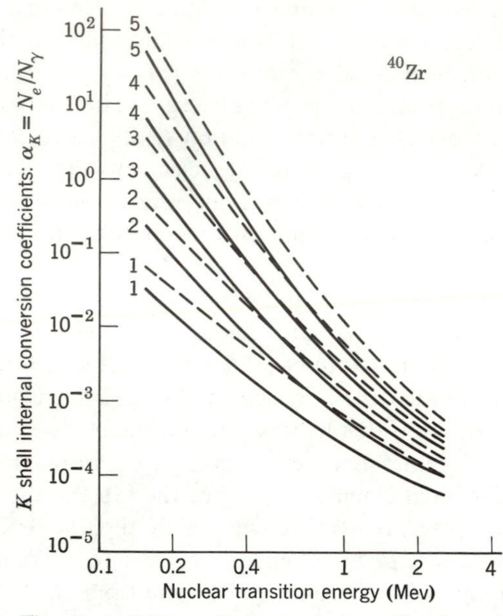

Figure 16–44. The *K* shell internal conversion coefficients. Solid curves, electric transitions; broken curves, magnetic transitions. From R. D. Evans, *The Atomic Nucleus*, McGraw-Hill Book Co., New York, 1955.

of the transition rate S_γ for gamma ray emission plus the transition rate S_{ic} for internal conversion. That is,

$$S_t = S_\gamma + S_{ic} \tag{16–86}$$

This can be written in terms of the internal conversion coefficients since

$$S_{ic} = (\alpha_K + \alpha_L + \alpha_M + \cdots)S_\gamma \equiv \alpha_t S_\gamma$$

where α_t is the total internal conversion coefficient. Thus

$$S_t = S_\gamma(1 + \alpha_t)$$

If the initial state can decay only to a single final state, as is usually true for isomeric transitions, equation (13–118) shows that the lifetime τ of the initial state is

$$\tau = \frac{1}{S_t} = \frac{1}{S_\gamma(1 + \alpha_t)} \tag{16–87}$$

As α_t can be accurately calculated, the experimental values of τ can be compared with the values predicted by the S_γ of (16–82) and (16–82′).

Figure (16–45) shows such a comparison for a group of isomeric transitions that have been identified as magnetic sixteenpole ($\Delta i = 4$, parity change). The dots are experimental and the curve is theoretical. For these transitions, we see that the predictions of (16–82) and (16–82′), which are based on the shell model, are reasonably accurate. But inspection of figure (16–23) will demonstrate that in the shell model all $\Delta i = 4$, parity change

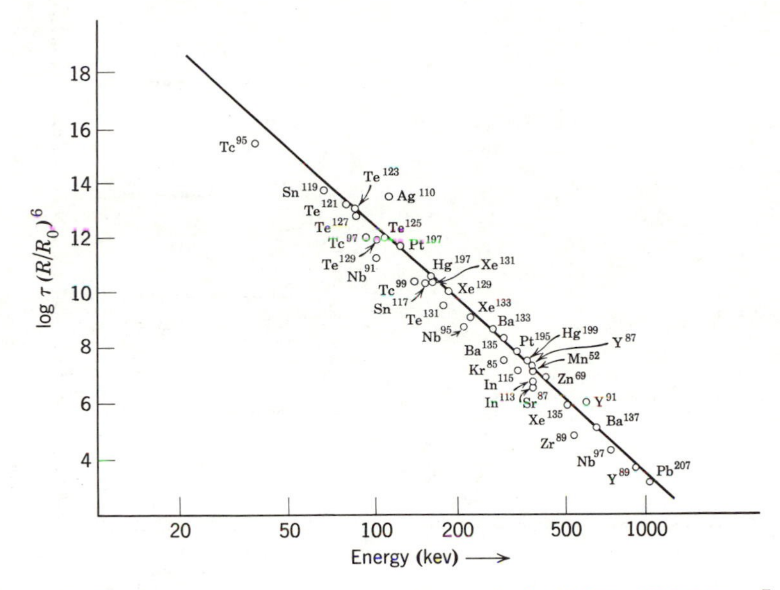

Figure 16–45. The lifetimes for a group of magnetic sixteenpole transitions. From M. Goldhaber and A. Sunyar, *Phys. Rev.*, **83**, 906 (1951).

transitions are between states quite near those filled at the magic numbers,† and this is exactly where the model should be accurate. For transitions of a different character, the shell model predictions are not so accurate.

However, the shell model does explain one striking feature of gamma ray transitions from nuclear states of excitation energy of several Mev or less. In transitions from these states, which are the states excited by beta decay, electric dipole transitions are almost never observed. The selection rules show that electric dipole transitions require $\Delta i = 1$, parity change. But figure (16–23) shows that there is a particularly large energy difference between any two states for which the transitions satisfy this condition.

† The parity of a shell model state is even if l is even [cf. equation (13–112)]. An example of a $\Delta i = 4$, parity change transition would be the transition between $5g_{9/2}$ and $3p_{1/2}$.

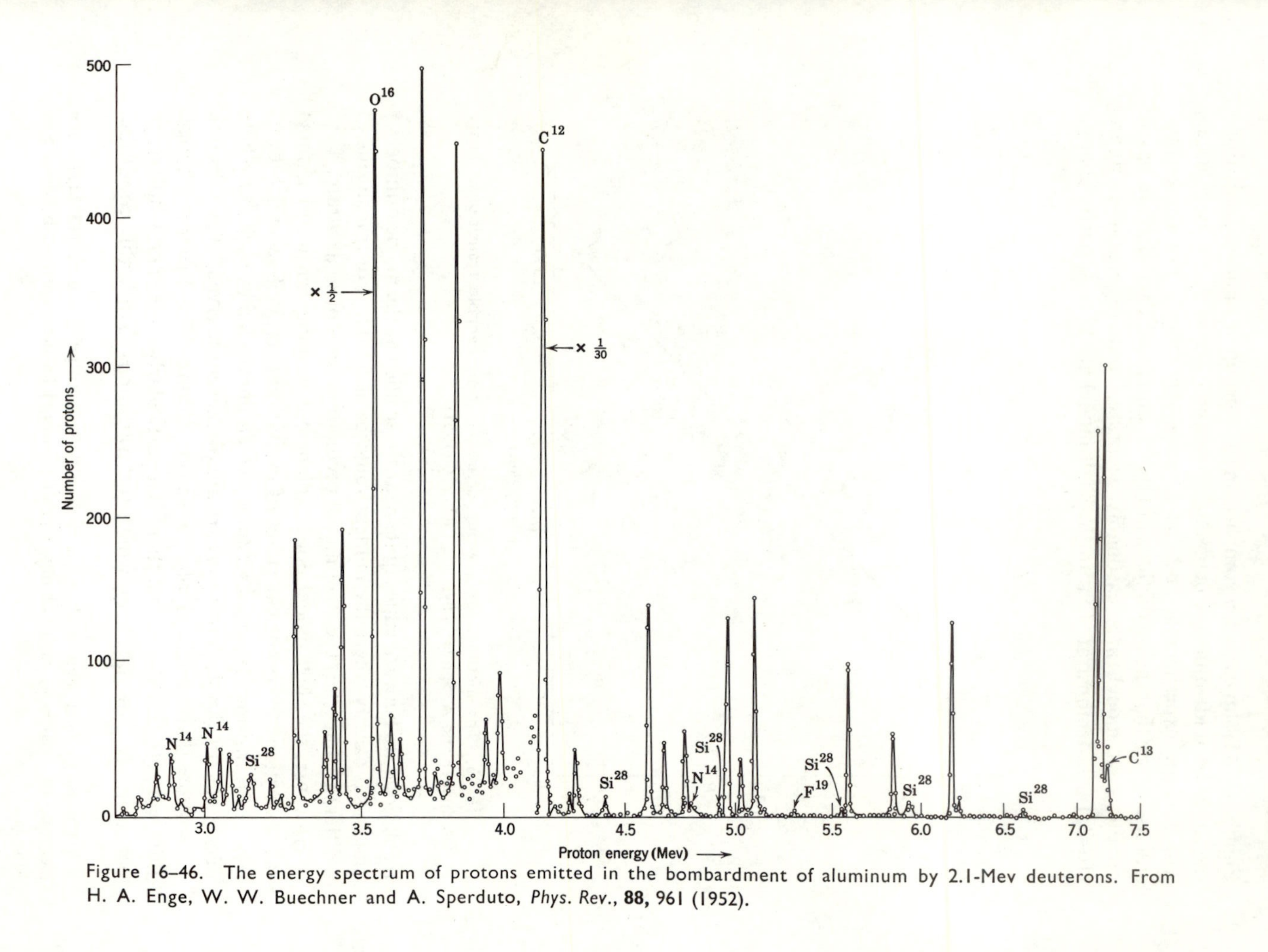

Figure 16–46. The energy spectrum of protons emitted in the bombardment of aluminum by 2.1-Mev deuterons. From H. A. Enge, W. W. Buechner and A. Sperduto, *Phys. Rev.*, **88**, 961 (1952).

This energy is almost always greater than several Mev, and so it is energetically impossible for a state which can decay by an electric dipole transition to be excited by beta decay. However, when enough excitation energy is available, electric dipole transitions are found, and they have transition rates which are not in disagreement with theory.

13. Properties of Excited States

In addition to the information about the excited states of nuclei obtained from the study of internal conversion, gamma ray emission, and alpha and beta decay to excited states, much information is obtained from the study of nuclear reactions. Consider the general case (16–16), which is written

$$a + A \rightarrow B + b$$

If the energy of the bombarding particles a is held constant, and the energy distribution of the product particles b is measured, this energy distribution is usually found to consist of a number of discrete groups. Figure (16–46) shows an example—the energy spectrum of protons produced in the reaction ${}^{1}H^{2} + {}^{13}Al^{27} \rightarrow {}^{13}Al^{28} + {}^{1}H^{1}$ for a bombarding deuteron energy of 2.1 Mev. (A nucleus of the hydrogen isotope ${}^{1}H^{2}$ is called a *deuteron.*) The labeled groups can be identified as originating from impurities in the target foil, but all others are proton groups from reactions on the target nucleus ${}^{13}Al^{27}$. When a proton is emitted in the highest energy group, the residual nucleus ${}^{13}Al^{28}$ is left in its ground state; when a proton is emitted in a lower energy group, the ${}^{13}Al^{28}$ is left in one of its excited states. Thus the spectrum of excited states in the residual nucleus is immediately obtained from the energy spectrum of the product particles [after using equation (16–20) to make small corrections for the recoil of this nucleus]. This is shown in figure (16–47).

Excited states can also be located by measuring the cross section for reactions between a bombarding particle and a target nucleus, while varying the energy of the bombarding particle. With low A nuclei and low bombarding energies, the cross section is usually found to exhibit *resonances*, or certain energies at which the probability of reactions taking place is unusually large. As an example, figure (16–48) shows the relative cross section for producing reactions by the bombardment of ${}^{13}Al^{27}$ with ${}^{1}H^{1}$.

Resonances occur at energies that excite the system consisting of the bombarding particle plus the target nucleus, which is called the *compound nucleus*, to one of its excited states. This is because at these energies the

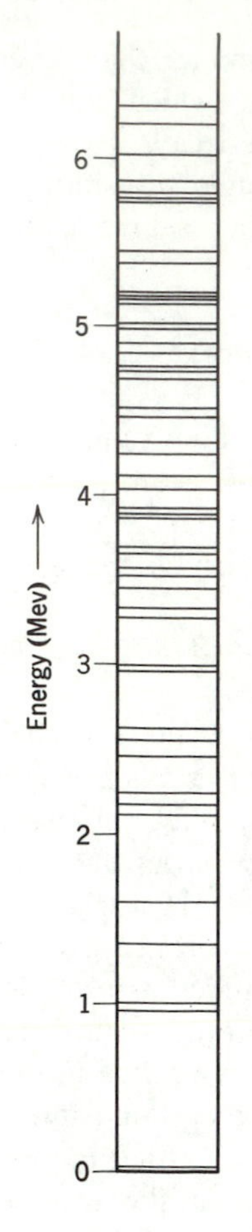

Figure 16–47. The energy levels of the nucleus $^{13}Al^{28}$. From H. A. Enge, W. W. Buechner, and A. Sperduto, *Phys. Rev.*, **88,** 961 (1952).

eigenfunction for the bombarding particle has a particularly large value in the region of the nucleus, and thus the cross section is greatly increased. These resonances are analogous to the resonances discussed in section 6, Chapter 15, and the excited states of the compound nucleus are analogous to the virtual states mentioned there. This is a very interesting situation. Since the compound nucleus is obviously energetically unstable to the re-emission of the bombarding particle, or some other particle, it might be expected that no well-defined excited states would be observed—if a particle is unbound, its allowed energies form a continuum. We shall see in the next section that well-defined excited states can actually exist for sufficiently long times to be observable, even though the total excitation

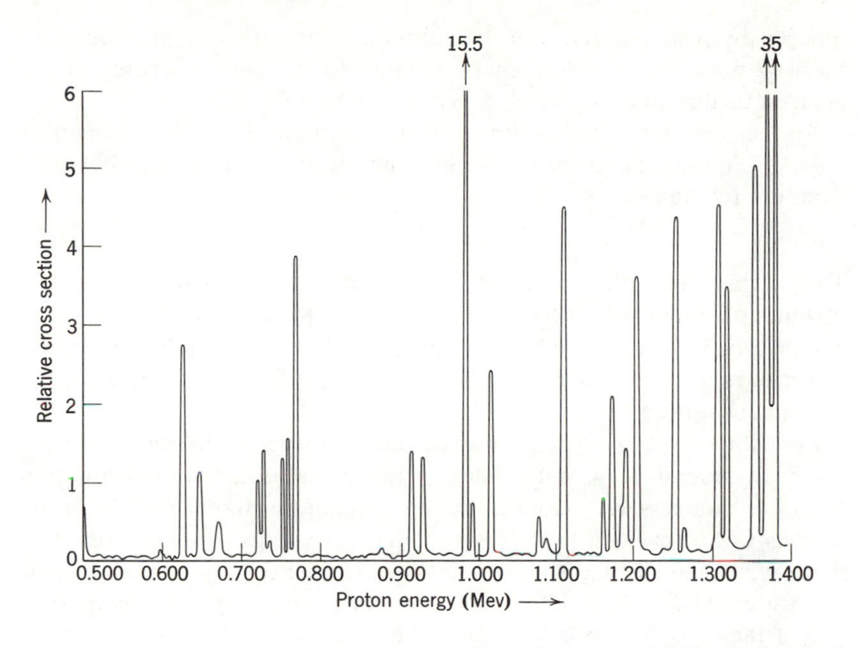

Figure 16–48. Resonances in the cross section for producing reactions by the bombardment of aluminum with protons. From K. J. Brostrom, T. Huus, and R. Tangen, *Phys. Rev.*, **71,** 661 (1947).

energy is great enough to liberate a particle if it is all concentrated on that particle, because the energy is usually divided among many particles of the compound nucleus, and also because reflection at the nuclear surface inhibits particle emission.†

The data of figure (16–48) locate states of the compound nucleus $_{14}Si^{28}$ in a range of excitation energies starting at 0.5 Mev above the 11.59-Mev binding energy of a proton to that nucleus. Lower energy protons cannot be used because they have negligible probability of getting through the Coulomb barrier of the target nucleus. However, in reactions

† These very sharp many particle resonances are definitely not the same as the very broad single particle resonances, mentioned in section 3, that are observed in the cross sections of nuclei for several Mev nucleons. The distinction will be made apparent in the next section. In atoms the broad single particle resonances are just the Ramsauer effect resonances in the cross section for scattering electrons, but the sharp many particle resonances in the cross section for capturing electrons have never been observed. This is due, presumably, to difficulties in obtaining adequate electron energy resolution ($\sim 10^{-6}$ ev?). However, sharp many particle resonances probably do exist because the *cut-off* potential for the interaction of a neutral atom and a free electron is qualitatively similar to that in the nuclear case, in that reflection from the "surface" should inhibit electron emission.

induced by neutrons states of the compound nucleus can be located all the way down to excitation energies equal to the binding energy of the neutron to that nucleus, which is typically 6 to 8 Mev.

By measuring the full width at half maximum Γ of the resonances observed in these experiments, the lifetimes of the states may be obtained from the relation

$$\tau = \hbar/\Gamma \tag{16–88}$$

This is the same relation as (13–125), which applies to the widths and lifetimes of states excited by the resonant absorption of quanta (including gamma rays). Its applicability to states excited by the resonant absorption of particles has been justified by Breit and Wigner. This will be discussed in the next section.

For either proton or neutron experiments using bombarding energies less than several Mev, only states of the compound nucleus with spin quantum numbers very close to the spin quantum number of the target nucleus will be excited. The reason is that the energy of the bombarding particle is so small that it must interact in a state of low orbital angular momentum ($KR \lesssim 1$) and, therefore, angular momentum can be conserved only if the spin change is low. In higher energy nuclear reactions, more states can be excited. Furthermore, experimental and theoretical studies of the angular distributions of the product particles can sometimes give information about the spins and parities of the states of the residual nucleus that are left excited after the emission of these particles.

Now that we have some feeling for the procedures used in studying the energies, lifetimes, spins, and parities of the excited states, let us consider the results of these studies. Briefly put, there seems to be little rhyme or reason in the results. That is, few of the regularities, which are so characteristic of the excited states of an atom, are found in the excited states of a nucleus. There are only three well-established exceptions:

1. In the first few excited states of nuclei, particularly those with N or Z near magic numbers, certain shell model characteristics are often found. An example, which we discussed in the previous section, is the absence of electric dipole transitions between these states.

2. In the first few excited states of even N, even Z nuclei, particularly those with N and Z far from magic numbers, certain characteristics of the rotational motion predicted by the collective model have been found. For a rigid body rotating about a point at the center of its axis of symmetry, the eigenvalues are

$$E = \frac{\hbar^2}{2I}\, i(i+1), \qquad i = 0, 2, 4, 6, \ldots \tag{16–89}$$

where I is the moment of inertia about the axis of rotation, and the parities of all the eigenfunctions are even.† Every even N, even Z nucleus has an $i = 0$, even parity ground state. In many of them the first excited state is $i = 2$, even parity, and the second excited state is $i = 4$, even parity. In some cases the third excited state is $i = 6$, even parity. Furthermore, the energies of these states agree with equation (16–89) for reasonable values of I.

3. In certain sets of light nuclei having common values of A, the first few excited states are found to have corresponding spins and parities. After corrections are made for the different Coulomb repulsion energies of the different nuclei of a set, the energies of the corresponding states are approximately equal. This is particularly true if the set consists of two mirror nuclei.

Beyond the first few excited states, even these modest beginnings of regularity are no longer observed, and the energies, lifetimes, spins, and parities seem to be essentially random. But this should not be surprising—the nucleus is much more complicated than the atom. In the Hartree theory of the atom, the net potential $V(r)$ is the dominant factor in determining the behavior of the electrons, and all the other effects are small perturbations. In the shell model of the nucleus there is also a net potential $V(r)$, but the other effects are not small perturbations. Furthermore, because it is only a model instead of a theory, there are additional effects (the collective effects, and probably a number of others) that the shell model leaves out, and these effects are also not small perturbations. Apparently in a real nucleus there are a number of different effects of comparable importance that govern the properties of the excited states and, as a result, it is not possible to give a description of the *individual* properties of every excited state, except by listing them in a table.

However, it is possible to give a simple description of certain *average* properties of the excited states. With respect to their energies, this is done by specifying the *average energy spacing* D between adjacent states. Some values of D are given in table (16–4). Because of the large fluctuations from one case to the next, and also because of large uncertainties, these average spacings must be taken cum grano salis. Nevertheless, they do indicate the general trends, and these trends are easy to understand.

In the shell model, at very low excitation energies only one of the nucleons carries the total excitation energy of the nucleus, and so the spacing

† Cf. results of exercise 6, Chapter 10. The $i = 1, 3, 5, \ldots$ states are absent because in these states the parity is odd, whereas only even parity states are possible since the symmetrical system to which equation (16–89) applies is transformed into itself in the parity operation; that is, the system is completely symmetrical with respect to reflection in the origin and must therefore be described by eigenfunctions of even parity.

TABLE (16–4)
Average Spacings between Excited States of Nuclei

Excitation energy E	$A \sim 20$	$A \sim 200$
~0 Mev	~10^6 ev	~10^5 ev
~6 to 8 Mev	~10^3 ev	~10^0 ev
~16 Mev	~10^2 ev	

of the excited states is the relatively large spacing of the low lying states of the shell model. In the collective model, the low lying excited states are mainly rotational, and their spacing is also relatively large. At higher excitation energies no known model is adequate, and the nucleus is perhaps best thought of as a complicated system of many particles all of which can participate in various ways in the excitation of the system. As there are a large number of divisions of excitation energy between the many particles that lead to approximately the same total excitation energy, it is evident that the spacing between the higher excited states of the system will be small. It is also evident that this spacing will decrease with increasing total excitation energy E, and with increasing A.

An *approximate* description of the energy dependence of the average spacings that are observed can be obtained by using the Fermi gas model for the highly excited nucleus. Simple thermodynamic arguments, starting from the relation

$$E = a(kT)^2 \tag{16–90}$$

between the total excitation energy E of a Fermi gas and its temperature T, lead to the following expression for the *density of excited states* $\rho \equiv 1/D$:

$$\rho = \rho_0 e^{2\sqrt{aE}} \tag{16–91}$$

The quantity $a = \pi^2 \mathcal{N}/4E_f$, where $\mathcal{N}$ is the number of Fermi particles in the system and E_f is its Fermi energy (12–38). Treating the neutron gas and the proton gas separately, and evaluating $\mathcal{N}$ and E_f in terms of N, Z, and the nuclear volume as in equation (16–35), we find that

$$a \simeq A/10 \tag{16–91$'$}$$

where a is measured in Mev^{-1}.

The lifetimes of the excited states are usually expressed in terms of their *widths* Γ. As we have indicated, the relation (16–88) is valid for all excited states. Thus any excited state with lifetime τ has a full width at half maximum Γ, where

$$\tau = \hbar/\Gamma \tag{16–92}$$

We have seen that for excitation energies up to several Mev the excited states can decay only by gamma ray emission and internal conversion, and that the lifetimes range from about 10^{-15} sec to as long as 10^8 sec. Since $\hbar \sim 10^{-15}$ ev-sec, this means that Γ ranges from about $\Gamma \sim 10^{-1}$ ev to $\Gamma \sim 10^{-23}$ ev. For excitation energies somewhat above this range, but less than the binding energy of a nucleon, the evidence indicates that Γ is not strongly dependent on E or A. It varies from $\Gamma \sim 1$ ev near the neutron binding energy in nuclei with $A \sim 20$, to $\Gamma \sim 10^{-1}$ ev near the neutron binding energy in nuclei with $A \sim 200$. This comes about because in this energy range there are very many lower energy states to which the excited state can decay, so there will always be states available whose spins and parities are such as to allow decay of the excited state by electric dipole transitions with lifetimes of the order of 10^{-15} sec.

As the excitation energy goes above the binding energy of the last nucleon in the nucleus, typically 6 to 8 Mev, decay by the emission of a nucleon becomes energetically possible. In some nuclei, proton emission becomes energetically possible first, and in others neutron emission becomes possible first, but in all, except those of very small Z, the Coulomb barrier effectively inhibits proton emission until the excitation energy is quite high. When nucleon emission is energetically possible, the total rate for transitions from the excited state will be larger than that due to transitions involving gamma ray emission and internal conversion alone. This decreases the lifetime and increases the width. In fact, since widths are inversely proportional to lifetimes and lifetimes are inversely proportional to transition rates, widths are proportional to transition rates and we can write

$$\Gamma = \Gamma_\gamma + \Gamma_n \tag{16-93}$$

because transition rates are additive. In this expression, Γ_γ means the width due to all possible gamma ray and internal conversion transitions, and Γ_n means the width due to neutron emission (or nucleon emission if proton emission is important).

It is easy to see what the general form of the energy dependence of the neutron width Γ_n must be. Imagine using Golden Rule No. 2 to calculate the rate for transitions from an initial state consisting of the nucleus in an excited state of energy E to a final state consisting of the nucleus in its ground state plus a nucleon moving off with kinetic energy $(E - E_b)$, where E_b is its binding energy. The matrix element for the transition will be a slowly varying function of E, and the energy dependence of the transition rate comes in mainly through the density of states for the nucleon. From equation (12–37), we see that this density of states is proportional to $(E - E_b)^{1/2}$. Therefore we expect

$$\Gamma_n \propto (E - E_b)^{1/2} \tag{16-94}$$

at least for a restricted energy range. This is the energy dependence exhibited by the average results of the measurements. When $\Gamma_n \gg \Gamma_\gamma$, then $\Gamma \simeq \Gamma_n$, and these measurements simply involve finding the widths of the resonances in data such as those given in figure (16–48). Since $\Gamma_\gamma \propto$ constant, while $\Gamma_n \propto (E - E_b)^{1/2}$, Γ_n will always become large compared to Γ_γ when $(E - E_b)$ becomes larger than some value. For nuclei with $A \sim 20$, this occurs for $(E - E_b) \sim 1$ ev, and at this value $\Gamma_n = \Gamma_\gamma \sim 1$ ev. For nuclei with $A \sim 200$, $\Gamma_n = \Gamma_\gamma \sim 10^{-1}$ ev at $(E - E_b) \sim 10^3$ ev.

When $(E - E_b)$ is not too large, there is an interesting empirical relation between the average neutron widths Γ_n of the excited states of a nucleus and the average spacings D between these states. It is

$$\Gamma_n \simeq CK_2D \tag{16–95}$$

where

$$K_2 = p_2/\hbar = \sqrt{2M(E - E_b)}/\hbar$$

The quantity p_2 is the momentum of the emitted nucleon, and M is its mass. For almost all nuclei, the constant C is within a factor of 5 of

$$C = 0.5 \times 10^{-13} \text{ cm} \tag{16–95′}$$

although the values of Γ_n vary by many orders of magnitude.

From (16–95), or from (16–91) and (16–94), we see that Γ_n/D increases with increasing excitation energy. At the point where $\Gamma_n/D = 1$, the excited states begin to *overlap* and form a *continuum*. As an indication of the general trend, it might be said that for nuclei with $A \sim 20$ the continuum begins at $(E - E_b) \sim 10$ Mev, and for nuclei with $A \sim 200$ it begins at $(E - E_b) \sim 10^{-1}$ Mev.

14. Nuclear Reactions

Nuclear reactions provide much useful information about the properties of excited states. In turn, the properties of excited states provide the key to the understanding of nuclear reactions. In this section we shall use this point of view to give a simplified version of the present theoretical description of nuclear reactions. This description is due to Bohr (1936), Breit and Wigner (1936), Weisskopf (ca. 1950), Butler (ca. 1953) and many others.

Before the discovery (1936) of very sharp resonances in the cross sections for reactions induced by low energy neutrons, it had been assumed that nuclear reactions involve a direct transition from an initial state consisting of the bombarding particle and the target nucleus to a final state consisting of the product particle and the residual nucleus. However, the discovery of resonances replaced this *direct interaction* theory by the *compound nucleus* theory. Bohr proposed the latter theory because he realized that in reactions exhibiting resonances there must be an intermediate state of relatively long lifetime, which he called the compound nucleus. This state is necessary because such reactions are essentially measurements of the incident neutron energy, since the cross sections are very rapid functions of energy. Therefore the uncertainty principle demands that the duration of these measurements be at least $\tau = \hbar/\Gamma$, where the resonance widths Γ are their energy resolutions. The compound nucleus theory was quite successful, and for a number of years it was thought that it could explain all nuclear reactions. Recently, however, it has been found that some reactions actually are most easily explained on the basis of a direct interaction theory.

We shall describe first the compound nucleus theory. It is convenient to distinguish between two cases: (a), $(E - E_b)$ is small enough that the compound nucleus is excited into a region in which its states are well separated since $\Gamma_n/D \ll 1$; (b), $(E - E_b)$ is large enough that the compound nucleus is excited into a region in which its states strongly overlap since $\Gamma_n/D \gg 1$. However, in both cases the initial step in the nuclear reaction is essentially the same. This is the step described by the optical model. Consider a beam of particles bombarding a target of nuclei. According to the optical model, a particle striking a nucleus will either be scattered out of the incident beam or removed from that beam by absorption. If it is absorbed it will initiate a nuclear reaction, but the optical model cannot tell us anything about the details of the reaction because it only describes the particle moving through the absorptive nucleus in a single particle state of the complex potential that represents the initial particle-nucleus interaction. Because the potential is fairly absorptive for all positive values of $(E - E_b)$,† this single particle state will have a very short lifetime, or very large width. In fact, since it immediately follows from (15–106) that, if the imaginary part of the complex potential has a constant value $-W$, the width of the single particle state is $\Gamma = 2W$, and since figure (16–10) shows that $2W \sim 10$ Mev for the typical case $(E - E_b) \sim 10$ Mev, the lifetime of the single particle state is typically $\tau = \hbar/\Gamma \sim 10^{-15}$ ev-sec/10^7 ev $= 10^{-22}$ sec.

The very short lifetime single particle state rapidly decays into the longer lifetime many particle states that correspond, within the limits of the uncertainty principle, to the same total excitation energy of the system consisting of the bombarding particle plus the target nucleus. This is the

† Here we use $(E - E_b)$ for the external kinetic energy of the bombarding nucleon. It equals the E of figure (16–10).

system we call the compound nucleus, and the many particle states are precisely the higher excited states that we were discussing in the last section. These many particle states are certainly different from the single particle state. One difference is that in the cases we are discussing the width of the single particle state is measured in Mev while the widths of the many particle states are measured in ev. Another is that the single particle state and the many particle states do not describe the same physical situation, even though they correspond to the same total excitation energy. In the single particle state all the excitation energy is carried by a single particle, but in any of the many particle states it is distributed over a large number of particles. The excitation of the many particle states by the single particle state takes place through a process in which energy is transferred from one to the other in multiple collisions of the bombarding particle with the particles of the nucleus.†

Of course, the many particle states of the compound nucleus will eventually decay with an average lifetime $\tau = \hbar/\Gamma \sim 10^{-16}$ sec, where we take $\Gamma \sim 10$ ev as a typical many particle state width. We know that, in general, this decay must be by nucleon emission since, except at small $(E - E_b)$, $\Gamma_n \gg \Gamma_\gamma$ for many particle states. But it is perhaps not apparent how this can happen. The point is that in the many particle states the total excitation energy E is distributed among many nucleons, and in order for any nucleon to escape from the nucleus the nucleon must have an excitation energy at least equal to its binding energy E_b. Thus a large fraction of the excitation energy must somehow become concentrated again on a single nucleon if it is to escape. Some idea of the way this comes about, and also some idea of the process in which the many particle states are initially excited, can be obtained from the following argument.

Let Ψ be the wave function for the compound nucleus, and ψ_k be the eigenfunctions for the many particle states that are excited in that nucleus. Assume that the spacing between all adjacent eigenvalues E_k of these many particle states is a constant equal to their average spacing D, so that $E_k = E_0 + kD$, where E_0 is a constant and $k = 1, 2, 3, \ldots$. Then the wave functions Ψ_k for the many particle states are $\Psi_k = \psi_k e^{-iE_k t/\hbar} = \psi_k e^{-iE_0 t/\hbar} e^{-ikDt/\hbar}$. Next write Ψ as a linear combination of the Ψ_k. That is,

$$\Psi = \sum_k a_k \Psi_k = e^{-iE_0 t/\hbar} \sum_k a_k \psi_k e^{-ikDt/\hbar} \tag{16–96}$$

where the summation is taken over the many particle states that are excited. In this expression $t = 0$ is the time at which they are initially excited, and

† This gives rise to a simple relation between the sharp many particle resonances and the broad single particle resonances—the single particle resonances define the average *strength*, Γ_n/D, of the many particle resonances.

$a_k = A_k e^{-i\phi_k}$, where the real quantities A_k are their amplitudes and the real constants ϕ_k are phase factors that allow the t in all the Ψ_k to have the same zero. Now note that the summation appears to be a perfectly periodic function of t, with repetition period

$$T = 2\pi\hbar/D \tag{16–97}$$

At $t = 0$, Ψ actually describes a single particle state, since at that time the compound nucleus is just being formed and all the excitation energy is carried by the bombarding particle. If the summation is perfectly periodic with period T, then Ψ will again describe a single particle state at $t = T$ (and also $2T$, $3T$, etc.) In fact, it will describe exactly the same single particle state that was initially formed, and all the total excitation energy will be concentrated on the single particle. This particle will then escape the nucleus, providing it is not reflected by the change in potential at the surface of the nucleus.

Let us consider case (a), in which the quantity $(E - E_b)$ is small. Here reflection at the nuclear surface is very important. For a proton, the Coulomb barrier makes the probability of escape negligible. For a neutron, the probability of escape, P, can be estimated from equation (8–46). It is

$$P = 4K_1K_2/(K_1 + K_2)^2 \tag{16–98}$$

where

$$K_1 = p_1/\hbar \simeq \sqrt{2MV_0}/\hbar$$

and

$$K_2 = p_2/\hbar = \sqrt{2M(E - E_b)}/\hbar$$

The quantity p_1 is the momentum of the neutron of mass M when inside the nucleus. We equate this to the momentum it would have, approximately, when moving in an optical model potential with real part of depth V_0. The quantity p_2 is the momentum of the neutron after escaping. It is related to the external kinetic energy $(E - E_b)$ as shown. Equation (16–98) is based on the approximation that the potential at the surface of the nucleus changes rapidly in a distance of the order of 1 de Broglie wavelength. This approximation is not bad for small $(E - E_b)$. Furthermore, with small $(E - E_b)$, $K_1 \gg K_2$ and so

$$P \simeq 4K_1K_2/K_1^2 = 4K_2/K_1 \tag{16–99}$$

Now the probability per second that a neutron will be emitted from the nucleus is equal to T^{-1}, the number of times per second that Ψ repeats its original single particle form, multiplied by the probability P that any one of these times the neutron will escape. Thus the lifetime for neutron emission is

$$\tau = 1/T^{-1}P = T/P \tag{16–100}$$

and the neutron width Γ_n is $\Gamma_n = \hbar/\tau = \hbar P/T$. From (16–97) and (16–99) this is

$$\Gamma_n = \frac{\hbar 4K_2}{K_1}\frac{D}{2\pi\hbar} = \frac{2K_2 D}{\pi K_1}$$

or

$$\Gamma_n = \frac{2\hbar}{\pi\sqrt{2MV_0}} K_2 D \tag{16–101}$$

where

$$K_2 = \sqrt{2M(E - E_b)}/\hbar$$

Taking the value $V_0 = 50$ Mev that is indicated by the optical model, the coefficient $2\hbar/\pi\sqrt{2MV_0} = 0.39 \times 10^{-13}$ cm, and we see that equation (16–101) is in excellent agreement with (16–95). Note from equation (16–100) that the lifetime τ of the many particle states is much longer than the repetition period T, when $(E - E_b)$ is small, simply because the escape probability P is small—the wave function Ψ for the compound nucleus repeats its original single particle form many times before the particle is emitted.

This is not true for case (b), in which $(E - E_b)$ is large, because in these circumstances the escape probability $P \simeq 1$. To handle this case we must correct a weak point in the argument. The quantities A_k, which are the amplitudes of the a_k in equation (16–96), are not constants, but on the average decrease in time according to the law†

$$A_k(t) = A_k(0)e^{-t/2\tau_k} \tag{16–102}$$

where $\tau_k = \hbar/\Gamma_k$ is the lifetime of the many particle state of width Γ_k. This can be ignored for the purpose of evaluating the repetition period T in case (a), since all the $\tau_k \gg T$ and the $A_k(t)$ are very nearly constant over any repetition period. But it cannot be ignored in case (b), since the condition $\tau_k \gg T$ does not obtain. Now, if all the τ_k had exactly the same value τ, and all the $A_k(t)$ had the same t dependence $e^{-t/2\tau}$, this term would factor out of the summation (16–96) and appear on the outside as a term describing the decay of the compound nucleus. This would leave the summation a perfectly periodic function of t. But the τ_k do not all have exactly the same values. Instead they have a spread of values, and thus the summation is not a perfectly periodic function of t. Even so, it will be an approximately periodic function if the spread in the τ_k is not too large. Then the wave function Ψ for the compound nucleus should, within a certain time, return to a form that has an approximately single particle character, and the particle is emitted on the first occasion on which this

† As the probability of finding the nucleus in the many particle state k is $a_k^* a_k = A_k^2(t)$, equation (16–102) is simply the usual exponential decay law.

happens. However, the particular approximately single particle wave function to which Ψ returns should have no necessary connection with the exact form of the initial single particle wave function. This will certainly be true *if* a large number of overlapping states are excited in the compound nucleus, and *if* the width of each of these states is not strongly correlated with the widths of the others. In these circumstances, *the decay of the compound nucleus will be independent of the exact details of its formation.*

This fundamental result of the theory of compound nucleus reactions also holds for case (a). This can be thought of as a consequence of the fact that the values of $A_k(t)$ actually change slightly during one repetition period, so Ψ deviates slightly from being a perfectly periodic function of t. Thus Ψ returns repeatedly to a single particle wave function that is slightly different from the previous single particle wave function and, in the large number of cycles that occur before the particle is finally emitted, Ψ loses all memory of the exact details of the initial single particle wave function. In both cases, this result is only strengthened by deviations from the assumption that the eigenvalues E_k of the many particle states are evenly spaced. But, in any circumstances, the decay of the compound nucleus must be independent of the exact details of its formation for case (a). The reason is that energy conservation within the limits of the uncertainty principle demands that only many particle states of excitation energy within Γ_n of the energy E be excited by the time $t \sim \tau = \hbar/\Gamma_n$ when the compound nucleus has existed long enough to decay.† For case (a), $\Gamma_n/D \ll 1$, so there can only be one many particle state within these limits. Therefore, in case (a) the decay of the compound nucleus *must necessarily* depend only on the properties of that many particle state and cannot depend on the details of how it was excited.

After this somewhat esoteric description of compound nucleus reactions, it might be well to recapitulate in the kind of language used in the original papers of Bohr. When a particle enters a nucleus, it collides with the particles of the nucleus and very rapidly gives up its kinetic and binding energy to those particles. The bombarding particle is thereby captured, and an excited compound nucleus is formed in a state, somewhat reminiscent of a boiling liquid, in which the excitation energy is distributed at random among all its particles. Each particle then has an excitation energy considerably less than its binding energy. Occasionally, statistical fluctuations cause enough energy to be concentrated on a single particle to make it energetically possible to escape. If the circumstances are such that

† Of course, at an earlier time many particle states whose excitation energies fall within broader limits can be excited, as we have assumed. For example, when only 10^{-22} sec has elapsed (the lifetime of the initial single particle state), many particle states over a range of $\sim$10 Mev can be excited. Cf. section 5, Chapter 9.

reflection at the nuclear surface is important, this must happen a number of times before a particle is finally emitted. But, in any circumstances, the compound nucleus has a relatively long lifetime and during this long lifetime it forgets everything about how it was originally formed (except that it must remember to conserve energy, angular momentum, etc.). Thus the final decay process is independent of the exact details of the formation process. This means, for example, that the ratio of the probability that the compound nucleus will decay by nucleon emission to the probability that it will decay by gamma ray emission and internal conversion is equal to the ratio Γ_n/Γ_γ (since transition rates are proportional to widths), where Γ_n and Γ_γ are evaluated for the state or states excited in the compound nucleus, and where this is true independent of exactly how these states were excited. If in a certain reaction a particular compound nucleus is formed with a particular excitation energy, and in a different reaction (different bombarding particle and different target nucleus) the same compound nucleus is formed at the same excitation energy, both compound nuclei should decay in the same way. At small values of $(E - E_b)$, Γ_γ can be comparable to Γ_n, and there can be an appreciable probability that the compound nucleus decays by gamma ray emission and internal conversion to form a completely stable nucleus. At higher values of $(E - E_b)$, $\Gamma_n \gg \Gamma_\gamma$, and the compound nucleus always decays by particle emission. If enough excitation energy is available, the compound nucleus can decay either by neutron emission or by proton emission. For quite high excitation energies, more than one particle can be emitted.

Figure (16–49) shows the measured cross sections for the set of reactions

$$
{}^{1}\mathrm{H}^{1} + {}^{29}\mathrm{Cu}^{63} \rightarrow \begin{cases} {}^{30}\mathrm{Zn}^{63} + {}^{0}n^{1} \\ {}^{30}\mathrm{Zn}^{62} + {}^{0}n^{1} + {}^{0}n^{1} \\ {}^{29}\mathrm{Cu}^{62} + {}^{0}n^{1} + {}^{1}\mathrm{H}^{1} \end{cases}
$$

and also for the set

$$
{}^{2}\mathrm{He}^{4} + {}^{28}\mathrm{Ni}^{60} \rightarrow \begin{cases} {}^{30}\mathrm{Zn}^{63} + {}^{0}n^{1} \\ {}^{30}\mathrm{Zn}^{62} + {}^{0}n^{1} + {}^{0}n^{1} \\ {}^{29}\mathrm{Cu}^{62} + {}^{0}n^{1} + {}^{1}\mathrm{H}^{1} \end{cases}
$$

at quite high bombarding energies. The abscissa is the kinetic energy of the bombarding proton for the first set of reactions, and the kinetic energy of the bombarding alpha particle in the second set of reactions; the relative displacement of the two abscissas is equal to the difference in E_b for the two bombarding particles. The ordinate is the cross section for the various reactions, and each reaction is labeled according to a common notation. This figure provides a test of the prediction that the decay of a compound nucleus is independent of its formation, in case (b), where

$(E - E_b)$ is large enough that the compound nucleus is excited into a region in which there are presumably many overlapping states whose widths are not strongly correlated, because for both sets there is the same compound nucleus $^{30}Zn^{64}$, and because the abscissas have been adjusted

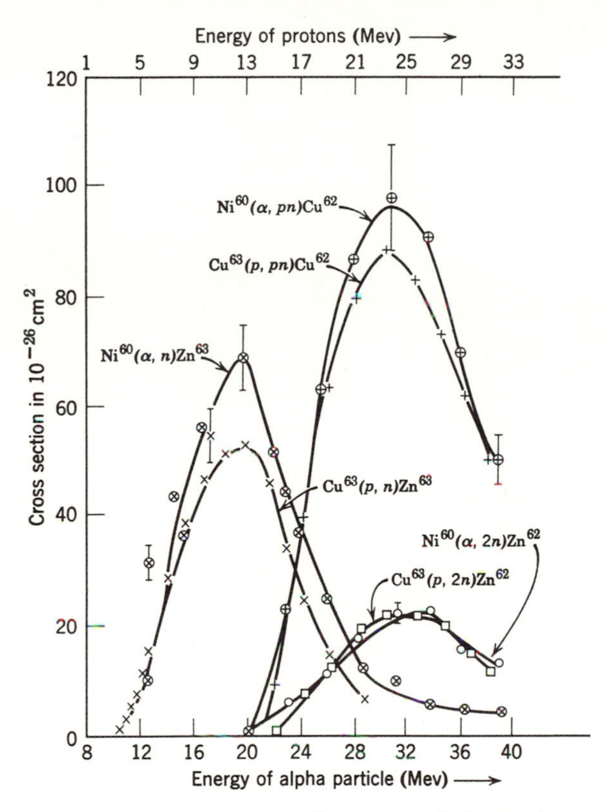

Figure 16–49. Cross sections for a set of reactions which test the predictions of the compound nucleus theory. From S. N. Ghoshal, *Phys. Rev.*, **80**, 939 (1950).

to equalize its excitation energy. We see that the prediction is well verified because for both sets there are the same relative probabilities for the various decay modes of the compound nucleus.

It has not been possible to make such a test for case (a), where $(E - E_b)$ is small enough that the states of the compound nucleus are well separated and only a single state is excited; and such a test is hardly necessary. But other predictions of the compound nucleus theory can be tested for this case. For instance, Breit and Wigner have shown that the energy dependence of the cross section for absorbing an S-wave neutron ($K_2R \ll 1$) to form a compound nucleus excited to a state of energy E_0 and width $\Gamma = \Gamma_n + \Gamma_\gamma$, which subsequently decays by gamma ray emission and internal conversion, is essentially identical with the energy dependence (13–124) of the probability for absorbing a quantum to form an excited state of an atom, which subsequently decays by emitting another quantum. They found

$$\sigma(E) = \frac{\pi}{K_2^2} \frac{\Gamma_n \Gamma}{(E - E_0)^2 + (\Gamma/2)^2} \frac{\Gamma_\gamma}{\Gamma} \tag{16–103}$$

where

$$K_2 = \sqrt{2M(E - E_b)}/\hbar$$

The term π/K_2^2 is the maximum possible cross section for absorbing an S-wave neutron (cf. equation 15–98 with $\eta_0 = 0, \eta_1 = \eta_2 = \eta_3 = \cdots = 1$). The term $\Gamma_n\Gamma/[(E - E_0)^2 + (\Gamma/2)^2]$ gives the resonance its shape; it contains a factor Γ_n which, being proportional to the probability for emitting a neutron, must also be proportional to the probability for absorbing one (cf. equation 13–84 and the associated discussion of microscopic reversibility), and it also contains a factor Γ that comes from normalization. The term Γ_γ/Γ is the ratio of the probability that the excited compound nucleus will decay by gamma ray emission and internal conversion to the probability that it will decay by any process. The energy dependence of (16–103) has been accurately verified in many different experiments. Granting its validity, it may be used to obtain values of Γ, Γ_n, and Γ_γ from measurements of $\sigma(E)$. Breit and Wigner also found that equation (16–88), which is identical with (13–125), applies to the resonant absorption of particles in any situation covered by case (a).

Although the predictions of the compound nucleus theory seem to be verified for reactions that proceed through a compound nucleus process, it has been found in recent years that there are certain situations in which reactions proceed through a direct interaction process. Figure (16–50) presents some data illustrating one of these situations. These are energy distributions of protons inelastically scattered (scattered with energy loss) at five different angles in the LAB system from a target of ^{50}Sn nuclei bombarded by a beam of protons of external kinetic energy 31 Mev. The abscissa is the external kinetic energy of the inelastically scattered protons, and the ordinate is the cross section for scattering a proton into a unit energy and solid angle range. Thus the data also give the angular distribution of the inelastically scattered protons. Note that in the energy range between 10 and 20 Mev the cross section does not depend very strongly on energy but increases rapidly with decreasing angle. On both points, this is exactly the opposite of what would be expected from the compound nucleus theory. The theory predicts that in the long-lived compound nucleus essentially all sense of direction would be lost, and that the angular distribution of the protons which are finally boiled off should be isotropic. It also predicts that the protons boil off above the 10-Mev high Coulomb barrier of the $Z = 51$ compound nucleus with a strongly energy dependent Maxwellian-like energy distribution $N(\varepsilon) \propto \varepsilon e^{-\varepsilon/kT}$, where ε is their external kinetic energy and kT is obtained from equation (16–90) with $E = (31 + E_b)$ Mev $= 38$ Mev. That is, the theory predicts $N(\varepsilon) \sim 0$, $\varepsilon < 10$ Mev, and $N(\varepsilon) \propto \varepsilon e^{-\varepsilon/1.7\,\text{Mev}}$, 10 Mev $< \varepsilon <$ 31 Mev.

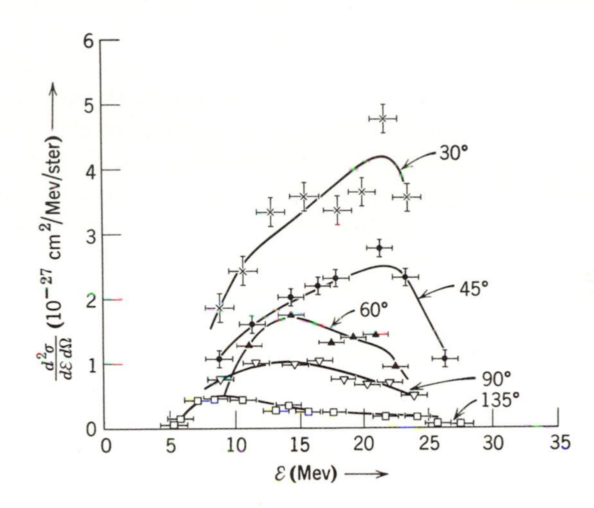

Figure 16–50. Energy and angular distributions for a reaction which proceeds through a direct interaction process. From R. M. Eisberg and G. Igo, *Phys. Rev.*, **93,** 1039 (1954).

For neutrons, there is no Coulomb barrier, and their predicted energy distribution is $N(\varepsilon) \propto \varepsilon e^{-\varepsilon/1.7\,\text{Mev}}$ over the entire range of ε. In the compound nucleus theory almost all the emitted nucleons would be neutrons, since only a few protons in the tail of the Maxwellian with $\varepsilon > 10$ Mev can escape, and it would be expected that considerably less than 1 percent of the reactions induced by the incident 31-Mev protons should lead to the emission of inelastically scattered protons. But, on this third point, the data also contradict the predictions, because the magnitude of the cross section indicates that about 15 percent of the reactions lead to the emission of inelastically scattered protons.

This does not mean that the compound nucleus theory is wrong. Instead, it means that inelastic proton scattering reactions on intermediate and high Z nuclei proceed through some process, other than the compound nucleus process, which has a cross section for these reactions so much larger than the small compound nucleus cross section that the compound nucleus effects are completely hidden. The process that is involved is thought to consist of a direct interaction in which a bombarding proton collides with a nucleon in the diffuse rim of the target nucleus, and either the bombarding proton or the struck nucleon (or both) immediately escapes.

Consider a 31-Mev proton incident upon a ^{50}Sn nucleus. As the de Broglie wavelength of the proton is several times smaller than the nuclear radius, it is legitimate to think of the collision as being well enough localized that a distinction can be made between head-on collisions and collisions with the diffuse rim. Now the optical model shows that attenuation lengths in nuclear matter at 31 Mev are also several times smaller than the nuclear radius. Thus, in any head-on collision with the nucleus, the bombarding proton will soon collide with the nucleons of the nucleus, these nucleons will in turn collide with other nucleons, and so the kinetic energy of the bombarding nucleon will rapidly be shared with all the nucleons of the nucleus. This is the first step in the compound nucleus process that will eventually lead to the emission of low energy neutrons. However, in a collision with the diffuse rim of the nucleus, the bombarding proton could collide with a nucleon in such a manner that either it or the struck nucleon (or both) would immediately escape, carrying away a large fraction of the kinetic energy. As the optical model shows that the thickness of the diffuse rim in the ^{50}Sn nucleus is about 10 percent of its radius, its projected area is about 20 percent of the projected area of the entire nucleus. This indicates that the cross section for this direct interaction process could be of the order of magnitude observed experimentally. The other characteristics of this process are such that they could also be in agreement with experiment. The process would lead to a slowly varying energy distribution of high energy inelastically scattered protons, and to an angular distribution of these particles which would increase rapidly with decreasing angle (if it were not for the initial momentum of the struck nucleon, and refraction in the nuclear potential, conservation of momentum and energy would require the angular distribution to be concentrated entirely in the forward 90° of the LAB system).

A number of semi-quantum mechanical calculations based on this direct interaction process have shown that it is capable of a very adequate fit to the 31-Mev inelastic proton scattering reaction, and to many similar reactions. Butler (ca. 1953) and others have made completely quantum mechanical calculations for this process for reactions in which it is possible to measure the angular distribution of a single group of inelastically scattered protons that leave the residual nucleus excited in a particular low lying excited state. The calculations are basically the same as the Born approximation calculations for elastic scattering studied in Chapter 15, but the details are somewhat more complicated—very much more complicated if the scattered particle imparts angular momentum as well as energy to the nucleus. Instead of giving them in the form of a long equation, the results of these calculations for a particular case are shown as the solid line in figure (16–51). The points are experimental data. This is the angular distribution for the inelastic scattering of alpha particles, of external kinetic energy 31 Mev, from a target of $^{12}Mg^{24}$. The alpha particles

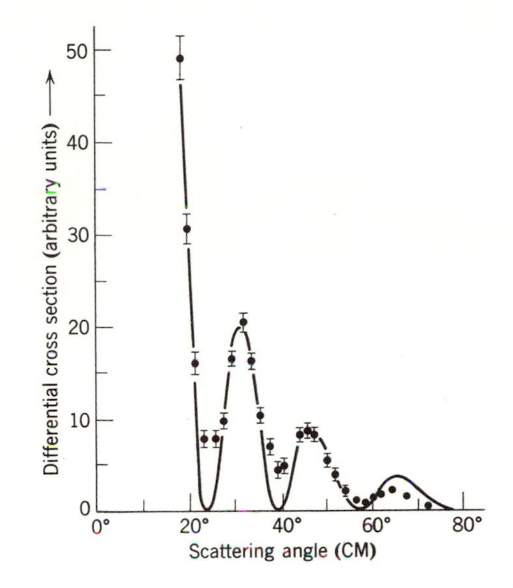

Figure 16–51. Experimental and theoretical angular distributions for a direct interaction leaving the residual nucleus in a single excited state. From S. T. Butler, *Phys. Rev.*, **106,** 272 (1957).

measured are the highest energy inelastically scattered group, and they leave the residual nucleus in its first excited state at 1.37 Mev. Note that, in addition to the general increase in the angular distribution with decreasing angle, there are very pronounced oscillations that are reminiscent of the elastic scattering angular distributions. This is typical of what is found in these cases. The locations of the oscillations in the theoretical results turn out to be a very sensitive function of the orbital angular momentum transferred from the inelastically scattered particle to the nucleus, and thus provide a powerful tool for studying the spins and parities of the excited states of the nuclei involved.

There may be direct interaction processes other than those involving nucleon-nucleon collisions in the rim of the nucleus. In fact, recent evidence indicates that in some cases the highest energy inelastically scattered particles come from a process in which the incident particle collides with the rim of the nucleus but, instead of interacting with a single nucleon, the particle interacts coherently with the entire nucleus and puts it into one of its low lying rotational states. The particle immediately moves off, having suffered only a small energy loss and angular deflection.

Direct interactions seem to play an important role in all reactions in which a fairly high energy product particle is emitted; this includes reactions where the product particle is different from the bombarding particle. Important examples are reactions where a deuteron ${}^1H^2$ is the bombarding particle, and a proton ${}^1H^1$ or a neutron ${}^0n^1$ is the product particle. These reactions apparently go through a process in which one of the nucleons of the deuteron is captured, while the other continues on.

Recently there have been attempts to use the formalism of the compound nucleus theory to describe direct interactions. This is done by dispensing with the condition that the widths of the overlapping states excited in the intermediate system are not strongly correlated. By adjusting these correlations, which is of course permissible since the widths are not directly measurable when the states overlap, it is possible to make the intermediate system do almost anything from decaying immediately in a manner dependent upon its formation (direct interaction) to decaying after a long time in a manner independent of its formation (compound nucleus). With regard to direct interactions, this might be criticized as an overly complicated way of describing a simple process. However, it has the advantage of providing a single unified theory of all nuclear reactions, including those which are compound nucleus processes but are in the awkward range, lying between our cases (a) and (b), where $\Gamma_n \sim D$.

15. Nuclear Forces

In the preceding sections we have concentrated on describing the properties of the system of nucleons called the nucleus and have largely avoided discussing the properties of the forces acting between the nucleons of the system, even though these properties are clearly the origin of the properties of the system. Nevertheless, we have learned something about nuclear forces.

1. Nuclear forces must be attractive, at least in regard to their net effect on the average, because otherwise nuclei would not exist.

2. Nuclear forces must be of short range because the spatial distribution of the nuclear interaction potential exerted by a nucleus on a passing nucleon extends beyond the spatial distribution of the nucleons in the nucleus by a distance which is only of the order of 10^{-13} cm.

3. Nuclear forces between two neutrons in particular quantum states must be nearly the same as nuclear forces between two protons in the same quantum states, not counting the Coulomb repulsion of the protons. This follows from the fact that the total ground state binding energies of each member of a set of mirror nuclei are equal, after correction is made for the different total Coulomb repulsion energies of the two members, and from the fact that for one member the number of neutron-neutron interactions between nucleons in various quantum states is equal to the number of proton-proton interactions between nucleons in those quantum states for the other member, and vice versa (cf. figure 16–42). As the number of neutron-proton interactions in various quantum states is the same for both members of a set, mirror nuclei do not allow a comparison between neutron-proton interactions and neutron-neutron or proton-proton interactions.

4. Nuclear forces cannot be such that there is nearly the same purely attractive interaction potential between all pairs of nucleons in all quantum states. This follows from the fact that the total ground state binding energy ΔE is approximately proportional to A for all nuclei with $A \geqslant 4$, and from the following argument due to Heisenberg (1933). Consider some nucleus and fix attention on a particular nucleon in its interior, say the ith nucleon. Now assume that there *is* a purely attractive interaction potential, of range a, acting between this nucleon and all the other nucleons of the nucleus. Then the average potential energy $\bar{V}_i$ of that nucleon is proportional to the number of other nucleons found within a sphere of radius a, centered on that nucleon, and this energy is negative. That is,

$$\bar{V}_i \propto -\rho \tag{16–104}$$

where ρ is the density of nucleons. If this were the only effect, the radius R of the nucleus would collapse to the range a of the interaction potential, since this would make $\bar{V}_i$, as well as the average potential energies of all the other nucleons, as negative as possible and, therefore, lead to the largest total binding energy and the most stable arrangement. But there is an effect which tends to oppose such a collapse. It is the increase in the kinetic energy of the nucleons demanded by the uncertainty principle if the volume in which they are confined decreases. For a nucleus of radius R, the uncertainty Δp_i in any component of the linear momentum of the ith nucleon must be at least

$$\Delta p_i \sim \hbar/R$$

The magnitude of the linear momentum vector is

$$p_i \sim \Delta p_i$$

and the uncertainty arises because its direction is unknown. Thus the average kinetic energy of the ith nucleon is

$$\bar{T}_i \sim p_i^2/2M \propto (\Delta p_i)^2 \propto 1/R^2$$

But

$$\rho \propto 1/R^3$$

so†

$$\bar{T}_i \propto +\rho^{2/3} \tag{16–105}$$

From (16–104) and (16–105) we find that the ρ dependence of the total binding energy of that nucleon is

$$\Delta E_i \equiv -E_i = -(\bar{T}_i + \bar{V}_i) = -\alpha\rho^{2/3} + \beta\rho$$

where E_i is the negative total energy of the ith nucleon, and where α and β are constants. The total binding energy of the nucleus is

$$\Delta E = \sum_{i=1}^{A} \Delta E_i = -\gamma\rho^{2/3} + \delta\rho \tag{16–106}$$

where γ and δ are other constants. If this equation is valid, the nuclear radius R would collapse to the range a of the potential because it would lead to a large ρ and maximize ΔE.‡ In these circumstances, all the nucleons of the nucleus are within the range of each other's interaction potential, and the average potential energy of the ith nucleon is proportional to the total number of nucleons in the nucleus. That is,

$$\bar{V}_i \propto -A$$

As the same is true for any other nucleon, the total average potential energy of the nucleus is

$$\bar{V} = \sum_{i=1}^{A} \bar{V}_i \propto -A^2 \tag{16–107}$$

But, as the average kinetic energy of any of the nucleons would not depend on A, the total average kinetic energy is

$$\bar{T} = \sum_{i=1}^{A} \bar{T}_i \propto +A \tag{16–108}$$

With a collapsed nucleus, $\bar{V}$ is much larger than $\bar{T}$, and this is particularly true for large A. Thus

$$\Delta E = -(\bar{T} + \bar{V}) \simeq -\bar{V} \propto +A^2 \tag{16–109}$$

† Note that equations (12–38) and (12–39) show that the same result is obtained for a Fermi gas in which the exclusion principle is explicitly taken into account.

‡ The second term of equation (16–106) must be larger than the first term, as ΔE must be positive. Therefore ΔE increases with increasing ρ since the second term varies faster than the first term.

We see that the total binding energy is proportional to A^2, despite the effect of the uncertainty principle, under the assumption that there is the same purely attractive interaction potential between all pairs of nucleons in a nucleus. But this is not true, and neither is it true that all nuclei have the same radius $R = a$. Therefore the assumption is false. It must be that the interaction potentials are different between pairs of nucleons in different quantum states, being repulsive in some states, or it must be that there is some kind of impenetrable repulsive core in the interaction potentials, or else it must be that there is some combination of both these effects. One of these alternatives is absolutely necessary in order to keep the nucleons far enough apart that at any instant one nucleon interacts only with a *limited* number of other nucleons. Then the binding energy of each nucleon will be independent of the total number of nucleons in the nucleus and the total binding energy of the nucleus will be proportional to A, as observed. This requirement is described by saying that the nuclear forces must lead to *saturation*.

Most of the information about nuclear forces that can be obtained from the study of systems as complex as a *saturated nucleus* ($A \geqslant 4$) is contained in the four items listed above. More detailed information is obtained by studying simpler systems where the effects of the nuclear forces can be analyzed without the complications that characterize a saturated nucleus. The simplest of these is the *deuterium* nucleus $^1H^2$, or *deuteron*, in its ground state. This system consists of a neutron and a proton bound together by the interaction potential for the neutron-proton force. Its observed properties are

$$\begin{aligned} M_{^1H^2} &= 2.014732 \text{ amu} \\ \Delta E &= 2.22 \text{ Mev} \\ i &= 1 \\ \mu_{i_z} &= +0.8574\mu_N \\ Q &= +2.74 \times 10^{-27} \text{ cm}^2 \end{aligned} \tag{16–110}$$

The deuteron has a small quadrupole moment Q, which means that its probability density is not completely spherically symmetrical. This immediately shows that the neutron-proton interaction potential is complicated because the ground state would be a spherically symmetrical S state† if the interaction potential could be written $V(r)$, where r is the magnitude of the vector from one nucleon to the other. [We have seen that this is true in the case of the H atom and its Coulomb potential

† The symbols, and the quantum numbers they represent, are taken over directly from the theory of the electronic structure of atoms, except that i is often used for j as in the shell model.

$V(r) \propto -1/r$. It is not difficult to prove it for any attractive potential $V(r)$ by showing that the lowest eigenvalue of the Schroedinger equation (10–27) is always found for $l = 0$.] Thus the neutron-proton interaction potential is not simply of the form $V(r)$. A form which is consistent with the existence of a quadrupole moment is

$$V(x, y, z) = V(r) + V'(r)\{3(\mathbf{S}_1 \cdot \mathbf{r})(\mathbf{S}_2 \cdot \mathbf{r})/r^2 - \mathbf{S}_1 \cdot \mathbf{S}_2\} \quad (16\text{–}111)$$

The first term is the dominant spherically symmetrical potential. The second term gives rise to small departures from spherical symmetry because its value depends on the angles between the vector **r** and the nucleon spin angular momentum vectors $\mathbf{S}_1$ and $\mathbf{S}_2$; the quantity $\mathbf{S}_1 \cdot \mathbf{S}_2$ is subtracted so that the average of the entire second term over all directions of **r** vanishes.†

It is conventional to call the second term the *tensor potential.* This nomenclature is somewhat unfortunate as it is not a tensor but a scalar, just like the first term. Neither term changes sign in the parity operation, so both are scalars and $V(-x, -y, -z) = V(x, y, z)$. The small non-spherically symmetrical term mixes a state of higher l value in with the dominant S state to form the ground state of the deuteron. But, since $V(-x, -y, -z) = V(x, y, z)$ and the ground state is not degenerate, the eigenfunction for the system must have a definite parity (cf. footnote on page 465), and the state which is mixed in must have the same parity as the S state. As the parities of these states are even if l is even and odd if l is odd (cf. equation 13–112), this state can be a D state, a G state, etc. The D state is the only one of these which is possible because for all the others l is too large to lead to the spin quantum number $i = 1$. Thus the ground state is assumed to be a mixture of S and D states. In order to obtain $i = 1$, these must be the ${}^3S_1$ and ${}^3D_1$ states. Calculations of probability densities, and corresponding values of quadrupole moments, show that the observed quadrupole moment is obtained with the mixture

$$96\%\ {}^3S_1 + 4\%\ {}^3D_1 \quad (16\text{–}112)$$

where these percentages give the probability of finding the deuteron in either of the states.

This is confirmed by comparison with the observed magnetic moment μ_{i_z}. *If* the ground state of the deuteron were a pure ${}^3S_1$ state, then

$$\mu_{i_z} = (\mu_{i_z})_p + (\mu_{i_z})_n \quad (16\text{–}113)$$

† This term has the same directional dependence as the magnetic interaction potential for two magnetic moments. However, it must not be thought of as due to magnetic interactions between the magnetic moments associated with the spins, because the strength of those interactions is several orders of magnitude smaller than what is required to explain the quadrupole moment.

where $(\mu_{i_z})_p$ and $(\mu_{i_z})_n$ are the z components of intrinsic magnetic moments of the proton and neutron that arise from their spins. There would be no orbital contribution from the motion of the charged proton since $l = 0$, and the z components of the intrinsic magnetic moments would add since the two nucleons are in the triplet state with their spins "parallel." Evaluating $(\mu_{i_z})_p$ and $(\mu_{i_z})_n$ in equation (16–113) gives $\mu_{i_z} = +2.7896\mu_N - 1.9103\mu_N = +0.8793\mu_N$, which is 2.6 percent greater than what is observed.† The general agreement confirms the

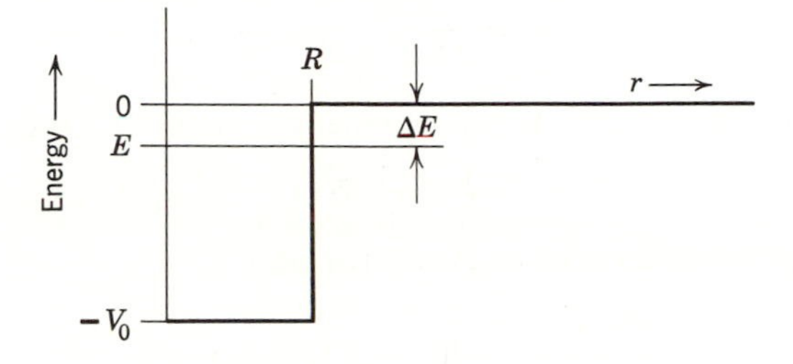

Figure 16–52. An attractive square well potential.

conclusion that the ground state is almost a pure $^3S_1$ state and the discrepancy confirms the conclusion that it does have a bit of some other state mixed in. When the magnetic moment is calculated for the mixture (16–113), agreement with the observed value is better than the accuracy of the observation. The fact that the neutron-proton interaction potential is not completely spherically symmetrical is interesting but, to a first approximation, it can be ignored at low energies because the asymmetry is small. In most of our discussion we shall ignore it and, therefore, take the ground state of the deuteron to be a pure $^3S_1$ state.

Consider, then, the spherically symmetrical interaction potential $V(r)$, and assume that it is in the form of a square well of radius R and depth V_0, as in equation (16–114) and figure (16–52).

$$V(r) = \begin{matrix} -V_0, & r < R \\ 0, & r > R \end{matrix} \qquad (16\text{–}114)$$

We would like to establish a relation between R, V_0, and the measured deuteron ground state binding energy $\Delta E = 2.22$ Mev. As might be guessed, it will turn out that we shall not be able to evaluate separately both R and V_0 from the single known quantity ΔE, and we shall certainly not be able to evaluate the *shape* of the potential well from this single

† The signs give the relative orientations of the magnetic moment and spin angular momentum vectors.

quantity. Therefore any shape is equally appropriate and, to simplify the calculation, we might as well take the square well. The time independent Schroedinger equation that we must solve for this problem is (10–27), with $l = 0$ since we are considering an S state. It is

$$\frac{1}{r^2}\frac{d}{dr}\left(r^2\frac{dR(r)}{dr}\right) + \frac{2\mu}{\hbar^2}[E - V(r)]R(r) = 0$$

where μ is the reduced mass for the system of two nucleons. That is,

$$\mu = \frac{MM}{M + M} = \frac{M}{2}$$

with M the nucleon mass. It is convenient to make the substitution

$$u(r) = rR(r) \qquad (16\text{–}115)$$

This leads to the differential equation for $u(r)$,

$$\frac{d^2u(r)}{dr^2} + \frac{2\mu}{\hbar^2}[E - V(r)]u(r) = 0 \qquad (16\text{–}116)$$

which is (15–76) for $l = 0$. Equation (16–116) is identical in form with the one dimensional square well equation we solved in Chapter 8, and we could adapt that solution to our present purposes by redefining the terms. But it is just as easy to solve it again. Write it as

$$\frac{d^2u}{dr^2} + K_{\text{I}}^2 u = 0, \qquad r < R \qquad (16\text{–}117)$$

where

$$K_{\text{I}} = \sqrt{2\mu(V_0 - \Delta E)}/\hbar \qquad (16\text{–}117')$$

and

$$\frac{d^2u}{dr^2} - K_{\text{II}}^2 u = 0, \qquad r > R \qquad (16\text{–}118)$$

where

$$K_{\text{II}} = \sqrt{2\mu\,\Delta E}/\hbar \qquad (16\text{–}118')$$

The general solution to equation (16–117) is

$$u = A \sin K_{\text{I}}r + B \cos K_{\text{I}}r, \qquad r < R \qquad (16\text{–}119)$$

Since the eigenfunction $R(r)$ never diverges, equation (16–115) shows that $u = 0$ at $r = 0$. This requires setting $B = 0$ in (16–119). Then

$$u = A \sin K_{\text{I}}r, \qquad r < R \qquad (16\text{–}120)$$

The general solution to equation (16–118) is

$$u = Ce^{-K_{\text{II}}r} + De^{K_{\text{II}}r}, \qquad r > R \qquad (16\text{–}121)$$

To prevent this from diverging as $r \to \infty$, we must set $D = 0$. Then equation (16–121) becomes

$$u = Ce^{-K_{||}r}, \qquad r > R \tag{16–122}$$

Now we must equate u and du/dr for both (16–120) and (16–122) at $r = R$. This gives

$$A \sin K_{|}R = Ce^{-K_{||}R} \tag{16–123}$$

and

$$AK_{|} \cos K_{|}R = -CK_{||}e^{-K_{||}R} \tag{16–124}$$

Dividing equation (16–124) by (16–123) gives

$$K_{|} \cot K_{|}R = -K_{||} \tag{16–125}$$

or

$$\frac{\sqrt{2\mu(V_0 - \Delta E)}}{\hbar} \cot \left(\frac{\sqrt{2\mu(V_0 - \Delta E)}}{\hbar} R \right) = -\frac{\sqrt{2\mu\,\Delta E}}{\hbar} \tag{16–125$'$}$$

This is essentially (8–82), and it can be solved by the same graphical method used for that equation. The solution only establishes a relation between those values of V_0 and R that will lead to the observed value of ΔE. However, we shall soon see that there is evidence from scattering experiments that can be used to evaluate R separately, so that V_0 can be determined. It is found that $V_0 \gg \Delta E$. This being the case, it is a good approximation to neglect ΔE in the term $(V_0 - \Delta E)$ and write equation (16–125′) as

$$\frac{\sqrt{2\mu V_0}}{\hbar} \cot \left(\frac{\sqrt{2\mu V_0}}{\hbar} R \right) \simeq -\frac{\sqrt{2\mu\,\Delta E}}{\hbar}$$

or

$$\cot \left(\frac{\sqrt{2\mu V_0}}{\hbar} R \right) \simeq \sqrt{\frac{\Delta E}{V_0}} \simeq 0$$

The first root (corresponding to the ground state) is

$$\frac{\sqrt{2\mu V_0}}{\hbar} R = \frac{\pi}{2}$$

so

$$V_0 R^2 \simeq \pi^2\hbar^2/8\mu = \pi^2\hbar^2/4M \tag{16–126}$$

If we take $R = 2 \times 10^{-13}$ cm, a value which is indicated by the scattering experiments, we obtain $V_0 \simeq 25$ Mev and confirm the consistency of our approximation.

From this information concerning the V_0R^2 value for the interaction potential that fits the ground state of the deuteron, it is not difficult to calculate the locations of the excited states. The calculation predicts that there are no bound excited states, and this is verified by experiment.† Figure (16–53) indicates the assumed square well potential $V(r)$ with its single bound state, and also indicates the associated eigenfunction. There are no bound excited states of the neutron-proton system because the interaction potential is only barely strong enough to bind the ground state. For the same reason, the ground state eigenfunction has a long exponential tail extending into the region $r > R$, and there is a probability higher than 50 percent that the separation between the two nucleons of a deuteron is greater than the range of their interaction potential.

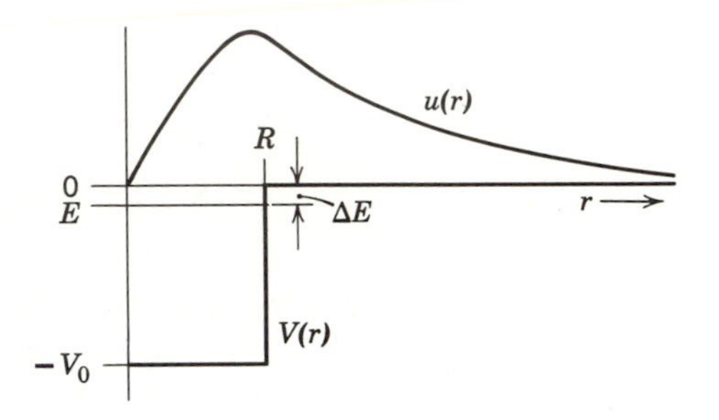

Figure 16–53. The assumed potential and the single bound eigenfunction for the deuteron.

We turn now from the bound state ($E < 0$) of the neutron-proton system to the unbound states ($E > 0$) of that system. They are investigated experimentally by measuring the cross section for scattering neutrons from protons, using techniques similar to those used for scattering nucleons from nuclei, and are analyzed theoretically by the procedures of Chapter 15. As we saw in equations (15–73) and (15–74) of that chapter, the scattering cross section is specified by the phase shifts δ_l. Furthermore, equation (15–78) showed us that the δ_l will have appreciable values only for values of l through $l = l_{\text{max}}$, where

$$l_{\text{max}} \simeq KR = \frac{\sqrt{2\mu E}}{\hbar} R \tag{16–127}$$

Taking the radius of the interaction potential as $R \simeq 2 \times 10^{-13}$ cm, and evaluating the total energy E which gives $l_{\text{max}} = 1$, we find $E \simeq 10$ Mev. Thus, at energies small compared to this value, only δ_0 should have an

† Compare this with the H atom. This simplest of atomic systems has an infinite number of bound excited states whose properties provide almost all the information that was needed in developing the theory of atomic systems. One of the reasons why the development of the theory of nuclear systems is difficult is that the deuteron provides little information because it has no bound excited states.

appreciable value, and isotropic S-wave scattering should be observed.† As isotropic angular distributions actually are observed up to energies of about 5 Mev, it is certainly safe in treating scattering in this energy range to use the equations (15–80) and (15–81) that apply to S-wave scattering. The latter equation reads

$$\sigma = \frac{4\pi \sin^2 \delta_0}{K^2} \tag{16–128}$$

where

$$K = \sqrt{2\mu E}/\hbar \tag{16–128′}$$

Still assuming the square well potential (16–114), we may immediately take over any useful results from the calculation for the identical potential (15–82). However, we shall not just write down the expression (15–91) for δ_0 because it gives δ_0 in terms of V_0, and we prefer an expression in terms of the known quantity ΔE.

An approximate expression of this type is easy to obtain if we realize that the form of the eigenfunction in the region $r < R$ is not going to change very much as the total energy E goes from the small negative value $-\Delta E$ to the small positive values we are considering here since, for all these values, $|E| \ll V_0$, and the form of the eigenfunction in the region of the well is determined almost entirely by its depth V_0. Thus we use equation (16–120) inside the potential well. That is,

$$u \simeq A \sin K_{|}r, \qquad r < R \tag{16–129}$$

Outside the potential well we use equation (15–88), which is

$$u = F \sin (Kr + \delta_0), \qquad r > R \tag{16–130}$$

Matching u and du/dr at $r = R$ gives

$$A \sin K_{|}r \simeq F \sin (KR + \delta_0) \tag{16–131}$$

and

$$AK_{|} \cos K_{|}r \simeq FK \cos (KR + \delta_0) \tag{16–132}$$

Dividing (16–132) by (16–131), we obtain

$$K_{|} \cot K_{|}R \simeq K \cot (KR + \delta_0) \tag{16–133}$$

But equation (16–125) shows that the left side of (16–133) equals $-K_{||}$, where

$$K_{||} = \sqrt{2\mu\, \Delta E}/\hbar \tag{16–134}$$

Therefore

$$K \cot (KR + \delta_0) \simeq -K_{||} \tag{16–135}$$

† It is to be understood that in this section all statements refer to the CM system, and that all energies are measured in that system. The relation between the CM energy E and the LAB energy is $E = \frac{1}{2}E_{\text{LAB}}$.

Since we are going to use the S-wave scattering equations that are valid only for $KR \ll 1$, in (16–135) we may as well drop KR in comparison with δ_0, and write the equation as

$$K \cot \delta_0 \simeq -K_{||} \tag{16–136}$$

A little work with trigonometric identities will convince the reader that (16–136) is identical with

$$\sin^2 \delta_0 \simeq \frac{K^2}{K^2 + K_{||}^2} \tag{16–137}$$

Evaluating (16–128) from (16–137), and using (16–128′) and (16–134), we obtain an expression for the low energy neutron-proton scattering cross section,

$$\sigma \simeq \frac{4\pi}{K^2 + K_{||}^2} = \frac{4\pi}{2\mu E/\hbar^2 + 2\mu\, \Delta E/\hbar^2} = \frac{2\pi\hbar^2}{\mu} \frac{1}{E + \Delta E} \tag{16–138}$$

in terms of the scattering energy E and the known deuteron binding energy $\Delta E = 2.22$ Mev. This expression was first obtained by Wigner in 1933. It is in fairly good agreement with experiment for E from about 1 Mev up to 5 Mev. However, the expression considerably underestimates the observed values of σ when E is small compared to 1 Mev. For example, (16–138) predicts that σ approaches the constant value $\sigma \simeq 2\pi\hbar^2/\mu\, \Delta E \simeq 2.4 \times 10^{-24}\ \text{cm}^2 = 2.4$ bn, for $E \simeq 0$.† But the experimental value of σ for $E \sim 1$ ev is $\sigma = 20.4$ bn. Although (16–138) was obtained from certain approximations, calculations in which these approximations are not used show that it can hardly be off by as much as a factor of 2. Thus the factor of 10 discrepancy between the predicted and measured values of σ shows that something is definitely wrong with the theory.

In 1935 Wigner suggested that the point at which the theory might be wrong is the assumption that the interaction potential used to fit the deuteron binding energy can also be used to explain the scattering of a free neutron by a proton. In the deuteron, the neutron and proton must always interact in the $^3S_1$ state because the spin quantum number is $i = 1$, whereas a free neutron and proton can interact in the $^1S_0$ state as well as in the $^3S_1$ state. In fact, for an unpolarized beam of neutrons incident upon an unpolarized target of free protons, $\frac{1}{4}$ of the scatterings will take place in the singlet $^1S_0$ state where the spins are "antiparallel," and $\frac{3}{4}$ of the scatterings will take place in the triplet $^3S_1$ state where the spins are "parallel." The 1 to 3 ratio occurs because there are 1 singlet state and 3 triplet states (cf. section 4, Chapter 12). If the target contains free protons,

† The *barn* (bn) is a unit of cross section commonly used in nuclear physics; 1 bn = 10^{-24} cm^2.

each scattering will be incoherent with all the others, and the observed scattering cross section σ will be an average of the singlet scattering cross section σ_s and the triplet scattering cross section σ_t, weighted according to the relative probabilities $\frac{1}{4}$ and $\frac{3}{4}$. That is,

$$\sigma = \tfrac{1}{4}\sigma_s + \tfrac{3}{4}\sigma_t \tag{16–139}$$

The triplet scattering cross section is what was really evaluated in (16–138). The singlet scattering cross section will have the same general form as (16–138), but the quantity ΔE will have a different value if the neutron-proton interaction potential in the singlet state is different from what it is in the triplet state. Allowing for this possibility, Wigner wrote

$$\sigma \simeq \frac{2\pi\hbar^2}{\mu}\left[\frac{1}{4(E + \Delta E_s)} + \frac{3}{4(E + \Delta E_t)}\right] \tag{16–140}$$

where $\Delta E_t = 2.22$ Mev. By equating this to the observed value $\sigma = 20.4$ bn at $E \sim 1$ ev, ΔE_s may be determined. The result is $\Delta E_s \simeq 0.06$ Mev. As this is very different from ΔE_t, it is apparent that the interaction potential is different in the singlet and triplet states. Since the former corresponds to the spins "antiparallel" while the latter corresponds to the spins "parallel," we see that the interaction potential of a neutron and proton depends on the relative orientation of the spins of the two nucleons. This is described by saying that *the interaction potential is spin dependent.*

The experimental value of σ at $E \sim 1$ ev does not determine the sign of ΔE_s because, for $E \ll |\Delta E|$, the expression

$$\sigma \simeq \frac{2\pi\hbar^2}{\mu}\frac{1}{E + |\Delta E|}$$

can be shown to be as valid if $\Delta E < 0$, as (16–138) is if $\Delta E > 0$. The phase shift changes sign if ΔE changes sign, but the scattering cross section (16–128) is independent of the sign of the phase shift. Thus the singlet state could be either a stable bound state with binding energy $\Delta E_s > 0$, or it could be an unstable *virtual* state with binding energy $\Delta E_s < 0$ (cf. section 6, Chapter 15). The observation that there are no stable systems containing two neutrons or two protons, for both of which the exclusion principle demands that the lowest energy state be the singlet state $^1S_0$, suggests that the singlet state of the system containing a neutron and a proton is also unstable, or virtual. The fact that this state actually is virtual is proved in measurements of the scattering of low energy neutrons from a target of molecular hydrogen at very low temperatures. The hydrogen molecule can either have the two protons "parallel" (ortho-hydrogen) or "anti-parallel" (para-hydrogen). The latter type has the lowest energy state, and so at very low temperatures essentially all the molecules will eventually be converted to that type by collisions and other processes. For neutrons of de Broglie wavelength comparable to the internuclear spacing of the molecule, the total wave scattered from the molecule will be a coherent combination of the waves scattered from

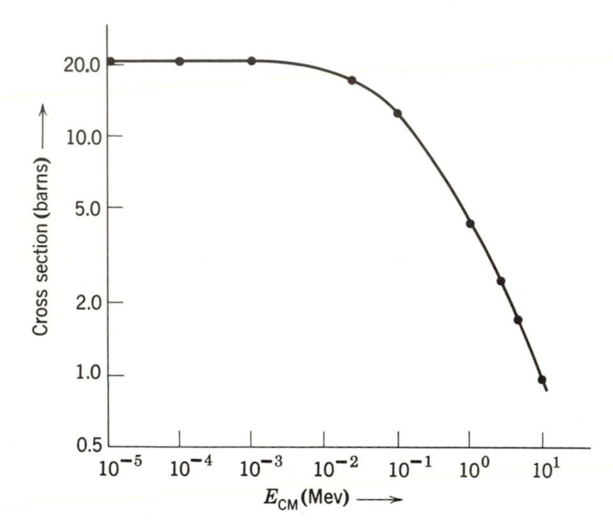

Figure 16–54. Experimental cross sections and theoretical fits for the scattering of neutrons from free protons in the S-wave energy range.

the two protons. Since in one of these scattering processes the neutron and proton are in the singlet state while in the other they are in the triplet state, the total scattering cross section is sensitive to the relative signs of the singlet and triplet phase shifts. An analysis by Teller (1937) of the measurements, which were made at his suggestion, shows that these phase shifts are of the opposite sign and, therefore, that the singlet state of the neutron-proton system is virtual.

The points of figure (16–54) shows some measured values of the cross section for scattering neutrons from free protons in the S-wave energy range. To within their accuracy, these data can be fitted perfectly by an equation basically the same as (16–140), but including corrections for the approximations we made in its derivation. This is indicated by the solid line. These corrections make the exact equation more dependent on the characteristics of the neutron-proton interaction potential than (16–140) and thus allow the radius and depth of both the triplet and singlet interaction potentials to be determined by fitting the data.† (It is also necessary to bring in the data obtained in the scattering of neutrons from low temperature molecular hydrogen.) The values so determined are

$$\begin{aligned} R_t &= 2.05 \times 10^{-13} \text{ cm} \\ V_{0_t} &= 33.9 \text{ Mev} \\ R_s &= 2.9 \times 10^{-13} \text{ cm} \\ V_{0_s} &= 11.7 \text{ Mev} \end{aligned} \tag{16–141}$$

Note that the interaction potential certainly is spin dependent, and that the singlet potential is considerably weaker than the triplet potential for the neutron-proton interaction.

Let us turn now to the proton-proton interaction. For these identical particles with $s = \frac{1}{2}$, the exclusion principle plays an important part in ruling out certain states because it demands that the complete eigenfunction for the system of two particles be antisymmetric with respect to an exchange of their labels. Thus the spin part of the eigenfunction must be antisymmetric if the spatial part is symmetric, and vice versa (cf. section 4, Chapter 12). The symmetry of the spatial part is easily evaluated if it is realized that, for this system of two particles, exchange of their labels is the same as changing the signs of the coordinates because the origin of coordinates is at the CM of the two particles. Therefore the spatial part of the eigenfunction is symmetric if its parity is even and antisymmetric if its parity is odd, and the parity is even if l is even and odd if l is odd (cf. equation 13–112). Thus the S, D, G, . . . states correspond to symmetric spatial parts and require the antisymmetric triplet spin parts. The opposite is true for the P, F, H, . . . states. The possible states in which two protons can interact are, then, the ${}^1S$, ${}^3P$, ${}^1D$ ${}^3F$, ${}^1G$, ${}^3H$, . . . states only.

In scattering experiments at energies somewhat less than 10 Mev, KR is small enough that the two protons interact only in the ${}^1S$ state, which is the ${}^1S_0$ state. The exclusion of the ${}^3S_1$ state simplifies the analysis of the proton-proton scattering experiments, and this certainly helps because the analysis is complicated by the existence of a Coulomb as well as a nuclear interaction between the two protons. Because of the Coulomb interaction, the overall angular distribution is not isotropic in the CM system, even though it is S-wave scattering as far as the nuclear interaction is concerned. Figure (16–55) shows the differential scattering cross sections for proton-proton scattering measured at several different low energies, and also shows theoretical fits to these measurements. Consideration of figure (15–2) shows that $d\sigma/d\Omega$ must be symmetric about 90° because, if

† For example, the radius R appears separately in the term which was dropped between equations (16–135) and (16–136). Cf. exercise 20 at the end of this chapter.

one proton is scattered at the angle θ, the other one must be scattered at $180^\circ - \theta$. This is verified experimentally, and it is sometimes used as an experimental check. At angles smaller than about 30°, $d\sigma/d\Omega$ rises very rapidly. In this region it is following the rapid angular dependence of Coulomb interaction scattering (cf. equation 15–94) which dominates the nuclear interaction scattering. Near 90°, the S-wave nuclear interaction

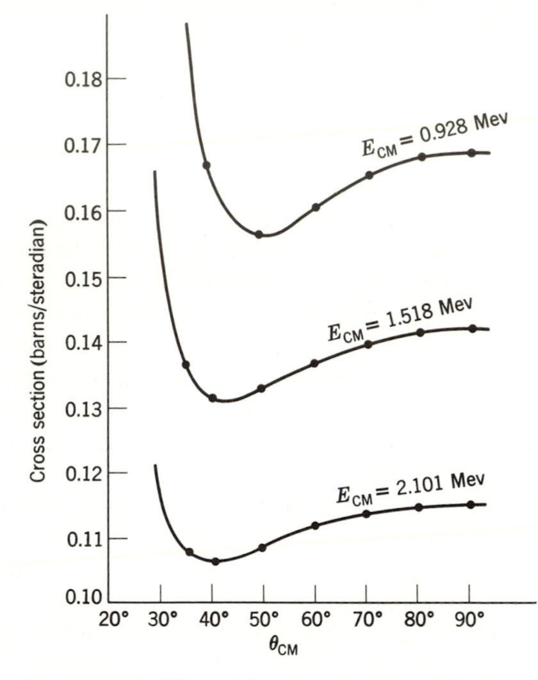

Figure 16–55. Experimental differential cross sections and theoretical fits for proton-proton scattering in the S-wave energy range. From G. Breit and R. L. Gluckstern, *Annual Review of Nuclear Science*, Vol. 2, Annual Reviews, Stanford, 1953.

scattering dominates and $d\sigma/d\Omega$ is almost isotropic. In between, there is a region in which the nuclear and Coulomb interactions interfere destructively. This interference allows a determination of the sign as well as the magnitude of the nuclear interaction phase shift. It is found that, for energies up to about 5 Mev, the phase shift can be fit with a square well proton-proton interaction potential with radius and depth exactly the same as R_s and V_{0_s} of (16–141), the radius and depth that are found for the $^1S_0$ state neutron-proton interaction potential. Thus at low energies the neutron-proton and proton-proton interaction potentials appear to be the same in the $^1S_0$ state, which is the only state in which a comparison can be made. Also recall that mirror nuclei indicate that the neutron-neutron and proton-proton interaction potentials are the same in the low energy range characterizing the relative energies of two nucleons in a nucleus. From all of this it is concluded that *at low energies the interaction potentials between all pairs of nucleons are the same.* It should be remembered, however, that this does not necessarily mean that the shapes of the interaction potentials are the same, but only that the phase shifts they produce are the same. None of the low energy measurements can give detailed

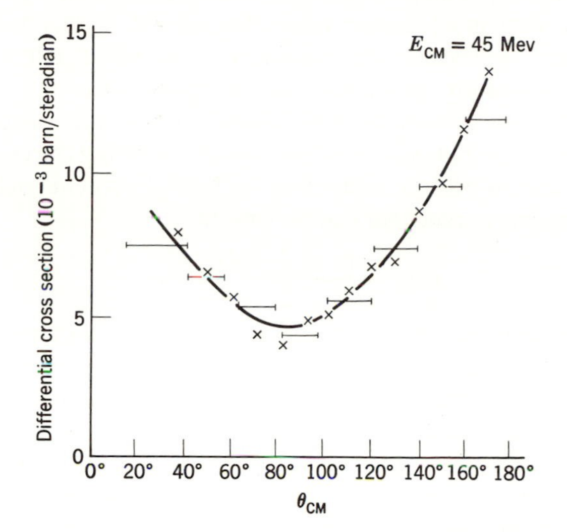

Figure 16–56. Experimental differential cross sections and theoretical fits for neutron-proton scattering at a high energy. From R. Christian and E. W. Hart, *Phys. Rev.*, **77**, 441 (1950).

information about the shapes of the interaction potentials since, when $KR < 1$, the shapes are not important in determining the scattering cross sections.

In order to learn more about the interaction potentials it is necessary to measure and analyze nucleon-nucleon scattering at high energies, where $KR \gg 1$. According to statements made in the previous chapter, it would be expected that such measurements should be able to give information about the shapes of the interaction potentials. Since, for $KR \gg 1$, partial waves with several l values will be involved, the measurements should also be able to tell whether or not the interaction potentials have an explicit dependence on l.

Figure (16–56) shows the neutron-proton differential scattering cross section measured at $E = 45$ Mev, and a theoretical fit to these data. At other energies from about 15 to 150 Mev, $d\sigma/d\Omega$ has much the same appearance except that its magnitude has an approximately $1/E$ behavior. Even the qualitative form of this $d\sigma/d\Omega$ indicates that there is an l dependence in the neutron-proton interaction potential, as can be seen from the following argument. If there were no l dependence the potential could be written as $V(r)$, and its high energy differential scattering cross section could be evaluated, at least qualitatively, by the Born approximation. But, in section 4, Chapter 15, we saw that for $KR \gg 1$ it seems to be a general characteristic of all potentials $V(r)$ that $d\sigma/d\Omega$ is a highly asymmetric function of θ, with much larger values for $\theta \sim 0°$ than for $\theta \sim 180°$. This is in strong contrast to the symmetry about 90° that the $d\sigma/d\Omega$ of figure (16–56) evidently possesses, and it shows that the interaction potential cannot be written as $V(r)$. Serber (1948) has proposed an l dependent interaction potential which is capable of fitting the experimental data for the neutron-proton interaction, at least up to energies of about 150 Mev. This *Serber potential* can be written

$$V_l(r) = \frac{[1 + (-1)^l]}{2} V(r) \qquad (16\text{–}142)$$

Since $V_l(r)$ vanishes for all odd l, it gives zero phase shift for all these l values. This, in turn, gives a $d\sigma/d\Omega$ which has symmetry about 90°, because only even l terms will remain in the general phase shift expression (15–73) for $d\sigma/d\Omega$, and because all the Legendre polynomials $P_l(\cos\theta)$ for even l have symmetry about 90° (cf. 15–53). Note that the Serber potential helps the nuclear forces lead to saturation by removing the nucleon-nucleon interaction potential in half of the possible quantum states. However, it does not do enough. For nuclear forces actually to lead to saturation, it is necessary that there actually be repulsions in certain circumstances.

The study of high energy proton-proton scattering indicates where the repulsions arise. Figure (16–57) shows $d\sigma/d\Omega$ for proton-proton scattering measured at $E = 170$ Mev. In the energy range from about 50 to 200 Mev, $d\sigma/d\Omega$ has almost the same form—a long flat region where $d\sigma/d\Omega \simeq 4 \times 10^{-3}$ bn per unit solid angle, and an abrupt rise at small angles to follow the Coulomb interaction $d\sigma/d\Omega$ that dominates at these angles. When these measurements were first performed, it was thought that, because $d\sigma/d\Omega$ is isotropic in the angular region where nuclear interaction scattering dominates, this was somehow a case of pure S-wave scattering even though $KR > 1$. However, the maximum possible value of $d\sigma/d\Omega$ for S-wave scattering of two protons is†

$$d\sigma/d\Omega = (2/K)^2$$

† Except for the factor of 2, this is equation (15–80) for $\sin^2\delta_0 = 1$. The factor of 2 arises because there are two identical particles in proton-proton scattering.

At $E = 170$ Mev this gives the upper limit $d\sigma/d\Omega \simeq 2.5 \times 10^{-3}$ bn per unit solid angle, which is less than what is observed. Thus it must be that several partial waves contribute to the scattering and happen to mix together in such a way as to yield an isotropic $d\sigma/d\Omega$.

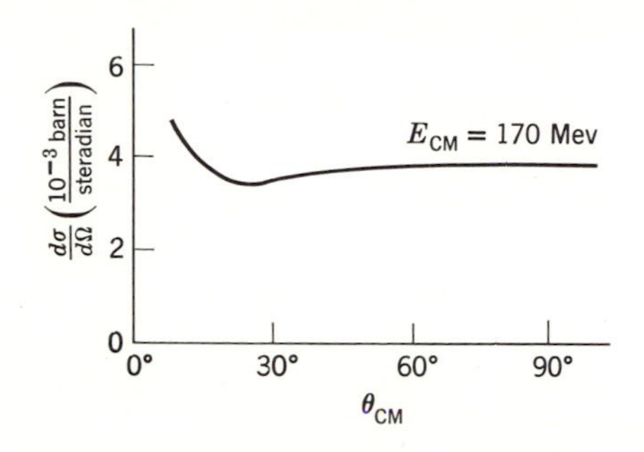

Figure 16–57. The measured differential cross section for proton-proton scattering at a high energy.

Jastrow (1951) has shown that this implies the existence of a *strong repulsive core* in the interaction potential, i.e., that the potential has a form of the type indicated in figure (16–58). The detailed analysis which led to this conclusion is complicated, but the essential points can be seen

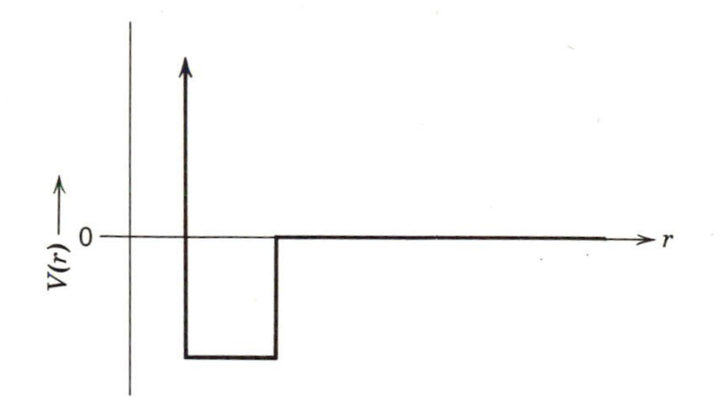

Figure 16–58. A potential with a strong repulsive core.

from a simple argument. Consider proton-proton scattering at $E = 170$ Mev, and assume that, as for neutron-proton scattering, the interaction is described by the Serber potential. Since $KR \simeq 2$ at this energy, only S-, P-, and D-wave phase shifts can be appreciable. But the P-wave phase shift is zero for the Serber potential, so the differential scattering cross section can be obtained from (15–73) by setting all phase shifts zero except δ_0 and δ_2.† As $P_0(\cos\theta) = 1$, this gives

$$\frac{d\sigma}{d\Omega} = \frac{1}{K^2}\{e^{-i\delta_0}\sin\delta_0 + 5e^{-i\delta_2}\sin\delta_2 P_2(\cos\theta)\}$$
$$\times \{e^{i\delta_0}\sin\delta_0 + 5e^{i\delta_2}\sin\delta_2 P_2(\cos\theta)\}$$

$$\frac{d\sigma}{d\Omega} = \frac{1}{K^2}\{\sin^2\delta_0 + 25\sin^2\delta_2 P_2^2(\cos\theta)$$
$$+ 10\sin\delta_0\sin\delta_2[e^{i(\delta_0-\delta_2)} + e^{-i(\delta_0-\delta_2)}]P_2(\cos\theta)\}$$

$$\frac{d\sigma}{d\Omega} = \frac{1}{K^2}\{\sin^2\delta_0 + 25\sin^2\delta_2 P_2^2(\cos\theta)$$
$$+ 20\sin\delta_0\sin\delta_2\cos(\delta_0-\delta_2)\,P_2(\cos\theta)\} \tag{16–143}$$

Now $P_2(\cos\theta) = (3\cos^2\theta - 1)/2$, so $P_2(\cos 0°) = P_2(\cos 180°) = 1$, and $P_2(\cos 90°) = -\frac{1}{2}$. Thus the second term of (16–143) is larger at 0° and 180° than at 90° and, if both δ_0 and δ_2 are positive, the third term will be positive at 0° and 180° and negative at 90°. This will make $d\sigma/d\Omega$ much smaller at 90° than at 0° and 180°, and certainly not isotropic as observed experimentally. To achieve isotropy it is necessary for δ_0 and δ_2 to have opposite signs. Now, for a purely attractive potential of the strength used to fit the low energy data, both δ_0 and δ_2 will be positive (cf. section 5, Chapter 15). But, for a potential of the form indicated in figure (16–58), δ_0 will become negative at energies high compared to the depth of the attractive region since the effect of the attractive region becomes negligible and only the repulsive core, which produces a negative phase shift, is important. The repulsive core is, however, not important in determining the phase shift δ_2 until the energy becomes extremely high because the r^l dependence of the eigenfunctions, for $r \to 0$, makes the $l = 2$ partial wave smaller in the region of the repulsive core than in the attractive region; at $E = 170$ Mev, and similar energies, δ_2 will be positive. By fitting the high energy proton-proton scattering data, it is found that *the radius of the repulsive core is about* 0.5×10^{-13} cm. The attractive region can then be adjusted to give agreement with the low energy data. If its strength is made somewhat greater than that of the simple attractive square well potential described by the parameters (16–141), the correct positive value for δ_0 can be achieved at low energies because the positive phase shift produced by the attractive region becomes larger than the negative phase shift produced by the repulsive core at these energies.

† As this is proton-proton scattering, only the ^1S and ^1D states are possible. Therefore there is only one phase shift for $l = 0$ and one phase shift for $l = 2$.

Recent work of a number of investigators shows that by using a *repulsive core* in a *spin dependent potential*, which is similar to the *Serber potential* in the sense that it is weak for odd l, an interaction potential can be found which gives a good fit to both neutron-proton and proton-proton data at all energies up to more than 100 Mev, providing the *tensor potential* and a *spin-orbit potential* are included. The spin-orbit potential is proportional to $\mathbf{S} \cdot \mathbf{L}$ and is attractive if this quantity is positive and repulsive if it is negative.† One reason for its inclusion is to help fit experiments which find that the spin angular momentum vectors of two nucleons are partially polarized by the process of scattering from each other. We avoid quoting the parameters for the very complicated interaction potential that fits all the data, because these parameters have not been determined uniquely. In fact, it is felt by many that it will not be possible to find a unique potential, and that it is better to express what is known about nuclear forces in terms of sets of phase shifts describing the nucleon-nucleon interactions in various quantum states at various energies. There are even ambiguities concerning the values of these phase shifts.

However, there is, at present, little doubt concerning the existence of repulsive cores (at least in the singlet states, which are the ones most important in the proton-proton interaction). For one thing, repulsive cores obviously produce nuclear forces that lead to saturation, although it was originally thought that they would not lead to saturation at the right density because the radius $\simeq 0.5 \times 10^{-13}$ cm of the repulsive cores is considerably smaller than the mean center-to-center spacing $\simeq 1.2 \times 10^{-13}$ cm of the nucleons in a saturated nucleus. Recent calculations indicate, though, that repulsive cores should actually produce nuclear forces that lead to saturation at densities and binding energies close to those observed in nuclei.

These calculations are based on a *theory of nuclear structure* due to Brueckner (ca. 1953) and others. The theory starts from the nucleon-nucleon interaction potential described in the next to last paragraph, and then goes through a self-consistent procedure similar to that of the Hartree theory of the atom. However, for the nucleus the procedure is very much more difficult than for the atom, because there is no single dominant interaction analogous to the electron-nucleus interaction of the atom, and because the nuclear forces which must be treated in the theory are so complicated. Although the theory is still in its infancy, some results have been obtained, and they are quite encouraging. In addition to finding nuclear densities and binding energies which are close to what is observed,

† Note that the spin-orbit interaction between two nucleons is similar to the spin-orbit interaction between a nucleon and a nucleus (cf. section 8). The two effects are probably related.

fairly good values are found for V_0 and W_0, the real and imaginary parts of the optical model potential. Furthermore, the theory gives some feeling for what is actually happening in a nucleus. For instance, the self-consistent procedure indicates that the nucleons interact primarily two at a time, and not three or more at a time. This is the justification for assuming that the nucleon-nucleon potential, obtained from analysis of experiments involving the interaction of two free nucleons, applies to the interactions between the nucleons of a nucleus. It is also found that the features of the interaction potential which are the most important in determining nuclear structure are: (a), there are repulsive cores; (b), the attractive regions of the S state interaction potentials are just barely strong enough to overcome the repulsive cores and give a net binding for the neutron-proton system; and (c), the entire P state interaction potentials vanish or are very weak.† Of course, the exclusion principle is also extremely important. Despite the uncertainties in the present state of knowledge concerning nuclear forces, progress has been made in the theory of nuclear structure because the features of nuclear forces which are apparently the most important are fortunately those which are the least uncertain.

16. Mesons

In the preceding section we presented a description of certain features of nuclear forces that are observed in experiment. Although theory was used in the description, it was used only to correlate the experimental observations, and not to explain them. There is, however, a theory that attempts to explain how the properties of nuclear forces are based on more fundamental attributes of nature. This is the *meson theory*, which originated with the work of Yukawa in 1935.

Let us imagine two nucleons in interaction. Yukawa proposed that there is associated with this system a "field" which describes the interaction, and that the actual mechanism of the interaction involves the exchange of a "quantum" of the field between the two nucleons (i.e., in interacting, one nucleon emits a "quantum" and the other subsequently absorbs it). This is pictured schematically in figure (16–59). In such an interaction the momentum carried by the "quantum" is transferred from one nucleon to the other, and this is the origin of the force acting between the nucleons.

In making this proposal, Yukawa was guided by analogy with the case of the Coulomb interaction between a system of two charged particles.

† States with l values higher than those of the P state are not very important at the energies characterizing the motion of nucleons in nuclei.

According to the well-developed and highly successful theory of quantum electrodynamics (cf. footnote on page 462), associated with this system there is an electromagnetic field which describes the interaction, and the interaction actually arises from the exchange of a quantum of that field between the two charged particles. The quanta of the electromagnetic field are the zero rest mass entities which throughout this book we have called simply quanta.

Quantum electrodynamics shows that the long range of the Coulomb interaction is a result of the zero rest mass of the quanta that are exchanged in the interaction, and that the range would not be long if it had happened

Figure 16–59. A schematic representation of a meson exchange process.

that quanta had finite rest mass. Yukawa adapted the theory to the case of a system of nucleons, interacting with a short range nuclear force, by assuming that the "quanta" of the "field" describing the interaction have finite rest mass. These "quanta" are called *mesons*, and the "field" is called the *meson field*. We shall not attempt to go beyond the first steps of a quantitative development of the meson theory because that would rapidly carry us into some difficult mathematics. However, there are certain features of the theory which are not at all difficult to describe.

For instance, it is easy to estimate the meson rest mass m_π required to lead to the observed range R of nuclear forces. Consider the process represented in figure (16–59). In this process one nucleon emits a meson of rest mass m_π, the meson travels a distance of the order of R, and finally the meson is absorbed by the other nucleon. While the meson is in flight, the conservation of total energy is violated because the total energy of the system equals two nucleon rest mass energies in the initial and final state, and equals two nucleon rest mass energies plus at least one meson rest mass energy in the intermediate state. But the uncertainty principle shows that violation of energy conservation by an amount

$$\Delta E \sim m_\pi c^2$$

is not impossible if it does not happen for a time longer than $\Delta\tau$, where

$$\Delta\tau\, \Delta E \sim \hbar$$

because such a violation would be immeasurable. The velocity of the meson can be no greater than c, so its flight time $\Delta\tau$ is at least

$$\Delta\tau \sim R/c$$

These three equations give

$$m_\pi c^2 \sim \hbar/\Delta\tau \sim \hbar c/R$$

or

$$m_\pi \sim \hbar/Rc \tag{16–144}$$

Setting $R \sim 2 \times 10^{-13}$ cm, this predicts $m_\pi \sim 200m$, where m is the electron rest mass. It is worth while to restate the argument in the following

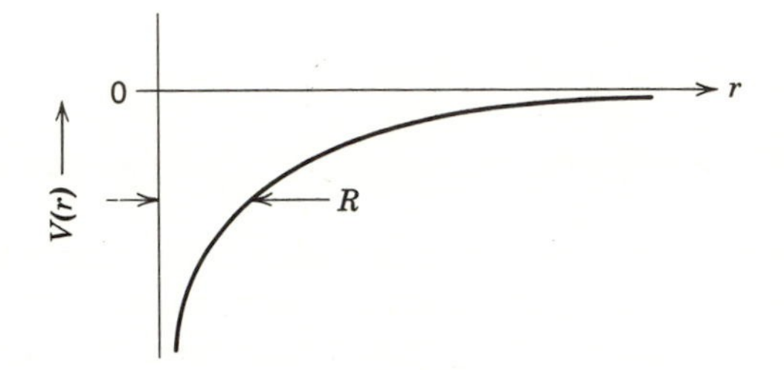

Figure 16–60. The Yukawa potential.

way. A meson of rest mass $m_\pi \sim \hbar/Rc$ leads to an interaction of range $\sim R$ because, if the nucleons are separated by a distance much larger than R, there could be no interaction as its flight time would be so long that the uncertainty principle would allow an accurate enough measurement of total energy to detect the violation of energy conservation.

A detailed treatment, using relativistic quantum mechanics, leads to a detailed prediction of the interaction potential arising from meson exchange. This is the *Yukawa potential*,

$$V(r) = -\frac{q^2}{r}e^{-r/R} \tag{16–145}$$

where

$$R = \hbar/m_\pi c \tag{16–145$'$}$$

which is plotted in figure (16–60). It is not difficult to follow the steps leading to this potential. A relativistic wave equation for the mesons is obtained by writing (1–25),

$$E^2 = c^2p^2 + m_\pi^2c^4$$

where

$$p^2 = p_x^2 + p_y^2 + p_z^2$$

making the substitutions (7–77),

$$p_x \to -i\hbar \frac{\partial}{\partial x}$$

$$p_y \to -i\hbar \frac{\partial}{\partial y}$$

$$p_z \to -i\hbar \frac{\partial}{\partial z}$$

$$E \to i\hbar \frac{\partial}{\partial t}$$

and then multiplying the resulting operator equation into the equality

$$\psi = \psi$$

This gives

$$-\hbar^2 \frac{\partial^2 \psi}{\partial t^2} = -c^2\hbar^2 \nabla^2 \psi + m_\pi^2 c^4 \psi$$

or

$$\nabla^2 \psi - \frac{1}{c^2} \frac{\partial^2 \psi}{\partial t^2} = \frac{m_\pi^2 c^2}{\hbar^2} \psi \qquad (16\text{–}146)$$

which is called the *Klein-Gordon equation*. For $m_\pi = 0$ it reduces to the wave equation

$$\nabla^2 \psi = \frac{1}{c^2} \frac{\partial^2 \psi}{\partial t^2}$$

for the quanta of the electromagnetic field. This has a static solution of the form

$$\psi = -e^2/r, \qquad r > 0$$

as can easily be verified by substitution, using the relation

$$\nabla^2 \psi = \frac{1}{r^2} \frac{d}{dr}\left(r^2 \frac{d\psi}{dr}\right)$$

for $\psi = \psi(r)$. For $m_\pi \neq 0$ the Klein-Gordon equation has a static solution of the form

$$\psi = -\frac{q^2}{r} e^{-r/R}, \qquad r > 0$$

where

$$R = \hbar/m_\pi c$$

as can also easily be verified by substitution. Since the solution to the wave equation for zero rest mass quanta gives the Coulomb interaction potential for the electromagnetic field, the solution for non-zero rest mass quanta is assumed to be the interaction potential for the meson field, that is, the Yukawa potential (16–145).

The constant q^2 of (16–145) determines the strength of the potential. Although this potential is obviously not capable of explaining the existence of the short range repulsive core in the nucleon-nucleon interaction, it

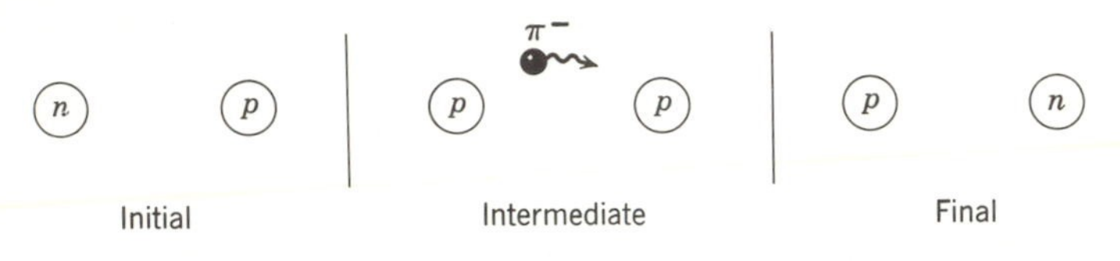

Figure 16–61. The exchange of a π^- meson.

can explain the attractive region outside the core. Fitting it to this region fixes the value of q^2. In terms of the dimensionless constant $q^2/\hbar c$, it is

$$q^2/\hbar c \simeq 15 \tag{16–147}$$

In the original form of the meson theory, it was assumed that all mesons carry the electric charge $-e$ or $+e$. The theory then predicts that a typical interaction process between a proton and a neutron would lead to the situation pictured schematically in figure (16–61). In this figure n represents a neutron, p represents a proton, and π^- represents a π^- *meson* of charge $-e$. We see that, as a result of the interaction, the neutron is changed into a proton and the proton is changed into a neutron.

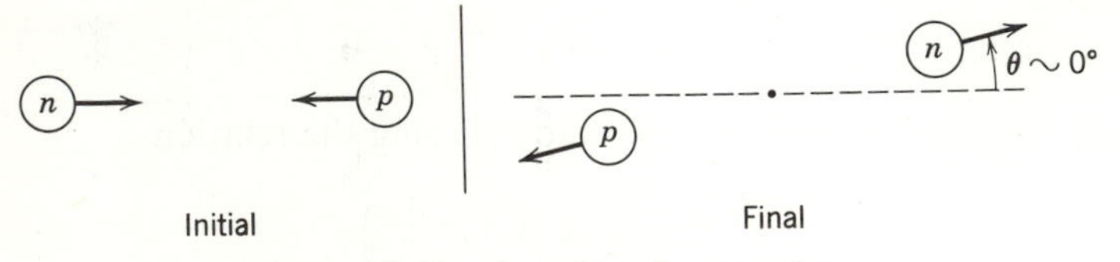

Figure 16–62. A small angle scattering.

This might seem like an unreasonable prediction, but there is very direct experimental evidence that it really can happen. Consider the high energy neutron-proton differential scattering cross section curve of figure (16–56). We have interpreted its symmetry about 90° as evidence for the Serber potential (equation 16–142) because an ordinary interaction potential $V(r)$ would lead primarily to scattering at $\theta \sim 0°$, as indicated in figure (16–62). However, we could also interpret the symmetry about 90° as evidence for the conclusion that there is a probability of $\frac{1}{2}$ that in the

scattering the neutron is changed into a proton, and vice versa, by the exchange of a π^- meson emitted from the neutron, or by the exchange of a π^+ *meson* of charge $+e$ emitted from the proton. Figure (16–63) indicates such a scattering for a case in which a π^- meson is exchanged. An inter-

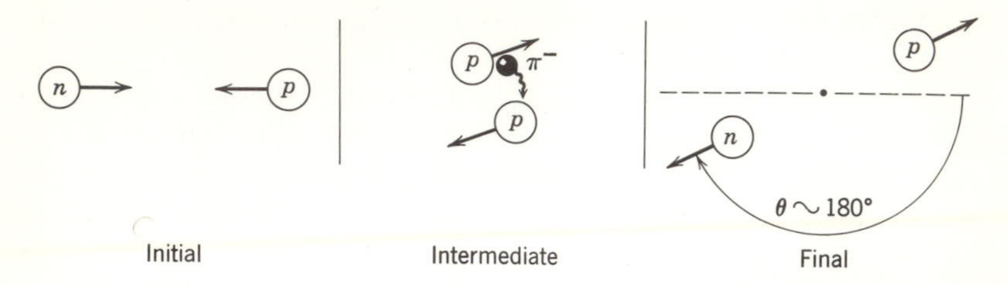

Figure 16–63. A small angle scattering with the exchange of a π^- meson.

action potential, that would lead to scatterings in which there was a probability of $\frac{1}{2}$ that such processes occur, can be written formally as

$$\frac{[1 + P]}{2} V(r) \tag{16–148}$$

where P is an *exchange operator* that exchanges the labels of the two particles. This potential always comes into formulas for the cross section through the matrix element

$$\psi^* \frac{[1 + P]}{2} V(r)\psi$$

so it is of interest to consider the quantity

$$\frac{[1 + P]}{2} V(r)\psi_l = \tfrac{1}{2}V(r)\psi_l + \tfrac{1}{2}V(r)P\psi_l \tag{16–149}$$

where we have specified the eigenfunction by its quantum number l. In the paragraph following (16–141), we found that exchanging the labels of the particles is the same as changing the signs of the coordinates. In this operation, ψ_l will be unchanged if l is even, and will change sign if l is odd, because its parity is even in the first case and odd in the second. Therefore

$$P\psi_l = \psi_l, \qquad l \text{ even}$$

$$P\psi_l = -\psi_l, \qquad l \text{ odd}$$

or

$$P\psi_l = (-1)^l\psi_l$$

and equation (16–149) can be written

$$\frac{[1+P]}{2}V(r)\psi_l = \tfrac{1}{2}V(r)\psi_l + \tfrac{1}{2}V(r)(-1)^l\psi_l = \frac{[1+(-1)^l]}{2}V(r)\psi_l$$

We see, then, that (16–148) is completely equivalent to the Serber potential (16–142), and that the meson theory provides a qualitative explanation of the origin of that potential.

The high energy neutron-proton differential scattering cross section shows, however, that the original assumption that all mesons are charged cannot be correct. If it were, processes of the type indicated in figure (16–63) would take place with a probability of 1, instead of with the

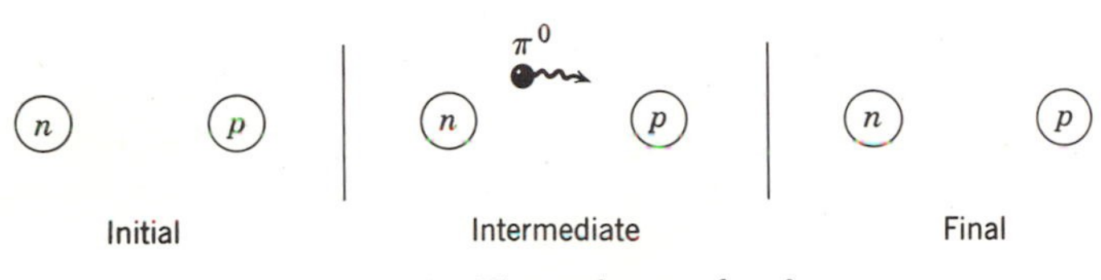

Figure 16–64. The exchange of a π^0 meson.

probability of $\frac{1}{2}$ that is required to fit the data. Thus there must also be uncharged mesons which allow interactions of the type pictured schematically in figure (16–64), where π^0 represents an uncharged π^0 *meson*. Furthermore, the π^0 mesons are necessary in order to allow nuclear interactions between two nucleons of the same charge.†

Meson theory also provides a qualitative explanation of how the neutron can have an intrinsic magnetic moment even though its net charge is zero. According to the theory, a neutron can spontaneously emit a π^- meson and become a proton, if the proton reabsorbs the meson within a time $\Delta\tau \sim \hbar/m_\pi c^2$ and becomes a neutron again. As a result of such processes, there are internal currents associated with the neutron, and these currents produce the intrinsic neutron magnetic moment. The theory predicts that protons also spontaneously emit and reabsorb mesons. Thus meson theory pictures nucleons as being surrounded by a "cloud" of mesons, which have a transient existence and are bound to the nucleon "core" by the requirements of energy conservation.

At the time of Yukawa's proposal, there were no known particles of rest mass between the electron rest mass m and the proton rest mass $1836m$. The π^+ and π^- mesons were first observed by Powell and collaborators in 1947 as a component of the cosmic radiation. Shortly after,

† At the time of Yukawa's original proposal it was thought that nuclear forces acted only between neutrons and protons.

these mesons were produced artificially at a large particle accelerator in collisions between high energy (~300 Mev) nucleons and nucleons in a target. Presumably the cosmic radiation mesons are also initially produced in high energy collisions. Measurements show that the π^+ and π^- mesons have the same rest mass,

$$m_{\pi^\pm} = 274m \tag{16–150}$$

This is certainly close enough to Yukawa's prediction (16–144). The π^0 mesons were first observed by Moyer and collaborators in 1950 as products of high energy collisions. Their rest mass is found to be

$$m_{\pi^0} = 264m \tag{16–151}$$

According to the theory, the free π mesons, which are observed in these experiments, are liberated from the clouds surrounding the colliding nucleons by the energy made available in the collision. It is assumed, then, that these are the same as the mesons discussed in the meson theory. Of course, their discovery constituted a striking verification of the qualitative validity of the theory.

Unfortunately, there has been little opportunity to test the quantitative validity of the theory. The problem is that the usual calculational techniques cannot be applied to meson theory because the dimensionless constant (16–147) is so large. This constant, $q^2/\hbar c \simeq 15$, plays a part in meson theory which is analogous to that played by the fine structure constant, $\alpha \equiv e^2/\hbar c \simeq 1/137$, in quantum electrodynamics. It is possible to make very accurate calculations in the latter theory by the use of perturbation techniques, because it turns out that the perturbations associated with electromagnetic interactions are always proportional to powers of $e^2/\hbar c$ and are therefore very small.† Perturbation techniques cannot be used in meson theory because the perturbations are proportional to powers of $q^2/\hbar c$ and are therefore very large. Quantitative calculations in meson theory await the development of more powerful techniques.

Mesons are unstable. The π^+ and π^- mesons decay spontaneously in free space according to the scheme

$$\pi^\pm \rightarrow \mu^\pm + \nu \tag{16–152}$$

with lifetime

$$\tau_{\pi^\pm} = 2.54 \times 10^{-8} \text{ sec} \tag{16–153}$$

The symbol μ^+ represents a μ^+ *meson* of charge $+e$; μ^- represents a μ^- *meson* of charge $-e$; and ν represents a neutrino. The π^0 mesons

† As a particular example, note that the spin-orbit perturbation (11–51) is proportional to $\alpha^2 = (e^2/\hbar c)^2$.

decay according to the scheme

$$\pi^0 \begin{array}{l} \nearrow \gamma + \gamma, \qquad\qquad 99\% \\ \searrow e^+ + e^- + \gamma, \qquad 1\% \end{array} \tag{16–154}$$

with lifetime

$$\tau_{\pi^0} \sim 10^{-15} \text{ sec} \tag{16–155}$$

The symbol γ represents a gamma ray; e^+ represents a positron; and e^- represents an electron. The μ^+ and μ^- mesons also decay. Their decay scheme is

$$\mu^\pm \rightarrow e^\pm + \nu + \nu \tag{16–156}$$

and their lifetime is

$$\tau_{\mu^\pm} = 2.22 \times 10^{-6} \text{ sec} \tag{16–157}$$

There are no uncharged μ mesons. All values quoted for lifetimes are values in frames of reference in which the mesons are at rest. In a frame in which a meson is moving, its lifetime appears to be lengthened because of the relativistic time dilation effect described by equation (1–7). Measurements of the observed values of $\tau_{\pi^\pm}$ and $\tau_{\mu^\pm}$, made in both rest frames and moving frames, give direct experimental verification of time dilation.

The μ mesons have no part in Yukawa's meson theory of nuclear forces, although this was not appreciated until some time after their discovery in 1936 by Anderson and Neddermeyer. These investigators found the particles as components of the cosmic radiation and showed that their rest mass is intermediate between the rest mass m of an electron and the rest mass $1836m$ of a proton. We now know that they are probably produced in the cosmic radiation from the decay of π mesons. But, in 1936, π mesons had not been discovered, and it was naturally assumed that the μ mesons were Yukawa's mesons. An ever increasing accumulation of evidence showed, however, that the interaction of μ mesons with matter is very weak. For instance, the μ mesons of the cosmic radiation can penetrate great thicknesses of solid matter with little attenuation since they can be detected in deep mines. This being the case, μ mesons can hardly be the particles responsible for the strong interactions of nuclear forces, despite the fact that their rest mass

$$m_{\mu^\pm} = 207m \tag{16–158}$$

is quite close to the value predicted by Yukawa.

This situation was the source of considerable confusion in the ten years before the discovery of the π mesons. But, after their discovery, it was immediately assumed that the π mesons are Yukawa's mesons since the early evidence indicated that their interaction with matter is strong. This was confirmed by later measurements which showed that the cross section for interaction of a π meson with a nucleus is close to its maximum possible value, the projected geometrical cross sectional area πR^2 (cf. equation 15–103). The interaction is also particularly violent; when a π meson interacts with a nucleus, most of its rest mass energy goes into splitting the nucleus into several fragments which fly apart with considerable energy. It is now known that π mesons are closely associated with nucleons and interact with the strong nuclear interaction, while μ mesons are closely associated with electrons and interact with the very weak beta decay interaction. One of the pieces of evidence for the association between μ mesons and electrons is that they both have spin quantum number $i = \frac{1}{2}$, antisymmetric symmetry character, and magnetic moments corresponding to the spin g factor $g_s = 2$. The π mesons have $i = 0$, symmetric symmetry character, and no magnetic moments.

Within the past few years, more than a dozen other unstable particles have been discovered. These particles fall into two classes: the *K mesons* with rest masses around $1000m$, and the *hyperons* with rest masses around $2500m$. At present, not too much is known about their properties. However, some of them seem to have strong interactions, and so probably contribute to nuclear forces. In fact, the relation (16–144) makes it tempting to associate those that interact strongly with the short range repulsive core of the nucleon-nucleon potential, because the relation shows that the range of the force is inversely proportional to the mass of the meson. This may be true, but it is probably also true that part of the effects leading to the core arise from the simultaneous exchange of several π mesons. We shall not even attempt to list the K mesons and hyperons that are presently known, since such a list would undoubtedly soon be obselete. For up to date information concerning this rapidly developing field, and the important part which it surely will play in the fundamentals of modern physics, the reader must consult current research literature and review articles.

BIBLIOGRAPHY

Bethe, H. A., and P. Morrison, *Elementary Nuclear Theory*, John Wiley and Sons, New York, 1956.

Blatt, J. M., and V. F. Weisskopf, *Theoretical Nuclear Physics*, John Wiley and Sons, New York, 1952.

Elton, L. R. B., *Introductory Nuclear Theory*, Isaac Pitman and Sons, London, 1959.

Evans, R. D., *The Atomic Nucleus*, McGraw-Hill Book Co., New York, 1955.

Halliday, D., *Introductory Nuclear Physics*, John Wiley and Sons, New York, 1955.

EXERCISES

1. Show that the minimum energy quantum that will eject a 4.5-Mev proton from a thin hydrogenous layer is $E_\gamma \simeq 50$ Mev, as stated in section 2.

2. Evaluate, in Mev, the energy of gravitational attraction, and the energy of Coulomb repulsion, for two protons separated by a distance of 5×10^{-13} cm.

3. Prove equation (16–3). *Hint:* Use Gauss' law, which is stated in section 2, Chapter 13.

4. Verify equations (16–11), (16–11′), and (16–11″). Also show that (16–11′) reduces to (16–12).

5. Use the appropriate classical expressions to derive (16–20).

6. Use (16–20) to calculate the energy in the LAB frame of the protons emitted in the forward direction, and at 30° from the forward direction, in the reaction ${}^2He^4 + {}^7N^{14} \rightarrow {}^8O^{17} + {}^1H^1$ induced by 30-Mev alpha particles. For this reaction $Q = -1.18$ Mev.

7. Using the data of table (16–1), evaluate, in Mev, the energy released in the fusion of two ${}^1H^2$ nuclei to form a ${}^2He^4$ nucleus. Also evaluate, in Mev, the height of the Coulomb repulsion barrier which must be overcome before there is a large probability that the two nuclei can get close enough together for fusion to take place, assuming that the radius of ${}^1H^2$ is 2×10^{-13} cm.

8. Evaluate the terms of the semi-empirical mass formula for the ${}^{92}U^{238}$ nucleus. Use the results to calculate the predicted values of total binding energy ΔE and average binding energy per nucleon $\Delta E/A$, and compare with table (16–1). Calculate the fraction which each of the terms f_1 through f_5 contributes to ΔE.

9. Look up the ground state spin quantum numbers of all the odd A nuclei, and compare with the predictions of the shell model and figure (16–23).

10. Assuming that ${}^{71}Lu^{175}$ is an ellipsoid of revolution, find the lengths of the semiminor and semimajor axes that lead to the value of quadrupole moment given in figure (16–24), and also to a volume consistent with normal nuclear densities.

11. Show that the semi-empirical mass formula correctly predicts that ${}^{92}U^{238}$ is stable to the emission of ${}^0n^1$, ${}^1H^1$, ${}^1H^2$, or ${}^2He^3$, but that it is unstable to the emission of ${}^2He^4$. Explain why this is so. *Hint:* Take the masses of the emitted particles from table (16–1) because the semi-empirical mass formula is not accurate for very small A.

12. An unstable nucleus whose probability per second of decay is λ is being produced by artificial means in a particle accelerator at a constant rate of I nuclei per second. If the production process commences at $t = 0$, calculate the number of these nuclei which will be present at $t = t'$ by setting up and solving an equation similar to one of the equations (16–46).

13. Use the semi-empirical mass formula to estimate how the data of figure (16–25) would extrapolate to $A = 190, 180, 170, 160$.

14. Use the semi-empirical mass formula to evaluate points on a figure similar to (16–36) for $A = 77$. From it predict the energies available for electron and positron decay. Then look up the data and compare.

15. Use the shell model to predict the spins and parities of the ground states of several sets of odd A nuclei in the region $N = 50$ to $N = 82$. Then predict which of the beta transitions between these ground states will satisfy the Fermi selection rules (16–74), and which will satisfy the Gamow-Teller selection rules (16–77).

16. Evaluate the attenuation length Λ for neutrinos passing through matter of density 10 gm/cm^3, using the typical interaction cross section $\sigma = 10^{-44}$ cm^2 predicted by (16–80).

17. Taking the predictions of the shell model and figure (16–23) for the spin quantum numbers and parities of the ground and first excited states of odd A nuclei in the regions of the magic numbers, where the model should be valid, use the selection rules (16–83) and (16–83′) to predict the character (L value and whether electric or magnetic) of the gamma decay transitions from the first excited state to the ground state in these nuclei. Also determine the predictions of the collective model for even N, even Z nuclei in the regions between magic numbers, where the model should be valid, by consulting section 13.

18. Compare the value of the $\sigma(p, pn)$ cross section for $^{29}Cu^{63}$ and 25-Mev protons, quoted in figure (16–49), with the projected geometrical area for a nucleus of radius $R = 1.25A^{1/3} \times 10^{-13}$ cm^2. Discuss the implications of this comparison in terms of the optical model. Also calculate the probability that a 25-Mev proton would make such a reaction in traversing a foil 0.001 inch thick and containing the normal mixture of ^{29}Cu isotopes.

19. Solve equation (16–125′) graphically, but without approximation, for $R = 2 \times 10^{-13}$ cm. Compare with the predictions of (16–126).

20. Derive the low energy neutron-proton scattering cross section formula without making the approximation of using the bound state eigenfunction for $r < R$ in (16–129), and without dropping the KR in (16–135). Show that this gives $\sigma = [4\pi/(K^2 + K_{||}^2)](1 + K_{||}R)$ instead of (16–138). Taking $R = 2 \times 10^{-13}$ cm, evaluate the error in (16–138) at $E = 1$ ev.

21. Given the fact that at 1 ev the isotropic neutron-proton scattering cross section is $\sigma = 20.4$ bn in the CM frame, use equation (15–17) to evaluate $d\sigma/d\Omega$ in the LAB frame. Plot the results.

22. A π meson comes to rest in the LAB frame and then decays into a μ meson and a neutrino according to the scheme (16–152). Show that the kinetic energy of the μ meson in the LAB frame is $(m_\pi - m_\mu)^2c^2/2m_\pi$, where m_π and m_μ are the π and μ meson rest masses, respectively. Evaluate this kinetic energy, in Mev, from the rest mass values quoted in (16–150) and (16–158).

Index

Absorption, of electromagnetic radiation, 453–7, 505–12
Einstein's coefficient for, 459
physical picture of, 461–2
transition rate for, 457
of nucleons by nuclei, 577–8
of particles, 555
by absorbing sphere, 556–7, 573
in complex potential, 557–8
in Schroedinger theory, 554–8
Absorption spectra, 113, 121
Absorptivity, 47
Actinide elements, 141, 411
Age of earth, 615
Alkali atoms, 418–25
energy levels of, 418–24
fine structure in, 421–4
quantum numbers for, 421
selection rules for, 424–5
spectra of, 112, 424–5
Alkali elements, 413
Allowed transitions, 634
Alpha decay, 89–90, 236–9, 609–16
barrier penetration probability, 238, 610–2
energy, 609–10
equilibrium decay in, 614–5
fine structure in, 641
Geiger-Nuttall law for, 612–3
lifetime, 613
long range components in, 641
motion of alpha particle before, 612
nuclear radii and, 239, 612
Alpha decay, potential, 237, 611
radioactive series and, 613–6, 620–1
rate, and magic numbers, 609
and semi-empirical mass formula, 609
with several optical electrons, 425–8
Alpha particles, discovery and nature of, 89–91
emission through barrier, 236–9, 610–2
Alpha particle scattering, 91–2
and atomic charge, 106–7
and discovery of nuclei, 99–100
and nuclear radii, 107–8
by Rutherford atom, 100–6
by Thomson atom, 92–8
Ampere's law, 339
Angular momentum, 313–4
addition of, in *JJ* coupling, 442
in *LS* coupling, 430–2
in Paschen-Bach effect, 447
grand total, 450
operators, 314–5
orbital, 323; *see also* Orbital angular momentum
spin, 336–9, 449; *see also* Spin angular momentum
total, 346–9, 429–30, 442
total orbital, 429; *see also* Orbital angular momentum
total spin, 428; *see also* Spin angular momentum
Annihilation, 516–7

Anomalous Zeeman effect, 447
Antiparticle, 517
Antiproton, 517
Antisymmetric eigenfunctions, 363–73
and exchange force, 373
and exclusion principle, 366–7
for interacting identical particles, 368
for non-interacting identical particles, 363–4, 367
singlet, 370
and spin, 369
triplet, 370
Associated Laguerre functions, 302, 304–5
Associated Legendre functions, 301, 304–5
Atomic bombs, 619
Atomic energy states, 124–8
Atomic form factor, 500
Atomic mass formula, 590–2
Atomic mass unit, 582
energy equivalent of, 584
Atomic model, of Rutherford, *see* Rutherford's model of atoms
of Thomson, *see* Thomson's model of atoms
Atomic number (Z), 87, 560
Atomic radii, in Bohr theory, 116, 133
in Hartree theory, 403
macroscopic estimate, 78–9
in Thomas-Fermi theory, 396
Atomic reactor, 619
Atomic spectra, absorption, 113, 121
of alkali atom, 112, 424–5
emission, 113
fine structure of, *see* Fine structure
of hydrogen, *see* One-electron atom
hyperfine splitting of, 357–8, 449
of JJ coupling atom, 443
lines of, 111
of LS coupling atom, 440–1
measurement of, 110–1
of one-electron atom, 111, 119, 425
optical, 417
series of, 112
series limit of, 112
spark, 120–1
Zeeman, 447
Atomic theory, of Bohr, *see* Bohr's theory of atoms
of Hartree, *see* Hartree's theory of atoms
of Schroedinger, *see* One-electron atom
of Thomas and Fermi, *see* Thomas-Fermi theory of atoms
Atomic weight (A), 78, 560
Atoms, charge of, 87–8, 106–7, 488–9
charge density of, 395–6, 400–1
energy levels of, *see* Energy levels, of atoms
excitation of, 119, 416–8
mass of, 78, 585
stability of, 108–9, 114–5, 152, 187, 414, 467
Attenuation, exponential law for, 511–2
of nucleons, 577–8
of X-rays, 493–512
Attenuation coefficient, 517
Attenuation length, for nucleons, 577–8
for X-rays, 512
Avogadro's number, 78
determination of, 481
Axial vectors, 637

Balmer series, 113, 120
Barn (bn), 684
Barrier penetration, and alpha particle emission, 236–9, 610–2
for barrier potential, 233–9
by electromagnetic radiation, 235–6
for step potential, 226–7
and uncertainty principle, 239
and particle-wave duality, 235, 239
Barrier penetration probability, for barrier potential, 233–4
for nuclear potential, 610–2
in three dimensions, 610–1
Barrier potentials, 231–9
alpha particle emission and, 236–9
boundary conditions for, 232
classical behavior of, 231
eigenfunctions for, 232
energy levels for, 232
Ramsauer effect and, 234–5
transmission coefficients for, 233–4
transmission resonances for, 234–5

Bessel functions, spherical, 538
Beta decay, 89–90, 620–40
  coupling constant, 628, 635, 640
  Coulomb effects in, 631
  density of final states for, 628–30
  energy, 621–4
  energy conservation in, 624–5
  Fermi selection rules for, 633–4
  Fermi theory of, 627–33
  $F\tau$ values for, 633–6
  Gamow-Teller selection rules for, 636
  Kurie plots for, 630
  lifetime, 632–3
  matrix element, 628, 633–8
  neutrino induced reactions in, 638–40
  neutrino recoil in, 638
  neutrinos and, 626
  origin of, 621
  parity non-conservation in, 636–8
  pseudo-scalar matrix element for, 636–8
  shell model and, 634–5
  spectra, 624
Beta particles, 89–90, 599
Binding energies, of electrons in atoms, 128, 373, 412–3
  of nuclei, 585–6
    constancy of, 586, 590
    and fission, 587
    and fusion, 587–8
    and saturation of nuclear forces, 675–7
Black body radiation, 47–9
  Planck theory of, 63–6
  Rayleigh-Jeans theory of, 51–7, 62–3
Black body spectrum, 48–9
  and spontaneous emission, 458, 460
Bohr magneton, 329
Bohr's microscope, 157–8
Bohr's theory of atoms, 110–37
  angular momentum of, compared to quantum mechanics, 313
  energy levels of, compared to quantum mechanics, 303
  and magnetic moments, 327–9
  postulates of, 114–5, 123, 128–30
  postulates of, de Broglie's interpretation, 151–2
  predictions of, 115–24
Bohr's theory of atoms, radii of orbits, compared to quantum mechanics, 308
  Sommerfeld's relativistic modification of, 131–4, 355–6
Boltzmann probability distribution, 61
  applications of, 121, 458–60
  derivation of, 57–61
  relation to quantum probability distributions, 388–90
Boltzmann's constant, 61
Bombarding particle, 563
Born approximation, 524–30
  applicability of, 534, 554
  for Coulomb potential, 554
  for Gaussian potential, 530–2
  for nuclear direct interactions, 672–3
  for relativistic electrons, 569
  for square well potential, 532–3
Born's interpretation of wave functions, 172–5, 294–5
Bose probability distribution, 389–90
Boundary conditions, for eigenfunctions, 178, 300
    of barrier potential, 232
    of one-electron atom, 300–2
    of simple harmonic oscillator potential, 258–9, 261
    of step potential, $E < V_0$, 223–4
    $E > V_0$, 228
  of square well potential, $E < V_0$, 241
    $E > V_0$, 250
  periodic, *see* Periodic boundary conditions
    for Thomas-Fermi theory, 394–5
Box normalization, 220
  applications of, 252, 270, 525–6, 629
Brackett series, 113, 120
Bragg's law, 479
Breit-Wigner formula, 670
Bremsstrahlen, 489; *see also* X-ray continuum spectra
Broadening of spectral lines, *see* Line widths
Brueckner theory of nuclear structure, 693–4
Bucherer's experiment, 37, 75–6

Cascades in gamma decay, 642
Cathode rays, 71
Cathode ray tube, 71
Center of mass, 122
separation of time independent Schroedinger equation for, 295–8
Center of mass (CM) frame, 39, 519
relation to laboratory frame, 519–22
Centrifugal potential, 404–5
Chain reactions, 618–9
Charge, of atoms, 87–8, 106–7, 406, 488–9
of electrons, 74
measurement of, 73–5
of nuclei, 106, 560
Charge density, of atoms, 339–400
of nuclei, 570–2
Charge independence of nuclear forces, 675, 688–9
Classical limit of quantum theory, 206–9
and correspondence principle, 135–6
and identical particles, 364–5
and orbital angular momentum, 324
and simple harmonic oscillators, 68, 264–5
and spin angular momentum, 337–8
and square well potentials, 248–50
and time dependent perturbation theory, 289–90
Classical physics, 1
Cloud chamber, 513
Collective model of nuclei, 605–8
and electric quadrupole moments, 606–8
and low lying energy levels, 658–9
and magnetic dipole moments, 606
and liquid drop model, 605
and shell model, 605
Collision broadening of spectral lines, 470
Compound nucleus, 655
and resonances, 655–7
Compound nucleus reactions, 662–71
angular and energy distributions of, 670–1
Breit-Wigner formula for, 670
excitation and decay of, 664–7
experimental evidence for, 668–70
Compound nucleus reactions, independence of formation and decay in, 667
random width assumption of, 667
uncertainty principle and, 663
Complex conjugate, 172
Complex potentials, 557–8, 572–80
Compton scattering, 81–5, 502–5
cross section for, 503–5
from protons, 564
quantum theory of, 82–5, 503
Compton wavelength, 85
Conduction electrons, 382, 386–7
Configuration, 409
Conservation, of charge, 584, 644
of energy, in beta decay, 624–5
and limitations of uncertainty principle, 289, 667, 695–6, 700
in relativity theory, 35–6
and nuclear reactions, 582–5
of probability, 175; *see also* Normalization, of wave functions *and* of eigenfunctions
and complex potentials, 557–8
Conservation equation, 174–5
Continuity of eigenfunctions and their derivatives, 178, 321
Continuum of energy levels, atomic, 128, 184
nuclear, 662
Contraction of length, *see* Lorentz contraction
Correspondence principle, 135–6; *see also* Classical limit of quantum theory
Cosmic radiation, 513
Coulomb cross section, 554
Coulomb potential, 295; *see also* One-electron atom
Coulomb scattering, 100–6
angular distribution, 106
cross section, 554
Coulomb's law, 42
and quantum electrodynamics, 694–5, 696–8
Coulomb repulsion energy of nuclei, 566
Coupling constant, beta decay, 628, 635, 640

Coupling constant, comparison between magnitudes of, 640
  Coulomb field, 701
  meson field, 698, 701
Coupling of angular momenta, *JJ*, 441–3
  LS, 428–41
  orbital, 426–7
  restrictions of exclusion principle, 435–9, 442
  spin, 426
  spin and orbital, 428
  strong field, 447
Cross product, 313
Cross section, 495, 504–5, 522
  absorbing sphere, 557
  Born approximation formulas for, 529–30
  Breit-Wigner, 670
  CM and LAB, 522–4
  Coulomb, 554
  differential, 498, 504
  Gaussian potential, 531
  geometrical interpretation of, 496
  hard sphere, 548
  Klein-Nishina, 503–4
  neutron-proton, 685–6, 689
  pair production, 508–9, 510
  phase shift formulas for, 544, 556
  photoelectric, 507, 510
  proton-nucleus, 574
  proton-proton, 688–91
  square well, 532, 551, 684
  S-wave, 549
  Thomson, 495, 498
  X-ray, 509–10
Current loops, 327–8

Daughter nucleus, 609, 613
Davisson-Germer experiment, 147–9
De Broglie's postulate, 139–41
  experimental confirmation of, 146–51
  relation to Schroedinger's theory, 165–6
  relation to uncertainty principle, 153
De Broglie wavelength, 146
  typical values of, 146–7, 150
Degeneracy, 278
  in atoms with several optical electrons, 425–6
Degeneracy, and effect of perturbations, 278–9
  and exchange of identical particles, 363
  in Sommerfeld theory, 133
Degenerate perturbation theory, *see* Perturbation theory for degenerate case
Density of energy levels for free particle, 527–8
Density, of atoms, 79
  of nuclei, 560, 572, 580
  of final states, in beta decay, 628–30
    in nucleon emission, 661
    in time dependent perturbation theory, 290
Deuterium, *see* Deuteron
Deuteron, 677–82
  eigenfunction for, 682
  electric quadrupole moment of, 677–9
  excited states of, 682, 685
  ground state of, 678
  potential for, 679, 687
  properties of, 677
  time independent Schroedinger equation for, 680
  virtual state of, 685
Differential cross section, 498, 522; *see also* Cross section
Differential operators, 201–5
  for angular momentum, 314–5
  and associated dynamical quantities, 203
  and development of Schroedinger equations, 293–4
  and eigenvalue equations, 318–22
  and relation to Schroedinger's theory, 204, 321
  and relativistic wave equations, 697
Diffraction, of electromagnetic radiation, 140, 148, 149–50, 160–2, 467–80
  of electrons, 147–50, 567–8
  of helium atoms, 150
  of protons, 573–5
  of quantum mechanical waves, 533, 548–9, 557
Diffraction gratings, 110, 148, 149, 467–80

Dilation of time, *see* Time dilation
Dirac theory, of quantum mechanics, 338
  and fine structure, 356–7
  and pair production, 513–7
  of vacuum, 514–6
Direct interactions in nuclear reactions, 662–3, 670–4
  angular and energy distributions for, 670–1, 673
  Born approximation and, 672–3
  excitation and decay of, 672
  experimental evidence for, 670–1, 672–3
Discrete energy levels, 183
  finite widths of, 470
Doppler broadening of spectral lines, 469
Dot product, 331
Doublets, of energy levels, *see* Fine structure
  of spectral lines, 424
Dynamical quantities, 202
  and associated differential operators, 203
  fluctuations in, 318–21

Earth, age of, 615
Effective charge, 401, 488
Ehrenfest's theorem, 206–9
Eigenfunctions, 185, 320
  antisymmetric, 363–4, 367–8
  for barrier potential, 232
  in beta decay, 627
  conditions of acceptability, 178, 300, 321
  for deuteron, 682
  finiteness of, 178, 321
  for free particle, standing wave, 217
    three dimensional, 526, 540
    traveling wave, 213
  and gamma decay transitions, 647
  for helium atom, 375–8
  for incident and scattered waves, 535
  for infinite square well potential, 253, 254
  for interacting identical particles, 368–9
Eigenfunctions, linear combinations of, 189–90, 191, 269, 272–3, 276–7, 538
  for *LS* coupling, 432–3
  mathematical properties of, 184–92
  for multi-electron atoms, 397–8
  for non-interacting identical particles, 363–4, 367–8
  normalization of, *see* Normalization, of eigenfunctions
  for one-electron atom, *see* One-electron-atom eigenfunctions
  orthogonality of, *see* Orthogonality of eigenfunctions
  for simple harmonic oscillator potential, 263–4
  singlet, 370–1
  for spin angular momentum, 370
  for square well potential, $E < V_0$, 243, 247
    $E > V_0$, 250–1
  for step potential, $E < V_0$, 224, 225
    $E > V_0$, 229
  for string, 196
  symmetric, 363–4, 368–9
  triplet, 370–1
Eigenstates, 187
  and emission of radiation by atoms, 461–2
Eigenvalue equations, 318–22
  for angular momentum, 323, 538
  for energy, 320–1
  for linear momentum, 321–2
Eigenvalues, 185, 320; *see also* Energy levels
Einstein's coefficients, 459
Einstein's photoelectric theory, 79–81
Einstein's relativity postulates, 15–6
Einstein's mass-energy relation, 34
  experimental verification of, 585–6
  and nuclear binding energy, 585–6
  and nuclear reactions, 582–5
Einstein's quantum postulate, 80
  and theory of radiating atoms, 455, 462
Einstein's spontaneous emission theory, 458–60
Electric dipole matrix elements, 455, 460–1

Electric dipole transitions, 454, 467, 644
Electric quadrupole moment, 606, 646
and collective model, 606–8
of deuteron, 677–9
of nuclei, 451–2, 607
and shell model, 608
Electric quadrupole transitions, 468, 644
Electromagnetic radiation, absorption of, 453–7, 461–2
classical theory of, 42–6
dual nature of, 85–6, 139, 161–2
energy density of, 43
energy flux of, 45
induced emission of, 457, 461–2
rate of energy emission, 46
spontaneous emission of, 453, 460, 461–2
Electromagnetic wave equation, 697
Electron, capture in beta decay, 623–4
diffraction, 147–50, 567–8
discovery of, 70–3
emission in beta decay, 623–4
scattering by nuclei, 567–70
Electron charge, 74
measurement of, 73–5
Electron mass, 74
measurement of, 73–5
Electron-proton mass ratio, 124
Electron spin, *see* Spin angular momentum
Electron volt (ev), 125
Emission spectra, 113
Emission theories, 15
Emissivity, 46
e/m ratio for electrons, 73
Energy, equipartition of, 61
relativistic kinetic, 33
rest mass, 34
total relativistic, 34
Energy conservation, in nuclear reactions, 582–5
in relativity theory, 35–6
and uncertainty principle limits, 289, 667, 695–6, 700
Energy level diagram, 64
Energy levels, of atoms, for alkali atoms, 418–24
Energy levels, of atoms, for helium, 374, 376
for hydrogen, *see* One-electron atom
in *JJ* coupling, 441–2
in *LS* coupling, 433–4, 439–40
for multi-electron atoms, 403–5, 408–9, 411–2, 415, 425–8, 433–4, 439–40, 441–2
for one-electron atoms, 117, 134, 303–4, 308–9, 403–5
Zeeman splitting of, 446–7
of bound particle, 184
continuum, 128, 184
discrete, 183
of Fermi gas, 383
of free electron, 128, 513–7
of free particle, 312
of infinite square well potential, 255
of nuclei, average spacing, 659–60
in collective model, 658
continuum of, 662
density of, 660
excitation of, 641
in Fermi gas model, 595
measurement of, 655–8; *see also* Beta decay; gamma decay
in shell model, 601–2, 658–9
widths of, 660–2
of rigid rotator, 138, 325
of simple harmonic oscillator, 64, 131, 262
of step potential, 222
of square well potential, $E < V_0$, 244–5
$E > V_0$, 240, 250
of unbound particle, 184
X-ray, 485–6
Equilibrium decay, 614–5
Equipartition of energy, 61
Ether, 7
Ether drag, 14
Ether frame, 8
Even-odd effects in nuclei, 589, 591
and Fermi gas model, 596–9
Exchange, of labels for identical particles, 362–3
of mesons, 694–5
and repulsive core, 703

Exchange, of mesons, and Serber potential, 698–700
Exchange degeneracy, 363
in helium atom, 375
Exchange force, 372–3
in helium atom, 377–8, 380
Exchange integral, 380
and nuclear Coulomb energy, 566
Exchange operator, 699
Excitation, of atoms, 119, 416–8
of nuclei, 641
Excited states, 119; *see also* Energy levels
Exclusion principle, 366–8
and angular momentum of holes, 485
and antisymmetric eigenfunctions, 367
and capacity of subshells, 408
and conduction electrons, 386–7
and experimental evidence, 381
and Dirac theory of pair production, 514
and Fermi gas, 384, 386–7, 595
and Fermi probability distribution, 390
and independent particle motion in nucleus, 595
and *JJ* coupling, 442
and *LS* coupling, 435–9
and properties of atoms, 414
and proton-proton scattering, 687
and theory of nuclear structure, 694
Expectation values, 201–5
general formula, 205
order of terms in integrals, 204, 211
relation to perturbation theory, 272
Exponential attenuation, 511–2
Exponential decay, 468–9
of compound nucleus states, 666

Fermi beta decay theory, 627–33
Fermi energy, 385, 387
of conduction electrons, 386
of nuclei, 596
Fermi gas, 381–9
applications of, 382
conduction electrons and, 386
energy distribution of, 383, 388
energy level density of, 383
energy levels of, 383
Fermi gas, exclusion principle and, 384
Fermi gas model of nuclei, 594–9
and depth of nuclear potential, 595–6
and even-odd effects, 596–9
and optical model, 595–6
and semi-empirical mass formula, 596–8
and shell model, 595, 599
Fermi probability distribution, 388–90
Fermi selection rules, 633–4
Field, electromagnetic, 694–5
meson, 694–5
Fine structure, 131, 334–5, 338–9, 355–7
in alkali atoms, 421–4
in *JJ* coupling, 441
in *LS* coupling, 433–4, 441
Fine structure constant, 134
and quantum electrodynamics, 701
and spin-orbit energy, 352
Finitenesse of eigenfunctions and their derivatives, 178, 321
Fission of nuclei, 616–20
chain reactions in, 618–9
energy diagram for, 616, 618
energy emitted in, 587
fragment distribution, 619–20
fragments, 616
induced, 617–8
neutron emission in, 618
origin of, 616
spontaneous, 617
Fluctuations in dynamical quantities, 318–21
Flux, classical, 174–5
incident, 504–5, 522
probability, 175; *see also* Probability flux
scattered, 505, 522
Forbidden transitions, 635
Form factor, 500
Fourier analysis, 154, 158, 191, 200
Franck-Hertz experiment, 124–8
Free particle, 212–20
classical behavior of, 212
eigenfunctions, standing wave, 217
three dimensional, 526
traveling wave, 213
energy levels, 128, 213

Free particle, normalization of wave functions for, 215, 219
  probability density, standing wave, 218
    traveling wave, 213
  probability flux, standing wave, 218
    traveling wave, 214
  time independent Schroedinger equation, 212
  wave functions, standing wave, 217
    traveling wave, 213
$F\tau$ values, 633–6
Fusion of nuclei, 587–8

Galilean transformation, 5
  and mechanics, 3–9
Gamma decay, 640–55
  cascades, 642
  electric and magnetic transitions, 644–8
  energy of gamma rays in, 642
  internal conversion and, 650–3
  isomeric transitions, 643
  lifetimes, 652–3
  matrix elements, 649–50
  measurement of energy, 642–3
  measurement of lifetime, 643
  selection rules, 648
  transition rates in shell model, 647–8
Gamma rays, 89–90, 559–60
Gamow-Teller selection rules, 636
Gaussian potential, 531
Gauss' law, 393
Geiger counter, 563–4
Geiger-Nuttall law, 612–13
General theory of relativity, 38
g factor, Lande, 446
  meson, 703
  nuclear, 450
  orbital, 329
  spin, 336–7, 357
Golden Rule No. 2, 291
  and beta decay, 627–30
  and Born approximation, 527–9
  and widths of nuclear energy levels, 661–2
Gravitational forces compared to nuclear forces, 566
Ground state, 119
Groups of wave functions, 142–5, 153–4, 159–60, 163, 215–6
Group velocity, 142, 144, 166

Halogen elements, 413
Hamiltonian, 321
Hartree's theory of atoms, 396–405; *see also* Multi-electron atoms, in Hartree theory
  and nuclear shell model, 600
  and theory of nuclear structure, 693
  and verification by X-ray measurements, 400, 501
Heisenberg's theory of quantum mechanics, 164–5, 337
Hermite equation, 259
  solution of, 259–62
Hermite polynomials, 263
Holes, and pair production, 515–7
  properties of, 437, 516
  quantum numbers for, 485
  selection rules for, 486
  subshells with, 437
  X-ray spectra and, 484–5
Hydrogen, *see* One-electron atom
Hyperfine splitting, 357–8, 449
Hyperfine structure, 449–52
  and electric quadrupole interaction, 451–2
  and magnetic dipole interaction, 448–51
Hyperons, 703

Identical particles and indistinguishability, 360–1
Independent particle motion in nucleus, 593–5
Indistinguishability, 360–1
Induced emission, 457
  Einstein's coefficient for, 459
  physical picture of, 461–2
  transition rate for, 457
Induced fission, 617–8
Inertial frame, 4
Infinite square well potential, 251–6
  applications of, 252
  boundary conditions for, 252–3
  eigenfunctions, 253, 254
  energy levels, 255

Infinite square well potential, normalization of eigenfunctions for, 254
parity of eigenfunctions for, 253–4
time independent Schroedinger equation, 253
zero point energy of, 255–6
Interaction strengths, comparison between, 640
Interference of eigenfunctions and wave functions, 148, 151–2, 153–4, 365
Interferometer, 11
Internal conversion, coefficients, 651–2
energetics of, 650–1
gamma decay and, 650–3
Interval rule of Lande, 434, 441
Ionization energy, 128, 413
Isomeric transitions, 643
Isotopes, 581
abundance of, 582, 588–9
hyperfine structure and, 449

JJ coupling, in atoms, 441–3
in nuclei, 603

Kirchhoff's law, 47
Klein-Gordon equation, 697
Klein-Nishina cross sections, 503–5
K mesons, 703
Kurie plots, 630

Laboratory (LAB) frame, 39, 519
relation to center of mass frame, 519–22
Laguerre functions, associated, 302, 304–5
Lamb shift, 357
Lande *g* factor, 446
Lande interval rule, in atoms, 434, 441
in nuclei, 600
Lanthanide elements, 411, 414
Laplacian operator, in rectangular coordinates, 294
in spherical coordinates, 298
Larmor frequency, 332
Larmor precession, 331–2
of net magnetic moment, 444–5
Last nucleon in nucleus, 592–3
Legendre functions, associated, 301, 304–5
Legendre polynomials, 537
Lifetime of quantum states, 469
and widths of energy levels, 472
Limiting velocity in relativity, 17, 27, 29, 38, 44
and range of nuclear forces, 696
Line, spectral, 111
of stability, 618
Linear combinations, *see* Superposition of wave functions
of eigenfunctions, 189–90, 191, 269, 272–3, 276–7, 538
of wave functions, 186, 192, 285
Linearity of Schroedinger equation, 166–7, 185–6
Line widths, 469–72
and collision broadening, 470
and Doppler broadening, 469
and natural broadening, 469–70
and uncertainty principle, 469–70, 472
Liquid drop model of nuclei, 589–91
and collective model, 605–8
and fission, 617
and semi-empirical mass formula, 590–1
and shell model, 605
Lorentz contraction, 14, 18, 21–2
Lorentz transformation, 26
derivation of, 24–7
LS coupling, 428–41
energy levels, 433–4, 439–40, 441
and exclusion principle, 435–9
fine structure splitting, 433–4, 441
of orbital angular momenta, 429
quantum numbers, 429–33, 435
selection rules, 440–1
spectra, 440–1
of spin angular momenta, 428
of total angular momenta, 429
Lyman series, 113, 120

Magic numbers, 592–4
and alpha decay energy, 609
and electric quadrupole moments, 608
and fission fragment distribution, 620

Magnetic dipole moment, of deuteron, 677–9
of electron, 336–7
of filled subshell, 438–9
net, 443
of neutron, 679
and meson theory, 700
in non-uniform magnetic field, 333–4
orbital, of electron, 327–9
of positron, 516
and precessional motions, 331–2, 444–5
of proton, 679
quantization, 329, 336
spin, of electron, 336–9
Magnetic dipole moments of nuclei, 335–6, 357–8, 449
and collective model, 606
and electrons in nuclei, 562
and shell model, 603–4
Magnetic dipole transitions, 468
Magneton, Bohr, 329
nuclear, 450
Mass, of deuteron, 677
of electron, 74
measurement of, 73–5
of μ meson, 702
of π meson, 701
of neutrino, 626
of neutron, 585
of nuclei, 585
measurement of, 580–5
of positron, 516
of proton, 585
relativistic, 32
rest, 32
Mass deficiency of nuclei, 585–6
Mass density of nuclei, 572
Mass-energy relation, 35–6
and nuclear reactions, 565
Mass spectrometry, 580–2
Mass unit, atomic, 582
energy equivalent of, 584
Matrix element, 273, 286
for beta decay, 628, 633–8
electric dipole, 455, 460–1
for gamma decay, 649–50
nuclear, *see* Beta decay
Maxwell's equations, 7
Mechanics, classical, 3–4
relativistic, 30–7
Mesons, 694–703
charge of μ, 701
charge of π, 698, 700
decay of μ, 702
decay of π, 701–2
discovery of μ, 702
discovery of π, 700–1
hyperons, 703
interaction of μ, 702
interaction of π, 702–3
K mesons, 703
lifetime of μ, 702
lifetime of π, 701–2
lifetime and time dilation, 702
magnetic dipole moment of μ, 703
magnetic dipole moment of π, 703
mass of μ, 702
mass of π, 696, 701
and range of nuclear forces, 695–6
spin of μ, 703
spin of π, 703
symmetry character of μ, 703
symmetry character of π, 703
Meson theory, 696–700
and role of perturbation theory, 701
Michelson-Morley experiment, 9–13
Microscopic reversibility, 458
Mirror nuclei, 565
beta decay of, 634–5
binding energies and nuclear radii, 565–6
Models, 88, 559
and theories, 604–5
Modern physics, 1
Multi-electron atoms, in Hartree theory, 396–405; *see also* Alkali atoms; Atoms, with several optical electrons; Periodic table of elements
effective charge for, 401
eigenfunctions for, 397–8
net potentials for, 397, 400–1
probability density for, 397–401, 404–5
quantum numbers for, 398
self-consistency in, 397
shells in, 402–3

Multi-electron atoms, in Hartree theory, subshells in, 403–5, 408–9, 411–2, 415
time independent Schroedinger equation for, 396
in Thomas-Fermi theory, 393–6
boundary conditions for, 394–5
net potential for, 394–5
probability density for, 394–5, 401
self-consistency in, 393
Multiple scattering, 92–8
angular distribution for, 98
Multiplet, 434

NaI scintillation detector, 537
Natural broadening of spectral lines, 469–70
Natural width of spectral lines, 470
Net potential, in Hartree theory, 397, 400–1
in multi-electron atoms, 392
in Thomas-Fermi theory, 395
Neutrino, beta decay and emission of, 626–33
meson decay and emission of, 701–2
induced reactions, 638–40
properties of, 626
recoil in beta decay, 638
Neutron, 562
beta decay of, 630–1
discovery of, 562–5
magnetic dipole moment of, 679
and meson theory, 700
mass of, 585
spin angular momentum of, 562
symmetry character of, 562
Newton's laws, 4
for angular momentum, 331
Noble gas elements, 412–3
Normalization, of eigenfunctions, 187–8, 190
examples, 254, 367–8, 525–6
of probability density, 175
of wave functions, 187, 190–1, 215, 219–20
Nuclear bombs, 619
Nuclear collective model, *see* Collective model of nuclei
Nuclear Coulomb repulsion energy, 590–1
Nuclear even-odd effects, 589, 591
and Fermi gas model, 596–9
Nuclear Fermi gas model, *see* Fermi gas model of nuclei
Nuclear fission, *see* Fission of nuclei
Nuclear forces, 674–94
angular momentum dependence of, 690
attractive character of, 556, 674
charge independence of, 675, 688–9
and deuteron, 677–82
and eigenfunction for deuteron, 682
and meson theory, 694–703
and neutron-molecular hydrogen scattering, 685–6
and neutron-proton scattering, high energy, 689–90
low energy, 682–7
and nuclear structure, 693–4
and potential for deuteron, 679, 687
and proton-proton scattering, high energy, 690–3
low energy, 687–8
range of, 579, 674, 687, 692
compared to other forces, 579
and meson theory, 696, 697
repulsive cores in, 691–3
and meson theory, 703
and saturation of, 693–4
saturation of, 675–7, 690, 693
and Serber potential, 690, 691–2, 693, 699–700
spin dependence of, 685, 693
and spin-orbit potential, 693
and stability of nuclei, 566
strength of, compared to other forces, 556, 579–80, 640
and tensor potential, 678–9, 693
and virtual state of neutron-proton system, 685–6
Nuclear fusion, *see* Fusion of nuclei
Nuclear liquid drop model, *see* Liquid drop model of nuclei
Nuclear magic numbers, *see* Magic numbers
Nuclear magneton, 450

Nuclear mass formula, 590–2
and alpha decay energy, 609
and beta decay energy, 621–2
Nuclear matrix element, 628, 633–6
Nuclear model of atoms, *see* Rutherford's model of atoms
Nuclear optical model, *see* Optical model of nuclei
Nuclear potential, 572–80; *see also* Shell model of nuclei
composition of total real part, 579–80
Nuclear radii, for charge distribution, 565–72
comparison with potential distribution radii, 578–9
electron scattering measurements of, 567–72
mirror nuclei binding energy measurements of, 565–6
values of, 571
for potential distribution, 572–80
alpha particle scattering and decay measurements, 100, 107–8, 239, 612
comparison with charge distribution radii, 578–9
neutron cross section measurements of, 572–3
proton scattering measurements of, 573–9
values of, 575–6
Nuclear reactions, 662–74
compound nucleus, *see* Compound nucleus reactions
direct, *see* Direct interactions in nuclear reactions
energy balance in, 582–5
energy level measurements and, 655–8
history of, 563–5
notation for, 563
Q value of, 583
Nuclear reactors, 619
Nuclear shell model, *see* Shell model of nuclei
Nuclear structure, theory of, 693–4
Nuclei, binding energies of, *see* Binding energies, of nuclei
charge density of, 570–2
Nuclei, composition of, 561–5
electric quadrupole moments of, *see* Electric quadrupole moments
energy levels of, *see* Energy levels, of nuclei
magnetic dipole moments of, *see* Magnetic dipole moments of nuclei
mass of, 585
mass deficiency of, 585–6
mass density of, 560, 572, 580
number of protons and neutrons in, 565
spin of, *see* Spin angular momentum, of nuclei
stability of, 587–8, 609, 615–6, 622
symmetry character of, 561
Nucleons, 565
Numerical integration of differential equations, 180–2, 268, 276–7

Old quantum theory, 136
critique of, 136–7
One-electron atom, 295–325, 350–8
and role in theory of multi-electron atoms, 402–3
One-electron-atom angular momenta, expectation values, 317
precessional motion of, 323–4
quantization of, 323
One-electron-atom eigenfunctions, behavior near origin, 309
degeneracy of, 303–4, 350–1
cause, 308–9, 404–5
normalization of, 307
orthogonality of, 307
parity of, 466
quantum numbers for, 302–4, 337, 349–50
One-electron-atom energy levels, degeneracy of, 303–4, 350–1
cause, 308–9, 404–5
with fine structure, 355–7
in old quantum theory, 117, 134
in Schroedinger theory, 303
One-electron-atom probability densities, angular dependence of, 309–13
behavior near origin, 309
measurability of, 311–2

One-electron-atom probability densities, radial dependence of, 306–9
relation to angular momenta, 312–3
One-electron-atom time independent Schroedinger equation, complete, 295
for relative motion, 298
separation of complete equation, 295–8
separation of relative equation, 298–9
solution of relative equation, 299–302
Operators, *see* Differential operators
Optical excitation of atoms, 417, 484
Optically active (optical) electron, 418
Optical model of nuclei, 572–80
and Fermi gas model, 594–6
and initial stage of nuclear reaction, 663
and shell model, 594
values of parameters, 575–7
Orbital angular momentum, 323
of filled subshell, 438
of one-electron atom, *see* One-electron atom
precessional motion of, 323–4
quantization of, 114, 129–30, 323
total, 429
in atoms with several optical electrons, 426–7
quantization of, 429
Orbital *g* factor, 329
Orbital magnetic dipole moment, 327–9
quantization of, 329
Ordinary integral, 380
Orientational potential energy, of magnetic dipole moment, 330–1
of net magnetic dipole moment, 445
Orthogonality of eigenfunctions, 188–90
in continuum, 189
when degenerate, 189–90

Pairing energy of nuclei, 598
Pair production, 508–9, 513–7
cross section for, 508–9, 510
Parent nucleus, 609, 613
Parity, 247, 464–6
and atomic transition selection rules, 467
and beta decay selection rules, 634, 636
of deuteron ground state, 678
of eigenfunctions, for infinite square well potential, 253–4
for one-electron atom, 466
for simple harmonic oscillator, 263
for spherically symmetrical potential, 466
for square well potential, 247–8
and form of potential, 248
and gamma decay selection rules, 648, 650
non-conservation in beta decay, 636–8
of nuclear energy levels, 653, 658–9
measurement of, 658; *see also* Beta decay; Gamma decay
origin of, 465
and perturbation theory, 276
and solution of time independent Schroedinger equations, 248, 253–4
Partial wave analysis, 534–44, 554–6
for absorbing sphere scattering, 556–7
applicability of, 542, 549, 554
cross section formulas for, 544, 556
for hard sphere scattering, 546–9
for neutron-proton scattering, 682–4
for proton-proton scattering, 691–2
for S-wave scattering, 549–52
Partial waves, 537–8, 555
orbital angular momentum of, 538
and plane waves, 540–2
Particle-wave duality, 85–6
and barrier penetration, 235, 239
and uncertainty principle, 160–2, 226–7
Paschen-Bach effect, 447–8
Pauli principle, *see* Exclusion principle
Periodic boundary conditions, for electromagnetic radiation, 52
in relativistic range, 629
for wave functions and eigenfunctions, 220, 525–6

Periodic table of elements, 406
actinide elements, 411, 414
alkali elements, 413
halogen elements, 413
Hartree theory interpretation of, 407–14
lanthanide elements, 411, 414
noble gas elements, 412–3
rare earth elements, 411, 414
transition group elements, 413–4
transuranic elements, 414
Perturbation, 269
and effect on degeneracies, 278–9
Perturbation theory, and meson theory, 701
and quantum electrodynamics, 701
Perturbation theory for degenerate case, 278–83
and atoms with several optical electrons, 425–8
classical example of, 283–5
formulas for, 279–81
and helium atom, 373–81
and *JJ* coupling, 442
and *LS* coupling, 428–9, 432–4
and relativistic corrections to Schroedinger's equation, 353–4
simplified treatment of, 282–3, 350–1
and spin-orbit energy, 350–3
and Zeeman effect, 445–6
Perturbation theory for non-degenerate case, 268–73
accuracy of, 277–8
formulas for, 272
use in degenerate case, 282–3
and helium atom, 373–6
and V-bottom potential, 274–8
Perturbation theory for time dependent case, 285–91; *see also* Golden Rule No. 2
and absorption of electromagnetic radiation, 453–7
and Einstein's quantum postulate, 455
formulas for, 286
and Golden Rule No. 2, 286–91
and induced emission of electromagnetic radiation, 457
Pfund series, 113, 120
Phase shifts, 542, 555–6
for absorbing sphere scattering, 557
and cross sections, 544, 556
energy dependence of, 553
for hard sphere scattering, 547
interpretation of, 544–6
for $l > KR$, 547
for Ramsauer effect, 552
for resonance, 551–2
for S-wave scattering, 551
Photoelectric cell, 76
Photoelectric effect, classical theory of, 78–9
cross sections for, 507–8, 510
early experiments on, 76–8, 81
quantum theory of, 79–81, 507
X-ray line spectra and, 507
Photoelectrons, 77
Pilot waves, 140; *see also* Wave functions
Planck's black body spectrum theory, 63–6
Planck's constant, 66
and uncertainty principle, 158
and X-ray continuum spectra, 482
Planck's quantization postulate, 64
comments on, 66–8, 128–31, 262–3
Polarization, of scattered nucleons, 602, 693
of scattered X-rays, 476
Polar vectors, 637
Positronium, 517
Positrons, 513–7
annihilation of, 516–7
Dirac theory of, 514–7
discovery of, 513
emission of, in beta decay, 623–4
holes and, 516
properties of, 516
Potential, alpha particle-nucleus, *see* Alpha decay
barrier, *see* Barrier potentials
constant, *see* Free particle
Coulomb, *see* Coulomb potential; One-electron atom
deuteron, *see* Nuclear forces
Gaussian, *see* Gaussian potential

Potential, infinite square well, *see* Infinite square well potential
net, *see* Net potential
nucleon-nucleon, *see* Nuclear forces
nucleon-nucleus, *see* Optical model of nuclei; Shell model of nuclei
simple harmonic oscillator, *see* Simple harmonic oscillator
square well, *see* Square well potential
step, *see* Step potential
V-bottom, *see* V-bottom potential
Yukawa, *see* Yukawa potential
Potentials, and forces, 178
radial dependent, general treatment, 298–9
and spatial quantization, 324–5
relative coordinate dependent, and separation of center of mass motion, 298
time dependent, general treatment, 285
general treatment, 176, 294
Precession of angular momenta, Larmor, 331–2, 444–5
and uncertainty principle, 324, 430
Probability conservation, 175
Probability density, 171–2, 172–5, 294–5
for atom emitting radiation, 461–2
for free particle, standing wave, 218
traveling wave, 213
for general wave function, 186–7
for eigenstate wave function, 187
for multi-electron atom, 397, 401, 404–5
for one-electron atom, *see* One-electron atom
radial, 307
in multi-electron atom, 401
in one-electron atom, 306–9
for simple harmonic oscillator, 264
for square well potential, $E < V_0$, 248
$E > V_0$, 251
for step potential, $E < V_0$, 226
for two non-interacting identical particles, 363–5
Probability flux, 175
for free particle, standing wave, 218
traveling wave, 214
for incident and scattered waves, 536
for step potential, $E < V_0$, 225
$E > V_0$, 229
Probability in quantum mechanics, 160
Product particle, 563
Proper length, 22
Proper time, 21
Proton, 560
Pseudo-scalars, 636
and beta decay, 636–8

Quadrupole moment, *see* Electric quadrupole moment
Quantization, 64; *see also* Energy levels
of atomic energy states, 117, 124–8
of energy in Schroedinger theory, 178–84
of grand total angular momentum, 450
of nuclear spin angular momentum, 449
of orbital angular momentum, 114, 323
of orbital magnetic moment, 329
of simple harmonic oscillator, 64
of total angular momentum, 347, 429–30
of total orbital angular momentum, 429
of total spin angular momentum, 429
and uncertainty principle, 157–8
and Wilson-Sommerfeld rules, 128–32
Quantum, of electromagnetic field, 80, 694–5
of meson field, 694–5
Quantum defects, 420
Quantum electrodynamics, 462
and origin of Coulomb interaction, 694–5, 696–8
and perturbation theory, 701
and spin *g* factor, 357
and spontaneous emission of radiation, 458, 462
Quantum numbers, 64, 185
assignment of, 448–9

Quantum numbers, for atoms, with *JJ* coupling, 442
with *LS* coupling, 429–33, 435
with one electron, 302–4
with one optical electron, 421
with several optical electrons, 425
for nuclear shell model, 600
spectroscopic notation, for *l*, 408
for *n*, 486
for *s*, 422–3
for X-ray levels, 485
Quantum states, 64
lifetime of, 469, 472
Quantum statistical mechanics, 390
Quantum theory, 41
Q value of nuclear reaction, 583

Radial probability density, 307
in multi-electron atoms, 401
in one-electron atoms, 306–9
Radiation by accelerated charges, 42–6
Radii, of atoms, *see* Atomic radii
of nuclei, *see* Nuclear radii
Radioactive decay, 89; *see also* Alpha decay; Beta decay; Gamma decay
Radioactive series, 613–6, 620–1
branching in, 621
equilibrium in, 614–5
Ramsauer effect, 234–5, 552, 657
Random walk, 97
Range, of charged particles, 563
of nuclear forces, 579, 674, 687, 692
compared to other forces, 579
and meson theory, 696, 697
Rare earth elements, 411, 414
Rayleigh-Jeans theory of black body radiation, 52–7, 62–3
spectral distribution of, 62
Recursion relation, 260, 268
Reduced mass, 123, 519
in time independent Schroedinger equation for CM motion, 297
Reflection coefficient, for step potential, $E < V_0$, 225
$E > V_0$, 229–31
Relativistic energy, 33–4
Relativistic mass, 34–5
Relativistic mass-energy relation, 34–5
Relativistic mass-energy relation, and binding energy of atomic electrons, 421, 580
and binding energy of nuclei, 585–6
and energy balance in nuclear reactions, 565, 582–5
experimental verification of, 585
Relativistic mechanics, 30–7
Relativistic transformation, of momentum and energy, 36–7
of position and time, 26
of velocity, 29
Relativistic wave equation, 697
Relativity theory, general theory, 38
experimental verification of, 37–8
special theory, 3–38
Relativity theory applications, *see* Relativistic mass-energy relation
Compton effect, 83
corrections to Born approximation, 569
corrections to Schroedinger equation, 353–4
non-radiation of uniformly moving charge, 43
pair production, 513–7
Sommerfeld theory of fine structure, 133
spin angular momentum, 338
Thomas precession, 340–4, 359
Repulsive cores in nuclear forces, 691–3
and meson theory, 703
and saturation, 693–4
Residual Coulomb interactions, 426–8
Residual nucleus, 563
Resonances, 551–2
many particle vs. single particle, 657, 664
in nuclear reactions, 655–7
in nuclei, 577
in square well potentials, 251
for transmission through barrier potentials, 234–5
and virtual states, 552–3
Rest mass, 32
Rest mass energy, 34
Rigid rotator, energy levels of, 138, 325

Russell-Saunders coupling, *see* LS coupling
Rutherford scattering experiments, 91–2, 98–9, 106–8; *see also* Coulomb scattering
Rutherford's model of atoms, 99–100
  and Bohr's theory, 114
  predictions of, 100–6
  stability of, 108–9, 114–5, 152, 187, 461
  verification of, 106–7
Rydberg constant, for hydrogen, 112
  for infinitely heavy nucleus, 119, 123
  for nucleus of mass M, 123
Rydberg series, for alkalis, 112, 420, 424
  for hydrogen, 112

Saturation of nuclear forces, 675–7, 690, 693–4
Scalars, 636
Scattering, 555
  of alpha particles by nuclei, 91–108
  of electrons, by atoms, 400
    by nuclei, 567–70
  of neutrons, by molecular hydrogen, 685–6
    by protons, 682–7, 689–90
  and nuclear physics, 518
  of protons, by nuclei, 572–80
    by protons, 687–8, 690–3
  of X-rays, 493–505
Schmidt lines, 604
Schroedinger's equation, 170, 294
  and classical wave equation, 193
  complex character of, 170
  development of, 165–70, 293–4
  and differential operators, 204
  linearity of, 166–7, 185–6
  and Newton's laws, 206–9
  time independent, *see* Time independent Schroedinger equation
Schroedinger's theory of quantum mechanics, and de Broglie's postulate, 165–6
  and energy quantization, 178–84
  postulates of, 165–6, 321
Scintillation counter, NaI, 573
  ZnS, 92
Selection rules, for alkali spectra, 424
  for beta spectra, 634, 636
  example of evaluation of, 463–7
  for gamma spectra, 648
  initial and final eigenfunctions and, 463
  for JJ coupling spectra, 443
  for LS coupling spectra, 440–1
  in old quantum theory, 134, 135–6
  for X-ray spectra, 486
  for Zeeman spectra, 447
Self-consistency, 392
  in Hartree theory of atom, 397
  in nuclear structure theory, 693–4
  in Thomas-Fermi theory of atom, 393
Semi-empirical mass formula, 590–2
  and alpha decay energy, 609
  and beta decay energy, 621–2
Separation constant, 176
Serber potential, 690, 691–2, 693
  and meson theory, 698–700
Series, atomic, 112
  Balmer, 113, 120
  Brackett, 113, 120
  Lyman, 113, 120
  Pfund, 113, 120
  Rydberg, 112, 420, 424
Series limit, 112
Series solution of differential equations, 259–60, 301–2
Shadow scattering, 548–9, 557
Shell model of nuclei, 599–604
  and beta decay, 634–5
  and collective model, 605–8
  and electric quadrupole moments, 608
  and Fermi gas model, 594–5
  and gamma decay, 647
  and Hartree theory, 600
  and JJ coupling, 603
  and liquid drop model, 605
  and magnetic dipole moments, 603–4
  and optical model, 594
  and spin angular momentum, 603
  and spin-orbit interactions, 600–2
  theoretical basis of, 593–5

Shells, in multi-electron atoms, 402–3
in nuclei, 602
in one-electron atoms, 307–8
Shielding, *see* Effective charge
Simple harmonic oscillator, 256–66
applications of, 256
boundary conditions for, 258–9
classical behavior of, 68, 256
eigenfunctions for, 263–4
energy levels of, 262
in old quantum theory, 64, 131
parity of eigenfunctions for, 263
probability densities for, 264
quantization of, in old quantum theory, 64, 130–1
selection rules for, 136
time independent Schroedinger equation for, 257–8
zero point energy of, 262
Simultaneity, 16–8
Singlet eigenfunctions, 370–1
and exchange force, 372–3
quantum numbers for, 372
Singlet state of deuteron, 684
Single-valuedness of eigenfunctions and their derivatives, 178, 300, 321
Slater determinant, 367
Smoothed step potential, 231
Solid angle, 497
Sommerfeld's relativistic theory of atoms, 131–4
and fine structure, 131, 338
and quantum mechanics, 355–7
Spatial quantization, 324
experimental verification of, 335
and form of potential, 324–5
Spectra, analysis of, 448–9
atomic, *see* Atomic spectra
black body, 48–9
beta, 624–33
optical, 417
simple harmonic oscillator, 136
X-ray, continuum, 482–3, 489–92
line, 483–9
Spectral line, 111
widths of, *see* Line widths
Spectroscopic notation, for l, 408
origin of, 424
for n, 486
Spectroscopic notation, for s, 422–3
Spherical Bessel functions, 538
Spin angular momentum, of deuteron, 677
eigenfunctions for, 300, 337
of electron, 336–9
of filled subshell, 438
of μ meson, 703
of π meson, 703
of neutrino, 626
of neutron, 562
of nuclei, and beta decay selection rules, 634, 636
and electrons in nuclei, 562
and gamma decay selection rules, 648, 650
measurement of, 451, 658; *see also* Beta decay; Gamma decay
quantization of, 449
and shell model, 603
and symmetry character, 369
values of, 451
of positron, 516
of proton, 562
quantization of, for $s = \frac{1}{2}$, 336
quantum numbers for, for $s = \frac{1}{2}$, 336
and symmetry character, 369
total, 371, 428
in atoms with several optical electrons, 426
quantization of, 371, 429
Spin dependence of nuclear forces, 685, 693
Spin g factor, 336–7, 357
Spin-orbit interaction, in atoms, 339–46, 351–3
for alkali atoms, 421–4
formulas for, 345, 353
for JJ coupling, 442
for LS coupling, 433–4
magnitude of, 345
in nuclei, 600–3
and JJ coupling, 603
magnitude of, 602
and nuclear properties, 600–4
and nucleon-nucleon scattering, 693
and nucleon-nucleus scattering, 602

Spontaneous emission, by atoms, 453, 458–60
  Einstein's coefficient for, 459
  justification for treating only electric dipole interactions, 468, 644–6
  physical picture of, 461–2
  transition rate for, 460, 463
 by nuclei, *see* Gamma decay
Spontaneous fission, 617
Square well potential, 239–51
 applications of, 240–1
 boundary conditions for, $E < V_0$, 241
  $E > V_0$, 250
 classical behavior of, 240
 eigenfunctions, $E < V_0$, 243, 247
  $E > V_0$, 250–1
 energy levels, $E < V_0$, 244–5
  $E > V_0$, 240, 250
 infinite, *see* Infinite square well potential
 parity of eigenfunctions for, 247–8
 probability density, $E < V_0$, 248
  $E > V_0$, 251
 transmission coefficient, $E > V_0$, 240, 251
Stability, of atoms, 108–9, 114–5, 152, 187, 414, 467
 of nuclei, 587–8, 609, 615–6, 622–3
Standing waves in cavity, 52–7
 number of allowed frequencies for, 57
Stationary state, 187; *see also* Eigenstates
Stefan's law, 46
Step potential, 221–31
 applications of, 221
 barrier penetration of, $E < V_0$, 226–7
 boundary conditions, $E < V_0$, 223–4
  $E > V_0$, 228
 classical behavior, $E < V_0$, 222
  $E > V_0$, 227
 eigenfunctions, $E < V_0$, 224–5
  $E > V_0$, 229
 energy levels, $E < V_0$, 222
  $E > V_0$, 227
Step potential, probability density, $E < V_0$, 226
 probability flux, $E < V_0$, 225
  $E > V_0$, 229
 reflection coefficient, $E < V_0$, 225
  $E > V_0$, 229–31
 time independent Schroedinger equation, $E < V_0$, 222
  $E > V_0$, 227
 transmission coefficient, $E > V_0$, 229–31
 wave function, $E < V_0$, 224
Stern-Gerlach experiment, 334–5
 and electron spin, 335
 on hydrogen atoms, 335
Stokes' law, 74
Strength function, 664
Subshells, 405
 in alkali atoms, 419–20
 in multi-electron atoms, 405
 ordering according to energy, 408–9, 411–2, 415
 spherical symmetry of filled, 398–9, 412
Superposition of wave functions, 143, 144, 148, 167; *see also* Linear combinations
Surface tension energy of nuclei, 590
S-wave scattering, 549–52
 and limits to proton-proton scattering cross section, 690
 in neutron-proton scattering, 682–7
 in proton-proton scattering, 687–8
Symmetric eigenfunctions, 363–4, 368–9
 for interacting identical particles, 369
 for non-interacting identical particles, 363–4, 368
 and spin angular momentum, 369
Symmetry character of eigenfunctions, for alpha particles, 368
 and behavior of atoms, 368, 414
 for electrons, 367
 and exclusion principle, 367
 for μ mesons, 703
 for π mesons, 703
 for neutrons, 368
 for protons, 368

Symmetry character of eigenfunctions, and spin angular momentum, 369
Synchronization of clocks, 16–7, 23–4, 27

Target nucleus, 563
Tensor potential, 678–9, 693
Thermal emission of electrons, 124–5
Thermal radiation, 42
  and quantum theory, 41
Thomas-Fermi theory of atoms, 393–6
Thomas frequency, 343
Thomas precession, 340–4, 359
Thomson scattering, 493–8, 505
  cross sections, 495, 498, 505
Thomson's model of atoms, 88
  and alpha particle scattering, 92–8
Time dependent perturbation theory, *see* Perturbation theory for time dependent case
Time dilation, 19, 20–1
  for mesons, 38, 702
Time independent Schroedinger equation, 176–7, 294
  acceptable solutions of, 178, 300
  and classical wave equation, 194–5, 235
  for deuteron, 680
  and energy quantization, 178–84
  for Fermi gas, 382
  for free particle, 212
  for infinite square well potential, 253
  for multi-electron atom, 396
  for one-electron atom, *see* One-electron-atom time independent Schroedinger equation
  real character of, 177
  for simple harmonic oscillator, 257–8
  solution, by numerical integration, 180–2, 268, 276–7
    by series method, 259–60, 301–2
  for step potential, $E < V_0$, 222
    $E > V_0$, 227
Total angular momentum, 346–9, 429–30, 442
Total cross section, 510, 556; *see also* Cross section
Total internal reflection, 235–6
Total relativistic energy, 34
Transformation, of cross sections, CM to LAB, 523–4
  Galilean, 5
  Lorentz, 26
  of momentum and energy, classical, 36
    relativistic, 37
  of velocity, classical, 521
    relativistic, 29
Transition group elements, 413–4
Transition probability, 290
Transition rate, 291
  for absorption, atomic, 453–7
  for induced emission, atomic, 457
  and old quantum theory, 137, 291
  for spontaneous emission, atomic, 453, 458–60, 463
    nuclear, 647–8
Transmission coefficient, for barrier potential, 233–4
  for square well potential, 240, 251
  for step potential, $E > V_0$, 229–31
Transmission probability of nucleons through nuclear surface, 665
Transmission resonances, for barrier potential, 234–5
  in Ramsauer effect, 552
Transuranic elements, 414
Triplet eigenfunctions, 370–1
  and exchange force, 372–3
  quantum numbers for, 372
Triplets, of energy levels, *see* Fine structure
  of spectral lines, 424
Triplet state, absence of, in ground state of helium, 381
  of deuteron, 684

Ultraviolet catastrophe, 63
Uncertainty principle, 155, 156, 159
  and barrier penetration, 226–7, 239
  and compound nucleus reactions, 663
  and de Broglie's postulate, 153–5
  and energy definition in nuclear reactions, 667
  and indistinguishability, 360
  and line widths, 469–70, 472
  and meson rest mass, 695–6

Uncertainty principle, and particle-wave duality, 160–2, 226–7
physical basis of, 157–8
and quantization of angular momentum, 323
and quantization of electromagnetic radiation, 157–8
and saturation of nuclear forces, 675–7
and spontaneous emission and absorption of mesons, 700
and spreading of wave groups, 159–60
and time dependent perturbation theory, 288–9
and zero point energy, 255–6
Uniform translation, 4
Unsold's theorem, 312

V-bottom potential, 274–8
eigenfunctions, 276–7
energy levels, 276–7
Vector diagrams for addition of angular momenta, 348–9, 431
Velocity of light, 8
Velocity transformation, classical, 521
relativistic, 27–9
Virtual level, 553
Virtual state, 553
of deuteron, 685–6
of neutron-proton system, 685–6

Wave equation, electromagnetic, 235, 697
quantum mechanical, *see* Schroedinger's equation
relativistic, 697
for string, 193
Wave functions, 165, 185; *see also* Pilot waves
complex character of, 170
for compound nucleus reactions, 664
for free particle, standing wave, 217
traveling wave, 213
groups of, 142–5, 153–4, 159–60, 163, 215–6
for identical particles, 364
interference of, 148, 151–2, 153–4, 167
Wave functions, interpretation of, 205, 206, 214, 217, 294, 295, 318–21, 321–2
general, 170–5
linear combinations of, 186, 192, 285, 288
mathematical properties of, 141–6, 184–92
normalization of, 187, 190–1, 206
for step potential, $E < V_0$, 224
superposition of, 143, 144, 148, 167; *see also* Linear combinations, of wave functions
time dependence of, 191–2
Wave number, 112
Wave-particle duality, *see* Particle-wave duality
Waves in string, 192–200
Wave velocity, 141, 142, 144
Widths, of nuclear energy levels, 660–2
and Breit-Wigner formula, 670
measurement of, 658; *see also* Alpha decay; Beta decay; Gamma decay
and predictions of compound nucleus theory, 666
and random width assumption of compound nucleus theory, 667
of quantum states, 470–2
and lifetimes, 472
and uncertainty principle, 470, 472
of spectral lines, *see* Line widths
Wien's law, 50–1
Weizsacker's atomic mass formula, *see* Semi-empirical mass formula
Wilson-Sommerfeld quantization rules, 128–9
applications of, 129–31
de Broglie's interpretation of, 152

X-ray attenuation, 493
attenuation coefficient, 512
attenuation length, 512
cross section for, 509–11
exponential law for, 511–2
X-ray energy levels, 485–6
X-ray excitation of atoms, 417, 483–4
X-ray continuum spectra, classical theory of, 490–1

X-ray continuum spectra, origin of, 489–92
  Planck's constant and, 482
  properties of, 482–3
  quantum theory of, 491–2
X-ray line spectra, and Bohr theory, 488
  and determination of atomic charge, 488–9
  and holes, 484–5
  and Moseley experiments, 487–9
  origin of, 483–7
  photoelectric effect and, 507
  properties of, 483
  quantum numbers for, 485
  selection rules for, 486
X-ray pair production, 508–9, 513–7
  cross sections for, 508–9, 510
X-ray photoelectric effect, 505–8
  cross sections for, 507–8, 510
X-rays, diffraction of, 476–80
  discovery of, 475–6
  measurement of spectra, 476, 480–1
  properties of, 476
X-ray scattering, 493–505
  classical theory of, 493–501
  coherent, 476–80, 499–501
  coherent cross sections for, 499–501
  Compton, 81–5, 502–5
X-ray scattering, Hartree theory and, 400, 501
  incoherent, 498–9
  Klein-Nishina cross sections for, 503–5
  measurement of atomic structure by, 400, 501
  quantum theory of, 501–4
  Thomson, 493–8, 505
  Thomson cross sections for, 495, 498, 505
  total cross sections for, 510

Yukawa potential, 696–8

Zeeman effect, 332, 443–7
  anomalous, 447
  energy level splitting in, 446–7
  Lande *g* factor and, 446
  net magnetic moment in, 444–5
  normal, 447
  selection rules, 447
  spectra, 447–8
Zero point energy, 255
  and electrons in nuclei, 561–2
  for simple harmonic oscillator, 262
  and spontaneous emission of radiation, 462
  and uncertainty principle, 255–6
ZnS scintillation detector, 92